146080

#12.99
sierl

AIR Enthusiast

VOLUME ONE

William Green	Managing Editor
Gordon Swanborough	Editor
W R Matthews	Modelling Editor
Dennis Punnett	Contributing Artist
John Weal	Contributing Artist

Doubleday & Company, Inc
Garden City, New York

ISBN 0-385-08171-5

© 1971 Pilot Press Limited

First Published in the United Kingdom, 1972, by Fine-Scroll Limited, De Worde House, 283 Lonsdale Road, Barnes, London SW13 9QW, United Kingdom

Printed by G. A. Pindar & Son Ltd., Scarborough

VOLUME SUMMARY

Volume 1 Number 1   June 1971

THE AIR ENTHUSIAST is a monthly journal for people having an interest in aircraft in particular and aviation in general, and is unashamedly concerned with aeronautical history. The Oxford English Dictionary defines history as "The continuous methodical record of . . . events", and the editorial task that we have set the AIR ENTHUSIAST is that of providing such a record. History, in this sense, does not concern only the events and artifacts of bygone eras; it also concerns the contemporary scene. Nor need it mean a dry-as-dust recital of facts, as we hope that the contents of this and subsequent issues of the AIR ENTHUSIAST will reveal. Aviation *can* make interesting reading for many whose work-a-day connection with the subject is remote or non-existent, and the AIR ENTHUSIAST is aimed at the tyro, the well-informed layman and the professional alike.

The serious historian, in his efforts to achieve complete accuracy in recording events of any kind, is confronted with problems and pitfalls, but few aspects of modern history afford more difficulties than that of aviation. The reasons are several. The pace at which aircraft development proceeds has often meant that events have outpaced the written record. Two world wars and numerous more localised conflicts have temporarily imposed measures of secrecy on the activities of aircraft manufacturers, their products and the air arms to which they were supplied, many records being lost in part or in total before the end of hostilities rendered their publication harmless. In more pacific times, national or commercial interests have frequently motivated against publication of detailed information; aircraft development, more than most fields of technology, has been greatly influenced by political factors and even personal whims which have led to the publication of distorted facts, half-truths, and even pure fabrications, and finally, the passage of time has dimmed the memories of those most closely involved with events, and has seen the destruction, deliberately or by accident, of the priceless records of many an aircraft manufacturer, operator, official department or individual.

The AIR ENTHUSIAST will endeavour to add much to the record that has hitherto gone unrecorded, and brings together what we believe to be a skilled, professional and highly-experienced team of editors, illustrators and contributors, backed by unrivalled resources for research. A primary editorial objective is to ensure that each feature appearing in our pages is both detailed and accurate, and our success or failure in achieving this objective will be adjudged in the course of time by you, the reader. There will be occasions when statements appearing in the AIR ENTHUSIAST will contradict what has been published in the past and accepted as fact, and opinions expressed may, from time to time, be controversial. Readers' views on these and other matters will be welcomed for inclusion in our correspondence column *Talkback*.

We will endeavour to maintain a high standard of completeness and accuracy in our monthly summary of aviation news which appears under the heading *Airscene,* and two other regular features that we consider to be of importance are *In Print*, which will provide news of important new books of general aviation interest, and *Model Enthusiast*, furnishing the modeller with information and illustrations that we hope he will find to be of assistance, as well as reviews of the latest plastic model aircraft kits. The *Veteran and Vintage* column offers news and illustrations, of older aircraft types surviving in museums, private collections or in flying trim, and readers will have access to the information resources of our editorial offices through the monthly *Plane Facts* in which specific requests for information will be answered in print.

AIRSCENE

MILITARY AFFAIRS

ALBANIA
The Albanian People's Army Air Force is now reportedly in process of receiving 25-30 examples of the Sino-Communist copy of the **MiG-21 interceptor** to boost its air defence capability. Albania denounced the Warsaw Pact in September 1968, having played no part in the alliance for some nine years as a result of ideological differences, and has been the recipient of Chinese military aid since 1964 when the Sino-Communist government furnished MiG-15 spares, Chinese-manufactured RD-45FA turbojets, and a number of *Fong Chou* (Chinese-built An-2) utility transports. From 1965, China supplied the Albanian air arm with MiG-17F fighters, copies of the SB06 (*Atoll*) AAM and *Wu Ming* AGMs, following these during 1966-67 with Chinese-built MiG-19 fighters and Mi-4 helicopters. The MiG-21s will presumably replace the MiG-19s in the Albanian air arm's two intercept squadrons, the MiG-19s being assigned to the four ground attack squadrons operating the ageing MiG-17.

ARGENTINA
The logistic support capability of the *Servicios de Transportes Aéreos Militares,* the primary air transport component of the *Fuerza Aérea Argentina* (FAA), is expected to be augmented by four **Canadair CC-106 Yukons,** the last of which was scheduled to be retired from the Canadian Armed Forces in April. One CC-106 has reportedly already reached the FAA which has an option on a further three transports of this type.

AUSTRALIA
The Army Aviation Corps anticipates receiving the first of 12 **OH-58A Kiowa** light observation helicopters being supplied by Bell Helicopter during the next three months. The OH-58A Kiowa has been selected as the Australian Army's standard LOH, and the co-production agreement signed earlier this year between the Australian government and Bell Helicopter covers a further 179 helicopters of the Model 206A Jet Ranger type of which some 40 per cent will be for defence purposes. The Army has a requirement for 75 LOHs of the OH-58A Kiowa type, and the Australian

Navy has a requirement for a further nine, current plans calling for the service requirements to be fulfilled at a rate of one helicopter per month from Australian production commencing next year. Commonwealth Aircraft is the prime Australian licensee, and Hawker de Havilland and the Government Aircraft Factory are participating in the eight-year programme. Other new Bell helicopters scheduled for operation by the Australian armed forces include 11 Bell AH-1G HueyCobra gunships to be delivered to the RAAF by 1974 at a cost of $A12·4m (£5·75m), inclusive of spares and support equipment, and a further 35 UH-1H Iroquois (seven Iroquois of this model already being in service with the RAAF) for which approval has been given with the placing of orders awaiting the conclusion of a satisfactory "offset order opportunities" agreement.

BELGIUM
The long-delayed choice of a successor for the aged Fairchild C-119Gs serving with the 15ème *Wing de Transport* of the *Force Aérienne Belge* (FAéB) was finally announced on 10 March when a Defence Ministry spokesman stated that 12 **Lockheed C-130H Hercules** transports are to be purchased at a cost of $55m (£22·9m), approximately 60 per cent of this sum being offset over a period of 10 years by contracts to be placed with Belgian industry. The first six Hercules for the FAéB are scheduled to enter service during the second half of next year with the remainder of the order being completed during the course of 1973. Meanwhile, FAéB transport capability is being augmented by the lease from SABENA of two Boeing 707-320s.

CAMBODIA
A major US-assisted **re-equipment programme** is now being undertaken by Cambodia's air arm, National Khmer Aviation, in an attempt to increase its effectiveness in anti-Communist operations. Until recently, National Khmer Aviation (NKA) possessed the most cosmopolitan aircraft inventory of any SE Asian air arm as an inheritance of the "neutralist" regime of Prince Sihanouk, its equipment including aircraft of French, Chinese, Soviet and US origin. At the time of the 22 January attack by Viet Cong and North Vietnamese forces

on the major Cambodian base of Pochentong, Phnom Penh, much of the inventory was unserviceable owing to spares shortages, and the NKA, commanded by Colonel So Satto, comprised six squadrons as follows: one fighter-bomber sqdn operating a mix of MiG-15s, MiG-17s and CM 170-1 Magisters; one ground attack sqdn with T-28Ds and A-1D Skyraiders; one transport sqdn with C-46s, C-47s and Il-14s; one AOP sqdn with Cessna O-1 Bird Dogs, one helicopter sqdn with Alouette IIs and IIIs and Iroquois, and one training sqdn with SOCATA GY-80 Horizons. In addition, a liaison flight operated DHC-2 Beavers, Antonov An-2s and Cessna 170s. Most of these units could muster only a nominal strength. The NKA is currently receiving an infusion of additional T-28Ds, UH-1 Iroquois and Cessna U-17s. Most Soviet- and Chinese-supplied aircraft have been phased out, and it is likely that the MiG-15s and -17s will be replaced by Cessna A-37s and, possibly, Northrop F-5s.

CANADA
Deliveries to the CAF began in April of 25 **Beech Musketeers** which are now replacing the DHC-1 Chipmunk in the primary training rôle. The Musketeer emerged as the winning contender in a Chipmunk replacement feasibility study concluded by the CAF late in 1969.

FEDERAL GERMANY
The *Aufklärungsgeschwader* 51 'Immelmann' at Bremgarten is now in process of working up with the **McDonnell Douglas RF-4E Phantom,** the first examples of which were received by the *Luftwaffe* on 20 January, and *Aufklärungsgeschwader* 52 at Leck is scheduled to commence conversion to the RF-4E later this year. Each unit will comprise two 15-aircraft *Staffeln,* and of the remainder of the 88 RF-4Es being delivered to the *Luftwaffe,* 24 will be held in reserve against anticipated attrition over a 10-year period, two will serve in the training rôle and two will be employed for trials purposes.

INDIA
At least eight squadrons of the Indian Air Force (Nos 1, 4, 8, 28, 29, 30, 45 and 47) are now fully operational on the limited all-weather MiG-21FL, and the earlier MiG-21F day interceptors previously operated by Nos 28 and 45 Squadrons have been phased out. The Sukhoi Su-7 serves with Nos 3, 31, 32, 221 and 222 Squadrons, and the attrition of the IAF's Canberra-equipped tactical bombing component (Nos 5, 16 and 35 Squadrons) has been partly made good by the purchase from New Zealand of the 10 Canberra B(I) 12s made redundant when the RNZAF acquired A-4K Skyhawks.

Current **HAL activities** include the investigation of the possible application of the Rolls-Royce/Turboméca Adour turbofan to the HF-24 Marut to endow this indigenous strike fighter with Mach 2 capability, and long-term studies for the design and production of a twin-turbofan multi-purpose fighter in the general category of the F-4 Phantom to succeed the HF-24 in the 'eighties. It is anticipated that the new fighter will initially be powered by proven engines of foreign origin but will eventually receive power plants of Indian design. In addition to original design activity, HAL has concluded an agreement with the Soviet

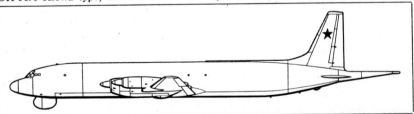

Illustrated above and below is the maritime patrol version of the Ilyushin Il-18 transport known by the NATO code-name of "May". These illustrations, reproduced by courtesy of Herkenning, *show the lengthened fuselage terminating in an MAD tail "sting" and the radome beneath the fuselage nose.*

Union for the licence manufacture of an advanced version of the MiG-21 for 1973-4 delivery to the IAF. Although no details of this model, referred to as the MiG-21M, have been revealed, it is claimed to be superior to the current MiG-21FL in all respects — speed, range, weapon load and manoeuvrability — and is presumably powered by an uprated version of the Tumansky RD-11 turbojet. The first MiG-21FL manufactured entirely in India was formally handed over to the IAF at HAL's Ojhar (Nasik) factory on 19 October, but both the airframe programme and the production at Koraput of the RD-11-300 turbojet are reportedly well behind schedule, and production of the HF-24 Marut fighter displays no signs of gathering momentum (this type currently equips only two IAF squadrons, Nos 10 and 210). Work is progressing in Bangalóre on a new HAL factory specifically for the manufacture of helicopters and light aircraft, the former being initially of Aérospatiale design. HAL has manufactured 80 SA 316 Alouette III helicopters under licence for the IAF and Indian Navy, is currently fulfilling Aérospatiale orders for Alouette components, and is initiating production of the SA 315 Alouette II. HAL is also setting up its own helicopter design team with the object of producing helicopters of indigenous design within the next decade.

Believed to be illustrated for the first time in the West, the MiG-21 depicted above is one of the most recent production variants of this Soviet fighter, referred to by NATO as the "Fishbed-J". This model embodies a re-contoured rear fuselage, repositioned pitot head, and two additional underwing strong points. For the intercept rôle two "Atoll" AAMs are normally carried in combination with a 23-mm NR-23 cannon housed by a centreline pod, and for the attack rôle up to four UV-16-57 pods each containing sixteen 55-mm rockets, four 550-lb (250-kg) bombs, or four 220-mm or 325-mm rockets may be carried on wing pylons.

IRAN

The Iranian Imperial Air Force (IIAF), or *Nirou Havai Shahanshahiyé Irân*, currently engaged in a **major expansion programme**, expects to have increased its personnel strength to some 30,000 by the end of the year, and to 35,000 by the end of 1972. The combat element is now provided by 125 Northrop F-5As, RF-5As and F-5Bs, and 32 McDonnell Douglas F-4D Phantoms, with 32 F-4E Phantoms currently on order, and the IIAF has a requirement for up to 64 F-4E(F) interceptors or up to 100 F-5E interceptors for 1973-4 delivery. The capacity of the IIAF transport component is being more than doubled by the orders placed late in November for 30 Lockheed C-130H Hercules (which will supplement 26 C-130E Hercules already delivered) and 14 Fokker F.27 Friendships, the latter being intended to replace the service's Douglas C-47s. Current helicopter procurement calls for the delivery of 70 AB 206 Jet-Rangers, 45 AB 205 Iroquois, five AB 212 Twin Two-Twelves, 16 Meridionali-built Boeing-Vertol CH-47C Chinooks, and 10 Agusta-built Sikorsky SH-3s.

JAPAN

At the time of closing for press, the first of two **Mitsubishi XT-2** supersonic trainer prototypes was about to be rolled out of Mitsubishi's Komaki factory with initial flight testing scheduled for later this month or early July. The first production order calling for 50 T-2 trainers is expected to be placed during the 1972 Fiscal Year to enable the first two squadrons to be formed early in 1975, the total Air Self-Defence Force requirement being for 100 T-2s. Development of a strike fighter variant, currently designated SF-X, will be undertaken in parallel with the T-2 development programme, and three SF-X prototypes (modified T-2s) will be ordered during the 1972 Fiscal Year, with an initial production order for 50 machines for 1975-6 delivery being placed in the following Fiscal Year. It is anticipated that 150 SF-X strike fighters will be ordered during the Fourth Five-year Defence Programme (1972-6), and a tactical reconaissance version of the SF-X is currently under consideration.

The first squadron to be equipped with the **NAMC C-1** transport is now scheduled to be formed in 1973, current plans calling for the purchase of 30 aircraft of this type during the 1972-6 Defence Programme. The first XC-1 prototype was flown on 12 November and the second on 16 January.

The Air Self-Defence Force will acquire 158 licence-built **F-4EJ Phantoms** during the 1972-6 Defence Programme, and of these 34 were ordered in March last year, and a further 48 form the largest single item of expenditure in the 1971 Defence Budget. It is anticipated that the ASDF will also acquire 20 RF-4Es during the 1972-6 programme, and these will also be licence-built by Mitsubishi except for the forward fuselages which will be imported from the parent company. The first of two pattern F-4EJs built by the parent company was flown at St Louis on 15 January and is expected to be ferried to Japan during the next two months.

NETHERLANDS

The first of nine **Breguet Atlantic** maritime patrol aircraft for the Dutch *Marineluchtvaartdienst* (MLD) to be built at the new Breguet factory at Toulouse-Colomiers was scheduled to be delivered to the Netherlands in April. Nine Atlantics (designated SP-13A by the MLD) were ordered in July 1968 with completion of deliveries originally being scheduled for March of this year. The first aircraft built against this contract did not, in fact, fly until 30 January, but No 321 Squadron, formed on 8 August 1969 to operate the Atlantics, has been working up with four Atlantics diverted from *Aéronavale* contracts.

SOUTH VIETNAM

Plans have been formulated to establish an indigenous aircraft industry in South Vietnam on 19 July, the Republic's National Day. The aircraft, which is to be built at Bien Hoa, the South Vietnamese Air Force's main supply and maintenance base, is the side-by-side two-seat **Pazmany PL-2**. This is almost identical to the PL-1 trainer being built under licence on Taiwan but has a redesigned and simplified structure.

SPAIN

It has been unofficially reported that the Spanish air arm, the *Ejército del Aire*, will now receive six **Lockheed C-130 Hercules** transports and two KC-130 Hercules tankers instead of the two C-130s originally mentioned as part of the new US aid package. This package will now also include 25 helicopters and funding for the modernisation of the air defence system.

TANZANIA

According to sources in Washington, China is furnishing the Tanzanian Defence Force Air Wing with sufficient **MiG-17** fighters to equip two squadrons, which, some observers in the US State Department say, will give Tanzania air superiority over neighbouring Rhodesia. As long ago as mid-1968, the leader of the Tanzanian National Assembly announced that some defence capability would be provided the Air Wing by the addition of a jet fighter component, and it is understood that a number of Tanzanian pilots, originally Canadian trained, have undergone jet conversion courses in China. The MiG-17s will probably be based at Mikumi, near Dar-es-Salaam.

YUGOSLAVIA

The Yugoslav government is understood to be negotiating with the Soviet Union for the supply of 12 late-model MiG-21 interceptors, but has also expressed some interest in the possibility of purchasing a number of F-5E light interceptors from the USA. Apart from a relatively small number of MiG-21Fs acquired several years ago, the entire interceptor component of the Yugoslav Air Force comprises F-86 Sabres.

ZAMBIA

The Zambian Air Force has recently been supplied with a nucleus of a **combat component** by Yugoslavia with the delivery of four Soko Jastreb single-seat light strike-aircraft and two Soko Galeb two-seat jet trainers to Mbala (formerly Abercorn) on the Tanzanian border. A small Yugoslav mission is currently training Zambian personnel on the new aircraft, and the Zambian Air Force is scheduled shortly to take delivery of 12 Aermacchi MB 326-GB strike trainers from Italy.

AIRCRAFT AND INDUSTRY

CANADA

Preliminary details have been published of the **Canadair CL-246**, a project for a large STOL transport. Basis of the design is a conventional 70-seat fuselage, with a high-mounted tilt wing similar in concept to that used on the Canadair CL-84s now in flight test. Four 1,900 hp Lycoming T53-10A turboprops would be spread along the wing to produce a slipstream over the whole span. The wing would have whole-span Krüger flaps and slotted trailing edge flaps, the outer portions of which would also function as ailerons. Gross weight of the project is 53,030 lb (24 055 kg) and the payload is about 14,330 lb (6 500 kg).

FRANCE

A "third-level" airline version of the Falcon biz-jet has been proposed by Dassault as the **Falcon 20-T.** Details and a full scale mock-up were being presented at the Paris Air Show in May/June. The Falcon 20-T makes use of the wings, tail unit, undercarriage and engine installations of the Falcon 20-F, with a new fuselage of greater diameter and length which will accommodate, in a typical three-abreast layout, 24 passengers. Comparative data for the Falcon 20-T and Falcon 20-F are: fuselage diameter, 7 ft 9 in/6 ft 8 in (2,36 m/2,04 m); overall length 60 ft/56 ft 3 in (18,30 m/17,15 m); gross weight 29,100 lb/27,560 lb (13 200 kg/12 500 kg). The Falcon 20-T will be powered by two 4,315 lb (1 957 kg) thrust General Electric CF-700 2D2 engines and a prototype is expected to fly early in 1972.

Reims-Aviation claims a 40% improvement in the take-off and landing performance of its new military STOL version of the Cessna 337 Skymaster which is now under-going flight testing. Principal modification is the addition of high-lift flaps and four wing strong points to carry rocket pods or bombs. The aircraft is known as the **Reims FTMA Milirole,** the initials standing for "France/Turbo/Militaire/ADAC", and the all-up weight is 4,430 lb (2 010 kg). Take-off distance is 722 ft (220 m) and the take-off speed 48.5 mph (78 km/h); the approach speed is 51 mph (82 km/h), landing distance 732 ft (223 m), and landing roll not more than 165 ft (50 m).

Flight testing of two new engines has started in France. A Turboméca/SNECMA **Larzac 01** has been mounted beneath a Super Constellation for trials at Istres prior to installation of this engine type in the Dassault-Breguet/Dornier Alphajet, the Dassault Falcon 10 and Aérospatiale Corvette. Turboméca has modified a North American Hawk Commander to take two **Astafan** turbofans in underwing nacelles. Both these test-beds are expected to be on show at the Paris Air Show in May/June. Meanwhile, the **SNECMA M-53** turbofan of about 12,350 lb (5 600 kg) thrust, has completed more than 150 hours of ground running and the possibilities of using a Mirage F1 or a Caravelle as a flying test-bed are being investigated.

INTERNATIONAL

While assembly of the first **Airbus A-300B** proceeds on schedule, with delivery of the first fuselage section from MBB in Hamburg to Aérospatiale in France having been made during March, studies of a series of variants continues. The basic A-300 is the B-1; this version has 49,000 lb (22 226 kg) thrust CF6-50A engines, 162 ft (49,38 m) fuselage length and 291,000 lb (132 000 kg) gross weight, and can carry 259 passengers. The B-2 has 51,000 lb (23 132 kg) CF6-50C engines, 167 ft 6 in (51 503 m) fuselage length, 304,240 lb (138 000 kg) gross weight, and can carry 279 passengers. The B-3 is as the B-1 with CF6-50C engines and 321,875 lb (146 000 kg) gross weight. The B-4 is as the B-2 with 321,875 lb (146 000 kg) gross weight and reduced max payload for greater ranges. The B-5 is an all-freight version of the B-1 with a 70,500 lb (32 000 kg) payload. The B-6 is another specialised cargo version which may have a different fuselage diameter; it is based on the CF6-50C engine and has a projected gross weight of 322,140 lb (146 120 kg). The B-7 was the version proposed for BEA last September with a fuselage length of 174 ft 2 in (53,09 m), a gross weight of 321,875 lb (146 000 kg) and accommodation for 296 passengers; this variant could have CF6-50C engines or (now highly unlikely) Rolls-Royce RB.211-62s. The latter engines were also proposed as an option in the B-3, and the earlier RB.211-56s were at one time considered for the B-2 and B-4 variants.

JAPAN

As a successor to the YS-11 (sales of which now total 153), NAMC has projected the Y-X or, in its present phase, the **N-B-X,** indicating NAMC-Boeing X. Discussions with Boeing centre upon joint development of a 200-seat transport, although the Y-X was originally somewhat smaller, and the possibility of BAC and/or Fokker-VFW joining the programme is also under study. Choice of powerplant continues to be a major difficulty; after three Rolls-Royce Trents had been considered in the very early stages, NAMC next studied two P & W STF 19s but these did not meet the required timescale, and an alternative scheme to use two JT3D-7s was not favoured by the Japanese domestic airlines who represent a major market for the project. Late in 1970, the use of two 27,000 lb (12 250 kg) thrust Rolls-Royce RB.211Js was under consideration, but the collapse of Rolls-Royce has again left the position wide open. Meanwhile, NAMC is con-tinuing to study the prospects for a jet-powered YS-11, called the **YS-11J** and to be powered by 7,500 lb (3 400 kg) thrust SNECMA M-45H-01s or two 7,910 lb (3 590 kg) thrust M-45H-04s. A recent market survey showed a prospective demand for 90 aircraft in the USA, for Piedmont (20), Frontier (30-40) and Texas International (25-30). Deliveries could begin in 1974 at $2.3m (£958,300).

SWEDEN

A more powerful version of the **Saab MFI-15A** two-seat trainer has been under test since 26 February. The 160 hp Lycoming IO-320 engine originally installed has now been replaced with a 200 hp Lycoming IO-360. The aircraft will be available with either nosewheel (MFI-15A) or tailwheel (MFI-15B), and with empty and loaded weights of 1,213 lb (550 kg) and 1,984 lb (900 kg) respectively, it has a max speed of 165 mph (265 km/h) at sea level, cruises at 150 mph (245 km/h) at 8,000 ft (2 440 m), and has an initial climb rate of 1,375 ft/min (7,0 m/sec).

UNITED STATES OF AMERICA

Two new variants of the range of Piper aircraft have been announced. The **Cherokee Flite Liner** is a simplified version of the two-seat Cherokee 140 intended primarily for use at the Piper operated flying schools known as Piper Flite Centers. With a 150 hp Lycoming engine, the Flite Liner operates at a gross weight of 1,800 lb (815 kg) but is structurally approved for the normal Cherokee 140 gross weight of 2,150 lb (975 kg). The **PA-36 Pawnee II** is offered with 285 hp or 320 hp engines, and comple-ments the PA-25 Pawnee I, which is still in production with 235 hp or 260 hp engine. Pawnee II has larger overall dimensions, with a span of 39 ft (11,88 m) and length of 27 ft 8 in (8,44 m), and the gross weight is increased to 3,800 lb (1 724 kg). Fuel capacity is also increased.

Newest engine test-bed to fly in the USA is a Boeing B-47 carrying a **General Electric TF34** high by-pass turbofan under the port wing, between the inboard and outboard nacelles. First flight was made on 21 January. The TF34-GE-2 is under develop-ment as the power plant for the Lockheed S-3A ASW aircraft which is scheduled to fly in 1972, and is in the 9,000 lb (4 082 kg) thrust class. It is also specified for the Boeing E-3A, the AWACS derivative of the 707, which will have eight TF34s, and for the two Fairchild Hiller YA-10A (AX) prototypes which will each have two.

The USAF is currently studying the possi-bility of adapting the 24 F-111Cs built for the RAAF, and which, at present stored at Fort Worth, have been neither accepted nor rejected by the Australians, as **EF-111** electronic countermeasures aircraft to satisfy the service's need for an airborne tactical jamming system.

Turbine-engined conversions of two Sikor-sky helicopters have now been certificated in the USA and deliveries have begun. The **Sikorsky S-58T,** engineered by Sikorsky, is a basic S-58 re-engined with a Pratt & Whitney (UACL) Twin-Pac, comprising two PT6T-3s coupled through a single gear box. Orders and options to date are for 14 S-58Ts converted by Sikorksy and 14 con-version kits; included in this total are five conversion kits for Air America and two S-58Ts and a kit for Okanagan Helicopters Ltd. Delivery of conversion kits began in

————————Continued on page 56

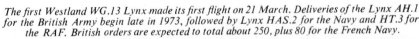

The first Westland WG.13 Lynx made its first flight on 21 March. Deliveries of the Lynx AH.1 for the British Army begin late in 1973, followed by Lynx HAS.2 for the Navy and HT.3 for the RAF. British orders are expected to total about 250, plus 80 for the French Navy.

THE FIGHT FOR THE SKIES

Fighter Spectrum Part 1 – Air Superiority

by Roy Braybrook, BSc, CEng, AFRAeS

IT IS a strange reflection on Western fighter design that, despite the plethora of types in service, the market is now completely dominated by two aircraft which have reached the top in rôles for which they were never intended.

The F-4 Phantom was developed for naval all-weather interception, but found fame as the only type suited to both carrying out and protecting strikes with iron bombs in the hostile environment over North Vietnam. Starting with a spectacular mission in which this aircraft brought the MiG-21s to battle by simulating an F-105 raid out of Thailand (the MiG lead pilot's last-minute recognition of his mistake is on tape in USAF archives), the Phantom proved the value of a two-man crew, long-range radar, and a mix of missiles. Against this, the MiG-21's only advantages were in presenting a smaller target area and in better manoeuvrability.

The F-105 "Thud" had previously borne the brunt of the fighting, but its design was based on all-weather strike with an internal nuclear store. It is better equipped than the Phantom for accurate bomb delivery and has exceptional airspeed limits, but in combat with MiG-21s it lost out on both wing loading and thrust/weight ratio. When bounced, the "Thud" driver could generally only clear the bomb-racks, push the nose down, and hit the burner, which (given sufficient warning) left even the MiG-21 trailing behind in a cloud of part-burned kerosene. The Phantom was found to be much better in a dogfight, especially given the M-61 cannon of the F-4E, and also had the advantage that any losses could be replaced from continuing production. Over 4,000 Phantoms have now been manufactured.

Europe's own success is, of course, the Mirage III and its derivatives, for which orders have already exceeded 1,000. Designed to defend point targets in France against high altitude bombers, it has achieved international status as a fighter-bomber combining genuine Mach 2 capability with a price (for a model combining full intercept and strike capability) of the order of £750,000–£800,000 ($1·8m–$1·92m), and proven superiority over the MiG-21. This is no mean achievement for an aircraft which (although openly admired by the RAF) has been described in Britain as one which combines an airframe inspired by the Fairey FD2 with an engine basically designed by Germans, and a radar owing a great deal to Ferranti's Airpass. The fighter business is full of sore losers!

By comparison with the Soviet product, which from certain airframe aspects is a superior design, the Mirage is better powered and armed, and in Middle East encounters may have been better flown. Its later models of Thomson-CSF Cyrano radar are outstanding multi-mode equipments, reportedly equal to Hughes' TARAN and Autonetics' NASARR F-15A. Nevertheless, the inferiority demonstrated by the MiG-21 in 1967 dogfights may still be reversed by engine and weapon developments.

Combat philosophy

The fact that today's main fighter successes have been the results of air superiority operations has resulted in overwhelming importance being attached to this aspect of design in several current military requirements. The danger of this philosophy is that in striving to guarantee that a fighter can outclass anything the opposition can put into the air, the end product may cost so much that numbers are limited; the aircraft becomes too valuable to risk in anything less than vital operations, and funds for other developments are severely restricted. These risks are particularly applicable to the USAF's Tactical Air Command, whose insistence on ruling the skies deep over enemy territory is both unique and extremely expensive.

One of the problems in this field is to assess the relative combat merits of various designs. This can be difficult enough for aircraft already flying: the comparison of such widely differing types as the Phantom and Harrier requires a

Manufacturer Type		Lockheed CL-1200-1 Lancer	CL-1200-2 Lancer	Vought Super V-1000	McDonnell Douglas F-4E(F) Phantom	Grumman F-14A Tomcat
Power Plant		1 × J79-GE-19	1 × TF30-P-100	1 × J79-GE-17	2 × J79-GE-17	2 × TF30-P-412
Maximum thrust	lb (kg)	17,900 (8 120)	25,130 (11 400)	17,900 (8 120)	17,900 (8 120)	22,500 (10 205)
Span	ft (m)	29·2 (8,9)	29·2 (8,9)	35·76 (10,9)	38·38 (11,7)	64·12 (19,57)[2]
Length	ft (m)	56·76 (17,3)	57·09 (17,4)	54·79 (16,7)	62·99 (19,2)	61·85 (18,85)
Height	ft (m)	15·1 (4,6)	15·4 (4,7)	15·88 (4,84)	16·27 (4,96)	16 (4,88)
Wing area	sq ft (m²)			374·58 (34,8)	530 (49,2)	
Empty weight[1]	lb (kg)	17,700 (8 030)	18,630 (8 450)	18,080 (8 200)	36,828 (16 705)	36,000 (16 330)
Internal Fuel	lb (kg)	8,510 (3 860)	8,510 (3 860)	6,480 (2 940)		
Ordnance (max)	lb (kg)	9,215 (4 180)	8,995 (4 080)	6,000 (2 720)		
Normal loaded	lb (kg)	25,000 (11 340)[3]	25,800 (13 100)[3]	26,000 (11 790)[3]	46,526 (21 102)	53,000 (24 040)
Max speed	Mach	2·16	2·4	1·7	2·1	2·5
Unstick speed	mph (km/h)	200 (322)	205 (330)		196 (315)	
Take-off run	ft (m)	2,500 (762)	1,850 (565)	1,970 (600)	1,970 (600)	
Landing run	ft (m)	2,180 (665)	2,280 (695)		2,800 (854)	
Ceiling	ft (m)	60,000 (18 290)	60,000 (18 290)	55,000 (18 045)	60,000 (18 290)	60,000 (18 290)
Initial climb	ft/min (m/sec)	38,140 (195,6)	55,295 (289)		35,000 (177,8)	
Radius of action	mls (km)	335 (540)	330 (530)	315 (510)	215 (346)	
Armament		1 × 20-mm M61 2 × AIM-9	1 × 20-mm M61 2 × AIM-9	2 × 20-mm M39 2 × AIM-9	1 × 20-mm M61 4 × AIM-7 and 4 × AIM-9	1 × 20-mm M61 6 × AIM-7 and 4 × AIM-9

[1]) Equipped for intercept rôle. [2]) With minimum sweep (reducing to 33·2 ft/10,12 m with max sweep). [3]) Normal intercept mission.

lengthy flight programme even for the basic step of optimising tactics. When the fighters under discussion are still at the project stage, the task may be verging on the impossible.

The only satisfactory approach appears to be a combat simulator in which two pilots can "fly" the opposing types from initial contact through to the final "kill". By repeating the process many times and by changing the pilots, the relative merits of the two designs can be assessed with reasonable confidence. Unfortunately such simulators are too costly to permit widespread use. The writer knows only of examples owned by LTV and McDonnell Douglas.

In these circumstances low-cost combat simulation has become a thriving pseudo-art, although the simplifications involved can be so sweeping that it is difficult to take the results seriously. What remains is the traditional approach of merely comparing performance parameters for acceleration, climbing, turning, deceleration, etc. There have been attempts to combine these in a single quantity, giving an overall measure of an aircraft's worth in a dogfight, but so far without success.

Lacking such an all-embracing parameter, one of the favourite substitutes is SEP (specific excess power), which at least combines acceleration and climb capability. This is computed from the power in hand after the engine has overcome drag, customarily in level flight. When related to aircraft weight this excess gives the rate at which energy (in the form of height or speed) can be increased.

To illustrate the numbers involved, subsonic ground

attack aircraft have peak SEPs of 200–300 ft/sec (60–100 m/sec), while interceptors may reach 500–800 ft/sec (150–250 m/sec), and the F-14, F-15, and MiG-23 will presumably be up around 1,000–1,200 ft/sec (300–350 m/sec). All these figures represent the best attainable values at sea level, although high performance fighters also have a comparable peak approaching Mach 2 at medium altitude.

Demand for Dragmasters

SEP also has to be computed for various g-loadings, since aircraft differ markedly in lift-dependent drag. A Mirage III at really high incidence is about the best airbrake in the business! Deceleration capability can be useful in forcing the opponent to overshoot into your field of fire during a scissors manoeuvre, hence sales jargon has taken on board "specific power deficit". Unusually good values can be achieved by such aircraft as the A-7 and Harrier, due respectively to an outsize airbrake and to the momentum drag of the Pegasus turbofan when its jets are directed downwards.

Maximum (lift-limited) and sustained (thrust-limited) turning performance may be specified in terms of g, turn radius, or turn rate. It is debatable which of these is most significant, but g is slightly easier to compute, and radius naturally wins if the only consideration is to turn inside your opponent. Radius is also critical in ground attack, to keep a target in sight or to maintain air-surface communications. Northrop are thumping the tub for turn rate, pre-

America's first response to the threat posed by the new generation of Soviet fighters is the variable sweep Grumman F-14 Tomcat, selected by the US Navy to replace the abortive F-111B programme. Shown here at take-off, the first of 12 development F-14As made its initial flight on 21 December 1970, but crashed on its second flight on 30 December after a complete flying control failure. The second F-14A should have flown by the time this issue of AIR ENTHUSIAST is published.

	Northrop		Dassault
P-530 Cobra		F-5E	Mirage F1E
2 × GE-15	2 × J85-GE-21		1 × Atar 9K-50[4]
12,000 (5 440)	5,000 (2 268)		15,870 (7 200)
33 (10,05)	26·5 (8,08)		27·56 (8,4)
51 (15,55)	48·4 (14,76)		49·21 (15)
	13·15 (4,0)		14·75 (4,5)
400 (37,16)	186 (16,28)		269·1 (25)
17,000 (7 710)	8,500 (3 855)		16,314 (7 400)
7,200 (3 265)	4,570 (2 073)		
12,000 (5 443)	6,200 (2 810)		7,055 (3 200)
25,000 (11 340)	15,500 (7 030)		24,030 (10 900)
2·0	1·6		2·2
	190 (306)		200 (322)
1,800 (550)	2,050 (625)		1,475 (450)
			1,640 (500)
60,000 (18 290)	55,000 (18 045)		65,600 (20 000)
44,000 (223,5)	35,200 (179)		
400 (645)	400 (645)		350 (530)
1 × 20-mm M61	2 × 20-mm M39		2 × 30-mm DEFA
4 × AIM-9	2 × AIM-9		2 × Matra 530
			2 × AIM-9

[4]) Offered as alternative to SNECMA M-53 or J79-GE-19.

·sumably because the F-5's straight wing has a better transonic turning performance than the swept-wing competition, which may hit a smaller radius at low speeds. However, it is possible that turn rate is the best criterion only in special situations, for instance in turning to deny IR acquisition to an aircraft coming up from astern.

Turning performance now receives more attention than sheer speed, although Mach 2 capability is often an incidental fall-out from SEP demand, and may be essential in high altitude interception. One result of the new emphasis has been a growing interest in manoeuvring flaps, not only to increase maximum lift, but also to lower fuselage attitude in a turn and thus reduce drag. The alternative approach is to design for turning, by building in more camber, washout, and a higher wing incidence on the fuselage, although these measures can only be applied at the expense of cruise economy, maximum speed, and landing performance.

The classic air superiority aircraft is small, cheap, well-powered, manoeuvrable, and carries both cannon and short-range missiles. Provided it is not also required to intercept high-flying aircraft, there may be no need for good ceiling or Mach 2 capability, since recent combat experience has been at relatively low level and below Mach 1·3. Surface-launched missiles are said to have ruled out tactical operation at the altitudes used by the F-86 Sabre and MiG-15 over Korea (ie, heights at which Mach 2 dashes might now be used), although clearly this only applies to some theatres.

In the light of these considerations, it is hardly surprising that the Northrop F-5-21 was selected as the IFA (International Fighter Aircraft) to be supplied by the US to South Vietnam, Thailand, Taiwan and South Korea. Some 325 of these aircraft will be manufactured as the F-5E, at a unit price of about $1·3m, (£550,000) compared with $750,000 (£310,000) for the "cooking" F-5A, of which over 1,000 have already been produced for 16 countries.

The normal problems of any lightweight fighter are that, with external loads, it takes forever to get the rubber off the road, and there is little space for fuel. In this latest variant of the F-5, Northrop has improved take-off (not to mention SEP and max speed) by modifying the fuselage to take J85-21s of 5,000 lb (2 280 kg) reheat thrust, and internal fuel has been boosted by over 500 lb (230 kg). A two-position nose gear improves unstick without spoiling landing roll, a fix that may be required by the centre-line store being ahead of the *cg*. Best speed with two Sidewinders is Mach 1·5, and turning is enhanced by leading- and trailing-edge flaps. Internal fuel tanks have reticulated foam fillings to prevent explosions due to combat damage. External stores weighing up to 6,200 lb (2 820 kg) can be carried, including three 230 Imp gal (1 050 l) tanks. It is significant that even in SE Asia

a search and track radar is felt necessary for air superiority, and the F-5E will have a lightweight radar from Emerson Electric.

Since F-5A pilots claim they already have more manoeuvrability at low level than either the F-4 or F-104, the F-5E should prove a cost-effective answer to the MiG-17 and -21 at the altitudes now in fashion, assuming Sidewinder developments are supplied to match any improvement in the Soviet K-13 air-to-air missile. However, an aircraft with a basic weight of about four tons must be limited in ground attack, and this may explain why the USAF originally opted for LTV's V-1000.

The Capped Crusader
The V-1000 proposal is a lighter, improved, and more easily maintained variant of the F-8D Crusader, with a J79-GE-17 of 17,900 lb (8 150 kg) in place of the heavier but equally powerful J57-P-20. The F-8 may be limited to Mach 1·7, but it is reputed to be the most manoeuvrable fighter in US service. In fact minimum radius was even better on the F-8C (ie, before the heavy provisions for underwing stores). Rear hemisphere view benefits from the high wing which blocks off less sky than the low wing of the F-4 Phantom II.

The V-1000 has been lightened by deleting wing fold and flight refuelling, and by reducing fuel capacity. Two USAF M-39 cannon are used in place of the standard four Navy Mk 12s. External stores up to 6,000 lb (2 730 kg) can be carried, including two 250 Imp gal (1 140 l) tanks. The cost of restarting production of this relatively large fighter may have proved decisive in the IFA contest, but LTV are nonetheless promoting the so-called Super V-1000 in countries demanding better attack capability. Not so limited by budget-initiated constraints imposed on IFA contenders, the Super V-1000 basically features more advanced equipment and improved performance.

The F-104 Starfighter was designed on the basis of Korean War dogfights, to carry an M-61 gun and two Sidewinders, and to excel in performance at high altitudes. Pouring on the power, an F-104G can have an SEP of over 500 ft/sec (155 m/sec) as it hits Mach 2, and very few aircraft will stay with it under such conditions!

On the other hand, fighters must also operate at low speeds and are supposed to be straightforward to fly. The trouble with the F-104 is that you have to look closely to find the wing, and the tailplane is in the wrong place. However, pitch-up was cured by APC (auto pitch control) which pushes the nose down before things get out of hand. Secondly, landing speeds were cut on the C and G models by flap blowing, although this has the drawback that a flame-out can mean loss of both onwards and upwards arrows. Italy's F-104S is a Mach 2·2 air superiority development of the F-104G, with more wing area, a J79-GE-19 of 17,900 lb (8 150 kg) (ie, 13 per cent more thrust), and the addition of two Sparrow AAMs. Ventral fins boost its directional stability at high speeds.

The proposed Lockheed CL-1200 Lancer may also be powered by the J79-GE-19, but differs in representing a major structural redesign. Wing span is increased to 29 ft (8,85 m), or 33 per cent up on the F-104G, and full span flaps are used. Despite the lowering of the tailplane to the fuselage, Lockheed has still had to raise the wing to a shoulder position to give the desired stability improvement. The acute angle below the wing root is filled with a ramp-shaped fairing, which may penalise top speed. Compared with the F-5E and V-1000, the Lancer requires more research and development but is clearly attractive to countries which have built or are still building the F-104, such as Italy. A modified Lancer has been proposed to the USAF under the research designation X-27. This variant would accept the TF30, F100

Above, the Northrop F-5E, recently selected as the IFA (International Fighter Aircraft) to be supplied by the US to foreign powers through military aid programmes. Below, the McDonnell Douglas F-15A, the USAF's new all-weather interceptor. Scheduled to fly next year, the F-15A, unlike the F-14, has a fixed wing.

or F401 engine, and is predicted to be capable of Mach 2·6. Two USAF F-104 airframes would be modified into X-27s if the proposal is accepted.

In broadly the same performance category is the Israeli development of the Mirage, code-named *Salvo*, which began flight trials last October. Based on the IIICJ, it has greater fuel capacity and a J79-GE-17 in place of the standard Atar, giving an impressive 35 per cent thrust boost. Motivation for this particular programme was at least in part to overcome the shortage of Atar spares and to make continued operation of Mirages independent of the French arms embargo.

Homesick Angels
Stepping up to the major league, one yardstick against which new designs are measured is the MiG-23, a Mach 3 fighter first shown to Western observers in 1967. Unfortunately, Soviet statements on its record-breaking flights have revealed nothing of its engines. The MiG-23 appears to weigh about 30 tons for interception, and perhaps over 40 tons for strike, but there is no straightforward way to estimate its SEP with confidence.

More likely to be met in dogfights are Mikoyan's 20-ton duo, *Faithless* and *Flogger*, which get STOL performance respectively by means of two 5-ton lift engines and by a

Below, the proposed Lockheed CL-1200 Lancer, a model of which is shown here undergoing wind-tunnel testing, is a major structural redesign of the F-104G and therefore of special interest to European users of the Starfighter.

variable sweep wing. When these aircraft appeared in 1967, it was evident that they were competitive experiments based on a common fuselage, tail and a single propulsion engine of about 14 tons thrust. The MiG selection will presumably be augmented by the double-delta Sukhoi *Flagon B*, a 25-ton interceptor with two 10-ton propulsion engines and three lift engines.

America's first response to the new Soviet generation is the Grumman F-14 Tomcat, a two-seat fighter using variable sweep to provide long endurance and easy carrier stowage. The first F-14A flew late in 1970 fitted with two TF30-P-412s of 22,000 lb (10 000 kg). Deliveries are scheduled for 1973, despite the loss of the first aircraft after a flying control failure reportedly caused by a titanium hydraulic line fracture. The second of an initial development batch of 12 F-14As was to have flown by the time these words appear in print.

By 1974 this variant will be superseded by the F-14B, with Pratt & Whitney F401s of about 29,000 lb (13 200 kg). Take-off weight will range from 52,000 lb (23 750 kg) for air superiority to 63,000 lb (28 700 kg) when carrying six long-range Phoenix AAMs, four AIM-9H Sidewinders, and two drop tanks. The programme originally called for 67 F-14As and 406 F-14Bs, but the US Navy now talks of 722 aircraft at a unit cost of $11·5m (£4·8m), including some F-14Cs which differ in having uprated avionics.

The single-seat McDonnell Douglas F-15 is scheduled to fly next year and enter USAF service in 1974. In order to give superiority at long range, it will gross more than 40,000 lb (18 000 kg), and be fitted with a new Hughes attack radar and two Pratt & Whitney F100s of 25,000 lb (11 400 kg). In this case the wing is fixed, presumably because of the aircraft's ground attack commitment. Two of the 20 test and evaluation aircraft will be two-seaters. The initial follow-on order will be for 107 aircraft, but eventual total procurement is expected to be 500–700 aircraft at a unit cost around $12m (£5m).

Both F-14 and F-15 will initially have a 20-mm M-61 multi-barrel gun, but this is to be replaced by a 25-mm cannon from General Electric or Philco-Ford, firing caseless ammunition. There were originally separate Navy (Agile) and USAF (AIM-82A) programmes for dogfight missiles, but the latter has been cancelled, and Agile continues only as a back-up to the AIM-9X Super Sidewinder and AIM-7F Sparrow. There is speculation that economy measures will lead to the F-14B fulfilling the Phantom replacement rôle for both services, but it would fall short of USAF demands both for SEP and ground attack capability.

The writer recently discussed fighter costs with a Grumman representative somewhere over the Pacific (another tax-deductible holiday!) and was told that nobody could build an air superiority fighter below $10m (£4·2m). This may be true in American terms, but somebody will have to protect the skies of Europe for a lot less than that!

A possible competitor for the Mirage F1 is the P-530 Cobra project which will weigh in at slightly more and may be expected to obtain a better thrust/weight ratio by virtue of two 12,000 lb (5 400 kg) thrust GE15s. In terms of weights and areas, this is virtually a 2·3 times scale-up of the F-5, although possessing no family resemblance, having a high-mounted wing and the fashionable twin vertical tail arrangement. Seven pylons provide for external loads up to 12,000 lb (5 400 kg), giving a gross weight of some 39,000 lb (17 690 kg). Northrop has been trying to persuade various countries to sponsor and participate in the Cobra's development as an air superiority replacement for the Mirage III and F-104G (which represented a market of 250 aircraft for the *Luftwaffe* alone until that service's

————————————————————*Continued on page 38*

End of an Era . . .

Polikarpov's Chaika

AT 11.00 HOURS on a mid-July morning in 1939, at the height of what the Japanese referred to as the "Nomonhan Incident", nine small fighter biplanes, dust and dry grass churned up by their airscrew blades billowing behind them, took-off from a makeshift airfield just to the west of the Khalkhin-Gol. Tucking up their undercarriages as they climbed through the heat haze, and forming up into loose three-aircraft *zvenos,* the grey-painted fighters turned a few degrees towards the front line and steadily climbed into the clear air above the barren expanse of the Barga Steppe straddling the nearby and ill-defined Manchukuoan-Mongolian border. A hundred or so kilometres to their north, the Wall of Ghenghis Khan was out of sight, but as they levelled-off at about 9,850 ft (3 000 m) and turned southeast to fly along the Soviet side of the front line, the pilots of the fighters could see Ta Hingan Ling, a mountain range rising from the steppe to port.

The front had been extremely active during the previous two weeks. The Japanese forces, having failed in their attack at Bain-Tsagan, had struck from the north in strength, and had succeeded in wresting from the Soviet and Mongolian forces their foothold on the east bank of the Khalkhin. Then, on 5 July, the Japanese thrust had been checked and the Japanese Army forced back across the river. Aerial warfare throughout this phase of the "incident" had been sporadic; but there was evidence that the Imperial Japanese Army was markedly increasing the combat aircraft units deployed in support of its forces committed to the area, and the local V-VS RKKA (Soviet Air Forces) Headquarters at Tamtsak-Bulak, some 40 miles (65 km) south of the Buyr Nur — a large lake in the vicinity of which much of the aerial combat had taken place — had repeatedly stressed in its reports to Moscow the importance of increasing the V-VS commitment in this undeclared but nevertheless full-scale conflict.

On that particular morning, the first *zveno* was led by Lt Col Sergei Gritsevets, a highly-experienced fighter pilot, a veteran of the Spanish Civil War, and already a Hero of the Soviet Union. The other *zvenos* were led by Nikolai Viktorov

and Aleksandr Nikolayev who, too, were no novices to combat. As the fighters cruised steadily along the front line, the ground haze lifted, and on the slopes of one of the foothills, the Khamardaba, the pilots could see a white marker laid out by Soviet troops to indicate to the patrolling V-VS fighters the direction in which an advanced post had reported by radio the appearance of Japanese aircraft.

Within seconds of sighting the marker, Gritsevets' deputy Major Smirnov, flying as his port wingman, waggled the wings of his fighter to attract the formation leader's attention, and then pointed in the direction of the Uzur Nur, a lake on the Japanese side of the line. Some 5,000 ft (1 525 m) above the lake a gaggle of Nakajima Army Type 97 fighter monoplanes was climbing towards the Soviet *zvenos*. The Army Type 97, or Ki.27, was a truly formidable warplane of which five *Sentais,* or Groups (the 1st, 11th, 24th, 59th and 64th), comprising some 80 aircraft, had been available to the Imperial Japanese Army in the Khalkhin-Gol area from the outset of the fighting between Soviet and Japanese forces on 4 May. Supremely manoeuvrable, the Type 97 had immediately taken the measure of the ageing I-152 fighter biplane, but had found a stouter opponent in the I-16 Type 10, the outcome of combat between the two monoplane types depending largely on the capabilities of their respective pilots.

Somewhat to Smirnov's surprise, Gritsevets, instead of turning towards the enemy force, merely changed course a few degrees to take the Soviet formation further away from the front line. It was unlike Gritsevets to avoid combat, whatever the odds, but his motives were soon obvious to his companions. Their wily CO was expecting the Japanese pilots to mistake the Soviet fighters for I-152s which they could master with ease; he was luring them steadily deeper into Soviet air space. Discarding their usual caution in anticipation of easy prey, the Japanese gradually overhauled Gritsevets' formation which maintained course and speed, giving no hint that the Soviet pilots were aware of the existence of their pursuers.

The Soviet pilots cocked their machine guns, rotated their

(Above and below) One of the I-153 (alias I-15ter) prototypes photographed during the winter of 1938-9. This M-25V-powered aircraft was a production prototype, and the radio masts on the starboard wing leading edge and tailfin, and bomb shackles beneath the interplane struts, are noteworthy.

gun triggers to the "fire" position, and patiently awaited Gritsevets' signal. The Type 97s steadily closed the gap between the two groups of fighters, their pilots having apparently failed to notice their quarries' "lack of under-carriages". When the gap had shrunk to some 2,200 yards (2 000 m), Gritsevets made a circling motion above his head with one gloved hand, and immediately the nine Soviet bi-planes, with engines at full boost, turned to face their pursuers. This sudden volte-face took the Japanese totally by surprise, and they broke formation and scattered. The hunters were now the hunted! Victims of the withering hail of fire from the Soviet fighters' highly-effective batteries of ShKAS machine guns, two of the Type 97s had plunged into the steppe below, another had broken up in the air, and a fourth had spiralled down trailing smoke before the surviving Japanese fighters reached the comparative safety of their own air space!

The I-153 had been blooded in combat and the first lines in the closing chapter of the story of the fighter biplane had been written. A few days after this encounter between the I-153 and the Type 97, the Japanese newspaper *Yomiuri* published a despatch from a correspondent attached to the Headquarters of the Kwantung Army. This despatch record-ed the operational début of the new Soviet fighter, stating that (*sic*) ". . . although flown by veritable devils it had been bested by fighters of the Imperial Japanese Army which had accounted for no fewer than *eleven* of this new warplane". In fact, all *nine* of Gritsevets' I-153s participating in the

skirmish had safely regained their base, and only two of these had sported bullet tears in their fabric skinning as evidence that they had *been* in combat!

Demise of a theory

The initial success of the I-153 over the Khalkhin-Gol area was destined to be short-lived. On 6 July 1939, when Major Smirnov received orders to transfer the I-16 fighter mono-planes of his *eskadrilya* (squadron) to another commanded by a Major Zherdev, and proceed to the railhead at Chita, in the Chitinskaya Oblast, to take delivery of the first batch of 20 I-153s fresh from the *Zavod* 156 assembly line, the fighter biplane still possessed, in theory, a place in the in-ventory of the modern air arm. The I-153 was destined to provide Soviet protagonists of warplanes of this configura-tion with the disillusionment that sounded the death knell of the fighting biplane.

The good fortune enjoyed by the I-153 in its first skirmish with the highly respected Nakajima monoplane resulted in the greatest urgency being attached to the conversion of Sergei Gritsevets' entire fighter regiment to the new biplane, but the element of surprise employed so successfully during the I-153's first combat sortie could not be sustained for long. Despite the superlative handling qualities and manoeuv-rability of the I-153, the effectiveness of its quartette of ShKAS machine guns, and the élan with which it was flown by such pilots as Gritsevets*, Korobkov, Nikolayev, Pisanko, and Smolyakov, the *eskadrilii* flying the biplane were soon experiencing serious combat attrition.

As soon as the Japanese pilots appreciated the fact that they had to deal with a bird of somewhat different feather to that of the I-152 they changed their tactics accordingly. Some Soviet sources subsequently claimed that I-153 pilots adopted an artifice to regain a measure of surprise, approach-ing the combat area with undercarriage extended in the hope that their opponents would assume their aircraft to be lower-performance I-152s. Once the Japanese pilots had committed themselves, the pilots of the I-153s would then retract their undercarriages, pull their fighters round in a tight 180-deg turn, and attack head on. This ruse was claimed to have proved effective on more than one occasion, but it would seem improbable that the story has any foundation in fact, for, while the I-153 could undoubtedly have lowered its undercarriage after climbing to operational altitude, it is unlikely that any pilot would have possessed the dexterity to pull his fighter round in a high-speed turn and simul-taneously operate the handcrank to retract its wheels!

Whatever the truth of the matter, the V-VS was forced to admit tacitly that the I-153 could give a reasonable account of itself only if operated in concert with I-16 monoplanes. If unaccompanied, the biplanes suffered a disastrous mauling from the Nakajima fighters. In so far as the V-VS was concerned, the fighter biplane was no longer viable except under conditions of air supremacy.

The Soviet Air Forces were among the last major air arms to accept the demise of the biplane, and the I-153 was virtually the last such warplane to enter service†. It was not

Sergei Gritsevets had by this time been named Hero of the Soviet Union for the second time as a result of an exploit on 26 June. During a large battle between 50 Soviet and 60 Japanese aircraft over the Buyr Nur, the CO of the 70th Fighter Regiment, Major V M Zabaluev, had been forced to bale out some 20 miles (30 km) inside enemy territory. Gritsevets had promptly landed his I-16 and, with Zabaluev on his lap, had taken-off under a hail of fire from Japanese ground forces.

†The last fighter biplane to be introduced into service was, in fact, the Fiat C.R.42 Falco (Falcon) which, flown as a prototype some six months after the I-153, entered service with the Regia Aero-nautica *within a couple of weeks of the I-153 joining the V-VS, and equipped three* Stormi *(eighteen squadriglie or squadrons) at the outbreak of WW II.*

that the V-VS was unaware of contemporary fighter technology or lacked imagination. On the contrary, with the I-16 the V-VS had been the first of the world's air arms to adopt the fighter of low-wing cantilever monoplane configuration with retractable undercarriage. Thus, the service's apparent reluctance to relinquish the fighter biplane seemed something of an anomaly. The continued development of fighters of biplane configuration in the Soviet Union alongside fighters conforming to the *nouvelle vogue* stemmed directly from experience gained during the Civil War in Spain.

From a relatively early stage in the Spanish conflict, pilots flying the I-16 had found some difficulty in combating the extremely nimble Fiat C.R.32 as the pilots of the Italian fighter persisted in employing the classic dog-fighting tactics in which the I-16 did not excel. Strong factions in the *Narkomavprom* and *Narkomat Oborony,* the People's Commissariats for the Aviation Industry and for Defence, and within the V-VS itself, had been loath to give the fighter monoplane precedence over the tried and tested fighter biplane, and early experience in Spain led to vigorous renewal of their demand that development of fighters of biplane configuration be continued. This question of fighter development was made the subject of a special meeting between leading members of the Central Committee of the Party, senior staff of the V-VS, and representatives of the *Narkomavprom,* Iosip Stalin himself taking part in the discussion. The final decision concerning future lines to be followed in fighter development lay, of course, with the Politbureau of the Central Committee, and this decision was that work on fighter biplanes should be reinstated.

Nikolai N Polikarpov, still considered the doyen of Soviet fighter designers, was assigned the task of evolving a more potent fighter biplane and, in turn, allocated this to one of his principal team leaders, Aleksei Ya Shcherbakov. To accelerate the development of the new fighter, Shcherbakov was instructed to base his work on the I-15 and I-152 (alias I-15bis), and the basic structure of the earlier fighters was retained although completely restressed and extensively refined. The project was examined and received official approval on 11 October 1937, at which time it was allocated the designation I-153 (or I-15ter).

Shcherbakov elected to revert to the "gull" upper wing configuration of the original I-15 which had been discarded by the I-152, but retained the Clark YH wing profile adopted for the latter. The fabric-covered wings were of wooden construction, the upper wing carrying metal-framed ailerons, and areas were marginally reduced by comparison with the I-152, the upper wing having an area of 153·8 sq ft (14,29 m²) and the lower an area of 84·5 sq ft (7,85 m²), giving a total of 238·3 sq ft (22,14 m²). The fuselage was of KhMA chrome-

molybdenum steel alloy tube with light alloy formers of L-section, the aft section having fabric skinning to a point level with the cockpit windscreen and D1 dural skinning forward. All tail surfaces had D6 dural frames and fabric skinning, and bell cranks and rigid rods were used for aileron and elevator actuation, cables and pulleys being employed for the rudder.

Cables were used to retract the main undercarriage members, the process being effected by turning a handcrank, and each unit comprised a 27·5-in (700-mm) wheel and three legs forming an inverted tripod when extended, the wheels being raised vertically and turning to lie flat in wing-root wells. The semi-castoring tailwheel was fixed. The pilot was provided with a bucket-type seat adjustable for height only, and some aft protection was given by an 8-mm shaped armour plate attached to the bulkhead aft of the cockpit. Power was provided by an M-25V nine-cylinder radial air-cooled engine which drove an AV-1 two-bladed metal two-pitch airscrew, fuel being drawn from a 46 Imp gal (210 l) tank immediately aft of the firewall. The M-25V was a licence-built derivative of the Wright Cyclone SR-1820-F-3 which had been progressively improved by a team led by Arkadii D Shvetsov. This gave 775 hp for take-off at 2,200 rpm and a maximum output of 750 hp at 9,515 ft (2 900 m).

Armament comprised four 7,62-mm ShKAS machine guns mounted in vertically-staggered pairs in each side of the fuselage, the lower weapons firing through long blast tubes which passed through the engine to emerge between the adjustable cooling shutters of the cowling. The ShKAS (*Shpitalny-Komaritsky Aviatsionny Skorostrelny,* or Fast-firing Aircraft Gun) provided the I-153 with a substantial increase in weight of fire over that of the I-152 which was fitted with four PV-1 (*Pulemot Vozdushny,* or Machine Gun — Air) weapons of similar calibre. Weighing 32 lb (14,5 kg), the PV-1 had been evolved by A Nadashkevich from the Maxim gun, and had a rate of fire of only 780 rpm, whereas the ShKAS, developed by B G Shpitalny and I A Komaritsky, weighed only 22 lb (10 kg) and offered a fire rate of 1,800 rpm. With a muzzle velocity of 2,706 ft/sec (825 m/sec), the ShKAS was believed by the V-VS to be the best aircraft machine gun in the world.

Committed to production

Unlike the production models that were to follow, the I-153 prototypes were fitted with single-channel RSI-3 receivers. Promptly dubbed unofficially the *Chaika* (Gull), the prototypes completed their State Acceptance Tests in the autumn of 1938, and preparations were immediately initiated for series manufacture to commence in *Zavod* 156, the responsibility for preparing the new fighter for production being assigned to a young and then virtually unknown engineer,

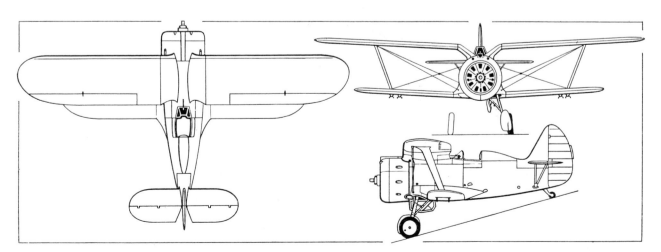

The Polikarpov I-153

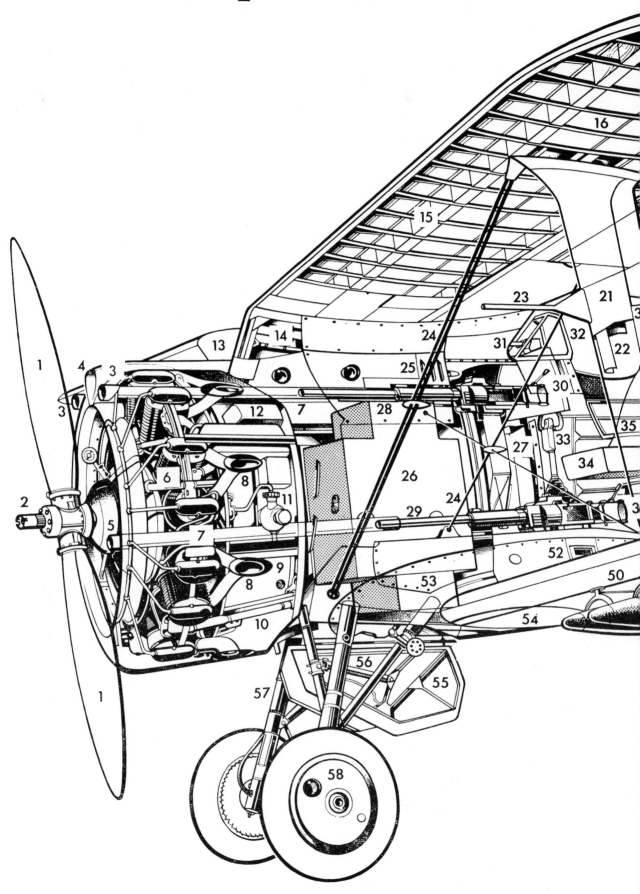

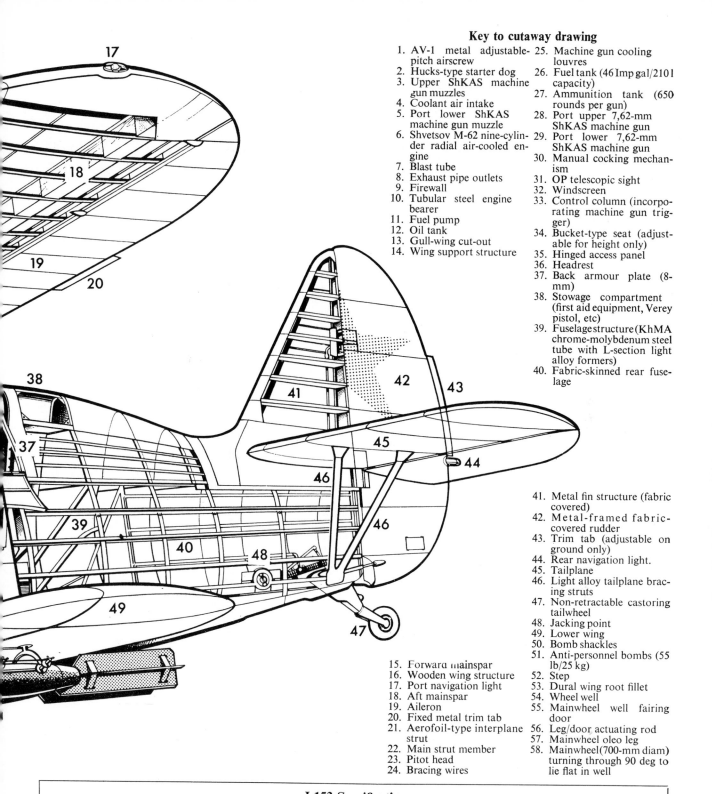

Key to cutaway drawing

1. AV-1 metal adjustable-pitch airscrew
2. Hucks-type starter dog
3. Upper ShKAS machine gun muzzles
4. Coolant air intake
5. Port lower ShKAS machine gun muzzle
6. Shvetsov M-62 nine-cylinder radial air-cooled engine
7. Blast tube
8. Exhaust pipe outlets
9. Firewall
10. Tubular steel engine bearer
11. Fuel pump
12. Oil tank
13. Gull-wing cut-out
14. Wing support structure
15. Forward mainspar
16. Wooden wing structure
17. Port navigation light
18. Aft mainspar
19. Aileron
20. Fixed metal trim tab
21. Aerofoil-type interplane strut
22. Main strut member
23. Pitot head
24. Bracing wires
25. Machine gun cooling louvres
26. Fuel tank (46 Imp gal/210 l capacity)
27. Ammunition tank (650 rounds per gun)
28. Port upper 7,62-mm ShKAS machine gun
29. Port lower 7,62-mm ShKAS machine gun
30. Manual cocking mechanism
31. OP telescopic sight
32. Windscreen
33. Control column (incorporating machine gun trigger)
34. Bucket-type seat (adjustable for height only)
35. Hinged access panel
36. Headrest
37. Back armour plate (8-mm)
38. Stowage compartment (first aid equipment, Verey pistol, etc)
39. Fuselage structure (KhMA chrome-molybdenum steel tube with L-section light alloy formers)
40. Fabric-skinned rear fuselage
41. Metal fin structure (fabric covered)
42. Metal-framed fabric-covered rudder
43. Trim tab (adjustable on ground only)
44. Rear navigation light.
45. Tailplane
46. Light alloy tailplane bracing struts
47. Non-retractable castoring tailwheel
48. Jacking point
49. Lower wing
50. Bomb shackles
51. Anti-personnel bombs (55 lb/25 kg)
52. Step
53. Dural wing root fillet
54. Wheel well
55. Mainwheel well fairing door
56. Leg/door actuating rod
57. Mainwheel oleo leg
58. Mainwheel(700-mm diam) turning through 90 deg to lie flat in well

I-153 Specification

Power Plant: One Shvetsov M-62 nine-cylinder radial air-cooled engine rated at 1,000 hp at 2,200 rpm for take-off, 850 hp at 2,100 rpm at 5,020 ft (1 530 m) and 800 hp at 2,100 rpm at 13,780 ft (4 200 m). Maximum continuous cruise, 765 hp at 2,030 rpm.

Armament: Four 7,62-mm ShKAS belt-fed machine guns in fuselage with 650 rounds per gun. Alternative external loads comprising six 82-mm RS-82 unguided air-to-air rocket missiles, or two 110-lb (50-kg) bombs, or four 55-lb (25-kg) bombs.

Weights: Empty equipped, 3,201 lb (1 452 kg); fuel, 331 lb (150 kg); oil, 42 lb (19 kg); maximum loaded (without external stores), 4,221 lb (1 960 kg); maximum overload (with two 22 Imp gal/100 l auxiliary tanks), 4,652 lb (2 110 kg).

Performance: (At 4,056 lb/1 885 kg) Maximum speed, 227 mph (366 km/h) at sea level, 280 mph (444 km/h) at 15,090 ft (4 600 m); normal cruise, 184 mph (297 km/h) at 6,560 ft (2 000 m); range at normal cruise, 292 mls (470 km); range with two 22 Imp gal (100 l) auxiliary tanks, 547 mls (880 km) at 168 mph (270 km/h); time to 3,280 ft (1 000 m), 0·85 min, to 9,840 ft (3 000 m), 3·0 min, to 16,400 ft (5 000 m), 5·3 min, to 22,965 ft (7 000 m), 8·3 min, to 29,530 ft (9 000 m), 13·2 min; maximum ceiling, 35,105 ft (10 700 m); landing speed, 68 mph (110 km/h); take-off distance, 116 yds (106 m).

Dimensions: Span, 32 ft 9½ in (10,00 m); length, 19 ft 3 in (6,17 m); height (on wheels with tail down), 9 ft 2¼ in (2,80 m); wing area, 238·31 sq ft (22,14 m²).

A standard production M-62-powered I-153 on skis. This aircraft was doped high-gloss forest green but the engine cowling remained natural metal and lacked the usual adjustable cooling shutters.

Artem I Mikoyan, Shortly afterwards, tooling was also begun for series manufacture of the I-153 in *Zavod* 1 at Vnukovo, and initial deliveries of the fighter to the V-VS began during the late spring of 1939.

The M-25V-powered I-153 had empty equipped and loaded weights of 2,972 lb (1 348 kg) and 4,098 lb (1 859 kg) respectively, wing loading being 17·2 lb/sq ft (84 kg/m²) and power loading being 5·467 lb/hp (2,48 kg/hp). Maximum speed was 258 mph (415 km/h) at 9,840 ft (3 000 m), and in normal loaded condition an altitude of 3,280 ft (1 000 m) could be attained in 1·1 min, 9,840 ft (3 000 m) in 3·3 min, 16,400 ft (5 000 m) in 5·8 min, and 22,965 ft (7 000 m) in 8·5 min. A full 360 deg turn could be accomplished in 11·4–12·4 seconds, maximum practical ceiling was 33,040 ft (10 700 m), and landing speed was 68 mph (110 km/h).

Soon after its arrival in the Manchukuoan-Mongolian border region, the armament of the I-153 was augmented by the application of 82-mm RS-82 air-to-air rocket missiles. The RS-82 had been developed at the RNII (Rocket Scientific Research Institute) by Ivan T Kleimenov and Georgi A Langemak, and had been first tested in combat on 20 August 1939 over the Buyr Nur when missiles of this type had been launched with some success from I-16 Type 10 fighters. Simple racks for up to six RS-82 missiles were attached to the lower wing, but hostilities between the Soviet Union and Japan terminated on 15 September before the I-153 could make any extensive operational use of this new weapon.

Although the shortcomings of the I-153 were by now clearly seen, two factories were fully tooled-up and were delivering fighter biplanes of this type in substantial quantities during the closing months of 1939. The development of more advanced fighters had the highest priority, but until these were ready to be placed in production, no alternative existed but to continue the I-153 manufacturing programme. However, some effort was expended on endeavouring to improve the I-153's capabilities, and the first change was the replacement of the M-25V by a more powerful derivative of this engine, the M-62.

By comparison with the M-25V, the cubic capacity and compression ratio remained unchanged, but a two-speed supercharger replaced the single-speed supercharger, take-off power being increased to 1,000 hp at 2,200 rpm, maximum rated outputs being 850 hp at 5,020 ft (1 530 m) and 800 hp at 13,780 ft (4 200 m). Loaded weight was raised by only some 104 lb (47 kg), and the additional power provided a useful increment in speed at altitude and climb rate, although sea level maximum speed remained virtually unchanged at 227 mph (366 km/h). The increase in fuel consumption had an adverse effect on range, and therefore wet points were introduced in the lower wing to permit two 22 Imp gal (100 l) auxiliary tanks to be carried.

The next change introduced was in the basic armament of the fighter, the 7,62-mm ShVAK machine guns giving place to 12,7-mm BS (*Berezina Skorostrelny*, or Berezina Fast-firing Gun) weapons. Developed by a team led by M Berezina, this larger weapon had a 1,000 rpm rate of fire and a muzzle velocity of 2,820 ft/sec (860 m/sec), and its bullets were considered as effective as 20-mm Oerlikon cannon shells, so good were its ballistic characteristics. Produced in smaller numbers was the I-153P (*Pushechny*, or Cannon-armed) which mounted two 20-mm synchronised ShVAK cannon in place of the normal quartette of machine guns, but the larger calibre weapon was in short supply, and few could be spared for the I-153 production line.

By August 1940, the first MiG-1s were following closely on the heels of the last I-153s on the *Zavod* 1 assembly line, and *Zavod* 156 was also in process of phasing out the little biplane, but when the last *Chaika* was rolled out during the closing weeks of 1940, no fewer than 3,437 fighter biplanes of this type had been manufactured, and despite its recognised obsolescence, the I-153 was numerically the second most important fighter in the V-VS inventory. A small number had been supplied to the Chinese government, and

An M-62-powered I-153 that had crash-landed at its base after combat during the opening phase of Operation "Barbarossa" (small-calibre bullet holes can be seen immediately aft of the engine cowling and in the lower wing/fuselage fillet) and was subsequently found by German forces when the airfield was overrun.

the *Chaika* thus once again found itself in combat with the Nakajima Army Type 97, enjoying even less success than had attended its operation over the Khalkin-Gol.

When German forces launched Operation *Barbarossa* on 22 June 1941, the I-153 still equipped even a number of the regiments of the IAP-VO (*Istrebitelnaya Aviatsiya Protivo-vozdushnoi Oborony,* or Fighter Aviation of the Anti-aircraft Defence), crack units responsible for the defence of primary targets, these including some of the 6th Fighter Aviation Corps, such as the 120 IAP commanded by Major Aleksandr S Pisanko at Alferovo, which was assigned the task of defending Moscow. As might have been expected, the *Luftwaffe* had a field day and the I-153-equipped *eskadrilii* were decimated.

Those I-153s that survived the initial attacks on their bases and succeeded in getting into the air stood little chance against the Messerschmitt Bf 109s that their pilots attempted to oppose. Flown by a highly-skilled and experienced pilot capable of taking full advantage of the I-153's superlative manoeuvrability, the *Chaika* could prove a troublesome opponent even to a first-class pilot flying a Bf 109F, but unfortunately for the I-153-equipped *eskadrilii,* few of their pilots *were* highly-skilled and experienced! Although the *Chaika* was to linger on in the first-line operational inventory of the V-VS until mid-1943, the swan-song of the fighter biplane was being sung by the Soviet "Gull" in the skies over Belorussia, the Ukraine and Lithuania during those fateful summer weeks of 1941.

Before, in 1941, the I-153 achieved the dubious distinction of proving once and for all that the fighter biplane no longer possessed a rôle in aerial warfare other than as prey for fighters of more modern concept, several experimental versions had been tested. Aleksei Shcherbakov had evolved a simple pressurised cockpit which was applied to an M-62-powered I-153 tested in 1939 as the I-153V. Also known as the I-153GK (*Germeticheskaya Kabina,* or Pressurised Cabin), this aircraft had a pressurised cockpit of the "cold wall" type comprising a welded light alloy shell and heavily-framed plexiglass windscreen and canopy, rubber packing pieces sealing the joints around the canopy and rotary movement of the main controls at their points of egress from the cockpit reducing the risk of leakage. Working on the air regeneration principle, the entire cockpit capsule weighed only 99 lb (45 kg).

A second I-153V, fitted with a pressurised cockpit of the so-called "soft" type developed by Nikolai Polikarpov, had an M-63 engine with a pair of TsIAM TK-3 turbo-superchargers. The nine-cylinder M-63 was essentially similar to the M-62 but had a higher compression ratio and operated at higher rpm, affording 1,100 hp at 2,300 rpm for take-off, 1,000 hp at 4,595 ft (1 400 m), and 930 hp at 14,765 ft (4 500 m). This version of the I-153V attained 275 mph (443 km/h) at 19,685 ft (6 000 m), but in all other respects performance was inferior to that of the standard M-62-powered model, and further development was therefore discontinued.

Another power plant fitted experimentally in an I-153 airframe during the course of 1939 was the 14-cylinder M-88V two-row radial developed by Sergei K Tumansky. This gave 1,100 hp for take-off at 2,300 rpm and possessed a maximum rating of 1,100 hp at 2,375 rpm at 13,120 ft (4 000 m). Redesignated I-190, the M-88V-powered model at 3,759 lb (1 705 kg) attained a speed of 304 mph (490 km/h) at 20,340 ft (6 200 m), and climbed to 16,400 ft (5 000 m) in five minutes. However, the only example built crashed early in its flight test programme, and proposals to install the 1,500 hp 18-cylinder M-90 two-row radial as the I-195 were abandoned.

The I-153 also participated in trials with ramjet engines which were considered as auxiliary power plants for short-

(Above and below) *The I-153 used to test Merkulov's DM-4 ramjets during August–October 1940.*

period boosting of fighter performance. These ramjets were developed by I Merkulov, and the initial version, the DM-2 (*Dopolnitelny Motor,* or Auxiliary Motor), had been successfully tested beneath the lower wing of an I-152 from 25 January 1940 until July of that year. In August 1940, improved ramjets of the DM-4 type were mounted beneath the lower wing of an I-153. The DM-4 was of 19·7-in (500-mm) diameter (as compared with the 15·75-in/400-mm of the DM-2) and ran on petrol with an ethyl-alcohol additive. Each DM-4, complete with mount, weighed 66 lb (30 kg). On 3 October 1940, Petr E Loginov achieved a speed of 267 mph (430 km/h) at an altitude of 6,560 ft (2 000 m) with the aid of two DM-4 ramjets, this representing a 26 mph (42 km/h) increase over the maximum speed attainable at that altitude without the auxiliary engines. This performance was improved upon on 27 October when 273 mph (440 km/h) was reached at the same altitude.

The I-153 marked the end of the era of the fighter biplane in the V-VS. The heyday of the biplane, with its struts and bracing wires, had passed while the *Chaika* was still on the drawing boards, and such concessions to modernity as a retractable undercarriage could not seriously affect the issue. Had the *Chaika* been introduced two years earlier it might have earned for itself a place among the truly great warbirds. Instead it was an anachronism appearing in service at a time when the biplane had long since been eclipsed by the monoplane. □   WG

The I-153V (also known as the I-153GK) employed in 1939 for the testing of Aleksei Shcherbakov's pressurised cockpit.

THIRTY SECONDS OVER SARGODHA

The short-lived Indo-Pakistan war of 1965 brought British and American aircraft into direct conflict. Of the many combats that ensued, none was more remarkable than that in which a PAF Sabre pilot destroyed four Indian Hunters in less than a minute. The pilot describes the action here in an exclusive interview with John Fricker

WING COMMANDER Mohammed Mahmood Alam of the Pakistan Air Force is a scrap of a man who appears almost lost in the none-too-roomy cockpit of a Sabre. Yet during the 1965 conflict with India, this Pakistani pilot established a combat record which is probably unequalled throughout the long history of air fighting.

Many pilots have scored several air victories in one sortie. A few have equalled or exceeded Alam's achievement of shooting down five enemy aircraft of superior performance within a few minutes. But nobody, in the annals of air combat, is believed to be able to match his record of destroying four opponents — Hawker Hunters of the Indian Air Force — within the space of somewhere around thirty seconds.

Unbelievable? On the face of it, yes. And yet this encounter was witnessed at close range by a number of Alam's fellow pilots. Nearest of these observers was his wingman, Flying Officer Masood Akhtar, who, protecting his leader's tail, clung like a leech throughout the action. Another section of PAF Sabres, led by Flt Lt Bhatti, was attempting to engage the Hunters but Alam (at that time a Squadron Leader) got there first. Flying top cover in an F-104A Starfighter was Sqdn Ldr M Arif Iqbal, who, with intense frustration, watched the brief combat admiringly. And, finally, there were the gun camera films to provide irrefutable confirmation of these eye-witness reports.

When, at the end of August 1965, the dispute between Pakistan and India over Kashmir threatened to explode into full-scale hostilities, the PAF, equipped with American aircraft supplied through the Mutual Aid Program, was confronted by an adversary operating greatly superior numbers of primarily British combat aircraft. The PAF order of battle comprised 92 F-86F Sabres in service, plus eight undergoing repair and overhaul at Maintenance Units; 12 Lockheed F-104 Starfighters, of which two were unarmed F-104B two-seat trainers; 25 Martin B-57s,

including several fitted out for reconnaissance tasks, and a dozen Lockheed T-33 jet trainers adapted for ground attack and tactical reconnaissance. These 141 combat aircraft were deployed at only four widely scattered airfields, and were supported by about 30 assorted trainers, transports and helicopters.

Corresponding IAF strength at the time was estimated to be at least 775 aircraft in the course of building up to a planned combat strength of some 45 squadrons. More than 500 of the IAF aircraft were first-line jet types, disposed in 27 fighter and three bomber squadrons, each with 16 aircraft on establishment. They included the first 10 Mach 2 MiG-21s, 118 British-supplied Hunter 56s, 80 Indian-built Gnats, 80 Dassault Mystères, 56 Dassault Ouragans, 132 obsolescent Vampires, 53 Canberra B(I)58 bombers, and seven Canberra PR57 reconnaissance aircraft.

Thus, in combat aircraft the PAF was outnumbered by nearly four to one, and if the IAF support element of 191 transport aircraft and 48 helicopters was included against the corresponding total for Pakistan, the ratio became closer to five to one. Even then the comparison was between the numbers of aircraft estimated to be in actual *service* with the IAF against the overall totals in the PAF inventory. For example, although India received its first six MiG-21s from the Soviet Union as early as February 1963, followed by a further six by the end of 1964 and another 18 in 1965, only 10 of the initial 12 were estimated by the PAF to be operational at Chandigarh by the end of August 1965. But this Mach 2 force exactly matched Pakistan's total of 10 F-104As at Sargodha.

Similarly, the PAF's 100 F-86F Sabres were balanced against the 118 Hunters operated by the IAF, although there were some doubts in Pakistan that the older North American fighter would be able to cope effectively with the more powerful and more heavily armed fighters of British origin. Most of the PAF Sabres had been modified to F-86F-40-NA

About six squadrons of Sabres still serve with the Pakistan Air Force, forming a significant part of its fighter force. Provision has been made for the Sabres each to carry two Sidewinder AAMs, and these weapons were credited with destroying at least eight Indian aircraft during the 1965 war.

(Left) Sqdn Ldr Mohammed Alam stands beside his Sabre which displays Indian flags representing 11 "kills" during the Indo-Pakistan war. (Right) A gun camera shot of one of Sqdn Ldr Alam's IAF Hunter victims going down in flames after the wing tanks had been ruptured by a burst from the Sabre's Browning guns.

standards, although there were still a few F-86F-35s in service. The main difference between the two versions was that the -35 Sabre had the so-called "hard" wing, without leading-edge slats, whereas the -40 had both slats and wingtip extensions for improved manoeuvrability and high-altitude performance. There were also minor differences in such details as the elevator actuator and rear fuselage sections. The uprating of the Sabres to the later standard was to prove a significant factor in future combat with the formidable Hunter.

Against this preponderance of Indian air power, the PAF could offer only a high standard of leadership throughout all its echelons, a determination springing from the realisation that nothing less than national survival was at stake, and a sound background of training and experience. From the technical point of view, the PAF had perhaps only one trump card to play against the IAF: the Sidewinder infra-red homing air-to-air missile. Less than 25 per cent of the Sabre force — in fact, only 22 aircraft — was equipped to carry Sidewinders, but in the heat of combat, the IAF would have to assume that every F-86F it encountered was so equipped.

One of the main effects of the Sidewinder in so far as the PAF was concerned was its ability to remedy in part the deficiencies in speed and climb which the Sabre suffered by comparison with the Hunter. With their superior performance, the IAF Hunters could normally expect to disengage from combat at will, but the two-mile (3,2-km) reach of the Sidewinder tended to eliminate the profitability of this manoeuvre. If the Hunters stayed they had to fight on the F-86Fs' terms and thus permit the latter to make use of their unrivalled low-speed manoeuvrability.

Into battle

This, then, was the situation on 7 September 1965 — the second day of the Indo-Pakistan war — when 32-year-old Mohammed Alam was commanding No 11 Squadron, flying Sabres from the main PAF operational base of Sargodha, in the Punjab. With virtually all of its fighter strength concentrated on this one first-line airfield, the PAF was fully alert to the dangers of IAF air attack — especially as it was the Pakistan pilots who had opened the airfield offensive at dusk on the previous day with strikes against the opposing Indian bases.

Night brought the B-57s and Canberras in harassing attacks against each other's airfields. From its own successes on the previous day, however, the PAF was under no doubt that the main danger lay in fighter strikes in daylight. This was why the half-light before dawn on the morning of 7 September found Sqdn Ldr Alam and some of his pilots already strapped at readiness in the cockpits of their Sabres,

awaiting the signal to scramble immediately warning of approaching IAF aircraft was received through the Pakistani radar and observer network.

With low-flying jets, any warning at all is a luxury to a front-line airfield. Although Sargodha already had its initial protective fighter patrols airborne, the first intimation of the arrival of the IAF was the sight of six Mystère IVA fighter-bombers pulling out of the sunrise from their tree-top approach to deliver their attack. Their navigation was good, and the great airbase — packed with aircraft, fuel bowsers and other rewarding targets — lay sprawled beneath them. While many of the PAF fighters were spread around in camouflaged dispersals, others were lined up, through military necessity, on the Operational Readiness Platform (ORP) at the end of the runway. Sargodha was well defended with light ack-ack, but it had only three fighters airborne and the Mystères, enjoying the advantage of virtually complete surprise, had the Pakistani base at their mercy.

The PAF Sabre and Starfighter pilots, strapped helplessly in their aircraft on the ground, watched with disbelief as the Mystères pulled up to about 1,000 ft (305 m), still maintaining tight and unwieldy echelon formation, and sprayed at random with their cannon and rockets the empty tarmac areas during a single high-speed pass. First they fired their underwing rockets, although several simply jettisoned the pods* with the rockets still in them, and then they opened up with their twin 30-mm cannon, but apparently without taking aim, and made off at high speed towards the south-west, leaving Sargodha unscathed. The Indian aircraft themselves were less fortunate however. One was shot down by the anti-aircraft defences during its high-speed pass, while a further two fell to the 20-mm Vulcan cannon of a patrolling Starfighter.

Immediately after this initial IAF daylight attack, Sqdn Ldr Alam and his No 2, Flg Off Masood, were scrambled for an airfield patrol at about 15,000 ft (4 570 m). Within five minutes they were directed by ground control to intercept an incoming Indian raid, but after flying eastwards for 10–15 miles (16–24 km) they were told to return as IAF fighters were over Sargodha.

"As I flew back," recalls Wing Cdr Alam, "I picked up four Hunters diving to attack our airfield. So I jettisoned my drop tanks to dive after them through our own ack-ack. Almost simultaneously I saw two more Hunters about 1,000 ft (305m) to my rear, so forgot the four ahead and pulled up to go after the pair behind. The Hunters broke off

---

*Some of these battered pods are still on display at the PAF Museum at Peshawar.

their attempted attack on Sargodha, and the rear pair turned into me. I was flying much faster than they were at this stage — I must have been doing about 500 knots (925 km/h) — so I pulled up to avoid overshooting them and then reversed to close in as they flew back towards India."

For all his apparently frail physique, Wing Cdr Alam has the incisive speech and mannerisms of a born leader: "I took the last man and dived behind him, getting very low in the process. The Hunter can out-run the Sabre — it's only about 50 knots (92 km/h) faster, but it has much better acceleration so can pull away rapidly. Since I was diving, I was going still faster, and as he was out of gun range, I fired the first of my two GAR–8 Sidewinder missiles at him." The GAR–8 homes on the heat radiations from the target aircraft's jet engine but is sometimes affected by ground returns if fired in a dive at low altitudes. "In this case, we were too low and I saw the missile hit the ground short of its target.

"This area east of Sargodha has lots of high-tension wires, some of them as high as 100–150 ft (30–45 m), and when I saw the two Hunters pull up to avoid one of these cables I fired my second Sidewinder. The missile streaked ahead of me but I didn't see it strike. The next thing I remember was overshooting one of the Hunters, and when I looked back its cockpit canopy was missing and there was no pilot in the aircraft. He had obviously pulled up and ejected, and just then I spotted him coming down by parachute. This pilot, Sqdn Ldr Onkar Nath Kakar, commander of one of the Indian Air Force Hunter squadrons, was later taken prisoner.

"I had lost sight of the other five Hunters, but I pressed on thinking that maybe they would slow down." Alam was under the impression, of course, that there were still only two Sabres pitted against the remaining five IAF aircraft. "I had lots of fuel so I was prepared to fly 50–60 miles (80–95 km) to catch up with them. We had just crossed the Chenab river when my wing man called out contact. I picked them up at the same time — five Hunters in absolutely immaculate battle formation. They were flying at about 100–200 ft (30–60 m), at around 480 knots (890 km/h), and just as I reached gunfire range they saw me. They all broke in one direction, climbing and turning steeply to the left, which put them in close line astern. This, of course, was their big mistake. If you are bounced, which means a close range approach by an enemy fighter within less than about 3,000 ft (915 m), the drill is to call a break. This is a panic manoeuvre to the limits of the aircraft's performance, splitting the formation, getting you out of the way of an attack and freeing you to position yourself behind your opponent. However, in the absence of one of the IAF sections initiating a break in the other direction to sandwich our attack, they all simply stayed in front of us.

"It all happened very fast. We were all turning very tightly — pulling in excess of 5g, or just about on the limits of the Sabre's very accurate A–4 radar ranging gunsight. I think that before we had completed more than about 270 deg of the turn, at around 12 deg per second, all four Hunters had been shot down! In each case, I got the pipper of my sight around the canopy of the Hunter for virtually a full deflection shot. Almost all our shooting throughout the war was at very high angles off — seldom less than about 30 deg. Unlike some of the Korean combat films I had seen, nobody in our war was shot down flying straight and level."

Accurate shooting is difficult enough at the best of times from a jet fighter travelling not far below the speed of sound, and the difficulties of flying with such precision as to be able to shoot down four opponents in a few seconds are immensely increased in a turn of such crushing force that the pilot's weight is quintupled. Throughout the short 1965 war, in which Alam destroyed nine enemy aircraft in only three encounters, he never had to fire more than twice at an opponent.

"I developed a technique of firing very short bursts — around a half-second or less. The first burst was almost a sighter, but with a fairly large bullet pattern from six machine-guns, it almost invariably punctured the fuel tanks so that they streamed kerosene. During the battle on 7 September, as we went round in the turn I could just see in the light of the rising sun the plumes of fuel gushing from the tanks after my hits. Another half-second burst was then sufficient to set fire to the fuel, and as the Hunter became a ball of flame, I shifted my aim to the next aircraft. The Sabre carries about 1,600 rounds of ammunition for its six 0·5-in (12,7-mm) guns, or sufficient for some 15 seconds continuous firing. In air combat this can be a lifetime. Every fourth or fifth round is an armour-piercing bullet, and the rest are HEI — high-explosive incendiary. I'm certain after this combat that I brought back more than half of my ammunition, although we didn't have any time to waste counting rounds.

"My fifth victim of this sortie started spewing smoke and then rolled on to its back at about 1,000 ft (305 m). I thought he was going to do a barrel roll, which, at low altitude, is a very dangerous manoeuvre for the pursuer if the man in

A Sidewinder-armed "slick-winged" F-86F-40-NA of the Pakistan Air Force.

front knows what he is doing. I went almost on my back and then realised that I might not be able to keep with him, so I took off bank and pushed the nose down. The next time I fired was at very close range — about 600 ft (180 m) or so — and his aircraft literally blew up in front of me. None of these four pilots ejected, and all were killed."

How could a formation of jet fighters flown by senior and experienced pilots (three were Squadron Leaders and three Flight Lieutenants) allow itself to be shot down by numerically fewer aircraft of inferior performance? "Hunter pilots will not believe it," comments Wg Cdr Alam, "and I have flown Hunters myself in England, and I know it to be a very manoeuvrable aircraft, but I think the F-86 is better!

"Actually, the Sabre has a fantastic turning performance. Although the normal stalling speed with flap is around 92 knots (170 km/h) you can fly it round in a steep turn down to as little as 80 knots (148 km/h) or less in a decending scissors manoeuvre. Between 100 and 120 knots (185–222 km/h) is quite a normal speed range to rack the Sabre round in combat. If you applied aileron at that speed in a Hunter you would flick out the other way. The F-86F is almost faultless on the controls, but it's a pity it doesn't have a bit more engine thrust. Provided that you see your opponent in time, you can never be out-turned or out-fought in a Sabre.

"I think, too, that the 0·5-in (12,7-mm) Browning machine gun is the best possible weapon for use against fighters in close combat. But if the Hunters hadn't all broken in the same direction I could have been in trouble. As it was they left themselves no initiative. The sixth and last Hunter disappeared from view, but we later heard that the pilot had ejected — because, he said, of engine trouble.

"All six Hunters in the Indian formation were therefore destroyed. This must have been a very serious blow to IAF morale."

Sabre versus Hunter

Not everyone shares Wg Cdr Alam's views concerning the relative merits of the Sabre versus the Hunter. Even in the PAF, where many pilots have flown Hunters with the RAF, opinions are divided, but there is a strong element in favour of the British aircraft, which after all, flew for the first time four years later than the Sabre. Wg Cdr Alam is prepared to defend his beloved Sabre with supporting technicalities, which he quotes with authority. "In a turn," he says, "the Hunter slows down more quickly than the F-86 for the same application of g. For one thing, it has a much higher aspect ratio, and induced drag increases in inverse propor-

Camouflage netting was used by the PAF to hide its aircraft.

THE EYE-WITNESS: Flt Lt Imtiaz Ahmed Bhatti, who had shot down two IAF Vampires on the first day of the 1965 war with India, was among the PAF pilots who observed Sqdn Ldr Alam's brief combat with the half-dozen enemy Hunters and on which he made the following report:

"On 7 September, four F-86s and an F-104 were scrambled from Sargodha after the first Mystère raid. We were sent here and there without contact until we had a call on the guard channel from our lookout on top of the Air Traffic Control building. He told us that six Hunters had pulled up from the end of runway 14 and were strafing the airfield. We immediately headed back to Sargodha, and my No 2 and I arrived at about 7,000 ft (2 134 m), with Sqdn Ldr Alam and his wingman slightly below us and to our left. I saw one of the Hunters, but when I tried to jettison my wing tanks, only one released. Nevertheless, I got my sights on the Hunter and was just about to open fire when the nose of an F-86 appeared from below. I broke to one side and saw two F-86s chasing the Hunter, so I looked for the other enemy aircraft. I had no visual contact, but as I knew we were outnumbered, I stayed behind Sqdn Ldr Alam with my wingman.

"The next thing I saw was a Hunter pulling up and exploding after Sqdn Ldr Alam had released his GAR-8. We continued the chase, and my No 2 called, 'Bogies at 3 o'clock'. I turned towards that direction without making contact, but I saw Alam break to the right and then shoot down two Hunters in quick succession. My No 2 then called 'Reverse left', and as I did so, I saw Alam shoot down the other two Hunters. The game was over, so we came home. It all happened very quickly."

tion to the square of the velocity. In other words, the lower the speed, the higher the induced drag."

He readily acknowledges the far better thrust/weight ratio of the Hunter, which has 10,000 lb (4 536 kg) of thrust from its Rolls-Royce Avon 203 engine for a clean aircraft weight of 19,000 lb (8 618 kg), compared with 6,000 lb st (2 721 kgp) output from the General Electric J47 turbojet of the 15,000 lb (6 804 kg) F-86F Sabre. In his opinion, however, this is more than offset in combat by the higher induced drag of the Hunter. "This means," adds the Wing Commander, "that the Hunter loses speed faster than the Sabre in a turn because of its bigger drag rise which the extra thrust cannot counter. So in the turn I steadily closed up on the Hunters which rapidly decelerated from about 450 knots (834 km/h) down to around 240 knots (448 km/h), and would have had to have pulled about 7g to get away from me. As it was they just slid back into my sights, one by one."

In the final reckoning, could it be the man in the cockpit that counts more than the aircraft? "Certainly," concedes Alam, "flying skill matters a great deal. When I went to war I had about fourteen hundred hours on the Sabre, which is a lot of experience and, with another pilot, I held the record for the highest gunnery scores in the PAF with an average of around 70 per cent. PAF air combat standards are as high as any in the world. Many of our pilots go on exchange postings to England and elsewhere, and come back proudly claiming that in air combat practice they have matched their skill with the best. I don't know whether by nature we Pakistanis are more aggressive, but we lay a lot of stress on aerobatics, gunnery, and air combat itself."

By the end of the 22-day war with India in 1965, the PAF was claiming local air superiority throughout the battle area, despite being greatly outnumbered, as well as a combat loss ratio of around five to one in its favour. Estimates of IAF aircraft losses as a result of air combat totalled 35, according to PAF intelligence sources, compared with only seven F-86s. A further dozen PAF aircraft were lost in accidents and from other causes during the period. □

IN PRINT

THE FLOW of books and publications of interest to the aviation enthusiast appears to be in inverse proportion to the contraction of the aircraft industries in Britain and elsewhere. Certainly, in the last year or two, there has been no reduction in the steady flow of new titles or re-issues to match the decline in aircraft production. The great majority of aviation books can be classed of general interest to the enthusiast, and as many of these as possible will be recorded in the monthly "In Print" column, indicating briefly the scope of the contents. Extended reviews will appear from time to time of selected titles considered to be of special importance.

Books published by British companies can normally be obtained through local booksellers or ordered direct from the publisher. In the case of books published abroad, these can often be obtained through one or other of the recognised aviation booksellers and dealers in Britain. The AIR ENTHUSIAST cannot supply copies of the books noted in this column.

This illustration, which appears in "Bristol Aircraft Since 1910", a second edition of which has recently been published, shows one of the least-known of all British aeroplanes of the inter-war periods, despite the fact that it was flown regularly at Filton from 1925 to 1928. It is the Bristol Type 92, also known as the Laboratory biplane, built and flown under Air Ministry contract to investigate engine cooling and drag problems. The photograph, one of only two known to exist, shows the Type 92 before installation of the Jupiter VI engine.

"Sopwith — the Man and his Aircraft"
by Bruce Robertson
Air Review Ltd, Letchworth, England, £4.00
244 pp, 11 in by 8½ in, illustrated
THIS is another volume in the familiar Harleyford series, despite the change of publisher's name. The contents fall into three main parts: a narrative account of the founding of the Sopwith company by T O M (now Sir Thomas) Sopwith and its activities until it went into voluntary liquidation in 1920; details of the Sopwith designs, including serial numbers and specifications, whether built by the parent company or under licence elsewhere, and three-view line drawings to 1/72nd scale of 29 Sopwith designs.

"Off the Beam"
by Robert Chandler
David Rendel, London, £1.80
184 pp, 5½ in by 8½ in
HERE IS a radio operator's view of his flying life — interesting if only because few books have been written by men of this calling. Of particular value historically are his accounts of the operations of the Atlantic Ferry Organisation set up by A V M Don Bennett during the war, and the subsequent birth of British South American Airways by the Bennett "team".

"Warplanes of the Third Reich"
by William Green
Macdonald & Co (Publishers) Ltd, London, £10.00
672 pp, 8½ in by 11 in, illustrated
FEW BOOKS can be more truly described as "definitive" in their particular subject than this one. The period covered is that of the Third Reich in Germany, from 1933 to 1945; the term "warplanes" embraces those types specifically designed to wage war and those used to support the *Luftwaffe*'s combat operations, such as troop transports and troop-carrying gliders. Each aircraft which qualifies for inclusion under these terms is the subject of a detailed account tracing its development and operational deployment. The many hundreds of photographs depict virtually every variant of every type, and these illustrations are supplemented by line three-views plus extra side views (by Dennis Punnett). The more significant aircraft are also shown in the form of cutaway drawings and three-view tone drawings which show typical service markings. Eight pages of full-colour drawings (by John Weal) show the major wartime camouflage schemes and typical unit emblems. A complete reference library to the military aircraft of Germany's Third Reich in a single volume.

"Jane's All the World's Aircraft 1945–6"
edited by Leonard Bridgman
David & Charles, Devon, £9.45
716 pp, 12¼ in by 7¾ in, illustrated
A GOOD QUALITY litho reprint of one of the more significant volumes in the Jane's series, including much material which became available with the end of World War II. Subsequent research, reflected in other publications, reveals all too clearly the gaps in this coverage, but many enthusiasts will welcome the chance to add this title to their collection.

"British Racing and Record-Breaking Aircraft"
by Peter Lewis
Putnam & Co Ltd, London, £6.30
496 pp, 5½ in by 9 in, illustrated
THE TITLE is a little misleading, for this volume is concerned more with the events and personalities than with the aircraft involved, although there is no shortage of illustrations of the latter. The author describes each event from the balloon races at the start of the present century to the England-Australia race at the end of 1969, as well as the major record-breaking flights by individuals. Lists are included of the aircraft, registration and pilot of all entrants in the major events. Within the space available, the text is necessarily a somewhat tiring recital of facts: names, dates, times, places, registration markings, all are here, but not much of the excitement of the times comes through. A useful reference volume, nevertheless.

"Bristol Aircraft Since 1910"
by C H Barnes
Putnam & Co Ltd, London, £4.20
416 pp, 5½ in by 9 in, illustrated
SECOND EDITION of one of the well-known Putnam company histories, with some new material at the end of the book to round off the story up to the time that the Bristol design office ceased to function as an independent unit. Of special interest are the brief details and an illustration of the Bristol 204, submitted to the same specification as the TSR-2.

"Pictorial History of the RAF"
Volume Three, 1945–69
by John W R Taylor and Philip J R Moyes
Ian Allan, Shepperton, £2.00
208 pp, 6 in by 9 in, illustrated
COMPLETING the trilogy of RAF pictorial histories, this volume covers the post-war ups and downs of the Service. As in the other two volumes, John Taylor's text covers the subject thoroughly but concisely. For the most part, the record is factual, but the author has made no attempt to hide his feelings about the political interference which has so bedevilled RAF planning in recent years. Detailed captions add to the historical value of the excellent selection of illustrations. □

Dassault Flies a Kite...

To "fly a kite" is, in the parlance of financial circles, to raise money on credit, but when we say that Avions Marcel Dassault is flying a kite we hasten to assure the reader that we use this idiom in *quite* a different connotation: that of "seeing how the wind blows" in the world export market for a new version of the increasingly-ubiquitous Mirage, which, appropriately enough, if for an entirely different reason, has been named "Kite", or rather its French equivalent of "Milan".

The ornithologist knows the kite as a bird of prey of the family *Falconidae*, which, while capable of high-speed flight, can virtually hover under full control. The Mirage is certainly a true predator among warplanes — which the aggressive Dassault sales force is ensuring remains in constant demand — and the Milan retains all the proven qualities of the Mirage. If it does not exactly add to these the ability to hover, it at least complements them with very acceptable low-speed controllability. Hence the appellation of Milan.

Aircraft design exercises inevitably call for compromise since no single configuration yet conceived is the optimum for the entire performance envelope of any aeroplane category. The wider the performance envelope demanded, the greater the compromise dictated. From no aspect is this compromise more readily apparent than in the high-speed fixed-geometry warplane of today; the higher the speed desired, the more difficult the design team's task of restricting take-off and landing speeds to figures compatible with the field lengths within which the aircraft must operate. Running out of runway has a deleterious effect on a pilot; it has a disastrous effect on his aircraft!

The delta wing planform offers an excellent case in point.

A "plain" delta, such as that used for the wing of the Mirage or MiG-21 to obtain a Mach 2 performance, possesses major shortcomings from the aspect of low-speed handling. High angles of incidence must be assumed to maintain lift at low speeds (hence, for example, the need to droop the nose of the Concorde and Tu-144 during take-off and landing), and since there is no horizontal tail surface, elevons on the wing trailing edge must be selected "up" to keep up the nose. In other words, elevons work in the opposite sense to conventional trailing-edge flaps which would normally be "down" to increase lift during take-off and landing. Denied this increment of lift, the aircraft with a delta wing tends to burn up rather more runway than its tailplane-equipped equivalent.

Examples of the compromises that this situation produces are afforded by the Saab Viggen and North American's XB-70A Valkyrie whose basic delta configuration is augmented by nose trimming surfaces, but a more interesting solution on the same theme is that now being offered by Dassault with its Milan, which, aerodynamically, is a Mirage to which have been added retractable foreplanes. The effect of foreplanes in all these cases is similar, but Dassault's success in evolving a satisfactorily moustached Mirage is no mean achievement when it is considered that the characteristics of its delta mainplane were already established and unalterable other than by a major redesign, and the effect of the foreplanes on pilot visibility and intake efficiency had to be watched with the utmost care. In each case, the foreplanes provide means of raising the nose of the aircraft without applying up-elevon, and, in fact, some down-elevon is needed to counteract the nose-up pitch produced by

forward lifting surfaces, the elevons thus providing some conventional flap effect.

Although the principle is clear, its application demands great care in the design of the foreplanes, which have to maintain a constant lift throughout the angle-of-attack range of the aircraft. Unless this continuity of lift is achieved, the pilot is forced to make elevon adjustments of variable magnitude throughout the range of aircraft incidence in order to counteract the varying effect of the foreplanes.

Swiss contribution

The Swiss requirement for a ground attack aeroplane to replace the Venom was a major factor in development of the Milan. The Swiss Air Force already operated the Mirage IIIS as an interceptor and the Mirage 5 close-support aircraft was of obvious possible interest in the Venom-replacement category. However, an improvement in low-speed control would, equally obviously, render Dassault's contender more competitive with the other types selected for Swiss evaluation, as well as enhancing sales prospects in other parts of the world.

Various methods of achieving a nose-up moment without using the elevons were studied jointly by Dassault and the Swiss Federal Aircraft Factory (*Fabrique Fédérale d'Avions*) in the latter half of 1967, and tested in the wind tunnel at Emmen in the FFA facility. Possibilities of modifying the wing leading edge profile or using a leading-edge flap were eventually discarded, and, by February 1968, the canard

layout had been selected as that offering a most satisfactory solution. While wind-tunnel testing continued, a mock-up was built to check the effect on pilot visibility, and a Mirage III was fitted with fixed canard surfaces for flight development. A series of 11 flights was made with this initial canard configuration during September and October 1968, including two by a Swiss test pilot.

Meanwhile, further wind-tunnel tests and the flight-trial results allowed Dassault to achieve an optimised design for the foreplanes — actually, Stage 7 in the design studies. Optimisation was concerned with achieving the constant nose-up pitch at various angles of attack. The final choice was derived from the St Cyr 156 aerofoil section, modified to have double the usual camber, and incorporating a fixed slot; a detached leading-edge section was added ahead of this aerofoil, forming with it another slot. Although these canard surfaces — soon dubbed *moustaches* — had only 1·7 per cent of the area of the wing, they were thus highly sophisticated in design. Flight testing of the Stage 7 surfaces continued throughout 1969, and included further tests by Swiss Air Force pilots. During the course of these tests the fixed angle of incidence relative to the aircraft datum was increased from 15 deg to 19 deg, the dihedral angle being 15 deg.

Having proved that the *moustaches* achieved their primary purposes, Dassault then designed a system whereby they could be retracted, since they contributed nothing to the performance of the aeroplane at high speed. For initial trials, a set of retractable foreplanes was built into the nose of a Mirage IIIR (the longer nose of the reconnaissance version providing adequate stowage space for the surfaces when retracted), and this aircraft flew for the first time on 24 May

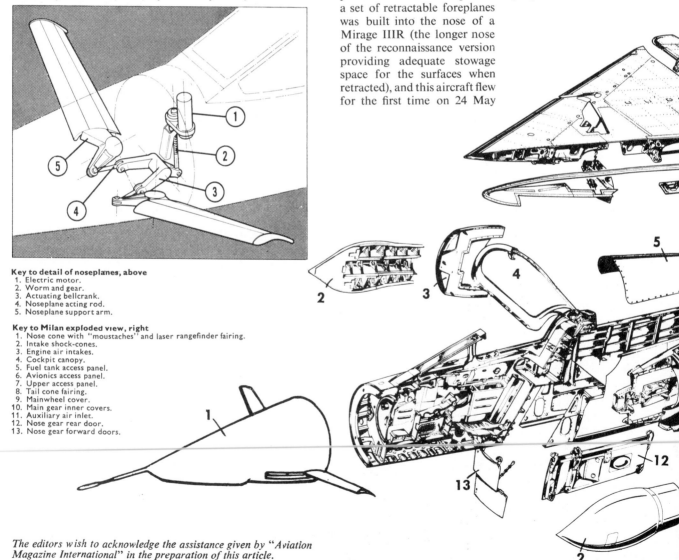

Key to detail of noseplanes, above
1. Electric motor.
2. Worm and gear.
3. Actuating bellcrank.
4. Noseplane acting rod.
5. Noseplane support arm.

Key to Milan exploded view, right
1. Nose cone with "moustaches" and laser rangefinder fairing.
2. Intake shock-cones.
3. Engine air intakes.
4. Cockpit canopy.
5. Fuel tank access panel.
6. Avionics access panel.
7. Upper access panel.
8. Tail cone fairing.
9. Mainwheel cover.
10. Main gear inner covers.
11. Auxiliary air inlet.
12. Nose gear rear door.
13. Nose gear forward doors.

The editors wish to acknowledge the assistance given by "Aviation Magazine International" in the preparation of this article.

1969, appearing in public two weeks later at the Paris Air Show. An electrically-actuated retraction system was employed, the surfaces being hinged to retract forwards through slots in the aircraft nose, full extension taking 6–7 seconds, and the total assembly weighing about 110 lb (50 kg).

The definitive Milan

Offered for delivery in 1972, the definitive Milan has certain other new features compared with the Mirage III, apart from the retractable foreplanes. These features include an up-rated Atar engine with some associated changes in the air intakes and fuel system, and the introduction of a new nav/attack system. All these features are incorporated in the Milan S-01, which is basically a Mirage IIIE airframe (on loan from the *Armée de l'Air*) configured to represent the proposed production Milan in general and the aircraft offered to Switzerland in particular, as indicated by the suffix letter "S" (for *Suisse*) in the designation. This aircraft made its first flight on 29 May 1970, and achieved Mach 2 on its seventh flight. Evaluation by Swiss pilots began on 16 July, with the 32nd flight.

To power the Milan, Dassault adopted the SNECMA Atar 9K-50, an engine with dry and reheat ratings of 11,023 lb (5 000 kg) and 15,875 lb (7 200 kg) respectively, and which is also used in the Mirage F1 and Mirage G8. The size of the air intakes is marginally increased to cater for the greater air mass flow of this engine, and electrical starting is employed, the aircraft batteries providing built-in starting capability. Drawing upon experience with the Mirage F1, Dassault has improved the fairing around the variable-area reheat nozzle where it joins the tail parachute housing.

The Milan S-01 photographed during the recent weapon trials at Cazaux. It is carrying two Sidewinder AAMs and five 990-lb (450-kg) bombs, four of the latter being suspended from two 110 Imp gal (500 l) external wing tanks.

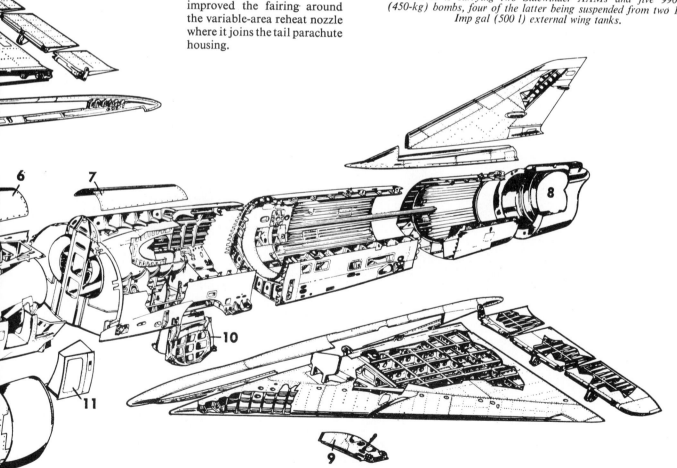

The Swiss Evaluation

The Swiss Air Force (*Schweizerische Flugwaffe*) has a requirement to replace, by 1976, the de Havilland Venom Mk 50s which currently equip 13 squadrons (*fliegerstaffeln*) in the ground attack role. The search for a suitable replacement began in 1966 when the Defence Ministry set a limit of SFr 1,300m (equivalent to about £108m at that time) on the cost of the programme.

In an effort to avoid a repetition of the circumstances surrounding the selection in 1960 of the Dassault Mirage III which now equips three *Flugwaffe* squadrons, the Defence Ministry directed that computer analyses should provide the basis for the new evaluation. The Mirage procurement had led to accusations of "gross negligence" and "incredible frivolity" being made against various of the individuals concerned with that contract; the commander of the *Flugwaffe* was dismissed and the Chief of the General Staff resigned, while cost escalation reduced the number of Mirages acquired from a planned 100 to a total of 57 aircraft.

Types included in the initial computer analysis comprised the Dassault Mirage 5, the BAC/Breguet Jaguar, the FFA P-16, the Saab Draken and Viggen, the Fiat G-91Y, the Northrop F-5, the Vought A-7 and the McDonnell Douglas A-4. Final detailed evaluation was between the A-7 and the Fiat G-91Y, and early in 1970 the Vought A-7G with Rolls-Royce Spey engine was finally announced as the primary technical choice. However, the funds earmarked for the programme would buy only about 40 A-7Gs equipped to the standard specified by the *Flugwaffe*, which had said it required a minimum of 60 aircraft. The Minister of Defence refused to sanction an increase in funds, and, on 15 July 1970, ordered a new study.

Listed officially as alternatives at this stage were the Fiat G-91YS, the Douglas A-4M and the Dassault Milan, the last having meanwhile been evolved to Swiss requirements in place of the Mirage 5 offered originally. Naturally, other manufacturers also took the opportunity of the reopened investigation to make new proposals. BAC tried again with the Jaguar; Saab proposed a version of its Saab 105; Hawker Siddeley offered a series of alternatives embracing refurbished or Super Hunters and/or Harriers, and Vought Aeronautics proposed various ways of rendering the A-7 less expensive.

All these aeroplanes have been available for flight evaluation during 1970–1. The Saab 105XT prototype, representing the proposed Saab 105XH, was inspected and flown by an 18-man team from Switzerland in August 1970, and Saab subsequently modified the 105XT to have the new wing leading edge intended to improve the manoeuvrability of the 105XH in tactical operations. Fiat completed a G.91YS prototype, which made its first flight on 16 October 1970, this aircraft differing primarily from the Italian Air Force version in having a Saab BT9R bombing computer. Both this type and the Milan S-01 have been flown by Swiss evaluation teams.

Meanwhile, the purchase of 30 HS Hunters by the Swiss Air Force was confirmed early in 1971. These aircraft, refurbished by Hawker Siddeley, are for delivery between 1972 and 1974, and will supplement the existing force of Hunters. Delivery of the Venom-replacement type is to begin in 1975, and a decision is expected to have been taken by the time these words appear in print, although definitive contracts will not be placed until mid-1972. The Vought A-7, in a less expensive form than that selected last year, is reported to be still the first choice, with the Fiat G-91YS second and the Milan third. However, the results of the recent Milan bombing trials and the fact that Switzerland has previously produced the Mirage may influence the final decision.

Modifications of the fuel system include the addition of two fuel tanks in the fuselage with a capacity of 154 Imp gal (700 l), introduction of a new leading-edge fuel tank, deletion of the provision for fuel tanks in place of the cannon — two 30-mm DEFA guns with 125 rpg are fitted for all missions — and deletion of the fuel system associated with the rocket engine beneath the rear fuselage. This rocket unit is a feature of some Mirage IIIs but is not used by Switzerland's Mirage IIIS version which, instead, can use an eight-bottle JATO unit. The Milan has the same provision.

The Milan S is fitted with a nav/attack system based upon work done for the Jaguar, although in its completeness the Milan system is closer to that designed for the British Jaguar S than that of the less comprehensively equipped French Jaguar A. Among the new items in the Milan are a projected map display (Thomson-CSF LC 102), a head-up display (Thomson-CSF Type 121RS) and a laser rangefinder (Thomson-CSF T-102 or CGE TAV 34). The navigation computer is a Crouzet 91S and the bombing computer is a Thomson-CSF 31DS; other equipment includes EMD-Decca RDN 72S Doppler, SFIM 251 twin gyro platform with SFIM CG111 standby heading system and Jaeger 21

Continued on page 55

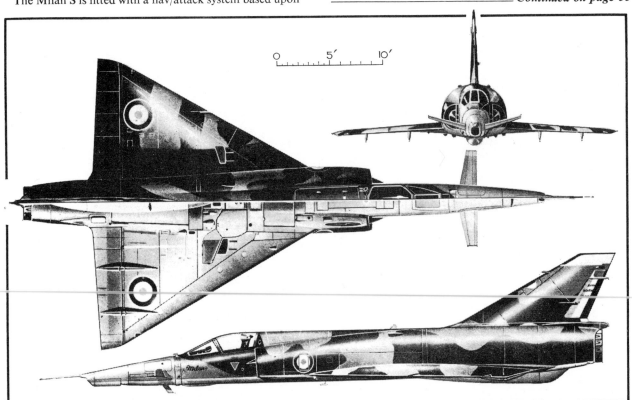

OIL WELL TOP COVER

SIXTY YEARS OF RUMANIAN MILITARY AVIATION

THE YEAR 1971 sees the sixtieth anniversary of the creation of military aviation in Rumania; three score years in which the switchback of growth and decline in the effectiveness of the Rumanian air arm has reflected the fluctuating fortunes of this southeast European country through two world wars, with their numerous border changes, and internal turmoil accompanying a succession of regimes.

Sixty years ago saw the formation in the Rumanian Army of an Aviation Group, or *Grupul Aeronauticii,* with a half-dozen aeroplanes; 30 years later, in 1941, a Rumanian air arm the *Fortelor Aeriene Regal ale România,* with a first-line strength of more than 500 combat aircraft, was girding itself for the assault on the Soviet Union in concert with Germany's *Wehrmacht* which had effectively taken control of Rumania. Today, a further 30 years on, Rumania's air arm is known as the Air Forces of the Rumanian Socialist Republic (*Fortele Aeriene ale Republicii Socialiste România*), a somewhat grandiloquent title for what is almost the smallest air arm of the Warsaw Pact forces.

The Rumanian Army had first begun to display interest in aviation in 1910 when Aurel Vlaicu built a parasol mono-plane of original design in the Army Arsenal in Bucharest. This, the Vlaicu I powered by a 50 hp Gnôme rotary, flew successfully for the first time on 17 June 1910, a date that is still celebrated in Rumania as National Aviation Day. In the following year, the interest generated by demonstrations of Vlaicu's monoplane and other aircraft of French design germinated in the decision to build under licence Henry Farman biplanes in a workshop at an airfield established at Chitila, and to form an Aviation Group which, in addition to the Farmans, acquired a two-seat and three single-seat

Blériot monoplanes by the end of 1911. The first group of military student pilots — six officers from the Corps of Engineers — had begun flying training at Chitila in April of that year, and the first Rumanian Army pilot had obtained his brevet in June.

In the meantime, another Rumanian national, Henri Coandă, had been invited to join the British and Colonial Aeroplane Company at Filton, and had begun the develop-ment and improvement of the Prier monoplanes. Son of General Coandă, the Rumanian Minister of War, Henri Coandă enjoyed considerable success, and Prince Serge Cantacuzene, who headed a Rumanian Army contingent that received flying training at Larkhill, on Salisbury Plain, in 1912, and subsequently organised a military flying school at Bucharest, ordered four tandem two-seat and three side-by-side two-seat Bristol-Coandă monoplanes for the Rumanian government. Another Coandă monoplane with extra fuel and oil tanks was flown to Bucharest by Prince Cantacuzene in September 1912, and was demonstrated with considerable success during manoeuvres of the Rumanian Army, leading to 10 aircraft of this type being ordered*.

The Aviation Group meanwhile obtained two Morane Type F monoplanes, and had intended to acquire the second aircraft designed and built by Aurel Vlaicu, the Vlaicu II, with which he had taken first place at an international aviation meeting held at Aspern, Vienna, in 1912, the trials having included a bombing contest. However, while on a

The first of these suffered an accident in which the port wing collapsed on 5 March 1913, and as a result the monoplanes were fitted with a strengthened wing. This was found to result in a marked reduction in performance, and the entire batch was converted as biplanes before delivery, this modification proving satisfactory.

cross-country flight in August 1913, the aircraft crashed near Ploesti, and Vlaicu was killed. One of the Bristol-Coandă biplanes had been fitted with rudimentary bomb racks at Filton, and delivered to Rumania in October 1913, and three months earlier, in July, when the Rumanian Army had crossed the Bulgarian frontier with the beginning of the Second Balkan War, occupying Southern Dobruja and advancing on Sofia, some limited réconnoitring had been undertaken by the *Corp al aeronautilor permanenti,* or Permanent Corps of Aeronauts as the Aviation Group had become on 20 April 1913. The further expansion of this nucleus of an air arm was delayed, however, by the outbreak of World War I in July 1914 which prevented the purchase of additional aircraft abroad.

Initially, Rumania maintained a state of armed neutrality, but after the Battle of the Marne, the body of Rumanian opinion in favour of intervention on the side of the Western Powers steadily increased. This was assiduously fostered by the latter, and in 1915, France supplied Rumania with 44 military aircraft for the equipment of the Rumanian Air Corps, or *Corpul Aerian Română,* which was created on 15 September 1915. The initial establishment of the Air Corps comprised three mixed groups and one independent squadron as follows:

Grupul 1 with one reconnaissance squadron, or *escadrilă reconoastere,* equipped with six Henry Farman F.20s, one artillery observation squadron with four Caudron G IIIs, and one fighter scout squadron with two Morane-Saulnier LA monoplanes.

Grupul 2 with one bomber squadron, or *escadrilă bombardement,* operating eight Voisin LAs, one artillery observation squadron with four Caudron G IIIs, and one fighter squadron with two Morane LAs.

Grupul 3 possessed a similar establishment to that of *Grupul* 1, its component units operating the same aircraft types, and the independent reconnaissance squadron attached to the Air Staff (which later became *Grupul* 4) had six Bleriot XI monoplanes.

On 17 August 1916, the Western Powers guaranteed Rumania the Banat, Transylvania, the Hungarian plain as far as the Tisza, and the Bukovina as far as the Prut, in return for an immediate declaration of war against the Central Powers. Ten days later, on 27 August, Rumania declared war on Austria-Hungary, and Rumanian troops immediately began to cross the passes into Transylvania. At this time, Rumania possessed 97 trained military pilots but only 57 of these were included in the active Air Corps personnel strength, and, in any case, the number of serviceable aircraft had shrunk to 19 — four Morane-Saulnier LAs and 11 Farman F.20s — distributed between five squadrons. Spares and ammunition were virtually non-existent, and the Air Corps could no longer be considered as an effective force.

On the ground the war began to go badly for Rumania. General Erich von Falkenhayn and the German 9th Army had thrown back the Rumanian forces, while August von Mackensen, up from the Salonika front with the German-reinforced Bulgarian Danube Army, was driving north through the Dobruja. Nieuport 11 *Bebe* single-seat fighters and Farman MF.11 reconnaissance aircraft were hurriedly ferried into Rumania from Salonika by Royal Naval Air Service pilots, and with their arrival the Air Corps was reorganised. The *Grupul* 1 now consisted of two squadrons with a combined strength of two Farman MF.11s, one Morane-Saulnier LA and two Voisin LAs; *Grupul* 2 possessed one squadron with three Farman MF.11s and one Caudron G III; *Grupul* 3 comprised one fighter squadron, the *escadrilă Nieuport,* with eight Nieuport 11s, one squadron with two Farman F.20s, one Farman MF.11, and one Caudron G III, and one squadron with four Farman F.20s, two Farman MF.11s and two Voisin LAs, and *Grupul* 4 with two MF.11s.

The Air Corps was commanded by a French officer, *Colonel* de Verguette, who led a French mission which had been sent to Rumania, but the fledgeling air arm was comparatively ineffective in countering German aerial activities, confining itself largely to reconnaissance missions, and steadily disintegrated. The German forces had launched a bombing offensive against Rumanian targets from the beginning of hostilities, *Kampfgeschwader* 1 operating from Rasgrad, to the South of Ruschuk, in Bulgaria, and from Turtucaia, and a number of Zeppelin airships had also begun operations from Bulgarian bases. In fact, the first Zeppelin, LZ 101, appeared over Bucharest on the night of 28-29 August 1916. Mackensen's forces had, in the meantime, crossed the Danube, penning Rumanian General Alexandru Averescu into a salient. Averescu had attempted to envelop Mackensen's left flank, but Russian co-operation had been essential and this had not been forthcoming, resulting in the disastrous defeat of the Rumanians in the Battle of the Arges River on 1-4 December, the occupation of Bucharest on 6 December, and the retreat of the Rumanian forces north into Russia.

Sheltered by Russian troops, the Rumanian forces began to regroup and re-equip in Moldavia during the early months of 1917, and the Air Corps was once again reorganised and supplied with French equipment. The initial composition of the reorganised Air Corps was as follows:

Grupul 1 comprising the 1° Fighter Squadron with Nieuport 17s and the 2° and 6° reconnaissance squadrons with Farman MF.11s.

Grupul 2 consisting of the 3° Fighter Squadron with Nieuport 11s and the 4° Reconnaissance Squadron with Farman MF.11s.

Grupul 3 with the MF.11-equipped 5° Reconnaisance Squadron and the 10° Fighter Squadron with Nieuport 17s.

Grupul 4 comprising the 7° Reconnaissance Squadron with Farman MF.11s and the 8° Bomber Squadron with Bréguet-Michelin BM 4s.

Each squadron varied in strength from two to eight aircraft, and by the time Rumanian forces launched an offensive in July 1917, the 4° Squadron had been transferred to *Grupul* 1, two more fighter squadrons, the 11° and the 14°, had been formed on Nieuport 17s, being assigned to *Grupul* 3 and 2 respectively, and a second bomber squadron, the 12°, had been established with Caudron G IVs and assigned to *Grupul* 2, the *Grupul* 4 having been disbanded and its component squadrons re-assigned, the 7° Squadron going to *Grupul* 2 and the 8° Squadron going to the *Grupul* 3.

The Russian collapse left General Mackensen free to throw all his forces against the Rumanian Army, which was rendered incapable of further resistance after the prolonged struggle of Mǎşǎreşti in August 1917. The disposition of the Rumanian Air Corps at this time was:

Grupul 1 based in the vicinity of Bacǎu in support of the 2nd Rumanian Army with 1° Squadron at Borzesti, 2° Squadron at Onesti, and 6° Squadron at Girbovana, while its 4° Squadron was deployed in support of the 4th Russian Army at Domnesti.

Grupul 2 based in the vicinity of Tecuci with its 3° and 14° Squadrons at Cioara, and the 7°, 9° and 12° Squadrons at Calmatui, these supporting the 1st Rumanian Army.

Grupul 3 based in the Galati area in support of the 6th Russian Army, with the 5° Squadron at Folteşti and the 8° and 11° Squadrons at Galati.

The situation deteriorated rapidly, the Russian Army disintegrated into pillaging bands, and the Rumanian forces` had no recourse but to suspend operations, an armistice following on 6 December 1917. However, the Rumanian Air Corps was not disbanded, although it had suffered heavy attrition and was short of spares for its remaining aircraft.

Re-establishing military aviation

On 8 November 1918, when defeat of the Central Powers was assured and German troops had begun the evacuation of Rumania, General Coandă was called to power, and an attempt was made to reorganise the armed forces, but this process had barely begun when, on 21 March 1919 the pro-Bolshevik Béla Kun seized power in neighbouring Hungary and established a Communist dictatorship. Three weeks later, on 10 April, Rumanian forces invaded Hungary to forestall threatened Hungarian efforts to regain Transylvania. With Hungary torn by an anti-Communist counter-revolution, Rumanian troops advanced rapidly, but the Rumanian Air Corps suffered a severe mauling from the Hungarian Red Airborne Corps, the *Vörös Légjárócsapat*, which quickly established air supremacy, and few airworthy aircraft remained in the inventory of Rumania's air arm by the time Rumanian forces occupied Budapest on 4 August.

The collapse of the Hungarian communist regime provided the Rumanian air arm with a very necessary infusion of equipment, some 87 former-Hungarian warplanes being absorbed, including Fokker D VII and Hungarian Aviatik (Berg) D I fighters, and Phönix and UFAG versions of the Hansa Brandenburg C I reconnaissance biplane. Small numbers of SPAD S.VII and S.XIII fighters and 20 Bréguet XIV bombers were also obtained from France for what, in 1920, became officially Rumanian Military Aviation, or *Aeronautica Militara România*. Supervised by the Ministry of War, or *Ministerul de Război*, the service was organised as a number of separate groups each possessing one task, the first line groups being:

Grupul 1 *reconoastere* (reconnaisance) based at Iași.
Grupul 2 *bombardement* (bomber) based at Pipera-București.
Grupul 3 *vinâtoare* (fighter) based at Galati.
Grupul 4 *instructie* (training) based at Tecuci.
Grupul 5 *reconoastere* based at Cluj.

Other components of the *Aeronautica Militara* comprised the *Grupul de exploatatie* (transport and liaison) based at Baneasa, the *Grupul de aerostatie* (balloons) at Pantelimon-București, the *Escadrila de hidroaviatie* (naval squadron) at Constanza, the *Arsenalul Aeronauticii* (Aviation Arsenal) at Cotroceni-București, and the *Depozitul central de materiale* (Central Materials Depot) also at Cotroceni-București. The Aviation Arsenal was initially confined to the overhaul and repair of aircraft but expanded its activities in 1922 when series production was launched of the UFAG 269 version of the Hansa Brandenburg C I, a two seat-observation biplane powered by a 200 hp Austro-Daimler engine. Seventy-two aircraft of this type were to be built by the Arsenal for the *Aeronautica Militara* between 1922 and 1925.

By 1923, when an Air Force Command was created as the

Among the first military aircraft of indigenous design in Rumanian service was the Astra-Sesefsky observation biplane (above), a tandem two-seater powered by an Astra-manufactured Benz engine.

Comandamentul trupelor de Aeronautica (this was later to be redesignated *Divizia I-a aeriana,* or 1st Air Division), the inventory of the *Aeronautica Militara* included approximately 250 aircraft, although a substantial proportion of these were obsolescent. Some effort was expended to increase Rumania's self-sufficiency in aircraft manufacture, and in 1923, the Astra company, which produced railway rolling stock, established an aeronautical department which, during the course of the year, built 25 Proto 1 training biplanes, following these in 1924 with 22 Proto 2 trainers and commencing series production of the Astra-Sesefsky observation biplane intended to supplement the UFAG 269.

The *Societatea pentru exploatári technice* (Technical Development Society), or SET, also established an aircraft manufacturing section in 1923 in București, subsequently building the SET 3 two-seat advanced training biplane to the designs of Grigore C Zamfirescu, and on 1 December 1925, the *Industria Aeronautica Romana* (Rumanian Aeronautical Industry), or IAR, was created, and construction of a new aircraft factory began at Brașov. While preparations for quantity production of military aircraft in Rumania got under way, however, the *Aeronautica Militara* was in sore need of modern combat aircraft. Fifty Fokker D XI fighters were ordered from the Netherlands in 1925, but upon arrival in Rumania these were relegated to the advanced training role, and it was planned to purchase 70 Armstrong Whitworth Siskin V fighters from Britain, this proposal being abandoned when one of the Siskins crashed at Whitley with

One of the few Rumanian combat aircraft of indigenous design to see any extensive service was the IAR 80 (above and below), both photographs illustrating the initial production version, the IAR 80A.

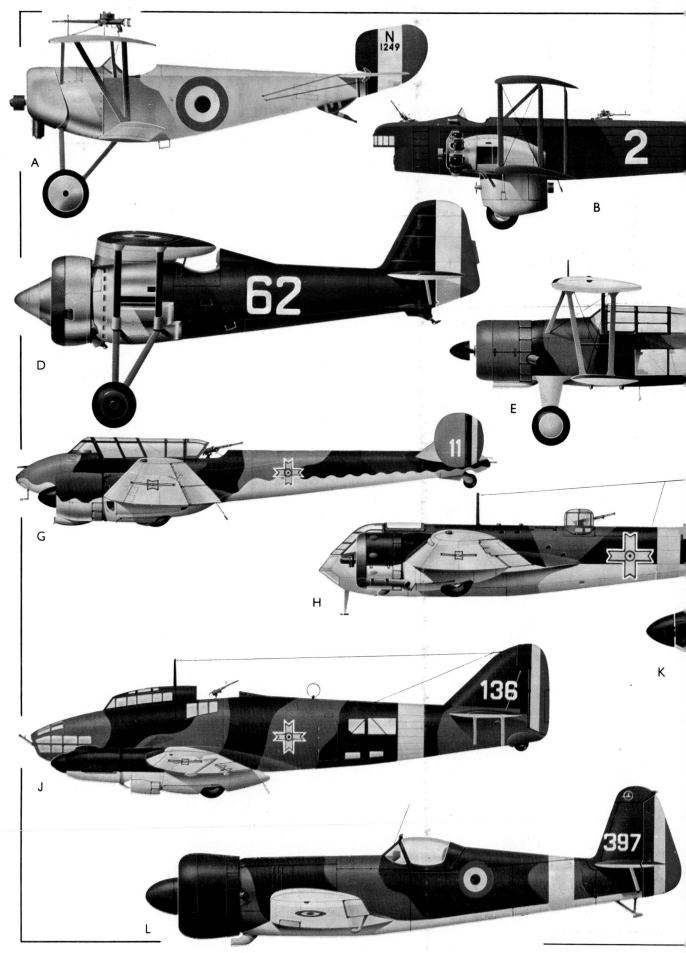

A

B

D

E

G

H

J

K

L

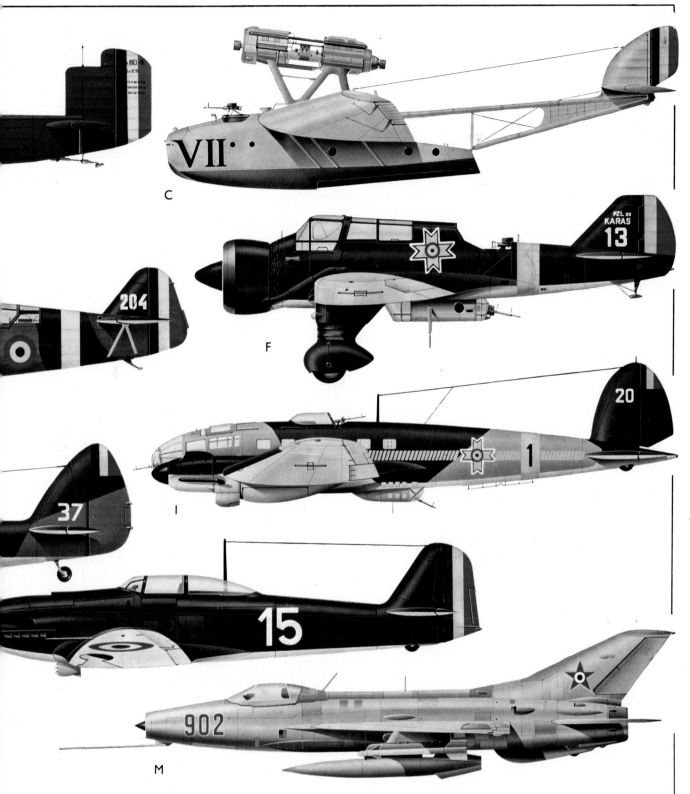

A: Nieuport 11 "Bebe" of the 3rd Fighter Squadron at Cioara attached to the 1st Rumanian Army, September 1917.

B: Liore et Olivier LeO 20 of the Night Bomber Squadron of the Air Combat Flotilla, Pipera-Bucureşti, 1929.

C: Savoia-Marchetti S.M.55 of the Naval Flotilla, Constanza, 1934.

D: PZL P.11b of the Fighter Flotilla, Pipera-Bucureşti, 1935.

E: IAR 39 employed in the army co-operation rôle, autumn 1945.

F: PZL P.23 Karas of the 3rd Bomber Flotilla, Focşani, June 1941.

G: Potez 633B of the 2nd Bomber Flotilla, Râmnicu Sărat, June 1941.

H: Bristol Blenheim I of the 3rd Bomber Flotilla operating on the Soviet Front, 1942.

I: Heinkel He 111H-3 of No 5 Group, Zaporozh'ye area of the Ukraine, early 1943.

J: IAR-built Savoia-Marchetti S.M.79-JR of the 3rd Air Corps, 1943.

K: Heinkel He 112B-1, Bessarabia, September 1941.

L: IAR 81c of the 2nd Fighter Group, summer 1945.

M: MiG-21F of an Interceptor Regiment, 1967.

a pilot of the Rumanian mission at the controls. The fighter requirement of the *Aeronautica Militara* was finally fulfilled by the purchase from France during 1926-7 of 100 Blériot-SPAD 61-2s.

Other purchases from France included Potez XVs, XXVs and XXVIIs for the reconnaissance role, supplemented by SET 7K reconnaissance biplanes of indigenous design and construction; a large number of Bréguet Br XIX B2s and Br XIX.7s for the light bomber task; seven Lioré et Olivier LeO-20 BN3 twin-engined night bombers which were delivered in September 1928, and 50 Loire-Gourdou-Leseurre LGL-32 fighter monoplanes which were ordered in the same year. The IAR factory at Braşov had begun operations in February 1927 with licence manufacture of the Morane-Saulnier MS 35 trainer and Potez XXV reconnaissance aircraft, following these with production of the Dewoitine D.27 fighter monoplane and simultaneously initiating its own line of single-seat fighter development under the leadership of Ing-Dr Elie Carafoli, the aero engine division producing Lorraine and Gnôme-Rhône engines under licence.

In 1929, despite the serious economic problems being faced by Rumania, the expansion of the *Aeronautica Militara* was accelerated, and a number of organisational changes made. Whereas previously the largest formation had been the group, the operational components of the air arm were now divided into *Flotile*, each *Flotila* consisting of two or more groups. The largest of these was the *Flotila aviatie de luptâ* (Air Combat Flotilla) which, based at Pipera-Bucureşti, embodied almost the entire combat element, including all fighter groups, day bomber groups, and the heavy night bomber group. There were three reconnaissance flotillas, *Flotila* 1 at Iasi, *Flotila* 2 at Cluj, and *Flotila* 3 at Galati, and a naval flotilla, the *Flotila de hidroaviatie* at Suit-Ghiol, Constanza, plus the *Flotila de aerostatie* (balloon flotilla) at Pantelimon-Bucureşti. An experimental squadron was established at Pipera-Bucureşti for the testing and evaluation of new foreign aircraft and indigenous prototypes, and at the same time a Training Command the *Comandamentul scolilor si centrelor de instructie,* was created, this supervising the flying training school at Tecuci, the gunnery and bombing school at Mamaia, the advanced training centre at Buzau, and other training establishments.

The 1920 Treaty of Trianon had forbidden Hungary any form of military aviation, but in 1932 it became obvious that a small clandestine air arm was being created by the Hungarians. Rumania immediately saw this as the first step on the part of Hungary towards achieving her ambition of recovering Transylvania, and one of the results was the dissolution of the Air Combat Flotilla and its replacement by a *Flotila vinatoare* (Fighter Flotilla) comprising six squadrons based at Pipera-Bucureşti, and a *Flotila bombardement* (Bomber Flotilla) with three squadrons based at Braşov. The other operational components were now three mixed flotillas: *Flotila aviatie N° 1* with two fighter squadrons and five reconnaissance squadrons at Iasi, *Flotila aviatie N° 2* with two fighter squadrons and four reconnaissance squadrons at Cluj, and the *Flotila de hidroaviatie* which, in addition to a squadron with a mix of Savoia-Marchetti S.M.55, S.M.59 and S.M.62 flying boats, now included five bomber squadrons and two fighter squadrons.

The SET factory had begun licence production of the Fleet F-10-G trainer and was also producing the SET 10 single-seat fighter training biplane, while the IAR had built a small number of IAR 14 Lorraine-powered single-seat fighter monoplanes which had been assigned the advanced training role, and was producing limited batches of IAR 15 and IAR 16 fighter monoplanes, these being powered by the IAR-built Gnôme-Rhône 9Krsd and the Bristol Mercury IVS respectively. However, the capabilities of the IAR fighters were adjudged inferior to those of some foreign contem-

poraries, and in 1934 the Polish PZL P.11 fighter monoplane was selected for the re-equipment of the fighter squadrons, 50 examples being ordered from the parent company with IAR-built Gnôme-Rhône engines (as the P.11b) and a manufacturing licence being acquired. IAR commenced tooling for the PZL fighter during the following year, and was to manufacture 70 during 1936-7.

Preparations for a new war

With the creation of a Ministry of Aviation and Marine (*Ministerul aerului si Marinei*) in 1936, the Rumanian air arm, now known as the *Fortelor Aeriene Regal ale România* (Royal Air Forces of Rumania), or FARR, gained a greater degree of autonomy. Administered by the *Comandamentul Fortelor Aeriene* (Air Forces Command) which was directly responsible to the Ministry, the FARR was split between the three Air Regions into which Rumanian territory was now divided: *Regiunea 1-A Aeriană* with Headquarters at Iasi, *Regiunea 2-A Aeriană* with Headquarters at Cluj, and *Regiunea 3-A Aeriană* with Headquarters at Bucureşt. At the same time, the number of *Flotile* was increased from six to ten — three fighter, three bomber, three reconnaissance and one naval — and, by 1938, the FARR possessed a personnel strength of 5,900 officers and men, and a first-line aircraft inventory of 340 warplanes, including 25 naval aircraft. There were also 260 training and reserve aircraft, and thus, numerically, the FARR was a substantial force, although serviceability gave serious cause for concern, this being primarily the result of the wide diversity of types* in FARR service.

From 1936 greater emphasis had been placed by the FARR on bombing capability, and in that year eight twin-engined Potez 543 bombers had been purchased from France. An order had also been placed in Poland for 24 examples of the LWS-4 Zubr (Bison) bomber, but with the crash of the LWS-1 prototype on 7 November 1936 and the loss of two members of the FARR purchasing commission, this contract had been cancelled, and, instead, an order placed in France for a similar quantity of Bloch 210s, delivery of the French bombers commencing on 4 June 1937 and continuing until midsummer 1938. Twenty-four S.M.79B bombers each powered by two Gnôme-Rhône radials were purchased from Italy during the course of 1938, and the year also saw the placing of an order for 20 Potez 633 light bombers, deliveries of which began before the end of the year.

In 1936, the PZL P.24E had been selected as a successor to the P.11 as the standard Rumanian single-seat fighter, six examples being purchased from the parent company in 1937, together with a manufacturing licence, and IAR producing a further 50 during 1938-9. Other foreign aircraft purchases at this time included a number of C.R.D.A.Cant Z.501 flying boats for the re-equipment of the *Flotila de hidroaviatie*, some Nardi FN.315 and Focke-Wulf Fw 44 Stieglitz trainers, and five Potez 65s for use as paratroop transports. The IAR factory at Braşov had meanwhile completed the prototype of the IAR 37 light reconnaissance-bomber biplane which had been ordered into production for the FARR.

In the meantime, Rumania had vacillated between a policy of friendship towards the democratic countries and a growing reliance upon Italy and Germany, while endeavouring to form an alliance with Poland for the purpose of establishing a common defensive neutrality bloc between Germany and the Soviet Union. Internal strife had resulted, on 10 February 1938, in King Carol establishing a personal dictatorship, but by the end of the year, Rumania found herself in a most hazardous situation, facing the revisionist claims of Hungary, Bulgaria, and the Soviet Union for areas of Rumanian

---

*Between 1936 and 1940, the FARR operated no fewer than 77 different types of aircraft — French, Polish, German, British, Italian and Rumanian — with 58 different types of engine!

territory, and coveted by Germany on account of her oil, wheat and strategic position. On 24 March 1939 a five-year economic treaty was concluded with Germany, and the interpretation of this treaty was to form a point of dispute between the two countries for many months. In an attempt to offset the effects of the German treaty, Rumania concluded a trade treaty with Britain on 12 May, and Britain and France jointly guaranteed Rumania's territorial integrity in the event that she should have to face German aggression.

In so far as the FARR was concerned, these treaties meant changes in its sources of aircraft supply, and the variety of aircraft in its inventory was now further expanded by the arrival of a number of German aircraft. Initially the quantities were insignificant, comprising nine Messerschmitt Bf 108B liaison aircraft, a few Focke-Wulf Fw 58 Weihe crew trainers, 24 Heinkel He 112B fighters, and some Czech-built Bloch M.B.200s. A licence was acquired by the IAR for the Junkers Jumo 211Da engine, a further 24 S.M.79B bombers were ordered from Italy with these engines, and a manufacturing licence for the airframe was purchased, the Jumo-engined S.M.79B subsequently being built by the IAR as the S.M.79-JR.

Britain, by now thoroughly alarmed by the extent of the German penetration of Rumania, agreed, in something of a diplomatic gamble to persuade Rumania into the allied fold or, at least, to remain neutral in the struggle that was now so obviously imminent, to contribute 12 Hawker Hurricane I fighters and 34 Bristol Blenheim I bombers to the FARR inventory. The first of the Hurricanes was, in fact, despatched to Rumania on 28 August 1939, only a few days before war broke out in Europe, and the first batch of 13 Blenheim Is followed in November. A second batch of 20 Potez 633s had been withheld by the French government, but in mid-September 1939, the FARR received an unexpected windfall when the surviving aircraft of the Polish Air Force flew in to Rumanian airfields. These comprised 39 PZL P.37 Loś bombers, 30 PZL P.23 Karaś light reconnaissance bombers, 17 Lublin R.XIII and 11 RWD-14 Czapla army co-operation aircraft, and 38 P.7 and P.11 fighters, as well as several commercial transports. These aircraft were promptly absorbed by the FARR which simultaneously increased both its numbers and its maintenance problems.

Before the end of the first month of hostilities in Europe, Rumania's position had become extremely precarious. The Rumanian government had relied upon German-Soviet antagonism and its own friendship with Poland, but Germany and the Soviet Union had suddenly become allies and had divided Poland between them. Rumania now found herself powerless to resist the pressures that were being exerted upon her by her neighbours. On 27 June 1940, Soviet forces occupied Bessarabia and northern Bukovina. The Rumanian government renounced the British guarantee, and the *Wehrmacht* immediately began to infiltrate the country, ostensibly to protect the Rumanian oil fields. On 30 August, Germany and Italy forced Rumania to cede northern Transylvania to Hungary, and on 6 September the dismemberment was completed when southern Dobruja was ceded to Bulgaria.

On the credit side, Rumania, losing most of her minorities, became ethnically united; her armed forces were still intact, and her natural resources, and in particular the Ploesti oil wells (which Winston Churchill was later to refer to as the "taproot of German might"), gave her some bargaining power. General Ion Antonescu now took over the government, placing Rumania squarely in the Axis camp and, on 23 November, signing the Axis Tripartite Pact.

Rumania goes to war
During the winter of 1940-1, a substantial *Luftwaffe* training mission arrived in Rumania, together with a number of

Representative of the wide variety of different combat aircraft with which the Rumanian air arm went to war are the PZL P.23 Karaś (above) and the Bristol Blenheim I (below).

Heinkel He 111H-3 bombers, Junkers Ju 87B dive-bombers, and Junkers Ju 52/3m transports, and on 22 June 1941, when Rumanian forces attacked the Soviet Union in concert with the *Wehrmacht*, the FARR comprised the following: Three fighter *flotile* with a total of 12 squadrons (five equipped with the P.11, four with the P.24, one with the Hurricane, and two with the He 112B); three bomber *flotile* with a total of 20 squadrons (two with Bloch 210s, two with Potez 633s, three with P.37s, three with Blenheim Is, three with He 111H-3s, four with S.M.79Bs, and three with Ju 87Bs); three reconnaissance *flotile* with 18 squadrons (three with P.23s and the remainder with IAR 37s, 38s or 39s), and one naval *flotile*

On the Russian Front: (above) the crew of a Potez 633 preparing to take off on a support mission late in 1941, and (below) Heinkel He 111H-3s being "bombed up" in the Zaporozh'ye area of the Ukraine in the winter of 1942-3.

(Above) An S.M.79-JR medium bomber, which, together with the He 111H-3, provided the backbone of the FARR bombing component throughout most of the war, and (below) a Junkers Ju 87D-3 of Grupul 6 Picaj, the dive-bombing component of the Corpul 1 Aerian.

with two squadrons (one equipped with Z.501 flying boats and the other with He 114B-2 floatplanes). The statutory strength of each fighter squadron was 12 aircraft and that of other squadrons was nine aircraft. The FARR order of battle thus comprised 504 aircraft excluding reserves.

The FARR was deployed—under the overall command of *Luftflotte 4*—in support of the Rumanian 3rd and 4th Armies' thrust into Bessarabia, and, subsequently, the Ukraine, which culminated in the capture of Odessa on 16 October. During this initial stage, at least 12 of the bomber squadrons (S.M.79Bs, Potez 633s, Ju 87Bs, He 111Hs, P.37s and Blenheims) were committed, together with virtually the entire reconnaissance and light bombing force, and four fighter squadrons (one with He 112Bs and the remaining three with P.24s), together constituting more than 10 per cent of all aircraft employed in the initial assault on the Soviet Union. Although aerial opposition encountered was modest, attrition suffered by the FARR was inordinately high, largely as a result of the multiplicity of aircraft types being operated and the maintenance difficulties in consequence. By the end of 1941, the Rumanian forces had suffered 70,000 dead and 100,000 wounded, the number of Rumanian divisions in the Soviet Union had been reduced from a peak of twenty-two to six, and almost the entire FARR element had been pulled back for re-equipment and regrouping.

The total personnel strength of the FARR had now risen to more than 13,000 (including personnel of the anti-aircraft artillery component), and during the first months of 1942 additional equipment was forthcoming from Germany. Sixty-nine Bf 109E-4 fighters arrived to replace the surviving He 112B and P.24E fighters of *Flotila 1 vinâtoare; Junkers Ju 88A-4s replaced Potez 633s in the *Flotila 2 Bombardement,* the latter being relegated to the advanced training role in which they were supplemented by refurbished ex-*Armée de l'Air* Potez 63.11s supplied by the Vichy government in exchange for oil, and a second batch of 12 He 114 floatplanes was delivered to the *Flotila de hidroaviatie.*

By July 1942, the number of Rumanian divisions in the Soviet Union had once more risen to 22, these advancing with German forces on Stalingrad where the advance bogged down. By September the number of Rumanian divisions committed had increased to 34, but the FARR forces deployed in their support were weaker than in the previous year. The FARR component deployed in the Soviet Union was designated the *Corpul 1 Aerian,* or 1st Air Corps, this comprising two fighter groups each with two squadrons of Bf 109Es, two bomber groups each of three squadrons, one operating the S.M.79-JR and the other operating the He 111H-3, one dive bomber group with two squadrons of Ju 87Bs, and several light bomber and reconnaissance groups operating the Blenheim, the IAR 37, the IAR 38 and the IAR 39.

The Soviet offensive in the Stalingrad area began on 19 November 1942, the Rumanian 3rd Army taking the initial impact and losing 75,000 men within four days. The Soviet forces then concentrated on the Rumanian 4th Army to the south, armoured spearheads being followed up by great bodies of cavalry, and the Rumanian and German forces reeled back as the Soviet pincers closed on Stalingrad. Having suffered a severe mauling, the Rumanian forces in the Soviet Union were reduced to 10 divisions, but the *Corpul 1 Aerian* was greatly increased in strength.

Operating as a component of the I *Fliegerkorps,* the *Corpul 1 Aerian* received 70 Bf 109G-6s and G-8s to re-equip its four-squadron (*Escadrile 45, 46, 47 and 48*) fighter component based at Mariupol (Zhdanov) in the Ukraine; Ju 87D-3s and D-5s supplanted the Ju 87B-2s of the *Grupul 3 picaj,* the three-squadron (*Escadrile 84, 85 and 86*) dive bomber group; sufficient Ju 88A-4s had been obtained to equip all three squadrons (*Escadrile 75, 76 and 77*) of *Grupul 5 bombardement;* Ju 88D-1s had been made available to equip a long-range reconnaissance squadron (*Escadrila 2*), and a ground

The first deliveries of the IAR 80 fighter to the FARR began in March 1942, entering service with the Corpul 1 Aerian during the following summer and seen here at a field in the vicinity of Stalingrad in September 1942.

attack unit, *Grupul 8 Asalt,* had been formed with three squadrons (*Escadrile* 41, 42 and 43) of Henschel Hs 129Bs. Other components of the *Corpul 1 Aerian,* which was operating primarily in the Zaporozh'ye area of the Ukraine, were two transport squadrons (*Escadrile* 112 and 118) operating the Ju 52/3m, four liaison squadrons (*Escadrile* 111, 113, 114 and 115) with the Fieseler Fi 156 Storch, and three bomber squadrons (*Escadrile* 78, 79 and 80) equipped with the He 111H-3, the original deliveries of this aircraft from Germany having by now been supplemented by licence-built He 111H-3s from SET. Finally, there were four squadrons (*Escadrile* 11, 12, 13 and 14) operating the indigenous IAR 80 and IAR 81 which combined the tactical reconnaissance task with fighter-bomber and close-support duties. One other FARR unit operating in the Soviet Union under I *Fliegerkorps* but not attached to the *Corpul 1 Aerian* was the *Grupul 6 picaj* with three squadrons (*Escadrile* 81, 82 and 83) equipped with the Ju 87D-1, D-3 and D-5. Thus, in the summer of 1943, the FARR was committing some 350 aircraft to the Russian Front.

By this time, much more attention was being given to the air defence of Rumania herself, and particularly to the defence of her oil wells and refineries from which came more than half of the petroleum and lubricants used by the German armed forces. A new Air Corps, the *Corpul 3 Aerian,* was created specifically for home defence. The IAR at Braşov had produced some 120 IAR 80 fighters during 1942-3*, enabling a number of additional squadrons to be formed, including *Escadrile* 59, 61, 62, 63, 64, 65 and 66. Apart from new German combat aircraft, the FARR had received 20 Bloch M.B.151s and 152s, and 150 Dewoitine D.520s for the advanced training role, and various other training aircraft had been obtained, including Bücker Bü 131s and Arado Ar 96s. In fact, by late 1943, the FARR had more than 1,300 training aircraft of more than *eighty* different types! The two naval squadrons (*Escadrile* 101 and 102) at Constanza and Odessa had replaced their He 114s with Arado Ar 196s, but the supply of Rumanian-built aircraft had dwindled to a trickle as the factories were choked with aircraft in need of repair.

On 1 August 1943, the Ploesti refineries had received their first visit from B-24 Liberators of the US 15th Air Force, this attack, given the codename Operation *Tidal Wave,* proving disastrous for the participants who had to make a 2,300-mile (3 700 km) round-trip. German intelligence had allegedly cracked the Allied code and tracked the B-24s all the way from their bases in the vicinity of Benghazi. The lead B-24 carrying the mission navigator crashed in the Mediterranean, the lead wave consequently taking a wrong heading just short of the target, and *Tidal Wave* disintegrated into chaos. Some B-24s followed the lead wave on its wrong heading, and others that succeeded in reaching the target area flew into an intense barrage of anti-aircraft fire and then ran the gauntlet of large numbers of FARR and *Luftwaffe* fighters. Of the 163 B-24s that reached Rumania no fewer than 54 were shot down. Nevertheless, the capacity of the refineries in the Ploesti area was temporarily reduced by 40 per cent at an extremely critical time for the Axis.

By early 1944, although the number of Rumanian divisions deployed on the Russian Front had once again risen to 25, the FARR contingent had declined considerably in strength, and from February only one FARR Bf 109G squadron remained, and this, *Escadrilă* 49 based at Saki on the Crimea, possessed a strength of only some five serviceable aircraft. The other Bf 109G squadrons of the *Corpul 1 Aerian* had

The IAR 80 and its development, the IAR 81, were phased out of production at Braşov during the course of 1943 to enable IAR to concentrate on manufacture of the Bf 109G-6, but after the assembly of 30 fighters of this type from German-supplied sub-assemblies and components, only 16 more Bf 109G-6s had been completed by IAR before, on 6 May 1944, that Braşov plant was largely gutted by a US 15th Air Force bombing attack.

An FARR "ace", Captain Alexandre Serbanesco, being congratulated by ground staff after adding to his score (summer 1944) while flying a Messerschmitt Bf 109G-6/R6.

been pulled back for home defence duties, these operating independently while the similarly-equipped *Escadrile* 51 and 52 at Tepes-Voda and Mamaia respectively were operating under the *Jagdabschnittsführer Rumänien* (Fighter Sector Leader Rumania) which included the Bulgarian Bf 109G-equipped 6 *Polk* (6th Regiment) as well as *Luftwaffe* units. The only major FARR units still in the Soviet Union were *Flotila 1 bombardiere* (consisting of *Grupul 5 bombardement* with two squadrons of Ju 88As, *Grupul 3 picaj* and *Grupul 6 picaj* each with three squadrons of Ju 87Ds, *Grupul 8 asalt* with three squadrons of Hs 129Bs, and two Ju 88D-equipped reconnaissance squadrons), and *Flotila 2 bombardiere* (comprising *Grupul 1 bombardement* with three squadrons of He 111H-3s and *Grupul 2 bombardement* with a similar number of S.M.79 squadrons).

A change of sides

On 10 April 1944 Odessa was evacuated by Rumanian and German forces, and Soviet forces entered Moldavia and swept forward, being halted only 11 miles (18 km) from Iaşi. The situation inside Rumania deteriorated rapidly; friction between Rumanians and Germans escalated, and there were numerous incidents of sabotage. Nevertheless, on 20 August, when Soviet forces launched their offensive across the Prut River, the FARR remained a substantial force, with 37,196 personnel, 508 first-line aircraft distributed between 58 squadrons, and 1,131 reserve and training aircraft. The *Corpul 1 Aerian* had 249 serviceable aircraft in 30 squadrons, and the *Corpul 3 Aerian* had 191 (including four Gotha Go 242 gliders) in 22 squadrons.

There were three fighter *Flotile* with a total of 195 IAR 80s and Bf 109Gs distributed between 20 squadrons; three bomber *Flotile* comprising eight dive bomber squadrons with 64 Ju 87Ds and IAR 81s, eight medium bomber squad-

(Below) An He 114C-1 float seaplane of the Flotila de hidroaviatie *during the summer of 1943.*

rons with 51 Ju 88As, He 111Hs and S.M.79s, and three assault squadrons with 32 Hs 129Bs; three army co-operation *Flotile* with 11 reconnaissance and observation squadrons flying 94 Ju 88Ds, IAR 38s, IAR 39s and SET 7Ks, two transport squadrons with 17 Ju 52/3ms and four Go 242s, and four liaison squadrons with 33 Fi 156s, and one naval *Flotila* with 22 Ar 196s, He 114s and Cant Z.501s.

As the Soviet armour rolled towards Bucharest, King Michael led a plot which overthrew Ion Antonescu, and General Johannes Friessner was suddenly faced with the fact that half of his forces – Rumania's 25 divisions – had allied themselves with the Soviet forces opposing the 26 divisions of the German 6th and 8th Armies that remained to him. Rumania capitulated on 23 August, most of the German forces were trapped, and on 24 August Rumania declared war on Germany, the Rumanian armed forces receiving orders to attempt to disarm their former comrades-in-arms. Serviceability in the FARR had dropped alarmingly by 7 September when the *Corpul 1 Aerian* was subordinated to the 5th Soviet Air Force (Second Ukrainian Front), strength having also diminished and now comprising 2,915 personnel and 18 squadrons with fewer than 200 serviceable aircraft. From 21 September, the FARR initiated operations against the German forces, its 18 squadrons possessing 197 aircraft (81 fighters, 24 assault aircraft, 12 dive bombers, 16 medium bombers and 17 reconnaissance aircraft, plus liaison and transport aircraft). By 26 October, the number of FARR squadrons had been reduced to 15 and the first-line strength to 174 aircraft.

In the meantime, those Rumanian aircraft factories that had escaped total annihilation had resumed production and repair work, these being crammed with 360 damaged machines (132 FARR aircraft and 228 confiscated from the *Luftwaffe*), but the task of making good FARR attrition from these damaged aircraft was complicated by the fact that the Rumanian mechanics were now required also to maintain Soviet Air Force aircraft. Thus, by the end of 1944, the IAR had succeeded in completing only 46 aircraft that had been on the assembly line, and had repaired 23 Rumanian and 66 Soviet aircraft. The ASAM repaired 12 S.M.79-JR bombers during the course of October, and SET repaired 41 reconnaissance aircraft. Three large workshops were also engaged on the repair of damaged aircraft.

With the repaired aircraft, the FARR had succeeded in increasing its squadrons to 20 by the beginning of 1945, by which time the service was engaged on operations in

When Rumania declared war on Germany in 1944, the cross insignia gave place to the pre-war roundel, seen here on an IAR 81c fighter-bomber (above) and an IAR 39 army co-operation aircraft during the spring of 1945.

Slovakia. The number of aircraft on strength totalled 239, these including 88 fighters, 44 bombers and dive bombers, and 54 reconnaissance aircraft. Between 23 August 1944 and 12 May 1945, the FARR flew a total of 4,400 sorties (2,578 of these between 21 December 1944 and 12 May 1945) against the German and Hungarian* forces, dropping a total of 1,360 tons (1 382 tonnes) of bombs in the process, and claiming the destruction (including the claims of Rumanian anti-aircraft batteries) of 101 enemy aircraft. Aircraft losses during this period had ranged between 10 and 60 per cent of unit strengths except in the case of the dive bomber squadrons which lost 95 per cent of their strength.

The FARR saw the end of hostilities with only 10 combat squadrons remaining. These were divided between the Fighter Command (*Comandament Vinatoare*) and the Bomber Command (*Comandament Bombardement*), the former comprising *Grupul* 1 (*Escadrile* 63 and 64) and *Grupul* 2 (*Escadrile* 65 and 66) equipped with the IAR 80, and *Grupul* 9 (*Escadrile* 47 and 48) equipped with the Bf 109G, while the latter consisted of the *Grupul Bombardement* (*Escadrile* 72 and 82) operating a mix of He 111Hs, S.M.79-JRs and Ju 88As, and the *Grupul Asalt-Picaj* with 16 Hs 129Bs (*Escadrila* 41) and nine Ju 87Ds (*Escadrila* 74).

With the end of World War II the Rumanian air arm was run down rapidly. Few of the surviving squadrons were able to maintain their statutory strength, cannibalisation and normal attrition rapidly reduced the number of available aircraft, and training was reduced to a minimum. The service continued to operate a miscellany of wartime aircraft types for which – as the IAR and other aircraft factories had been demilitarised by the Soviet occupation authorities – no replacements were available. Under the terms of the Peace Treaty signed in Paris in 1947, Rumania was not permitted to maintain an air arm of more than 150 aircraft with a personnel strength in excess of 8,000 men. In the meantime, elections held on 19 November 1946 had resulted in victory for the Communist bloc, and on 30 December 1947 King Michael was obliged to abdicate. With the declaration of a People's Republic, the designation of the Rumanian air arm became *Fortele Aeriene ale Republicii Populare România* (Air Forces of the Rumanian People's Republic), or FR-RPR.

Administered by the Armed Forces Ministry, or *Ministerul Fortelor Armate*, the FR-RPR continued to operate obsolescent Bf 109G and IAR 80 fighters, the IAR 39 and the Fi 156 Storch for army co-operation and liaison tasks, the Ju 52/3m and converted S.M.79-JR bombers for the transport role, and a miscellany of types, such as the Nardi FN 305, the Fleet F-10-G, and the Fw 58 Weihe for training. The Sovrom-Tractor plant at Braşov (as IAR had become) undertook the conversion of a few IAR 80 fighters as tandem two-seat advanced trainers, and subsequently, after the signing of the "Treaty of Friendship, Co-operation and Mutual Assistance" between the Soviet Union and Rumania in 1948, undertook licence manufacture of the Yak-11 trainer. At the same time, the reorganisation and re-equipment of the Rumanian armed forces was initiated under close Soviet supervision.

The FR-RPR adopted Soviet training techniques, and its organisation was now based broadly on that of the Soviet Air Forces. The largest formation was the *divizie aeriană* (Air Division) which was sub-divided into *corpuri aeriene* (Air Corps), each *corp aerian* consisting of two or more *regimente de aviatie* (Aviation Regiments) roughly equivalent to RAF wings. The basic unit remained the *escadrilă* (Squadron) of 12-15 aircraft, three *escadrile* usually comprising one *regiment*. Po-2s and UT-2s were supplied for primary training, Lavochkin La-7s and Yakovlev Yak-9s supplanted the Bf 109Gs and IAR 80s, the ageing Ju 52/3m and S.M.79 trans-

The Rumanian government had declared war on Hungary on 6 September 1944.

The MiG-15bis (seen above) began to enter service with the FR-RPR during the early 'fifties, but has long since been relegated to the armament training rôle.

ports were supplemented by Li-2s, and Ilyushin Il-10s were provided to equip a Ground Attack Regiment.

In the early 'fifties the FR-RPR received its first jet aircraft in the shape of the Yak-23 which equipped one *regiment*; the transport component was expanded by the provision of Ilyushin Il-14s and Antonov An-2s, permitting the creation of an airborne battalion for which a parachute training school was established. In line with Soviet policy in southeast Europe during the early 'fifties, the FR-RPR was steadily expanded and modernised, Yak-18s replacing the older primary trainers, and MiG-15 interceptors and Il-28 tactical bombers being supplied. The process of expansion and modernisation was accelerated after 14 May 1955 when

Continued on page 43

AIRCRAFT THAT HAVE SERVED WITH THE RUMANIAN AIR ARM

Type	Rôle	Year of Intro	Quantity	Type	Rôle	Year of Intro	Quantity	Type	Rôle	Year of Intro	Quantity
H Farman Type 1909	T	1911	4(1)	Savoia-Marchetti S.M.55	R	1934	—	Polikarpov I-16(7)	T	1941	3
Blériot	T	1911	4	Savoia-Marchetti				Heinkel He 111H-3(2)	B	1942	30
H Farman Type 1910	T	1912	4(2)	S.M.59	R	1934	—	Heinkel He 114C-1	R	1942	12
Bristol-Coanda(3)	T	1912	7	Savoia-Marchetti				Junkers Ju 88A-4	B	1942	
Bristol-Coanda(4)	R	1913	10	S.M.62	R	1935	—	Junkers Ju 88D	R	1942	}50
Morane Type F	R	1913	2	IAR 36	Tpt	1936	5	IAR 80	F	1942	127
H Farman F.20	R	1915	12	PZL P.11b(2)	F	1936	70	Potez 63.11	T	1942	63
Morane Type LA	F	1915	6	Potez 543	B	1936	8	Fieseler Fi 156	L	1942	45
Caudron G III	O	1915	12	Avia BH 25	Tpt	1937	2	Arado Ar 196	R	1943	28
Voisin LA	B	1915	8	PZL P.24E	F	1937	6	Henschel Hs 129B	A	1943	62
Blériot XI	R	1915	6	Bloch 210	B	1937	24	Junkers Ju 87D	DB	1943	115
Farman F.27	R	1916	37(5)	PZL P.24E(2)	F	1938	50	Messerschmitt Bf			
Nieuport 11	F	1916	28	Savoia-Marchetti				109G	F	1943	135(8)
Bréguet-Michelin				S.M.79B	B	1938	24	Bloch 151-152	T	1943	20
BM 4	B	1917	10*	C.R.D.A. Cant Z.501	R	1938	—	Dewoitine D.520	T	1943	150
Nieuport 17	F	1917	—	Nardi FN.315	T	1938	20*	IAR 81	DB	1943	40
Farman F.40	R	1917	20*	Fleet F-10-G(2)	T	1938	200	Consolidated B-24	B	1943	3
Caudron G IV	B	1917	—	Potez 65	Tpt	1938	5	Messerschmitt Bf			
Sopwith 1½ Strutter	F	1919	—	Potez 633	B	1938	20	109G(2)	F	1944	16
Aviatik (Berg) C I	R	1919	—	IAR 37-38-39	R	1938	325	Gotha Go 242	G	1944	4
Aviatik (Berg) D I	F	1919	8	Klemm Kl 35	T	1938	—	Polikarpov Po-2	T	1948	—
Phönix H-B C I	R	1919	}18	Focke-Wulf Fw 44	T	1938	—	Yakovlev UT-2	T	1948	—
UFAG H-B C I	R	1919		Messerschmitt Bf 108	L	1939	30	Zlin 181	T	1948	40
Fokker D VII	F	1919	6	Focke-Wulf Fw 58	T	1939	—	Avia C-2 (Ar 96)	T	1948	—
SPAD S.VII	F	1919	—	Bloch 200	T	1939	25	Lavochkin La-7	F	1948	40*
Bréguet XIV	B	1919	20	Heinkel He 112B	F	1939	24	Yakovlev Yak-9	F	1948	100
SPAD S.XIII	F	1919	—	Savoia-Marchetti				Ilyushin Il-10	A	1949	120
D.H.9	R-T	1922	65	S.M.79-JR	B	1939	24	Aero 45	L	1950	—
UFAG 269(2)	O	1922	72	Heinkel He 114B-2	R	1939	12	Mraz M-1 Sokol	T	1950	—
Astra Proto 1	T	1923	25	Hawker Hurricane I	F	1939	12	Li-2	Tpt	1950	50*
Astra Proto 2	T	1924	22	Bristol Blenheim I	B	1939	34	IAR 811	T	1950	100
Astra-Sesefsky	O	1924	—	PZL P.23 Karaś	A	1939	30	Yakovlev Yak-11(2)	T	1950	—
SPAD 33	T	1925	4	PZL P.37 Łoś	B	1939	39	Yakovlev Yak-23	F	1950	40*
SPAD 46	F	1925	—	PZL P.7	F	1939	10	IAR 813	T	1951	—
Fokker D XI	T	1925	50	PZL P.11	F	1939 (6)	28	MiG-15	F	1951	450*
Potez XV	R	1925	80*	Lublin R.XIII	R	1939	17	Yakovlev Yak-18	T	1951	—
Potez XXV	R	1926	—	RWD-14 Czapla	R	1939	11	Ilyushin Il-28	B	1953	—
SET 7K	R	1926	100*	RWD-8	T	1939	30	IAR 814	L	1953	10
Blériot-SPAD 61-2	F	1926	100	Lockheed 10	Tpt	1939	—	MiG-17F	F	1954	—
Bréguet XIX B2	B	1926	108	Lockheed 14	Tpt	1939	—	Antonov An-2	Tpt	1954	—
Potez XXVII	R	1927	—	Bücker Bü 131	T	1940	150*	Ilyushin Il-14	Tpt	1955	—
Junkers F 13	Tpt	1927	—	Arado Ar 96	T	1940	—	Mil Mi-1	H	1955	—
Bréguet XIX.7	B	1928	12	Junkers Ju 87B	DB	1940	50*	Mil Mi-4	H	1955	—
Loire-Gourdou-				Heinkel He 111H-3	B	1940	35	MiG-17PF	F	1956	—
Leseurre LGL-32	F	1928	50	Junkers Ju 52/3m	Tpt	1940	31	MiG-19S	F	1961	—
Lioré et Olivier				Douglas DC-3	Tpt	1941	—	MiG-21F	F	1962	—
LeO-20 BN3	B	1928	7	Savoia-Marchetti				L 29 Delfin	T	1965	—
SET 31	T	1928	30*	S.M.79-JR(2)	B	1941	16	L 60 Brigadyr	L	1965	—
IAR 15	F	1930	5	Messerschmitt Bf				L 200 Morava	L	1966	—
IAR 16	F	1930	—	109E	F	1941	69	MiG-21PF	F	1967	—
SET 15	F	1930	—	Junkers W 34	T	1941	—	WSK-Swidnik Mi-2	H	1967	—
Dewoitine D.27(2)	F	1931	—	Polikarpov Po-2(7)	L	1941	—	Mil Mi-8	H	1968	—
ICAR-Universal	T	1931	—	Yakovlev UT-2(7)	T	1941	—	MiG-21UTI	T	1968	—
PZL P.11b	F	1934	50					Ilyushin Il-18	Tpt	1968	1

* Approximate.
Rôles: *A, Assault; B, Bomber; DB, Dive Bomber; F, Fighter; G, Glider; H, Helicopter; L, Liaison; O, Observation; R, Reconnaissance; R-T, Reconnaissance-Trainer; T, Trainer; Tpt, Transport.*
Notes: 1) *Three built in Rumania;* 2) *Rumanian-built;* 3) *Four tandem two-seaters and three side-by-side two-seaters;* 4) *Built as monoplanes but modified as biplanes;* 5) *Including six ex-British aircraft;* 6) *Ex-Polish Air Force.* 7) *Captured in Soviet Union;* 8) *Includes those taken over from Luftwaffe in August 1944.*

THE NORTHROP P-61 BLACK WIDOW IN USAAF SERVICE, 1944-45

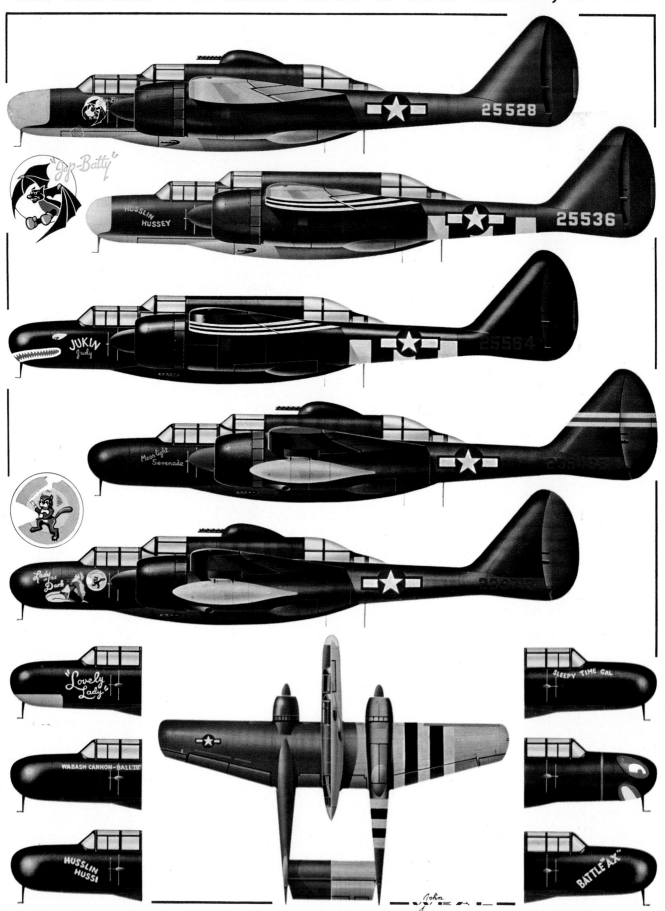

"Jap-Batty"

25528

HUSSLIN HUSSEY

25536

JUKIN Judy

25564

Moonlight Serenade

Lady of the Dark

"Lovely Lady"

SLEEPY TIME GAL

WABASH CANNON-BALL IV

HUSSLIN HUSSI

BATTLE AX"

John

MODEL ENTHUSIAST

IN THIS, the first issue of the AIR ENTHUS-IAST, it is fitting that we establish the primary purpose of this monthly column devoted to the interests of the members of the modelling fraternity among our readers: it is that of aiding the modelling art. Modelling may be defined as the representation in three dimensions of some material object, showing the proportion and arrangement of its parts, and, in so far as this column is concerned, the object is the aeroplane represented in miniature.

Our policy will be to provide our readers with a critique of each important new aircraft model kit as it is issued, but we would stress that our aim is to be fair to purchaser and manufacturer alike, an aim which perhaps demands some explanation. There are two distinct approaches to the art of aircraft modelling, those of what may be termed the *artistic* realist and the *engineering* realist. The former applies primary effort to the final appearance of his model; his aim is the creation of an overall impression, and he endeavours to simulate in miniature the minutiae of the surface detail of the full-scale original of the aircraft that he is modelling. He has a penchant for the more exotic in marking and finish; the finer points of outline detail are of limited importance and working parts of dubious value.

In direct contrast, the engineering realist will devote an infinite amount of time to the actual component parts of the aircraft kit, researching the subtlest detail of internal and external physical shape. His *bête noire* can be what to the artistic realist represents the most minor inaccuracy in the kit of his choice, and when it comes to external finish he will select the simplest and most readily available. Possessing some degree of knowledge of both sides of the modelling fence, we are all too well aware that what, to the engineering realist, may represent an infuriating and "unnecessary" deficiency in accuracy of the kit that he has purchased, may have been forced upon the manufacturer by production or economic necessity. Again, it may be the result of careless research, and for this there is no excuse.

(Opposite) Top to bottom: P-61A-1 (Serial No 42-5528) "Jap-Batty" of the 6th NFS, Saipan, 1944; P-61A-5 (Serial No 42-5536) "Husslin Hussey" of 422nd NFS, Scorton, 1944 (also illustrated by plan view at foot of page); P-61A-5 (Serial No 42-5564) "Jukin Judy" of 422nd NFS, Scorton, 1944; P-61B-1 (Serial No 42-39468) "Moonlight Serenade" of 550th NFS, Morotai, 1944; P-61B-15 (Serial No 42-39713) "Lady in the Dark" of 548th NFS, Ryukyu, August 1945. All nose details from 9th Air Force P-61As (based Britain and/or Normandy, summer/autumn, 1944, and featuring invasion striping as per 42-5564) with exception of nose illustrated in the centre of the righthand column, this belonging to a P-61B-1 (Serial No 42-39405) of the 6th NFS at Kagman Field, Saipan, 1944. This particular machine was fitted with a dorsal barbette.

We will be outspoken in our criticism, and if a kit is, in our opinion, a poor kit, we will convey this opinion to our readers. If a kit possesses shortcomings stemming from factors other than carelessness on the part of the manufacturer, we will endeavour to explain to our readers the reasons for these deficiencies.

From time to time, we will include in this column tips which we trust will be of value to both the skilled and the less skilled modeller for improving the accuracy and external finish of the product of his labours, and suggestions for the conversion of standard kits into *non*-standard models, but while we will welcome letters from our readers with suggestions for subject matter for this column, we cannot undertake to answer such letters by post.

Northrop's nocturnal prowler

A component part of this column will, each month, comprise a page in full colour devoted to one specific aircraft type of which one or more kits are available. We would stress, however, that this page is *not* intended as a main course satisfying the appetite of the modeller for external finishes of the chosen subject. Owing to limitations on space, the items depicted in colour should be considered more in the nature of *hors-d'œuvres,* their task being to titillate the appetite, encouraging the modeller to provide the bulk of the repast with original research of his own.

The subject of this month's colour page is perhaps one of the least colourful and certainly least publicised of any fighter that saw widespread use during World War II, the Northrop P-61 Black Widow. Its choice is appropriate as elsewhere in this issue of the AIR ENTHUSIAST will be found the first detailed account of the development of this mighty warplane.

There are currently three kits of the P-61 Black Widow generally available from the stockists, those from Airfix and Frog being to 1/72nd scale and that from Aurora to 1/48th scale. All three represent the P-61A in outline, the nose of the P-61B being a mere 8 in (20,32 cm) longer — no great amount on an aeroplane of this size but, nevertheless, noticeable. Both the 1/72nd scale kits possess virtues and faults, and both are generally accurate in outline. The tailbooms of the Airfix "Widow" are about a tenth of an inch too short, and the tip wing flap segments, which could be operated differentially with the spoiler-type ailerons, are incorrect in shape, while the principal faults of the Frog "Widow" are a foreshortened nose of too slender contour, an excessively-pointed aft transparency, too much taper on the engine cowlings, oversize vertical surfaces, and too angular wingtips. Most of these errors are, however, readily corrected. Airfix provides a beautiful revolving gun barbette, the weapons of which may be elevated, but Frog has moulded the turret as a *transparency* and omitted the guns! The wheels of the Frog model are too small and the undercarriage legs oversimplified, but

the arrangement of the well doors is more accurate that that of the Airfix model. Airfix provides plenty of interior detail while that provided for the Frog model is minimal. and while the surface detail of Airfix's model includes numerous rivets, Frog sticks to straight lines, which in a kit of this scale, we consider preferable. On balance, the Airfix kit is slightly the better of the two, but the best solution to the problem of creating a really accurate model of the "Widow" is to combine the Airfix fuselage nacelle, engines, engine cowlings, airscrews and undercarriage with the Frog wings, tailbooms, tail surfaces, and wheel doors, finishing the model with Airfix decals.

While on the subject of decals, Airfix provides two sets, one including the serial number 42-5558 which was a P-61A-5 and which, if used, should *not* be accompanied by the dorsal barbette, The other decals, for "Lady in the Dark", include the serial number 42-39773 which was allocated to the 16th Vega-built B-17G-1 Fortress! The Frog decals are for a P-61A-10 "Double Trouble" serial number 42-5565 which was not, of course, fitted with the barbette, while accompanying the larger Aurora kit is the serial number 42-39728 which was one of the last of the batch of P-61A-15s, yet the nose is definitely that of a P-61A. Again, for some inexplicable reason, the Aurora decals omit the blue surround of the insignia.

The 1/48th scale Aurora kit is now 11 years of age, and is understandably, therefore, somewhat less sophisticated. Nevertheless it is generally accurate apart from the tailbooms which are about three-quarters of an inch too long, and the vertical surfaces and engine cowlings which are inaccurate in contour. Fortunately, the thickness of the plastic from which the Aurora "Widow" is manufactured enables these faults to be corrected without undue difficulty by simply cutting and filing. As with most Aurora kits, surface detail is sparse, but the beautifully clear transparencies reveal detailed cockpits, and, as usual with kits from this manufacturer, the component parts fit with precision.

A giant from Revell

Hitherto, Revell has confined its highly-successful 1/32nd scale series of kits to single-engined aircraft, and therefore this company is breaking new ground with its recently-released kit of the Lockheed P-38J Lightning. Though a comparatively small aeroplane by comparison with, say, the P-61 Black Widow, the Lightning spanned a respectable 52ft (15,85 m) and results in a model which, by miniature standards, has the equally respectable span of just over 19 inches (48,26 cm). This must be close to the limit of practicality from the viewpoint of storage, although anyone who can accommodate a 1/72nd scale Lancaster should be able to accept Revell's Lightning.

From most aspects this is a very good kit indeed, and despite its size and apparent complexity is easily assembled, due in no small part to the really superb fit of the components. The wing sections engage the fuselage nacelle and booms by hooks, a system resulting in a very strong model and one of the select few whose parts fit so well that they may be largely assembled without cement. The kit is accurate, apart from some minor points such as the wing tips which are too blunt of contour and the radiator intakes that demand opening out slightly. It incorporates no working parts, but there is a mass of cockpit detail and an excellent miniature Allison engine, the latter being exposed by removal of the upper section of the port cowling. The decal sheet provides the markings of one of the aircraft flown by Major Richard Bong, who scored all 40 of his "kills" while flying Lightnings in the Pacific Theatre.

Where the standard of this kit falls is in tooling detail. The rivet detail is decidedly heavy, the injection gates are large and thick, many parts suffer heavy mould marks, and there is a good deal of flash. These faults, although of a comparatively minor nature, are unfortunately obvious, and prevent us from including this kit among the "true greats", a category in which, from all other aspects, it is clearly entitled to be included. Nevertheless, it would be idle to suggest that Revell's Lightning is not excellent value for money at £1·40, but oh, how we wish that some manufacturer would remember that there *were* variants of the Lightning before the P-38J.

A mammoth from Airfix

It is a long established fact in the aircraft model kit business that one thing more than any other ensures a satisfactory level of sales of a particular model: the aircraft that it portrays must be familiar to the schoolboy, enthusiast and modeller in general. It is also a fact that, other things being equal, the larger the model the greater the contribution that it makes to its manufacturer's profitability. The odds in the success stakes are therefore stacked in favour of a very large Spitfire, and that is exactly what Airfix has produced — a 1/24th scale kit of a "Battle of Britain" Spitfire IA.

If a kit manufacturer has the temerity to select so large a scale for his kit, he is faced with certain obligations. Not only must a great deal of detail be included but this detail must be executed to the highest possible standard. This is a challenge and one which there is no doubt in our mind that Airfix has met handsomely. Many years ago the writer was, for his sins, a rigger on Spitfires, and the assembly of this huge model brought back a profusion of memories. Everything is in the right place, and by exercising a little imagination, one can conjure up the odours so characteristic of the cockpits of British fighters of the period.

The detail of the exterior is to the same high standard as that of the interior — all rivets are in the right places, the ailerons are of the correct section and function properly, and the simulated fabric skinning of the tail control surfaces is absolutely realistic. Even the instruments are properly glazed! The Merlin engine is as faithful a miniature replica of the original as is the rest of the model, and, moreover, is large enough to conceal an electric motor (not included in the kit, incidentally) to rotate the airscrew.

As a kit this is a *beautiful* piece of work. The 150 component parts assemble with precision, little trimming being necessary anywhere. Provided with vinyl rubber tyres, the undercarriage retracts accurately and efficiently, and all control surfaces operate smoothly, although it seems a pity that the elevators could not be linked together. The canopy slides, but it has not proved possible to hinge the access flap in the cockpit side. Apart from the cockpit canopy, the outline of the model is completely accurate, and the fact that the canopy is not correctly shaped is hardly surprising as it is virtually impossible to reproduce in miniature that almost-unique bulged shape of the Spitfire's hood.

The decal sheet, providing alternative markings for two aircraft, is first class, being well printed and matt finished, although the shade of grey used for the code letters may be a little on the dark side. It is a pity, however, that the instructional booklet accompanying the kit attains a lower standard. The booklet accompanying our kit omitted all reference to a vital part (No 5A — the airscrew shaft bearing). Airfix informs us that booklets with this omission should not have been released. This is as may be, but it was only by sheer luck that we avoided ruining the engine of our model. We are doubtful of the wisdom of using instructions of the all-drawing variety on relatively complex models of this type, the drawings not always being easily followed. Furthermore, the instructions make no reference to the correct colouring of such small items as knobs and engine details which are very obvious on a model of this size. Revell's drawings-and-text approach, with the colouring of individual parts clearly indicated, is, to us, infinitely to be preferred.　　　　　W R MATTHEWS

FIGHTERS————————————*from page 8*

choice of the F-4E(F) Phantom). Time scale obviously favours the Mirage F1, but the Cobra offers the prospect of local design participation.

European fighter manufacturers have for some time been frightened by stories that the market was about to be flooded with stripped Phantoms, selling at below £1m ($2·4m) each. The threat has now materialised, but not at the throat-cutting price that had been feared. Germany has announced an intention to purchase a batch of 175-220 F-4E(F)s at a reported flyaway price of £1·5m ($3·6m) each, as an interim replacement for the Lockheed F-104Gs of the *Luftwaffe* in the 1974-8 period.

The standard J79-GE-17 engines and Westinghouse APG-120 multi-mode radar of the standard USAF F-4E are

This full-scale mock-up shows salient features of the proposed Northrop P-530 Cobra, including the unusual location of air intakes for the General Electric GE15 engines.

retained in this new model, but some major items are deleted and other changes made to improve the SEP and reduce cost. These changes include deletion of the rear cockpit, boundary layer blowing, leading edge flaps, the Sparrow system, rear fuselage tank and flight refuelling installation. Allowing for the addition of manoeuvring slats on the wing leading edge, the overall weight saving in the F-4E(F) is 3,300 lb (1 500 kg.). This new version can maintain a sea level Mach 0·6 turn with a radius of 2,900 ft (885 m), compared with a minimum of 3,900 ft (1 180 m) achieved with the basic F-4E version of the Phantom II.

An attractive multi-rôle fighter could have been built around a single reheated Spey, but there is no room for such an aircraft in British planning. However, the Mirage F1 will enter this category when fitted with the M53 Super Atar of 20,000 lb (9 000 kg). With a clean gross weight 20 per cent more than this, it will fall far short of the F-15 in SEP, but if costs can be held down, the F1 may nonetheless repeat the *succèss fou* of the Mirage III.

Curtains for Knuckleheads?

To operate a fighter intensively for ten years is now reckoned to cost more than twice its purchase price. Coupled with the escalation in initial cost, this is making control of the skies an astronomically expensive business. One way out may be to give part of the task to pilotless aircraft which could be far smaller and cheaper, and in peacetime would only require periodic ground checks rather than 200–300 hours per year in the air. In the USA such studies are already in hand, and simulated combats between a modified Firebee drone and conventional fighters are scheduled for late this year. Let's face it, those "knuckleheads" have been hogging the credit for far too long! □

HS141 THE AIRPORT SHRINKER

IN NEW YORK, development of a badly-needed fourth airport has been blocked for years by the objections of local residents who, not unnaturally, take unkindly to the idea of having several square miles of concrete for a backyard. In England's rural Bedfordshire, the inhabitants are manning the barricades to prevent the coming of London's third airport to peaceful Cublington. In a hundred and one cities around the world, noise and pollution make life daily less tolerable for the citizens living close to busy airports. But at Hatfield, near London, designs are already taking shape on the drawing boards for a new airliner which, within 10 years, could reverse this trend towards more and mightier airports.

At a time when the "social manners" of an aircraft are of the greatest importance, great hopes centre upon aeroplanes such as the Hatfield-designed Hawker Siddeley HS.141, which combine vertical or very short take-off-and-landing capability with standards of comfort and speed expected of any new airliner in the 'seventies.

The most obvious advantage of a V/STOL airliner is that it can operate from smaller airports than conventional aircraft, which, to obtain a safe flying speed, must first rush headlong down a broad concrete swath at least two miles in length, their engines at full blast. Not so long ago, it was being assumed that V/STOL operation would enable airliners to be brought closer in to the centres of population, thus reducing the frustrations of travellers whose journey times are often more than doubled by time consumed on the ground travelling to and from airports. Efforts during the past 20 years to develop viable commercial helicopters and scheduled helicopter routes have been motivated primarily by this requirement, but have enjoyed only limited success.

It is now becoming clear, however, that development of V/STOL airliners may be spurred on by the means they offer of reducing annoyance at *existing* airports. This could well become the most urgent reason for pressing ahead with an aeroplane such as the HS.141, even if the total "infrastructure" for an inter-city V/STOL transport system appears unobtainable. For many cities, it is probably already too late to establish special STOLports or VEEports much closer to their centres than existing airports, and travellers will thus have to continue to endure frustrating journeys to and from these airports. But a new generation of V/STOL airliners could at very least stop the trend towards mammoth new airports, sited at ever greater distances from the cities they are supposed to serve, by increasing the capacity of existing airports and simultaneously reducing the noise and inconvenience inflicted on the neighbourhood.

If this view *is* correct, then it is axiomatic that the money needed to develop a third London Airport — wherever it is sited — could be better spent developing, in the same timescale, a genuine V/STOL airliner. Of several designs in the project stage in the UK, the HS.141 is among the most promising, as well as being at a relatively advanced stage of definition.

The term V/STOL is used to indicate an aircraft which

possesses vertical (V) take-off-and-landing capability but would probably be used also in the STOVL mode — that is, making a short take-off run but landing vertically, and thereby achieving better economics. These economies are obtained because an aircraft with installed thrust adequate to enable it to rise vertically at a given weight, can take-off at a greater weight if a short forward run allows the wings to develop some lift. Many of the advantages of a V/STOL type can be claimed for a STOL aircraft (ie, one which does not possess the capability to operate vertically), but it is Hawker Siddeley's view that any new project in this category intended for service towards the end of the decade should have at least the potential of being developed for full V/STOL, even if only STOL-operated initially. This is the underlying philosophy of the HS.141.

Variety of configurations

Many configurations for V/STOL aircraft have been projected and even built in prototype form during the past few years. While some of these configurations may be rejected as impractical for further development, several alternative lines of development remain open to designers. Broadly, these may be considered to fall into three groups: those deriving their lift primarily from rotors, winged types with tilting propellers, and winged types with fans. All make use of powered lift in some form. There is yet a further group based on aerodynamic lift but they can achieve only STOL performance unless aided by some form of powered lift such as the deflection of the slipstream from the propellers.

Hawker Siddeley designers have studied a wide range of configurations with full "V" capability, including the compound helicopter, tilt-wing, convertible rotor and circulation-controlled rotor types, as well as the fan-lift type as represented by the HS.141. The broad parameters of the last-mentioned configuration were defined by a specification issued in 1969 by the Transport Aircraft Requirements Committee, an official body responsible for co-ordinating work on new civil aircraft in the UK. The Outline Requirement then issued requested studies for a 100-seat VTOL airliner with a range of 450 st miles (725 km). Hawker Siddeley's comparative studies led to choice of the fan-lift principle after some early doubts about noise levels had been allayed by Rolls-Royce's work on the RB.202 project. This power plant (see accompanying information box), or a similar type, holds the key to the achievement of acceptable noise levels by a fan-lift type. But it is not only the low noise of the installed engines that is making the V/STOL airliner look increasingly more attractive. It is also the high rates of ascent and descent that it can offer, and which will minimise the area over which noise constitutes a nuisance.

Having decided that fan-lift was the preferred method of achieving vertical capability, Hawker Siddeley designers went on to study a variety of possible airframe configurations, both for the TARC Outline Requirement and for other aircraft categories. Out of these studies the HS.141 emerged as economically the most viable, although location of the lift engines in fairings alongside the fuselage produces some complication by comparison with the use of engine pods, as on the Dornier Do 31-type arrangement.

As would be expected at this stage in a project design, many details of the aircraft have still to be finalised. The description and illustrations appearing in these pages relate to what can be called the "datum" aeroplane — the basis of the submission to the TARC in January 1970. Since that time, studies have been made in which such things as the fuselage cross-section shape and width, the main undercarriage retraction geometry, and the overall dimensions have been varied. Large question marks also hang above the

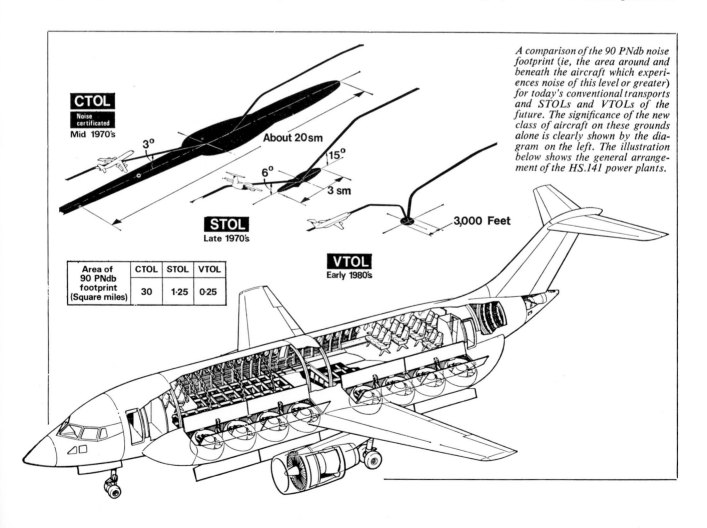

CTOL
Noise certificated
Mid 1970's
3°
About 20 sm
15°
6°
3 sm
STOL
Late 1970's
3,000 Feet
VTOL
Early 1980's

Area of 90 PNdb footprint (Square miles)	CTOL	STOL	VTOL
	30	1·25	0·25

A comparison of the 90 PNdb noise footprint (ie, the area around and beneath the aircraft which experiences noise of this level or greater) for today's conventional transports and STOLs and VTOLs of the future. The significance of the new class of aircraft on these grounds alone is clearly shown by the diagram on the left. The illustration below shows the general arrangement of the HS.141 power plants.

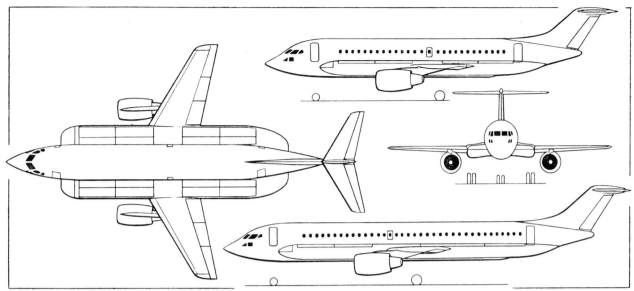

General arrangement of the "datum" HS.141 and, bottom right, the stretched variant with a fuselage length of 135 ft 2 in (41,2m).

choice of power plant, in view of the present Rolls-Royce situation, and the exact power ratings which may eventually be available, either from Rolls-Royce engines or those of other manufacturers. Variations in the number of lift engines required by the HS.141 are therefore likely.

The engine installation in an aircraft such as the HS.141 is of much greater consequence to the airframe design than in a more orthodox airliner. Up to 15 per cent by weight and 35 per cent by cost of the final aircraft may be taken up by power plants and their installation. Upon the engines such aircraft depend not merely for thrust but for control and safety when in the VTOL mode.

Hawker Siddeley's answer

The HS.141 is a conventional low-wing monoplane with modest sweep on wings and tail unit. Its only unusual feature, in so far as structural design is concerned, is the provision of sponsons or fairings running the length of the fuselage on each side to house the lift-fan engines. In the datum aircraft, eight engines are located on each side of the fuselage, four ahead and four aft of the wing. A proposed stretched version to seat up to 150 passengers has 20 engines (10 each side), while STOL variants have been studied with four, six, eight and 12 engines.

Propulsion engines are carried in conventional underwing pods, and in the datum aircraft these power plants each require a thrust of about 27,000 lb (12 250 kg). A Rolls-Royce project in this power bracket is the RB.220, while a French alternative might be the proposed SNECMA M.56 "10-tonne" commercial engine.

The HS.141 variant illustrated here has a fuselage width defined for five-abreast seating, but wider bodies for six-abreast arrangements have been studied. With five abreast, the basic cabin accommodation provides for 102 passengers at 32-in (0,81 m) pitch, rising to 119 at a seat pitch of 28 in (0,71 m). Baggage containers would be tailored to fit the underfloor space, sliding on side runners.

Whilst the mode of operating an aircraft such as the HS.141 will be the subject of much further study and research in the years to come, Hawker Siddeley suggests that, for minimum noise levels around the airport, maximum lift thrust would be used at take-off only to a height of 250 ft (76,2 m). To avoid exceeding the noise limit at the community boundary, power would then be reduced to 83 per cent for the remainder of the climb-out.

Forward transition from the vertical mode would commence at 1,000 ft (305 m) by tilting the fan engines, and the

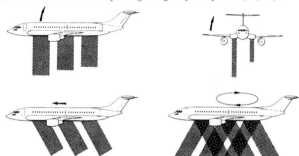

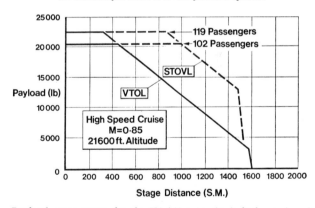

Differential thrust and tilt of the lift-fan engines would be used by the HS.141 for control at low forward speeds.

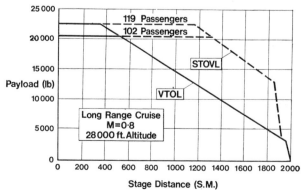

Payload-range curves for the HS.141 using (top) high-speed and (bottom) long-range cruise procedures. The advantage of using a short take-off procedure is clearly illustrated.

aircraft would continue to accelerate as it climbed to 2,000 ft (610 m), aided by the thrust of the cruise engines. At the latter altitude a forward speed of 168 kt (310 km/h) EAS should be reached, and the aircraft would then be fully wing-borne, the lift engines being shut down and inlet and exhaust doors closed to fair off the top and bottom sides of the sponson.

Similarly, for the approach to landing, the lift-fans would be started at a height of 2,000 ft (610 m), at a distance of 4 miles (6.4 km) from the landing point. Allowing one minute for starting and checking the engines, deceleration transition would begin at an altitude of 1,000 ft (305 m) at a distance of 2,650 ft (808 m). Height would then be maintained to almost overhead the landing pad, the let down commencing at 800 ft (244 m) with the vertical speed being reduced from an initial 35 ft/sec (10,7 m/sec) to 10 ft/sec (3,05 m/sec) when the HS.141 was 100 ft (30 m) above the pad. The transition and landing would occupy one-and-a-half minutes.

Control problems

The most unorthodox and most challenging aspect of the HS.141 design lies in the control system. Whilst the aircraft has conventional ailerons, elevators and rudder, these lose effectiveness as speed drops off below what would be the

---

Some Hawker Siddeley V/STOL Projects

D.H.129 1961–2 design studies to NATO NBMR 4 specification, with co-operation from Nord Aviation (France) and Bell (USA). High wing configuration with lift engines in wing pods plus cruise engines in underwing pods.

HS.133 Fan-lift 100-seat civil transport with ogival delta wing and RB.202 engines in fuselage sponsons.

HS.139 Fan-lift 100-seat civil transport using similar layout to D.H.129.

HS.140 Five-seat general purpose and research aircraft with single lift fan.

HS.141 Fan-lift 100-seat civil transport, low swept-back wing and RB.202 engines in fuselage sponsons.

HS.145 Executive transport with lift fan engines.

H.S.681 V/STOL military transport intended to support the RAF Harrier/HS.1154 force. High wing, four engines in underwing pods with vectored thrust capability.

HS.803 HSA Woodford study of civil airliner using NGTE-developed circulation-controlled rotors.

HS.807 HSA Woodford studies of application of RB.162 lift engines to an HS.125-type.

HS.810 HSA Woodford design for a 100-seat civil airliner to TARC Outline Requirement, using circulation-controlled rotor at tip of each wing, and driven by RB.162 engines in wing-tip pods.

An impression of the HS.803 project.

The Rolls-Royce RB.202

Design studies were started by Rolls-Royce Ltd. at Derby about five years ago with a view to developing a quiet lift engine. The company already had experience of lift jet engines through its original work on the two "Flying Bedsteads" and construction of the RB.108 and RB.162. These two engines were pure turbojets designed to operate in the vertical mode when installed in such aircraft as the Short SC.1, Dassault Mirage III-V, Dornier Do 31 and VAK 191B. This line of development has been continued in a collaborative programme with Allison, funded by the US government. Few details of this joint engine are available; it is designated the RB.198 or XJ99 and one possible application was in the projected US-German AVS fighter. Little further Anglo-American work on the programme is likely.

The RB.202 differs from the earlier engines in being a turbofan, designed with low noise as the primary objective, coupled with high reliability and low installed weight. The engine has a gas generator section similar to that of the RB.162, with a single-stage HP turbine, and a single stage fan driven by a three-stage LP turbine. The layout produces a very squat engine, diameter being greater than height.

Velocity of the jet exhaust has an important bearing on noise levels and in the RB.202, this velocity is 640 ft/sec, compared with 2,170 ft/sec (661 m/sec) for the RB.162. The 13,000 lb (5 900 kg) thrust RB.202-31 has an overall diameter of 75·1 in (1,90 m) for a height of 45·4 in (1,15 m), and a by-pass ratio of 9·5:1. The thrust-to-weight ratio of 15:1 compares with 16:1 for the RB.163-81.

The engine has reached the project design phase but further work now depends on the outcome of the recent Rolls-Royce crisis. The company had previously set 1976 as the target date for engines to be available for STOL prototypes.

---

minimum control speed in conventional wing-borne flight. For vertical and very slow forward flight the lift-fan engines are used to provide control, serving the same functions as the "puff-pipe" controls used by the Harrier. This requires differential operation of the engines, both in terms of thrust and thrust vector.

For pitch control, differential thrust is used between forward and aft engines, while roll control comes from differential thrust between port and starboard engines. Control in yaw calls for differential tilt of the engines, those on one side vectoring their thrust forward while the others are tilted to give a rearward thrust. As already noted, for acceleration or deceleration, all the engines are tilted together, rearward or forward. A further consequence of the use of the engines for control purposes is that very rapid response must be available to demands for changes of thrust.

Allowing for the fact that various of the control modes will be required in combination, a complex system is clearly required to convert pilot demands into the correct engine response. Hawker Siddeley proposes to use a manoeuvre demand autostabilisation system with electrical signalling in normal operation, and reversion to a direct mechanical linkage following failure of any autostabilisation channel. Outputs from the control system feed into the fuel control units of three groups of the lift-fans on each side, and into the multiplex tilt actuators on opposite sides of the aircraft.

A further complication arises from the need to design against possible engine failures. Failure of any one lift-fan during the critical take-off and landing stages would require immediate shut-down of its opposite number on the other side of the aircraft to avoid assymetric forces. Such forces might well be aggravated by any differential which was being used for control purposes at the time of the engine failure. Because of the speed of response needed in a situation of this kind, an automatic system is likely to be required, this sensing any engine failures and taking appropriate action to retain control.

Because of the need to design for "two out", the required excess of thrust over weight for vertical operation has to be based on the output of only 14 of the lift-fans in the datum HS.141. Hawker Siddeley assume a thrust of 10,300 lb (4 670 kg) for the RB.202-25, although this engine could be scaled to give anything from 10,000 to 22,000 lb (4 535–9 070 kg), and Rolls-Royce were referring to it as a 13,000 lb s t (5 900 kgp) power plant. At the assumed Hawker Siddeley rating, the HS.141 has a thrust-to-weight margin of 1·16:1 with two failed engines.

Whilst the VTOL operating mode has set the limits within which the HS.141 is designed, the structure will be stressed for operations at weights about eight per cent above the maximum for VTOL. This will allow better economics when the aircraft is operated STOVL. The airframe will be designed for a high cruising speed of Mach 0·85 at 21,600 ft (6 583 m) giving a range of 450 miles (724 km) from a vertical take-off with 102 passengers. Somewhat better ranges are possible using a long-range cruise procedure at Mach 0·8 at 28,000 ft (8 535 m).

As part of the evolution of the HS.141, Hawker Siddeley has conducted extensive wind tunnel and simulator studies, and these are continuing. A 1/10th scale powered model used in the wind tunnel investigation has individually-powered fans and instrumentation to measure both engine performance and aerodynamic characteristics over the whole speed range. Work is continuing on a privately-funded basis, but

for the aircraft to reach production status government aid would be needed. The company believes the HS.141 could technically be prepared for service by 1978–9, and the cost of launching the airframe plus its lift-fan engines has been estimated at about £200m. This excludes the cost of any necessary infrastructure, but as previously suggested, the concept of a V/STOL airliner now appears to be worth pursuing even if operations are restricted initially to existing airports. The decision to go ahead should therefore not need to depend upon setting up a completely new system of STOLports, VEEports or other "downtown" airport sites, and if taken, the HS.141 could prove itself a true "airport shrinker" in the course of time. ☐

HS.141 Specification

Power plant: Two 27,000 lb (12 250 kg) thrust Rolls-Royce RB.220 cruise turbofan engines and 16 10,300 lb (4 670 kg) thrust Rolls-Royce RB.202 lift fans.

Dimensions: Wing span, 75 ft 0 in (22,86 m); wing area, 1,060 sq ft (98,47 m²); aspet ratio, 5·3:1; sweepback, 28 deg on quarter-chord line; overall length, 120 ft 2 in (36,63 m); max fuselage diameter, 11 ft 3 in (3,43 m); max internal cabin width, 10 ft 8 in (3,25 m); overall height, 29 ft 10 in (9,00 m); tail-plane span, 32 ft 10 in (10,00 m); tailplane and elevator gross area, 239 sq ft (22,20 m²); fin and rudder gross area, 145 sq ft (13,47 m²).

Weights: Design take-off weight, 134,200 lb (60 872 kg); max VTO weight, 124,200 lb (56 336 kg); design landing weight 118,000 lb (53 524 kg); max zero fuel weight, 110,300 lb (50 031 kg); fuel weight, 33,500 lb (15 195 kg).

Performance: Design Vc, 375 kt (695 km/h); design Mc, 0·85; design VD, 435 kt (806 km/h); design MD, 0·92; range with max payload (VTOL), 400 st mls (644 km); range with max payload (STOVL) 1,200 st mls (1 931 km).

RUMANIA ───────────────────────── *from page 35*

Rumania signed the Warsaw Pact, and in the following year the fighter element of the FR-RPR began to receive day and all-weather versions of the MiG-17, and additional Il-28s to permit conversion of the Il-10-equipped Regiment. The Rumanian air arm now possessed two fighter regiments, or *regimente vinâtoare,* each possessing 60-70 MiG-15s and MiG-17s, a *regimente bombardiere* with 30-35 Il-28s, a liaison and transport group, a naval group and an airborne battalion, personnel strength being 10,800 men.

By 1960, a third fighter regiment had been created, the bomber regiment had been increased in strength to some 50 Il-28s, several independent interceptor squadrons had been formed, the transport capability of the FR-RPR had been markedly increased, and personnel strength had risen to 16,000 men. Early in the 'sixties a number of MiG-19S fighters were acquired, but with the withdrawal of the last Soviet forces from Rumania in 1964, the expansion of the FR-RPR stopped, the status-of-force agreement concluded with the Soviet Union in 1957 was allowed to lapse, and the Rumanian armed forces stopped participating in Warsaw Pact manoeuvres. In 1965, in which year the FR-RPR began to receive L 29 Delfin jet trainers from Czechoslovakia, the service had phased in its first MiG-21F interceptors and included some 220 first-line combat aircraft in its inventory, but personnel strength had been reduced to fewer than 10,000 men.

Today, Rumania, despite being the second largest Warsaw Pact nation in both area and population, after the Soviet Union herself, maintains one of this alliance's smallest air arms. The *Fortele Aeriene ale Republicii Socialiste România,* or FR-RSR, as the air arm was renamed in 1966 when Rumania was declared a Socialist Republic, is solely a defensive force, and the larger part of its first-line strength consists of interceptor fighters which, together with the V750VK (Guideline) surface-to-air missile batteries operated by the Army, are assigned to the Warsaw Pact air defence

system directed by the P-VO *strany* in Moscow. The Rumanian government, possibly as a result of the events of August 1968 in Czechoslovakia, has now apparently abandoned its tendency to adopt a neutralist policy, having given the Soviet Union assurances that Rumania will fulfil all her obligations to the Warsaw Treaty.

Possessing some 10,000 personnel, representing approximately six per cent of the regular Rumanian armed forces, the FR-RSR is equal in status to the Armour and Infantry components. Its interceptor force comprises three fighter regiments equipped with the MiG-21F and -21PF, and the remainder of the operational element reportedly comprises two ground support regiments and several independent ground attack squadrons which, having a secondary intercept role, operate the MiG-19SF, -19PM, and -17F, although a small number of Sukhoi Su-7s have allegedly been delivered.

Logistic support is provided by the so-called *divizie transport aeriană,* which possesses only regimental strength with some three squadrons each with 8-10 Il-14s and Li-2s. There is a small helicopter component equipped primarily with the Mi-4, but including a number of Polish-built Mi-1s and augmented in recent years by small quantities of Mi-2s and Mi-8s.

The training organisation consists of two flying training schools which employ an all-through jet training syllabus using the L 29 Delfin and the MiG-15UTI, but all pupils have received some *ab initio* training on such types as the Zlin 226 or Yak-18 before induction by the FR-RSR. Conversion to the MiG-21 interceptor is usually undertaken at unit level with the two-seat MiG-21UTI. For the utility role small numbers of such types as the An-2, L 60 Brigadyr and L 200 Morava are employed, and the current inventory of the FR-RSR comprises some 360 aircraft of which about 220 are combat types. ☐

The editors wish to acknowledge the assistance of Ferenc-Antal Vajda and Georges Zagonesco in the compilation of this feature.

The 'WIDOW'

It is one of the paradoxes of warplane development that some of the most successful combat aircraft in the annals of military aviation have found their true métier fulfilling tasks unforeseen at the time of their conception. Nocturnal interception was a role that was to provide several outstanding examples of this paradox. From the birth of the *true* night fighter — the aircraft carrying airborne intercept radar enabling it to seek out its quarry in the night skies without the aid of searchlights, flares or moonlight — until the end of World War II, all aircraft used for nocturnal interception were, with but one exception, the products of improvisation. For the first time, AIR ENTHUSIAST tells the full story of this one deviation from the pattern of improvisation, the Northrop P-61 Black Widow.

Hitherto, the historian has largely neglected Northrop's P-61; its virtues have rarely been extolled and its successes have gone unsung. Aesthetically the P-61 could lay little claim to distinction. Indeed, it was perhaps the ugliest fighter of its generation. It was by no means the heaviest if marginally the largest fighter to be committed to the conflict. But the P-61, which made its operational début on both sides of the world almost simultaneously, enjoyed the unique distinction of being the first warplane designed from the outset for the nocturnal intercept role to attain service status. The P-61 possessed its share of shortcomings, as, indeed, has every warplane ever designed, but these were outnumbered by its attributes. It was an imaginative answer to a highly-demanding requirement.

IN THAT last full summer of World War II, to the Londoner with a taste for the grape seeking that most efficacious remedy for depression and festive wine *par excellence*, champagne, mention of the "Widow" would have meant but one thing: a bottle of Veuve Clicquot. To the occupants of the hotter regions of Latin America, the "Widow" signified *Latrodectus Mactans*, a highly unpleasant spider the virulence of whose poison caused intense pain and possibly death. At the airfields of Scorton and Hurn in the UK and at bases in the Central and South West Pacific Areas, however, this appellation had taken on another meaning, for it was borne by an entirely new combat aircraft which USAAF night fighter squadrons were just committing to operations.

The wraps had *officially* been taken off this new warplane, the Northrop P-61 Black Widow, some six months earlier, on 9 January 1944, when it made its public début before 75,000 spectators at the Los Angeles Coliseum, although the external appearance of this immense fighter had become familiar months before as a result of its inclusion in a comic strip! Conceived three-and-a-quarter years prior to its arrival at operational bases, the P-61 had certainly made an impressive sight over the Coliseum. As large as a medium bomber and sporting a shiny black finish, the P-61 possessed all the squatness of appearance that characterises most venomous spiders, and the name "Black Widow" thus appeared singularly appropriate.

The finish of the P-61 was not the least interesting of the features of the new nocturnal predator. Intended to prowl at night in enemy air space, it was important that the P-61 should be rendered as inconspicuous as possible if illuminated by the beam of a searchlight. The first P-61s off the

assembly line were sprayed dull black and subsequent aircraft were given an olive drab-and-grey finish, but both paint schemes glowed white in the beams of searchlights, the dull black being barely less conspicuous than the olive drab and grey. Shortly before deliveries of the P-61 to the USAAF began, the National Defense Research Committee had submitted proposals that a new high-gloss black finish be tested, and these tests had been conducted at Eglin Field, Florida, in the late autumn of 1943. Three aircraft were used, only one of which had a high-gloss black finish, and these were flown at night over Fort Barrancas where searchlight crews awaited them.

The aircraft sporting dull black and olive drab-and-grey finishes were picked up immediately, presenting bright silhouettes as they made their runs through the glare of the searchlights. The aircraft to which the high-gloss black finish had been applied did not put in an appearance — or so it seemed to the searchlight crews. The pilot was contacted and it was only then discovered that he had made his runs through the lights on schedule. The gloss black had refracted the rays of the searchlights! Therefore, from the beginning of January 1944, the high-gloss black finish had been applied to P-61s as standard, but a long period of gestation had elapsed at Hawthorne, California, between the conception of Northrop Aircraft's contribution to the genus *Latrodectus* and the roll-out of the first production "Widow" in its definitive warpaint.

Development of airborne radar for use by fighters at night had begun in Britain in 1936, and within weeks of the commencement of hostilities in Europe, operational trials with A.I.Mk III radar were being performed from Manston with a trio of Blenheim IF fighters. From the outset it was apparent that, in order to create a fully effective night fighter, *design* rather than *adaptation* of the aircraft for nocturnal combat was mandatory. The exigencies of the times precluded such a development in Britain, however, and for the RAF there was no alternative *but* improvisation. The Beaufighter was the only aircraft in production capable of carrying the operational A.I.Mk IV radar without serious sacrifice of endurance or firepower, and it was therefore adapted as a night fighter.

In the USA, the Army Air Corps Material Command, after studying the report on British night interception experience compiled by General Emmons and his team of observers who had visited Britain during the late autumn of 1940, concluded that conversions of existing aircraft for the nocturnal role were necessary as emergency measures but that the design of a specialised night fighter should be initiated immediately. The broad parameters of the Army Air Corps' requirement were presented to John K Northrop on 21 October 1940, and two weeks later, on 5 November, Northrop and his assistant, Walter J Cerny, were at the Material Command Headquarters at Wright Field, Dayton, with their outline proposals.

As a result of the discussions at Wright Field, Northrop Specification NS-8A was submitted to the Material Command on 5 December and, after some revisions, re-submitted on 17 December, the project having by this time been allocated the designation XP-61. A fixed-price contract was prepared and signed by Northrop on 11 January 1941, and this, covering a total cost of $1,167,000, received official approval on 30 January. In addition to two XP-61s in accordance with Specification NS-8A, the contract included the tunnel testing of an 8 ft (2,44 m) model of the XP-61 in the Northrop wind tunnel, the provision of a model of the XP-61 suitable for testing in Wright Field's 5 ft (1,52 m) tunnel, and a free-spin model for testing by the NACA Laboratory.

A spider emerges

The XP-61 was an extremely advanced design from several aspects, apart from being a giant among fighters. Of twin-boom configuration with a central nacelle housing the three crew members, the intercept radar and much of the armament, the XP-61 was intended to be powered by two Pratt & Whitney R-2800-A5G Double Wasp 18-cylinder radials cooled by means of ducted spinners. Each boom was to house a 225 Imp gal (1 023 l) fuel tank between the wing spars, four 20-mm cannon were to be mounted in the wings, and the potency of this sting was to be augmented by a pair of remotely-controlled power-operated barbettes mounting a total of six 0·5 in (12,7 mm) machine guns with 500 rpg — four in the dorsal barbette and two in the ventral barbette.

Soon after approval of the XP-61 contract had been given, it became apparent that some increase in fuel capacity was desirable, and Northrop was asked to investigate the feasibility of raising the fuel load. Calculations indicated that it would be possible to provide a further 100 Imp gal (455 l) to bring the total to 550 Imp gal (2 500 l), and this change was duly incorporated in the specification, but on 14 March the Northrop team was faced with what promised to be its first serious setback — an Army-Navy Standardisation Committee decision to standardise on up-draft carburettors for the R-2800 engines. The XP-61 was being designed around engines with down-draft carburettors, and a change at that stage would have necessitated scrapping some 75 per cent of all work completed on the power plant installation, and a delay of some two months in the schedule! Fortunately, a week after taking this decision the Standardisation Committee performed a *volte-face*, enabling engines with down-draft carburettors to be furnished.

But Northrop's difficulties had only just begun. On 2 April the Hawthorne factory was visited by the mock-up board from Wright Field, this comprising Majors F C Wolf, R C Wilson, and M S Roth, and Lt Col J G Taylor. The mock-up board concluded that, in general, the configuration of the XP-61 was satisfactory, but declared its opinion that the 20-mm cannon should be moved from the wings to the lower portion of the fuselage nacelle, the ventral barbette being discarded. This change, it was alleged, would simplify maintenance and simultaneously improve airflow over the wing, but, unfortunately, it dictated major modifications including several structural changes, these necessitating revision of the stress analyses and the reworking of many of the drawings already completed under Northrop's accelerated development programme. Nevertheless, the change was made at the cost of one month to the schedule and $38,137 to the contract.

Late in October 1941, Northrop was notified that the model supplied for testing in the Wright Field wind tunnel did not comply with the specification, but since the deviations from the specification were of a relatively minor nature, it was agreed between Northrop and the Material Command that the model be accepted at reduced cost. At the same time, it became apparent that General Electric could not furnish a dorsal barbette for the first XP-61 prior to the initial test flight unless this item was allocated an A-1-A priority. At that time experimental aircraft were being procured under an A-1-B priority, and for this reason it was decided to mount a mock-up of the barbette on the first XP-61 to provide the necessary external contours for determining drag during flight testing.

On 16 December, the Material Command was notified by Northrop that the XP-61 project had been delayed a further three weeks as a result of the high priority that had been given to work on the Vultee A-31 (Vengeance I) dive bomber*. Immediately after the Japanese attack on Pearl Harbour, action was taken to obtain the greatest number of combat aircraft possible by 31 December, but the directive was *not* intended to cover Northrop's production of the A-31, and was promptly cancelled as soon as the matter was brought to the attention of the Material Command. Nevertheless, more valuable time had been lost.

Meantime, such was the urgency attached to the XP-61 project that, on 1 September 1941, an initial order had been

---

*In 1940 the British Purchasing Commission had placed orders for the Vultee V-72 Vengeance which was assigned the designation A-31 for procurement purposes. Contracts had been placed with both Vultee and Northrop, the latter setting up a production line at Hawthorne for 400 dive-bombers which were delivered between January 1942 and April 1943.

(*Above and below*) The first XP-61 (41-19509) which, initially flown on 26 May 1942, was fitted with a mock-up of the dorsal barbette.

(Below) The first YP-61 (41-18876) as completed in August 1943 with four-gun dorsal barbette, and (above) the second YP-61 (41-18877) after delivery to the AAF with reinforced barbette and inboard guns removed.

placed for 150 production aircraft, this being followed on 12 February 1942 by an order for a further 410, and the first XP-61 had still to make its maiden flight. Yet a further problem had to be faced before this event. On 10 February, Northrop was informed that the Curtiss Electric airscrews intended for production aircraft would not be available, and was requested, therefore, to ascertain the suitability of Hamilton Standard airscrews. These were to be fitted to the first XP-61 but were to prove unsatisfactory owing to severe vibration. An investigation was to be initiated to discover means of eliminating this problem but discontinued when it was finally determined that sufficient Curtiss Electric airscrews *could* be made available.

The "Widow" gets airborne

The first XP-61 (Serial No 41-19509) took off on its maiden flight at 16.37 hours on 26 May 1942 with Northrop test pilot Vance Breese at the controls. The prototype remained airborne for 15 minutes and, in general, its characteristics were considered satisfactory, but subsequent testing revealed the need for a number of changes, the first of these being the replacement of the original R-2800-25S engines by -10 engines. This modification was recommended by Pratt & Whitney in order to eliminate the trapping of oil and gas in the crankcase and resultant loss of power.

On 19 June, Northrop decided to design and fit new horizontal tail surfaces in order to improve longitudinal stability. The XP-61 had been flown initially with so-called Zap flaps but owing to the close manufacturing tolerances demanded by such flaps for their satisfactory operation, they were considered to be impracticable for production aircraft, and the new horizontal surfaces were designed to complement the aerodynamic characteristics of the conventional full-span slotted flaps which were to be coupled with retractable spoiler-type ailerons on production P-61s.

Two engine failures were experienced during the initial test phase, the starboard engine failing on 9 September and the port engine on 17 September, both failures being attributed to lubrication faults which were corrected by Pratt & Whitney by the introduction of oil jets within the power plants. Early in October, Northrop ascertained that the rear spar caps were some three per cent under strength, and although this shortcoming was not considered sufficiently serious to prevent preliminary diving tests, it was adjudged unsafe for final demonstration dives and pull-outs. The time factor in the test programme motivated against any attempt

to strengthen the spar caps, and, in consequence, the XP-61 had to be restricted to safe speeds and accelerations.

In view of the experimental nature of the radar (SCR-720) to be used by the aircraft and its secret status, it was decided that this equipment would be installed after delivery of the aeroplane to the Material Command at Wright Field. It was also decided that Northrop should redesign the cannon installation, this including the replacement of the hydraulic gun chargers by manual chargers.

The second XP-61 (Serial No 41-19510) joined the flight test programme on 18 November 1942, and initial reports were satisfactory with the exception of the rudders, the operation of these demanding an excessive amount of energy from the pilot. This fault was overcome by modifying the shape of the rudder leading edges. Gun firing trials began on 18 December, but difficulties were encountered with the forward and aft canopies, the life raft hatch and the nose-wheel doors as a result of blast effect, and these structures had to be reinforced. During February 1943, there was some indication that the spoiler-type ailerons tended to jerk in and out in turbulent air, and the Material Command recommended that these be statically balanced. The Command also cautioned Northrop against flap flutter, which was, in fact, to be experienced on 4 April. At the time of this incident, the pilot's visual accelerometer indicated positive 9g and negative 6g, and although the aileron controls became inoperative as a result of the flap flutter, the pilot was able to make a safe landing. The flutter was attributed to faulty adjustment of the flap stops, but as a precautionary measure mechanical locks were fitted to hold the flaps in the retracted position when not in use in case of failure of the hydraulic system.

The second XP-61 was ferried to Wright Field for installation of the AI radar on 15 April 1943, and a month later performance tests were conducted by AAF pilots at Hawthorne with the first XP-61. On 8 July, the resident AAF representative at Hawthorne was notified that the demonstration flights of the first XP-61 were considered satisfactory, and that this aircraft should be accepted in accordance with the terms of the contract, and on the following day the second prototype was accepted at Wright Field.

Deliveries commence

Barely more than five weeks after the XP-61 contract had received official approval, a further contract calling for a static test airframe and 13 YP-61s for service test and evaluation had been approved, these aircraft (Serial Nos

(Opposite page) A P-61A-1 (42-5528) "Jap Batty" of the 6th Night Fighter Squadron operating from Saipan, autumn 1944. It should be noted that although this was the 44th P-61A-1 in serial number sequence it was fitted with the gun barbette which was featured by only the first 37 aircraft. This anomaly is presumably accounted for by the fact that the initial batch of aircraft were not completed in serial sequence. The "Jap Batty" is illustrated in colour on page 36.

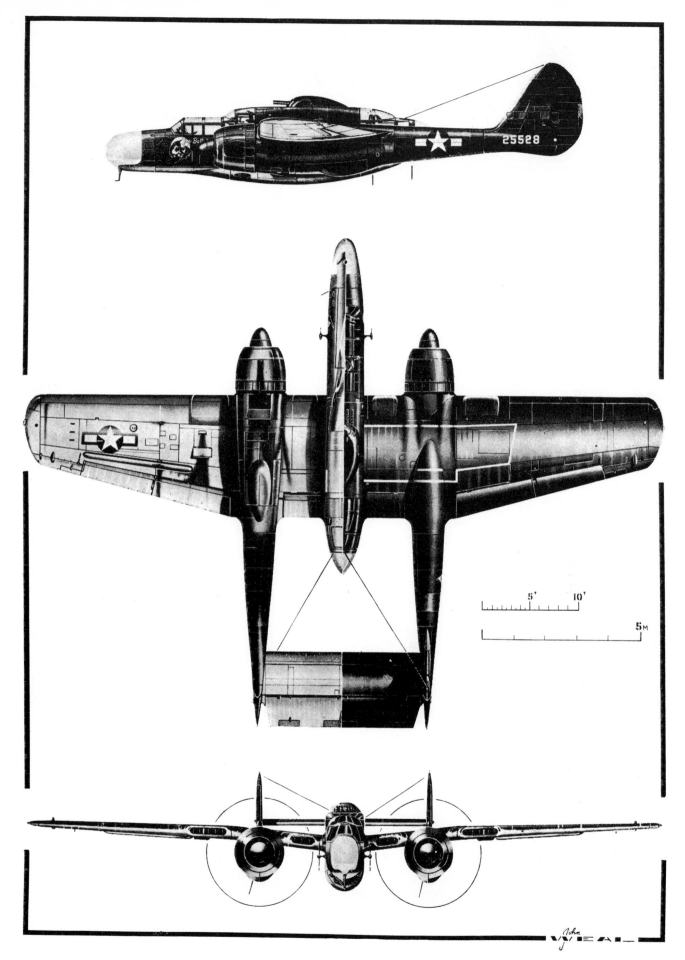

25528

John YEAL

41-18876 to -18888) rolling off the Hawthorne assembly line during August and September 1943. Their testing revealed serious buffeting when the four guns in the barbette were elevated or rotated in azimuth, and while a solution to the problem was being sought in the 20-ft (6,10-m) wind tunnel at Wright Field where flow tests were conducted with a complete YP-61 fuselage nacelle, the YP-61s were tentatively accepted with structurally-reinforced barbettes from which the two inboard guns had been removed.

Before the buffet problem was encountered, however, the first production P-61A-1 Black Widow night fighters had begun to leave the assembly line, deliveries commencing in October 1943, and the first 37 aircraft off the line (Serial Nos 42-5485 to -5522) retained the four-gun barbette, while the remaining 163 P-61As were completed as two-seat aircraft *without* the barbette* pending a solution to the buffet problem. Somewhat surprisingly, testing revealed that removal of the barbette resulted in a gain of only 3 mph (4,82 km/h) in maximum speed both at war emergency power at 17,600 ft (5 364 m) and at military power at 21,600 ft (6 584 m). Maximum sea level climb rate was increased from 2,610 ft/min (13,26 m/sec) to 2,925 ft/min (14,86 m/sec) and from 2,120 ft/min (10,77 m/sec) to 2,420 ft/min (12,29 m/sec) at war emergency and military power respectively, while an altitude of 20,000 ft (6 096 m) was attained in 10 min as compared with 11·7 min. Take-off gross weight with the barbette fitted was 29,240 lb (13 263 kg) and without the barbette was 27,600 lb (12 519 kg).

The first 45 Black Widows delivered were P-61A-1s with R-2800-10 engines each rated at 2,000 hp, but the 46th and following 34 aircraft were fitted with R-2800-65 engines as P-61A-5s (Serial Nos 42-5530 to -5564), these having war emergency ratings of 2,250 hp at 1,500 ft (457 m). Water injection was incorporated in the 100 P-61A-10s (Serial Nos 42-5565 to -5604, 42-5607, 42-5615 to -5634, 42-39348 to -39374, 42-39385 to -39386, and 42-39388 to -39397) that followed, and interspersed on the assembly line with these were 20 P-61A-11s (Serial Nos 42-5605 to 5606, 42-5608 to -5614, 42-39375 to -39384, and 42-39387) which had pro-

Some P-61As originally delivered without the dorsal barbette were retroactively fitted with the barbette after the buffet problem had been resolved.

vision for two 258 Imp gal (1 173 l) external fuel tanks or two 1,600-lb (726-kg) bombs.

Despite her awesome size, the Widow was found by AAF pilots to be an extremely docile aircraft in flight and surprisingly manoeuvrable, although overly conservative manufacturer's limitations resulted initially in the P-61 acquiring something of a reputation for being sluggish and unresponsive. Once service pilots had thoroughly familiarised themselves with the Widow's characteristics she received almost universal praise. All control surfaces were highly effective in all flight configurations, from limiting diving speed down to the stall. Forward visibility during taxying and take-off was extremely good apart from some minor distortion resulting from the curvature of the canopy panelling, but poor during the climb owing to the nose-up attitude of the aircraft.

For an aircraft of its size and weight, manoeuvrability was quite outstanding. Loops, Immelmann turns, barrel rolls and most other precision and semi-precision manoeuvres could be performed with ease, and the Widow could out-turn many single-seaters. The only manoeuvres prohibited because of its weight were outside loops, inverted flight, spins and snap rolls. The Widow was statically and dynamically stable about the normal and lateral axes and neutrally stable about the longitudinal axis. It handled satisfactorily at all altitudes up to its service ceiling, although it became somewhat sluggish above 30,000 ft (9 145 m). At a gross weight of 26,200 lb (11 884 kg) without gun barbette and with power off, a "clean" Widow stalled at 105 mph (169 km/h) IAS, and with undercarriage extended and flaps fully down, the indicated stalling speed was 85 mph (137 km/h).

Below 150 mph (241 km/h) IAS the controls became a little "mushy" and considerable movement of the control column was required, but adequate control could be maintained down through the stall. The recommended safe gliding speed on the approach with full flap was 120 mph (193 km/h) IAS. One feature of the Widow that occasioned some concern, however, was the position of the pilot who was seated almost directly in line with the plane of the airscrews. If a forced landing with wheels up proved necessary, or the undercarriage collapsed during take-off or landing, the pilot was in some danger from the starboard airscrew which usually broke up on contact with the runway, fragments sometimes penetrating the cockpit.

The central fuselage nacelle of the Widow was of stressed-skin, semi-monocoque aluminium alloy construction, and the wing was a stressed-skin two-spar cantilever structure composed of six panel assemblies — two inner panels, two outer panels and two tip panels — which were bolted together by internal fittings accessible via detachable panels. The engine nacelles, which also housed the main undercarriage members, were extended aft as tailbooms. On the XP-61 and YP-61 models these booms were manufactured from welded magnesium alloy, but severe manufacturing and quality

(Above) A P-61A-1 (42-5508) retaining the four-gun dorsal barbette. This, the twenty-third production "Widow", was among the first to reach the Pacific Theatre of Operations. (Below) One of the thirty-five P-61A-5s (42-5544) after arrival in the UK in 1944.

(Above) The fourteenth production P-61C-1 (43-8334). Note the broad-chord high-activity airscrew blades, the cheek-type carburettor air intakes and ventral turbo-supercharger air intakes. (Below right) This early P-61B-1 (42-39419) was unusual in having the dorsal gun barbette which was not generally reintroduced until the appearance of the P-61B-15.

control problems experienced by Northrop with small-gauge magnesium alloy sheet led to the redesign of the booms of the production Widow for aluminium alloy construction.

The "Widow" stings

The first unit to receive production Widows was the 348th Night Fighter Squadron, a component of the 481st Fighter Group in Florida which had the primary responsibility of training AAF night fighter crews, and eventually included the 349th and 420th Night Fighter Squadrons. Almost simultaneously, in March 1944, personnel of the first P-61 unit assigned to the European Theatre of Operations, the 422nd Night Fighter Squadron, arrived at Charmy Down airfield, moving on 9 May to Scorton to commence working up. A second P-61 squadron, the 425th NFS, reached Charmy Down in May 1944, proceeding to Scorton in June, and on 29 June six Black Widows from the two squadrons were attached to an RAF night fighter squadron at Hurn from where the first sorties were flown on the evening of 3 July.

Both the 422nd and 425th were equipped with barbette-less P-61A-5s and P-61A-10s, and their full complement was intended to be 18 aircraft and 16 crews per squadron, but they were handicapped by insufficient deliveries of P-61s throughout the summer and autumn of 1944, their average strength rarely exceeding 10 aircraft during this period. Their P-61s effected a number of successful night interceptions of FZG 76 flying bombs during July, and on their first intruder missions over France in August they claimed four "kills", including a Bf 110 which was stalked by the Widow for 23 minutes before finally being despatched.

It was in the nocturnal skies of the Central Pacific Area that the Widow first drew blood, however, for the initial

P-61 unit to achieve operational status in the CPA, the 6th NFS, claimed its first "kill" — a Mitsubishi G4M bomber — on the night of 6 July. This was followed on the next night by the initial "kill" of the first P-61 unit to join operations in the South-West Pacific Area, the 421st NFS, which destroyed a Mitsubishi Type 100 reconnaissance aircraft over Japen Island. The 421st NFS had received its first Widows in June as replacements for the unit's P-70s, being followed in the SWPA by the 418th NFS which converted from B-25G Mitchells to P-61s at Hollandia in August 1944.

The 14th Air Force in China received its first Widows on 6 October, these being flown into Chengtu by the 426th NFS, and the 10th Air Force in the India-Burma area added P-61s to its order of battle at about the same time with the arrival of the 427th NFS which, originally sent to Italy in August, had flown only five sorties before transfer. From December 1944 until May 1945, the Widows of the 427th were to be employed primarily in the ground attack rôle with rocket tubes beneath their wings.

Four night fighter squadrons sent to the UK in 1943, the 414th, 415th, 416th, and 417th (these subsequently transferring to North Africa and Italy), were equipped with Beaufighters in lieu of Widows, but shortages of P-61s were to delay their planned conversion until the final months of hostilities in Europe.

P-61B-20-NO Black Widow Specification

Power Plant: Two Pratt & Whitney R-2800-65 Double Wasp 18-cylinder two-row radial air-cooled engines each rated at 1,850 hp for take-off at sea level and 2,000 hp at 2,300 ft (700 m), with war emergency ratings of 2,250 hp at 1,500 ft (457 m), 2,060 hp at 13,000 ft (3 960 m), and 1,925 hp at 17,500 ft (5 335 m), and military ratings of 1,760 hp at 16,600 ft (5 060 m) and 1,600 hp at 21,600 ft (6 585 m).
Armament: Four 20-mm cannon in fuselage with 200 rounds per gun, and four 0·5-in. (12,7-mm) Colt-Browning machine guns in dorsal barbette with 560 rounds per gun. External loads comprising four 100-lb (45,3-kg), 250-lb (113,4-kg), 325-lb (147,4-kg), 500-lb (227-kg), 1,000-lb (454-kg), or 1,600-lb (726-kg) bombs.
Weights: Empty equipped, 23,450 lb (10 637 kg); normal loaded, 29,700 lb (13 471 kg); maximum overload, 36,200 lb (16 420 kg).
Performance: (At 29,700 lb/13 471 kg) Maximum speed at war emergency power, 330 mph (531 km/h) at sea level, 333 mph (536 km/h) at 5,000 ft (1 524 m), 352 mph (566 km/h) at 10,000 ft (3 050 m), 362 mph (582 km/h) at 15,000 ft (4 570 m), 366 mph (589 km/h) at 20,000 ft (6 095 m); maximum speed at military power, 316 mph (508 km/h) at sea level, 322 mph (518 km/h) at 5,000 ft (1 524 m), 331 mph (533 km/h) at 10,000 m), 348 mph (560 km/h) at 15,000 ft (4 570 m), 354 mph (570 km/h) at 20,000 ft (6 095 m), 356 mph (573 km/h) at 25,000 ft (7 620 m); range at maximum continuous power, 360 mls (579 km) at 300 mph (483 km/h) at 10,000 ft (3 050 m), 410 mls (660 km) at 315 mph (507 km/h) at 25,000 ft (7 620 m), at maximum cruising power, 785 mls (1 263 km) at 290 mph (467 km/h) at 10,000 ft (3 050 m); range at maximum continuous power at 32,100 lb (14 560 kg), 550 mls (885 km) at 305 mph (491 km/h) at 10,000 ft (3 050 m), 610 mls (982 km) at 339 mph (545 km/h) at 25,000 ft (7 620 m); range at longe-range cruising power at 32,100 lb (14 560 kg), 1,350 mls (2 172 km) at 229 mph (368 km/h) at 10,000 ft (3 050 m); initial climb rate at war emergency power, 2,550 ft/min (12,9 m/sec), at military power, 2,090 ft/min (10,6 m/sec); time to 5,000 ft (1 524 m), 2·7 min, to 10,000 ft (3 050 m), 5·6 min, to 15,000 ft (4 570 m), 8·6 min, to 20,000 ft (6 095 m), 12 min.

Dimensions: Span, 66 ft 0¾ in (20,14 m); length, 49 ft 7 in (15,11 m); height, 14 ft 8 in (4,47 m); wing area, 662.36 sq ft (61,53 m²).

This shortage of the Northrop night fighter was largely the result of delivery delays due to a host of minor modifications dictated by operational experience. The AI radar was particularly troublesome and during the first months of operations AI failures prevented the completion of a number of interceptions. Modifications introduced at early stages in the production run included the incorporation of an emergency release handle in the pilot's canopy that could be operated from outside; the provision of emergency release catches for the crew entry doors; the introduction of emergency downlock releases for the main undercarriage members, and a control device to prevent retraction of the mainwheels until after the struts had been fully extended. Stiffeners were added to reduce skin wrinkling; the tail cone was reinforced; shell case ejection chute doors were eliminated, and a shield was perfected to deflect exhaust gases from the air filter.

Months before the Widow first fired her guns in anger, the P-61A had given place on the assembly line to the P-61B of which 450 examples were built (Serial Nos 42-39398 to -39757 and 43-8231 to -8320). Powered by the R-2800-65 engine, the P-61B was similar to the P-61A apart from an 8-in (20,32-cm) nose cone extension which increased overall length from 48 ft 11 in (14,91 m) to 49 ft 7 in (15,11 m). Deliveries of the P-61B began in July 1944, and the two wing racks introduced by the P-61A-11 were standard on the P-61B-2, -5 and -11, while the P-61B-10 was the first model to introduce four wing racks for either 258 Imp gal (1 173 l) fuel tanks or 1,600-lb (726-kg) bombs. With the 201st B-series Widow, the first P-61B-15, the four-gun dorsal barbette was reintroduced, the buffeting problem having been largely resolved, and with the P-61B-20 a new General Electric barbette and a redesigned fire control system were incorporated.

Reports from the combat zones indicated two areas in which improvements in the Widow's performance envelope were desirable — speed and altitude. In an endeavour to improve the capabilities of the Widow in these respects Northrop evolved the P-61C. This differed from the P-61B primarily in having R-2800-73 engines equipped with General Electric CH-5 turbo-superchargers and driving fully-feathering hollow-steel Curtiss Electric airscrews. These engines possessed a war emergency rating of 2,800 hp, and despite an increase in normal gross weight to 32,200 lb (14 606 kg) it was calculated that maximum speed would be raised to 430 mph (692 km/h) at 30,000 ft (9 145 m), and that service ceiling would be increased to 41,000 ft (12 500 m).

While this increment of speed could be expected to reduce interception time, the greater closure rate would obviously increase the chances of overshooting the target. Therefore, to provide the pilot with more latitude in the control of his aircraft, the P-61C was fitted with interconnected air brakes above and below the wings. The maximum weight had by now risen to 40,300 lb (18 280 kg), and when flight testing commenced in July 1945 it was soon ascertained that above 35,000 lb (15 876 kg) the aircraft became sluggish, difficult to handle, and tended to "wallow" in flight. In addition, excessive ground roll was necessary before flying speed was attained, and AAF test pilots recommended that, at 40,000 lb (18 144 kg), a take-off should not be attempted unless at least three miles (4,82 km) of obstruction-free space was available. Nevertheless, production of the P-61C had already begun, and 41 (Serial Nos 43-8321 to -8361) had been completed by V-J day when a further 476 Black Widows of this version were cancelled.

Prior to the completion of the P-61C, testing of the XP-61D had begun. The XP-61D had been intended to offer an improvement in high-altitude performance, considered the most serious shortcoming of the Widow, and during February 1944, a P-61A-5 (Serial No 42-5559) had been selected for adaptation as the XP-61D by the installation of two Navy-type R-2800-14 engines with CH-5 turbo-superchargers. The R-2800-14 was essentially similar to the AAF's R-2800-77 and had been chosen as it was more readily available than the AAF power plant. As Northrop was working at full capacity, the Goodyear Aircraft Corporation, which was already building component parts of the Black Widow, was selected as the sub-contractor for the XP-61D, Northrop retaining responsibility for all preliminary design layouts and mock-up construction, and Goodyear assuming responsibility for all detail design and construction.

Northrop requested permission to select a second airframe for adaptation as an XP-61D to ensure that the test programme would not be delayed in the event of damage being suffered by the first aircraft. This request was granted and a P-61A-10 (Serial No 42-5587) was chosen as the second XP-61D. The first engine runs were conducted on 18 November 1944, and the first flight of the XP-61D followed a few days later. Initial testing proved encouraging but the programme was plagued by repeated engine failures which prevented the full performance envelope from being explored. In the meantime, the first P-61C had been completed, and further development of the XP-61D was abandoned.

In an effort to evolve a high-performance, long-range escort fighter, two P-61B-10 Black Widows (Serial Nos 42-39549 and 42-39557) were selected for modification as XP-61Es. This modification comprised cutting down the decking of the fuselage nacelle flush with the wing and adding a large blown canopy, that of the first XP-61E being hinged on one side while that of the second slid aft. The centre and aft sections of the fuselage nacelle housed additional fuel tanks raising the normal internal load to 964 Imp gal (4 382 l), and the AI radar in the nose was supplanted by four 0·5-in (12,7-mm) machine guns, the ventral battery of four 20-mm cannon being retained.

Possessing a normal loaded weight of 31,425 lb (14 254 kg), the XP-61E attained a maximum speed of 376 mph (605 km/h) at 17,000 ft (5 180 m), and the normal range of 2,250 mls (3 620 km) could be extended to 3,750 mls (6 035 km) by means of four 258 Imp gal (1 173 l) drop tanks. The second XP-61E was written off as a result of a take-off accident on 11 April 1945, and the surviving aircraft was eventually converted as the XF-15 photo-reconnaissance aircraft. The XP-61F was intended to be a similar conversion of a P-61C-1 (Serial No 43-8338), but modification of the airframe had not commenced when the contract was cancelled on 24 October 1945.

During the last year of WW II, the Widow served with distinction as the standard AAF night fighter, and P-61 squadrons employed in the war against Japan comprised (5th AF) the 418th, 421st, 547th, (7th AF) 6th, 548th, 549th, (13th AF) 419th, 550th, (14th AF) 426th and 427th. In the immediate postwar years, the Black Widow soldiered on in the USA with the 52nd (2nd and 5th Sqdns) and 325th (317th, 318th and 319th Sqdns) All-weather Fighter Groups, and in the Pacific with the 51st (16th, 25th and 26th Sqdns) and 347th (4th, 68th and 339th Sqdns) Groups.

The last two squadrons to operate the P-61 as principal equipment were the 68th and 339th which finally relinquished their Widows early in 1950, the P-61 being withdrawn from the USAF's active inventory without ceremony or publicity. The latter was hardly to be expected as the Widow, having largely confined her activities to sombre night skies, had seen little publicity *during* her career. Nevertheless, this immense Northrop fighter had proved herself a lady to her crews, had been feared by her opponents, and had enjoyed the unique distinction of having been the only fighter designed from the outset for the nocturnal rôle to have seen service during WW II. □

VETERAN & VINTAGE

THE RAF MUSEUM TAKES SHAPE

RECENT YEARS have seen a remarkable awakening of interest in the preservation of historic relics of all kinds, and not least is this true in the field of aviation. Already it is too late to preserve genuine examples of many famous aeroplanes which, in the span of a single lifetime have been conceived, produced, operated — and disappeared forever. But a great deal remains, ranging from complete aircraft which, despite being 50 or more years old, are still maintained in near-mint, flyable condition, to pathetic heaps of metal or wood structures which, at first glance, it is difficult to believe could ever have had any possibility of getting off the ground.

A few nations have both the resources and the inclination to make a positive and concerted effort to preserve the memorabilia of their aeronautical heritage in a coherent fashion. Both the USA and Canada, for example, can boast national aeronautical collections which, although not perhaps as complete as some would wish, do at least have an official status.

For many years, Britain has been less fortunate in this respect. A number of collections of considerable importance exist, including those in the Aeronautical Galleries of the Science Museum in London, with the Fleet Air Arm Museum at Yeovilton, and in the "flying museum" of the Shuttleworth Trust at Old Warden. Outstanding efforts by various individuals have also led to the creation of other smaller but important groups of historic aeroplanes being preserved in various parts of the country. Only in the last few years, however, have steps been taken to create anything approaching a national aircraft museum. This, in the shape of the RAF Museum, is now making good progress towards a hoped-for opening date sometime in 1972.

Official backing for such a museum dates from 1964, and the first appeal for public donations towards the capital costs was made as recently as April 1968. Since that time, about £600,000 has been donated or promised, but at least another £300,000 is

The Vickers F.B.5 Gunbus, RFC serial number 2345 (also registered G-ATVP) is one of two replicas included in the RAF Museum collection, the other being the Vickers Vimy.

needed to establish the Museum facilities which have been planned. Recurring costs will be met by the government, but the costs of providing buildings and other facilities — and, for that matter, of acquiring additional aircraft, have to be borne from the voluntary fund.

The site of the Museum, as most enthusiasts will know, will be the old (and historic) RAF airfield at Hendon, in the northern outskirts of London and adjacent to the M1 motorway. There could hardly be a better site, and the main exhibition hall actually comprises two of the original RAF hangars, thrown into one by the addition of a new roofed section. Also new are two- or three-level galleries along one side and one end of the main building, to provide such facilities as a library, archives and a cinema, as well as small exhibition rooms for the display of armament, radio and radar, aero-engines and general personalia. The side galleries also provide additional high-level viewing facilities into the main hall.

According to present plans — which are necessarily somewhat provisional — the main hall will contain some three dozen aircraft on permanent or semi-permanent display at the time of opening. They will represent a broad cross-section of aircraft used by the RFC and RAF from before World War I up to the present day, and they have been drawn from a variety of sources.

There has existed in the RAF for some time an "official collection" of historic aeroplanes and this collection has provided about half of the permanent exhibits for the museum. Others have been acquired by the RAF Museum itself during the last few years — the Hawker Hind which was returned recently from Afghanistan is one example, the Supermarine Stranraer returned from Canada another — while contributions have been made by the Royal Aeronautical Society from its Nash Collection, by Hawker Siddeley and by the Ministry of Technology.

The aircraft contributed by Hawker Siddeley, together with other Hawker types, are to be grouped in what is known as the Camm Memorial Hall, occupying the space between the two hangars and displaying a range of types designed by the late Sir Sydney Camm. This group includes the only non-RAF type in the Museum in the shape of the Hawker Cygnet biplane — but this is an interesting aeroplane in its own right and is justified by the fact that it was Camm's first design for the Hawker company. Logically, Camm's last design, the P.1127 (progenitor of the Harrier) can be expected to take its place in the Museum in due course.

The aircraft which are likely to be on show in the main hall initially include the following:

Morane-Saulnier Type BB   A301
Bleriot XI

Among the Hawker biplanes included in the Camm Memorial Hall at Hendon will be (left) this Hind, returned from Afghanistan and restored in full Royal Afghan Air Force markings, and (right) the Hart Trainer K4972 discovered in a barn in Cumberland in 1962.

Exhibits representing the World War II period include (left) a Sikorsky Hoverfly I, one of the small number of R-4Bs supplied to the RAF, and (right) this Boulton Paul Defiant, N1671, in night fighter finish.

Vickers FB 5 (replica)	2345
Caudron G III	
Sopwith Triplane	N5912
Sopwith Camel	F6314
SE 5A	F938
Avro 504K	E449
Vickers Vimy (replica)	H651
Supermarine Stranraer	920
Gloster Gladiator	K8042
D.H. Tiger Moth	NL985
Airspeed Oxford	MP425
Miles Magister	T9967
Vickers Wellington	MF628
Westland Lysander	R9125
Boulton Paul Defiant	N1671
Avro Lancaster	R5868
Bristol Beaufighter	RD253
D.H. Mosquito T.III	TW117
Supermarine Spitfire I	K9942
Supermarine Spitfire 24	PK724
Gloster Meteor III Specia	EE549
E.E.C. Canberra PR3	WE139
E.E.C. P.1B Lightning	XA847
Sikorsky Hoverfly I	KK995
Bristol Belvedere	XG474

The Camm Collection:

Hawker Cygnet	G-EBMB
Hawker Hart	J9941
Hawker Hart Trainer	K4972
Hawker Hind (Afghan)	
Hawker Hurricane	PZ865
Hawker Typhoon	MN235
Hawker Tempest	NV778
Hawker Hunter 5	WP185

go into the hall — the Blackburn Beverley XH124 and the Sunderland ML824. The Beverley, apart from any other claims to fame, is of interest as the last fixed-wing aeroplane to land at Hendon (on 19 June 1968), having been flown in to overcome the almost insurmountable problems of trucking it in disassembled. However, the possibility of future landings at the site, either by helicopters or by more advanced VTOL types, cannot be ruled out.

As already noted, the RAF Museum depends heavily upon public donations, but it also receives enormous practical help from volunteers within the Service itself. Teams of volunteers at various RAF stations have undertaken the restoration of several individual aircraft, and this has often involved a massive research effort followed by long hours of skilled labour to reassemble a machine to represent its original operational state.

Authenticity is the keynote in all that the Museum does. This applies even to obtaining the correct paint colours for interiors as well as exteriors, and it hardly needs to be said that great care is taken to apply the correct standards of national markings and unit markings. To prove the point, the Defiant in the collection, representing an aircraft of No 307 (Polish) Squadron, at present carries no unit markings since positive evidence has yet to be discovered to show exactly how the markings were applied (can any reader help?). Again, the Museum

has discovered that the blue in the roundels used by the Royal Canadian Air Force differed slightly from that used by the RAF, and is now obtaining actual colour samples from Canada to ensure that the Stranraer (which will have its original RCAF finish restored) is correct.

The full time staff of the Museum includes a Director (Dr Tanner) and his two assistants (J M Bruce and M P Sayer) who work from a London office; about six persons already working at Hendon, and another dozen (under R Lee) at RAF Henlow, where there are workshop facilities for stripping and rebuilding exhibits.

The library and archives of the museum will be of special importance to future historians and *bona fide* researchers. It seems likely that, in the long term, the RAF Museum will also apply its unique facilities to the preparation of publications of special relevance to both specific exhibits and to RAF subjects in general.

Readers may care to be reminded of one publication already on sale in support of the Museum funds. This is the booklet entitled "A Royal Air Force 75", written by J M Bruce and describing and illustrating 75 types of aircraft which have served with the RFC or RAF up to 1968. Copies are available in return for a minimum donation of 25p; correspondence in this connection should be addressed to the RAF Museum, Turnstile House, High Holborn, London WC2. □

For the future, the Museum can call upon a large residue of types in the official collection and it is hoped that provision will be made in due course, in a separate covered building, for small temporary exhibitions of special interest. For example, drawing only upon aircraft already preserved by the RAF, it would be possible to display a group of World War II enemy aircraft including German, Italian and Japanese types, or a group of significant British experimental aeroplanes including the Fairey FD-2, Saro S-R 53, BAC TSR-2, Short S.B.5 and Avro 707.

Changes may also be made, from time to time, in the permanent exhibits, although there are obvious difficulties in getting aircraft into and out of the large hall. The Lightning is represented at present by XA847, the first English Electric P.1B prototype, and although this was the first RAF aircraft to reach Mach 2 and is therefore of considerable interest, a genuine squadron-service Lightning will no doubt be preferred in due course.

Two aircraft at Hendon are too large to

Above, the hull of the Supermarine Stranraer recently returned from Canada and now at RAF Henlow to be restored before going to Hendon. Below, the last aeroplane to land at Hendon was this Beverley XH124, which will be exhibited in the open as it is too large for the hangars (seen in the background).

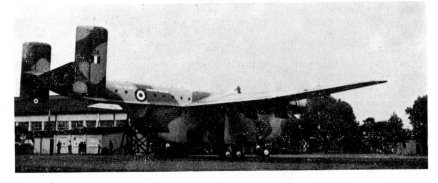

PLANE FACTS

This column is intended to provide an information service to readers requiring details relating to aviation subjects in general and aircraft in particular that may not be found in readily-available reference books. We ask readers to confine their requests to ONE subject or aircraft, and we will answer as many of these as space permits, but the editors reserve the right to use their discretion in selecting requests for reply in this column so that space is not taken up with facts that will already be well known to the majority of our readers. In no circumstances can we reply by post to requests for information nor supply copies of illustrations appearing in the column. Readers should not expect to see the information requested appearing in the issue of the AIR ENTHUSIAST immediately following the dispatch of their letter. There are two principal reasons for this: first, once a letter has been selected for reply the details or illustrations may not be on file and time may be consumed in research and, secondly, our printing leadtime means that when you, the reader, receive your copy of the AIR ENTHUSIAST, much of the next issue will have already been passed for press. Until a flow of readers' requests has been established, we will include on this page items that we believe to be of interest and hitherto unpublished. As requests for information are received, *Plane Facts* will expand to two or more pages each month.

Turboprop-powered fighter-bomber

Shortly before WW II, the aspiring Hungarian aircraft industry found the primary stumbling block in the development of world-standard combat aircraft of indigenous design to be the lack of a sufficiently powerful aero-engine. At this time, György Jendrassik, who had run an experimental 100 bhp gas turbine in 1937, was working on a turboprop at the Ganz Wagon and Engine Works, Budapest. Designated Cs-1, the turboprop was an ingenious unit, incorporating many design features to be seen in aero-engines today. These included a rigid compressor-and-turbine rotor assembly carried on front and rear bearings only, an annular reverse-flow combustor, extended-root turbine blades to reduce the flow of heat to the turbine discs, and air-cooled turbine discs.

First run in August 1940, the Cs-1 comprised an annular air intake surrounding a front-mounted reduction gearbox of 0·119:1 ratio, a 15-stage axial compressor, an annular combustor, an 11-stage axial turbine, and an annular exhaust duct. The initial rating of the Cs-1 was expected to be 1,000 bhp at 13,500 rpm, and such was its promise that, before bench running commenced, the design of a fighter-bomber intended specifically for a pair of Cs-1 turboprops was begun. Design development was led by László Varga at the Technical Institute for Aviation (*Repülö Müszaki Intézet*), and the prototype airframe, which was known both as the X/H and RMI 1, was built at the Sóstó Repair Works. The X/H was a low-wing cantilever monoplane of all-metal stressed-skin monocoque construction, suitable for the heavy fighter, attack bomber and dive bomber rôles, and accommodating two or three crew members.

The X/H had a wing span of 48 ft 0⅛ in (15,70 m) and empty and loaded weights of the order of 9,920 lb (4 500 kg) and 14,330 lb (6 500 kg) respectively, and it was anticipated

(Above) The first photograph to be published of the definitive Ba 201.

that, with two 1,000 bhp Cs-1 turboprops, maximum speed would be of the order of 335 mph (540 km/h) at 13,125 ft (4 000 m). In the event, difficulties were experienced with the final assembly of the X/H owing to the Hungarian team's limited familiarity with the monocoque structural techniques, and, meanwhile, the Cs-1 power plant was suffering combustion problems which were limiting its output on the test bench to some 400 bhp. With the signing of the German-Hungarian Mutual Armament Programme in June 1941, and the decision to manufacture the Daimler-Benz DB 605 engine and the Messerschmitt Me 210 in Hungary, work on the Cs-1 was halted. Thus, what was probably the world's first turboprop engine to be built and tested never took to the air. Nevertheless, the X/H airframe was completed in 1942, and stored at Sóstó pending availability of a pair of sufficiently powerful engines. Two Jumo 210E engines salvaged from He 112B fighters were installed for taxying trials in September 1943, and during the following December, the X/H was taken to the Aircraft Research Workshops (*Repülö Kisérleti Mühely*) where DB 605B engines were fitted and further taxying trials performed. A delay resulted from the failure of an undercarriage leg

during these trials, but flight testing was scheduled to begin in June 1944. Shortly before this, the X/H was completely destroyed in a USAAF bombing attack.

Breda dive bomber

One of the most mysterious of Italian wartime experimental aircraft was the Breda Ba 201 which was at first thought to be a licence-built version of the Junkers Ju 87B. It was subsequently learned that the Ba 201 was an entirely original design, but hitherto, all photographs that have been published purporting to depict this single-seat dive bomber have, in fact, illustrated the highly-detailed mock-up in one or other of its forms. The accompanying photo is believed to be the first to illustrate the definitive Ba 201 as tested at Guidonia in 1941. Few details of the Ba 201 have ever been revealed, apart from a loaded weight of 8,598 lb (3 900 kg) and some dimensions, these including a span of 42 ft 7⅞ in (13,00 m) and a length of 36 ft 6¼ in (11,13 m). Power was provided by a Daimler-Benz DB 601 engine rated at 1,180 hp. Of all-metal construction, the Ba 201 emulated the Ju 87 in featuring an inverted gull wing, and a 551-lb (250-kg) or 1,102-lb (500-kg) bomb was to have been carried by a ventral crutch. □

(Left) The Jendrassik Cs-1 turboprop of 1940, and (right) a general arrangement drawing of the X/H fighter-bomber project with Cs-1 turboprop engines.

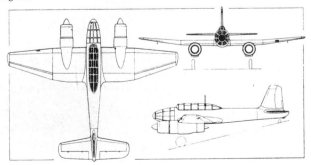

FIGHTER A TO Z

With this, its first issue, the AIR ENTHUSIAST commences an encyclopædia of fighter aircraft; a complete survey which, over the months ahead, will provide readers with a reference to every fighter type, whatever its nationality, be it prototype or production machine, built and flown during the past three-score years. The fighters will be arranged chronologically in alphabetical order of manufacturer, designer or design bureau, and all will be illustrated photographically, the more important or more interesting aircraft also being accompanied by a general arrangement drawing.

(Above and below) The A.D. Scout. Photograph depicts one of the two Blackburn-built aircraft, No 1536.

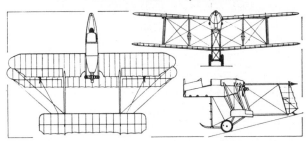

A.D. SCOUT (SPARROW)   GREAT BRITAIN

Designed by Harris Booth of the Admiralty's Air Department to meet a requirement for an anti-airship fighter, the single-seat A.D. Scout had a fabric-covered wooden airframe and a 100 hp Gnome Monosoupape 9-cylinder rotary engine. Four prototypes were ordered and built in 1915 (two by Hewlett & Blondeau and two by Blackburn), but no further development was undertaken. Known unofficially as the "Sparrow", the A.D. Scout was to have carried a single 0·303-in (7,7-mm) Lewis gun. Max speed, 84 mph (135 km/h). Endurance, 2·5 hr. Span, 33 ft 5 in (10,18 m). Length, 22 ft 9 in (6,93 m).

A.E.G. D I   GERMANY

The first fighter produced by the Allgemeine Elektrizitats-Gesellschaft (A.E.G.), the D I was primarily of steel-tube

(Below) The prototype of the A.E.G. PE ground attack fighter.

construction with fabric skinning, and was powered by a 160 hp Daimler D IIIa engine. The first prototype appeared in May 1917, and two further prototypes were built, these differing in having cheek-type radiators and revised exhaust pipes. Intended armament comprised twin synchronised 7,92-mm (0·312-in) Spandau machine guns. Max speed, 137 mph (220 km/h). Time to 3,280 ft (1 000 m), 2·5 min. Empty weight, 1,510 lb (685 kg). Loaded weight, 2,072 lb (940 kg). Span, 27 ft 10⅝ in (8,50 m). Length, 20 ft 0⅛ in (6,10 m).

(Above and below) The A.E.G. D I. The photograph depicts the third prototype and the general arrangement drawing illustrates the second prototype.

(Below) The sole prototype of the A.E.G. Dr I.

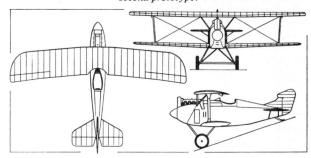

A.E.G. Dr I   GERMANY

Evolved from the D I, the Dr I was essentially a triplane derivative of the earlier single-seat biplane, with wings of extended span and additional interplane bracing struts. Having a similar eight-cylinder Daimler D IIIa engine, the Dr I appeared in October 1917, but development was not pursued owing to its poor performance. Max speed, 106 mph (170 km/h). Empty weight, 1,565 lb (710 kg). Loaded weight, 2,138 lb (970 kg). Span, 30 ft 10 in (9,40 m). Length, 20 ft 0⅛ in (6,10 m).

A.E.G. PE   GERMANY

The PE (*Panzer-Einsitzer*) was an experimental single-seat
ground attack fighter with a 195 hp Benz Bz IIIb eight-
cylinder engine which was flown early in 1918. Featuring
fabric-covered wings with dural spars and an armoured,
light alloy-covered fuselage, the PE had two synchronised
7,92-mm Spandau guns and could carry four light anti-
personnel bombs. Development was discontinued in favour
of the DJ I. Max speed, 101 mph (166·km/h). Empty weight,
2,606 lb (1 182 kg). Loaded weight, 3,113 lb (1 412 kg). Span,
36 ft 8⅞ in (11,20 m). Length, 20 ft 1⅜ in (6,60 m).

A.E.G. DJ I   GERMANY

Flown in September 1918, the DJ I armoured single-seat
ground attack fighter was of aerodynamically advanced
design. Powered by a 195 hp Benz Bz IIIb engine, it had no
interplane cables, all bracing being provided by aerofoil-
section struts. The armoured fuselage had sheet aluminium
skinning, and the fabric-covered wings had dural spars.
Armament was intended to comprise twin Spandau guns and
anti-personnel bombs, but development ended with the
Armistice. Max speed, 112 mph (180 km/h). Empty weight,
2,606 lb (1 182 kg). Loaded weight, 3,031 lb (1 375 kg). Span,
32 ft 9¾ in (10,00 m). Length, 21 ft 11⅛ in (6,69 m).

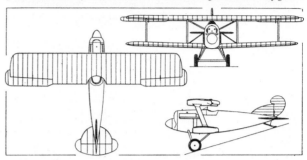

(Above and below) The A.E.G. DJ I armoured ground attack fighter.

MILAN————————————————— *from page 24*

air data computer. The complete system was installed in the
Milan S at the end of 1970 and weapon dropping trials were
conducted early in 1971 at Cazaux. These trials were made
with the particular object of demonstrating accuracy of bomb
aiming with typical Swiss weapon loads. Dassault claimed
that 990-lb (450-kg) bombs, dropped at a distance of 1·2–2·5
miles (2–4 km), fell with an accuracy of 65 ft (20 m) longitudinally and 13 ft (4 m) laterally.

The Milan S has an empty equipped weight exactly half
the maximum take-off weight, the figures being 15,430 lb
(7 000 kg), and 30,860 lb (14 000 kg) respectively. Allowing
for the pilot and the full internal fuel load of 8,320 lb (3 775
kg), then 6,600 lb (3 000 kg) remains for external loads
which can be disposed on four strong points under the wing
and three under the fuselage. The maximum external load,
with reduced internal fuel, is 9,240 lb (4 200 kg). The two
outboard wing positions are each stressed for loads of up to
330 lb (150 kg), such as a Sidewinder AAM or a light bomb.
The inner wing positions can each carry up to 3,300 lb (1 500
kg), possible loads including 2,200-lb (1,000-kg) bombs,
fuel tanks of up to 374 Imp gal (1 700 l) capacity, Napalm
containers of up to 88 Imp gal (400 l) capacity, rocket pods
containing up to 32 HVARs each, or the Matra combination
fuel tank/rocket pods or fuel tank/bomb carriers.

The strong point on the fuselage centre line is stressed
for loads of up to 2,645 lb (1 200 kg) and can therefore carry
most of the loads accommodated on the inboard wing
points, or an ASM such as the Nord AS.30. The lateral
fuselage points accommodate loads of 550 lb (250 kg) each.
There is no provision for flight refuelling, but a total external
fuel load of 1,034 Imp gal (4 700 l) can be carried to add to
the internal capacity of 830 Imp gal (3 775 l).

Flying a typical lo-lo-lo combat mission, the Milan has a
radius of action of 155 miles (250 km), with allowance for
5 minutes over the target, carrying a load of two AAMs and
six 990-lb (450-kg) bombs or 12 550-lb (250-kg) bombs.
These bombs are Swiss standard weapons, as used on the
Mirage IIIS. NATO standard bombs may also be carried.

The effect of the foreplanes and increased engine thrust is
to reduce the take-off speed for a given weight by 20–25 knots
(37–47 km/h), thereby reducing the take-off distance by
nearly 2,000 ft (600 m). Dassault attributes this remarkable
improvement in equal amounts to the *moustaches* and the
Atar 9K-50. For a given take-off distance, the maximum

take-off weight can be increased by 8 per cent, and the
landing approach speed is reduced by 25–30 knots (47–55
km/h), saving about 460 ft (140 m) on the roll-out. At
maximum weight, when operating from an airfield 1,640 ft
(500 m) amsl in a temperature of 82°C (ISA+20°), the Milan
has a take-off run of 5,250 ft (1 600 m).

Other benefits claimed for the Milan S include an increase
in the average approach angle from 3 deg to 4 deg, an
important factor when operating into restricted airfields,
and improved precision of touchdown; rate of climb increased by about 3,600 ft/min (18,3 m/sec); better low-speed
handling, with an increase of 1·3g in the turn, reducing the
radius by 820 ft (250 m), and better high-speed manoeuvrability (resulting from the increase in engine power alone) with
a gain of 0·5g reducing the turn radius by 492 ft (150 m).

The Milan nose section provides adequate space for the
additional equipment associated with the nav/attack system,
and also incorporates the window for the laser rangefinder.
The entire nose section is hinged forward of the slots into
which the *moustaches* retract, primarily to keep the dimensions within those required for accommodation in the
nuclear-resistant hangars built into the mountains of Central
Switzerland, where the Swiss Air Force maintains a number
of emergency operational bases. The wing span is 26 ft 11½ in
(8,22 m) and the overall length, 49 ft 3½ in (15,03 m).

Whether or not the Milan is purchased by Switzerland —
and indications at the time these words were written were
that it was rated third after the Vought A-7G and the Fiat
G.91YS — it has good prospects elsewhere. Two other
Mirage III users, South Africa and Australia, are reported
to have already held preliminary discussions with Dassault
concerning possible purchase of the Milan. □

*The Milan S-01 in ferry configuration, with two 375 Imp gal
(1 700 l) fuel tanks on the inboard wing pylons.*

AIRSCENE

Continued from page 4

January, but the Sikorsky-owned prototype was destroyed in an attempted auto-rotational landing on 22 January. Aviation Specialties Co started deliveries of the S-55-T, a conversion of the S-55 with a Garrett TSE 331 turboshaft engine, during February.

WEST GERMANY

Roll-out of the **VFW-Fokker 614 G1** (the first flying prototype) was made on schedule on 5 April at the company's Bremen works, but first flight is expected to be delayed for several months because of late delivery of the Rolls-Royce/SNECMA M.45H turbofans. Ground running units have already been delivered and will allow systems testing to begin. The company has announced options on 26 VFW 614s from the following airlines: Filipinas Orient, 2; Sterling Airways, 5; Bavaria-Fluggesellschaft, 3; General Air, 2; TABA (Argentina), 2; Spanish Ministry of Aviation, 1; Yemen Airlines, 3; Cimber Air, 3, and unspecified airlines, 5. The unannounced orders are believed to include an option from the Algerian domestic airline, STA.

CIVIL AFFAIRS

DENMARK

A new Danish domestic airline has been formed jointly by SAS, Maersk Air and Cimber Air. Known as **Danair A/S** (not to be confused with the British airline Dan-Air), it will operate 12 routes in Denmark and a service to the Faroe Islands. Initial equipment includes Caravelles and Convair 440s from SAS, an HS.748 and Fokker F.27 from Maersk Air and Nord 262s from Cimber Air.

FRANCE

Delivery was expected in April of the **Aero Spacelines Guppy 201** which has been purchased by Aérospatiale to support the Airbus A-300B programme. The aircraft is the second production Guppy 201 and it will be operated by UTA to fly components to the final assembly line at Toulouse from Chester (England), Hamburg, Bremen and Munich (Germany), and St Nazaire (France).

INDONESIA

Rapid expansion of local airline operations in Indonesia is reflected by the growing fleets of the new domestic operators. **Mandala Airlines,** after acquiring three Viscount 806s from BEA in 1970, has now purchased the two HS 748s from BOAC Associated Companies hitherto operated by Bahamas Airways, which went out of business recently. **Seulawah Air Service,** an associate of Mandala, has acquired five Convair 600s from Frontier to join two Convair 340s and five DC-3s. **Merpati Nusantara** has purchased four Viscount 828s from All-Nippon Airways and has followed this up by the purchase of two NAMC YS-11s from the same operator. A new domestic operator, named **Suryadirgantara,** has bought four Skyvans.

UNITED KINGDOM

Britten-Norman will deliver the first **Trislander** in July, following first flight of the first production example on 6 March. It will go to Aurigny Air Services, which has ordered three for services from the Channel Islands to Britain and France. Aurigny already has eight twin-engined Islanders in service; the Trislander, the prototype of which first flew on 11 September 1970, differs in having a 7 ft 6 in (2,29m) fuselage stretch and a third Lycoming O-540-E engine mounted on a new fin structure. The initial production batch includes eight for company distributors: Trans West, 1; Jonas, 3 and Alamo and Virgin Island Airways, 4.

UNITED STATES OF AMERICA

Delta Air Lines became the first Lockheed TriStar customer to react positively to the Rolls-Royce RB.211 situation, by placing an order for five **McDonnell Douglas DC-10 Srs 10s** during March. The move was an insurance against delivery delays or possible cancellation of the TriStar, but Delta did not immediately drop its order for 12 (plus 12 on option) of the Lockheed aeroplane. The Delta order increased the DC-10 backlog to a total of 124 firm orders plus 119 options; these totals reflect a number of recent new contracts and changes in earlier orders, including a cut of eight in the United contract. The order book is as follows: Series 10, United, 22, American, 25, Continental 8, National, 11, Delta, 5; Series 20, Northwest, 14; Series 30, Air Afrique, 3, Air New Zealand, 3, Alitalia, 4, Finnair, 2; KLM, 6, Lufthansa, 4, Swissair, 6, UTA, 4; Series 30CF, Overseas National, 3, Trans International, 2; Sabena, 2. In addition to options held by most of the above companies, other option holders for Series 30s include SAS, 8; Atlantis, 2; Paninternational, 1; and World Airways, 3.

CIVIL CONTRACTS AND SALES

Aérospatiale Caravelle: Sobelair, IT associate of Sabena, now has three ex-Sabena Caravelle VIs for the 1971 season. □ China Airlines has bought two from United. □ Transavia increased its fleet to nine with four from United.

BAC One-Eleven: Aviateca (Guatemala) leased one Srs 400 from Court Line for 1971, pending delivery of a new Srs 500 in March. □ Laker bought one, ex-Bahamas Airlines.

Boeing 707: Portuguese government ordered two -320C for military service to Angola, replacing DC-6s. □ In second-hand deals, Lloyd International acquired a second -320 from Pan Am, Dan-Air bought one from Pan Am, Paninternational (Germany) bought two from Pan Am and Britannia Airways bought one from World Airways; all have been delivered to their new owners. □ Caledonian/BUA is to buy an eighth -320C. □ Royal Air Maroc is leasing one -320C from Air France to inaugurate a Casablanca-New York service. □ Donaldson is getting two from Pan Am. □ United Irish Airways will lease three, source unknown, for transatlantic charters.

Boeing 720: MEA is acquiring a total of nine 720Bs from American Airlines by 1972, and will trade back five Convair 990As obtained from American in 1969. □ Boeing has sold five ex-Eastern 720s to Calair, a Danish IT operator. □ Belgian independent Trans European Airways is reported to have bought one 720.

Boeing 727: Three more -200s ordered by Air France, making a total of 20. □ Royal Air Maroc ordered its second -200. □ Tunis Air has placed its first order for a 727-200.

Boeing 737: Fourth aircraft for NZ National Airways Corporation is ex-Aloha. □ Pacific Southwest ordered its eleventh -200.

Boeing 747: Air France ordered three more for a total of 11 by 1973. □ BOAC received government permission to order another four, making 16 in all: regular services began mid-April. □ Japan Air Lines agreed to order four more. □ Eastern has leased three from Pan American, after earlier selling its four delivery positions to TWA; leased aircraft have been modified to 747A configuration with 735,000 lb gross weight and JT9D-3AW engines.

Convair 990A: Modern Air has acquired two more from American, ex-MEA lease. □ Spantax SA is adding three to its fleet of five, source unknown.

DHC Twin Otter: Air Madagascar is buying five for local service. □ Air Canada is expected to buy 10-12 for an intensive evaluation of "third level" STOL services. □ Royal Nepal Airlines ordered two.

Fokker F.28: Italian airline Turavia ordered two. □ Nigeria Airways has one, plus one on order, for domestic operations.

HSA Comet: Dan-Air bought three Comet 4s from East African and one from Kuwait to provide an operational fleet of 11 by summer 1971.

Ilyushin Il-18: United Arab Airlines received two (ex-Tarom) in March to replace grounded Comets, and six more are expected to follow; airline had previously acquired five Il-18s from the Soviet Union.

McDonnell Douglas DC-8: World Airways ordered three DC-8-63Fs (replacing cancelled Boeing 747s) and took delivery of the first in mid-March. □ Japan Air Lines received approval to order three more DC-8-62Fs and to buy two DC-8-61s from Eastern; the latter company is also leasing five -61s to JAL for one-three year periods. □ Air Spain bought one -20 from Eastern.

McDonnell Douglas DC-9: Garuda Indonesian has ordered a fourth, following delivery of its third in February. □ SAS is ordering five more.

NAMC YS-11A: Olympic Airways ordered three more, adding to five already delivered. □ Norwegian operator MayAir acquired two previously on lease to Olympic.

MILITARY CONTRACTS

Boeing 707: The Portuguese government has, with US governmental approval purchased two Boeing 707s for operation by the *Força Aérea Portuguesa* on routes to Portugal's African Territories.

Aérospatiale Caravelle: The Material Administration of the Swedish Armed Forces has purchased two Caravelles from SAS at a cost of SKr 8·8m (£708,000), including spares and training facilities. The first Caravelle was delivered in March and the second is scheduled to be delivered in September, and these aircraft are to be used as equipment test platforms by the National Defence Research Institute.

Lockheed C-130H Hercules: The *Fuerzas Aereas Venezolanas* (Venezuelan Air Forces) have placed a contract for four Lockheed C-130H Hercules transports, deliveries of which began on 25 March.

Vickers Viscount: Three Viscount 794s have been sold to the Turkish Air Forces by THY.

Volume 1 Number 2   July 1971

CONTENTS

AIRSCENE

MILITARY AFFAIRS

AUSTRALIA

During his visit to Australia earlier this year, General Yves Ezanno, the chairman of OFEMA (Office Française d'Exportation de Matériel Aéronautique), offered the **Dassault Mirage F1** to the Defence Department, including a proposal that this be licence-manufactured in Australia on the same basis as the Mirage IIIO for which it is offered as a successor. A year ago the RAAF circulated an Air Staff Target for a Mirage replacement for the 'eighties, this placing emphasis on the air superiority rôle. The AST propounded the dictum that multi-rôle capability invariably degrades overall performance in all rôles, but maintained that specialisation for the air superiority mission would, nevertheless, permit "satisfactory accomplishment of secondary missions". In addition to the Mirage F1, the Lockheed CL-1200-2 Lancer is being offered to meet the AST, Lockheed-California proposing production by a SE Asia consortium comprising Australia, Japan and Taiwan, but in assessing future requirements for a Mirage successor, the RAAF has to consider the question of an F-111C alternative if it is finally decided to reject the General Dynamics aircraft. In so far as the F-111C itself is concerned, the situation remains unresolved, and even if the F-111C *is* accepted for RAAF service, there now seems little hope that the aircraft could attain operational status before 1974.

It is not now expected that the RAAF will receive the first of the 12 **Boeing-Vertol CH-47C Chinook** helicopters ordered last year earlier than the late autumn of 1972 *if* the order is confirmed by the Australian government. The Chinook order, which is worth \$A37m (£17·25m), is frozen pending the outcome of intensive field trials of the CH-47C's Lycoming T55-L-11 turboshafts which have suffered serious heating problems. The trials with modified engines were nearing completion at the time of closing for press when it was hoped that the T55-L-11 would be cleared for service operation from next month onwards. If the Australian government does confirm its Chinook order, there is now no hope of meeting the original March 1972 schedule for delivery of the first helicopter of this type to the RAAF.

BRAZIL

At the time of closing for press, negotiations between the Brazilian government and a French mission for the purchase of an initial quantity of six **Breguet Atlantic** maritime patrol aircraft for the *Fôrça Aérea Brasileira* (FAB) had reportedly reached an advanced stage. The total FAB requirement for Atlantics is understood to be 12 aircraft to replace the survivors of the 14 P-2E Neptunes which have served with the 7° *Grupo de Aviação* at Salvador, Bahia, for some 12 years. French sources suggest that there is a potential Latin American market for between 30 and 40 Atlantics.

CAMBODIA

The Australian government has re-purchased six ex-RAAF **Douglas C-47** transports from Jetair Australia Limited, which acquired them in February at a cost of \$A275,000,

Delivery of the first of 50 Bell CUH-1N helicopters for the Canadian Armed Forces was made at Uplands Airport on 3 May. These helicopters will be assigned to Mobile Command tactical helicopter squadrons, replacing 10 CUH-1Hs which will be re-assigned to the local base rescue rôle.

and has supplied them, after overhaul, to the Cambodian air arm as part of the Commonwealth's foreign aid programme.

CANADA

The Canadian Minister of National Defence, Donald S Macdonald, recently announced the purchase of a fifth **Boeing 707-320C** for the Canadian Armed Forces at a cost of \$10m–\$11m, including spares. The Boeing 707s have replaced the Canadair CC-106 Yukons, the last of which was withdrawn at the end of April, and it is claimed that the annual operating cost of the five Boeings will total slightly less than \$Can 11·5m as compared with \$Can 16m for the dozen Yukons. Two of the Boeing 707s are currently awaiting conversion for use as tankers by the CAF CF-5 fighter-bomber force. The flight refuelling kits for these aircraft, developed by the Beech Aircraft Corporation, are scheduled to be delivered this month (July), each kit comprising two 16 ft 9 in (5,10 m) pods which, pylon-mounted at the wingtips, house flexible hoses, drogues, booms and associated controls. The booms swing down to bring the drogues some 15ft (4,57 m) below the wingtip vortices, and each tanker will be able to refuel two CF-5s simultaneously.

Deliveries to the CAF recently commenced of eight de Havilland Canada **CC-138 Twin Otters** for search-and-rescue and utility transport rôles. When deliveries are completed in August, five of the CC-138s will replace five C-47s currently employed in the search-and-rescue rôle, two being based at Namao and four at Trenton, one will replace the existing Caribou assigned for the support of United Nations' operations, and the remaining two will be used to provide the newly-created Northern Region Headquarters with utility transport support, for which purpose they will be based at Yellowknife.

The Canadian Armed Forces have formed three **new tactical helicopter squadrons** — No 427 at Petawawa, No 430 at Valcartier, and No 444 at Edmonton. These will operate the first of the 124 new helicopters of which deliveries to the CAF Mobile Command began recently. Early in May, the CAF started phasing in the CUH-1N Iroquois, 50 of which have been ordered, and was awaiting delivery of the first batch of 74 Bell COH-58A Kiowa LOHs which are scheduled to reach the CAF at a rate of five per month. The Kiowas are replacing the CAF's fleet of 44 miscellaneous small helicopters and fixed-wing aircraft.

CEYLON

The tiny Royal Ceylon Air Force (RCyAF) has received a substantial **infusion of equipment** as a result of the Ceylonese government's request for aid in its attempt to defeat the Che Guevarist revolutionaries, although it remains to be seen how much of this equipment will remain in Ceylon after the rebellion is suppressed. At the time the rebellion assumed serious proportions in April, the RCyAF comprised five Doves and four Herons employed for both coastal patrol and transport tasks, three Pioneer CC 2 STOL liaison aircraft, nine Chipmunk primary trainers, four Bell JetRanger helicopters, and an operational element of eight armed Jet Provost T 51s. During April, six additional JetRanger helicopters were delivered from the USA, funds for their purchase having been provided by the UK government, and two Alouette IIIs arrived from India, together with IAF personnel, although these were to be withdrawn early in May. On 20 April, An-12s landed at Katunayake with spares and support equipment for five MiG-17s, a two-seat MiG-15UTI and two Ka-26 helicopters which arrived on the following day aboard An-22s. Maj-Gen D S Attygalle, Army Chief-of-Staff, subsequently stated that the contingent of Soviet personnel accompanying the aircraft was a training mission, and added,"The MiGs will not be used in combat until *our* pilots are trained". Nevertheless, there is evidence that the MiG-17s *have* since flown operational patrols with Soviet pilots. On 14 May an agreement covering the supply of aircraft to the RCyAF was signed by the Soviet and Ceylonese governments, the actual types being unspecified.

CONGO (KINSHASA)

The air arm of the Congolese Republic, the *Force Aérienne Congolaise,* is currently receiving considerable foreign assistance in **reorganisation** and the upgrading of its training standards. Belgian, Italian, Israeli and US missions are assisting the FAC, and recent additions to its equipment inventory include the first of seven SA 330 Puma helicopters and three Lockheed C-130 Hercules transports. The latter were ordered under a \$17m contract specifically for airlifting the Congolese Army's élite paratroop force which has been trained by the Israeli Army and is intended as a counter-insurgency force for rapid deployment anywhere in the Congo.

CZECHOSLOVAKIA

The Czechoslovak Air Force anticipates

phasing the Aero **L 39 Super Delfin** into its training syllabus as a successor to both the L 29 Delfin and the MiG-15UTI from the autumn of next year. Three pre-production examples of the L 39 have now joined the prototypes in the development programme, and full production is scheduled to commence before the end of this year, current schedules calling for the delivery of 500 by the beginning of 1975, with 300 of these going to the Soviet Air Forces and the remainder being divided between the Czechoslovak and East German air arms. An afterburning version of the Walter Titan (AI-25V) turbofan affording some 4,410 lb st (2 000 kgp) is currently under development for installation in a light strike/reconnaissance version of the L 39 which will be available in both single- and two-seat variants, and major sales drives for the L 39 in Africa, Asia and Latin America are anticipated during 1972–3.

FRANCE
The first prototype **Dassault Mirage G8** variable-geometry aircraft made its initial flight at Istres on 8 May with Jean-Marie Saget at the controls. During its fourth flight, on 12 May, Mach 2·03 was reached, and the Mirage G8 was scheduled to be demonstrated at the Paris Air Show. The Mirage G8 is powered by two SNECMA Atar 9K-50s but will eventually receive M-53s.

GERMANY
Current schedules call for the *Luftwaffe* to receive the first **Panavia MRCA** in the battlefield interdiction configuration in March 1977, and some six *Geschwader* each of two *Staffeln* will progressively convert to the MRCA over the following five years, deliveries to the *Luftwaffe* being scheduled for completion by 1981–2. The *Marineflieger* is also expected to receive sufficient MRCAs to equip one *Geschwader*.

ISRAEL
The Israeli air arm, the *Heyl Ha'Avir*, is continuing to receive deliveries of **F-4E** and **RF-4E Phantoms** from the USA, approval having recently been given the supply of a further batch of 12 aircraft, bringing to 80 the number of Phantoms acquired by Israel since September 1969. The initial contract called for the supply of 50 F-4Es and six RF-4Es, these being followed by a further 12 F-4Es of which delivery was recently completed. Approval has also been given the supply to the *Heyl Ha'Avir* of 18 A-4E Skyhawks to supplement the initial batch of 48 A-4H Skyhawks delivered 1967–8. The new aircraft are to be equipped with an advanced Lear Siegler weapons delivery avionics system with which Skyhawks already in the *Heyl Ha'Avir* inventory will be retrofitted.

Replacement wing components for **Mirage IIICJ** fighters of the *Heyl Ha'Avir* are being manufactured in the USA by Metal Resources Incorporated of Gardena, California, under sub-contract from Israel Aircraft Industries.

ITALY
The funding of Lire 5,000m (£3·3m) for the **Fiat G.222** included in 1971–2 Italian defence appropriations supports reports current at the time of closing for press that an order will shortly be placed on behalf of the *Aeronautica Militare* for the production of some 40–50 transport aircraft of this type.

JAPAN
A formal order for a further 11 **Kawasaki P-2J** maritime patrol aircraft is expected to be placed before the end of the year for delivery during the course of 1973, this order raising the number of P-2Js ordered during the 3rd (1967–71) Defence Programme to the planned total of 46 aircraft. The purchase of a further 45 P-2Js is planned during the 4th (1972–6) Defence Programme, the last of these aircraft being scheduled for delivery during the 1978 Fiscal Year.

The third **Shin Meiwa PS-1** (the first pre-series) maritime patrol flying boat is scheduled to be delivered to the Maritime Self-Defence Force in September and will be followed by the second pre-series PS-1 before the end of the year. The MSDF anticipates having 30 PS-1 flying boats in service by 1976, average production rate being five per year from 1972. Shin Meiwa is currently engaged in the development of an air-sea rescue version, the SS-2, and plans call for the acquisition of three aircraft of this type by the MSDF to replace the service's five Grumman Albatross amphibians.

LIBYA
Despite French concern regarding the possibility of the use of Libyan Air Force **Mirages** against Israel in the event of a renewal of hostilities in the Middle East now that a federation has been formed by Libya, Egypt and Syria with a combined military command, the supply of Mirages to the Libyan air arm, and the training in France of a cadre of Libyan flying instructors and ground technicians is continuing. The Libyan order comprises 10 Mirage IIIB two-seat trainers, 32 Mirage IIIEs, 10 Mirage IIIRs and 58 Mirage 5s, the first four Mirage IIIBs arrived at Wheelus air base in March, two months ahead of schedule, and a total of 12 Mirages is expected to have passed into the Libyan inventory by the end of this year. France agreed to furnish Libya with Mirages on the grounds that the country was not a "battlefield participant" in the war with Israel, and the contract forbids transfer of the aircraft to a third country. French government spokesmen have insisted that France's attitude remains unchanged, and that an embargo will be imposed on the delivery of further aircraft and spares if those Mirages already delivered "seem to be changing their destinations", but it would seem at the time of closing for press that the combining of the Libyan, Egyptian, and Syrian military commands does not necessarily constitute transferring control of aircraft to either Egypt or Syria, both of which countries are covered by the French arms embargo. In addition to the Mirages, France is currently furnishing the Libyan air arm with nine Super Frelon and a number of Alouette helicopters.

NETHERLANDS
At the time of closing for press the selection of an F-104G **Starfighter** replacement by the *Koninklijke Nederlandse Luchtmacht* (KLu) was believed imminent. Two of the contenders for KLu orders, the F-4E(F) Phantom and the Vought Super V-1000, were eliminated from the contest earlier in the year, leaving a short list comprising the Dassault Mirage F1, the Lockheed CL-1200-2, the Northrop P-530 Cobra, and the Saab 37 Viggen. Approximately 100 aircraft are involved in the KLu requirement.

PERU
The *Fuerza Aérea del Perú* (FAP) was to receive the first of 16 **DHC-5 Buffalo** transports in May with completion of the order scheduled for February next year. The Buffalo are to replace the aged DC-3s and C-46s currently operated by the *Transportes Aéreo de Transportes Comerciales* (SATCO), the FAP-manned airline which operates services to 27 towns and cities.

SOVIET UNION
The Soviet Air Forces are reportedly deploying two squadrons of **MiG-23** *Foxbat* interceptors at Cairo West airfield and a similarly-equipped squadron at Mers-el-Kebir, Algeria. The MiG-23, which has not hitherto been seen outside the Soviet Union, allegedly enjoys a clear performance superiority over available western interceptors and, powered by two Tumansky turbojets with reheat ratings of the order of 24,250 lb (11 000 kg) thrust each, is believed to possess a short-period dash speed of Mach 3·2, or 2,100 mph (3 380 km/h) at high altitude, and maximum sustained speeds ranging from Mach 2·7, or 1,780 mph (2 865 km/h), at high altitude to Mach 1·3, or 975 mph (1 570 km/h) at 4,920 ft (1 500 m). Latest US estimates purportedly based on US and Israeli intelligence analyses of monitored test flights of the MiG-23 from Cairo West (presumably acceptance checks after re-assembly) indicate that the interceptor model can climb from sea level to 36,000 ft (10 970 m) in 2·5 min with full afterburning and in 7·8 min on full military power. Normal combat radius is of the order of 700 miles (1 125 km) but, depending on duration of supersonic flight, can vary between 460 and 805 miles (740 and 1 300 km) on 25,000 lb (11 340 kg) of fuel.

Deliveries of the Saab AJ 37 Viggen, the first production example of which is illustrated here, are scheduled to commence at the beginning of this month (July), and the first Flygvapen wing to convert from the A 32A Lansen to the AJ 37 Viggen will be F7 at Såtenäs. It will be noted that the production AJ 37 features a humped upper fuselage decking, presumably indicating an increase in internal fuel capacity. The Swedish government has recently authorised the Defence Material Administration to order continued development work on the S 37 reconnaissance version of the Viggen. Current production orders call for 150 AJ 37 all-weather attack aircraft and 25 two-seat SK 37 conversion trainers.

The US Marine Corps is evaluating two YOV-10D Broncos as night observation gunships, for which purpose they have been modified to carry a triple-barrel 20-mm gun in a remotely operated ventral turret, and a Hughes forward-looking infra-red (FLIR) aiming system in the extended nose, as shown in this photograph.

Empty equipped weight is calculated at some 34,000 lb (15 420 kg) and maximum gross take-off weight at 64,200 lb (29 120 kg). The MiG-23 is equipped to receive signals from the ground-to-air digital transmission system known to NATO by the code name *Markham,* which enables tracking radars to feed data directly to a cockpit display. This eliminates the need for any ground-to-air voice transmissions, the Israeli monitoring of which was a vital factor in the dogfight of 30 July last year when four of the *Fishbed-J* version of the MiG-21 flown by Soviet pilots were destroyed by F-4Es and Mirage IIICJs of the *Heyl Ha'Avir.*

A small number of *May* maritime patrol aircraft (see this column last month) of the Soviet Naval Air Force, the *Aviatsiya Voenno-morskovo Flota* (AV-MF), are now based at Matru airfield, near Cairo, for operations over the Mediterranean. A derivative of the Il-18 commercial transport, the *May* is now reported to be designated Il-38 in AV-MF service.

SOUTH VIETNAM
The South Vietnamese Air Force, which possessed 186 aircraft in 1962, currently has an inventory of 932 aircraft, and plans to have 1,299 aircraft of all types on strength by 1973. At Bien Hoa the complete overhaul and repair of some seven different types of fixed-wing aircraft and helicopters is now carried out entirely by Vietnamese personnel.

SPAIN
The *Ejército del Aire* has now taken delivery of the first of 36 McDonnell Douglas F-4C(S) **Phantoms** that it is receiving from the US as part of the payment for the use of Spanish bases by US forces. The F-4C(S) Phantoms, ex-US Navy aircraft with carrier equipment removed, are being overhauled at Getafe by Construcciones Aeronauticas SA (CASA), and six pilots of the *Ejército del Aire,* who will serve as F-4C(S) instructors, are currently undergoing training in the USA.

The aviation element of the Spanish Army, the *Aeronautica Ejército,* has recently taken delivery of a small batch of Agusta-Bell AB 206A **JetRanger** helicopters. These bring to 55 the number of helicopters supplied to Spain by Agusta, including AB 47Gs and AB 205s included on the strength of the *Aeronautica Ejército.*

UNITED STATES OF AMERICA
It is expected that the US Navy contract covering future purchases of the Grumman **F-14 Tomcat** shipboard fighter will be restructured before the end of the year in recognition of the cost escalation above that anticipated at the time the original contract

was signed in January 1969. At the present time, however, the US Navy has no legal recourse but to endeavour to hold Grumman to the $806m (£335·8m) price agreed in 1969 for the Fiscal 1972 procurement of 48 Tomcats. Grumman reportedly considers this sum "commercially impracticable under existing terms and conditions" and estimates that a further $35m (£14·5m) must be added if a loss is to be avoided. Proposals have been made to reduce the number of Tomcats purchased in Fiscal 1972 to 40 aircraft but such would increase the unit cost by some $1m (£416,000) to $12·5m (£5,208,000) per aircraft.

The USAF is further reducing planned purchases of the **General Dynamics F-111** by cutting from 24 to 12 the number of F-111Fs originally scheduled for procurement with authorised Fiscal 1971 funding. This reduces the F-111F order from 82 to 70 aircraft. The average flyaway cost of an F-111 at current cost figures is estimated by the USAF at approximately $7·6m (£3·16m), and the total USAF order now comprises 430 F-111s and 77 FB-111s. The US Defense Department has stated that the further reduction has been necessitated by general inflation in the economy coupled with "technical difficulties" in the development of the F-111F's Mk 2 avionics system. Following a fatal accident to an F-111E on 23 March during trials of the M61 cannon on a flight from Edwards AFB, all F-111s were again grounded by the USAF at the end of March.

AIRCRAFT AND INDUSTRY

CANADA
De Havilland of Canada has now delivered 300 **Twin Otters,** with another 30 on order. Of those delivered, 170 have been sold to the United States. The 300th aircraft was one of six for Pakistan International Airlines.

FRANCE
Construction of a second prototype of the **Aérospatiale** Corvette is continuing for a September roll-out despite the loss of the Corvette 01 in a fatal accident on 23 March. Compared with 01, Corvette 02 has the fuselage lengthened by 3 ft 3½ in (1,0 m), a higher gross weight and a small reduction in the area of the tail unit. The engines will be relocated so that the thrust line is parallel to the fuselage axis rather than being toed outwards. Production aircraft with 2,315 lb (1 050 kg) thrust JT15D-4 engines and gross weight of 13,450 lb (6 100 kg) are expected to be available at the end of 1973.

Flight testing of the Turboméca Astafan

turbofan began on 8 April in the **Astafan Commander** test-bed. This aircraft, converted from a Hawk Commander, has two Astafans in pods slung under the wing, replacing the turboprop engines previously fitted. For the initial series of trials, Astafan II engines have been fitted; this variant of the engine is derived from the Astazou XVI turboprop and has a rating of 1,570 lb st (712 kg) or 1,687 lb st (765 kg) with water injection. Other variants are the 1,367 lb st (620 kgp) Astafan I, derived from the Astazou XIV; the 1,764 lb st (800 kgp) Astafan III, derived from the Astazou XVIII and the 2,248 lb st (1 020 kgp) Astafan IV, derived from the Astazou XX.

INTERNATIONAL
After a period on the ground for modification of the air intake ramp mechanisms, both prototypes of the **Concorde** resumed flight testing during April, with 002 flying again on 3 April and 001 on 15 April. The latter flight was the 200th by Concorde, of which 001 had made 123 and 002 had made 77. On 17 April, Concorde 001 made the longest flight to date, with a duration of 3 hr 23 min. On April 22, the French and British Ministers responsible for Concorde development met for a regular review of progress and agreed to allow production of aircraft numbers 7 to 10 to proceed (the purchase of long lead time components for these aircraft had previously been sanctioned), and also approved the purchase of long-dated items for a further six production Concordes. Meanwhile Concorde 01, which was rolled out at Filton on 31 March, is now undergoing resonance testing prior to a first flight in August or September. This pre-production aircraft incorporates several significant modifications, including a revised transparent visor and a lengthened forward fuselage. The second pre-production Concorde, 02, was originally to have been completed to the same standard, but will now be further modified to bring it closer to the final production standard, with a lengthened rear fuselage cone and uprated Olympus engines with the new TRA (thrust reverser aft) nozzles.

A new version of the **Airbus A 300B** proposed primarily for the North American market is the A-300B-8. This would have General Electric CF6-6 engines, as used in the domestic versions of the DC-10, whereas the basic A-300B has the more powerful CF6-50 engines. Of the other Airbus variants (AIR ENTHUSIAST June 1971, p 4), principal interest now centres on the B-1 (259 passengers, short range), the B-3 (259 passengers, medium range) and the B-7 (290 passengers, short range). First flight is still expected at the end of summer, 1972. The work-force now engaged in four countries totals 8,600; more than 7,500 hours of wind tunnel testing have been completed and 12,000 production drawings issued.

NETHERLANDS
First flight of the **Fokker F-28 Mk 2000** was made at Amsterdam on 28 April. The aircraft, with fuselage lengthened by 7 ft 3 in (2,21 m) to accommodate up to 79 passengers, has been modified from the original F-28 Fellowship prototype.

JAPAN
Sales of the **Mitsubishi MU-2** now total 205, of which 168 are in North America, 30 in Japan, six in Europe and one in Australia. The MU-2J, developed from the MU-2G with a stretched fuselage and uprated TPE-331-6-251M engines, has recently been certificated in Japan, and a

new version, the MU-2K, has been reported under development for 1972. The latter would have the same engines as the MU-2J but would be similar in size to the MU-2F.

SPAIN
The first of two prototypes of the **CASA C.212 Aviocar** light STOL transport was flown for the first time on 26 March at Getafe, Madrid. The Aviocar has been designed primarily for operation by the *Ejército del Aire,* this service having a requirement for some 60 aircraft to replace its aged Douglas C-47s and CASA-built Ju 52/3m transports. It can accommodate a flight crew of two and 15 paratroops plus a jumpmaster or, in the ambulance rôle, up to 10 casualty stretchers, three sitting casualties and medical attendants. A planned commercial version for feederline operation will carry 18 passengers with an alternative high density arrangement for 21 passengers.

UNITED KINGDOM
Eventual production of the Handley Page Jetstream in North America is one possible result of the revived interest in this aircraft following the acquisition of design and production rights and a number of completed aircraft by Captain W Bright. To handle certification, marketing and support of the Jetstream, Capt Bright has set up **Jetstream Aircraft,** and the first delivery of a Jetstream Mk 1 by this company was expected in July, to Voyage Wasteels in France. Tropical trials of the more powerful Jetstream 200 with Astazou XVI engines were being made in the Trucial States during May, with certification of this version expected in June.

UNITED STATES OF AMERICA
The first manufacturer of business jet aircraft to pass the 300 sales mark is **Gates Learjet.** By the end of April, the company had sold 309 of its Lear Jet variants, including 99 of the original Model 23s, 140 Model 24s and 70 Model 25s.

Lockheed calculates that it will need $35m (£14·5m) funding in order to develop and build two prototypes of its proposed **X-27** research version of its CL-1200 Lancer air superiority fighter project, this figure assuming USAF supply of two F-104 Starfighter airframes and three J79 engines at no cost. The CL-1200 is essentially a modification of the F-104, and the highly-modified version proposed as the X-27 could also be marketed as an air superiority fighter. The X-27 will be capable of attaining Mach 2·6, and will employ manoeuvring tabs, both flaps and leading-edge slats operating as a function of angle of attack and g-forces applied to the aircraft. One of the two X-27 prototypes would be a single-seater and the other a two-seater, and first flight of the X-27 could take place within nine months of the contract go-ahead. Hughes Aircraft is proposing the use of a Dassault Aida — a 180-lb (82-kg) radar with a 23-in (58,4-cm) antenna dish — for the fighter version of the X-27, and as alternative power plants to the J79-GE-19, the aircraft will be capable of taking the Pratt & Whitney TF30-P-400 series turbofans or the advanced-technology F100-PW-100 or F401-PW-400.

Kaman's interim **LAMPS** (Light Airborne Multi-purpose System) helicopter for the US Navy, a modified twin-turbine H-2D SeaSprite redesignated SH-2D, was flown earlier this year at Bloomfield, Connecticut, and is currently undergoing testing with operational deployment scheduled for November. Ten SH-2D LAMPS helicopters

powered by General Electric T58-GE-8B engines of 1,250 shp each are to be produced during Fiscal 1971 by modification of existing SeaSprites, and a further 10 are expected to be delivered in Fiscal 1972. The SH-2D differs from the HH-2D search-and-rescue helicopter in having a radome under the nose housing the search radar antenna, and a pylon on the starboard side of the fuselage containing a winch for a magnetic anomaly detector. Auxiliary fuel tank mounts can be used as Mk 46 homing torpedo launchers.

Bell Helicopter has obtained FAA certification of the **JetRanger-II,** a higher-powered version of the basic Model 206. Principal change is the use of a 400 hp Allison 250-C20 engine in place of the 317 hp -C18. The new engine will maintain an output of 317 hp at altitudes up to 7,000 ft (2 134 m) in ISA, giving the JetRanger-II significantly improved lifting performance in hot or high conditions. For the same gross weight, the hovering ceiling is increased by about 4,000 ft (1 220 m) and the max weight for hovering at equivalent altitudes is increased by about 400 lb (181 kg.).

Production of the North American Rockwell **A-9 Sparrow Commander** and **A-9B Quail Commander** has been transferred to Aeronautica Agricola Mexicana SA from the Aero Commander plant in Albany, Georgia. The Mexican company, in which NR has a 30 per cent interest, has acquired design rights, tooling and production inventory for the two agricultural aircraft.

Fairchild Hiller Corporation has changed its name to **Fairchild Industries Ltd.** As a result, all aircraft produced by the company, including those of its Republic Aviation Division, will in future be known simply as Fairchild types.

CIVIL AFFAIRS

CAMEROUN
The former French territory of Cameroun, now independent, has decided to leave the Air Afrique consortium, and is in process of setting up its own airline to be known as **Air Cameroun.** Technical assistance is to be obtained from an airline in Britain or France, and suitable equipment is expected to be obtained either from Aérospatiale (Caravelles?) or from Hawker Siddeley (HS 748s?) to allow operations to begin in the latter half of 1971.

ITALY
The helicopter-operating subsidiary of Alitalia, **Elivie,** has ceased operation. The company operated scheduled services in the Naples area and undertook charter flights, carrying 82,000 passengers during 1970.

MALAYSIA
Malaysian-Singapore Airlines opened its **Singapore-London service,** via Bombay, Bahrein and Rome, early in June, using Boeing 707s. Two more 707-320Cs were purchased from Boeing earlier in the year to bring the total fleet to five. Meanwhile, discussions are continuing between the Malaysian and Singapore governments with a view to ending the present joint airline operation by the beginning of 1973. It is expected that Malaysia will concentrate upon domestic and regional services, and Singapore will expand the international routes, probably retaining the MSA identity under some such name as Metropolitan Singapore Airways. The regional airline will be known as Malaysia Airlines and its fleet of F.27s will be based at Kuala Lumpur.

THAILAND
Operations by a new Thai carrier, **Air Siam,** have been inaugurated with a DC-8-63 leased from Overseas National Airways. Initial routes link Thailand with the US. The company is backed by local Thai interests.

UNITED KINGDOM
Under the name of British Overseas Air Charter Ltd, BOAC has set up a subsidiary to operate non-scheduled services in competition with UK and US charter companies. The new company is expected to lease aircraft as required from BOAC (the initials being the same, no significant change in livery will be required) and initially the aircraft used are likely to be Boeing 707-420s. BOAC is also pressing IATA for massive cuts in excursion fares on scheduled flights across the North Atlantic (down from the present minimum of £113 to about £75).

UNITED STATES OF AMERICA
Several US airlines, suffering severe financial difficulties in the train of the recent business recession, are continuing to seek possible **merger partners.** A merger between American Airlines and Western Airlines has been approved by the two companies' stockholders, but is subject to CAB approval. American's merger with Trans Caribbean became effective on 8 March, but the projected mergers between Aloha

Now undergoing certification trials in Norway, this Larsen Special is an all-metal aerobatic two-seater designed and built by Captain Carl L Larsen, chief pilot of Braathens SAFE Air Transport Company. First flown on 20 October 1970, the aircraft carries the same LN-11 registration used by Capt Larsen on an earlier design which he built some 15 years ago. The new aircraft has a 100 hp RR-Continental O-200-A engine, wing span of 25 ft 5 in (7,75 m) and gross weight of 1,543 lb (700 kg). Maximum speed is 140 mph (225 km/h).

and Hawaiian, and between Northwest Orient and Northeast have been called off. Northeast is seeking an alternative merger agreement with Delta Airlines. Mohawk Airlines, after losing £11·9m (£5m) last year and then suffering an 18-week strike, is considering merger possibilities and has had discussions with Allegheny Airlines and North Central Airlines.

FAA approval has been obtained for operation of the **Boeing 747** in Category IIIA weather — a runway visual range of not less than 700 ft (213m) — and British approval by the ARB was expected to follow as this issue went to press. Operation down to these minima requires that the Boeing 747 — first US transport so certificated — be fitted with a triple-channel fail-operational version of the Sperry SPZ-1 auto-pilot/flight director, plus Bendix automatic throttle, AC Electronics Carousel 4 and radio altimeter which are part of standard Boeing 747 systems; in addition, the airport used must have a Cat 2 ILS. BOAC, which finally put its Boeing 747s into regular service on the North Atlantic on 25 April, is expected to be the first to use the aircraft down to Cat IIIA conditions; KLM has also ordered one 747 to be equipped to the necessary standards.

USSR
Domestic operation of the **Tupolev Tu-144** SST by Aeroflot is expected to begin during 1972 according to the latest reports from the Soviet Union. Likely route for the first service is Moscow-Khabarovsk. First international services may start in 1973, with Moscow-Calcutta the probable first route. Both dates are well ahead of the earliest possible Concorde operation, which will follow immediately after certification scheduled for the end of April 1974.

CIVIL CONTRACTS AND SALES

Aérospatiale Caravelle: Air Inter will acquire five Caravelle 12s on lease from Aérospatiale, from September 1973, pending delivery of Dassault Mercures already on option. ☐ One Caravelle has been bought for the personal use of the President of the Central African Republic. ☐ In another second-hand deal, Belgian International Air Services acquired one with an option on a second. ☐ Air Charter International, subsidiary of Air France, now has seven Caravelle IIIs in its own colours, of which four are owned and three loaned from Air France.

BAC One-Eleven: Germanair ordered a fourth Srs 500 and took delivery of it in April. ☐ Phoenix Airways, new Swiss IT operator, has one Srs 500. ☐ Air Malawi is now using one Srs 200, leased from Zambia Airways. ☐ Mohawk has bought another Srs 200 (the third) from Braniff.

Bell helicopters: New commercial sales include 11 JetRangers equipped for ambulance use, to Okanagan Helicopters; 17 JetRangers for Petroleum Helicopters Inc and four 47G-4As for Petroleum Helicopters Inc. ☐ Hong Kong International bought one Model 212 for scheduled services.

Boeing 707: Korean Air Lines ordered from Boeing one -320 for June delivery. ☐ Trans Mediterranean has added two more second-hand -320Cs to its first aircraft of this type. ☐ Two -120s are to be purchased by Air Manila from TWA. ☐ Cathay Pacific will acquire its first second-hand -320B in July,

two more by April 1972. ☐ Nigerian Airways has acquired a -320C to supplement one leased from Ethiopian. ☐ BWIA has bought the only four -220s in existence from Braniff. ☐ Aerolineas Argentinas has acquired two more. ☐ Transavia Holland bought one -320C from Braniff. ☐ Southern Cross Malaysian Airways acquired two -320s from Pan American.

Boeing 720B: Los Angeles Dodgers baseball team bought one from American Airlines.

Boeing 727: Lufthansa ordered one more -200. ☐ Braniff acquired three -100s from BWIA. ☐ TWA ordered two more -200s. ☐ Continental leased one -100C from Boeing.

Boeing 747: Air India ordered a fourth 747-200B, for delivery in March 1972. ☐ Condor ordered a second 747-100. ☐ SAA ordered two more, for a total of five.

Canadair Yukon: Canadian Armed Forces sold one to Aerotransportes Entre Rios.

Hawker Siddeley 748: Canadian local airline Air Gaspe ordered one. ☐ Dan-Air acquired one from Maersk Air.

Lockheed Electra: Pacific Western (Canada) has acquired two, modified to cargo configuration. ☐ Aerovias Condor de Colombia bought one from Los Angeles Dodgers, adding to four, ex-American Airlines, already in service. ☐ Air Manila has bought three, source unknown.

Lockheed Hercules: Saturn Airways ordered a fourth L-100-30; its first entered service early this year.

McDonnell Douglas DC-8: SAS ordered its sixth -63; eight -62s are also in service already. ☐ German charter airline Atlantis will acquire two -63s this year, adding to one in service. ☐ New Belgian charter operator Pomair bought one -33 from Pan American.

McDonnell Douglas DC-9: Iberia ordered 11 more DC-9 Srs 30s, including four in "rapid change" configuration with seats on pallets. Deliveries begin in January 1972. ☐ A third Srs 30 will be delivered to Atlantis shortly. ☐ Inex-Adria has bought a third DC-9, in the -30F convertible passenger-freight version.

McDonnell Douglas DC-10: National changed its contract to include two Series 30s plus nine Series 10s; the long-range Series 30s will be used on the Miami-London service in place of DC-8s.

NAMC YS-11: New operator Bouraq Indonesia Airlines took delivery of one from NAMC.

Saunders ST-27: First three customers for this turboprop conversion of the DH Heron are Millardair (Canada), for one; Air North (USA), for one; and Chicago and Southern, for two.

Short Skyvan: Gulf Aviation ordered a third, to fulfil a contract to support oil exploration activities by the Abu Dhabi Marine Corp. Two already delivered are under contract to Petroleum Developments (Oman) Ltd.

Sikorsky S-61: BEA Helicopters added one new (equipped with nose weather radar) and one second-hand to become largest S-61N operator in the world. ☐ All Nippon Airways bought one for oil rig support in

Sea of Japan. ☐ Carson Construction Co added one S-61L to fleet of five S-58s and S-61Rs. ☐ NYA acquired a fourth and put it into immediate service.

Swearingen Metro: First customer is Société Minière de Bakwanga (MIBA) of Congo Kinshasa; delivery was made earlier this year. ☐ Texas International has ordered five, for services starting 1 July.

Tupolev Tu-134: Six have been ordered by CSA of Czechoslovakia.

Vickers Vanguard: Two are now operated by Thor Flying Services of Iceland, ex-Air Canada acquired through Air Holdings (Sales).

Vickers Viscount: Alitalia sold a third to Somali Airlines. ☐ British Air Ferries leased one from Aer Lingus.

Business jets: HS 125 orders reached total of 239 with sale of one to Banco do Brasil ☐ Travelair GmbH of Bremen ordered three more Falcon 20 to make a total fleet of 16; total Falcon 20 sales are now 274, plus 55 Falcon 10s. ☐ Four Swearingen Merlin IVs (executive version of the Metro) have been delivered, including one to Air Charter in Hamburg, one in Brazil and two in the USA.

MILITARY CONTRACTS

Aérospatiale SA 330 Puma: Three Puma helicopters with 20-seat executive interiors have been sold to the Mexican government via Helair, the Mexican distributor for Aérospatiale.

Bell UH-1H Iroquois: The US Army has awarded Bell Helicopter a contract for 15 UH-1H Iroquois, supplementing a contract for a further 300 examples of this helicopter placed in January.

Bell 212 Twin Two-Twelve: The Colombian Air Force has taken delivery of a Bell Model 212 helicopter in executive configuration for use by the Colombian President.

Boeing 707-320C: The Canadian Armed Forces have purchased an additional Boeing 707-320C, raising to five the number of transports of this type in the CAF inventory.

De Havilland Canada DHC-6 Twin Otter: The Canadian Armed Forces are taking delivery of eight DHC-6 Twin Otters at a cost of $Can 9·65m, for search and rescue duties.

Fokker F.27-400: The Ivory Coast Air Force is scheduled to receive a second F.27 in November. The first, an F.27-600, was handed over in January, and the second, an F.27-400, is a convertible model for passengers or freight.

NAMC YS-11: The Japanese Defence Agency has signed a contract with NAMC for the supply of two additional YS-11s to the Self-Defence Forces, raising to 20 the total number of aircraft of this type ordered by the Defence Agency. One of the YS-11s will be delivered to the MSDF for the cargo rôle and the other is to go to the ASDF as an equipment test aircraft.

Sikorsky CH-53E: Sikorsky Aircraft has received a $1·7m (£708,000) cost-plus-fixed-fee contract from the US Navy to redesign the CH-53D to three-engined configuration as the CH-53E.

THE TURKISH AIR FORCES...

...NATO'S LINCHPIN OR ACHILLES HEEL?

This month Turkey provides the venue for the 1971 Allied Air Forces Southern Europe (AirSOUTH) Fighter Weapons Meet in which a Turkish team flying F-100 Super Sabres and a Greek team flying Northrop F-5As will represent SixATAF and will match their skill against two Italian Fiat G.91 teams representing FiveATAF and a combined USAF-USN team flying F-4 Phantoms and A-7 Corsair IIs. At a time when Turkey acts host to this important NATO event, the AIR ENTHUSIAST believes it is appropriate to survey the Turkish air arm and its position in the North Atlantic Treaty Organisation's defensive system. The feature is illustrated with photographs by Stephen Peltz.

WITH THE vast mass of her territory in Asia and a mere toehold in Europe, and thus the most easterly member of the North Atlantic Treaty Organisation, Turkey has for nearly two decades been considered protector of NATO's eastern flank. Yet Turkey is without doubt the most exposed of NATO nations. Her coastline includes the entire southern boundary of the Black Sea, an exposed frontier stretching over 800 miles (1 300 km); flanking the Black Sea to the west, Turkey shares 150 miles (240 km) of border with Bulgaria, the staunchest and militarily among the most reliable of the Soviet Union's allies; to the east Turkish and Soviet territory have a common border stretching 324 miles (525 km), the lengthiest frontier between NATO and the Soviet Union, while to the south-east Turkey borders on two of the most radical of the pro-communist Arab countries, Syria and Iraq, for a distance of more than 500 miles (800 km).

The 1,980-mile (2 750-km) defensive arc that is intended to be provided by Allied Forces Southern Europe, or AFSouth, terminates with Mount Ararat, whose 16,945-ft (5 165-m) peak towers over Turkey's eastern boundary, and it is the Turkish segment of this arc that provides NATO with what is probably this alliance's most persistent headache. With its rugged, mountainous terrain, snow-covered for six months of the year, Eastern Turkey affords good ground defensive positions, but other areas of Turkish territory offer less favourable aspects from the viewpoint of defence. For example, the plains of Turkish Thrace, the key to control of the Bosporus and the Dardenelles which provide access to the Aegean and eastern Mediterranean from the Black Sea, are easily approachable from Bulgaria.

The question of providing effective air cover for this strategically vital area is thus of the utmost importance to NATO, and NATO's answer is the 6th Allied Tactical Air Force (SixATAF), almost two-thirds of the muscle in which is furnished by Turkey's air arm. In recent years, however, doubt that SixATAF provides an *effective* answer to the Soviet bloc forces that it may one day be called upon to oppose has steadily increased. SixATAF is intended to defend the largest geographical area of any allied tactical air force in NATO; an area encompassing all of Greece and Turkey, and extending some 1,500 miles (2 400 km) from west to east — from the Ionian Sea to the Caucasus mountains — and 850 miles (1 370 km) from north to south. The bulk of this area of responsibility is provided by the 301,340 square miles (780 576 km²) of Turkey.

Along the northern periphery of SixATAF's tactical area, the Soviet bloc has created an interlocking network of more than 300 airfields capable of supporting some 3,000 combat aircraft, or more than six times the number of combat aircraft available to SixATAF. While these airfields are predominantly fighter, attack and strike aircraft bases, a number of them are capable of supporting medium bombers, and there are now more than 1,500 interceptors protecting potential SixATAF target areas, as well as protective screens of surface-to-air missile batteries. The air defence capability of Soviet bloc forces facing SixATAF has more than doubled over the past few years, and offensive capability has increased commensurately, yet SixATAF in general and its Turkish component in particular have declined numerically over the same period, and much of the hardware on their squadron flight lines today is made up of the same aircraft that they flew in the early years of the *last* decade!

NATO concedes vital numerical *and* qualitative edges in

(Above) Convair F-102A Delta Dagger all-weather interceptors of No 114 Squadron at Murted, near Ankara, and (below, left) a North American F-100F Super Sabre used for conversion at squadron level.

(Below) One of the ex-Luftwaffe T-33As employed for advanced training at Cigli which is also the commercial airfield for Izmir.

some weaponry, and particularly aircraft, to the Soviet forces facing SixATAF, a fact stressed on several occasions by AirSOUTH (Allied Air Forces Southern Europe) commander, Lt Gen F M Dean. The SixATAF is one of the two component tactical air forces of AirSOUTH, and Lt Gen Dean has stated unequivocally that the combat aircraft under his command would be placed at a numerical disadvantage of between four and five to one in any encounter with Soviet bloc forces. Furthermore, it is tacitly admitted that AFSouth has now been outflanked and virtually encircled as a result of the steady Soviet Mediterranean incursion.

AirSOUTH comprises two subordinate commands, Five-ATAF which is responsible for the planning and operation of assigned air elements in the Italian area, and consists almost entirely of units provided by Italy's *Aeronautica Militare,* and SixATAF with similar responsibilities in the area of Greece and Turkey, and consisting primarily of units from the Greek and Turkish air arms, the *Elliniki Aeroporia* and the *Türk Hava Kuvvetleri.* It is the upgrading of Six-ATAF's defensive and retaliatory capabilities that presents NATO with one of its most serious challenges to the viability of the alliance. But the problem posed is not confined to the provision of more effective combat equipment; it is compounded by Greek and Turkish national differences which impose major restrictions on co-operation between the forces of the two countries. A half-dozen or so Greek Air Force personnel are attached to SixATAF Headquarters in Sirinyer, a suburb of Izmir, but there is no interchange of combat units, and the respective national components of SixATAF confine themselves to their individual, non-overlapping defence zones. Furthermore, widespread disapproval of the current Greek régime, and the struggle for power between the Turkish armed forces and Turkey's political leaders earlier this year, have hardly helped to loosen US Congressional purse strings on the Military Assistance Program on which Greek and Turkish air arms are dependent.

There is close parity between Turkish and Greek defence

expenditure as percentages of their respective gross national products, although, with her substantially smaller population, Greece spends virtually two-and-a-half times as much *per capita,* but neither nation's budget permits national purchase of modern, high-performance combat aircraft in quantity. The problems facing the two air arms concerning the renewal of hardware differ fundamentally only in scale, the Turkish air arm being roughly twice the size of that of Greece, but the proportion of ageing aircraft in the first-line strengths of both services is closely comparable.

Elderly combat aeroplanes preponderate and neither Greek nor Turk faces with equanimity the prospect of increasing hardware senility with, in consequence, a steady downgrading of the operational capabilities of their air arms. Both have repeatedly stressed their need for modern warplanes if their contribution to the effectiveness of SixATAF is not to be drastically reduced; both have constantly demanded air superiority fighters which, in their view, can meet the MiG-21 on even terms, and both attach considerable importance to increasing their retaliatory capability. Both are reliant on the US Military Assistance Program and some military aid from Federal Germany — the latter, in the case of Greece, having now *officially* terminated — and neither considers such aircraft as the F-84 Thunderstreak and F-100 Super Sabre, excellent warplanes though they undoubtedly were when the 'sixties were born, to give their air arms the muscle that they need for the 'seventies. Yet the ageing Republic and North American fighter-bombers are apparently destined to soldier on for years to come with many of the Greek and Turkish units that would come under SixATAF command in the event of hostilities.

Turkey's contribution to NATO air strength

The Turkish Air Forces, or *Türk Hava Kuvvetleri* (THK), with Headquarters in the Inönu Bulvari, Ankara, have been primarily tactical since their systematic reorganisation and modernisation under USAF tutelage which began in the late 'forties, and which, within a decade, resulted in a modern,

The Republic RF-84F Thunderflash remains standard THK tactical reconnaissance equipment and currently equips one squadron in each Tactical Air Force.

Attrition suffered by Republic F-84F Thunderstreak squadrons over the past dozen years has been made up, in part, by the delivery of ex-Luftwaffe aircraft referred to as F-84FQs.

efficient and numerically substantial arm of 18 squadrons, mostly of the standard NATO 25-aircraft strength, with expanding early warning and communications back-up, transport support, an excellent training organisation, and a personnel strength of some 35,000 men. During the past decade, however, the THK has suffered steady erosion in combat aircraft modernity, although personnel strength has increased some 40 per cent and the 18 combat squadron level has been maintained, albeit with a reduced establishment in many cases of 18 aircraft.

The THK possesses two tactical air forces committed entirely to NATO, the 1st Tactical Air Force (*1nci Taktik Hava Kuvveti*) with Headquarters at Eskisehir, and the 3rd Tactical Air Force (*3nci Taktik Hava Kuvveti*) with Headquarters at Diyarbakir. Both air forces place emphasis on strike capability but also possess air defence rôles, some all-weather intercept potential being provided by surplus USAF Convair F-102A Delta Daggers transferred to the THK inventory in 1969. These were introduced into the 4th Wing (*4nci Üs*) last year at Murted, near Ankara, initially equipping the 1st Tactical Air Force's No 114 Squadron (114 *Filo*). This unit was subsequently split into two 18-aircraft squadrons, one being incorporated into the 3rd TAF.

The 1st TAF, largest of the two THK operational components, possesses, in addition to its F-102A all-weather intercept squadron, two 18-aircraft squadrons of F-104G Starfighters, also based at Murted, which serve primarily in the strike rôle with ground attack and air defence as secondary and tertiary missions respectively. Earlier this year, on 7 January, the THK lost almost 10 per cent of this Starfighter component when three F-104Gs collided in mid-air, crashing near Balikesir, their pilots losing their lives.

Two squadrons of Northrop F-5As, based at Bandirma and Balikesir, each possess a similar 18-aircraft complement, and have a primary air defence commitment with close support as a secondary mission. The F-5A, which began to enter the THK inventory early in 1966 with about 140 having now been delivered to Turkey, is popular with THK aircrew and ground staff alike, but the Turks, while admitting the qualities of the little Northrop warplane, consider it inadequate in the counter air rôle if opposed by aircraft of the calibre of the MiG-21, and while the THK would undoubtedly accept the newer F-5E model, the service has been pressing for the supply of McDonnell Douglas Phantoms for several years, and would like to obtain the F-4E(F) for the air superiority rôle.

The balance of the 1st TAF is made up of somewhat older combat aircraft, with three squadrons of F-100 Super Sabres operating primarily from Eskisehir and Bandirma in the tactical strike rôle, a squadron of F-84F Thunderstreaks at Eskisehir, and a tactical reconnaissance squadron of RF-84F Thunderflashes at the same base. The Super Sabre first entered THK service in 1958, 260 F-100C, D and F models being supplied to provide the service with its principal combat aeroplane, but the inevitable attrition of more than a dozen years of service (during a proportion of which the Super Sabre was in much demand for service in SE Asia, thus preventing MAP deliveries of surplus USAF aircraft as replacements) has taken a steady toll of the number of F-100s in the THK inventory. This erosion was partly arrested in 1969 by the transfer from the USAF of 40 F-100Ds which, having served their time in Vietnam, supplemented the surviving Super Sabres with the five THK squadrons operating this type, their delivery being accompanied by an outburst in the Turkish popular press which decried the supply of well-worn "hand-me-downs" by the USA to Turkey's air arm. The establishment of the THK Super Sabre squadrons varies between 20 and 25 aircraft.

The F-84F Thunderstreak entered Turkish service at much the same time as the F-100 in the late 'fifties, and the three remaining THK Thunderstreak squadrons were restored to full 25-aircraft strength some time ago with the arrival of a number of ex-*Luftwaffe* fighter-bombers of this type. These aircraft differed in equipment from Thunderstreaks already serving with the THK, and are referred to by the designation F-84FQ to distinguish them from the F-84Fs that survived from the 1958–60 MAP deliveries.

The 3rd Tactical Air Force, like the 1st, suffers a preponderance of obsolescing combat hardware. Apart from its single all-weather intercept squadron of F-102As and an F-5A squadron flying from Merzifon in the air defence rôle, its primary strength is provided by two squadrons of F-100 Super Sabres, two squadrons of F-84F Thunderstreaks, and a single squadron of RF-84F Thunderflash tactical reconnaisance aircraft. The defence of Turkish air space, originally invested in the 2nd Tactical Air Force which was disbanded and its North American F-86 Sabres retired, is the responsibility of the F-102A and F-5A squadrons of the 1st and 3rd Tactical Air Forces, and comes within the control of the Air Defence Command (*Hava Savunma Komutanligi*), the interceptors being backed by six batteries of THK-operated Nike-Hercules SAMs. The air defence system includes an extensive radar and communications network, and the NATO tropospheric forward scatter radio facility,

F-104G Starfighters of No 110 Squadron which, with the F-102As of No 114 Squadron, form No 4 Wing at Murted (No 4 Base), near Ankara.

known as the ACE High system, includes 17 stations manned by THK personnel and distributed over thousands of square miles of the rugged and remote Anatolian plateau in central and eastern Turkey.

Transport support and training

Transport support is provided by the four squadrons of the Transport Group, two of which are equipped with aged Douglas C-47s, one having a mix of C-47s, Beech 18s, a trio of Douglas C-54s and, from earlier this year, three ex-THY Viscounts, and the remaining squadron operating the four survivors of the five Lockheed C-130E Hercules supplied under the MAP in the mid 'sixties. This Transport Group, which has Headquarters at Etimesğut, Ankara, has extensive commitments, including the provision of transport for the Army's two paratroop battalions, and the THK possesses an urgent requirement for more modern transport aircraft to increase its logistic support capability. It is possible that Turkey may, like Greece, receive some ex-*Luftwaffe* Nord 2501D Noratlas transports, but there is a stronger likelihood that the THK will be provided with up to 15 Transall C.160Ds surplus to *Luftwaffe* requirements.

The Training Command (*Hava Egitim Komutanligi*) with Headquarters at Izmir, is responsible for all THK training activities. Grading and primary instruction is provided at Gaziemir, near Izmir, on CCF-built Beech T-34 Mentors, future transport and support aircraft pilots receiving basic instruction on the T-6J Harvard and twin conversion on the C-47. From Gaziemir future fighter pilots are sent to Konya for basic jet conversion on the Cessna T-37C, which is followed by advanced training on the Lockheed T-33A at Cigli. The latter base, handed over to the THK by the USAF last year, is also the commercial airfield for Izmir, and the original batch of 24 Canadair-built T-33A-N Silver Stars delivered to Turkey at the beginning of the 'fifties and supplemented over the years by additional MAP deliveries has more recently been augmented by a number of ex-*Luftwaffe* T-33As. Operational training is conducted at squadron level on TF-102As, TF-104Gs, F-100Fs and F-5Bs.

Apart from a handful of Sikorsky UH-19s, the THK currently possesses no vertical lift capability, the bulk of the helicopters available to the Turkish armed forces being included in the inventory of the *Kara Ordusu Havaciligi*, the aviation element of the Turkish Ground Forces Command. First formed in the late 'forties and initially confined to air observation post tasks with fixed-wing light aircraft such as the Piper L-18B, this force is steadily expanding its rotary wing element. Entirely independent of the THK and possessing its own flying training and maintenance organisations, the Turkish Ground Forces Command has a Central Army Aviation Establishment in Ankara and a major base at Euvercinlik, a few miles from Etimesğut.

Flights of the Turkish Army's aviation component are

Beech T-11 Kansans (above) and CCF-built Beech T-34 Mentors (below) remain in the THK training aircraft inventory.

deployed at corps and divisional levels for air observation post, liaison and communications, aeromedical and support tasks, and for administrative purposes are controlled by the Artillery Command. Fixed-wing equipment introduced in recent years has included the Cessna U-17 and the Dornier Do 27, which have, in part, replaced the substantial fleet of some 200 Piper L-18B and L-18C Super Cubs. Emphasis on helicopters has increased steadily since 1966, when, after six Turkish Army pilots had completed helicopter conversion courses in the USA, a small number of Bell Model 47Gs were supplied under the MAP. These have seen supplemented since 1968 by substantial numbers of Agusta-Bell AB 205 Iroquois and AB 206 JetRanger helicopters.

Apart from instruction on fixed-wing aircraft and helicopters, Army establishments provide instrument flying training with Dornier Do 28B-1s, training in aircraft maintenance, etc, and the current inventory includes a number of MAP-supplied DHC-2 Beavers and two Do 28D Skyservants, but the ratio of fixed-wing aircraft to helicopters is expected to change in favour of the latter during the next few years, and requirements exist for small numbers of medium transport helicopters and gunship helicopters.

The magnitude of the infrastructure programme in Turkey is evidenced by the amount of money spent by NATO in the country during the period 1965-7 alone, this totalling some $67m — the largest slice of infrastructure funds allocated to any NATO country during this three-year period. Apart from MAP aid, Turkey had received DM 200 million ($54.8m) in Federal German military aid up to mid-1970, about half of this in the form of new military equipment and the remainder in surplus material. The Turkish defence estimates for 1970-1 of 4,700 million Turkish liras (£167m or $401m) represents by far the lowest *per capita* expenditure of any NATO country, although as a percentage of the gross national product it is among the highest of NATO nations.

The bulk of this defence budget is absorbed merely in maintaining the Turkish armed forces, and any substantial modernisation of the THK inventory can only result from US military aid. In this connection, THK Headquarters personnel have expressed their opinion that the continued denial of advanced-performance aircraft in the category of the Phantom will inevitably result in a downgrading of the combat rôle of the THK during the early 'seventies, coupled with a progressive reduction in the service's combat strength as attrition makes inroads into the inventory of ageing combat aircraft. There can be no doubt that the cost to NATO of assisting Turkey to maintain a powerful, modern tactical air arm is high, but Turkish territory, forming the eastern flank of NATO, is among the most vital of the entire alliance, and NATO can ill afford *not* to underwrite this cost if it is to remain a viable defence organisation.

There is a strong school of thought in the upper echelons of the THK that US refusal to furnish even a nucleus of really modern combat aircraft may reflect a gradual shift in US strategy. It is argued that the basic mission of the THK is to "keep up with the potential enemy", and thus the best air superiority fighter available should be supplied to Turkish squadrons against the day they may have to oppose the MiG-21 or its more advanced successors. This is not to suggest that Turkey is disenchanted with NATO, but the Turkish High Command has qualms concerning the capability of some of the combat aircraft that the THK is forced to fly. Meanwhile, as NATO air space, Turkish skies witness some strange contrasts these days. While USAF Tactical Air Command Phantoms fly into Incirlik, near Adana, on rotation from bases in the UK and Germany, MiG-15UTIs, MiG-17s and various support types en route from the Soviet Union to Syria, Iraq, and other Middle Eastern countries, are staged through Diyabakir and other Turkish 3rd Tactical Air Force bases! □ WG

THE ASIATIC THUNDERBOLT

As THE US 37th Division battered its way south into the city of Manila, capital of the Philippines, and the 1st Cavalry Division swung in an arc around the city to link up with the 11th Airborne Division in their effort to surround Rear Admiral Mitsuji Iwafuchi's garrison during the dramatic February weeks of 1945, a technical intelligence team following in the wake of the advancing US troops made an interesting discovery. Among a number of Japanese Navy aircraft abandoned among the trees alongside the Dewey Boulevard, on the outskirts of Manila, was a single example of a barrel-like little fighter that the Allies had had no previous opportunity to examine — the Mitsubishi J2M3 Raiden (Thunderbolt).

The Raiden had been encountered in combat over the Marianas eight months earlier, during the Battle of the Philippine Sea, when small numbers of fighters of this type had operated from Guam. The Technical Air Intelligence Command (TAIC) was conversant with this new warplane's general characteristics, a technical manual on the J2M2 Raiden having been discovered after the capture of Saipan, and the Allied code-name of "Jack" had been assigned, but the interrogation of Japanese Navy personnel had suggested that this fighter was being reserved primarily for the defence of Japan's home islands, a fact that was being borne out by B-29 crews of the 21st Bomber Command that had begun flying against Japan from their new bases in the Marianas. The TAIC was therefore most anxious to evaluate the Raiden in flight.

The Raiden abandoned on the Dewey Boulevard, although somewhat dilapidated, was intact apart from airscrew and armament. It was carefully disassembled and trucked to nearby Clark Field, where, after undergoing repair by TAIC personnel, it was ready to be flown within two months, although by this time testing of the Raiden was of little more than academic interest; the USAAF and the USN having by then clearly established air superiority over the heart of Japan. Nevertheless, the captured fighter was tested by a senior pilot attached to TAIC who had flown virtually every Allied fighter, as well as the Messerschmitt Bf 109, the Focke-Wulf Fw 190, and such Japanese types as the Nakajima Ki-43 Hayabusa and Ki-84 Hayate, the Kawasaki Ki-45 Toryu and the Mitsubishi A6M Zero-Sen. After logging three hours and twenty minutes in the Raiden he stated unequivocally that it was the best Japanese fighter that he had flown, offering a good performance, good stability, good stalling characteristics, and good take-off and landing qualities!

A change in Japanese thinking

Unbeknown to the team evaluating the captured Raiden at Clark Field, this fighter was noteworthy from two aspects: it reflected a radical change in Japanese Navy fighter design thinking and it provided an outstanding example of mismanagement and vacillation. In the Raiden the Japanese Navy acquired for the first time a point defence interceptor; a fighter whose design placed speed and climb rate paramount in importance, and relegated manoeuvrability to a secondary position. It took Navy pilots — for long firm adherents of the classic dog-fighting style of combat and excelling in the manoeuvres demanded by close-in conflict — some time to adjust to the new fighter concept as represented by the Raiden, and when they finally recognised the qualities

WARBIRDS

The J2M3 Raiden found by US forces on Dewey Boulevard, on the outskirts of Manila, during February 1945. It was subsequently repaired and flown.

of this warplane they were involved in the final hectic mêlée over the Japanese home islands with which WW II was to end. The Raiden possessed most of the attributes of a highly successful warplane, but its development was plagued by teething troubles with which the small and overburdened Mitsubishi fighter design team, initially led by Dr Jiro Horikoshi, could not adequately cope; troubles that produced indecision in the Navy procurement department at a critical time in the development cycle of the Raiden and, fortunately for the Allies, denied it the chance to play an important rôle in the Pacific air war.

From several aspects, the Raiden was a brilliant design, as might have been expected from Dr Horikoshi who had had primary responsibility for the A6M Zero-Sen (or Reisen), and the Imperial Navy's decision to place emphasis on speed and climb rate was a bold departure from its accepted practice. The requirement for such a fighter was established as part of the 14-*Shi* programme (14th year of the Showa reign or 1939), and the first discussions concerning this type actually took place between Dr Horikoshi and the Bureau of Aeronautics of the Imperial Japanese Navy in October 1938, but preoccupation with the A6M Zero-Sen necessitated the shelving of the project for nearly a year. Thus, it was not until September 1939 that the official specification was drawn up for the 14-*Shi* interceptor fighter. The performance parameters called for by this specification were a maximum speed of 373 mph (600 km/h) at 19,685 ft (6 000 m); the ability to attain this altitude within 5·5 minutes; an endurance of 45 minutes at full rated power; a take-off run at overload weight and in nil-wind conditions not exceeding 985 ft (300 m), and a landing speed no greater than 81 mph (130 km/h). Armament was to comprise two 20-mm cannon and two 7,7-mm machine guns, and for the first time some protection was requested for the pilot in the form of an armour plate behind the seat.

In so far as a power plant was concerned, Dr Horikoshi's choice lay between the Aichi Ha-60 Atsuta derivative of the Daimler-Benz DB 601A inverted-vee 12-cylinder liquid-

The third J2M1 Raiden, the Kana character Ko used as a prefix in the tail identification letter-and-numeral sequence indicating that the aircraft was included in the inventory of the Koku Gijutsu Sho (Air Technical Arsenal). This should not be confused with the Kanji character Ko which was applied as a suffix to aircraft, power plant etc, designations to indicate a modification.

cooled engine offering 1,185 hp at 2,500 rpm for take-off and war emergency power of 1,205 hp at 1,380 ft (420 m), and the Mitsubishi Ha-32 Kasei Model 13 radial 14-cylinder air-cooled engine rated at 1,440 hp at 2,450 rpm for take-off and giving a war emergency power of 1,530 hp at 8,400 ft (2 560 m). Although the Aichi Tokei Denki KK (Aichi Clock and Electric Co Ltd) promised an increase of at least 15–20 per cent in the war emergency power of the Atsuta at all altitudes and as much as 25–30 per cent in the military power, Horikoshi elected to use the more powerful Kasei despite its greater weight (1,697 lb/770 kg dry compared with 1,378 lb/ 625 kg), higher fuel consumption and larger frontal area. This choice was to prove singularly unfortunate for the future Mitsubishi fighter.

Dr Horikoshi, assisted by Yoshitoshi Sone and Kiro Takahashi, began detail design during the early weeks of 1940 of what by this time bore the Service Aeroplane Development Programme number M-20. Basically, the new fighter was an all-metal cantilever low-wing monoplane, the wing having a single mainspar at 35 per cent chord and smooth flush-riveted stressed skinning, the fuselage being an oval-section semi-monocoque, and the tail surfaces having metal-framed fabric-covered movable surfaces. In order to minimise the drag of the Kasei engine an extension shaft was introduced to drive the 10 ft 6 in (3,20-m) diameter three-bladed airscrew, this allowing the cowling to be finely tapered, with an engine-driven fan sucking cooling air through a narrow annular intake. The wing was of low aspect ratio and employed a laminar flow aerofoil section, and in a further attempt to reduce drag an extremely shallow, curved windscreen and canopy were adopted. Combat flaps were provided, to increase lift with the minimum of drag when extended and thus improve manoeuvrability, and to simplify manufacture forged components were used wherever possible.

The higher priority allocated the development of the A6M series and teething troubles suffered by the engine cooling system delayed the completion of the prototypes, and the first of these did not leave the experimental shops until February of 1942, by which time Dr Horikoshi, suffering from overwork, had relinquished the post of chief designer to Kiro Takahashi. Designated J2M1, the first prototype made its initial flight on 20 March 1942 at Kasumigaura with Mitsubishi test pilot Katsuzo Shima at the controls. The initial flight testing revealed stability and controllability to be excellent, but Shima expressed his opinion that forward view was totally unacceptable and that the curvature of the windscreen seriously distorted vision, this being particularly dangerous during landing. It was discovered that the undercarriage would not retract at speeds in excess of 100 mph (160 km/h), and while handling proved good throughout the entire speed range of the aircraft, the ailerons tended to stiffen up at speeds above 323 mph (520 km/h).

Apart from some modifications to the undercarriage retraction mechanism, it was decided to transfer the prototype without changes to the Suzuka Naval Air Base for initial service trials. The anticipated criticisms regarding visibility from the cockpit were duly voiced by the Navy test pilots, and the test schedule was frequently disrupted by the unsatisfactory operation of the airscrew pitch-change mechanism. It was soon patently obvious that the specified performance could not be attained with the Kasei Model 13 engine, measured speed and climb indicating a maximum

(Opposite page) A Mitsubishi J2M3 Raiden Model 21 of the 302nd Kokutai engaged in defence of the home islands (autumn 1944). Dark green upper surfaces and pale grey under surfaces, tan airscrew spinner, yellow wing leading-edge stripes, red Hinomarus with white outline on upper wing surfaces and fuselage sides, and off-white tail identification markings. (Drawing reproduced by courtesy of BPC Publishing Limited.)

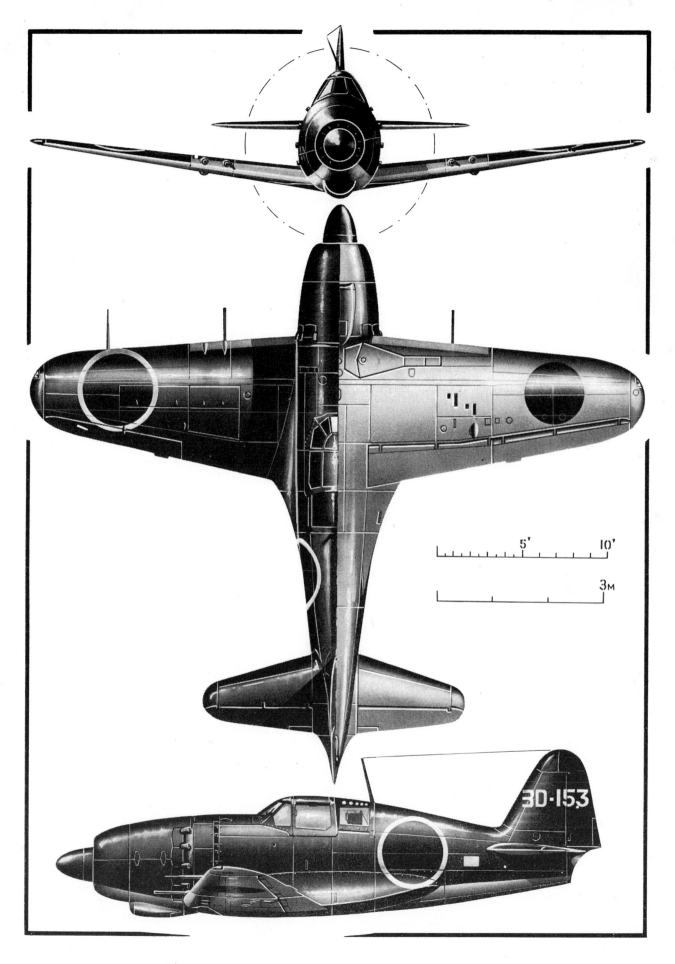

3D-153

5' 10'

3M

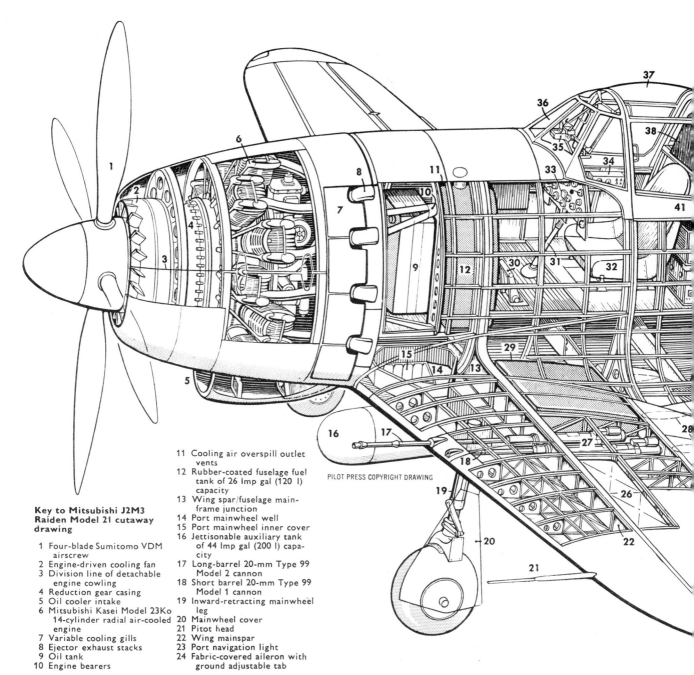

PILOT PRESS COPYRIGHT DRAWING

Key to Mitsubishi J2M3 Raiden Model 21 cutaway drawing

1 Four-blade Sumitomo VDM airscrew
2 Engine-driven cooling fan
3 Division line of detachable engine cowling
4 Reduction gear casing
5 Oil cooler intake
6 Mitsubishi Kasei Model 23Ko 14-cylinder radial air-cooled engine
7 Variable cooling gills
8 Ejector exhaust stacks
9 Oil tank
10 Engine bearers
11 Cooling air overspill outlet vents
12 Rubber-coated fuselage fuel tank of 26 Imp gal (120 l) capacity
13 Wing spar/fuselage mainframe junction
14 Port mainwheel well
15 Port mainwheel inner cover
16 Jettisonable auxiliary tank of 44 Imp gal (200 l) capacity
17 Long-barrel 20-mm Type 99 Model 2 cannon
18 Short barrel 20-mm Type 99 Model 1 cannon
19 Inward-retracting mainwheel leg
20 Mainwheel cover
21 Pitot head
22 Wing mainspar
23 Port navigation light
24 Fabric-covered aileron with ground adjustable tab

speed of 359 mph (578 km/h) at 19,685 ft (6 000 m), 7·8 minutes being required to attain that altitude. The J2M1 weighed 4,830 lb (2 191 kg) in empty equipped condition and 6,307 lb (2 861 kg) fully loaded, this providing a wing loading of 29·2 lb/sq ft (142,7 kg/m²) and a power loading of 4·4 lb/hp (2,0 kg/hp).

Premature production order

Satisfied with the taxying and ground handling, the steep climb angle, the stability and general controllability of the fighter, the Imperial Navy's Bureau of Aeronautics was confident that the shortfall in performance could be rectified by adaptation of the fuel-injection Kasei Model 23 for fan cooling and installation in the aircraft. Furthermore, this engine would offer sufficient power to compensate for any increase in drag resulting from the application of a deeper, more conventional windscreen. Indeed, so confident was the Navy of the efficacy of these changes — which it had

stipulated should be introduced on the fourth of the prototype and evaluation aircraft (which thus became the J2M2) — that the new Mitsubishi warplane was accepted for production as the Navy Interceptor Fighter Raiden Model 11 within a few days of 13 October 1942 when the J2M2 made its initial flight, and before any thorough evaluation of the fighter was possible. Tooling was begun immediately at Mitsubishi's No 3 (Airframe) Plant at Nagoya, but little time was to elapse before it was to be discovered that the production order was somewhat premature.

The modified engine, the Kasei Model 23Ko, drove a four-bladed Sumitomo hydraulically-operated VDM metal airscrew of 10 ft 10 in (3,30 m) diameter, and made provision for the injection of water-methanol into the supercharger below the rated altitude. A marginally shorter extension shaft permitted a reduction in the overall length of the fighter from 32 ft 5¾ in (9,90 m) to 31 ft 9⅞ in (9,69m); individual exhaust stacks were introduced which provided

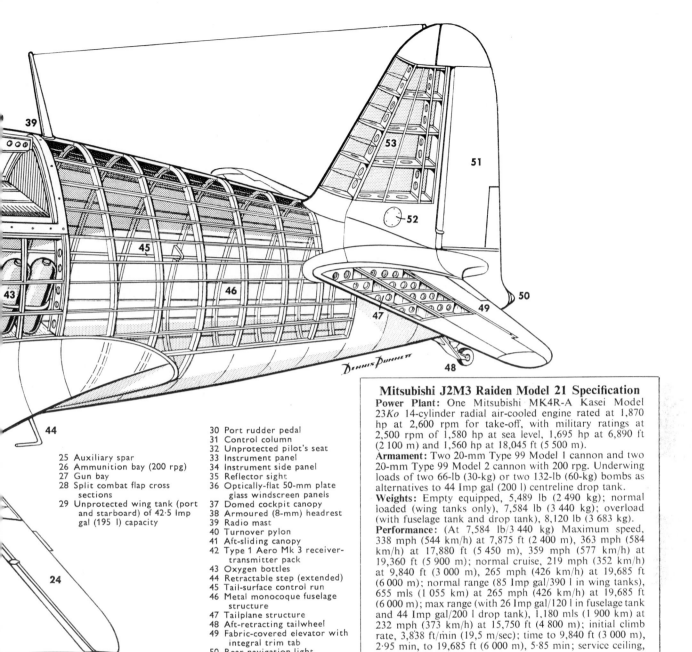

25 Auxiliary spar
26 Ammunition bay (200 rpg)
27 Gun bay
28 Split combat flap cross
   sections
29 Unprotected wing tank (port
   and starboard) of 42·5 Imp
   gal (195 l) capacity

30 Port rudder pedal
31 Control column
32 Unprotected pilot's seat
33 Instrument panel
34 Instrument side panel
35 Reflector sight
36 Optically-flat 50-mm plate
   glass windscreen panels
37 Domed cockpit canopy
38 Armoured (8-mm) headrest
39 Radio mast
40 Turnover pylon
41 Aft-sliding canopy
42 Type 1 Aero Mk 3 receiver-
   transmitter pack
43 Oxygen bottles
44 Retractable step (extended)
45 Tail-surface control run
46 Metal monocoque fuselage
   structure
47 Tailplane structure
48 Aft-retracting tailwheel
49 Fabric-covered elevator with
   integral trim tab
50 Rear navigation light
51 Fabric-covered horn-balanced
   rudder with integral trim
   tab
52 Control linkage acess panel
53 All-metal fin structure

Mitsubishi J2M3 Raiden Model 21 Specification

Power Plant: One Mitsubishi MK4R-A Kasei Model 23*Ko* 14-cylinder radial air-cooled engine rated at 1,870 hp at 2,600 rpm for take-off, with military ratings at 2,500 rpm of 1,580 hp at sea level, 1,695 hp at 6,890 ft (2 100 m) and 1,560 hp at 18,045 ft (5 500 m).
Armament: Two 20-mm Type 99 Model 1 cannon and two 20-mm Type 99 Model 2 cannon with 200 rpg. Underwing loads of two 66-lb (30-kg) or two 132-lb (60-kg) bombs as alternatives to 44 Imp gal (200 l) centreline drop tank.
Weights: Empty equipped, 5,489 lb (2 490 kg); normal loaded (wing tanks only), 7,584 lb (3 440 kg); overload (with fuselage tank and drop tank), 8,120 lb (3 683 kg).
Performance: (At 7,584 lb/3 440 kg) Maximum speed, 338 mph (544 km/h) at 7,875 ft (2 400 m), 363 mph (584 km/h) at 17,880 ft (5 450 m), 359 mph (577 km/h) at 19,360 ft (5 900 m); normal cruise, 219 mph (352 km/h) at 9,840 ft (3 000 m), 265 mph (426 km/h) at 19,685 ft (6 000 m); normal range (85 Imp gal/390 l in wing tanks), 655 mls (1 055 km) at 265 mph (426 km/h) at 19,685 ft (6 000 m); max range (with 26 Imp gal/120 l in fuselage tank and 44 Imp gal/200 l drop tank), 1,180 mls (1 900 km) at 232 mph (373 km/h) at 15,750 ft (4 800 m); initial climb rate, 3,838 ft/min (19,5 m/sec); time to 9,840 ft (3 000 m), 2·95 min, to 19,685 ft (6 000 m), 5·85 min; service ceiling, 38,385 ft (11 700 m).
Dimensions: Span, 35 ft 5¼ in (10,80 m), length, 32 ft 7½ in (9,94 m); height, 12 ft 11¼ in (3,94 m); wing area, 215,82 sq ft (20,05 m²).

a measure of thrust augmentation, and the Kasei Model 23*Ko* offered 1,870 hp at 2,600 rpm for take-off, military power ranging from 1,580 hp at sea level to 1,695 hp at 6,890 ft (2 100 m) and 1,560 hp at 18,045 ft (5 500 m). However, engine dry weight at 1,896 lb (860 kg) had risen some 14 per cent and thus, coupled with increases in airscrew and accessory weights, necessitated a reduction in fuel tankage in order to maintain the cg position of the fighter, capacity being reduced from 156 Imp gal (710 l) to 120·5 Imp gal (550 l).

The fuel was distributed between two 46 Imp gal (210 l) tanks in the wing roots and a 28·5 Imp gal (130 l) tank in the fuselage immediately aft of the engine firewall. A carbon dioxide fire extinguishing system was provided which could be operated by the pilot by means of a selector switch, but apart from a thick rubber coating enclosing the fuselage tank, the fuel was unprotected. The new windscreen, which impaired the lines of the fighter but greatly improved visi-

bility from the cockpit, embodied optically-flat shatter-proof panels of 50-mm plate glass, the only protection for the pilot being provided by a small piece of 8-mm armour plate just aft of the pilot's head and protecting only the base of his neck.

Armament comprised two 7,7-mm Type 97 machine guns in the upper fuselage decking with 550 rpg and two wing-mounted Type 99 Model 2 cannon of 20-mm calibre with 200 rpg. The smaller-calibre Type 97 Model 3*Otsu* installed in the J2M2 had a muzzle velocity of 2,460 ft/sec (750 m/sec), a rate of fire of 1,000 rpm, and was based on the Vickers gun, and the larger-calibre Type 99 Model 2 was a version of the Oerlikon with a muzzle velocity of 2,490 ft/sec (760 m/sec) and, in its *Hei* version, a fire rate of 490 rpm. Some attempt was made to provide the J2M2 with a measure of buoyancy in the event of alighting on the sea, two watertight compartments being introduced outboard of the wing guns, and a flotation bag being located aft of the cockpit.

The Raiden immediately began to experience what were to prove to be protracted troubles with the Kasei Model 23*Ko* engine. The initial flight test programme revealed strong vibration at maximum power, and the engine ran roughly, emitting a considerable amount of smoke. The adjustment of the fuel and water-methanol injection systems reduced the smoke emission to acceptable standards and largely rectified running roughness, but the vibration problem was not so simple of solution. At certain engine rpm both power plant and airscrew oscillation assumed critical proportions, and some months elapsed before it was discovered that the problem could be resolved by increasing the rigidity of the airscrew blades and the resilience of the engine-mount shock absorbers.

By the end of March 1943, only 11 J2M2 fighters had left the assembly line, but production was gathering momentum when, on 16 June, the second J2M2 crashed. Shortly after becoming airborne, the fighter inexplicably nosed down suddenly from an altitude of some 65 ft (20 m) and crashed into a barn, the fuel exploding on impact. A subsequent investigation failed to reveal the cause of the accident, but a possible explanation was provided some weeks later when a Navy acceptance pilot, after taking-off in the 10th J2M2, raised the undercarriage and found that, simultaneously, the control column was wrenched forward. Fortunately, he had gained some altitude and had the presence of mind to lower the undercarriage immediately, whereupon the controls responded normally. After landing, it was found that the tailwheel shock strut *could* press against the elevator torque tube during retraction, and it was assumed that this

had caused the crash of the second J2M2. A modification was promptly made to the elevator control run, and the lack of any further incident of this nature suggested that the assumption was correct.

Despite the fact that all the teething troubles of the Raiden had by no means been resolved, in December 1943 the Navy finally accepted the first batch of J2M2s, and these were flown to Toyohashi, south-east of Nagoya, for service with the 381st *Kokutai* (Air Corps — roughly equivalent to an RAF Group). In January 1944, within two weeks of its arrival at Toyohashi, the 30th J2M2 disintegrated over the airfield immediately after its pilot had made a firing pass at a target streamer. This accident was never satisfactorily explained, but it was believed that a violent oscillation was set up when an engine attachment point broke, causing a secondary airframe failure. An alternative explanation was that an engine cowling panel had been incorrectly fastened, had broken away and had hit the tail. The engine attachment points were reinforced and the cowling fasteners were strengthened, but there were to be other incidents in which Raiden fighters disintegrated in mid-air without any satisfactory explanation.

Progressive development
The Kasei 23*Ko* engine-Sumitomo VDM airscrew combination proved an endless source of trouble, and although it had been believed that the oscillation problem was resolved, recurrence of the trouble stressed the fact that the vibration was more complex in nature than had been thought. A variety of high-rigidity thick airscrew blades were tested and their characteristics studied, but the problem was never to be resolved entirely satisfactorily. A series of crankpin failures occurred, this problem being eliminated by raising the oil pressure. This, in turn, gave rise to an increase in oil temperature, necessitating the adoption of a new and enlarged oil cooler with an external intake which was standardised on the J2M3 Raiden Model 21, but the poor reliability of the Kasei 23*Ko* engine was to continue to plague the Raiden, and only 155 fighters of this type (including the trio of J2M1s and the first J2M3s) had been completed by the end of March 1944.

Weighing 5,176 lb (2 348 kg) in empty equipped condition, and possessing a clean loaded weight (with full wing tanks only) of 7,077 lb (3 210 kg), the J2M2 attained 371 mph (597 km/h) at 17,880 ft (5 450 m) and climbed to 19,685 ft (6 000 m) in 5·7 min. By comparison, the similarly-powered J2M3 weighing 7,584 lb (3 440 kg) was marginally

J2M3 Raiden 21 interceptors (above and below) based at Kasumigaura (indicated by the Kana character Ka*). The 1000-series identification number signified a land-based interceptor fighter.*

slower, attaining 363 mph (584 km/h) at 17,880 ft (5 450 m) and reaching 19,685 ft (6 000 m) in 5·85 min, but it carried a more potent armament, the fuselage-mounted 7,7-mm weapons being discarded and the wing weapons being augmented by a pair of 20-mm Type 99 Model 1 cannon. The Model 1 cannon was a shorter-barrel weapon than the Model 2 that it complemented in the wing of the Raiden, possessing a lower muzzle velocity and fire rate, but it substantially increased weight of fire and improved the potency of the fighter as a bomber interceptor.

The additional wing guns had dictated some local strengthening of the wing structure and necessitated slight reduction in the size of the wing tanks, their capacity being reduced to 42·5 Imp gal (195 l) each. The fuselage tank, which was considered as an auxiliary cell, was also slightly reduced in capacity to 26 Imp gal (120 l), but a 44 Imp gal (200 l) drop tank could be carried beneath the fuselage centreline as an alternative external load to a pair of 132-lb (60-kg) bombs beneath the wings. Owing to the different ballistic characteristics of the two versions of the cannon, this mix was not considered ideal, and 21 aircraft were, in fact, completed under the designation J2M3*Ko* Raiden Model 21*Ko* with the Model 1 cannon deleted and replaced by an additional pair of Model 2 cannon mounted in underwing gondolas. Although the quartet of Model 2 weapons proved a more effective armament, the drag of the underwing gondolas had an adverse effect on performance, and this arrangement was not standardised.

The first J2M3 had actually appeared in October 1943, some time before the J2M2 was delivered to the 381st *Kokutai* at Toyohashi. This model was adopted in succession to the J2M2 and placed in production at Mitsubishi's Nagoya and Suzuka factories, the first production J2M3s being completed at the beginning of February 1944, the last (and 155th) production J2M2 rolling off the Nagoya assembly line at the beginning of May. In the following month, however, the Imperial Navy's Bureau of Aeronautics, disenchanted with the Raiden because of its protracted teething troubles and low mechanical reliability, decided to adopt the Kawanishi N1K1-J Shiden (Violet Lightning) as its primary interceptor. Furthermore, it was pronounced that Raiden manufacture should continue at a *reduced* tempo only until the availability of the A7M Reppu (Hurricane) permitted its complete production phase out.

These decisions did not, in the event, sound the death knell for the Raiden. On the contrary, within weeks of their being taken the war situation had deteriorated to such an extent that the Raiden, now adjudged one of the most potent B-29 Superfortress interceptors, had its production priority reinstated, and the Koza Kaigun Kokusho (Koza Naval Air Arsenal) was instructed to join in Raiden production.

Meanwhile, the development of further Raiden variants had continued unabated, and the J2M4 Raiden Model 34 with a turbo-supercharged Kasei Model 23*Hei* was flown in August 1944. The mechanically-driven turbo-supercharger, which was installed on one machine by Mitsubishi and on another by the Naval Air Research and Development Centre, permitted rated power of 1,420 hp to be maintained up to 30,180 ft (9 200 m), at which altitude a speed of 363 mph (584 km/h) could be attained. Apart from the turbo-supercharged Kasei 23*Hei* engine, the J2M4 mounted an additional pair of cannon, these being installed in the fuselage to fire forward and upward at an angle of 70 deg from the horizontal in a similar fashion to the German *schräge Musik* (Oblique Music or Jazz) installation. Unfortunately for the Imperial Navy, the turbo-supercharger system was highly complex, and after several failures and with little prospect of rectifying the faults quickly, the J2M4 was abandoned, only the two experimental models having been tested.

A further high-altitude variant of the Raiden, the J2M5

A J2M3 Raiden 21 of the 302nd Kokutai taking-off from its base at Yokosuka.

Raiden Model 33, had actually preceded the J2M4, having flown in May 1944, and this variant proved rather more successful. It featured a Kasei Model 26*Ko* engine with a mechanically-driven three-stage supercharger and an enlarged intake manifold, and achieved speeds of 382 mph (615 km/h) at 21,600 ft (6 585 m) and 375 mph (603 km/h) at 26,250 ft (8 000 m). The J2M5 was immediately ordered into production at the Koza Naval Air Arsenal, although only 34 had been completed because of delays in deliveries of the Kasei 26*Ko* engine when Japan surrendered.

Both J2M4 and J2M5 had been preceded chronologically by the J2M6 Raiden Model 31, but this differed from the standard production J2M3 in relatively minor respects, the most noteworthy change being a new domed cockpit canopy which improved visibility. In fact, only one example of the J2M6 was completed, and the new canopy was fitted on some J2M3 and J2M3*Ko* Raidens. The impetus placed behind Raiden production when its qualities as a bomber interceptor were finally appreciated proved belated, and only 281 J2M3 and J2M3*Ko* fighters had been completed when hostilities terminated. The production of Raiden fighters of all types (including prototypes and experimental models) totalled 476 machines, these being operated primarily by the Genzan (Korea), Tainan (Formosa), 302nd, 332nd, 352nd and 381st *Kokutais*, and were too few to exert any marked effect on the air war in the Pacific Theatre.

The Raiden in the air
The Allied field grade evaluation pilot who tested the J2M3 Raiden Model 21 found near Manila after its restoration to flying condition made the following report on its characteristics:

"Taxying and ground handling in general are excellent, visibility from the cockpit on the ground being much better than that offered by the Nakajima Hayate due to the wide cockpit canopy and more forward position of the seat. Good taxying brakes are fitted on a British-type rudder bar, but the rudder is not very effective during taxying and no tailwheel lock is fitted. Take-off is normal with little tendency to swing, but the tail does not come up readily owing to the small elevators. When the tail is raised forward vision is good. Take-off run is short and the aircraft leaves the ground readily at 100 mph (160 km/h). There is practically no change of trim with gear retraction. Climb angle is steep and the rate of climb rapid, and the cooling of engine and oil appears to be excellent. Handling and control are good at all speeds from the stall up to 325 mph (523 km/h), although the ailerons are heavy at normal cruise speeds and become exceptionally so above 325 mph (523 km/h). The elevators are on the light side at all speeds up to 325 mph (523 km/h). The rudder is satisfactory at all speeds.

"Rudder and elevator trim tabs are controlled by parallel wheels in longitudinal plane on the port side of the cockpit, and the rudder control tab is ineffective, but little change in

continued on page 103

THE FIGHT FOR THE GROUND
Fighter Spectrum Part 2 – Ground attack

by Roy Braybrook, BSc, CEng, AFRAeS

FIGHTERS mainly concerned with surface targets may for practical purposes be categorised as close support, ground attack, and strike aircraft, although there are no precise and agreed definitions of these terms. Strike fighters (or interdictors) are 20-30 ton, high-performance aircraft intended to penetrate enemy territory to the order of 500 nm (900 km) HI-LO, while ground attack aircraft are generally just as fast, but half as heavy and penetrate roughly half as far. Pure close support aircraft are taken to be relatively low performance designs (and perhaps not really fighters), which provide special capabilities of value in a more permissive environment.

On these definitions a fighter will often operate in duties below its own category, but cannot move up the scale. Thus the Su-7 ground attack aircraft may be used in close support (one operator insists that that is *all* it's good for), but on the other hand the A-X close support aircraft would not be expected to graduate to targets defended by swarms of fighters.

The existence of these three distinct categories is not accepted by all operators. One air force has been shopping for a ground attack aircraft, yet specifies bombload and range figures that sound more like a Tu-22! Another claims that its fighter-bomber requirements can be met by a simple close support aircraft, although its potential adversary has been operating supersonic fighters for many years.

Close support means assisting ground forces in contact with the enemy. It involves co-ordination with these forces, and often the target is marked either by them or a FAC (forward air controller) aircraft. At its best, this form of attack is more spectacular than the movies. There were instances in Korea of aircraft strafing Chinese on one side of a road when UN troops were entrenched on the other! However, air support has often arrived too late, been inaccurate, short-lived, and seriously limited by weather and darkness. Soldiers have long memories for fighters that failed to find the target, or attacked *across* the battle-line, or tried to bomb through jungle canopies with cluster weapons that merely supplied the VC with thousands of booby-traps.

Direct support is not a particularly rational use for aircraft, despite their superiority in mobility over ground-based fire power. Their other advantages are in delivering certain area weapons and in having a psychological effect on the recipient, who can see his assailant but can do little to hit back.

For best results, this type of attack needs special training and radio fits, and aircraft designed specifically for the job, although until Vietnam it was left to standard ground attack fighters. Of these one of the most effective was the F4U, ie, THE Corsair. The F-51 Mustang was also used in Korea, but suffered high losses due to the vulnerability of its liquid-cooled engine. The F4U was followed by the A-1 Skyraider, and in due course by the A-4. Although considerably slower than today's F-4, the A-4s achieve better accuracies in pairs bombing due to the smaller separations associated with this low speed.

Over the same timescale Britain had its own successes, the best known probably being the Typhoons' operations over Normandy. Looking back, how anyone hit tanks with those 3-in (7,6-cm) "drainpipes" is a mystery. Post war, the Hornet (one of the finest-looking twins ever built), Sea Fury, Meteor, Sea Hawk, and Hunter excelled in this rôle. All have been used in earnest, although it can be argued that only the Meteor F.8 was proved in a shooting match of any duration.

The progress of fighter design to suit higher speeds has produced a divergence from the form required for close support. Many air forces are consequently ill-equipped to help ground forces, having fighters which carry little ordnance, are poor weapon platforms, have a turn radius that takes in most of Texas, fatigue lives measured in mere hundreds of hours, and sometimes tailplane-runaway problems that make low flying suicidal. They usually lack armour, self-sealing tanks, and the radio frequencies needed to talk with the men on the ground.

While fighters were becoming less suitable, demands for close support escalated with the insurgency problem, and this led to inter-service friction. Armies are now demanding a breakthrough in air support or the transfer of that function to their own service. Air forces respond that only they know how to control air power, and that the men on the ground would "fritter it away in penny packages" (a piece of airman's dogma that seems to have come in with the Wright Brothers). They simultaneously admit that close support comes low on their priorities, a list which appears to be topped by fighting the Navy!

Concepts of close support differ considerably between individual services. The USAF tends to allocate most of its resources to large pre-planned missions with secondary and tertiary targets, and then respond to calls for air support by diversions. One peculiarity of this system is that it seems impossible to hit the enemy with anything less than twelve aircraft! Criticism that reaction times are often over 30 minutes draws the response that it takes almost that long for the Army to call off the artillery and mortars, and for the FAC to mark the target.

The USMC achieves quicker reaction by means of standing patrols, a practice reviled by the USAF as "loiter-and-jettison". The former also tends to work with pairs of aircraft, whereas the Air Force swamps the target with successive flights of four.

The close support conflict that has been festering between the American services is now coming to a head, with possibly far-reaching results. The US Army, which helped fund the P.1127 V/STOL fighter until shorn of its fixed-wing aircraft in 1966 (save for the Mohawk and utility types), has set the pace by pressing development of the armed helicopter.

The USAF is meantime responding to criticism that its F-100, F-105, and F-4 are unsuitable for this work, and that the B-57, A-1, A-7, A-26, A-37, and T-28 all have serious limitations. Funding has been approved to start work on A-X, which will excel in warload, endurance, low attrition, and short-field operation. For its part, the USMC is planning to cut the wastage of standing patrols by use of AV-8A Harriers, providing quick reaction from aluminium pads close up with the "mud marines".

Whether all three categories can survive the economists' scrutiny is doubtful. However, some form of armed helicopter will surely remain, as nothing can secure an assault landing in insurgent territory like the point-blank fire from a modern gunship. On the other hand nobody should rely on this system where targets have concentrations of automatic weapons, or are supported by MiGs. A fixed-wing complement is essential, but in a COIN situation there is not much to choose between the V/STOL concept and an air loiter type that can carry twice the bombload. The only overwhelming difference arises when these aircraft have to be used in a more serious conflict.

Costly egg-beater

Production of the AH-56A Cheyenne was cancelled in 1969 because it had a serious rotor vibration problem, was overweight, and down on speed. Lockheed has since increased the stiffness and damping of the rotor, modified the flight control system, and doubled the fixed wing chord. The US Army has meanwhile relaxed its weight limit, and is calling for only 212 kt (390 km/h). An order for the AH-56B is thus possible, although unit cost is expected to be about $3·5m (£1·45m).

Cheyenne equipment includes laser ranging, Doppler

(Above) The Fairchild Hiller A-10A and, below, the Northrop A-9A — the two very different designs which have been selected by the USAF for prototype construction and a "fly-off" in the A-X competition.

radar, auto terrain avoidance, a ventral turret with a 30-mm cannon, and a nose turret with a 40-mm grenade launcher or 7,62-mm Minigun. Using STO, the internal ammunition load of 3,400 lb (1 550 kg) can be augmented by 4,500 lb (2 050 kg) of 7-cm rockets and TOW anti-tank missiles, carried on six pylons.

Another Army option is the Sikorsky AH-3 (S-67) Blackhawk, a less ambitious winged helicopter with (at last!) an aerodynamically clean rotor head. It weighs 10,900 lb (4 950 kg) empty, compared with 12,200 lb (5 550 kg) for the AH-56B, and has two 1,500 shp T58-GE-5s, rather than the single 3,435 shp T64-GE-16. Clean dash speed is 173 kt (320 km/h), and cruise speed with stores is 163 kt (300 km/h).

One of these types may be selected later this year to replace the Bell AH-1G Hueycobra, which costs a mere $500,000 (£210,000), weighs 5,510 lb (2 500 kg) empty, and carries 2,200 lb (1 000 kg) of ammunition and stores. Its 1,250 shp T53-L-13 gives a clean maximum of 160 kt (295 km/h). The nose turret with 7·62-mm Minigun (replaced by a three-barrel XM-197 30-mm in the case of the AH-1J) can be augmented by four 19-round pods of 7-cm rockets.

A more specialised form of close support resulted from USAF adaptations of transport aircraft, fitted with side-firing weapons, various sensors, and batteries of flares, so that villages under insurgent attack can be protected throughout the night. Beginning with the AC-47 "Spooky" with three 7,62-mm Miniguns, this programme led to the

Barred from operating armed fixed-wing aeroplanes, the US Army is pressing forward with development of armed helicopters. Despite early problems, the Lockheed AH-56 Cheyenne (below right) is now showing promise, while Sikorsky is competing for Army funds with the proposed AH-3 (below left) based on the well-established S-61 design.

The Saab 105XT is one of the growing group of trainer/strike aircraft which provide an economic alternative to the more specialised ground attack aeroplanes for smaller countries.

AC-119K "Shadow" with four 7,62s and two 30-mm M-61s, and finally the AC-130 "Spectre" with the devastating total of four 7,62s and four 30-mm.

Flying Fortress

Arising from the lessons of South Vietnam, Fairchild-Hiller and Northrop are now funded to produce two prototypes each of their single-seat A-X designs, respectively the A-9A and A-10A, for fly-off assessment to be completed by 1973. Lack of a suitable high-bypass engine seems to have been a problem: the former company has chosen two 9,000 lb (4 100 kg) General Electric TF34s, being developed for the Lockheed S-3A and Boeing E-3A (Phase II AWACS), while the second type has new Lycoming engines of only about 5,500 lb (2 500 kg), derived from the T55s of the Chinook.

Northrop is using a conventional layout, reminiscent of the Cessna T-37, while F-H has chosen a Caravelle engine location and a main gear which retracts forward to leave room for ten pylons. Both designs will have a maximum speed of about 500 kt (925 km/h), a new 30-mm "tank-killing" cannon, with 1,300 rounds and fuel for two hours over the target with a bombload possibly as high as seven tons. To reduce the effects of ground fire, they will have armoured cockpits, duplicated flying control systems, and self-sealing tanks.

A-X was originally planned as a turboprop, and it may be that the switch to turbofans was made to give it some credibility in Europe, following the USA's withdrawal from SE Asia. However, in some respects the project now comes

(Above) The Mitsubishi XT-2 trainer is now at the start of its flight test programme, and a strike fighter derivative is to be developed.

(Below) The Dornier/Dassault-Breguet Alphajet, for which Germany and France have a requirement for 300, is to fly in 1972.

dangerously close to the A-7 Corsair II. Target cost is an optimistic $1·5m (£625,000) but one estimate already puts the fully equipped price at $3·5m (£1·45m).

The A-X is thus likely to come into competition with high-priced fighters, rather than the new breed of trainer/close support aircraft for which a much larger market has been predicted. This category springs partly from the cost explosion ruling out real fighters for many operators, and from the need to boost the performance of basic trainers, so as to minimise the expense of advanced flying training.

The concept of an "all-through trainer" with close support capability originated in Australia, where pilots go straight from the MB.326H to operational training on the Mirage IIID. The idea then caught on in Europe, and when the market analyses had been done, everyone with a sheet of aluminium and a bag of nails set out to build one. However, the market will probably be dominated by the Franco-German Alphajet, and the HS.1182 or BAC P.59. The USAF has scheduled a T-37 replacement designated TBAX, although this may be too late to export in quantity. Czechoslovakia's superb Aero L 39 appears somewhat underpowered, but since 1966 there have been reports of a twin-Viper Jugoslav design that may undercut Western projects.

Setting the standard is the Saab 105XT, which is already in service in Austria. Two 2,850 lb (1 300 kg) J85-GE-17s take it to a sea level maximum of 525 kt (970 km/h), despite unfashionable side-by-side seating. Empty weight is 5,550 lb (2 520 kg), but the gross limit of 15,430 lb (7 000 kg) on the XH version offered to Switzerland permits it to carry a 30-mm cannon, two Sidewinders, and four 1,000 lb (450 kg) bombs. Unit cost is about £300,000.

As with Viggen, the Saab 105 is an unglamorous design, but its performance and reported price are difficult to beat. The Alphajet, which is scheduled to fly in 1972, weighs roughly the same, and will have Larzac 04 turbofans of only slightly more thrust. Judging by press releases on the HS.1182 and rumours of the BAC P.59, the British design will have one unreheated Adour or a developed Viper 600, giving a thrust of 4,000-5,000 lb (1 800-2 300 kg).

Neither Alphajet nor HS.1182 show much sign of their ancestry. The former, while emanating from the Dornier/Dassault-Breguet ensemble, looks to be designed more for the Italian jet set than the fight game. The HS.1182 model shown at Farnborough suggested that ties with McDonnell Douglas had led to an exchange of project engineers: it was far more like an A-4 than a Gnat! At first sight it would appear that the Alphajet will have rather more speed, while the British design will be cheaper.

Hunter encore?

Production-wise, Hawker Siddeley might have another string to its bow in the Super Hunter offered for Swiss ground attack duties. Although the first Hunter interceptor flew 20 years ago, this type is now one of the best of ground attack aircraft, and there are many drivers willing to back its manoeuvrability and firepower against all comers at low level. The main restrictions to further development have been that the intake ducts go through the main spar (as on the Comet, now twice reamed out at mind-boggling cost), that the internal fuel capacity benefits climb performance rather than cruise, and that the outboard pylons are somewhat far aft for really heavy loads. However, by judicious uprating and new pylon positions, the Super Hunter would overcome these problems.

The Fiat G.91YS is one of its main competitors as a cheap substitute for the A-7D Corsair II, which was technical winner of both the Swiss AF Venom replacement cost-effectiveness studies. The G.91YS weighs only 8,600 lb (3 900 kg) empty, and has two reheat J85-GE-13As of

4,080 lb (1 870 kg). Six pylons carry a load of four 1,000 lb (455 kg) bombs and two Sidewinders and it has two internal 30-mm DEFA guns. Some 75 of the basic four-pylon version have been ordered for the Italian Air Force.

In contrast the A-7D has an empty weight of 19,276 lb (8 750 kg), a single unreheated TF41-A-1 Spey of 14,250 lb (6 500 kg), a built-in M-61 cannon, and can carry up to 15,000 lb (6 800 kg) of stores on eight pylons. Navigation is by Singer General Precision doppler-inertial mix, with noteworthy provision for in-flight alignment of the stable table. A Texas Instruments APQ-126 radar gives ground mapping, limited terrain following, and slant range for bomb release, which is claimed to result in only 60 per cent of the mean error of current systems lacking a range sensor.

In the contest to replace Swiss Venoms, the Dassault Milan has had the advantages of commonality with the Mirage IIIS and a useful secondary daylight interception capability. Dassault quote a 135 nm (250 km) LO-LO radius with six 1,000 lb (455 kg) bombs and two Sidewinders. Aerodynamically ingenious retractable foreplanes curb the Milan's runway demands, and (as for the other CTOL contenders) the Milan can be brought into the STOL category by JATO and arresters, a concept virtually unique to Switzerland.

The Hunter's V/STOL stablemate is now serving with the

Three of the most formidable close support fighters now in production are illustrated on this page. They are the McDonnell Douglas A-4 Skyhawk (two-seat TA-4K, top picture), the Vought Corsair II (USAF A-7Ds, above) and the BAC/Breguet Jaguar (British single-seat Jaguar S-07, below.)

RAF as the Harrier, and with the USMC as the AV-8A. The latter is modified to carry US weapons, although 30-mm Adens are retained rather than domestic 20-mm, so that "Now we can knock whole trees down, not just blow the leaves off!" The Harrier is still without a competitor, and it seems late in the day to start putting a wing on the VAK 191, although VFW-Fokker salesmen seem to be pushing the idea.

With the Harrier and its current Pegasus engine now established, interest centres on further powerplant development. A Pegasus 15 of 24,500 lb (11 200 kg) is scheduled to run early in 1972, and Rolls-Royce has a project designated Pegasus 16 which would give a thrust of 34,000 lb (15 500 kg).

Reports speak of British MoD studying a second generation V/STOL combat aircraft, while the US Navy has a study termed VAX-L or HIPAAS (High Performance Aerial Attack System), which has been described conceptually as a Super Harrier. Where the Soviet Union stands on V/STOL is difficult to establish, although they were at the preliminary design stage with a close support application in 1968, and the *Freehand* test vehicle has been operating on *Moskva*-class ships. However, the writer recently came across a Russian textbook on V/STOL design (luckily translated by the IBM machine at Wright-Patterson AFB), which included some of his own project studies!

The BAC/Breguet Jaguar has the mixed blessings of coming two or three years behind the Harrier, and being designed in two countries. It is much faster without external stores than the Hawker aircraft, and its design availability in metric units helps its chances of licence manufacture. Handling when clean has been compared with the Hunter, but its remarkable LO-LO radius of 310 nm (575 km) on internal fuel puts it more in the conceptual ballpark of a G.91 with supersonic capability. Its stalky landing gear results in a particularly useful centre-line pylon, adding to four wing positions to give a total load of 10,000 lb (4 550 kg).

Japanese Jaguar

Fighter development problems are costly, and therefore successful features are inevitably copied. However, in the case of the Mitsubishi XT-2, which was scheduled to fly early in July, the resemblance goes beyond accepted norms. The only fundamental differences between this and the European product are that the former has a shorter undercarriage, no wing fuel (a lesson learned from F-100 experience in Vietnam), and a 20-mm M-61 in place of two 30-mm Aden/DEFA. A strike-fighter variant is to be developed in parallel with the XT-2 as the SF-X.

Turning finally to strike fighters, an outstanding example of current practice is the USAF's F-105D, a Mach 2·1 aircraft with a useful bad-weather capability, and a basic

weight of 27,235 lb (12 400 kg). In a typical mission it grosses 50,000 lb (22 800 kg), of which 32 per cent is fuel in the normal internal and bomb-bay tanks, plus two drop tanks. In this configuration it can deliver six 750 lb (340 kg) bombs over 500 nm (925 km) in an unrefuelled HI-HI mission.

The US Navy's opposite number is the A-6, a slightly lighter subsonic attack aircraft with the most sophisticated avionics in its class. The British interdiction rôle is performed by the rather faster Buccaneer, which (like the F-4) weighs just over 30,000 lb (13 700 kg) empty, and provides an impressive strike capability at less than twice the cost of a light interceptor. As in the case of the Soviet Yak-28 (*Brewer*), its effectiveness has led to a virtual ban on exports. For South American operators, it appears that the long-awaited Canberra replacement is simply a refurbished Canberra, available at one-third the cost of a new Buccaneer.

The next strike fighter generation is typified by MRCA, a two-seat STOL aircraft grossing about 17 tons. This will be powered by two RB.199-34Rs of 13,700 lb (6 300 kg), have a new German cannon, and achieve a LO-LO radius of around 250 nm (460 km). Unfortunately, in attempting to meet tri-service requirements MRCA may have fallen between three stools. It is small for the RAF, who would still like TSR 2, and yet it threatens to be too expensive for the *Luftwaffe*, who have evidently set a ceiling of £1·7m in today's values. Hopes that it will meet Italian air superiority demands seem to be ridiculed by German talk of it needing an escort fighter, possibly designated PANNAP. It is incidentally conceivable that Saab will produce a twin J97 fighter that might similarly escort Viggen.

Design-wise the key question is whether MRCA will be a significant improvement over the Saab Viggen in the standard European scenario, where the full benefit of the former's VG wing cannot be exploited. On the Soviet side, the Yak-28 may be replaced by a strike version of either the VG *Flogger* or the jet-lift *Faithless*. This will be augmented by the MiG-23, more in the category of the F-111E, which has an empty weight of 47,500 lb (21 700 kg) and a gross in the range of 74,000-91,500 lb (33 700-41 500 kg).

Meantime India seems intent on developing her own strike fighter, despite the agonising HF-24 Marut experience. Reports have spoken of a first flight in 1973 and squadron service in 1976, but if the aircraft is really in the MRCA category, this schedule is as unrealistic as references to a self-contained Indian aircraft industry by 1975.

One cannot end without again referring to Marcel Dassault, the *éminence grise* of St Cloud, still printing his own money and trying everything twice, still switching designations and giving out information to confuse rather than clarify. According to his mouthpiece, just as the Mirage F1 will repeat the success of the Mirage III series, so the G-4/G-8 is going to take over from the F-4 Phantom.

All that we (and possibly Dassault) really know about the production derivative of the G series is that it will have variable sweep, two M53s of about 20,000 lb (9 100 kg), and will gross somewhere between 17 and 25 tons. Whether it will resemble the current G-8 (a simply-equipped version of the G-4) when it enters service in 1977, or even whether it will finally emerge as two distinct types (one interdictor, one for air superiority) is anyone's guess.

Old Marcel may be regarded by the British daily press as the current hot-shot of the fighter world, but it should be remembered that he has had only one big success, and that was with only Draken as competitor. Some of his other projects have been less than memorable, and on at least one occasion he has fallen flat on his face (remember the Cavalier?). Fortunately for Dassault, there is no need to hit the jackpot with both the F *and* G series; either will see his company well into the '80s, and survival is the name of today's game. □

Expected to fly in 1973, the Panavia MRCA (above) represents the next generation of strike fighters. Meanwhile, deliveries of the Saab Viggen (below) to the Swedish Air Force are about to start.

THE ARSENAL VB 10 . . .

. . . un chasseur vraiment unique

FROM virtually the earliest days of combat aircraft development, the centre-line thrust concept — two engines both delivering their power along the centre-line of the aeroplane — intrigued the designers of piston-engined fighters. The output of available power plants suited for fighter installation invariably imposed a limiting factor on the performance that could be attained by the aircraft, and, in consequence, the fighter designer was constantly demanding greater power than that afforded by contemporary engines. The use of two engines in an orthodox wing-mounted arrangement exacted penalties in the form of increased drag and reduced manoeuvrability, but the centre-line thrust concept held the tantalising prospect of overcoming these disadvantages, offering minimal frontal area, a clean wing, and the elimination of power asymmetry in the event of the failure of one or other engine.

The tandem fore-and-aft power plant arrangement, with the tail surfaces carried by booms extending from the wings, was the most obvious centre-line thrust arrangement; a configuration adopted in World War I for the experimental Fokker K I and Siemens-Schuckert DDr I, and between the wars by the Chernyshov-designed ANT-23 and the Fokker D XXIII. A variation on the theme was provided during WW II by the Dornier Do 335 in which the rear engine drove an airscrew aft of a cruciform tail, but there was yet a third and even more ingenious centre-line thrust scheme for fighter propulsion: tandem-mounted engines driving contra-rotating airscrews via concentric shafts. This scheme was embodied by the Arsenal VB 10; a fighter that was unique at the time of its conception and was to remain so.

The year 1938 saw the VB 10 in embryo, but so protracted was its gestation that 10 years were to elapse before it was to achieve production status, and by that time piston-engined fighters, however ingenious their design, were *passé*, the era of the turbojet-driven fighter having long since dawned.

During the late 'thirties, *Ingénieur-Général* Vernisse, director of the Arsenal de l'Aéronautique, became increasingly convinced that the answer to the successful development of a twin-engined fighter of centre-line thrust concept was to be found in the use of tandem-mounted

engines fore and aft of the pilot's cockpit driving contra-rotating co-axial tractor airscrews, the rear engine driving the front airscrew by means of a shaft passing through the axis and airscrew boss of the front engine. Once this scheme had crystallised, Vernisse, together with Jean Galtier, began the design of a wooden experimental airframe, the VG 10, which, intended to prove the tandem-engined concept, was to be the predecessor of the planned fighter.

It had been calculated that the VG 10 would have an all-up weight of 7,496 lb (3 400 kg) and a gross wing area of 279·86 sq ft (26 m2), but by the early months of 1939, events in Europe had led to the decision to accelerate the development of the planned tandem-engined fighter by discarding the VG 10 test-bed and proceeding immediately with work on a prototype of the fighter itself. Designated VG 20, this, too, was to be of wooden construction, but the wing was to be substantially larger to accommodate the armament and sufficient fuel, gross area being increased to 387·5 sq ft (36 m2). The initial engine installation was to comprise two Hispano-Suiza 12Y-29 12-cylinder vee liquid-cooled engines each rated at 910 hp, the intention being to replace these with more powerful HS 12Z engines at the earliest opportunity. In the final stages of bench testing at the time, the HS 12Z promised an initial rated power of 1,200 hp, and it was estimated that with two engines of this type, the VG 20 would attain a maximum speed of 404 mph (650 km/h), normal endurance being 1 hr 30 min.

Jean Galtier was, by mid-1939, becoming increasingly preoccupied with the VG 30 series of fighters, and further development of the VG 20 was therefore passed to another Arsenal engineer, M Badie, who, after a thorough appraisal of the design, concluded that wooden airframe construction was unsuited to this highly-original power plant installation. The airframe was redesigned for all-metal construction in consequence, the designation of the fighter being changed to VB 10 to reflect the change of design responsibility, and so promising was the calculated performance that, in May 1940, an initial batch of 30 production VB 10s was ordered off the drawing board for the *Armée de l'Air*. The decision to order the VB 10 into production without awaiting prototype trials was motivated by the fact that the characteristics of the wing and tail surfaces were known, these surfaces being identical, apart from scaling up and structure, to those of the VG 33 fighter.

The Armistice of 24 June 1940 automatically cancelled the production order for the VB 10 before any metal had been cut, but the Arsenal de l'Aéronautique, which had evacuated its premises at Villacoublay and had re-established itself in a spinning mill at Villeurbanne, in the suburbs of Lyon, continued design work on the radical fighter as the Atelier d'Etudes Aéronautiques. Early in 1942, the *Ministère de l'Air* of the Vichy government ordered work to commence on a prototype of the VB 10, but before construction could begin, German forces had entered the Unoccupied Zone of France. The *Luftwaffe* occupied the nearby airfield of Bron, and the work being undertaken by Vernisse's team at the Atelier d'Etudes Aéronautiques was carefully scrutinised by representatives of the *Reichsluftfahrtministerium*. Presumably with tongue in cheek, *Général* Vernisse assured the visitors from the RLM that the VB 10 was, in fact, intended

(Above) The forward and centre fuselage sections of the VB 10.02 mounted on a provisional mobile test-bench at Bron.

for an attempt on the world air speed record, and the Germans appeared content with this explanation, granting permission for prototype construction to begin. It seems improbable that they believed this story, possibly adopting the viewpoint that the features embodied by the VB 10 might prove of interest to the German aircraft industry if the Vernisse-Badie design proved successful.

Leisurely progress

The first metal was cut on the VB 10.01 early in 1943 and, meanwhile, *Général* Vernisse, anxious to obtain some flight experience with the power plant arrangement, had obtained authority for the Laté 299.01 shipboard torpedo-bomber prototype to be converted as a test-bed for tandem-mounted HS 12Y-31 engines at Latécoère's Toulouse factory. The entire forward fuselage of the aircraft was rebuilt to accommodate the engines which drove two three-bladed Ratier contra-rotating airscrews via concentric shafts. The pilot's cockpit was repositioned just aft of the wing trailing edge, and accommodation was provided aft and below the pilot for a flight test engineer.

While this conversion was being undertaken, work on the VB 10.01 proceeded at Villeurbanne in an intentionally leisurely fashion. The fuselage was a rigidly-braced electrically-welded steel-tube structure covered by detachable prefabricated panels of so-called *contreplaqué métallique,* a smooth dural sheet electrically-welded to a base of corrugated sheet. The fuselage panels were non-stressed, but the *contreplaqué métallique* skinning of the wings and tail surfaces was stressed. The wing was an all-metal two-spar structure in one piece and possessing constant taper in both chord and thickness from the root. The engines were disposed fore and aft of the cockpit and completely independent of one another. These were of the pre-series Hispano-Suiza 12Z (HS 89) type, rated at 1,200 hp and equipped with Turboméca superchargers, the forward engine driving the rear airscrew and the rear engine driving the forward airscrew via a shaft passing through the axis of its fellow.

The airscrews were of the Ratier three-bladed electrically actuated feathering type, and both power plants were cooled by a ventral Chausson radiator. One or other engine could be shut down for cruising flight, and provision was made to maintain the stationary engine at a suitable temperature to permit immediate starting. Fuel was housed by two tanks inserted between the coupling-shaft tunnel and the radiator,

(Above and below) The VB 10.02 which began flight trials at Villacoublay on 21 September 1946. The photograph below illustrates clearly the ease with which engines and accessories could be reached by means of large detachable panels.

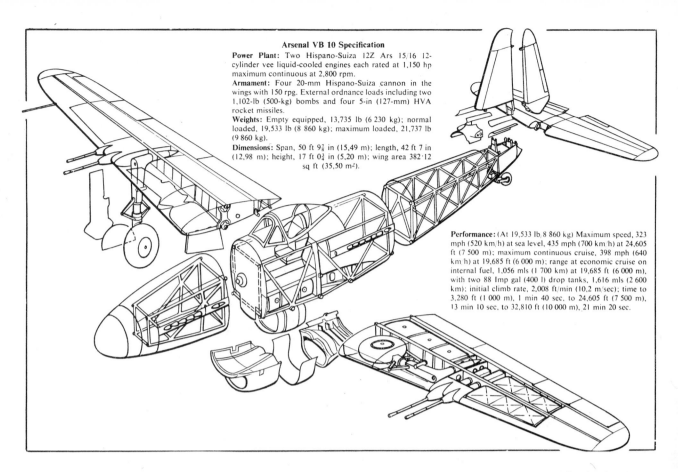

Arsenal VB 10 Specification
Power Plant: Two Hispano-Suiza 12Z Ars 15/16 12-cylinder vee liquid-cooled engines each rated at 1,150 hp maximum continuous at 2,800 rpm.
Armament: Four 20-mm Hispano-Suiza cannon in the wings with 150 rpg. External ordnance loads including two 1,102-lb (500-kg) bombs and four 5-in (127-mm) HVA rocket missiles.
Weights: Empty equipped, 13,735 lb (6 230 kg); normal loaded, 19,533 lb (8 860 kg); maximum loaded, 21,737 lb (9 860 kg).
Dimensions: Span, 50 ft 9⅞ in (15,49 m); length, 42 ft 7 in (12,98 m); height, 17 ft 0¾ in (5,20 m); wing area 382·12 sq ft (35,50 m²).

Performance: (At 19,533 lb/8 860 kg) Maximum speed, 323 mph (520 km/h) at sea level, 435 mph (700 km/h) at 24,605 ft (7 500 m); maximum continuous cruise, 398 mph (640 km/h) at 19,685 ft (6 000 m); range at economic cruise on internal fuel, 1,056 mls (1 700 km) at 19,685 ft (6 000 m), with two 88 Imp gal (400 l) drop tanks, 1,616 mls (2 600 km); initial climb rate, 2,008 ft/min (10,2 m/sec); time to 3,280 ft (1 000 m), 1 min 40 sec, to 24,605 ft (7 500 m), 13 min 10 sec, to 32,810 ft (10 000 m), 21 min 20 sec.

these having a total capacity of 77 Imp gal (350 l), and two 123 Imp gal (560 l) wing tanks aft of the forward mainspar and inboard of the gun bays.

Work on the VB 10.01 proceeded throughout 1943, and meanwhile, late in the year, the Latécoère test-bed, now designated Laté 299A, had flown with tandem-mounted HS 12Y-31 engines, each developing 830 hp for take-off and 860 hp at 10,500 ft (3 200 m). This aircraft had been ferried to Bron from where its flight test programme was conducted, but on 30 April 1944 the test-bed was destroyed in an Allied attack on the airfield. Fortunately, before its destruction, the Laté 299A had proven the practicality of the tandem-mounted engine arrangement.

With the German withdrawal from France, work on the VB 10.01 was accelerated, and construction of a second prototype, the VB 10.02, was begun. The first aircraft was finally completed in the summer of 1945, and, with the Arsenal chief test pilot, Vonner, at the controls, was flown for the first time on 7 July at Bron where the initial flight test programme was conducted. From Bron the VB 10.01 was flown on 13 September to the *Centre d'Essais en Vol* at Brétigny, the Arsenal de l'Aéronautique having meanwhile transferred to Châtillon-sous-Bagneux, together with the second prototype. The initial CEV reports on the VB 10 were sufficiently promising for the placing of contract 430/45 on 22 December 1945 which called for the production of no fewer than 200 aircraft which were to be manufactured by the SNCAN (Société Nationale de Constructions Aéronautiques du Nord).

By the time the VB 10.02 made its first flight at Villacoublay on 21 September 1946, financial difficulties had dictated drastic economies in France's military aircraft manufacturing programme. The VB 10 production contract had been cut back to 50 machines, and it had been decided to provide for the installation of two cameras in the aft fuselage to suit the aircraft for use by the 33e *Escadre de Reconnaissance* which then comprised one squadron with

Lightnings and one with Mustangs operating in the photographic rôle.

The VB 10.02 differed from the first prototype in having two 1,150 hp Hispano-Suiza 12Z-12/13 (HS 91) engines with direct fuel injection, an all-round vision cockpit canopy, revised undercarriage doors, and full armament comprising four 20-mm Hispano-Suiza cannon with 200 rpg, and six 0·5-in (12,7-mm) Browning machine guns with 600 rpg. The flight test programme was plagued by delays which largely stemmed from the poor reliability of various items of equipment and were the result of numerous petty economies that had been insisted upon by the authorities. For example, the generators used by the HS 12Z-12/13 engines had been adapted from those designed for the Gnôme-Rhône 14N series of air-cooled radials and were unsuited for their task, while budgetary restrictions had prevented the Ratier concern from providing airscrews with the full range of pitch changes demanded.

Meanwhile, the SNCAN had begun production of the VB 10, its Méaulte factory being assigned responsibility for the manufacture of the rear fuselage, the wings, and for final assembly; the works at Les Mureaux were allocated the task of producing the tail assembly; the Sartrouville plant was given the forward and centre fuselage sections, and the radiator tunnels were constructed at Le Havre, the factory at Villeurbanne (which had been retained by the Arsenal after its transfer to Châtillon-sous-Bagneux) producing the cast magnesium elements.

The first production VB 10 was flown on 3 November 1947, this differing externally from the second prototype primarily in having redesigned, taller vertical tail surfaces. The sextet of Browning machine guns was deleted from the armament which was reduced to the quartet of 20-mm cannon, but provision was made for underwing loads of two 1,102-lb (500-kg) bombs and four HVARs or two 88 Imp gal (400 l) drop tanks. The pilot was provided with an SF 1943B gunsight and TR 1143 radio, and his cockpit

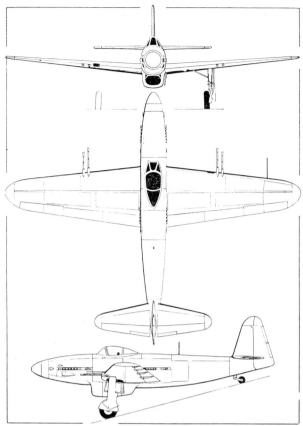

featured 7-mm side armour and 15-mm back armour, the windscreen comprising an 80-mm armourglass front panel and 40-mm side panels. The Hispano-Suiza 12Z Ars 15/16 each had a normal maximum continuous rating of 1,150 hp and drove Ratier 2035/2036 airscrews.

A bad reputation

On 10 January 1948, the front airscrew of the VB 10.02 suddenly changed its pitch from fine to coarse, several rods of the aft engine broke through the crankcase, and the oil promptly ignited. The test pilot, Decroo, stayed with the burning aircraft for sufficient time to ensure that it would crash in a sparsely-populated area, suffering severe burns in the process, and then baled out. Another SNCAN test pilot, Koechlin, was even less fortunate, losing his life on 15 September 1948 at Méaulte while testing the third pro-

duction VB 10. Eleven weeks earlier, on 30 June, a further prototype, the VB 10.03, had been flown, this being the first aircraft actually to have provision for camera installation, but the constant troubles suffered by the VB 10 coupled with the earlier crash of the second prototype had already endowed the fighter with something of an evil reputation.

The re-equipment of the 33e *Escadre*, for which the VB 10 was intended, was by now a matter of considerable urgency. From VE Day this unit had been engaged constantly in photographic survey work over France and North Africa, and its Lightnings and Mustangs were worn and weary. Indeed, until its transfer from Freiburg-im-Breisgau to Cognac late in 1949, the 33e *Escadre* succeeded in maintaining its equipment in airworthy condition only by dint of bargaining for spares from USAF dumps in Germany with cases of Champagne purchased from the unit's contingency fund! Nevertheless, it was becoming increasingly doubtful that continuation of VB 10 production to replace the aged Lightnings and Mustangs was justified, and with the crash of the third production aircraft, an immediate re-evaluation of the programme was ordered.

The Air Staff refused point blank to accept the VB 10, believing that much time would be consumed in eradicating the fighter's teething troubles, if, indeed, these *could* be eradicated, and that by the time it reached service status it would already be obsolete. Thus, on 21 September 1948, the entire VB 10 contract was cancelled. The first production aircraft had made only 13 test flights, the second had flown four times, and the fourth had been flown three times. More than 30 VB 10 airframes were in final assembly at Méaulte, and all of these were sentenced to be scrapped forthwith. Almost exactly a decade had elapsed since this truly unique fighter had begun to take shape on the drawing boards at Villacoublay.

At the time that development of the VB 10 was cancelled, several derivatives of the basic design were on the Arsenal drawing boards, including a version embodying boundary layer suction and blown tail surfaces, and the VB 15 which was to have featured modified forward and centre fuselage sections to accommodate a single 24-cylinder Hispano-Suiza 24Z engine affording 3,600 hp for take-off and 3,150 hp for climb and combat. There had been nothing *fundamental* wrong with the VB 10, and there would seem little doubt that, had sufficient funds and resources been made available, development would have been less protracted and the *Armée de l'Air* could have added to its inventory a multi-purpose fighter of inestimable value in the limited wars to which the service was committed in the 'fifties. □

(Below) The first SNCAN-built production VB 10, which flew for the first time on 3 November 1947. This featured redesigned and taller vertical tail surfaces, and armament was confined to four wing-mounted cannon.

A pictorial history of Turkish Military Aviation

As LIEUTENANT Ahmet switches on the gunsight of his F-100D Super Sabre, begins his run in on the target, and tenses himself to fire his quartet of Pontiac M-39E cannon, his thoughts will undoubtedly be centred on the contribution that he hopes to make to the Turkish team's score in the 1971 AirSOUTH Fighter Weapons Meet which takes place in Turkey this month. He will certainly spare no thought for the fact that some twelve hundred miles away in distance and sixty years earlier in time, perhaps to the very hour, a fellow countryman, Cavalry Captain Fesa, was perched in a Blériot XI at Pau, in the Basses-Pyrénées, tensing himself for his first solo flight which would mark the beginning of Turkish military aviation.

The Turkish General Staff had displayed an interest in the military potentialities of aviation from an early date, and had sent Capt Fesa and Lt Kenan to the Blériot Flying School at Pau for flying training. The use made of aircraft by the Italian Army in Tripolitania during the first months of the Italo-Turkish War, which began on 29 September 1911, accelerated Turkish plans to establish an Army Aviation Section, and on 15 March 1912 its first equipment arrived at Yesilköy in the shape of a single-seat REP monoplane and a two-seat Deperdussin monoplane. These were quickly joined by a second REP, and shortly afterwards, on 23 April, the two REP monoplanes were flown over the *Hürriyet Abidesi* (Liberty Monument) in Istanbul to coincide with the annual military parade.

A few Harlans, Aviatiks and Blériots were obtained during the Balkan Wars, but these were mostly operated on behalf of the Turkish forces by foreign personnel, as the Turkish Army possessed few pilots or aircraft mechanics, and the Army Aviation Section still remained little more than a nucleus when, on 29 October 1914, Turkey entered World War One as an ally of the Central Powers. Turkish military aviation expanded rapidly under the German aegis, such types as Albatros B IIs, C Is, C IIIs and D IIIs, and Halberstadt D Vs being provided, although these were largely manned by German personnel. Some Gotha WD 13 floatplanes were also supplied to Turkey for patrolling the Bosporus, the Dardanelles, and for operation over the Aegian Sea.

By the end of World War One, Turkish military aviation comprised 18 air companies, but both aircraft and crews were of indifferent quality, and serviceability was extremely low. The Treaty of Versailles forbade Turkey military aircraft, but before General Mustafa Kemal had established his new national government in Ankara, some 40 Turkish Army personnel were endeavouring to gather, by dint of cannibalisation, a small force of aircraft for use against the Greeks in Thrace and Anatolia. These few aircraft enjoyed some success, and in 1920, after the withdrawal of the Allies' support for Greek activities in Turkey, the Turkish Army received a number of British and French combat aircraft, which, by 1922, were having an important effect on the course of the Graeco-Turkish War, a Peace Treaty restoring to Turkey her Thracian territory finally being signed on 24 July 1923.

The newly-established Turkish Republic attached considerable importance to the creation of an effective air arm, and much effort was expended in encouraging air-minded-

(Above) Twenty Breguet XIXs were purchased in the late 'twenties, being supplemented in 1933 by 50 of the later Breguet XIX.7s.

(Above) When Turkey widened aircraft procurement in 1933, one of the first purchases was 24 Curtiss Hawk IIs which served until 1941.

(Above) In 1942 the Turkish fighter inventory was further extended by the delivery of Tomahawk IIBs (Hawk 81A-3) from RAF stocks.

(Above) The wide variety of military aircraft acquired by Turkey during the late 'thirties included a few Focke-Wulf Fw 58 trainers.

(Above) Between 1948 and 1954 the Turkish Air Forces operated the P-47D Thunderbolt fighter-bomber in some numbers.

(Below) One ANT-9 transport powered by two 500 hp M-17 engines was presented to Turkey and operated by the Turkish air arm.

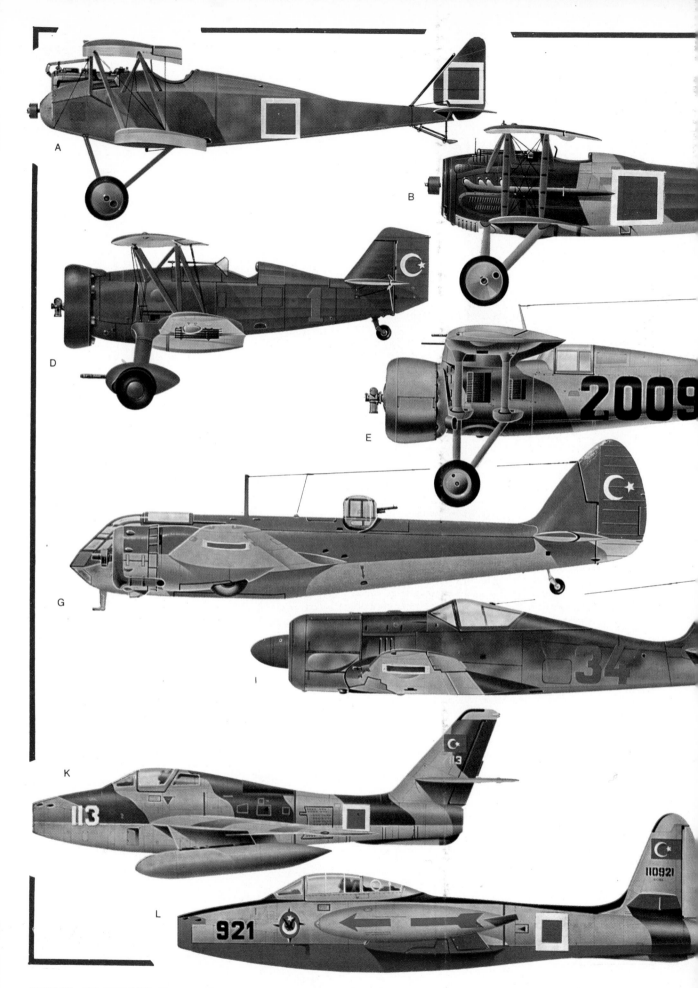

A: Halberstadt D V of the Turkish Army Air Service, early 1918.
B: SPAD S.XIII operational over Asia Minor in 1922 during the Graeco-Turkish War.
C: Breguet XIX.7 reconnaissance-bomber, 1935.
D: Curtiss Hawk II of the Fighter Battalion of the 3rd Regiment, Izmir, 1940.
E: PZL P.24C fighter at Yeşilköy, 1940.
F: Vultee V-11-GB of the 2nd Regiment, Diyarbakir, 1939.
G: Bristol Blenheim I of the Bomber Battalion of the 3rd Regiment, Izmir, 1939.

H: Curtiss Hawk 81A-3 (Tomahawk IIB), Eskişehir, 1942.
I: Focke-Wulf Fw 190A-3 of the 3rd Company, 5th Air Regiment, Bursa, 1944.
J: Westland Lysander II at Yeşilköy, 1940.
K: Republic F-84FQ Thunderstreak of the Turkish 1st TAF, Eskişehir, 1970.
L: Republic F-84G-16 Thunderjet of the Turkish 3rd TAF, Diyarbakir, 1954.
M: North American F-100D-15 Super Sabre of the Turkish Ist TAF, Bandirma, 1969.

(Above) A small number of Le Rhône-engined Nieuport 17s of 1916 vintage were acquired shortly after World War I and used for training.

(Above) The Turkish fighter force was strengthened in 1940 by the arrival of 45 Morane-Saulnier M.S.406s from France.

ness in Turkey. Personnel were sent abroad for training in modern techniques and, meanwhile, flying and technical schools were established in Turkey. The combat elements of the Turkish air arm continued to fly aircraft of WW I vintage into the second half of the 'twenties, purchases being confined primarily to training aircraft, exceptions being a half-dozen Rohrbach Ro III flying boats and a few Junkers A 20 reconnaissance-bombers. Extensive re-equipment took place from 1926, with the help of purchases made exclusively in France, apart from 16 Letov S-16T army co-operation aircraft, and an attempt was made to establish a domestic aircraft industry, Junkers assisting in creating a factory at Kayseri.

In the early 'thirties, the Turkish air arm widened its field of procurement with the purchase of 24 Curtiss Hawk II fighters and six Curtiss Fledgling trainers from the USA,

and in 1935 an order was placed for 40 PZL P.24C fighters in Poland, a licence to manufacture this type at Kayseri being acquired simultaneously*. By this time, the air arm, still an integral part of the Army, possessed some 160 aircraft of all types, and utilised 16 airfields and strips. With ominous signs of a new war on the horizon, Turkey began to create a bombing force, 20 Martin 139Ws being built in the USA for Turkey in 1937, and 24 Heinkel He 111F-1 bombers being ordered from Germany. Deliveries of the latter began in the summer of 1938, by which time the first 12 of 30 Blenheim Is had been taken on strength, deliveries of the British bomber being completed in February 1939. By November of the previous year an attack force had been created with the acquisition of 40 Vultee V-11-GBs, and the Turkish air arm, remodelled along British lines with RAF aid, was expanding rapidly.

When the first shots of WW II were fired, the Turkish inventory included some 370 aircraft of all types, and personnel strength had reached approximately 8,500 men of whom 450 were pilots. During the first weeks of the war, 15 Hurricane Is were sent to Turkey from the UK, although only one Spitfire I was delivered against a contract for 14 (two more being delivered a year later). These were followed by 29 Fairey Battles and, during the early months of 1940, 36 Westland Lysander IIs and 45 Morane-Saulnier MS 406s. Supplies of British military aircraft were not resumed until 1942, and in the meantime an arms agreement had been negotiated with the German government which resulted in the delivery of 75 Focke-Wulf Fw 190A-3 fighters from February 1942. The considerable variety of combat aircraft in the Turkish inventory was further expanded during WW II by the supply of Tomahawk IIBs, Hurricane IIs, Blenheim Vs, Baltimores, Spitfire Vs, and even Liberators.

By 1948, a USAF mission had been established in Turkey to guide the reorganisation and modernisation of the Turkish Air Forces, the tempo of this programme accelerating rapidly from 1952 when Turkey became a member of NATO. An account of the present status of the Turkish Air Force appears elsewhere in this issue. ☐

The first Kayseri-built P.24C flew on 27 May 1937, and limited production continued under Polish supervision until 1940.

AIRCRAFT THAT HAVE SERVED WITH THE TURKISH AIR FORCES

Type	Rôle	Year of Intro	Type	Rôle	Year of Intro	Type	Rôle	Year of Intro
REP (80 hp Le Rhône)	T	1912	Breguet XIX.7	R-B	1933	Supermarine Spitfire IX	F	1946
REP (80 hp Gnome)	R	1912	Curtiss Hawk II	F	1933	Supermarine Spitfire P.R.XIX	R	1946
Deperdussin (80 hp Gnome)	R	1912	Curtiss Fledgling	T	1933	De Havilland Mosquito T.III	T	1946
Bristol Prier-Dickson			Consolidated Fleet Model 7	T	1933	De Havilland Mosquito F.B.VI	F-B	1946
"Military" Monoplane	R	1912	Tupolev ANT-9	Tpt	1933	Douglas C-47	Tpt	1946
Harlan-Eindecker			Supermarine Southampton	R	1933	North American T-6D Texan	T	1947
(100 hp Mercedes)	R	1912	De Havilland D.H.84 Dragon	T	1934	Douglas B-26C Invader	A-B	1948
Aviatik (100 hp Mercedes)	T	1912	General Aircraft S.T.18 Monospar	T	1936	Republic P-47D Thunderbolt	F-B	1948
Bleriot XI (80 hp Gnome)	R	1913	PZL P.24C	F	1936	Beech T-11 Kansan	T	1948
Albatros B II	T	1916	Gotha Go 145	T	1936	Republic F-84G Thunderjet	F-B	1952
Albatros C I	R	1916	Martin 139 W	B	1937	Canadair T-33A-N Silver Star	T	1952
Albatros C III	R-B	1916	Heinkel He 111F-1	B	1938	Beech D18S	L	1952
Albatros D III	F	1916	Avro Anson I	R-B	1938	CCF Harvard Mk 4 (T-6J)	T	1952
Halberstadt D V	F	1916	Focke-Wulf Fw 58 Weihe	L	1938	Lockheed RT-33A	R	1952
Gotha WD 13	R	1916	Supermarine Walrus	R	1938	Canadair F-86E(M) Sabre	F	1954
Nieuport 17	F-T	1918	Bristol Blenheim I	B	1938	CCF T-34 Mentor	T	1954
De Havilland D.H.9	R	1920	Vultee V-11-GB	A-B	1938	M.K.E.K.4. Ugur	T	1955
Breguet XIV A2	R	1921	Curtiss CW-23	T	1939	Republic F-84F Thunderstreak	F-B	1957
SPAD S.XIII	F	1922	Hanriot 182	T	1939	North American F-100C &		
Breguet XIV B2	R-B	1923	Hawker Hurricane I	F	1939	F-100D Super Sabre	F-B	1958
Rohrbach Ro III	R	1923	Supermarine Spitfire I	F	1939	North American F-100F Super		
Junkers A 20	R-B	1924	Westland Lysander II	A-C	1940	Sabre	T	1958
Junkers F 13	L	1924	Morane-Saulnier M.S.406	F	1940	Lockheed T-33A	T	1958
Junkers G 23a	Tpt	1924	Fairey Battle	B	1940	Republic RF-84F Thunderflash	R	1958
Caudron C.27	T	1924	Airspeed Oxford	T	1941	Sikorsky UH-19D	H	1958
Caudron C.59	T	1924	Curtiss Hawk 81A-3			North American F-86D Sabre	F	1959
Savoia S.16ter	R	1925	(Tomahawk IIB)	F	1942	Douglas C-54	Tpt	1959
Morane-Saulnier M.S.35 (AR)	T	1925	Hawker Hurricane IIB & IIC	F	1942	Lockheed F-104G Starfighter	F-B	1963
Morane-Saulnier M.S.53	T	1926	Miles Magister	T	1942	Lockheed C-130E Hercules	Tpt	1965
Nieuport-Delage NiD.62	F	1926	Bristol Blenheim V	R-B	1943	Northrop F-5A	F-B	1966
Dewoitine D.21	F	1926	Focke-Wulf Fw 190A-3	F	1943	Northrop F-5B	T	1966
Potez 25 A2	R	1928	Consolidated B-24D Liberator	B	1943	Republic F-84FQ Thunderstreak	F-B	1966
Breguet XIX B2	R-B	1928	Martin Baltimore	B	1944	Convair F-102A Delta Dagger	F	1969
Letov S-16T	A-C	1929	Supermarine Spitfire V	F	1944	Convair TF-102A Delta Dagger	T	1969
Loire Gourdou-Leseurre LGL-32	F	1929	Bristol Beaufort	T	1946	Vickers Viscount 794	Tpt	1971
Morane-Saulnier M.S.147	T	1931	Bristol Beaufighter X	T-F	1946			

Rôles: A-B Attack Bomber; A-C Army Co-operation; B Bomber; F Fighter; F-B Fighter-Bomber; F-T Fighter-Trainer; H Helicopter; L Liaison; R Reconnaissance; R-B Recce-Bomber; T Trainer; Tpt Transport.

The Four-Pronged Trident Three

Fᴏʀ the past dozen years, the name Trident has served very well indeed for Hawker Siddeley's short-to-medium range airliner, with its three-pronged thrust from a trio of Rolls-Royce Spey engines clustered in its tail. With the service début of the 3B variant of the Trident, however, the name has become rather less appropriate, for this latest model has *four* engines in its tail — or, for that matter, five if the auxiliary power unit in the fin is counted.

Although there may be some doubt as to this new variant being truly tridental (should it, perhaps, be called the Quadrant — or even the Quintet?), there is *no* doubt that BEA is very happy indeed to have the Trident 3B to reinforce its fleet. The airline can, indeed, afford to view any quibbles about the continued suitability of the Trident name with equanimity, for, in the 3B, BEA certainly has another good revenue-earner.

With a cabin capable of accommodating 180 passengers at a time, the Trident 3B is undoubtedly destined to carry many millions of passengers before the 'seventies are through. The full fleet of 26 of these "stretched and boosted" Tridents which is scheduled to be plying the European routes by mid-1972 will have the capability of uplifting more than three-and-a-half million passengers a year. Even at typical BEA load factors and annual utilisation, they are likely to carry more than two million passengers in a 12-month period — a figure which places in perspective the importance of the purchase of these aircraft by BEA, whose entire fleet of 95 aeroplanes carried just over eight-and-a-half million passengers in 1970.

Whether these future passengers will be entirely happy with their experience of travelling in the Trident 3B is another question. One AIR ENTHUSIAST editor who travelled on a 3B from Paris to London during its introductory shake-down period was left with considerable misgivings on the score of comfort and convenience, when it took some 45 minutes to embark and seat a full cabin load of passengers! Limited aisle space and close pitching of the seats were largely responsible for this delay, with the added complica-

tion that individual seats had been allocated before embarkation began. Thus, unless passengers assigned to window seats happened to arrive at each seat row first — and the law of averages ensured that few did — those with centre and aisle seats had to crowd the aisle while waiting for the window-seat occupants to arrive.

With this happening at half-a-dozen or so adjacent seat rows at a time, confusion and frayed tempers were inevitable. Even when the passengers were seated, lack of stowage space for hand-held packages, cabin bags, brief-cases and so on, meant that few, if any, of the tourist-class passengers could relax, and attempts to recline the seats produced expletives in a half-dozen languages!

Criticism of this kind can no doubt be levelled at many contemporary aeroplanes operating on high-density, short-haul routes, and airlines that write the specifications for cabin layouts are more to blame than aircraft designers. Nevertheless, Hawker Siddeley includes, in its list of sales "plusses" for the Trident 3B, the fact that "close pitching of cabin windows ensures window vision at 28-inch seat pitching"— and with seat rows as close as *that* to each other, the shiniest new aeroplane in the world loses something of its gloss for the close-packed occupants. The BEA cabin arrangement, in fact, provides either 119 tourist-class seats at a pitch of 31-32 in (79-81 cm) plus 14 first-class seats, or a 140-seat all-tourist layout at about 32-in (81-cm) pitch.

Regular operations with the Trident Three (which is the BEA style of reference for what the manufacturers call the Trident 3B) began on 1 April on the routes from London to Paris and Lisbon. Other destinations added during the first week of April included Dublin, Palma, Glasgow, Milan, Åmsterdam and Madrid. Services to Brussels and Rome began in May. On these routes, the Trident 3B replaces the Trident 1 and 2; the Trident 1 first entered service with BEA exactly seven years previously, on 1 April 1964, and the Trident 2 joined the fleet three years ago in April 1968. At the time BEA inaugurated regular services with the Trident Three, only three aircraft had been delivered to the airline,

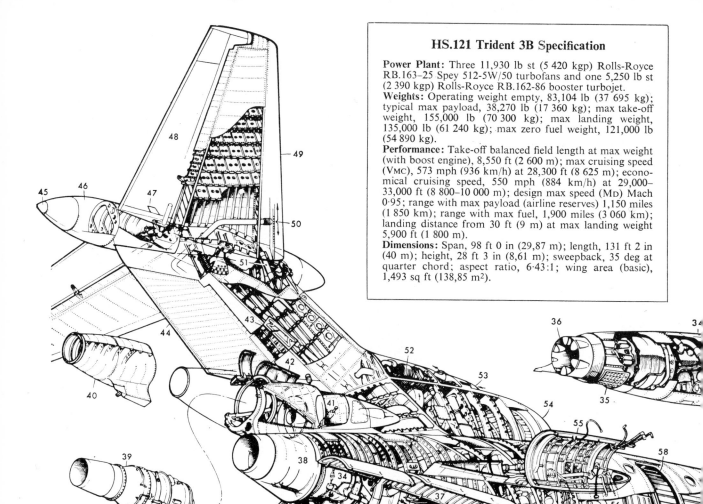

HS.121 Trident 3B Specification

Power Plant: Three 11,930 lb st (5 420 kgp) Rolls-Royce RB.163–25 Spey 512-5W/50 turbofans and one 5,250 lb st (2 390 kgp) Rolls-Royce RB.162-86 booster turbojet.
Weights: Operating weight empty, 83,104 lb (37 695 kg); typical max payload, 38,270 lb (17 360 kg); max take-off weight, 155,000 lb (70 300 kg); max landing weight, 135,000 lb (61 240 kg); max zero fuel weight, 121,000 lb (54 890 kg).
Performance: Take-off balanced field length at max weight (with boost engine), 8,550 ft (2 600 m); max cruising speed (V$_{MC}$), 573 mph (936 km/h) at 28,300 ft (8 625 m); economical cruising speed, 550 mph (884 km/h) at 29,000–33,000 ft (8 800–10 000 m); design max speed (M$_D$) Mach 0·95; range with max payload (airline reserves) 1,150 miles (1 850 km); range with max fuel, 1,900 miles (3 060 km); landing distance from 30 ft (9 m) at max landing weight 5,900 ft (1 800 m).
Dimensions: Span, 98 ft 0 in (29,87 m); length, 131 ft 2 in (40 m); height, 28 ft 3 in (8,61 m); sweepback, 35 deg at quarter chord; aspect ratio, 6·43:1; wing area (basic), 1,493 sq ft (138,85 m²).

and production is continuing at a rate of two aircraft per month at Hatfield to complete the present BEA order for 26 within the next year. The airline also holds options on 10 more Trident 3Bs, and a further order may be placed pending a decision on a future airbus type for BEA.

The second phase of Trident Three introduction will be to put the new aircraft on to the longer routes to Mediterranean holiday resorts, on which there is also a high level of traffic in summer months. It was the need to fly these routes non-stop from London, Birmingham or Glasgow — in effect a stage length of 1,200 miles (1 900 km) — that led to the introduction of a fourth engine in the Trident Three, this being one of the two most significant new features of the design. The other major change is, of course, the lengthened fuselage, but, in addition, numerous detail changes and improvements have been made.

Evolution of the stretched Trident

The Trident 3B was evolved to meet a specific BEA requirement, as were the other major Trident variants (see accompanying list). With European traffic growing steadily during the 'sixties, BEA projected the need for a 170-180-seat short-medium range aircraft to be in service around the

end of the decade, with better seat-mile costs than existing types.

A number of possible stretches of the Trident were studied in the Hawker Siddeley design offices during 1964-5, and in parallel, project designs were completed for a Trident derivative, the HS.132. This was itself studied in several versions, leading eventually to the HS.134 project, a more radical departure from the Trident design. The HS.132 was to have used a Trident wing with a new 158-seat fuselage and two rear-mounted 30,000 lb st (13 600 kgp) Rolls-Royce RB.178 engines, while the HS.134 was intended to have two

Key to Hawker Siddeley Trident 3B cutaway drawing

1. Captain's seat
2. Co-pilot's seat
3. Third crew member's seat
4. Folding seat for super-numerary crew member
5. Centre console
6. Control column
7. Rudder pedals
8. Nosewheel steering, port and starboard
9. Passenger windows
10. Forward cabin service door
11. Search radar
12. Upward hinged radome
13. Pitot heads
14. Offset nosewheel, retracting sideways
15. Unpressurised nose landing gear bay
16. Radio services bay
17. Static heads, port and starboard
18. Forward freight door, plug type
19. Forward freight hold
20. Seat attachment rail
21. Luggage racks housing passenger service panels
22. Emergency escape panels
23. Air conditioning duct
24. Leading edge Krüger flap
25. Centre passenger door
26. Main landing gear doors
  27. Main landing gear, starboard (turned through 90 deg for retraction)
  28. Main landing gear, port
  29. Centre torsion box
  30. Rear freight hold
31. Plug type rear freight door
32. Rear pressure dome
33. Equipment bay access door, with integral stairs and tail bumper
34. Rolls-Royce Spey engine
35. Thrust reverser cascades (outer engines only)
36. Exhaust silencers (outer engines only)
37. Centre engine access doors
38. Centre engine jet pipe fairing, fixed to engine
39. Rolls-Royce RB.162 boost engine
40. RB.162 jet pipe fairing
41. Air intake doors for RB.162
42. Access panels for RB.162
43. Access doors to rudder
44. One-piece rudder
45. Navigation lamp
46. Detachable tail cone
47. VHF aerials
48. Elevator
49. Variable-incidence tailplane
50. Tailplane anti-icing duct
51. Tailplane actuating jacks
52. AiResearch APU
53. Air inlet to APU
54. Air intake to centre Spey
55. Fixed portion of pod cowling
56. Rear toilet, port and stbd
57. Rear toilet servicing point
58. Rear cabin service door
59. Double slotted flaps, two sections each side
60. Lift dumpers, two sections each side
61. Main landing gear bays
62. Underwing access doors
63. Flap tracks (two per flap)
64. Air brake/spoiler (one each side)
65. No 1 fuel tank (port and stbd)
66. No 2 fuel tank (port and stbd)
67. Aileron
68. Retractable landing lamp
69. Kuchemann wingtip
70. Four-piece slat
71. Slat tracks
72. Slat screw jack
73. Telescopic supply pipes for slat anti-icing
74. Slat torque shafting
75. Leading edge anti-icing ducts

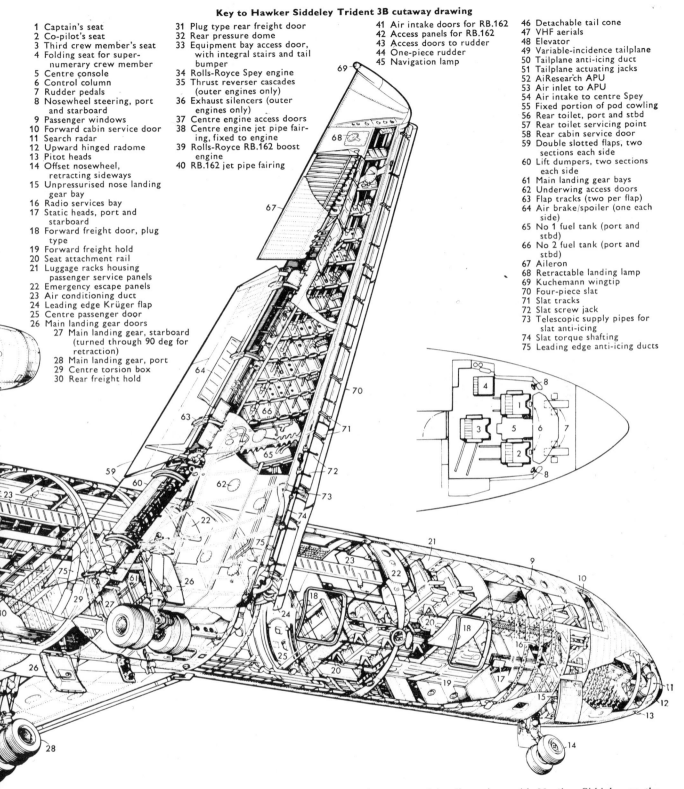

RB.178s in underwing pods, a 25-deg swept wing, and a 185-seat fuselage. Details of the HS.132 were first presented to BEA in June 1965 and it was offered for 1970 service; the HS.134 would have required at least a year more to develop. In the event, neither proceeded for the reasons that the RB.178 was not continued, and Hawker Siddeley failed to find a European partner interested in such projects, the latter being a prerequisite for obtaining government finance.

In mid-1966, BEA opted for a mixed fleet of Boeing 727s and 737s to meet its needs in the 1969-1975 period, but this plan did not receive governmental approval. The airline then resumed its discussions with Hawker Siddeley on the basis of stretched Trident variants, indicating before the end of 1966 that better range and airfield performance would be needed than previously outlined. This was the point at which Hawker Siddeley first offered the Trident 3B with a boost engine, in parallel with unboosted, stretched-fuselage Trident 3 projects. The BEA requirement for the new aircraft to operate on the longer-range routes to the Mediterranean meant that take-off performance had to be improved for the departures in high temperatures on return flights to the UK. To meet this specification, the Trident 3 as originally

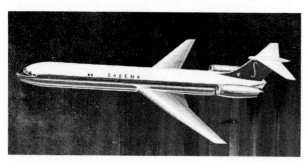

(Above) The HS.132 and (below) HS.134, two projected developments of the basic Trident design which were offered to BEA in 1965 but were eventually dropped in favour of the simpler and cheaper Trident 3B.

conceived needed more fuel. This meant greater take-off weight and, in turn, more power. To use either uprated Speys or completely new engines would have meant expensive airframe and engine development programmes, and the idea of using a small boost engine consequently had great merit. Both the Orpheus and the RB.162 were considered, the latter being adopted in its -86 version, this requiring development for horizontal installation as previous RB.162 versions had been lift-jets designed to operate vertically.

With a rating of 5,250 lb st (2 380 kgp), the RB.162 provides an increase of 15 per cent in total thrust available to the Trident 3B at take-off, for an engine weight penalty of only five per cent (less than 600 lb — 272 kg). Even more significant, in so far as certificated field lengths are concerned, is the 20 per cent increase in thrust available with one Spey out. Without this additional thrust, the runway length required for take-off (which must allow for a possible engine failure at the most critical point) would be excessive, and would mean that the aircraft would not be able to carry a commercially economic payload when tanked up for the longer stage lengths. The booster is intended to be used only for take-offs and noise abatement climb-outs, after which it is shut down, and is used in this way only when combinations of runway length and temperature would otherwise restrict the take-off weight of the Trident and prevent it from flying the required distance with a full payload.

Design changes

The RB.162 engine is located, in the Trident 3B, immediately above the centre Spey engine and beneath the rudder; this is the position occupied by the Artouste auxiliary power unit in the earlier marks of Trident. Air for the boost engine is taken in through doors on each side of the lower fin; when open, the doors form the inner wall of the intake ducts and when closed they form part of the external skin of the installation. The intake doors are pneumatically operated.

The auxiliary power unit, displaced from its position at the base of the fin, is now mounted forward of the fin and above the air intake to the centre engine. An Air Research unit is used in place of the Artouste employed previously, and this exhausts through a flush panel in the port side of the fin. Location of the APU in this position means that the fin fuel tank, used by the Trident 2, has been deleted. A supplementary fuselage tank between the undercarriage bays has also been deleted, with a total reduction of 780 Imp gal (3 545 l) in fuel capacity compared with that of the Trident 2, which has the longest range of all Trident versions.

Lengthening the fuselage of the Trident 3B is achieved by inserting three additional portions of constant cross section, one of 79 in (2,00 m) and one of 22 in (0,56 m) ahead of the wing, and one of 96 in (2,44 m) aft of the wing. Two additional servicing doors are provided in the starboard side of the fuselage but the main passenger doors, two ahead of the wings, are unchanged from earlier Tridents.

Significant changes have been made to the wing geometry of the Trident 3B, although these are difficult to distinguish

continued on page 94

THE TRIDENT VARIANTS

Airco D.H.121	Chosen by BEA in January 1958 as future short-haul jet. Three 13,790 lb st (6 255 kgp) RB.141/3 Medway engines, gross weight 150,000 lb (68 040 kg), range 2,070 mls (3 330 km) and up to 111 seats. Superseded by Trident 1.
HS Trident 1	Scaled down version of D.H.121 to meet revised BEA specification of 1959, with three 9,850 lb st (4 470 kgp) RB.163 Spey 505 engines, gross weight 105,000 lb (47 630 kg), range 930 miles (1 500 km) and up to 97 seats. BEA ordered 24 in August 1959; 21 now in service.
HS Trident 1A	Projected for American Airlines in 1960 with 10,700 lb st (4 850 kgp) Spey 510s, gross weight 120,000 lb (54 430 kg), increased wing area and fuel capacity.
HS Trident 1C	As Trident 1 with 1,000 Imp gal (4 546 l) more fuel in centre section tank. Gross weight 115,000 lb (52 165 kg) (same as Trident 1 as delivered).
HS Trident 1E	Improved Trident 1 with 11,400 lb st (5 170 kgp) Spey 511s, gross weight 128,000 lb (58 060 kg), increased wing area and same fuselage with up to 140 seats (six abreast). Three for Kuwait (two in service), three for Iraqi Airways, four for PIA (now with CAAC), two for Channel Airways, two for Northeast Airlines, one for Air Ceylon.
HS Trident 1F	Projected for BEA in 1963 with Spey 511s, 9 ft 3 in (2,82 m) fuselage stretch, gross weight 132,000 lb (59 875 kg) and up to 128 seats (five abreast). Superseded by Trident 2.
HS Trident 2ER	Chosen by BEA in August 1965 in place of planned purchase of 10 (plus option on 14) Trident 1Fs. Increased span and area, same fuselage, gross weight 142,500 lb (64 635 kg), 11,930 lb st (5 410 kgp) Spey 512s. BEA purchased 15, all in service; two for Cyprus Airways.
HS Trident 3	Projected for BEA in 1964, with modified Trident 2 wing and fuselage stretched by 16 ft 5 in (5,00 m). 12,000 lb st (5 445 kgp) Spey 512s, gross weight 143,000 lb (64 860 kg), up to 180 passengers.
HS Trident 3B	As Trident 3 with addition of one RB.162–86 boost engine. Gross weight 155,000 lb (70 305 kg). BEA ordered 26 in 1968, deliveries in progress.

IN PRINT

"The Observer's Book of Aircraft"
by William Green
Frederick Warne & Co Ltd, London, 45p
288 pp, $3\frac{1}{2}$ in by $5\frac{1}{2}$ in, illustrated
THE 1971 EDITION of this hardy little peren-
nial, now in its 20th edition, continues to
justify its unofficial appellation of the "Baby
Jane's", for, like its more splendiferous
contemporary, it maintains an impeccable
standard of accuracy and an unsurpassable
illustrative content. As a pocketable annual
reference to what's new in the air, the OBA
has no peer. In this new edition, the compiler has, wisely, supplemented Imperial
with metric measures, but in other respects
the formula is unchanged — concise
specifications, useful notes, beautifully-
reproduced three-view silhouettes, and a
carefully chosen selection of photographs,
these last including, incidentally, the first
that we have seen of a Sukhoi Su-7 in the
camouflage applied to most of these aircraft
serving with Middle Eastern air arms.

"Veteran and Vintage Aircraft"
by Leslie Hunt
Garnstone Press, London, £2·50
225 pp, $7\frac{1}{2}$ in by $9\frac{3}{4}$ in, illustrated
DETAILS of more than 5,000 of the world's
"oldest, rarest and most exciting" aircraft
are contained in this third edition of Mr
Hunt's well-known guide. The arrangement
of the information, which includes details
of the individual aircraft identity, is, as
before, on a geographical basis covering 70
countries, but an index has now been added
so that the location of any particular surviving aircraft type can be quickly turned up.
More than 1,000 different types are recorded,
and the book contains over 700 photographs.

The author does not give a definition of
what he means by "vintage" or "veteran",
but such modern types as Westland Scouts
and Wasps and Saab Safirs are included —
presumably being justified by the book's
sub-title of "The world's preserved aircraft". This also presumably accounts for
inclusion of details of 58 Hawker Hunters,
70 Lockheed F-80s and T-33s and over 100
North American Sabres, although all these
types are still serving in some quantities.
Information is included on all the exhibits
of the various aircraft museums and other
recognised collections, and all aircraft types
of pre-1945 vintage which are still flying
also appear to have qualified for inclusion
in this volume.

"The Lightplane"
by John W Underwood and George B Collinge
Heritage Press, Box 167, Glendale
California 91209, $4·95
100 pp, $8\frac{1}{2}$ in by 11 in, illustrated
THE SUB-TITLE of this volume is "A Pictorial
History, 1910–1969", but by no stretch of
the imagination can this be regarded as a
serious work of history. Essentially, it is a
scrapbook of lightplane photographs, printed in chronological order and provided
with captions containing variable amounts
of data. There is no narrative account save

*From the new edition of "The Observer's Book of Aircraft", noted on this page, this illustration
of the Tupolev Tu-28P (NATO code-name Fiddler) has not previously been published west of
the Iron Curtain.*

a short essay on Santos Dumont on the
first page; no explanation of the scope or
purpose of the book; and no index. Apart
from the 400-odd photographs, which are
well reproduced, the book contains 12
pages of three-view and cut-away drawings.

"Heinkel — an Aircraft Album"
by P St John Turner
Ian Allan, Shepperton, £2.00
128 pp, 9 in by 7 in, illustrated
IN THE SERIES of aircraft company histories
from this publisher, "Heinkel" is useful as
a primer for the relatively uninitiated, but
is not intended, and should not be regarded
as, a definitive record or one which con-
tributes much additional information to
that already in print elsewhere. The series
has merit for the new generation of aircraft
enthusiasts who may have limited funds to
expend on their hobby, but it is unfortunate
that many of the photographs in this book
have suffered from poor reproduction,
emphasising the deficiencies of the original
prints.

"World's Airliner Registrations"
by Gordon Swanborough
Ian Allan, Shepperton, England, 50p
176 pp, 5 in by $7\frac{1}{2}$ in, illustrated
PROVIDES complete production lists of the
35 types of turbine-engined airliners now in
service, and traces the history of each, from
delivery, through previous owners to
present operator or eventual fate. This third
edition is updated to the end of 1970 and
therefore covers two years' amendments and
additions since the previous edition appeared; regrettably, all illustrations have
been omitted, presumably to make room
for the 1,200 new entries.

"Aircraft of the Royal Air Force"
by Owen Thetford
Putnam & Co Ltd, London. £4·20
624 pp, $5\frac{1}{2}$ in by $8\frac{1}{2}$ in, illustrated
THIS was the very first title in the now well-
known series of Putnam aviation histories,
first published in 1957 and now appearing
in its fifth edition. The text has again been

extensively revised and updated, and some
of the anomalies of layout in the previous
editions have been overcome by grouping
together all the entries for the different
variants of each type. Despite the savings
in space which result, this edition is 100
pages longer than the first, and includes an
appendix covering the guided weapons
used by the RAF.

"Republic Thunderbolt"
by Roger Freeman
Ducimus Books Ltd, London, £1.25
72 pp, $7\frac{1}{2}$ in by 10 in, illustrated
FIRST in a new series entitled "Ducimus
Classics", this volume (in card covers) includes four pages of detailed line drawings
and four pages of colour tone drawings.
The text shows ample evidence of Roger
Freeman's concise knowledge of USAAF
operations during World War II, and does
full justice to the important rôle played by
the Thunderbolt in the ETO. Details of
production are lucidly set out and the
differences between each production block
are clearly explained. Strangely, in view of
all this erudition, the author fails to note
that the Thunderbolts still serving with the
USAF in 1948 were redesignated from P-47
to F-47D and F-47N.

"Supermarine Spitfire"
by Peter Moss
Ducimus Books Ltd, London, £1.25
72 pp, $7\frac{1}{2}$ in by 10 in, illustrated
ANOTHER "Ducimus Classic" (see above),
this volume is in the same format as that
on the Thunderbolt, but the text is much
more concerned with explaining the
differences between variants, and relatively
little space is given to the operational career
of the Spitfire. This is one of two Spitfire
volumes planned, and covers the Merlin-
engined types; the second will deal with
Griffon-engined Spitfires and will include
appendices relevant to both types. Meticu-
lous research into the history of individual
aircraft is shown by the text, but the colour
reproductions, one suspects, do less than
justice to the artist's originals. □

MODEL ENTHUSIAST

My wife doesn't understand me!

MODELLING aircraft is a peculiarly masculine pastime, as, indeed, is any form of modelling, if one ignores the common usage of the term in connection with the draping of clothes — or what in this day of permissive society can pass muster as apparel — on the female form for display purposes. In our experience, the female of *homo sapiens* is the true *rara avis* among aircraft modellers. That such *exist* we have no doubt, but these paragons are few and far between; the finer points of the modelling art rarely arouse enthusiasm in the female breast.

Many an unsuspecting member of the modelling fraternity has evinced surprise at the rapidity with which a professed interest in his pastime on the part of his fiancée has departed with her transition to wifehood. How *can* the ardent modeller convey to his spouse the pleasure that he derives from an accurate scale reproduction of an aircraft of which she has never even heard, let alone seen at full scale? Most wives do, to a greater or lesser extent, tolerate what they may consider a time-consuming and self-indulgent form of recreation on the part of their mate. Irrational though they may believe to be the pursuit of aircraft modelling, and strong though the temptation may be to banish plastic, paint and filler to garden shed or garage, they are usually wise enough to see that this diversion of their husbands is one solely *for* the boys, and better this than that they be indulging in "one *with* the boys"!

But there are *extreme* cases; examples of wives who, being unable to share in their husbands' enthusiasm for aircraft modelling, will go to any length short of physical violence in their endeavours to persuade their marital partners to change their chosen recreation to one of which they approve and understand. We can recall at least one highly-proficient modeller, who, having long since abandoned his attempts to persuade the mother of his children to mollify her philistine views, we encountered muttering plaintively, "My wife doesn't understand me", as he picked his way through the contents of the trashcan to which his wife had consigned vital components of a newly-acquired kit!

The answer to this problem of feminine apathy and, indeed, antipathy towards aircraft modelling *may* be the adroit use of psychological warfare. Rare is the skilful modeller who is not, if he sets his mind to it, adept at home decorating and the repair of household gadgetry. If he can reproduce a complex camouflage scheme in miniature he can surely repaint the face of a clock with equal skill; dexterity in filing sections to shape for kit conversions or modifying the complex undercarriage retraction mechanism of a working model can also be applied to fashioning a simple replacement part for an out-of-date vacuum-cleaner or repairing a food-mixer. Accompanying the completion of such tasks with the casual remark that they could not have been successfully undertaken without modelling expertise, or that modelling tools have been invaluable in performing the repairs, *could,* we repeat, *could* progressively lead to passive acceptance if not enthusiasm for modelling. But a word of caution: do *not* belabour the point. We have lost more married modellers that way . . . !

This month's colour

Although the large carnivorous feline quadruped known as the Jaguar is characterised by curvacious solidity whereas its namesake from the Anglo-French team of BAC and Breguet is noteworthy for its slender angularity, the use of this appellation is not inappropriate, for both the denizen of the wooded areas of America from Texas to Paraguay and the future strike fighter of the British and French forces are powerful predatory beasts. It will be next year before we are likely to see the *European* Jaguar sporting actual unit markings, but both Heller and Airfix have already marketed kits of this warplane to 1/50th and 1/72nd scale respectively.

Heller has displayed some enterprise in designing its Jaguar kit so that it may be issued in any version of the aircraft, single- or two-seat, land-based or shipborne, so far announced. We have seen this kit in its "A" and "M" single-seat versions for the *Armée de l'Air* and *Aéronavale* respectively, and we understand that it is also available in its "E" two-seat form, although we have not seen the last-mentioned version of the Heller kit at the time of writing. The Airfix kit represents the RAF's "S" single-seat strike variant, and the kits from both manufacturers are of excellent quality, accurate and finely detailed. The fit of their component parts is generally good, although the Airfix kit scores rather more points in this respect, Heller's model demanding some filling around the wing roots. On the other hand, the Heller kit scores by virtue of its larger scale which permits detail impracticable at 1/72nd scale. This is particularly noticeable on Heller's beautifully reproduced undercarriages, which, together with the decal sheets, provide the major differences between the "A" and "M" versions of the kit.

Naturally enough, the decal sheets provide the markings of prototype aircraft, and neither Airfix nor Heller is renowned for the quality of its decals, these kits providing no exception. The Airfix decal sheet is badly printed, with ragged edges to the serial numbers, while Heller's sheet offers off-centre and exceedingly glossy roundels. Both kits include camouflage pattern drawings on their instruction sheets, but Heller's is spoiled by a split planview which, in so far as the "A" version's camouflage pattern is concerned, is virtually useless. Airfix's kit, incidentally, offers a wide range of external stores whereas none is included by Heller. In short, both kits embody minor faults but both are basically good and well worth buying.

NATO warpaint

Appropriate to the Jaguar is the latest offering from Humbrol: a paint set consisting of six of the most commonly-used NATO standard aircraft colours. These are: dark green, dark sea grey, PR blue, medium sea grey, light aircraft grey, and extra dark sea grey. It should be noted that their shades are *very* different from those of their WW II namesakes.

These paints are of the usual Humbrol quality, and if correctly applied will give a superb finish to any model. Incidentally, applied directly from the tin these paints will give complete coverage in *one* coat, and re-coating should be avoided unless absolutely necessary! In other words, a camouflage scheme should be marked out in pencil before the application of paint which should then be applied directly to the appropriate areas. Painting the entire model in one colour and then applying the camouflage pattern over this will tend to result in brush marks which can seriously disfigure a model. Applied in one coat these paints dry rapidly to produce an absolutely even matt finish with a most realistic sheen. A model so finished should be handled with care as the paint readily shows finger-marks, but any modeller contemplating a NATO aircraft could not do better than to use these Humbrol Authentic Colours.

A Frog quartet

From the viewpoint of the modeller, one of the most neglected periods in the entire history of aviation — neglected, that is, by the kit manufacturer — is that immediately following WW II. The last generation of piston-engined combat aircraft had no opportunity to gain laurels comparable with those earned by its wartime predecessors, and thus, to the model kit manufacturer, lacked the glamour necessary for quantity sales. From time to time, of

Four different versions of the BAC/Breguet Jaguar have flown in prototype form so far, and all are shown in the colour drawings on the opposite page. The three French versions are shown in the side views, from the top of the page downwards: Jaguar E-02 two-seat trainer; Jaguar A-04 single-seat strike aircraft and Jaguar M-05 naval single-seat attack aircraft. The lower side view and the top and under side plan views show the Jaguar S-06, prototype of the British single-seat tactical strike aircraft.

course, the occasional kit of a warplane of the immediate post-war period *has* appeared, and we are happy to see that Frog has courageously issued a kit of one such aircraft, the de Havilland Hornet, which makes a truly worthy subject for a model and a fitting companion for this company's recently-issued Westland Wyvern and Gloster Meteor IV.

Frog's Hornet, which, in outline, represents the F Mk 1 version with the long dorsal fin added later, gives an overall impression of accuracy, but is at fault on a number of detail points. For example, the fuselage aft of the wing is too long, and this excessive length is exaggerated by an undersized fin. The tailplanes have too great a span and, in fact, represent those of the Sea Hornet F Mk 20, though the arrester hook and camera windows of this variant are not included in the kit, and the engine nacelles are inaccurately shaped, being too broad in planview at the wing leading edge, not extending sufficiently aft over the upper wing surface, and tapering too sharply forward to meet the slightly oversized spinners. All these errors can be corrected, but it would seem a pity that Frog did not check the outline drawings more carefully before cutting the tools for the kit.

The component parts are flash-free and assemble well, while the straight-line surface detail is commendably fine. The instruction sheet is satisfying as far as it goes, but suffers a rather important omission. The Hornet had handed airscrews, and these are provided perfectly correctly in the kit, but the instruction sheet fails to indicate which airscrew should be fitted to which nacelle. In fact, the airscrews rotated *inwards,* and part No 17 should therefore be fitted to the port nacelle and No 18 to the starboard. Incidentally, on the model illustrated by Frog's publicity photographs the airscrews are mounted the wrong way round! The decal sheet, which provides markings for aircraft of both Nos 19 and 41 Squadrons, maintains the usual high Frog quality, and the coloured illustration on the

boxlid provides other kit manufacturers with an outstanding example in the presentation of marking detail.

Released simultaneously by Frog were re-issues of three kits that have existed for some time but have now been redesigned for motorisation. To accompany them there is a set containing two tiny electric motors with the necessary wiring and copper connections. This is a first class idea for which we award full marks. The kits concerned are Bristol's Beaufort and Beaufighter, and the Junkers Ju 88A-4, and the motors are 0·8 in (2,0 cm) in length and 0·45 in (1,14 cm) in diameter, and thus slip easily into 1/72nd scale radial engine cowlings. Their shafts are offset, which may enable them to be used for some liquid-cooled engine installations, though nacelles such as those of the Hornet are too slim to accept them. The motors are driven by means of a penlight battery housed within the fuselage of the model (and accessible by means of a detachable panel in the fuselage underside) and are started and stopped simply by turning or stopping the airscrew. The price of the entire motorisation set is only 42½ pence in the UK, and we foresee a wide application for this clever device.

Turning briefly to the kits themselves, the Beaufighter is an excellent model, and has now been provided with the bulge forward of the windscreen that characterised Australian-built variants and an excellent new decal sheet for a Beaufighter 21. The Ju 88A-4 is a kit of older vintage, and suffers a number of inaccuracies, notably concerning the shapes of the wings and nacelles. The new decal sheet provides markings for examples of this aircraft that served both in North Africa and the Soviet Union. Oldest of the trio is the Beaufort, but it has stood the test of time well, and is generally a very good model if one throws away the atrocious engine cowlings and replaces them with suitable cowlings from another kit. Incidentally, despite the statement on the boxtop, the markings of neither aircraft provided by the decal sheet are those of a Beaufort II (W6476 was a

Taurus-engined Mk I while A9-408 was a Twin Wasp-engined Mk VIII). The cowlings of the two types of engine were noticeably different, and the box art is in error in showing the Australian machine with the forward collector rings and long exhaust pipes of the Taurus installation.

Yet another Draken
It would seem that the number of kits on the stockists' shelves of Saab-Scania's double-delta Draken continues to proliferate, but Airfix's recent 1/72nd scale addition to those kits of this Swedish warplane already available is by no means superfluous. In the first place, it is the only kit of the definitive version of the Draken for *Flygvapnet,* the J 35F, and secondly, it is an exceptionally good kit at a low retail price.

Airfix's Draken is accurate in outline and very neatly detailed. The component parts fit together with considerable precision, assembly is easy, and included in the kit are four Falcon AAMs and two drop tanks. Even the air brakes are separate parts and can be fitted in the open position. The decal sheet's standard is in keeping with that of the rest of this kit. In addition to the national insignia, it provides very gay yellow checkerboards that (it is alleged) may be applied to a J 35F of F 10, and offers alternative markings for an aircraft of F 13 — the minute lettering of the unit crest is perfectly legible through a lens, as is also the tiny stencilling.   W R MATTHEWS

RECENTLY ISSUED KITS			
Company	Type	Scale	Price
Airfix	TriStar	1/144	85p
Airfix	J 35F Draken	1/72	24p
Frog	Hornet	1/72	26p
Frog	Beaufighter (M)	1/72	36p
Frog	Ju 888A-4 (M)	1/72	36p
Frog	Beaufort (M)	1/72	36p
Aurora	DC-10	1/144	$2·50
Revell	OH-6A Cayuse	1/32	49p
Revell	Hurricane I	1/32	£1·05
VEB	Mil Mi-10K	1/100	—

TRIDENT——————————————*from page 90*
externally. The overall span remains the same as for the Trident 2, which, compared with the Trident 1, has extended, Kuchemann-style tips. The Trident 2 also introduced leading edge slats in place of the fixed droop on the leading edge, and these are retained on the Trident 3B, together with small sections of Krüger flap at the wing roots. The wing area has been increased, however, by extending the chord of the outer wing panels aft of the rear spar. In addition, the ailerons have been moved outboard and reduced in area, to allow more room on the trailing edge for the flaps. As a result, the flaps are both longer in span and greater in chord, having 26 per cent more total area. To compensate for the reduction in aileron area, greater reliance is placed on spoiler control, the maximum differential angle achieved being doubled, from 10 deg to 20 deg. The total area of the lift dumpers is increased 35 per cent by adding two extra sections outboard of the spoilers. Operation of the dumpers is automatic upon touchdown.

As the greater fuselage length imposes a limit on the angle of rotation which can be achieved during take-off, the wing is rigged at a 2 deg 33 min greater angle of incidence to the fuselage datum. The effect of all these changes to the wing, plus the boost engine power, is to allow the Trident 3B to

operate within similar field lengths to those of the Trident 2, despite the 12,500 lb (5 670 kg) increase in gross weight. The BEA aeroplanes operate at 150,000 lb (68 040 kg) gross, but the maximum approved weight of the 3B is 155,000 lb (70 305 kg). Structural changes have been made in the Trident 3B to match the increased weights, and systems have been modified to provide the necessary extra capacity for the greater cabin size.

The Trident 3B flew for the first time at Hatfield on 11 December 1969 and production, as previously mentioned, is now proceeding at a rate of two per month. Fairly obviously, the development history of the type is now approaching its end. The Trident is very much BEA's aeroplane, this airline having bought 65 of the 82 Tridents ordered to date. With excusable hyperbole, BEA calls its Trident Three "Europe's most beautiful, most advanced jetliner". If it isn't *quite* that, then it is at least a good, sound airliner, benefitting from many years of engineering development and with the added advantage of the world's most developed automatic landing system, which may render zero-zero landings commonplace before the end of the decade. The Trident Three certainly has a major rôle to play in keeping BEA "No 1 in Europe" throughout the 'seventies. □

The *Caproni* that Nearly Joined the RAF

O N THE TENTH of June 1940, *Il Duce,* finally convinced that France was tottering on the brink of defeat, summoned the courage to commit Italy to war as Germany's ally. Most of the consequences of this act, in so far as Benito Mussolini in particular and the Italian armed forces in general were concerned, have been thoroughly documented in the records of World War II, but one small facet of the effects of Italy's declaration of war is not to be found on the pages of the numerous works of reference devoted to that conflict; a side-light that has hitherto remained unknown to all but a few. The 10 June declaration automatically cancelled the largest single export contract that the Italian aircraft industry had ever obtained; a contract placed by an entirely new customer — the Royal Air Force!

Italy entered WW II with mixed feelings. While the Fascist followers of *Il Duce* rejoiced at the prospect of further expanding the *new* Roman Empire, many Italians did not consider the Germans to be their natural Allies. Among the latter was the pioneer Italian aircraft constructor, Count Gianni Caproni, who, controlling the largest component of the Italian aircraft industry, saw 10 June bring to nought weeks of secret negotiations and devious schemes aimed at enabling his organisation to evade a German embargo and

(At head of page) Caproni Ca 313S light reconnaissance and maritime patrol aircraft serving with Flottilj *11 of Sweden's* Flygvapen *at Nyköping during the summer of 1942.*

fulfil the substantial aircraft contract placed by the British government.

British interest in Caproni aircraft had been first aroused at the beginning of 1939. At this time, the Air Ministry, with the RAF immersed in an immense expansion scheme accelerated a few months earlier by the Munich crisis, was investigating the possibility of purchasing aircraft from any source that presented itself. On 2 January 1939, the Società Italiana Caproni had obtained from the Italian Air Ministry permission to receive a British Purchasing Mission at its main airframe works at Taliedo, Milan. The mission arrived in Milan two weeks later, and displayed some interest in the potentialities of the Ca 310 Libeccio (Southwest Wind) light twin-engined multi-purpose aircraft as a crew trainer for the RAF.

Although no contract was immediately placed for the Ca 310, British interest was perfectly genuine and had crystallised in a definite intention to purchase by the following December when Lord Hardwick arrived in Italy at the head of a further Purchasing Mission. On 12 December he informed Count Caproni that it was the British government's intention to purchase 200 Ca 310s as well as 300 examples of an improved and more powerful derivative of the basic design, the Ca 313. At that time, the Ca 313 prototype had still to make its first flight, an event which was to take place 10 days later, on 22 December, but Britain's Air Ministry had been kept fully informed of development work on the basic Ca 310 design by the Caproni Agency Corporation (England) Limited, which, occupying premises in Princes House, Piccadilly, was part of a highly active Caproni export sales organisation. Indeed, the British government was not the *first* customer for the Ca 313, despite the fact that this aircraft had still to commence its test programme. The French government had already ordered 200 Ca 313s, having placed the contract with Caproni on 26 September.

During the two weeks immediately following the arrival of Lord Hardwick's Mission in Italy, the planned purchase was subjected to several changes, the most important of these being the decision to buy 100 Ca 311s for the training rôle instead of 200 Ca 310s. The Ca 311, which had first flown on 1 April 1939, was a much improved development of the earlier aircraft, with uprated engines, a redesigned fuselage nose, and other modifications. The decision to purchase 300 Ca 313s remained unchanged.

An Italian team duly arrived at the Air Ministry in Harrogate, Yorkshire, early in January 1940 for contractual discussions which were conducted with Sqdn Ldr (later Air Commodore) N R Buckle and Caproni's representative in Britain, *Signor* Fronteras. All the drawings for installations and modifications specified by the RAF were passed to the Italian team which returned with them to the Taliedo factory. A few days later, Sqdn Ldr Buckle, in civilian clothes and carrying a passport describing him as an engineer, arrived in Milan to supervise the final details, this work

being done in the basement museum of the Taliedo factory. After a brief visit to the Isotta-Fraschini factory where the Delta R.C.35 engines for the Ca 313s were being manufactured, Sqdn Ldr Buckle was assured by Count Caproni that the work on the British contract would go ahead whatever action the Germans might take. Arrangements were made for the aircraft to be packed in boxcars bearing no distinguishing marks, and then sent by rail to Istres airfield, near Marseilles. Istres was being utilised by the RAF, and it was planned that the 400 Caproni aircraft would be assembled there and then flown to Britain. A week later, on 26 January, the Air Ministry officially confirmed the purchase.

Contract concluded

Although the negotiations between the Società Italiana Caproni and the British government had been conducted with the utmost discretion, as, indeed, had also those with the French government, and all foreign customers had been assigned "customer numbers" which were used for all correspondence and inter-office memoranda to avoid disclosure of the nationality of the customer, the German intelligence service was fully conversant with the details of the contracts, and "leaked" its knowledge to the Italian government. Somewhat embarrassed, the Italian Air Ministry hurriedly informed the German government officially of its dealings with Britain and France, enquiring if, in view of the Italo-German Alliance, Germany had any objection to Italy fulfilling these contracts. Surprisingly, the German government, in its reply of 8 March 1940, signified that it had no objection to the sale of the Italian aircraft! Nevertheless, a month later, on 6 April, the German government performed a *volte-face*, requesting that the contracts for the supply of aircraft to Germany's enemies should be terminated by the Italian government forthwith.

This was tantamount to a German embargo, and one which *Il Duce* could ill afford to ignore, but while the Italian government ostensibly acceded to the "request", it was loath to sacrifice the much-needed hard currency which would result from these contracts and there can be little doubt that it secretly connived with Count Caproni to fulfil the orders. It was thus, on 15 May 1940, with the German western offensive in its sixth day and *Il Duce* still straddling the fence of neutrality, that Count Caproni and Lord Hardwick finalised a scheme whereby the aircraft would apparently be purchased by Britain from Portugal, with a Portuguese subsidiary of the Società Italiana Caproni, the Sociedade Aeroportuguesa, fronting for the Italian company.

All this deviousness and subterfuge was to prove of no avail, however, for 26 days later Italy found herself at war with what had been potentially her largest customer for military aircraft, and the Caproni twins were destined never to enter the RAF's inventory. In view of subsequent Swedish experience with the Ca 313 and the incendiary

Type		Borea	Ca 309	Ca 310	Ca 311	Ca 312M
Power Plant Max power	hp	Alfa 115-I 200	Alfa 115-II 200	P.VII C.16 460	P.VII C.35 460	P.XVI R.C.35 650
Span Length Height Wing area	ft (m) ft (m) ft (m) sq ft (m²)	53·15 (16,20) 42·22 (12,87) 10·72 (3,27) 413·33 (38,40)	53·15 (16,20) 43·63 (13,30) 10·72 (3,27) 413·33 (38,40)	53·15 (16,20) 40·03 (12,20) 11·55 (3,52) 413·33 (38,40)	53·15 (16,20) 38·52 (11,74) 11·97 (3,65) 413·33 (38,40)	53·15 (16,20) 38·71 (11,80) 11·48 (3,50) 413·33 (38,40)
Empty weight Loaded weight	lb (kg) lb (kg)	4,134 (1 875) 5,952 (2 700)	4,409 (2 000) 6,607 (2 997)	6,702 (3 040) 9,270 (4 205)	8,020 (3 638) 11,063 (5 018)	7,496 (3 400) 11,464 (5 200)
Max speed at (altitude) Cruise at (altitude) Normal range	mph (km/h) ft (m) mph (km/h) ft (m) mls (km)	153 (246) 1,970 (600) 140 (225) 4,920 (1 500) 627 (1 010)	155 (250) 2,460 (750) 130 (210) 4,920 (1 500) 410 (660)	215 (347) 9,840 (3 000) 189 (305) 11,485 (3 500) 637 (1 025)	217 (349) 13,125 (4 000) 195 (314) 9,840 (3 000) 497 (800)	267 (430) 11,485 (3 500) 230 (370) 11,485 (3 500) 404 (650)
Armament Max bomb load	lb (kg)	Nil Nil	3 × 7,7-mm 440 (200)	3 × 7,7-mm 882 (400)	3 × 7,7-mm 882 (400)	3 × 7,7-mm 882 (400)

propensity that it was to reveal, this may well have been a stroke of *good* fortune for Britain's air arm!

The Borea line

The light Caproni twin-engined monoplanes that the British government had planned to acquire for use by the RAF had been derived from an eight-seat feederliner of wooden construction, dubbed the Borea (North Wind), which had appeared in 1935. The Borea had been evolved under the direction of Cesare Pallavicino of the Caproni Aeronautica Bergamasca at Ponte San Pietro, near Bergamo, and, powered by either two 205 hp Walter Major 6 or 200 hp de Havilland Gipsy-Six six-cylinder inline inverted air-cooled engines, was an aerodynamically clean and relatively efficient aeroplane. If it was to lay any claim on fame, however, it was to be for the somewhat dubious distinction of being sire to a long and *commercially* successful line of military aircraft.

The first development of the Borea, the Ca 309 Ghibli (Desert Wind) flown on 3 October 1936, set the pattern for all its successors in that it switched to mixed construction, the fuselage being a welded steel-tube structure with light alloy and fabric skinning. A light general-purpose aircraft in the category known between the wars as "colonial", the Ghibli retained the wing and tail surfaces of the Borea, marrying these to the new welded steel-tube fuselage which maintained the Borea's rectangular cross section, and a pair of 185 hp Alfa Romeo 115 six-cylinder inline inverted air-cooled engines. The trousered main undercarriage members gave place to streamlined spats, and provision was made for light defensive armament, bomb racks and cameras. The Ghibli was an immediate success, and was ordered in some numbers for policing duties in Italy's African territories by the *Aviazione Presidio Coloniale* and the *Aviazione Sahariana*.

Evolved in parallel with the Ghibli was the more powerful Ca 310 Libeccio which, flown on 9 April 1937, possessed an essentially similar airframe apart from some local strengthenin gand minor contour revision. Powered by two 460 hp Piaggio P.VII C.16 seven-cylinder radials and featuring retractable mainwheels which were raised backwards into the tails of the engine nacelles, the Libeccio quickly attracted export orders, and the output of the Ponte San Pietro factory was soon augmented by an assembly line at the main Caproni works at Taliedo. The commercial version obtained some publicity as a result of its success in the *III Raduno Sahariana* in 1938*, but only a limited order for the Ca 310 was placed by the *Regia Aeronautica* for the light reconnaissance-bomber variant, 16 being sent to Spain in July of that year for use by the *Aviazione Legionaria*.

The Caproni sales organisation was more successful in obtaining export orders for this new descendant of the Borea, and long before January 1939, when the British Purchasing Mission began to evince interest in the Ca 310, contracts had been obtained from several countries. A few had been delivered in 1938 to the Peruvian Air Force, and a dozen had been purchased by Yugoslavia. Hungary had ordered 36 Ca 310s with Piaggio P.VII C.35 engines, and these had been delivered in batches of a dozen machines in August, Septem-

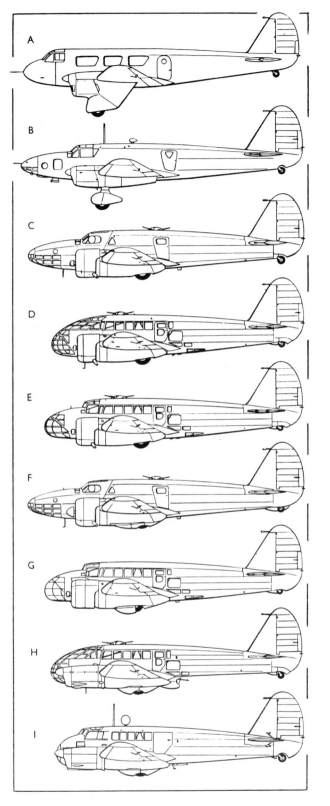

---

* *The III Raduno Sahariana, or Sahara Rally, was organised by the Aero Club of Tripoli from 18 to 28 February 1938, and the twin-engined Capronis swept the board. From among 25 competitors Ca 310s took first, second, third, fifth and sixth places and a Ca 309 Ghibli was fourth. Four other Ghiblis were directed to retire before the event ended in order to help search for a missing Italian competitor. The rally included three days of control and navigation tests at Gadames, followed by a four-stage flight round Libya, over the route Gadames-Brac-Cufra-Benghazi-Tripoli.*

ber and October of 1938, and the Norwegian government had placed contracts and acquired options on a total of 24. However, the Ca 310 soon proved incapable of meeting fully the brochure specification, and reliability left much to be desired.

In 1940, the 33 aircraft that survived in Hungarian service were returned to Italy where, after refurbishing by Caproni, they were passed to the *Regia Aeronautica* which issued them to the 50° *Stormo d'Assalto* as temporary replacements for the Breda 65. After taking delivery of four Ca 310s, the Norwegian government had, meanwhile, renegotiated the

The Ca 313 R.P.B.1 (above) was the first of the Caproni light twins to be Isotta-Fraschini-powered, and was the subject of export orders from France, Sweden and Britain.

One of the predecessors of the Ca 313-314 was the Ca 311, the first member of the Caproni family of light twins to have a fully-glazed unstepped nose. Britain's Air Ministry ordered 100 for use by the RAF as trainers but, in the event, the only Ca 311s to acquire RAF roundels were those that were captured in North Africa and Italy during the war (as below).

remainder of its contract on the basis that performance was substantially below guarantees, and in an attempt to meet Norwegian performance requirements, Caproni re-engined a Ca 310 airframe with 650 hp Piaggio P.XVI R.C. 35 radials, this flying on 7 December 1938 as the Ca 312. The Norwegian government agreed to accept this model but, in the event, none had been delivered by the time Norway was invaded, and the completed aircraft were taken over by the *Regia Aeronautica.*

During the course of 1938, a Taliedo-built Ca 310 *Serie 2°* with P.VII C.35 engines had been experimentally fitted with an entirely redesigned and extensively-glazed nose section not dissimilar to that of the Blenheim I and providing excellent visibility. Flown as the Ca 310bis, the modified aircraft met the requirements of the *Regia Aeronautica* for a light reconnaissance-bomber and observation aircraft and, with few modifications, was ordered into production as the Ca 311, the first genuine Ca 311 prototype flying on 1 April 1939. The Ca 311 began to supplant the Meridionali Ro 37 biplane in the *Gruppi Osservazione Aerea* during 1940, although extensive re-equipment with this type was not to take place until 1941. Apart from its new nose section, additional transparencies in the fuselage, and a repositioned Caproni-Lanciani Delta E gun turret, the Ca 311 was virtually identical to the Ca 310 *Serie 2°*, and the Ca 311M (*Modificato*) differed only in having a slightly lengthened nose with revised contours and a stepped windscreen. The Ca 311 was

first issued to the 61° *Gruppo* which took this type on operations in the Soviet Union with the *Corpo di Spedizione Italiano,* subsequently serving with all but two of the *Gruppi Osservazione Aerea.*

Boosting the breed

Apart from the contract from the British government, which considered it as a potential crew trainer, the Ca 311 was the recipient of no export order, largely as a result of its relatively low performance. The Società Italiana Caproni had proffered the Ca 312bis, which was essentially similar to the Ca 312 ordered by Norway apart from having a similar fuselage to that of the Ca 311, but foreign interest was, by the summer of 1939, focused on what promised to be a markedly improved development in the line of light twin-engined monoplanes that had stemmed from the Borea, the Ca 313.

Structurally, the Ca 313 differed in no major respect from the Ca 311, the principal change being the type of power plant installed. For a number of years the Caproni-controlled Fabrica Automobili Isotta-Fraschini had been developing a series of 12-cylinder inverted-vee air-cooled engines in the 700 hp category, and there was every reason to believe that the marriage of a pair of such engines with the soundly-designed and robust Ca 311 airframe would result in an outstanding aeroplane. That the calculated performance of the Ca 313 was impressive is indicated by the fact that, on 26 September 1939, before a prototype had flown, France's *Ministère de l'Air* had contracted for the supply of 200 Ca 313s plus 500 Isotta-Fraschini engines, confirming this order on 1 October; Britain's Air Ministry had ordered 300 Ca 313s on 26 January 1940, only five weeks after the maiden flight of the prototype, and Sweden promptly followed suit by ordering 54 for her *Flygvapen.*

The Ca 313 prototype (M.M.402), which made its initial flight on 22 December 1939, was a rebuilt Ca 310 *Serie 2°* fitted with two Isotta-Fraschini A.120 I.R.C.C.40 engines rated at 770 hp for take-off and 710 hp at 13,125 ft (4 000 m). These engines differed from the Delta R.C.35 I-D.S. engines mounted in the production model in having two-stage superchargers and different airscrew reduction gearing. The R.C.35 I-D.S. offered a maximum output of 730 hp at 2,600 rpm and an international rating of 700 hp at 11,485 ft (3 500 m), and drove a variable-pitch constant-speed fully-feathering Alfa Romeo H.365 three-bladed airscrew. The initial production model was the Ca 313 R.P.B.1 (*Ricognizione Piccolo Bombardamento*, or Reconnaissance and Light Bombing), and this was the subject of the French, British and Swedish orders. Five had, in fact, been delivered to France as the Ca 313F (the "F" suffix indicating *Francia*) during the month prior to Italy entering WW II, but the remainder of the contract was eventually taken over by the *Regia Aeronautica*. Sweden's *Flygvapen* thus became the first major operator of the Ca 313, a fact which the Swedish air arm was to have cause to regret.

By the time the first Ca 313S (the "S" indicating *Svezia*) reached Sweden on 5 November 1940, the Swedish order had been progressively increased to 70 and then to 84 aircraft. Had the Swedes, like the British, celebrated 5 November as *Guy Fawkes Day*, the date on which the first of the Caproni twins arrived in Sweden might, in retrospect, have been considered to be singularly appropriate, for the Ca 313 was soon to reveal a highly undesirable inflammability. Twenty Ca 313s reached Sweden during the course of November, these being ferried from Italy by air, and deliveries were completed during the first months of 1941, two being lost en route, and several having to make emergency landings at Bulltofta and elsewhere as a result of engine failures.

The Ca 313s supplied to Sweden evinced every sign of hurried manufacture and quickly acquired an unenviable reputation. The airframe itself was sound enough, and the handling characteristics of the aircraft were pleasant, but the low standard of component inspection at the Ponte San Pietro and Taliedo factories which had resulted from overly rapid expansion of production to cope with the wealth of orders had inevitable consequences. These were compounded by thoroughly unreliable engines which failed at the least provocation, poorly-designed hydraulic and electrical systems equally prone to failure, and the fact that the fuel lines were in close proximity to the exhaust pipes, so that the least fuel leak usually resulted in fire.

In service with *Flygvapnet* the Ca 313s were assigned to *Flottilj* (Wing) 7 at Såtenäs, fulfilling the bombing rôle as the B 16 until supplanted by the Saab 17, whereupon they were transferred to *Flottilj* 11 at Nyköping for reconnaissance and maritime patrol as the S 16. The Ca 313s were to serve with F 11 throughout the remainder of WW II, their most important task being the patrol of the Baltic coastline. Several were modified as the S 16A and S 16B, these modifications including some strengthening of the wing structure which permitted the take-off weight of the latter version to be raised by about 320 lb (145 kg), and two were adapted for the transport task as Tp 16s. Throughout its *Flygvapen* service, the Ca 313 was plagued with hydraulic, electric and power plant failures, and attrition was heavy, a total of 44 *Flygvapen* personnel losing their lives in Ca 313 crashes. However, despite these shortcomings, the Ca 313 was a remarkably sturdy aircraft, as was demonstrated on 14 May 1944 when an aircraft of this type was shot down by a Messerschmitt Bf 109G east of the island of Gotland. The three crew members rescued stated that the Ca 313 absorbed

a tremendous amount of punishment before finally being despatched by the Bf 109G.

Deliveries of the Ca 313 to the *Regia Aeronautica* began in March 1941, following completion of the Swedish order, the initial contract for aircraft built specifically for the Italian air arm calling for two batches of 60 examples of a modified version, the Ca 313 R.P.B.2, which embodied a further re-design of the nose section, reintroducing a stepped cockpit. In the event, only the first batch of 60 aircraft was completed as Ca 313 R.P.B.2s, the second batch being delivered as Ca 313 R.A.s (*Ricognizione Aerosiluranti*, or Reconnaissance and Aerial Torpedo). This designation was subsequently changed to that of Ca 314, by which time *Regia Aeronautica* orders had been increased to a total of 345 aircraft and, in addition to the Ponte San Pietro and Taliedo factories, the A.V.I.S. plant at Castellamare di Stabia, Naples, was being phased into the programme. The Ca 313 R.A. resulted from an experimental conversion of a Ca 313 R.P.B.1. (M.M. 12050) for torpedo-bombing trials at the request of *Generale*

(Above) A Ca 314 with the Caproni-Lanciani turret removed serving with the Italian co-belligerent air force, and (below) a Ca 314B with a 1,984-lb (900-kg) torpedo offset to starboard beneath the fuselage.

(Below) A Taliedo-built 1° Serie Ca 314A (M.M.12124) shortly after completion and prior to delivery to the Regia Aeronautica.

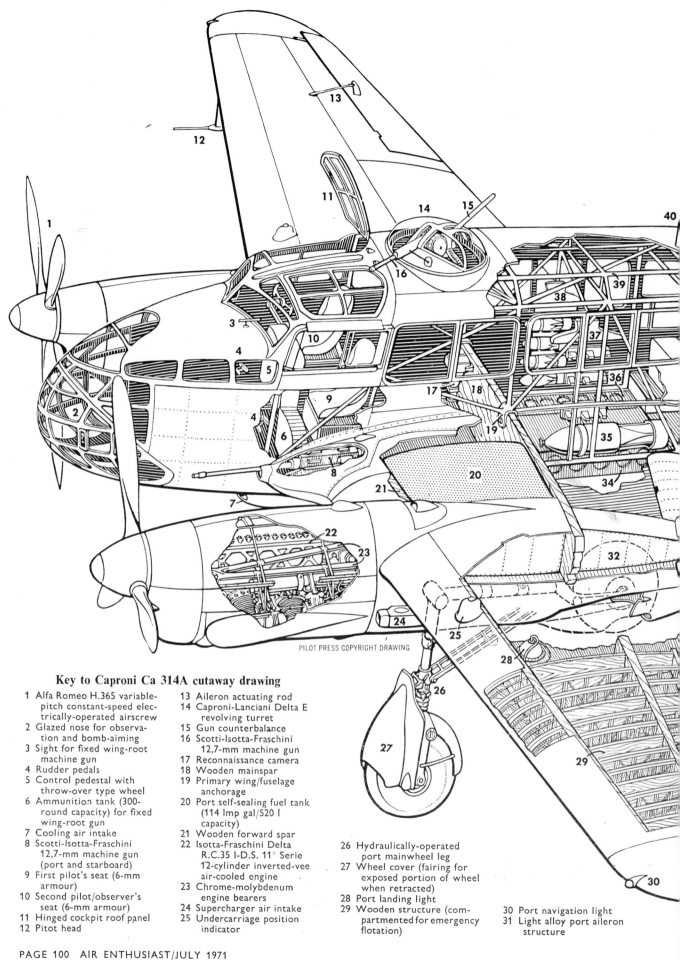

PILOT PRESS COPYRIGHT DRAWING

Key to Caproni Ca 314A cutaway drawing

1 Alfa Romeo H.365 variable-pitch constant-speed electrically-operated airscrew
2 Glazed nose for observation and bomb-aiming
3 Sight for fixed wing-root machine gun
4 Rudder pedals
5 Control pedestal with throw-over type wheel
6 Ammunition tank (300-round capacity) for fixed wing-root gun
7 Cooling air intake
8 Scotti-Isotta-Fraschini 12,7-mm machine gun (port and starboard)
9 First pilot's seat (6-mm armour)
10 Second pilot/observer's seat (6-mm armour)
11 Hinged cockpit roof panel
12 Pitot head

13 Aileron actuating rod
14 Caproni-Lanciani Delta E revolving turret
15 Gun counterbalance
16 Scotti-Isotta-Fraschini 12,7-mm machine gun
17 Reconnaissance camera
18 Wooden mainspar
19 Primary wing/fuselage anchorage
20 Port self-sealing fuel tank (114 Imp gal/520 l capacity)
21 Wooden forward spar
22 Isotta-Fraschini Delta R.C.35 I-D.S. 11° Serie 12-cylinder inverted-vee air-cooled engine
23 Chrome-molybdenum engine bearers
24 Supercharger air intake
25 Undercarriage position indicator

26 Hydraulically-operated port mainwheel leg
27 Wheel cover (fairing for exposed portion of wheel when retracted)
28 Port landing light
29 Wooden structure (compartmented for emergency flotation)

30 Port navigation light
31 Light alloy port aileron structure

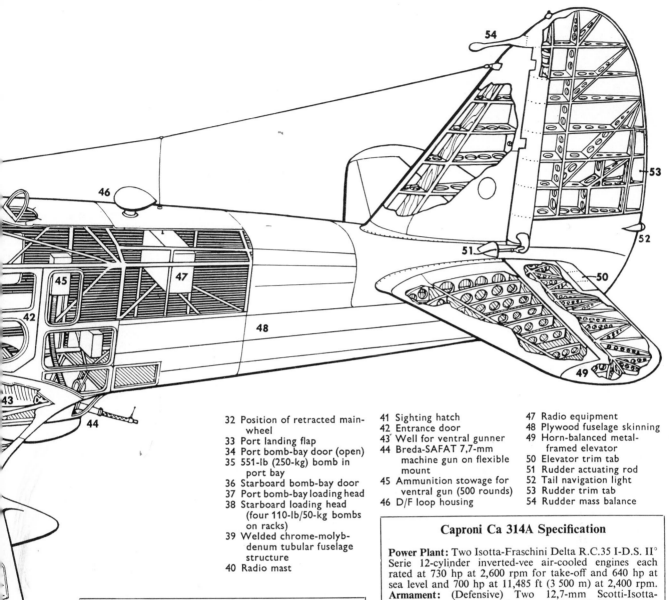

32 Position of retracted main-wheel
33 Port landing flap
34 Port bomb-bay door (open)
35 551-lb (250-kg) bomb in port bay
36 Starboard bomb-bay door
37 Port bomb-bay loading head
38 Starboard loading head (four 110-lb/50-kg bombs on racks)
39 Welded chrome-molybdenum tubular fuselage structure
40 Radio mast

41 Sighting hatch
42 Entrance door
43 Well for ventral gunner
44 Breda-SAFAT 7,7-mm machine gun on flexible mount
45 Ammunition stowage for ventral gun (500 rounds)
46 D/F loop housing

47 Radio equipment
48 Plywood fuselage skinning
49 Horn-balanced metal-framed elevator
50 Elevator trim tab
51 Rudder actuating rod
52 Tail navigation light
53 Rudder trim tab
54 Rudder mass balance

Caproni Ca 313 R.P.B.2 Specification

Power Plant: Two Isotta-Fraschini Delta R.C.35 I-D.S. II° Series 12-cylinder inverted-vee air-cooled engines each rated at 730 hp at 2,600 rpm for take-off and 640 hp at sea level and 700 hp at 11,485 ft (3 500 m) at 2,400 rpm.
Armament: (Defensive) Three 12,7-mm Scotti-Isotta-Fraschini machine guns — one in port wing root with 320 rounds, one in Caproni-Lanciani Delta E dorsal turret with 350 rounds, and one on flexible mount in ventral well with 500 rounds. (Offensive) Four 220-lb (100-kg), eight 110-lb (50-kg), 16 33-lb (15-kg), or 32 26·5-lb (12-kg) bombs.
Weights: Empty equipped, 9,480 lb (4 300 kg); normal useful load, 3,527 lb (1 600 kg); maximum loaded weight, 13,007 lb (5 900 kg).
Performance: Maximum speed, 220 mph (355 km/h) at sea level, 271 mph (436 km/h) at 11,485 ft (3 500 m); max cruise, 217 mph (350 km/h) at 70 per cent power at 11,485 ft (3 500 m); normal range, 497 mls (800 km); max range (with 59 Imp gal/270 l auxiliary tank) 746 mls (1 200 km); climb to 13,125 ft (4 000 m), 11 min 30 sec; service ceiling, 23,950 ft (7 300 m); absolute ceiling, 26,250 ft (8 000 m); take-off distance, 383 yds (350 m); landing distance, 415 yds (380 m).
Dimensions: Span, 54 ft 7½ in (16,65 m); length, 38 ft 8½ in (11,80 m); height, 12 ft 1⅓ in (3,70 m); wing area, 421·94 sq ft (39,20 m²).

Caproni Ca 314A Specification

Power Plant: Two Isotta-Fraschini Delta R.C.35 I-D.S. II° Serie 12-cylinder inverted-vee air-cooled engines each rated at 730 hp at 2,600 rpm for take-off and 640 hp at sea level and 700 hp at 11,485 ft (3 500 m) at 2,400 rpm.
Armament: (Defensive) Two 12,7-mm Scotti-Isotta-Fraschini machine guns in wing roots with 300 rpg, one 12,7-mm Scotti-Isotta-Fraschini machine gun in Caproni-Lanciana Delta E dorsal turret with 350 rounds (plus 300-round reserve), and one 7,7-mm Breda-SAFAT machine gun with 500 rounds in ventral well. (Offensive) Alternative internal loads of three 220-lb (100-kg), six 110-lb (50-kg), 16 33-lb (15-kg), or 32 26·5-lb (12-kg) bombs, or six containers for total of 252 4·4-lb (2-kg) fragmentation bombs. Without internal load, two 220-lb (100-kg) or 353-lb (160-kg) bombs could be carried externally.
Weights: Empty equipped, 10,053 lb (4 560 kg); normal loaded, 13,580 lb (6 160 kg); max loaded, 14,590 lb (6 618 kg).
Performance: (At 14,590 lb/6 618 kg) Maximum speed, 199 mph (320 km/h) at sea level, 245 mph (395 km/h) at 13,125 ft (4 000 m); normal cruise, 199 mph (320 km/h) at 13,780–14,765 ft (4 200-4 500 m); range with 132 Imp gal (600 l) fuel, 454 mls (730 km), with 229 Imp gal (1 040 l), 690 mls (1 110 km), with 288 Imp gal (1 310 l), 1,050 mls (1 690 km); initial climb, 797 ft/min (4,05 m/sec); climb rate at 13,125 ft (4 000 m), 827 ft/min (4,20 m/sec); time to 3,280 ft, 4 min, to 6,560 ft, 8 min, to 9,840 ft, 12 min, to 13,120 ft, 16 min; service ceiling, 21,000 ft (6 400 m); absolute ceiling, 21,980 ft (6 700 m).
Dimensions: Span, 54 ft 7½ in (16,65 m); length, 38 ft 8½ in (11,80 m); height, 12 ft 1⅓ in (3,70 m); wing area, 421·94 sq ft (39,20 m²).

Remondino. This, the Ca 313 R.P.B./S (*Silurante,* or Torpedo), being followed by a further six conversions.

Apart from its nose section, the Ca 313 R.P.B.2 was identical to the R.P.B.1. Its fuselage comprised a jig-built welded steel-tube framework built in two sections and bolted together, with fabric skinning over a light fairing structure; the wing was entirely of wooden construction with two box spars, former ribs and plywood skinning, the portion between the wings being watertight and divided into compartments for emergency flotation, and the tail surfaces were primarily of metal construction with plywood and fabric covering. Accommodation was provided for three crew members, the first and second pilot being seated side-by-side, the latter also performing the functions of observer and bomb-aimer, and operating the dorsal turret, with the radio-operator seated aft and responsible for the ventral gun.

Defensive armament comprised three 12,7-mm Scotti-Isotta-Fraschini machine guns, one being mounted in the port wing root with 320 rounds, one being flexibly-mounted in the tail of a ventral well with 500 rounds, and the third being installed in a Caproni-Lanciani Delta E dorsal turret with 350 rounds plus a reserve of 300 rounds. The bomb load comprised four 220-lb (100-kg), eight 110-lb (50-kg), 16 33-lb (15-kg) or 32 26·5-lb (12-kg) bombs in vertical racks, and all fuel was housed by two self-sealing tanks between the mainspars inboard of the engines, total capacity being 229 Imp gal (1 040 l), provision being made for an auxiliary tank in the aft fuselage with a capacity of 59 Imp gal (270 l).

During the course of 1940, the *Luftwaffe,* too, had evinced interest in the Ca 313 as a potential crew trainer, and three prototypes were delivered in 1941 as the Ca 313G (the "G" indicating *Germania*). The Ca 313G was basically similar to the Ca 313 R.P.B.2 apart from equipment and the fuselage nose, and in 1943 the *Reichsluftfahrtministerium* ordered 905 trainers of this type, although only 16 of these were to be delivered from the Ponte San Pietro and Taliedo factories before, in 1944, production was brought to a halt.

The last of the line

The three prototypes for the Ca 314 were the first three production Ca 313 R.P.B.2s (M.M.12051, 12054-4), these eventually becoming Ca 314As intended primarily for the convoy escort rôle. The Ca 314A differed from the Ca 313

R.P.B.2. in having Delta R.C.35 I-D.S. II° *Serie* engines affording 730 hp for take-off and having normal maximum ratings of 640 hp at sea level and 700 hp at 11,485 ft (3 500 m); revised armament comprising the introduction of a 12,7-mm machine gun in the starboard as well as the port wing root, and the replacement of the ventral 12,7-mm gun by a 7,7-mm Breda-SAFAT machine gun; provision for two 220-lb (100-kg) or 353-lb (160-kg) bombs mounted exernally beneath the fuselage; 6-mm armour protection for the radiators, and improved protection for the crew, 6-mm and 8-mm armour being provided for all three members.

During 1942, the Ca 313 and Ca 314 began to see extensive service with the *Gruppi Osservazione Aerea,* the former re-equipping the 64° and 67° *Gruppi* in North Africa, and being operated alongside Ca 311s by the 73° *Gruppo* in Italy, and the latter re-equipping the 65° *Gruppo* in Sardinia, but deliveries were erratic, production frequently being delayed by shortages of raw materials and equipment. Supplies of spruce used for the wings were soon exhausted after Italy entered the war, and red fur from Cadore in the Dolomites had to be used as a substitute. This entailed a great deal more work as the trees were not of very large girth and the laminations of the plywood therefore needed much joining. Completed airframes had frequently to be held in storage pending availability of engines, and machine guns were in such short supply that a new aircraft had usually to be fitted with one set of guns for tests, this set then being removed and installed in the next aircraft off the line.

Apart from the ex-French contract Ca 313 R.P.B.1s, the *Regia Aeronautica* took delivery of 120 Ca 313 R.P.B.2s, all other aircraft being completed to Ca 314 standards. Owing to its higher weights, the Ca 314A had a somewhat lower performance than that of the Ca 313 R.P.B.2, and a reduced bomb load. During 1942-43, Caproni's Taliedo factory delivered 224 Ca 314 1° *Serie* aircraft (comprising 70 Ca 314As, 80 Ca 314Bs and 74 Ca 314Cs) and 60 2° *Serie* aircraft (all Ca 314Cs). Simultaneously, the Ponte San Pietro factory delivered 40 1° *Serie* aircraft (Ca 314Cs) and 20 2° *Serie* aircraft (Ca 314Cs), and the A.V.I.S. factory at Castellamare di Stabia delivered 40 1° *Serie* and 20 2° *Serie* aircraft (all Ca 314Cs), total deliveries to the *Regia Aeronautica* thus being 404 aircraft. All 70 Ca 314As and the first 27 Ca 314Bs built at Taliedo had II° *Serie* engines, but all subsequent aircraft had III° *Serie* engines which provided

The Ca 314C (illustrated by the general arrangement drawing below) was essentially an attack bomber version of the basic design with heavier fixed forward-firing armament.

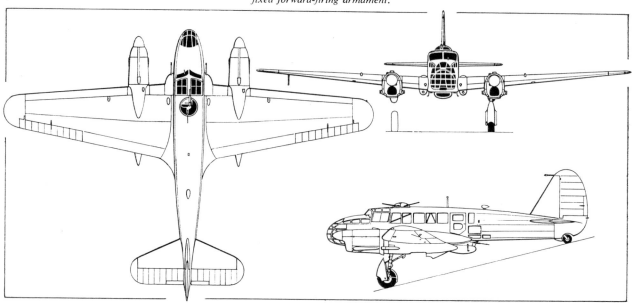

an additional 50 hp at all altitudes and increased the rated altitude from 11,485 ft (3 500 m) to 13,125 ft (4 000 m).

Whereas the Ca 314A was intended primarily for convoy escort and patrol tasks, the Ca 314B was produced for the torpedo-bombing rôle, carrying a single 1,984-lb (900-kg) S.I. (*Silurificio Italiano*) or W.I. (*Whitehead Italiano*) torpedo offset to starboard beneath the fuselage, alternative loads being one 1,102-lb (500-kg) or two 551-lb (250-kg) bombs. Empty and maximum loaded weights were increased to 10,240 lb (4 645 kg) and 15,432 lb (7 000 kg) respectively, fuel dumping facilities were provided, and 12,7-mm Breda-SAFAT machine guns could be installed in place of the normal Scotti-Isotta-Fraschinis. The Ca 314C was essentially an attack bomber, and fixed forward-firing armament was supplemented by an additional pair of Breda-SAFAT 12,7-mm machine guns with 300 rpg beneath the wing roots, internal offensive load comprising eight 353-lb (160-kg) or 220-lb (100-kg) bombs as alternatives to a single 1,102-lb (500-kg) bomb or two 551-lb (250-kg) or 353-lb (160-kg) bombs externally. An optional load consisted of 252 4·4-lb (2-kg) anti-personnel bombs.

Empty and loaded weights of the Ca 314C were 10,383 lb (4 730 kg) and 15,366 lb (6 970 kg) respectively, and performance with the IIIº *Serie* engines included maximum speeds of 211 mph (340 km/h) at sea level and 259 mph (417 km/h) at 11,485 ft (3 500 m), a normal range at 70 per cent power at 217 mph (350 km/h) at 14,765 ft (4 500 m) with an 882-lb (400-kg) bomb load of 485 miles (780 km), and a maximum range with auxiliary fuel tank of 746 miles (1 200 km).

After Italy capitulated on 8 September 1943, the Ca 313 and Ca 314 soldiered on in service with both the Italian co-belligerent air force and the air arm of the newly-proclaimed *Repubblica Sociale Italiana*, the *Aviazione della*

Some Ca 314s flew on into the postwar years with the Aeronautica Militare *in the liaison and training rôles, such as this example with aft fuselage deepened and transparencies reduced, photographed at Centocelli, Rome, during the summer of 1946.*

RSI, which continued to fight alongside the *Luftwaffe*. Production of the Ca 313G for the *Luftwaffe* was continued in Taliedo and Ponte San Pietro, but the aircraft took 23,000 man-hours to build and deliveries were little more than a trickle. The end of hostilities was not the end of the career of the Caproni twin, however, for the surviving Ca 313s and Ca 314s remained in the inventory of Italy's post-war *Aeronautica Militare* for a number of years and, during 1946–7, the Italian Air Ministry evinced interest in a blind-flying trainer version of the Ca 313, the Ca 313 A.V.S. A decision to reinstate the aircraft in production was postponed until, early in 1949, negotiations for the purchase of 20 aircraft of this type were initiated. In the event, these negotiations came to nought, and the career of the Italian warplane that so nearly saw service with the RAF finally came to an end. □

(The editors wish to acknowledge their indebtedness to Dr Rosario Abate for assistance in the compilation of this feature.)

RAIDEN—————————————— *from page 73*

rudder trim is required. The elevator tab is adequate and very effective, but it works in reverse of Allied procedure (ie, rolling the wheel backwards makes the aircraft nose heavy). The aircraft is dynamically and statically stable longitudinally and directionally, and neutrally stable laterally. In short, stability is excellent. The uneven use of fuel from either port or starboard wing tank quickly results in the aircraft becoming markedly wing heavy.

"The aircraft was stalled clean and dirty, and stalling characteristics are excellent except for lack of stall warning. The nose drops gently, either straight ahead or on either wing, and recovery is very rapid with very little loss of altitude. The oil and engine cowl flaps have no noticeable effect on stalling speed, and there is no tendency to spin. Rolls, Immelmans and turns are executed with ease at normal speeds, although ailerons are heavy at all operating speeds, and the aircraft cannot be rolled as rapidly as a P-51. Manoeuvre flaps of Fowler type are fitted, and are controlled by a safety switch and a trigger on the stick. These are extended only when the trigger is depressed and retract immediately the trigger is released, and their operation is superior to any used on our aircraft. The elevators are too light at normal and high speeds, and it is felt that the aircraft may be easily damaged by rough handling of the elevators.

"The engine was rough at cruising rpm in automatic mixture setting. This roughness was reduced as soon as the mixture was leaned out, but this caused the exhaust temperature to go above the limits. Vibration is not excessive, but the canopy fitted on this particular aircraft vibrates and makes considerable noise, and individual exhaust stacks also emit some objectionable noise. For a normal-sized pilot the aircraft is comfortable. There is ample headroom and body-room, and the cockpit enclosure is wide, permitting freedom

of movement of the head, and this improves vision. The ventilation system, which comprises forward and aft ventilators, is superior to any fitted to our fighters. The rigging of stick and rudder is satisfactory, and all controls are readily accessible. The cockpit layout is, in general, very satisfactory, and the engine and flight instruments are well grouped, although the airspeed indicator is too far from the rev counter and manifold pressure gauge (this is especially noticeable during the take-off run). Intermittent noise is the only real objection to comfort. Ground observers state that the noise emitted by the cooling fan is very noticeable, but this is not heard in the cockpit. Vision in climb, level flight and for landing is good, although rather poor for take-off until the tail comes up. Aft vision is good if the rear transparent panels are kept clean, but the metal framing of the windscreen tends to obstruct forward vision.

"The power plant is generally satisfactory. It is very easy to start, hot or cold, but, as already mentioned, runs roughly at cruising revs. Airscrew operation is satisfactory at normal revs but hunts at higher revs at about 10,000 ft (3 050 m), although it should be noted that airscrew control on this particular aircraft has been changed from hydraulic to electric so is non-standard. Oil and engine cooling are exceptionally good, and gave the impression that they would over-cool in cold weather. It was noted that there is considerable vibration for a short period when the engine is put into high blower at 12,000 ft (3 660 m) pressure altitude. In conclusion, it may be said that the favourable features of this fighter are: (1) Good stability; (2) Good stalling characteristics; (3) Comfort; (4) Good take-off and landing qualities; (5) Good performance; (6) Manoeuvre flaps. Its poor features are: (1) Brakes and rudder brake action; (2) Heavy ailerons and lack of manoeuvrability at high speed; (3) Low mechanical reliability; (4) Short range." □

Twenty-seven years ago, on the afternoon of 29 July 1944, a chance shell from a Japanese anti-aircraft gun exploding high above the industrial town of Anshan, Manchuria, started a chain of events that, indirectly, was to cost the North American tax-payer billions of dollars. This shell was to play a part in enabling the Soviet aircraft industry to take an immense leap forward in technology; the havoc it wrought on an example of the then new and revolutionary Boeing B-29 Superfortress — probably the most sophisticated and complex piece of movable machinery created to that time — forcing this warplane to seek sanctuary on Soviet soil. The damaged B-29, and others destined to follow it into Soviet air space, were to be the unwitting patterns for the Tupolev Tu-4; centrepiece of a fantastic crash programme aimed at raising Soviet combat air-craft manufacturing technology to that achieved in the USA — a programme embracing production tech-niques, materials, avionics, and pressurisation and remote-control armament systems.

The effects of this programme were destined to permeate throughout the entire Soviet aircraft industry. Around the bomber was to be created the second most powerful strategic air arm in the world, and from the Tu-4, by genealogical processes that were to take place over the next decade, was to be evolved the Tu-20 — the only Soviet bomber capable of performing strikes across the polar regions, and the sole factor justifying the billions of dollars spent on the immense chains of radar installations and interceptor bases intended to shield the North American continent from manned aircraft attack. Today, there can be no doubt that the seed from which the Tu-20 sprang was planted on Boeing's drawing boards at Seattle, Washington. The course of world events precluded any attempt on the part of this Soviet-built bomber of Boeing ancestry to strike the targets that it had been specifically built to attack. Its success took another form: it afforded the Soviet Union a major economic victory — a few hundred aircraft demanding a totally disproportionate counter-effort on the part of the USA and Canada.

North America paid dearly for the few B-29s that fell into Soviet hands, and, for the first time the AIR ENTHUSIAST tells the story of the line of strategic bomb-ers that evolved from the B-29s whose crews, perhaps unwisely but with little option, placed their trust in a country with which they believed themselves to be allied.

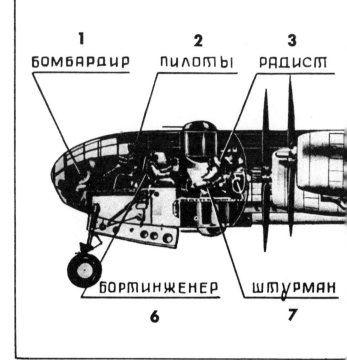

THE BILLION DOLLAR BOMBER

PART ONE

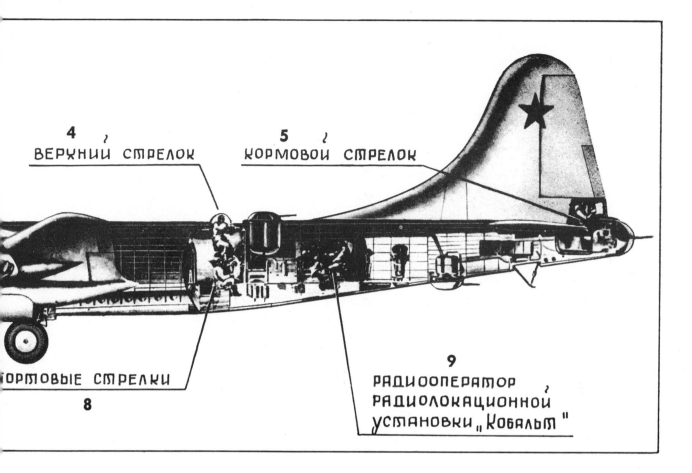

4 ВЕРХНИЙ СТРЕЛОК

5 КОРМОВОЙ СТРЕЛОК

9 РАДИООПЕРАТОР РАДИОЛОКАЦИОННОЙ УСТАНОВКИ „КОБАЛЬТ"

ОРМОВЫЕ СТРЕЛКИ
8

An official Soviet drawing of the Tupolev Tu-4 strategic bomber. The features annotated are: (1) Bombardir *(Bombardier); (2)* Piloty *(Pilot); (3)* Radist *(Radio Operator); (4)* Verkhniy strelok *(Dorsal gunner); (5)* Kormovoy strelok *(Tail gunner); (6)* Bortinzhener *(Flight engineer); (7)* Shturman *(Navigator); (8)* Bortovye strelki *(Beam gunners); (9)* Radiooperator radio-lokatsionnoy ustanovki 'Kobalt' *(Operator of 'Cobalt' radar equipment).*

Captain Howard R Jarrell and his crew completed their bombing run, closed the bomb doors of the B-29, and began a half-turn that would bring them on course for their advanced operational base of Kiunglai, in the vicinity of Chengtu, China, more than 1,500 miles (2 400 km) away. Early that morning, Capt Jarrell's B-29, together with other Superfortresses of the 771st Squadron of the 462nd Bombardment Group, normally based at Piardoba, India, had taken-off from Kiunglai, one of the complex of fields constructed around Chengtu specifically for use as advanced bases by the XX Bomber Command of the USAAF's 20th Air Force. The target was the Showa Steel Works at Anshan, Manchuria.

As a target Anshan was no sinecure. Apart from numerous mechanical teething troubles that the newly-delivered B-29s had still to overcome, and considerable distances involved in the mission, Anshan was heavily defended by anti-aircraft batteries and fighters. The first strike against the Showa Steel Works three weeks earlier — another high-altitude daylight precision attack — had hardly been an unqualified success. Of the 72 B-29s launched against Anshan only 60 had reached their target (one having crashed on take-off and the other 11 having suffered mechanical failures *en route* to Manchuria), and these had been greeted by heavy flak and fighters. Bombing conditions had been ideal, and the B-29s had begun their bomb runs at 25,000 ft (7 620 m) but, unfortunately, the bombs from the first wave of Superfortresses had produced a pall of smoke that had completely obscured the aiming points for the bombers following.

Capt Jarrell had just completed his half-turn when the B-29 was rocked by a tremendous explosion. An anti-aircraft shell had burst immediately beneath the starboard wing! Number three engine began to belch dense smoke and had to be cut out and its airscrew feathered, and number four engine was losing oil. The crippled bomber was

obviously incapable of the long flight back to Kiunglai, most of it over enemy-occupied territory. Jarrell had been told that the emergency procedure under such circumstances was to head for the Soviet Union which, at its nearest point, was some 450 miles (725 km) east-nor-east. The navigator gave Jarrell a new heading for Vladivostok, and, gingerly skirting the defensive perimeter of Anshan, the B-29 was carefully nursed on course for the Soviet Union.

Thus, in the early evening of 29 July 1944, B-29-5-BW Serial Number 42-6256, the 52nd production Superfortress to have rolled off Boeing's Wichita assembly line, was reported as an unidentified aircraft entering Soviet air space from the direction of Hunchun. Yak-9 fighters hurriedly scrambled from Mayke and Spassk Dalni were soon buzzing around the obviously crippled bomber like so many angry hornets, flying dangerously close and firing short bursts across its nose as the B-29 slowly circled Vladivostok Bay. Capt Jarrell and his crew had hardly expected so unfriendly a reception over a country which, if not at war with Japan, was at least allied in the war against Germany!

Eventually, one of the fighters drew alongside the bomber, and its pilot, by means of signs, succeeded in making Jarrell understand that he should land at a small airfield near Tavrichanka. Despite the inadequate dimensions of the field for an aircraft of the size of the B-29, Jarrell had no option but to obey, and with number four engine likely to die at any moment he could not risk an overshoot. Bringing the bomber in as low and as slow as possible, with full flap and

These *Tupolev Tu-4s* (220002 top and 2805103 immediately above) could, but for their national insignia, have been products of the WW II Wichita and Renton assembly lines rather than late 'forties progeny of factories on the far side of the Urals, the Boeing B-29 Superfortress and the Tu-4 being externally indistinguishable. Whereas the B-29 Superfortress normally mounted twin 0·5-in (12,7-mm) machine guns and a single 20-mm cannon in the tail, the Tu-4 replaced this combination with a pair of 23-mm NR-23 cannon, a similar pair of weapons being mounted in each of the four remotely-controlled barbettes. The photo below shows the tail position of the Tu-4, but the NR-23 cannon have been removed.

engines throttled right back, he touched down just within the boundary of the field, applied full brakes and, with smoking tyres, came to a standstill within yards of the sea.

In the event, Capt Jarrell was not to be faced with the problem of flying his bomber *out* of the postage stamp-sized airfield. Indeed, met by a heavily-armed group of grim-faced Russians as he clambered from the B-29, Capt Jarrell and his crew were never to see their aircraft again! Unbeknown to the 11 members of the B-29's crew, they had just delivered to the Soviet Union a "prototype" of the Soviet Air Force's first *true* strategic bomber. The reception group proved by no means friendly, hustling Jarrell and his crew into a waiting lorry which took them to the V-VS Headquarters in Vladivostok for interrogation. The puzzled USAAF group were being treated as prisoners. They had had some vague idea that the Russians would perform emergency repairs on their aircraft and then allow them to proceed on their way, but instead they had been ordered to surrender their sidearms and knives, and were soon to learn that the Soviet Union was sequestering their bomber and that they themselves were to be interned in a camp at Tashkent, in Turkestan.

Three weeks after arrival in the Soviet Union of Capt Jarrell and his B-29, on 20 August 1944, 61 B-29s of the XX Bomber Command attacked Yawata on the home island of Kyushu in a high-altitude daylight raid. In the mêlée that ensued with the arrival of the bomber force over Yawata,

one B-29 suffered a direct hit by flak and three were destroyed by Japanese fighters, one of them by ramming. Eight other B-29s were damaged by flak, and besides the four lost over the target no fewer than 10 failed to regain their bases in the Chengtu area. One of these 10 aircraft — actually the sixth Renton-built B-29A-1-BN (42-93829), piloted by Major Richard M McGlinn and belonging to the 395th Squadron of the 40th Bombardment Group, normally based at Chakulia, India, but operating for the Yawata mission from the advanced base at Hsinching — arrived in the Soviet Union. It was to prove of little use to the Soviet aircraft industry in the programme for which plans were already being formulated, however, as it crashed in the foothills of the Sikhote Alin Range, east of Khabarovsk, after its crew had bailed out, eventually finding themselves in the same Tashkent camp accommodating Capt Jarrell and his crew.

Nearly three months were to elapse before the Russians were to add to their inventory of intact B-29s, and then they were presented with two in quick succession. Both were B-29-15-BWs from the same squadron; both were returning from missions against Omura; one had followed the other from the Wichita assembly line by 10 days, and to carry coincidence still further, they arrived in the Soviet Union within 10 days of each other, though in the reverse order to that in which they had left the assembly line.

On the night of 10–11 November 1944, a force of 96 B-29s took-off from the Chengtu complex of bases to attack the aero engine and airframe manufacturing plants at Omura, on Kyushu. Two days earlier, a typhoon had been detected south of Japan, but the XX Bomber Command weather forecasters had predicted that it would move away from the proposed target in a south-easterly direction. Plans for the attack on Omura therefore went ahead and the B-29s were airborne by the time, somewhat belatedly, that the weathermen decided the typhoon *would* affect the target area. Sixty-seven of the bombers received instructions by radio to change course and attack Nanking, but the remaining 29 failed to receive the message and pressed on to Omura, bucking strong headwinds and, as they neared the target, encountering tremendous turbulence.

One of the bombers that reached Omura was the "General H H Arnold Special" (42-6365), which, piloted by Capt Weston H Price, belonged to the 794th Squadron of the 468th Bombardment Group operating from Penshan but normally based at Karagpur, India. The "Special" was on its 11th bombing mission, and after attacking Omura, insufficient fuel remained for it to regain Penshan. Capt Price was faced with several alternatives: he could head towards the newly-constructed B-29 bases on Saipan, hopefully calling air-sea rescue *en route* and ditching in mid-ocean; he could fly towards occupied China and then bail out, gambling on being found by friendly Chinese who would guide him and his crew to safety, or he could fly to Soviet territory. He opted for the last of the alternatives, and instead of heading south-west across the China Sea, turned the "Special" north towards Vladivostok.

As had Jarrell nearly four months earlier, Price arrived over Vladivostok Bay transmitting the international radio signal for "friend", but immediately anti-aircraft shells began to explode around the B-29. The crew of the bomber then saw fighters rising towards them, and, as the barrage ceased, these took up positions to port, starboard and ahead of the B-29. At least Price was permitted to land on the main airfield on the outskirts of Vladivostok, but from that point on — interrogation and the camp at Taskhent.

Ten days later, on 21 November, the Soviet Union obtained its last B-29 — at least, the last B-29 from an American *assembly* line! This, the "Ding How" (42-6358) from the 794th Squadron of the 468th Bombardment Group, had taken-off from Pengshan for an attack on Omura as part of a force of 109 B-29s, of which one had crashed on take-off. Omura was largely obscured by cloud, and most of the B-29s bombed by radar, Japanese fighters reacting strongly. Eight of the USAAF bombers failed to return to their Chinese bases, including "Ding How" piloted by Lieut William J Mickish who, knowing that he could not make Pengshan, had elected to head for Soviet territory.

Mickish's B-29 arrived over Vladivostok at an altitude of some 5,000 ft (1 525 m) at around 15.00 hours, with the weather closing in fast. Lieut Mickish circled slowly, looking for an airfield on which to land, until the bombardier pointed out another B-29 on the ground. It was, in fact, the "General H H Arnold Special", and Mickish promptly decided to put his aircraft down on the same field. With the B-29 at 500 ft (150 m) with the undercarriage lowered, the waist gunner reported that they were being fired on by anti-aircraft batteries, but Mickish continued his approach even though six Yak-9s, which had meanwhile taken-off, had begun shooting at the bomber. Mickish subsequently recounted: "When we were forced down it was broad daylight. First the anti-aircraft batteries started sending flak at us, although they could see that we were an American ship and that I had my flaps down for a landing. Then six fighters came up and the leader began spraying tracer bullets around our nose. My boys didn't like it. I had a hard time getting them to keep their trigger fingers off their guns."

Mickish's crew was puzzled by the attitude of the Russians, as had been the B-29 crews that had preceded it to the Soviet Union. The USAAF personnel had no specific directives on the Soviet situation, but they had assumed that their reception would be friendly and that their aircraft would be refuelled and allowed to depart. They were soon to discover otherwise, and when the US consul, O Edmond Chubb, put in an appearance, he confirmed their growing suspicion that they were unlikely to see their B-29 again. The Soviet government's *ostensible* reason for impounding the USAAF bombers was its neutrality pact with Japan; the *true* reason, which was to become patently obvious some three years later, was the plan already formulated to manufacture a copy of this remarkably advanced bomber which made all existing V-VS bombing aircraft appear relics of a bygone era!

The crew of "Ding How", provided with heavy clothing and assigned an interpreter, eventually found itself on a train, which, 24 hours later, deposited it in Khabarovsk. Here, the 11 men met up with Capt Price and his crew, and, on 6 December, the 22 USAAF personnel departed for the camp at Tashkent. In January, a Soviet-approved plan resulted in the occupants of the Tashkent camp being allowed to "escape" across the Iranian border to Teheran, but nothing further was heard of the B-29s. Strictly according to international law, the Soviet Union was within its rights to intern the bombers, but after declaring war on Japan it would have been a friendly gesture to have returned the B-29s to the United States; a gesture which the Soviet

Union would have been unable to make even had it deemed such to be necessary, for by this time the bombers had been broken down into sub-assemblies and components, and distributed throughout the Soviet aircraft industry.

The US government made no official comment on the sequestration of the B-29s by the Soviet Union, but WW II had been over little more than a year when, on 11 November 1946, the Berlin newspaper *Der Kurier,* quoting despatches from its Paris correspondent, declared that the Soviet Union was manufacturing a copy of the B-29 in factories in the Urals. This report was largely discounted as it was widely believed that the translation of so large and sophisticated an aeroplane back to its production breakdown was an engineering task of such magnitude that it was outside the capability of Soviet technology. However, when it became known that General Carl Spaatz, the USAF Chief of Staff, had disclosed in testimony before President Truman's Air Policy Commission that agents acting on behalf of the Soviet government had attempted during 1946 to purchase in the USA tyres, wheels and brake assemblies for the B-29, credence was given *Der Kurier*'s report.

Rumour and supposition was finally to take substance on 3 August 1947, during the Soviet Aviation Day parade over Tushino, Moscow. Three four-engined heavy bombers, the lead aircraft piloted by Chief Marshal of Aviation Golovanov, Commander-in-Chief of the Long-range Air Force (the *Dal'naya aviatsiya*), roared across Tushino at low level. Western observers had no difficulty in recognising them as B-29s! This, in itself, did not clinch the matter, for only *three* of the bombers participated in the parade over Tushino and these could conceivably have been the three B-29s acquired by the Soviet Union some three years earlier. What *did* settle all doubts was the appearance of a *fourth* aircraft which apparently differed from its fellows solely in having a redesigned fuselage for the transport rôle. The Soviet aircraft industry *was* producing a copy of the B-29, however fantastic this seemed; it had performed a feat unique in the annals of aviation.

The transport model was openly referred to as the Tu-70, an airliner accommodating 48–50 passengers and evolved for *Aeroflot* by a design collective led by the doyen of the Soviet aircraft industry, Andrei N Tupolev. No reference was made, needless to say, to the Boeing derivation of this aircraft, which had begun its test programme eight months earlier, on 27 November 1946. Even less information was forthcoming concerning its bomber counterpart. Western observers were totally unaware that the three bombers that flew over Tushino were, in fact, the first of no fewer than 20 pre-production aircraft that were being built for the test and development programme, and some time was to elapse before the appellation of Tu-4 that had been bestowed on the Soviet copy of Boeing's B-29 was to become common knowledge. □ WG

(To be continued next month)

The appearance of the Tupolev Tu-70 commercial transport derivative of the Boeing B-29 Superfortress over Tushino in 1947 provided proof positive that rumours of Soviet production of the American bomber were founded in fact.

FIGHTER A TO Z

AERFER SAGITTARIO 2   ITALY

Designed by Ing Sergio Stefanutti and built by Industrie Meccaniche Aeronautiche Meridionali AERFER, the Sagittario (Archer) 2 was an all-metal light fighter intended for clear-weather intercept and tactical support rôles. Powered by a 3,600 lb (1 633 kg) st Rolls-Royce Derwent 9, the sole prototype flew on 19 May 1956 and attained Mach 1·1 in a dive on 4 December 1956. Armament comprised two 30-mm Hispano-Suiza HDD-825 cannon. Max speed, 646 mph (1 040 km/h) at sea level and 634 mph (1 020 km/h) at 27,230 ft (8 300 m). Time to 39,370 ft (12 000 m), 10 min. Normal range, 475 mls (765 km). Empty weight, 5,070 lb (2 300 kg). Loaded weight, 7,275 lb (3 300 kg). Span, 24 ft 7¼ in (7,50 m). Length, 31 ft 2 in (9,50 m). Wing area, 156·08 sq ft (14,50 m²).

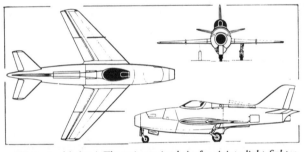

(Above) The sole example of the Aerfer Sagittario 2 fighter.

(Above and below) The twin-engined Aerfer Ariete light fighter.

AERFER ARIETE   ITALY

Evolved from the Sagittario 2, the Ariete (Ram) retained the wing of the earlier fighter, marrying this to a redesigned fuselage in which the 3,600 lb (1 633 kg) st Derwent 9 was supplemented by a 1,810 lb (820 kg) st Rolls-Royce Soar RSr 2 auxiliary turbojet to boost take-off, climb and combat performance, this engine drawing air through a retractable dorsal intake. The Ariete flew on 27 March 1958, and possessed an armament of two 30-mm HDD-825 cannon. Only one prototype was completed, and a progressive de-

(Above and below) Czechoslovakia's first fighter, the Ae 02.

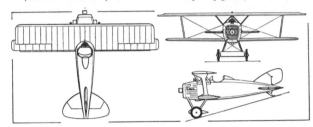

velopment, the mixed-power Leone (Lion), was abandoned. Max speed, 671 mph (1 080 km/h). Time to 39,370 ft (12 000 m), 4 min 20 sec. Empty weight, 5,291 lb (2 400 kg). Loaded weight, 7,793 lb (3 535 kg). Span, 24 ft 7¼ in (7,50 m). Length, 31 ft 5⅞ in (9,60 m). Wing area, 156·08 sq ft (14,50 m²).

AERO AE 02   CZECHOSLOVAKIA

The Ae 02, designed by A Husnik and A Vlasak, and built by the Aero Továrna Letadel, was the first fighter of Czechoslovak design. Built and flown in 1920, the Ae 02 was powered by a 220 hp Hispano-Suiza 8 Ba engine, the fuel tank for which was mounted between the undercarriage mainwheels, and armament comprised two Vickers machine guns. Only one prototype was built. Max speed,

The prototype Aero Ae 04 with initial (above) and definitive (below) engine cowling.

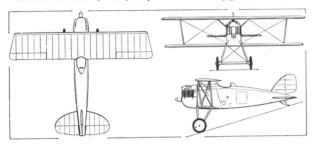

(Above and below) The first production Aero fighter, the A 18.

140 mph (225 km/h). Cruise, 118 mph (190 km/h). Time to 16,400 ft (5 000 m), 28 min 52 sec. Empty weight, 1,488 lb (675 kg). Loaded weight, 2,083 lb (945 kg). Span, 25 ft 3⅛ in (7,70 m). Length, 17 ft 10½ in (5,45 m). Wing area, 179·76 sq ft (16,7 m²).

AERO AE 04   CZECHOSLOVAKIA

Developed from the Ae 02, the Ae 04 was of similar mixed construction — steel-tube fuselage and wooden wings — and was powered by a 185 hp BMW IIIa six-cylinder liquid-cooled engine. Flown in 1921, it established a national altitude record of 20,869 ft (6 361 m) in that year. Armament consisted of two Vickers guns. As initially flown, the Ae 04 had a frontal radiator, and this and the definitive engine cowling are illustrated by the accompanying photographs. Only one prototype was built. Max speed, 140 mph (225 km/h). Cruise, 115 mph (185 km/h). Time to 16,400 ft (5 000 m), 14 min. Empty weight, 1,477 lb (670 kg). Loaded weight, 1,984 lb (670 kg). Span, 25 ft 3⅛ in (7,70 m). Length 18 ft 4½ in (5,60 m). Wing area, 157·15 sq ft (14,6 m²).

AERO A 18   CZECHOSLOVAKIA

Powered by a Walter-built BMW IIIa of 185 hp, the A 18 was flown as a prototype in March 1923, and 20 were built for the Czechoslovak Air Force. Armed with two synchronised Vickers machine guns, the A 18 was of mixed construction. The clipped-wing A 18B and the aerodynamically-refined, more powerful A 18C were one-off models which

(Below) The Aero A 20 was developed in parallel with the A 18.

participated in the national air races. Max speed, 142 mph (229 km/h) at 8,200 ft (2 500 m). Cruise, 121 mph (195 km/h). Time to 16,400 ft (5 000 m), 8 min 30 sec. Empty weight, 1,404 lb (637 kg). Loaded weight, 1,900 lb (862 kg). Span, 24 ft 11¼ in (7,60 m). Length, 19 ft 4⅓ in (5,9 m). Wing area, 171·15 sq ft (15,9 m²).

AERO A 20   CZECHOSLOVAKIA

Developed in parallel with the A 18, the A 20 was flown in 1923 but did not progress further than prototype status. Powered by a 300 hp Skoda-built Hispano-Suiza 8 Fb liquid-cooled engine, the A 20 carried two synchronised Vickers machine guns. Max speed, 140 mph (225 km/h). Cruise, 118 mph (190 km/h). Time to 16,400 ft (5 000 m), 14 min 10 sec. Empty weight, 1,728 lb (784 kg). Loaded weight, 2,381 lb (1 080 kg). Span, 31 ft 9⅞ in (9,70 m). Length, 21 ft 7¾ in (6,60 m). Wing area, 250·8 sq ft (23,3 m²).

(Above and below) The last pre-WW II Aero fighter, the A 102.

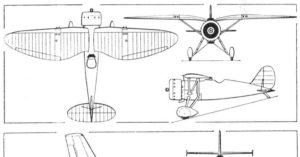

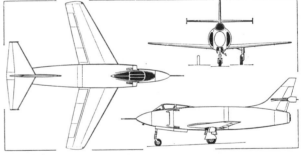

(Above and below) The NC 1080 with definitive tail.

AERO A 102   CZECHOSLOVAKIA

Flown in July 1934, the A 102 was powered by a 900 hp Walter-built Gnôme-Rhône Mistral Major 14 Kfs radial, and carried an armament of four 7,7-mm Model 30 machine guns in the wings. All-metal construction with fabric skinning were employed, and two prototypes were built, but landing characteristics were poor, and after suffering two accidents, the A 102 was abandoned. Max speed, 269 mph (434 km/h). Cruise, 221 mph (356 km/h). Time to 16,400 ft

(5 000 m), 5 min 46 sec. Empty weight, 3,258 lb (1 478 kg). Loaded weight, 4,488 lb (2 036 kg). Span, 37 ft 8¾ in (11,50 m). Length, 23 ft 11⅜ in (7,30 m). Wing area, 199·13 sq ft (18,5 m²).

AERÓCENTRE NC 1080   FRANCE
The NC 1080 was designed by M Pillon and built by the SNCA du Centre as part of an *Aéronavale* single-seat shipboard interceptor programme (competing designs being the Arsenal VG 90 and the Nord 2200). Powered by a 5,000 lb (2 268 kg) st Rolls-Royce Nene and intended to carry an armament of three 20-mm cannon, the sole prototype was flown on 29 July 1949, and underwent a number of changes to its tail surfaces before being destroyed in a crash early in 1950. Max speed, 608 mph (978 km/h) at 16,400 ft (5 000 m), 578 mph (930 km/h) at 29,530 ft (9 000 m). Range, 982 mls (1 580 km) at 522 mph (840 km/h). Initial climb, 5,512 ft/min (28 m/sec). Empty weight, 10,661 lb (4 836 kg). Loaded weight, 17,196 lb (7 800 kg). Span, 36 ft 0⅞ in (11,00 m). Length, 42 ft 2⅝ in (12,87 m). Wing area, 290·63 sq ft (27 m²).

AEROMARINE PG-1   USA
Designed by Isaac M Laddon for the Aeromarine Corporation, the PG-1 single-seat sesquiplane was evolved primarily for destroying armoured attack planes and for ground strafing, and three prototypes were built in 1923. Powered by a 330 hp Wright K-2 liquid-cooled engine, the PG-1 had a 37-mm engine-mounted cannon, a single 0·5-in (12,7-mm) machine gun, and ¼-in (6,3-mm) armour protection for the cockpit. Visibility from the cockpit was extremely bad, and development of the PG-1 was abandoned. Max speed, 130 mph (209 km/h). Time to 6,500 ft (2 130 m), 9 min 30 sec. Range, 195 mls (314 km) at max speed. Empty weight, 3,030 lb (1 374 kg). Loaded weight, 3,918 lb (1 777 kg). Span, 40 ft 0 in (12,19 m). Length, 24 ft 6 in (7,47 m). Wing area, 389 sq ft (36,13 m²).

(Above) The Aeromarine PG-1 armoured attack aircraft destroyer.

AERONAUTICA UMBRA T.18   ITALY
The T.18 single-seater was designed by Felice Trojani and built by Aeronautica Umbra SA at Foligno. Of all-metal stressed-skin construction, it was first flown on 22 April 1939 with a 1,030 hp Fiat A.80 RC 41 18-cylinder air-cooled radial. Armament comprised two 12,7-mm machine guns. Initial flights were performed with an NACA cowling, but in April 1940 a cowling of the type illustrated was introduced. Only one prototype was built. Max speed, 335 mph (540 km/h). Cruise, 278 mph (447 km/h). Range, 497 mls (800 km). Empty weight, 5,115 lb (2 320 kg). Loaded weight, 6,545 lb (2 975 kg). Span, 37 ft 8¾ in (11,50 m). Length, 28 ft 8¾ m (8,76 m). Wing area, 204·4 sq ft (19 m²).

AICHI TYPE H CARRIER FIGHTER   JAPAN
The designation Type H Carrier Fighter was applied to two examples of the Heinkel HD 23 single-seat shipboard fighter biplane designed and built by the Ernst Heinkel Flugzeugwerke at the request of the Aichi Tokei Denki KK and delivered to Japan in 1927. These prototypes were modified by Tetsuo Miki of Aichi to provide them with some flotation

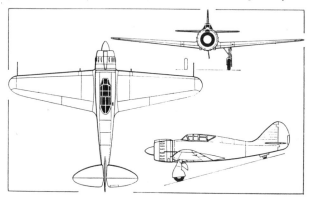

(Above and below) The Aeronautica Umbra T.18 in definitive form.

capability, but no production ensued (see Heinkel HD 23).

AIRCO D.H.2   GREAT BRITAIN
Designed by Geoffrey de Havilland of the Aircraft Manufacturing Company (Airco), the D.H.2 single-seat fighter was produced early in 1915, and the prototype was sent to France, falling into the hands of the Germans substantially intact on 15 August 1915. The standard engine was the 100 hp Gnome Monosoupape rotary, and armament comprised a free-mounted or fixed 0·303-in (7,7-mm) Lewis machine gun. Fabric-covered wooden construction was combined with tubular-steel tailbooms, and 400 were delivered. Max speed, 93 mph (150 km/h) at sea level, 77 mph (124 km/h) at 10,000 ft (3 050 m). Time to 5,000 ft (1 525 m), 8 min 25 sec. Empty weight, 943 lb (428 kg). Loaded weight, 1,441 lb (653 kg). Span, 28 ft 3 in (8,61 m). Length, 25 ft 2½ in (7,68 m). Wing area, 249 sq ft (23,13 m²).

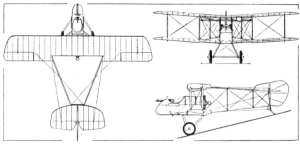

(Below) The prototype and (above) the production Airco D.H.2.

PLANE FACTS

This column is intended to provide an information service to readers requiring information that is not found in readily-available reference books. We would ask readers to confine their requests to ONE aviation subject or aircraft and we will answer as many of these as space permits. The editors reserve the right to use their discretion in selecting for reply in this column those requests which in their opinion will be of the widest general interest and which cannot be easily dealt with by a visit to the nearest local library.

Readers are asked to note that the editors cannot reply by post to requests for information, nor supply copies of illustrations appearing in the column. Until a flow of readers' requests has been established, items of particular interest have been drawn from our files for inclusion on this page. As sufficient requests for information are received, *Plane Facts* will be expanded to two or more pages each month.

The Yugoslav "Mosquito"

IN 1937, the Yugoslav designer Sima Milutinović initiated the design of a two-seat light bomber of wooden construction intended to carry an 880-lb (400-kg) bomb load over a distance of some 800 miles (1 300 km). It was to be equally suitable for the reconnaissance rôle, and, in some respects, was similar in concept to the D.H.98 Mosquito. It was proposed that the initial version should be powered by two 500 hp Walter Sagitta I-SR 12-cylinder air-cooled inline engines, and the Rogožarski concern began prototype construction in 1938.

Designated R-313, the prototype was completed late in 1939 and delivered early in the following year to a test unit of the Royal Yugoslav Air Force for evaluation. Apart from minor modifications dictated to improve longitudinal stability, the R-313 demanded few changes as a result of the official trials, and was considered an excellent aircraft from every aspect. During tests the R-313 attained a maximum speed of 234 mph (376 km/h) at sea level, 245 mph (394 km/h) at 3,280 ft (1 000 m), 258 mph (415 km/h) at 6,560 ft (2 000 m), 271 mph (437 km/h) at 9,840 ft (3 000 m), and 286 mph (460 km/h) at 13,120 ft (4 000 m). It could take-off fully loaded within 295

The only known photograph of the Vultee V-11-GB, alias PS-43, in Aeroflot service.

yards (270 m), climb to 3,280 ft (1 000 m) in 2 min, to 6,560 ft (2 000 m) in 3 min 54 sec, and to 9,840 ft (3 000 m) in 5 min 41 sec, and land within 837 yards (765 m) without application of brakes.

Armament comprised one fixed forward-firing 20-mm Hispano-Suiza cannon in the nose and a flexibly-mounted aft-firing 8-mm FN machine gun, and the R-313 could accommodate 882 lb (400 kg) of bombs internally. Overall dimensions included a span of 42 ft 7⅞ in (13,00 m), a length of 36 ft 1 in (11,00 m), a height of 8 ft 9½ in (2,68 m), and a wing area of 284·17 sq ft (26,4 m2). For the bombing rôle the empty equipped and normal loaded weights were

6,504 lb (2 950 kg) and 9,414 lb (4 270 kg) respectively.

For the structure of the fuselage a special sandwich material was evolved, comprising two plywood sheets with a limewood filling. This was covered by fabric to which a thick layer of lacquer was applied to achieve an exceptionally smooth finish. The wing was built in one piece, housed the fuel tanks, which had a total capacity of 150 Imp gal (680 l), and carried split flaps; the retractable main undercarriage members were developed in co-operation with Messier, and the Sagitta engines drove Ratier electrically-operated variable-pitch airscrews.

In view of its low-powered engines, the performance of the R-313 was considered outstanding, and consideration was given to the possibility of installing either Rolls-Royce Merlins or Daimler-Benz DB 601s but no definitive decision had been taken at the time of the German invasion. The prototype was used by an operational unit during the early months of 1941, and, during the German attack, was flown to Mostar. It was then decided to evacuate the prototype to Greece but, overloaded and flown by an inexperienced crew, its undercarriage collapsed during take-off from Nikšić airfield and the aircraft was abandoned.

Soviet Vultee V-11-GB

SOME MYSTERY has always surrounded the Soviet purchase in 1936 of a manufacturing licence for the Vultee V-11-GB attack bomber which, for a number of years, was believed to be in service with the Soviet Air Forces. The licence had, in fact, been acquired in order to provide the Soviet aircraft industry with experience in modern structural techniques, and during 1937–8, a total of 31 was built as the BSh-1 (*Bronirovanny Shturmovik,* or armoured assault aircraft). The BSh-1 was powered by a 750-hp M-62IR radial, but performance proved inadequate and the type was transferred to Aeroflot which used these aircraft as mail-carriers over the Moscow-Kiev-Tashkent route as the PS-43. During WW II the PS-43 served in the liaison rôle.

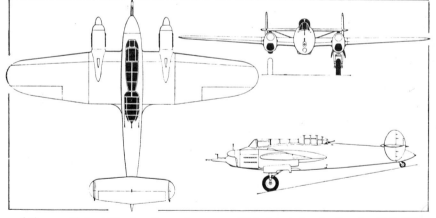

(Above and below) The first illustrations ever published depicting the Rogožarski R-313.

VETERAN & VINTAGE

The Royal Navy, Too

ALTHOUGH the Royal Navy does not boast so large a collection of historic aircraft as the Royal Air Force (see this column last month), the Fleet Air Arm Museum ranks as one of the most important of the air Museums in Britain. Located at the RNAS Yeovilton, Somerset, the Museum includes a dozen full-size aircraft, most of which relate to the World War II period or later. Special interest therefore attaches to the Sopwith Baby which was added to the collection last year as the first World War I aircraft represented.

This exhibit is also the only example of the Baby known to be extant. It was restored in three years of work at the RN Aircraft Yard Fleetlands. Painted as N2078 "The Jabberwock", which was a Blackburn-built Baby, the aircraft uses parts of two Babies which were loaned to Italy during the War and had been stored by the RAF after their return.

Various components for the restored aircraft had to be manufactured at Fleetlands, using working drawings prepared from the original plans. The propeller was found in Cyprus — with a clock in the boss! The Clerget engine has been rebuilt as far as possible, but still lacks pistons.

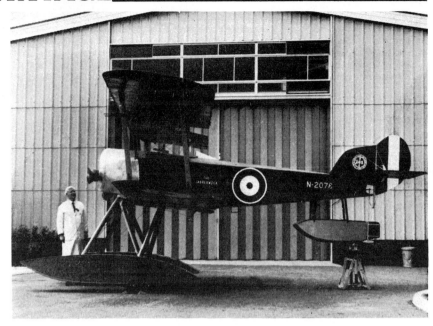

The restored Sopwith Baby photographed at Fleetlands before being transferred to the FAA Museum at Yeovilton. Standing alongside is Captain K R Hickson, a former commandant of the ETPS and CO of the test flying squadron at RAE Farnborough, who was in charge of the restoration as superintendent of the RN Aircraft Yard, Fleetlands.

Long-living Lightplanes

WHILST military aircraft tend to make up the bulk of the exhibits in the air museums around the world, and certainly attract the attention of the greater part of the visiting public, the preservation of vintage lightplanes in flying condition is a matter of concern and delight to another large group of enthusiasts.

Plenty of aeroplanes in this category survive from the 1930s and are now between 30 and 40 years old, although not so many of these have been airworthy for the whole of that period. Much more rare in flying condition are the aeroplanes of the 1920–30 period, now well over 40 years old. With spares virtually non-existent and works drawings just as hard to come by, keeping these veterans flying is a real labour of love.

We offer no prizes for finding the oldest survivor, but we do invite readers to submit details and photographs, for publication in this column, of the 40-year-plus lightplanes still in flying trim in their own localities. Letters should be addressed to "Veteran and Vintage Department", AIR ENTHUSIAST, PO Box 16, Bromley BR2 7RB, Kent.

A couple of American old-timers which were still flying recently are illustrated on this page, in photographs by Howard Levy.

(Top left) This Alliance Argo Model A was built in 1929 as the 11th of only 19 of the type built, and was restored a few years ago after being stored for some 30 years. Restoration, by Ed Cochran and Bob Sparling, took two years, including construction of a new fuselage coaming and substitution of a Warner Super Scarab for the original Hess Warrior engine. (Left) This Waco RBA dates from 1932 and was restored by Jim Loomis.

Volume 1 Number 3   August 1971

CONTENTS

114 AIRSCENE The monthly round-up of news and background to the news, specially extended in this issue to include news and illustrations from the Paris Air Show. Individual sections cover Military Affairs, Civil Affairs, Aircraft and Industry, and new Civil and Military aircraft contracts.

121 MiG-21—AIR SUPERIORITY SOVIET STYLE Roy Braybrook provides a penetrating analysis of the Soviet Union's foremost fighter, with new material on the distinguishing features of versions up to the Fishbed-J.

128 PIRATES ASHORE Marking the introduction into service with the USAF of the Vought A-7D, described in this article as one of the most potent and effective tactical strike aircraft yet produced.

134 THE CALAMITOUS 'COBRA The whole, never-before-told story of Britain's disastrous experience with the Bell Airacobra in 1941, when only four of the 675 Airacobras purchased were flown on actual operations by the RAF. Including colour profiles and an exclusive new cutaway drawing.

144 MODEL ENTHUSIAST The latest plastic aircraft model kits reviewed, with the Hawker Siddeley Harrier depicted in colour.

146 IN PRINT New aviation literature noted.

147 JUNGLE FLIERS OF THE RUBBER STATE A history of the Royal Malaysian Air Force, young in years but already old in experience of operating STOL aircraft and helicopters in rugged jungle environment. With colour profile drawings.

152 THE "HALFWAY-HOUSE" FOKKER Third in the AIR ENTHUSIAST "Warbirds" series, telling the history of the Fokker D XXI, with tone three-view and cutaway drawing.

160 THE BILLION-DOLLAR BOMBER The second part of the story describing how the Soviet Union put the Boeing B-29 into production as the Tupolev Tu-4, and then used it as the basis for a long line of improved bombers and transport designs.

164 FIGHTER A to Z Continuing AIR ENTHUSIAST'S encyclopaedia of the world's fighters, from the Airco D.H.5 to the Albatros D VI.

167 PLANE FACTS Little-known facts and illustrations from the AIR ENTHUSIAST files.

AIRSCENE

MILITARY AFFAIRS

AUSTRALIA

Although Australian defence officials were reportedly impressed by the capabilities of the Hawker Siddeley Nimrod during a recent presentation, it is anticipated that the **Lockheed P-3C Orion** will be selected as the successor to the aged Lockheed SP-2H Neptune currently operated by No 10 Sqdn from Townsville. Maintenance problems provided by the SP-2H are steadily escalating and will become serious by 1973, and the choice of a replacement is now assuming considerable importance. Although some factions in the RAAF prefer the Nimrod, the RAAF's other maritime patrol component, No 11 Sqdn, is operating the P-3B version of the Orion, and a measure of standardisation is likely to prove the decisive factor in the choice of the SP-2H replacement, the difficulties accompanying the introduction of yet another major type into the RAAF inventory motivating against selection of the British aircraft.

The Australian Defence Department is evincing a strong predilection for the SNECMA M-53-powered **Dassault Mirage F** as the RAAF's next single-seat fighter, according to Canberra sources. The RAAF requirement places emphasis on the air superiority rôle, and it is understood that the selected type will be licence-manufactured in Australia with the programme linked to a projected in-service time period of 1978-80. A definitive choice is not expected before 1974-5, but it is understood that the M-53-powered Mirage F meets the requirements of the current outline specification, having the advantage of being a firm programme whereas neither of its principal competitors, the Northrop P-530 Cobra and the Lockheed CL-1200-2 Lancer, is currently more than a project. The M-53-powered Mirage F is scheduled to commence flight trials late 1973, and the power plant, which attained a reheat thrust of 17,968 lb (8 150 kg) under test on 11 May, is expected to reach its planned reheat thrust of 18,650 lb (8 460 kg) during the course of this year.

DENMARK

The Danish Defence Ministry has recently requested authorisation to purchase 22 **CF-104 Starfighters** declared surplus by the Canadian Armed Forces. It is understood that the CF-104s are being offered to Denmark at a unit price of £125,000, and, if purchased, will be completely refurbished and modified to F-104G standards. At the present time, Denmark's *Flyvevåben* includes two Starfighter squadrons, ESK 723 and ESK 726, which operate from Aalborg each with 12 single-seat F-104Gs and two two-seat TF-104Gs, and the number of aircraft that it is proposed to purchase from Canada suggests that one of the existing *Flyvevåben* F-100D Super Sabre squadrons may convert to the Starfighter.

ETHIOPIA

Unofficial reports have suggested that the Imperial Ethiopian Air Force is displaying serious interest in the **Aerfer-Aermacchi AM.3C** battlefield surveillance and forward air control aircraft, although no order had been placed at the time of closing for press. At the present time, the Italian concern is building an initial batch of 40 AM.3C aircraft in anticipation that current negotiations with potential purchasers will crystallise in the form of firm orders. The Imperial Ethiopian Air Force has hitherto acquired the bulk of its equipment via the MAP, and comprises two fighter squadrons, one of which is equipped with the F-86F Sabre and the other with the Northrop F-5A; a bomber squadron with four Canberra B 2s; a light attack squadron operating North American T-28Ds, and transport support comprising five jet-augmented Fairchild C-119Ks, two Douglas C-54s, an Ilyushin Il-14, a dozen C-47s and two Hawker Siddeley Doves. A few Alouette II and III helicopters are included in the inventory, and training is provided on T-28As and T-33As.

FEDERAL GERMANY

Plans to replace the Alouette II as the standard LOH in the inventory of the German Army's air component, the *Heeresflieger,* from 1975 have now been delayed until 1978. Although the MBB Bo 105 helicopter is currently the leading contender as an Alouette II replacement, the revised time scale of the programme may result in competition with a design offered jointly by Dornier and VFW-Fokker.

INDONESIA

According to Marshal Suwoto Sukendar, CoS of Indonesia's air arm, the *Angkatan Udara Republik Indonesia* (AURI), negotiations are being conducted with France for the purchase of **Dassault Mirage** fighters. It is known that the Indonesian government has been offered the Northrop F-5, and according to Marshal Sukendar, Britain has also offered to supply the AURI with combat aircraft which are needed to replace the Soviet equipment furnished during the Sukarno régime and largely grounded since the mid 'sixties owing to lack of spares support. The quantities of Mirages being negotiated have been quoted variously as from 30 to 76 aircraft with deliveries commencing next year.

ITALY

The Italian naval air component, the *Aviazione per la Marina,* had taken delivery of 12 **Agusta A 106** *"vettore d'arma"* (weapon carrier) single seat helicopters by mid-year, the first production example having been accepted in July 1970. Intended primarily for anti-torpedo boat operations, the A 106 is being operated from the *Impavido*-class destroyers.

Italy's air arm, the *Aeronautica Militare Italiana* (AMI), has suffered a proportionately higher loss rate in its **F-104G Starfighter** force than has the *Luftwaffe,* although Italian losses have attracted less publicity. Of 125 F-104Gs and 28 TF-104Gs the AMI had allegedly lost more than 30 aircraft by the beginning of this year, or upwards of 20 per cent, some four per cent higher than the attrition suffered by the *Luftwaffe* to that time. The AMI Starfighter accident rate per flying hour is also allegedly worse than that of any other operator of the Lockheed-designed aircraft.

The Italian Army's air component, *Aviazione Leggera Esercito* (ALE), is not likely to introduce the **CH-47C Chinook** into its operational inventory before late 1972. The ALE order placed last year calls for 26 CH-47C Chinook helicopters, the bulk of which are to be licence manufactured by Elicotteri Meridionali. Included in the ALE order are four of the eight CH-47Cs being supplied by Boeing's Vertol Division, the first of which reached Elicotteri Meridionali in December with deliveries continuing until June 1972. In addition, two sets of certain assemblies and components are being supplied to the Italian company for the development of production tooling, and the first Italian-manufactured CH-47C is scheduled to be delivered to the ALE during the second half of next year.

IRAN

The first of 16 **CH-47C Chinook** helicopters

The BAC/Breguet Jaguar S06 (left) appeared at the Paris Air Show with underwing drop tanks, and bombs on the fuselage strong points. Prominent also in this view is the bulged rear fuselage incorporated since the prototype first flew. Meanwhile, Jaguar S07 has re-emerged at Warton with a redesigned nose (right) incorporating a Ferranti laser rangefinder; the production fin and computer-controlled navigation and weapon aiming sub-system (NAVWASS) have also now been fitted. Jaguar B08, the British two-seat prototype, was rolled-out at Warton on June 1 and has now joined the other Jaguars in flight test.

As revealed by the accompanying general arrangement drawing and photographs, the Dassault Mirage G8.01 differs in many respects from the Mirage G1 which was destroyed during its 316th flight on 13 January. Both the fuselage and the vertical tail surfaces have been substantially redesigned, and power is provided by two SNECMA Atar 9K-50 turbojets of 15,873 lb (7 200 kg) thrust with reheat which will eventually be supplanted by advanced-technology SNECMA M53 engines of 18,650 lb (8 460 kg) thrust. Since its first flight at Istres on 8 May, the Mirage G8 has made extremely rapid progress in its flight test programme and is to be equipped with the multi-mode Thomson-CSF Cyrano IV radar, the nav/attack system being based in other respects on that of the Jaguar A, and including laser range-finding, doppler and bombing computer elements. Maximum take-off is of the order of 40,500 lb (18 470 kg), and a second Mirage G8 prototype is scheduled to fly next year, and although no production funding is included in the current (1971-6) defence budget, both Dassault and the Armée de l'Air hope that some juggling of existing programmes will provide funds for production preparations with a view to an in-service date of 1978-9.

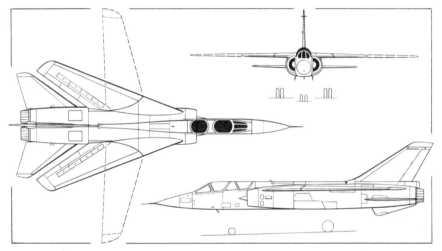

ordered from Elicotteri Meridionali for use by the Iranian Army Air Force Battalion, the *Gordan Havanirouz*, was handed over to the Iranian Imperial Government on 20 April. This helicopter is one of four Boeing-Vertol-built Chinooks that are being supplied to Iran prior to the delivery of the first licence-manufactured examples, and the bulk of the Iranian order will be fulfilled during the latter half of 1972 and early 1973. The *Gordan Havanirouz* is also receiving a substantial proportion of the 45 AB 205 Iroquois and 70 AB 206 Jet-Ranger helicopters ordered from Agusta by the Iranian Imperial Government.

JAPAN

The Japanese Defence Agency is seriously perturbed over forecast **Mitsubishi T-2** supersonic jet trainer cost escalation, the unit price being unofficially reported to have risen from the original figure of Yen 400m (£458,000) to Yen 1,500m (£1·75m) This increase is in part due to the doubling of the anticipated price of the Rolls-Royce/Turboméca Adour engine which is to be built under licence by Ishikawajima-Harima, and studies are currently being made of the alternative installation of a single General Electric engine derived from the basic GE1, the J79 or a pair of J85s. The Defence Agency has also requested submissions from Fuji, Kawasaki and Mitsubishi for the licence manufacture of the Northrop F-5

as a possible alternative to the indigenous T-2 trainer, the anticipated unit cost of the F-5 built in Japan being approximately Yen 792m (£916,000). The first XT-2 prototype was rolled out in April at Nagoya and was expected to commence its flight test programme in July. The second XT-2 was scheduled to be completed in July, and current schedules call for the first two pre-production T-2s to be rolled out in June and August next year.

The Air Self-Defence Force anticipates deploying an F-104J Starfighter squadron on **Okinawa** by the end of October 1972, the current schedule calling for the return of this territory to Japan in July of next year.

An air rescue wing has now been inaugurated within the Air Self-Defence Force. With Headquarters at Iruma, this wing comprises some 800 personnel and is equipped with seven S-62 and 14 V-107 helicopters, and 12 MU-2s and 15 T-34s. Detachments are deployed at Chitose, Matsushima, Hyakuri, Niigata, Hamamatsu, Ashiya, and Nyutabara.

NEW ZEALAND

Defence spending **economies** have dictated the withdrawal by the RNZAF of four of the service's 14 Iroquois helicopters and three of its 13 Bell Model 47G helicopters, these having been placed in storage. Three of its

Devon light twins were recently declared surplus and sold, and several of the nine Bristol 170 transports have been placed in storage.

SOUTH AFRICA

Unofficial reports stemming from Paris have referred to negotiations between the South African government and Avions Marcel Dassault for the purchase of 40 **Milan** multi-purpose fighters for the SAAF. Recent SAAF acquisitions from France have included a total of 20 SA 330 Puma helicopters, and four ex-South African Airways DC-3s have been added to the SAAF transport inventory, but on 26 May the entire complement of Hawker Siddeley HS 125s was lost when all three aircraft of this type crashed into Table Mountain while practising formation flying in preparation for Republic Day celebrations. The HS 125 was known in the SAAF as the Mercurius.

SWITZERLAND

Although the **Vought A-7G Corsair II** remained the aircraft favoured by the Swiss Defence Ministry as a successor to the ageing Venom after completion recently of the second evaluation of potential replacements, the McDonnell Douglas A-4M Skyhawk supplanted in second position the Dassault Milan S and Fiat G.91YS (which had previously shared second place), and in June a Swiss government and *Flugwaffe* team visited the McDonnell Douglas facility at St Louis for provisioning conferences on the A-4M. The Defence Ministry has said that under the £125·8m budget limitation set by the programme 61 A-7G Corsair IIs can be purchased as compared with 85 A-4M Skyhawks, and it has been unofficially estimated that if the selected aircraft is manufactured under licence by Swiss industry the number that can be procured within the stipulated sum will be reduced by approximately 30 per cent. A Vought Aeronautics spokesman at the Paris Air Show stated, however, that 68 A-7Gs could be supplied, and that this figure could be increased subject to certain proposals being accepted. Although the Federal Council was scheduled to decide on an aircraft during July, the final selection will not be made by

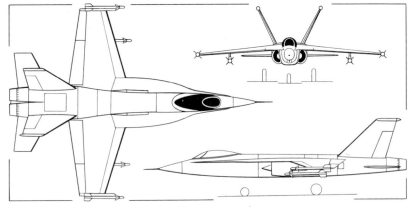

The Northrop P-530 Cobra advanced air superiority fighter, illustrated by the accompanying general arrangement drawing, has been based on F-5 concepts and philosophy but bears no resemblance to the earlier design. Northrop is currently endeavouring to enlist foreign financing for the development of the Cobra which is a purely private venture intended to form the basis of a collaborative manufacturing programme. Including research and development, the Cobra is expected to have a unit price of the order of $3m (£1·25m) and will be powered by two General Electric J101-GE-100 turbojets each having an augmented thrust of approximately 15,000 lb (6 804 kg). Development of the Cobra airframe and systems is expected to cost $100m (£41·6m), Northrop having already spent some $22m (£9·1m), and it is estimated that the development and flight testing of the J101 engine will cost about $150m (£62·5m.).

Parliament before September of next year, and a contract will not be signed before January 1973. Thus, the Federal Council choice can still be overruled, and Avions Marcel Dassault is continuing its efforts to gain acceptance of the Milan S, placed third in the evaluation. According to the Swiss Defence Ministry, only 70 Milans could be purchased within the financial limitations of the programme, but Dassault believes that it can supply 93 aircraft and remain within the budget limitation, and, furthermore, could subcontract 20 per cent of the contract value to the Swiss industry.

UNITED STATES OF AMERICA

The USAF has decided to scrap the **Convair B-58 Hustler** bombers currently in storage as too much time would be demanded to return the aircraft to combat-ready status in the event of an emergency. The B-58s will be stripped of their equipment and broken up before the end of this year.

AIRCRAFT AND INDUSTRY

FRANCE

Flight testing of the **Dassault Falcon 10-01** resumed on 7 May following five months on the ground during which time the prototype had undergone several modifications. The first series of nine flights was made between 1 and 18 December 1970, and indicated the need for some aerodynamic improvements. To overcome a tendency to Dutch roll, the wing has been given less dihedral and the angle of incidence has been changed. Wing sweepback has also been increased, to raise the Mc. The Falcon 10-01 is powered by 2,950 lb st (1 340 kgp) CJ-610 engines but the definitive 3,230 lb st (1 465 kgp) TFE-731 turbofans will be fitted in the 10-02, which is expected to fly in September, and in 10-03, to fly in mid-1972. SNECMA-Turboméca Larzac-02s will be fitted in 10-01 in due course, and the Larzac 03/06 will be offered as alternatives to the TFE-731 engines in production aircraft.

A STOL project developed in France as the **Aérospatiale A904** shares the characteristics of several other current STOL studies. The formula comprises a high, moderately-swept wing, four podded turbofan engines, large T-tail and accommodation in the 100-150

passenger range. Aérospatiale has adopted an externally-blown flap arrangement, depending primarily upon the large-area high-lift flaps operating in the jetstream of the turbofan engines to obtain the required lift at low forward speeds. The proposed engines are Rolls-Royce/SNECMA M45S/RB410s, derived from the M45H with a Dowty Rotol variable pitch fan replacing the standard by-pass fan at the front of the engine. With a gross weight of 112,500 lb (51 000 kg) the A904 has a projected range of 620 miles (1 000 km) with a payload of 20,720 lb (9 400 kg), and a max payload of 26,675 lb (12 100 kg). Cruising speed is quoted as 450 knots (830 km/h); the span is 101 ft 2½ in (30,85 m) and the length is 121 ft 8½ in (37,10 m).

Aérospatiale has bestowed the name **Lama** on the SA 315B version of the Alouette. This has the airframe of the Alouette II with the dynamic components of the Alouette III and an 870 shp Astazou IIIB turboshaft engine. It is intended particularly for "hot and high" operations.

Among the new light aircraft appearing at the Paris Air Show was the **CERVA CE-43 Guépard**, an all-metal four-seater which flew for the first time on 18 May. The Guépard is a joint product of the Siren and Wassmer companies, the former having been responsible for its structure and the

This model of the BAC P.59 basic trainer was displayed at the Paris Air Show. Submitted to meet the RAF specification, the P.59 is powered by a R-R Viper 600 variant.

latter for design and assembly; in effect, the aircraft is an all-metal version of the Wassmer 4/21. Marketing, starting next year, will be in the hands of Groupement d'Intérêts Economiques CERVA (Consortium Européen de Réalisation et de Ventes d'Avions). The French government is expected to order a small batch of CE-43s for use at State-aided flying clubs, and such an order is regarded as the only way to save the Wassmer company, which is in the hands of a receiver and has won a temporary reprieve with an order for nine Squale sailplanes. The CE-43 is powered by a 250 hp Lycoming engine, has a span of 32 ft 10 in (10 m), length of 25 ft 7 in (7,80 m), gross weight of 3,200 lb (1 450 kg) and cruising speed of 193 mph (310 km/h) at 75 per cent power at 6,560 ft (2,000 m).

INTERNATIONAL

Shortly before its appearance at Paris, the **Aérospatiale/Westland SA 341 Gazelle** set three new world speed records for helicopters, surpassing records held previously by the Hughes 500. The records, in the category for helicopters of 1,500–1,750 kg (2,200–3,860 lb) gross weight, were for speed in a straight line over 3 km and over 15–25 km at 310 km/h (192 mph), and speed in a closed circuit of 100 km (62 miles), at 295 km/h (183 mph). For the record flights, the Gazelle 01 had special streamlined fairings round the rotor head, and new streamlined struts carrying the landing skids, but only the latter will be featured on production helicopters. The prototypes 001 and 002 and pre-production examples 01 to 04 have together totalled more than 1,200 flying hours. The first production example is to be delivered from Marignane early in 1972, and present contracts are for 110 for the French and British armed forces, and 20 for Vought Helicopter Inc, to be sold as civil machines in the USA; the total French Army requirement by 1977 is 166. British versions will be designated Gazelle AH.I (Army and Royal Marines); Gazelle HT.2 (Royal Navy); Gazelle HT.3 (RAF) and Gazelle HCC.4 (RAF).

ITALY

First flight of the prototype **Agusta A 109C** helicopter was expected to be made during June, although plans to have this new type at the Paris Air Show were thwarted by labour problems earlier this year. The A 109C is a medium size, general purpose military or civil helicopter intended to fit between the Bell 206 JetRanger and the Bell 212, both of which are in production by Agusta. It is powered by two 400 hp Allison 250-C20 engines driving a single four-bladed rotor with a diameter of 36 ft 1 in (11,00 m). Gross weight is 5,290 lb (2 400 kg), max speed 149 mph (275 km/h) and range 380 naut miles (705 km) with a 1,322 lb (600 kg) payload. Four prototypes of the A 109C are being built and deliveries are expected to start in 1972.

JAPAN

Production of the **NAMC YS-11** is to end with a total of 182 aircraft built, the Japanese government having decided not to authorise any further expenditure on the programme. It is understood that the break-even quantity has not been reached. After production ends, NAMC will continue to provide after-sales support, but will not produce any more aircraft, development of the XC-1 jet transport having been transferred to Kawasaki.

NETHERLANDS

The **Fokker P301** project for a STOL transport derivative of the F-27 Friendship is regarded by the company as a "minimum step" towards STOL operation. Development costs are estimated to be about 200m guilders (£23m) and the company is seeking government support for a major share of this cost. The original Dutch government investment in the F-27 has now been fully repaid and with sales of the Friendship expected to continue at least until 1975, a profit is being shown by the government; it is hoped that some of this money might become available for the P301. Based on the fuselage of the F-27 Mk 600, the P301 has a substantially new wing and tail unit as different structural methods are used in its construction. It is powered by four 1,035 shp Pratt & Whitney PT6A-50 engines. High-lift double-slotted flaps, slats, spoiler-ailerons and lift dumpers are used to obtain the STOL performance, and controllability at low speed is enhanced by the use of vertical slats on the fin ahead of the rudder leading edge. With a gross weight of 43,500 lb (19 730 kg) the P301 could carry 48 passengers for 500 naut miles (925 km) from a 1,800 ft (550 m) runway in ISA + 19° at sea level, this runway length allowing for one engine failure at take-off.

UNITED KINGDOM

A model of the **BAC P59** exhibited at the Paris Air Show indicated the general appearance of the BAC submission for the new RAF basic trainer. Designated in the sequence of project numbers used in the Warton design office, the P59 is in fact a joint Warton/Weybridge study, marking the first military project at Weybridge since the TSR 2. The P59 is designed around a single Rolls-Royce Viper 600 military variant, and is a mid-wing aeroplane with slight sweepback, intended to fill a wide spectrum of training activities between the primary trainer and the operational conversion stage. Simplicity is being stressed, although the P59 has a Mach 0·9 speed and can carry up to 5,000 lb (2 270 kg) in external loads. The higher-performing P62, with a reheated Adour engine, is being soft-pedalled by BAC, but the competing Hawker Siddeley design, the **HS 1182**, has been offered with the latter engine. Hawker Siddeley is also understood to have offered to undertake HS 1182 production on a fixed-price contract at a unit price of about £570,000 for 175 aircraft.

Preliminary details of the STOL airliner project of BAC's Commercial Aircraft Division were released on the occasion of the Paris Air Show. Known simply as the **QSTOL** (the Q standing for "quiet"), the project is for a wide-body transport seating 108–140 passengers and with an externally-blown flap system to achieve STOL. Power is provided by four 14,525 lb st (6 600 kgp) Rolls-Royce/SNECMA M45S/RB 410 engines of the geared-fan type, these being based on the M45H gas generator with a Dowty Rotol variable-pitch fan added. At a STOL gross weight of 123,825 lb (56 170 kg) the QSTOL will carry 108 passengers for 600 miles (965 km) from 1,960 ft (600 m) field lengths in ISA + 20°C at sea level, and ranges up to 1,000 miles (1 610 km) can be achieved at higher weights and slightly increased take-off distances. The QSTOL has a span of 114 ft 6 in (34,90 m) and length of 110 ft 4 in (33,63 m).

An intensive campaign to sell the Jetstream executive and third level transport was started by **Jetstream Aircraft Ltd** during June with demonstrations of the aircraft at Le Bourget. The latter company acquired design and manufacturing rights in the

The world's largest and heaviest helicopter by substantial margins, the Mil Mi-12 (alias V-12) has been under flight test since the autumn of 1968. Developed by the Mil design bureau, headed by Marat N Tischenko since the demise of Mikhail L Mil in January last year, the Mi-12 is powered by four 6,500 shp Soloviev D-25VF engines and has a maximum gross take-off weight of 231,485 lb (105 000 kg), normal take-off weight being 213,848 lb (97 000 kg). A crew of six is carried of which the pilot, co-pilot, flight engineer and electrician are accommodated on the lower flight deck with the navigator and radio-operator housed on the upper deck. Maximum and cruising speeds are claimed to be 161 mph (260 km/h) and 149 mph (240 km/h) respectively, and range with a maximum payload of 78,000 lb (35 380 kg) is 310 mls (510 km) after a rolling take-off, maximum vertical lift capacity being 55,116 lb (25 000 kg). Rotor diameter is 114 ft 9½ in (35,00 m), overall length excluding rotors is 121 ft 4 in (37,00 m), overall width with rotors turning is 219 ft 9 in (67,00 m) and overall height is 41 ft 0 in (12,50 m).

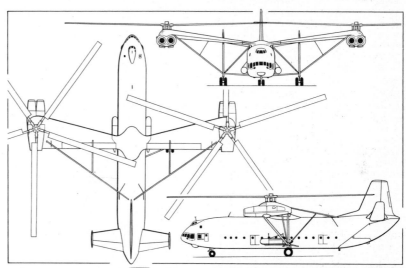

Jetstream about a year ago from the receiver of Handley Page Aircraft Ltd, together with the product support stores, bonded stores, and a number of Jetstreams in various stages of completion. Since then, five Jetstream Is have been completed (with Astazou 14 engines) and work is continuing to complete tropical certification of the Jetstream 200 with Astazou 16C-C2 engines. Both versions will be marketed by Jetstream Aircraft at prices respectively of $394,000 (£164,200) and $445,000 (£185,400) less avionics, flight instruments and interiors. Agreement has been reached with Scottish Aviation for the manufacture of most of the airframes and final assembly at Prestwick. The Jetstreams will then be flown to Leavesden for completion, and this is also the base for administration, sales, design and fatigue rig testing. The latter is running some 13 months ahead of the longest flying-time Jetstream.

Newest **Britten-Norman Islander** variant is a turbosupercharged version with Lycoming TIO-540-H engines providing improved performance at high altitudes. The first delivery of this version, which takes the place of the previously announced BN-2S, is scheduled for October to a customer in Central America. Another Islander variant is the **Defender,** intended for a wide variety of military rôles such as forward air control, logistic support, casevac, long range patrol, internal security, and search and rescue. It can carry a standard nose-mounted weather radar and has four underwing pylons with a total capacity of 2,000 lb (910 kg). With long-range wing tips, the Defender has a range of more than 1,200 miles (1 930 km) and endurance of over 10 hours. It can carry up to nine passengers or eight paratroopers.

Since rights to the original Beagle Aircraft **Bulldog** design were obtained by Scottish Aviation, the contract with Sweden has been renegotiated to provide for a delay of about 13 months in the original delivery date, but the revised contract date was beaten by a few days when the first production Bulldog for the Swedish Air Force

The accompanying photograph and general arrangement drawing of the Ilyushin Il-76 reveal the close similarity in configuration of this new Soviet heavy freighter to that of the Lockheed C-141A Starlifter.

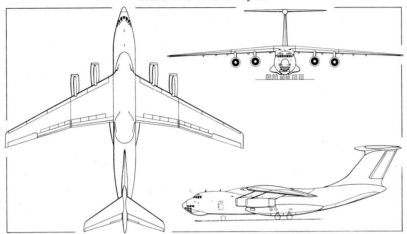

was handed over on 7 July. Production will have reached six a month by February 1972, to fulfil present orders for 58 for the Swedish Air Force, five for Kenya (deliveries starting in February 1972) and 15 for the Royal Malaysian Air Force (deliveries starting October 1971). One of the first developments of the basic design by Scottish Aviation will be the addition of a high-lift double-slotted flap to provide STOL performance. If this is successful, it is expected to be fitted to the 45 aircraft on option for Sweden to make them suitable for Army communication use. To a longer timescale, the company plans to develop civil derivatives of the Bulldog.

Slingsby Sailplanes of Kirby Moorside has completed and flown an experimental **Kestrel** sailplane with a carbon fibre mainspar. This is believed to be the first application of carbon fibre to the primary structure of an aircraft. The Kestrel, which first flew on 3 May, is a German Glasfugel design built by Slingsby under licence.

A prototype sailplane designed by J C Sellars and built by **Torva Sailplanes** made its first flight at RAF Driffield on 8 May and is the first British-designed glider of all glass-fibre construction. Two more prototypes of the 15-m (49·2-ft) design are being built and production is expected to start next year.

USA
Boeing is mounting an extensive effort to sell a short-range version of the 747 in Japan. Known as the **Boeing 747SR**, the proposed variant is structurally similar to the current Model 747-100 with a gross weight of 738,000 lb (334 750 kg) and would operate at weights of 520,000 lb–600,000 lb (235 870–272 155 kg) according to route; the structural integrity of the higher-weight version is needed to provide a 20-year fatigue life, including 53,000 landings. Accommodation would be provided for 537 passengers instead of 382 in the basic long-range model; also studied are various "stretches" of the design in which the upper deck behind the flight deck is extended aft and fuselage plugs are added to permit up to 1,000 passengers to be carried. An equivalent stretch of the long-range model would produce accommodation for 730 passengers, when the gross weight would run at about 840,000 lb (381 000 kg) and the wing span would be increased to give some 20 per cent more area. Growth versions of the Pratt & Whitney JT9D engine have been projected to match these Boeing studies, including a 65,000 lb st (29 480 kgp) JT9D-23, but the stretched 747SR would operate at about 750,000 lb (340 200 kg) gross weight and could be based on more readily available JT9Ds of about 50,000 lb st (22 680 kgp).

Boeing has announced details of an advanced **Model 727-200**, available for delivery from mid-1972, with increased range and improved sound suppression. Replacing an earlier advanced 727 proposal, the new model will now have 15,500 lb st (7 030 kgp) JT8D-15 engines as standard, and thicker wing skins allowing an increase in gross weight to 191,000 lb (86 636 kg). Fuel capacity will be increased by 1,332 Imp gal (6 055 l) to 9,780 US gal (44 460 l) and the range with typical payload will be about 800 miles (1 290 km) greater than that of the current 727-200.

Following announcement of a Boeing/Aeritalia agreement to join forces in development of a short-haul STOL transport, the Boeing Co has released brief details of one of the design studies made at Seattle for an aircraft in this category. Identified as the **Boeing Model 751**, this project makes use of the augmentor-type wing at present being pioneered by DHC. The Model 751 has a span of 119 ft 6 in (36,42 m) and an overall length of 147 ft 8 in (45,01 m). Similar designs are being studied with internally blown or externally blown flaps; all are high-wing aeroplanes with podded engines and T-tails, the latter being of large area to provide the required control moments at low forward speeds. The Boeing/Aeritalia STOL studies are for an aircraft to carry at least 100 passengers a distance of 500 miles (805 km) from a STOL strip (2,000 ft ± 500 ft; 610 m ± 152 m). Using longer take-off strips, the same aeroplane would probably have a range of about 1,000 miles (1 610 km). Other design parameters are an approach speed of 80 knots (148 km/h), a cruising speed of about Mach 0·8 at 30,000 ft (9 144 m), a 95 PndB noise footprint at 500 ft (152 m) distance and a relatively high wing loading. Engines would probably be turbofans in the 20,000 lb st (4 410 kgp) category.

While Lockheed and Rolls-Royce awaited the outcome of the US government's attempt to obtain Congressional approval of financial guarantees for Lockheed, the third **L-1011 TriStar** was flown for the first time on 17 May at Palmdale. At that date, the two other TriStars had totalled 160 hours in the air. Under the revised schedules agreed between Lockheed and Rolls-Royce, certification is expected in April 1972, but both British government funding for the RB.211 engine, and US banks' support to allow Lockheed to finance the remainder of the TriStar programme, are conditional upon the US government's guarantees, upon which final Congressional action was unlikely before August. The No 3 TriStar, in Eastern Air Lines markings, was demonstrated at the Paris Air Show from 1 to 5 June and visited Britain briefly during its return flight to the US, being displayed at East Midlands Airport for the benefit of Rolls-Royce management and employees.

In common with most of the other major aircraft companies, Lockheed has a growing

Flight testing of the Dassault Mercure short-haul twin-engined transport began on 28 May, and the aircraft was demonstrated at the Paris Air Show a week later. Powered by Pratt & Whitney JT8D-7 engines, this prototype will be followed by a second with the definitive JT8D-15 engines and higher operating weights.

interest in future STOL transport aircraft. This interest is centred upon versions of the **C-130 Hercules** for the short term and new designs in the longer term. Developments of the Hercules for STOL are believed to be of interest to the USAF as an interim solution to its need for a tactical STOL transport, and also to have commercial potential. The civil proposal is for a version of the L-100-20 with new flaps, increased vertical tail area, spoilers for roll control at low speeds, and 16 ft (4,87 m) diameter propellers for low noise levels. This aircraft would carry 80 passengers over 500 miles (805 km) from a 2,000 ft (610 m) runway. For military use, a derivative of the C-130H is proposed, with a gross weight of 130,000 lb (58 967 kg) and the ability to fly a 25,000 lb (11 340 kg) payload out of a 1,500 ft (457 m) strip in standard conditions. For the future, Lockheed is interested in turbofan-engined STOL types with externally blown flaps or an augmentor wing. Projects ranging in size up to a 150-seat 140,000-lb (63 500-kg) aircraft have been studied, but interest centres at present on the expected NASA request for proposals for a **STOL technology demonstrator.** Such an RFP would lead to selection of a manufacturer to build a prototype of a relatively small aircraft ("about JetStar-sized", say Lockheed) to prove the chosen technology.

The Piper Aircraft Corporation will enter its **PE-1 Enforcer** in the USAF's contest for a strike aircraft to be supplied to selected countries under the Military Assistance Program. The Enforcer is a newly-designed, newly-tooled aircraft based on the WW II North American P-51 Mustang and powered by a 2,535 hp Lycoming T55-L-9 turboprop. Originally designed by Cavalier Aircraft, the Enforcer was sold to Piper last year and is essentially a new aircraft, having no component commonality with the Mustang. Built-in armament consists of six 0·5-in (12,7-mm) machine guns, and there is provision for up to 10 wing ordnance hardpoints. Initially flown on 29 April, the Enforcer is expected to be offered at a unit price of $500,000 (£208,300).

Two new versions of the **General Electric CF6-6** engine are now on offer, in addition

(Above) An impression of the Aérospatiale A.904 and (below) the BAC QSTOL.

(Below) An impression of the Boeing 751.

The accompanying photograph of the CASA C.212 Aviocar STOL utility transport prototype was taken at Le Bourget just before an inadvertent asymmetric reverse thrust selection resulted in a heavy landing and a broken mainspar. The Aviocar had been under test since 26 March, and it is anticipated that a second prototype will shortly join the flight test programme, current plans calling for a pre-series of 12 aircraft with production deliveries commencing in 1973. Designed primarily for military operation, the Aviocar can accommodate 15 paratroops and a jumpmaster or 10 casualty stretchers, three seated casualties and medical attendants. The Spanish Air Force has a requirement for some 50 Aviocars to replace its aged Junkers Ju 52/3m transports.

to the basic 40,000 lb st (18 145 kgp) CF6-6D that has been certificated for use in the McDonnell Douglas DC-10 Srs10. The new variants are the 41,000 lb (18 600 kg) CF6-6D1, and the 43,000 lb (19 500 kg) CF6-6G. The latter is the highest-rated version of the -6 so far announced; also available are CF6-50 engines at ratings from 47,300 lb (21 455 kg) to 51,000 lb (23 135 kg).

First flight of the **Garrett TFE 731-2** turbofan engine was made on 19 May at Wichita. The engine was mounted in the starboard nacelle of a Learjet 25 which retained the standard CJ610 engine in the port nacelle. Plans to market a Learjet version with TFE engines are in abeyance pending a re-assessment of the market but the first two engines for the Dassault Falcon 10 are now being installed in pods by Grumman and are expected to fly in the French aircraft in September. The Garrett TSE-231 helicopter engine also began flight testing in a special test-bed on 19 May at Phoenix.

Under a US Navy-sponsored programme to investigate **rotating combustion engines** for possible use in light tactical aircraft, Cessna Aircraft Co has fitted a 185-hp Curtiss-Wright RC2-60-U5 engine of the Wankel rotary type in a Cardinal. With noise suppression as a primary goal, the aircraft has a three-bladed wide-chord propeller of 8 ft 4 in (2,54 m) diameter and an external exhaust and muffling system which ejects the exhaust above the aircraft. Flight testing is to extend from June to late summer.

Boeing Company's Vertol Division has been named winner of a competition to select a contractor to conduct Phase 1 studies of a tri-service **heavy-lift helicopter (HLH).** The decision eliminated Sikorsky, Gyrodyne, Hughes Tool and Kaman. The Boeing Vertol design has a tandem rotor layout, each rotor of 90 ft (27,43 m) diameter, and carries its 45,000 lb (20 410 kg) payload externally. A cabin is provided for up to 12 occupants in addition to the crew, and the helicopter has a rearward-facing station for the winch operator behind the cockpit. Phase 1 studies will cost $76m (£31·7m) and include design and testing of major components. Prototype construction depends on a Phase 2 contract being negotiated in due course.

A twin-engined 12-seat light transport and utility aircraft called the **Skytrader** is reported to be under development by

Dominion Aircraft Corp of Seattle and Vancouver. The Skytrader is a high-wing aircraft with two 400 hp Lycoming engines and a gross weight of 8,500 lb (3 855 kg).

General Electric's GE 15 engine, the powerplant for the projected Northrop P-530 Cobra, is now designated **YJ101-GE-100**, indicating USAF funding of development of the engine, afterburners and nozzle up to July 1972. The engine, in the 15,000 lb st (6 800 kg) class, is also available for international development and productionsharing, and the company has an eye on possible interest in European countries. First flight of a test engine is scheduled for mid-1973.

A USAF contract worth $2·1 million (£900,000) has been awarded to Martin Marietta company to develop and test-fly a **high-altitude reconnaissance drone.** Flight testing will begin early next year and the USAF evaluation of the aircraft will be in competition with the LTV Electrosystems Model 450F. Two prototypes of the latter have already flown, and this aircraft is reported to have a duration of over 24 hours and a ceiling of over 50,000 ft (15 240 m). The L450F was developed as a company-funded venture and it is being promoted as a communication relay and electronic reconnaissance vehicle. Israel is regarded as a potential customer.

Provisional type certification was granted to the **McDonnell Douglas DC-10** on 24 May. This step allows provisional certificates to be issued to individual DC-10s for such activities as airline pilot training and route proving, pending issuance of the final type certificate. Delivery of provisionally certificated DC-10s was to begin on 29 July. Up to mid-May, more than 1,000 flight hours had been logged by DC-10s, five examples of which were flying by that date.

After a brief attempt by the US House of Representatives to revive the **Boeing SST**, a second negative vote by the Senate on 19 May finally ended any possibility that the programme could be continued. The steps already taken by Boeing following the Senate's negative vote on 24 March have made it impossible for the SST to be resurrected without a sharp increase in costs, although none of the major jigs has been scrapped and all drawings, reports and other documents are being carefully filed. Under contract to the Department of Transportation, Fairchild Industries Corp has been studying the possibility of obtaining private finance for the programme, but clearly with little chance of success. Total expenditure on the Boeing SST up to cancellation, and including cancellation costs, is reported to have been $505m (£210m) of which Boeing had provided $150m (£62m), the government and customer airlines $334m (£140m) and General Electric $20m (£8·4m). Rather less than this total sum would have been required to continue the programme up to prototype flight testing. As a result of subsequent action by Congress, the money provided by the customer airlines is not to be refunded by the government.

USSR

Designation of a new wide-body airbus-type transport that is under development in the Soviet Union is **Ilyushin Il-86,** according to General Constructor Generi Vasilievich Novozhilov, who has succeeded Sergei V Ilyushin as head of the design bureau that bears his name. The new aircraft will seat about 350 passengers and will be powered by four developed Soloviev D-30 engines; no other details are forthcoming except that the first flight is not expected to take place before 1976.

Aleksandir S Yakovlev recently revealed the fact that the single-seat V/STOL aircraft of which two prototypes were seen at Domodedovo in July 1967 and which was subsequently dubbed *Freehand* by NATO is officially designated **Yak-36.** He would not comment on the current status of the aircraft, but the general consensus of opinion is that the Yak-36 was evolved essentially to evaluate V/STOL techniques and possesses no operational application.

Flown for the first time on 25 March and shown at Le Bourget during the Paris Show, the **Il-76** heavy freighter designed by a team led by Generi V Novozhilov was evolved primarily to meet a military requirement, and is one of the first Soviet aircraft to feature pylon-mounted engine pods (one of the few earlier Soviet examples of this type of engine mounting being provided by the Il-54 light bomber also from the Ilyushin bureau). The Il-76 is powered by four 26,455 lb (12 000 kg) thrust Soloviev D-30KP turbofans and has a maximum take-off weight of 346,122 lb (157 000 kg), maximum payload being 88,185 lb (40 000 kg). Overall dimensions include a span of 165 ft 8⅓ in (50,50 m), a length of 152 ft 10¼ in (46,59 m), and a height of 48 ft 5⅛ in (14,76 m). Maximum cruise and maximum range are quoted as 528 mph (850 km/h) and 3,107 mls (5 000 km) respectively. Generally similar in concept to the C-141A Starlifter which entered service with the USAF six years ago, but slightly larger, slightly more powerful and slightly heavier, the Il-76 employs a mechanised cargo handling system, a high-flotation undercarriage, the main members of which comprise four individual units each of four parallel-mounted wheels, and makes extensive use of high-lift devices to achieve short-field performance. It would seem unlikely that the Il-76 will join the transport inventory of the Soviet Air Forces much before 1973–4.

The recent admission by Professor Alexei Tupolev that only one prototype of the **Tu-144** SST has yet flown suggests that statements made in the Soviet Union to the effect that domestic operations with the Tu-144 will commence during 1972 were presumably intended to mean that *route proving* will begin next year. Professor Tupolev added that two additional Tu-144s will join the test programme before the end of this year, and it would seem, therefore, that the time scale for service introduction of the Soviet SST is much closer to that of the Concorde than has been suggested by some sources. New dimensions have recently been revealed for the Tu-144, these including an overall wing span of 90 ft 8½ in (27,65 m) and a length including nose probe of 194 ft 10½ in (59,40 m).

At the time of closing for press Aeroflot was scheduled to receive the first deliveries of the new high-capacity version of the **Yak-40.** This new model provides accommodation for 40 passengers in four-abreast seating, and has a thrust reverser on the centre engine, these features having been seen on the example displayed at Le Bourget in May-June. However, the production 40-seater is powered by Ivchenko AI-25T turbofans uprated to 3,858 lb (1 750 kg) thrust, and fuel capacity has been raised to 880 Imp gal (4 000 l), take-off weight being increased to 36,376 lb (16 500 kg).

Recently revealed was a modified version of the well-established **Ilyushin Il-62** long-range commercial transport, the Il-62M-200. The new model has 24,250 lb (11 000 kg) thrust Soloviev D-30KU turbofans in place of the 23,150 lb (10 500 kg) thrust Kuznetsov NK-8-4 turbofans used by the preceding version, and a small increase in passenger capacity achieved by the re-arrangement of the toilet and wardrobe areas at the rear of the cabin. Maximum high-density accommodation is provided for 198 passengers, with alternative arrangements for 186 tourist-class or 161 mixed-class passengers. Maximum take-off weight has been increased by 363,760 lb (165 000 kg) but the overall dimensions remain unchanged.

CIVIL AFFAIRS

PERU

Aerolineas Peruanas, the Peruvian airline with a fleet of two Convair 990s and a single DC-8, ceased operations at the beginning of May. Attempts to save the company through a government take-over appear to have been unsuccessful.

UNITED KINGDOM

Introduction of a network of "third-level" scheduled services in the UK by BEA subsidiary British Air Services is set for next January. The company was negotiating, at the time this issue closed for press, to purchase three 19-seat Short Skyliners, and plans to operate them in the markings of **InterSTOL**, the organisation set up by BAS to handle this particular operation. The Skyliner is a specially equipped version of the Skyvan, with slower-running propellers than usual to reduce noise levels.

Delivery of the first **Britten-Norman Trislander** to Aurigny Airlines was to be made shortly after this issue went to press. A British C of A was awarded to the Trislander on 14 May, by which time three examples had been completed and flown — the prototype (flown on 11 September 1970 and now grounded), a pre-production model (flown on 6 March) and the first production model for Aurigny (flown on 29 April). Jersey-based Aurigny Airlines has ordered three Trislanders, 15 of which are scheduled to be built by the end of 1971.

British Aircraft Corporation has formed a **Commercial Aircraft Division** to replace the former Weybridge and Filton Divisions. With effect from 1 June, the Filton, Hurn and Weybridge factories, the Fairford Flight Test Centre and the Wisley facility have been combined in the new division, of which Geoffrey Knight is the chairman.

USA

Southwest Airlines is the name of a new intrastate airline formed in Texas to operate scheduled services linking Love Field, Dallas, with Houston and San Antonio. Three Boeing 737s have been acquired for the operation, believed to be on lease from Boeing.

CIVIL CONTRACTS AND SALES

Aérospatiale Caravelle: The Series VI now operated by BIAS (this column last month) is ex-United Air Lines. ☐ Two purchased by China Airlines to replace YS-11s are ex-Swissair, not ex-United as previously reported. ☐ Sterling Airways is disposing of some Caravelle 10s as its larger 12s are delivered: one has gone to SATA and two to Syrian Arab Airlines.

BAC One-Eleven: Flamingo Airlines, an associate of Bahamas World Airways, may lease two ex-Bahamas Airways One-Elevens for inter-island service, subject to government approval. Court Airlines is also reported to be interested in acquiring the ex-Bahamas aircraft. ☐ Gulf Aviation has confirmed its order for a second Srs 400.

Boeing 707: Kuwait Airways ordered two -320Cs for early 1972 delivery.

Boeing 720: Monarch Airlines, British independent linked with Cosmos Tours, has contracted with Northwest for three 720Bs, the first for delivery in December. ☐ Trans Polar has added one, ex-Aer Lingus, to its fleet of two.

Boeing 727: Braniff has leased one from Executive Jet Aviation (EJA's last of this type). ☐ Olympic Airways has ordered another -200.

Boeing 737: Saudi Arabian Airlines reports that it will buy five -200s for deliveries starting second half of 1971, to replace DC-9s. ☐ A third -200 has been ordered by Air Algérie, for December delivery. ☐ Braathens-SAFE ordered one more. ☐ Nordair ordered its fourth -200.

Business Twins: Beech sold three BH125s to Sears Roebuck. ☐ McAlpine Aviation bought one HS 125. ☐ Modern Air bought its second Hansa Jet for European operations. ☐ Rousseau Aviation bought a Beech 58 for use as an instrument trainer and for business flying.

Canadair CL-44D: Airlift sold one to Transportes Aereos Rio Platenses in Argentina.

Convair 880-M: Boeing accepted four as trade-ins from Japan Air Lines, which has now disposed of its entire fleet of this type.

Convair 990: Modern Air has bought two from Varig, and two others from American Airlines, previously leased to MEA.

Fokker F.27: East-West Airlines ordered one more. ☐ New Zealand's DCA acquired two for airways calibration; they carry Civair markings.

continued on page 168

MiG-21

Air Superiority Soviet Style

In this article, Roy Braybrook, BSc, CEng, RAeS, discusses the effectiveness of the Soviet Union's most successful jet fighter and describes its many different variants

IT IS NOT necessary to push the imagination far to see that preliminary planning and design of Artem Mikoyan's MiG-21 was, as in the case of Clarence L "Kelly" Johnson's F-104A Starfighter, part of the fall-out of the air battles of the early 'fifties over Korea. Although forced to descend below their best altitudes to meet UNO aircraft, Soviet pilots on detachment to the ChiComAF had found their MiG-15s able to perform effectively against B-29 Superfortresses — 17 destroyed for 16 MiGs lost — and in forcing F-80 Shooting Stars and F-84 Thunderjets to jettison their ordnance loads. However, the MiG-15 had been outclassed in combat with the F-86 Sabre by virtue of the latter's better transonic handling — many MiGs engaging in combat with Sabres had simply spun in — and by the more accurate shooting that had resulted from the Sabre's radar ranging.

The MiG-21 first flew in 1955, and appeared in public over Tushino on 24 June 1956, almost exactly three years after the end of the fighting in Korea. By comparison, the first *production* F-104A Starfighter had flown on 17 February 1956, while the first pre-series Mirage IIIA was not to leave the ground until 12 May 1958. The MiG-21 entered service late in 1959, whereas deliveries of the F-104A had begun early in 1958, but the production Mirage, the IIIC, did not appear in quantity until 1961.

Design concept
Considerations behind the design of the MiG-21 undoubtedly centred on the provision of exceptional climb rate and effective firepower to deal with high-altitude bombers in

clear weather conditions. Also written into the requirement, however, was obviously the ability to counter whatever the West could produce to succeed the Sabre, the Hunter and Mystère IVA. In the year that saw the end of fighting in Korea, the MiG-17 — virtually a refinement of the MiG-15, with reheat on some sub-series — was entering service and, the Mach 1·4 twin-engined MiG-19 began prototype trials. Mikoyan's team had succeeded in keeping all of these fighters relatively light — in terms of clean gross weight, the MiG-15bis was a mere 13,500 lb (6 150 kg), and even the MiG-19 weighed only 16,700 lb (7 600 kg).

In coming to the next major project, there must have been considerable pressure to design a much heavier fighter, in line with the large missiles generally held necessary to destroy a heavy bomber. This philosophy would ultimately have led to production of the *Flipper*, as NATO dubbed the twin-RD-11 powered variant of Mikoyan's mighty Ye-166 record-breaker, which was seen over Tushino in June 1961, and was comparable in weight and performance with BAC's Lightning.

At least three considerations militated against this approach. Firstly, defence of the immense Soviet perimeter demanded large numbers of interceptors and, therefore, low unit cost. Secondly, the air superiority rôle favours aircraft that present small targets and are cheap to replace. Thirdly, during the Sino-Soviet technical collaboration programme there fell into Communist hands — or more accurately, lodged in the aft fuselage of a MiG-17 over the Formosa Strait — a Sidewinder lightweight air-to-air missile

(Head of page and below) The Fishbed-J, seen in service with the Czechoslovak Air Force, is the latest development in the MiG-21 series to see widespread service, and is a dual-rôle model embodying both structural and aerodynamic redesign.

that had the capability to destroy aircraft as large as B-52s.

We will probably never know the relative importance attached to these considerations, or, indeed, what other factors played a part in the decision to concentrate on the MiG-21. For example, the *Flipper* was evidently designed to carry two fully-active, two-stage missiles, and there is reason to believe that these proved too sophisticated to develop. If this failure left only air-to-air missiles that required the target illumination of the twin-engined Yak-28P's immense *Skip Spin* radar, then the *Flipper* programme was obviously doomed. Whatever the reason, the *Flipper* was abandoned, and the MiG-21 continued as mainstay of Soviet fighter forces.

This fighter, which received the somewhat ridiculous NATO appellation of *Fishbed*, has always been shrouded by a cloak of secrecy, and, indeed, for several years it was believed in the West that the swept-wing *Faceplate* rather than the delta-wing *Fishbed* had entered service as the MiG-21. Even when this misunderstanding was cleared up, no worthwhile photographs were released in the Soviet Union for some time, and the MiG-21 was generally thought to be quite a heavy fighter, partly because the six-ton thrust quoted for the power plant of the record-breaking Ye-66 version was taken to be a dry (military) rating rather than with afterburning.

Good photographs became available shortly before the security grading of the MiG-21 was relaxed to permit its export to Finland in 1963, removing any doubts as to the fighter's size. The MiG-21 now serves with no fewer than 23 air arms, including those of several non-aligned countries, but even this widespread use has not rendered the MiG-21 readily accessible to western observers! The Soviet Union may not be too concerned about who examines a MiG-17 or an Il-28, but insists on the use of secure airfields for the MiG-21, and remarkably few westerners have ever been inside the fighter's cockpit. This security screen holds true even in Indonesia, which could hardly be described as being on the best of terms with the Communist bloc!

Detail photographs became available after an Iraqi AF officer defected with his MiG-21 to Israel in August 1966, thus winning a bounty of £125,000. It is widely believed that the subsequent Israeli offer to allow US experts to examine the avionics on this aircraft was used as a lever to obtain clearance to purchase A-4 Skyhawks. Attempts made earlier that year to bribe an East German pilot to defect with his MiG-21 proved unsuccessful, but a break-through allegedly came during the 1967 Six-day War, when a flight of Algerian MiG-21s landed at El Arish airfield, their pilots being unaware that the field had been captured by the Israelis. Although this acquisition was never officially confirmed, reports at that time spoke of either two or three of these

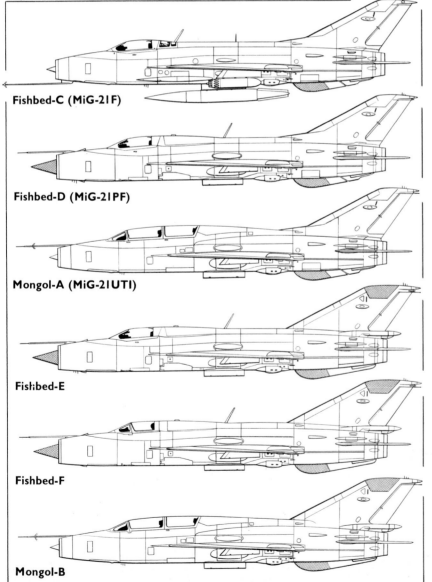

Fishbed-C (MiG-21F)

Fishbed-D (MiG-21PF)

Mongol-A (MiG-21UTI)

Fishbed-E

Fishbed-F

Mongol-B

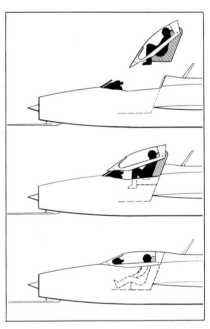

(Above) These drawings illustrate the novel semi-encapsulated escape system used by all Fishbed fighters prior to the Fishbed-F, to protect the pilot from blast effects until the seat has slowed down under the action of a drogue chute.

(Left) The accompanying drawings depict (from top to bottom) the Fishbed-C with centreline drop-tank which may be carried by all MiG-21 variants and UV-16-57 pods (containing 16 55-mm rockets) on the wings; Fishbed-D which introduced the enlarged nose, larger main undercarriage members, deepened dorsal fairing and redesigned forward speed brakes; Mongol-A which combined original intake with the larger main undercarriage members and a single centrally-mounted forward speed brake; Fishbed-E which combined repositioned parabrake with broader rudder; Fishbed-F which introduced a side-hinged canopy with fixed windscreen and quarterlights, and Mongol-B which had the broader rudder and repositioned parabrake, plus a slightly deeper dorsal fairing.

The Fishbed-D, or MiG-21PF (above) embodied a redesigned forward fuselage with enlarged air intake and fixed conical centrebody to house Spin Scan search and track radar. Other changes included enlarged mainwheels and revised canopy and aft fairing.

aircraft being broken down and flown out in C-130s for USAF testing at the Wright-Patterson Air Force Base, Dayton, Ohio.

As in the case of that flown by the Iraqi defector, these aircraft, if they exist, are probably of an early sub-series, and there is no known instance of a late model MiG-21 falling into western hands. Nevertheless, such an aircraft was presented for external examination at the display held in Domodedovo on 9 July 1967, and photographs of what is believed to be the latest production sub-series have been made available by Czechoslovakia (and are reproduced on these pages). By virtue of such photographs and the stories that circulate in odd corners of the fighter world, it is now possible to build up a relatively clear picture of the MiG-21, its systems, and development, although there remain many aspects that could be cleared up by access to possible WPAFB flight test reports, or pilot's notes (which the Israelis have) and other official Soviet publications.

NATO code-names

There is something of a gap in available information regarding the earliest forms of MiG-21 which NATO pundits, in their wisdom, considered to differ sufficiently one from another to warrant suffix letters being appended to their chosen code-name of *Fishbed*, but there can be no doubt that the *Fishbed-A* was supposedly the initial production model mounting a pair of 30-mm NR-30 cannon, while the *-B* was a variant — perhaps only an aerodynamic test-bed — that appeared simultaneously over Tushino in 1956, sporting a trio of large fences on the upper surfaces of each wing and twin splayed ventral fins. The first mass-produced model was dubbed *Fishbed-C*, and differed externally from the *-A* in having vertical tail surfaces of slightly broader chord and a fairing in place of the port NR-30 cannon, apparently to make space for avionics for two wing-mounted K-13 *Atoll* missiles that had meanwhile been evolved in the Soviet Union as counterparts of the Sidewinder.

By comparison with later variants, the *Fishbed-C* is easily recognised by its relatively small intake with translating three-shock centrebody — normally seen in the retracted position when the aircraft is on the ground or in airfield traffic patterns. The muzzle opening of the port NR-30 weapon mounted by the *Fishbed-A* is faired over as a continuation of what had previously been the blast tube covering, and although marginally broader than that of the first *Fishbeds*, its fin remains of relatively narrow chord and is distinguished by a highly-swept leading-edge root extension. Although this essentially simple day fighter model of the MiG-21 was progressively developed during its production life, external changes were negligible. Thus, no matter what modifications took place "under-the-skin", all versions of this day intercept model remain, in so far as NATO is

concerned, the *Fishbed*-C. For example, at a relatively early stage in the MiG-21's production life, the Tumansky RD-11 two-spool turbojet — which, according to non-western sources, originally had a military rating of 8,600 lb (3 900 kg) thrust boosted to 11,244 lb (5 100 kg) with afterburning — was uprated, the official Soviet designation for aircraft with the marginally more powerful engine being MiG-21F, the suffix letter indicating *Forsirovanny* or, literally, "boosted". All export day fighters have been F-series aircraft, a different number having been assigned to each recipient country and appended to the designation. The batch of aircraft purchased

(Above) The MiG-21F alias Fishbed-C with underwing UV-16-57 rocket pods, and (below) a Fishbed-F taking-off with the assistance of JATO rockets.

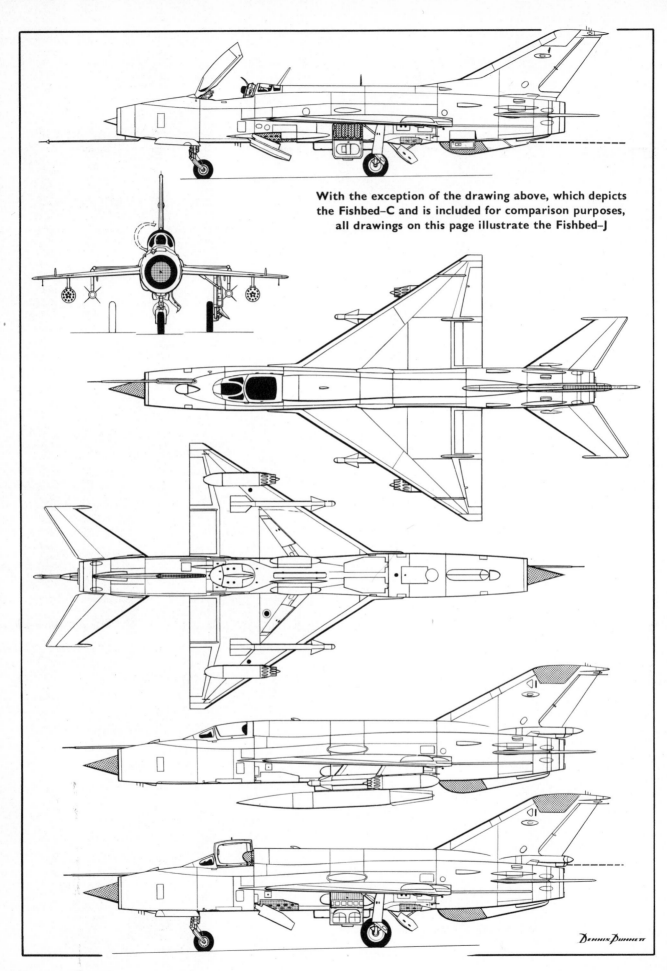

With the exception of the drawing above, which depicts the Fishbed–C and is included for comparison purposes, all drawings on this page illustrate the Fishbed–J

Dennis Punnett

A stage in the evolution of the MiG-21PF is illustrated by the aircraft above, presumably the Fishbed-E, with extended vertical fin accompanied by a new bullet-type parabrake housing, the latter being introduced on some late production Fishbeds-Ds.

by Finland were MiG-21F-12s, and that flown to Israel from Iraq (now generally referred to by its improvised serial of 007) was a MiG-21F-13.

The principal Soviet designation for later variants delivered to the Soviet Air Forces is MiG-21PF, the "P" indicating *Perekhvatchik*, or Interceptor, a term used in the Soviet Union in the *all-weather* intercept connotation*, and implying that search and track radar is fitted rather than the radar ranging of the day fighter MiG-21F. The export equivalent is the MiG-21FL which may differ in IFF fit. At the time of its début, the MiG-21PF was dubbed *Fishbed-D* by NATO, and the changes embodied by this all-weather version were comparable from some aspects with those that translated the F-104A day fighter into the all-weather F-104G.

By comparison with the *Fishbed-C*, the -D introduced a substantially enlarged intake — the lip diameter being increased from 2 ft 3 in (69 cm) to 3 ft (91 cm) — with a fixed conical centrebody for the *Spin Scan* search and track radar. The pitot-static boom, which evidently gave too much trouble in the underside position of the *Fishbed-C* (it tilted upwards on the ground to minimise eye-gouging), was moved to a position above the intake, and the attitude vanes were deleted. The canopy was revised to improve wave drag at the expense of rear view, leading into a fairing which extended halfway down the spine, the rear transparencies being deleted and the primary radio mast being moved aft, while the secondary mast was deleted. The removal of the fuselage cannon fairings permitted simplification in the design of the forward speed brakes, and substantially larger mainwheels were applied, with, presumably, some beefing-up of the legs, the new wheels dictating a considerable increase in the size of blister fairings over the wheel wells which protruded above the wing-roots.

The next stage in development involved moving the parabrake stowage from the lower port fuselage to a bullet fairing at the base of the fin, this bullet being formed by segments opening clamshell fashion to release the chute.

The *Fishbed-D*, like the -C, employed a somewhat primitive form of parabrake stowage, the chute being housed by a compartment alongside the ventral fin. On release, the twin doors opened and the cable was dragged from a channel in the fuselage and ventral fin to a tow-point at the aft end of the latter. It would seem that relatively few *Fishbed-D* aircraft had left the assembly lines after standardisation of the new parabrake housing before the vertical fin area was extended by carrying the leading-edge forward some 18 in (46 cm), thus eliminating the root fillet. This change was clearly intended to overcome the loss of directional stability which limited the high Mach performance of earlier models. It may be assumed that it was to this model with acorn brake fairing and broader fin that NATO applied the designation *Fishbed-E*.

Both *Fishbed-D* and -E have reportedly been employed in North Vietnam, and the latter was flown by the Soviet squadron that visited the Swedish Air Force base at Uppsala, north-west of Stockholm, in the summer of 1967. The next sub-series identified by NATO as *Fishbed-F* embodies all the progressive modifications featured by the -E to which have been added a conventional fixed windscreen and quarterlights associated with a simple ejection seat which supplants the earlier semi-encapsulated escape system, the canopy now hinging to starboard instead of tilting forwards. The dorsal spine appears to have been modified slightly to improve area distribution adjacent to the wing trailing edge. The *Fishbed-F* is the version of the MiG-21 currently being manufactured in India as the Type 77, and the Czechoslovak industry has developed its own short-field version (which has the Czech designation MiG-21SPS) with a bigger braking chute, RATO and, reportedly, blown flaps in place of the normal Fowler type.

One variant of the *Fishbed-F*, which, in view of its purely research nature, has somewhat surprisingly been assigned the NATO designation of *Fishbed-G*, is the STOL aircraft used as a low-speed testbed for lift engines, and shown at Domodedovo in 1967. Additional structure some 3·3 ft (1,0 m) in length had been inserted between the front and centre fuselage sections to provide space for two vertically-mounted lift engines. A scoop intake hinged at the rear ensured ram air for starting the lift engines in flight, and spring-loaded louvres in this door minimised the blockage

(Opposite page) The head-on, top plan and underside plan views of the Fishbed-J depict this dual-rôle fighter with K-13 Atoll air-to-air missiles on the inboard pylons and UV-16-57 rocket pods on the outboard pylons, and the sideview with undercarriage and speed brakes retracted and cockpit canopy closed illustrates the standard drop tank on the centreline pylon. The broken lines extending aft from the sideviews of the Fishbed-C (at head of page) and Fishbed-J (at foot of page) indicate the point of deployment of the parabrake cable. The hinged door of the Fishbed-C parabrake housing (above the forward portion of the ventral fin) is shown open, and the clamshell-type segments of the bullet-type parabrake housing of the Fishbed-J are shown open.

*The "P" suffix is applied to aircraft used for the all-weather intercept mission but not designed primarily for that rôle from the outset (eg, the Yak-28P "Firebar" all-weather interceptor is a derivative of the Yak-28 Brewer tactical strike and reconnaissance aircraft).

The Mongol-B tandem two-seater (above) embodies some of the refinements introduced by later Fishbeds, the initial training variant, the Mongol-A (below, left) being an adaptation of the Fishbed-C with the enlarged mainwheels first employed by the Fishbed-D.

when these engines were being run at low forward speeds.

External ducts took compressor bleed air to two reaction nozzles in the nose (probably duplicated in the tail) for longitudinal trim and control during transition. There was no sign of reaction controls at the wingtips. The main undercarriage members were fixed in the "down" position (and the fairing doors had been removed) since the space normally occupied by the retracted wheels was taken up by the diverted intake ducts to the main engine. The nosewheel was still retractable in order to avoid directional stability problems. A further experimental aircraft based on a Fishbed-F airframe has been used as the "Analogue" aerodynamic test vehicle for the modified-delta planform of the Tupolev Tu-144 SST, with emphasis on low-speed handling checks.

The Fishbed-H is allegedly a derivative of the -D with improved avionics and armament, and photographic capability, but there would seem to be some uncertainty concerning the external changes that are considered to have warranted the issue of a new suffix letter, and the latest known model, the Fishbed-J, is a dual-rôle aircraft easily recognisable by the fact that the dorsal fairing has been enlarged so that the depth is now almost constant back to the fin. The nose boom has been moved to the upper starboard side, and two further wing pylons have been added. The launching shoes for the K-13s remain on the inboard pylons, these pylons evidently being too close to the main gear to accept drop tanks. The centreline pylon is now understood to carry a 23-mm gun-pod as an alternative to the drop tank, and a camera pod may presumably be carried in this position.

Mach Two T-Bird

The two-seat operational training version of the fighter, the MiG-21UTI, has been allocated the name Mongol by NATO, and has appeared in two versions. The initial Mongol-A at first sight appeared to be a straight-forward adaptation of the Fishbed-C, with a second seat inserted behind the normal cockpit position at the expense of fuel

volume, but closer examination revealed that it employed the enlarged main undercarriage members first introduced by the Fishbed-D. Both canopies hinged to starboard, and were followed by a fairing which tapered down to a small spine leading in to the narrow-chord fin. The pitot-static boom was mounted above the nose but offset to starboard.

Like its single-seat counterpart, the two-seat Mongol was progressively developed, and the current Mongol-B embodies some of the refinements introduced by later Fishbeds, such as the broad-chord fin, and the bullet-type parabrake housing at the base of the vertical tail surfaces, while the dorsal fairing has been slightly enlarged at the fin junction. Both Mongol variants retain the relatively small air intake of the Fishbed-C with translating three-shock centrebody, and both may carry two K-13 missiles.

The principal dimensions of the basic MiG-21 day fighter, alias Fishbed-C, have recently been revealed by Soviet journals, but weights have not been published, possibly to give the Israelis the benefit of any doubt. These dimensions, which will differ only marginally in length and height for later variants of the MiG-21, include a span of 23 ft 5½ in (7,15 m) and a wheel track of 8 ft 9½ in (2,69 m). Overall length from the tip of the fin is 44 ft 2 in (13,46 m) to the intake lip and 51 ft 8½ in (15,76 m) to the end of the nose boom, while the wing, which is swept 53° at the leading edge, has an area of 247·57 sq ft (23 m²).

Eight years ago, the writer estimated a reheat thrust of 12,500 lb (5 670 kg) on the basis of the MiG-21F intake size, with potential for at least 13,200 lb (6 000 kg) on the MiG-21PF, and there still seems no reason to modify these figures. Basic weight is believed to be about 12,000 lb (5 450 kg), compared with 14,200 lb (6 450 kg) for the Mirage III and approximately 15,000 lb (6 800 kg) for the F-104G. This would give the MiG-21F a clean gross in the region of 15,500 lb (7 050 kg), increasing to 17,000 lb (7 750 kg) with the centre-line tank and two K-13s.

The intake of the MiG-21F is of relatively sophisticated design, with a centrebody made up of two conical sections, and a ring of vortex generators to prevent separations at the foot of the normal shock. Small spill doors on either side of the intake dump excess air. There are three longitudinal positions for the centrebody, automatically selected according to Mach number, and its location is indicated in the cockpit by a row of lights. Provision is made for manual override. In developing the MiG-21PF, the problems of moving the large Spin Scan radar and of making it look through a double-conical radome evidently outweighed engine performance considerations, and a simple two-shock intake using a fixed conical centrebody was chosen. In this case the centrebody boundary layer is removed by an annular slot at the foot of the normal shock, and sucked overboard through openings above and below the intake.

The most novel feature of the aircraft is the semi-encapsulated escape system used on *Fishbed-C*, *-D*, and *-E*. As far as can be established, the transparency assembly (which for normal cockpit access rotates about its forward end) when closed locks on to trunnions on the sides of the ejection seat. If the seat is fired, the transparency is drawn out from its forward attachment as it pivots about the seat trunnions, protecting the pilot from blast effects until the seat has slowed down under the action of a drogue chute. He can then free himself from both hood and seat, and open the main parachute.

This elaborate protection has proved unnecessary for a pilot wearing a helmet sealed by a faceplate, and with limbs firmly restrained. It also has the disadvantage that low level ejections — ie, below 1,000 ft (300 m) — are ruled out by the time wasted in complicated sequencing. The *Fishbed-F* (and all *Mongols*) therefore use a simple ejection seat. On all MiG-21 seats safety on the ground is ensured by a spring-loaded arm on top of the seat rather than the normal safety pin. Ejection cannot be initiated unless the hood is closed, and servicing accidents are thus avoided.

The MiG-21 cockpit is extremely simple, and has been compared favourably with that of the Hunter. Instrumentation includes the usual Soviet true airspeed indicator and an unspillable artificial horizon. The radio compass (ADF) is particularly admired. The most complicated dial in the cockpit is one combining displays of altitude and airspeed, from which the pilot judges which of two tailplane gearing ratios to select. Artificial feel for longitudinal control is provided by q (dynamic pressure) sensor, presumably used in combination with spring feel, which is the obvious choice for the ailerons. Cabin conditioning is poor.

Power by Tumansky

The engine has now been identified as a Tumansky RD-11-300 two-spool turbojet. Accessories are mounted on top of the compressor casing, because of the ventral droptank and the presence of the retracted mainwheels in the fuselage sides. Notwithstanding the aircraft's length of service, engine overhaul life is still only 250 hours (compared to 400-1,000 hr for Western fighter engines), and it is said that the turbine blades are thrown away after half this interval! The ceramic-coated jetpipe terminates in a petal-type variable convergent nozzle. Due to the extremely thin wing, all the internal fuel is contained in the fuselage, and it has been noted on many occasions in combat that strikes on MiG-21s have led to a tank explosion immediately aft of the cockpit. This particular tank is said to contain a volatile water-methanol mixture, which is presumably injected into the intake ducts to boost thrust, as on early F-4 Phantoms.

Whereas most Western fighters have fully duplicated hydraulic systems for the flying controls, with some form of back-up pump on one system, the MiG-21 appears to have a single system with only one pump on the engine. However, if this fails to provide pressure, an electrically-driven pump takes over, using power from the aircraft battery. If this back-up system also fails (for example, if hydraulic lines are fractured), the pilot can still trim the tailplane electrically, select the "long-arm" gear ratio, and land manually! Vulnerability is further reduced by armour plate both in front and behind the cockpit.

The MiG-21 established several international height and speed records under the designations Ye-66 (probably a souped-up *Fishbed-C*), Ye-66A with rocket boost, Ye-76 (*Fishbed-E*), and Ye-33 (*Mongol*). Of the absolute records, only the Ye-66A's 1961 altitude of 113,892 ft (34 714 m) remains unbroken, but it is more significant that as early as 1959 the Ye-66 reached 1,485 mph (2 376 km/h,) corresponding to approximately Mach 2·25. A more realistic speed for a standard MiG-21F is believed to be Mach 2 clean, with a diving limit of Mach 2·3, which speed the latest model

can probably reach in level flight. Early variants were reputed to be restricted to subsonic speeds at low level, but a recent unconfirmed report suggests Mach 1·07 can be attained with the stiffened wing of *Fishbed-J*.

"Goes nowhere, does nothing!"

Endurance and radius are severely limited. In normal supersonic interception carrying the ventral tank, the MiG-21F is only good for 35 minutes and a radius of 98 miles (155 km). The fuel in the dorsal fairing and drop tanks of the *Fishbed-J* will clearly enable it to improve substantially on these figures, but the MiG-21 remains suitable only for point defence and close support operations.

Flown clean, the MiG-21 has a marginally higher thrust/weight ratio than either the Mirage or F-104G, and can accelerate better transonically, although the maximum speed of *Fishbed-C* is exceeded by the Mirage at least. The Soviet aircraft's radius of turn is reportedly smaller than those of the Mirage, F-104G and F-4, and the MiG-21 is significantly better than the Mirage (with the possible exception of the Milan version of the latter) in terms of airfield performance. However, against these good flying qualities must be offset the fact that the MiG-21 is designed around an engine that is far too small, making it a lightweight in every sense. The basic layout, escape capsule, and variable geometry intake are evidence of ingenious design (in retrospect, perhaps *too* ingenious), but this is small consolation for the man sent to war with only six tons of thrust, one single-barrel gun and a couple of Sidewinder copies.

The availability of two extra pylons on the *Fishbed-J* give it a nominal ground attack capability, but this hardly makes it comparable with the dual-rôle Mirage. In reality the MiG-21 competes only by virtue of a heavily subsidised export price, making two aircraft available at the same cost as one from Avions Marcel Dassault. If Israel were to mate the Milan airfield and turning performance with the acceleration and climb provided by a late-model J79, then it would be no contest at all. Italy's F-104S Starfighter probably also has a clear lead.

The key issue is whether there is a Soviet powerplant in the J79 class available to re-engine the MiG-21, and on the basis of published information this currently appears unlikely. Nevertheless the 3,500 Warsaw Pact tactical aircraft in Central Europe, with the MiG-21 as principal type, still represent a powerful deterrent to their NATO counterpart, numbering only 1,500.

Whether Communist defences would have been more effective if the Soviet Air Forces had abandoned the *Fishbed* in favour of a man-sized design such as the *Flipper* is a matter for debate. Perhaps in the confused environment of European low-level operations, the search capability provided by sheer numbers more than offsets the MiG-21's incredibly pathetic armament. □

PIRATES ASHORE!

THE VOUGHT A-7 JOINS THE USAF

Twice in a decade, the United States Air Force has swallowed its pride in order to buy combat aircraft already. developed and in production for the US Navy, notwithstanding the vigour with which each service defends its right to initiate new aircraft designs tailored to its specific needs. The first of the USN types bought by the Air Force was the McDonnell Douglas F-4 Phantom II, destined to become one of the greatest of all fighters to serve with the USAF; the second is the Vought A-7D, only recently introduced into the Air Force inventory but already showing signs of repeating the Phantom's success. It also marks a significant "first" for the USAF, in that it has a pilot's head-up display incorporated in the advanced nav/attack system.

Selection of the A-7 by the USAF came in 1965 and was made at top levels in the Air Force and Department of Defense, despite some misgivings in lower echelons over the purchase of a subsonic aircraft. The A-7D, in fact, is the first subsonic jet-powered fighter aircraft type put into service by the USAF for nearly 20 years. The need to support the Army in Vietnam with a close-support aeroplane played an important part in the choice of the A-7, which was seen primarily as a replacement for the North American F-100 Super Sabre. At the time the A-7D was selected, the USAF had four fighter wings of F-100s operating in Vietnam, these aircraft being deployed almost exclusively in support of ground actions in South Vietnam, where no air opposition was encountered.

Intended to operate in a similarly permissive environment, the A-7D did not need to have the high supersonic performance of the F-4s and F-105s which were committed to strikes against targets in North Vietnam. Factors in its favour, so far as the Air Force was concerned, were that it was quickly available at a lower cost than would inevitably be required for a new type developed from scratch, and that it provided a good platform on which to mount a new nav/attack system designed to provide weapon delivery accuracy of an order previously impossible.

At the time the USAF decided in principle to buy the A-7, orders had been placed for two production versions, the A-7A and A-7B, for use by the Navy and Marine Corps. The Ling-Temco-Vought design, based on the F-8 Crusader, had won a Navy design competition (known as VAL) for a relatively inexpensive attack aeroplane, and had been ordered into production in mid-1963. First flight came on 27 September 1965, and the Navy selected the name Corsair II for its new type — although this name has not been adopted by the Air Force for the A-7D.

Just over two years were required to take the Corsair II from first flight to combat status over Vietnam, the first operational USN unit, VA-147 ("The Argonauts") sailing aboard the USS *Ranger* on 6 November 1967. In a change of policy, the US Marine Corps decided to defer procurement of the A-7 and has bought additional quantities of updated versions of the McDonnell Douglas A-4 Skyhawk instead. However, production of 199 A-7As and 196 A-7Bs for the US Navy proceeded as planned, while new and improved models were developed in parallel for the Navy and the Air Force, respectively as the A-7E and A-7D.

Air Force ideas

The USAF requested changes to the basic Corsair II design in five significant areas in order to produce an aircraft that would meet its specific requirements. These changes covered the powerplant, flight refuelling, armament, nav/attack system and protection from small arms fire. As a result of these and other more minor changes, commonality between the A-7D and A-7A was only about 25 per cent, being limited to the basic airframe structure and the hydraulic system. Much greater commonality is achieved between the A-7D and the Navy's A-7E, however, since the latter incorporates many of the changes specified by the Air Force.

The need to increase the engine power of the Corsair II quickly became apparent to the USAF. Power plant for the original A-7A was the Pratt & Whitney TF 30-P-6 rated at 11,350 lb st (5 150 kgp) and giving a nominal thrust-to-weight ratio of 0·3:1 — adequate for accelerated take-offs from the 250-ft (76-m) length of a steam catapult but likely to result in excessively long take-off runs for heavily laden aircraft operating in high temperatures such as encountered in Vietnam. As the Navy engine was derived from the afterburning TF30 used in the Air Force's F-111, the possibility of using reheat in the A-7D was considered, but discarded because of the difficulty of fitting the larger jet pipe. A suitable alternative was found in the Rolls-Royce RB-168 Spey, which was offered to the USAF in an American-built version by Allison Division of General Motors, designated TF41-A-1. Rated at 14,250 lb st (6 465 kgp), this engine provides the A-7D with a thrust-to-weight ratio of 0·34:1 despite the fact that the overload gross weight is increased by 4,000 lb (1 814 kg) compared with that of the A-7A and B.

A change from probe-and-drogue to boom-in-flight refuelling was a natural consequence of the different philosophies adopted by the US Navy and Air Force. The Navy Corsair IIs have a retractable probe in the starboard fuselage side just below the cockpit, giving the pilot an excellent view when making contact with the drogue trailed by the tanker aircraft. On the A-7D, this feature is omitted, and in its place there is a receptacle for the "flying boom" of the tanker, located in the top fuselage decking just aft of the cockpit. For production convenience, the first 16 A-7Ds for the USAF were completed with the Navy refuelling system.

Experience in Vietnam had demonstrated, by 1966, the effectiveness of the 20-mm General Electric Vulcan gun, with a rate of fire of 6,000 rpm from its six rotating barrels. This gun was therefore considered an essential item for the A-7D, and was adopted in place of the two 20-mm single-barrel cannon fitted in the A-7A and A-7B. The same basic provision for fuselage and underwing loads was retained, that is to say, two fuselage-side mountings for 500 lb (227 kg) each; two inner wing pylons for 2,500 lb (1 134 kg) and four centre and outer wing pylons for 3,500 lb (1 588 kg) each. The fuselage mounts can carry only air-to-air missiles or rocket pods, while the wing pylons variously can accommodate missiles, gun pods, rocket pods, bombs or (inner and outer points only) fuel tanks. Fully loaded, the A-7D can carry a maximum external load of 15,000 lb (6 804 kg), leaving 7,510 lb (3 406 kg) for internal fuel. The basic weight of 19,490 lb (8 840 kg) includes an allowance for 1,000 rounds for the M 61 gun — enough for 10 seconds' firing. Gross weight is 42,000 lb (19 050 kg).

Although the USAF Chief of Staff, General John P McConnell, had referred to the need for A-7s "for the purpose of providing close air support to the ground forces in a permissive environment", this did not mean that they would not be subjected to enemy fire. In fact, aircraft operating in the ground attack rôle in Vietnam had proved uncomfortably vulnerable to small arms fire from the ground, and this led the Air Force to specify a higher standard of protection for airframe and systems than the Navy had originally required. Whereas only the main sump fuel cell was protected in the A-7A, all fuel tanks in the A-7D were provided with self-sealing capability, vital controls were duplicated and protection provided for other controls. The extra engine power more than compensated for the increased weight brought about by these changes.

The nav/attack system

As already noted, the A-7D is the first USAF fighter to feature a Head-up Display of navigation and attack information. The British-supplied equipment was specified as part of the extensive redesign of the total avionics package for the USAF. This package is a fully integrated nav-attack system built round an IBM digital computer — IBM also serving as avionics system integrator for the A-7D under contract to LTV.

The computer is a System/4 Pi, Model TC-2 (AN/ASN-91), with a 16,384 word memory which stores trajectory in-

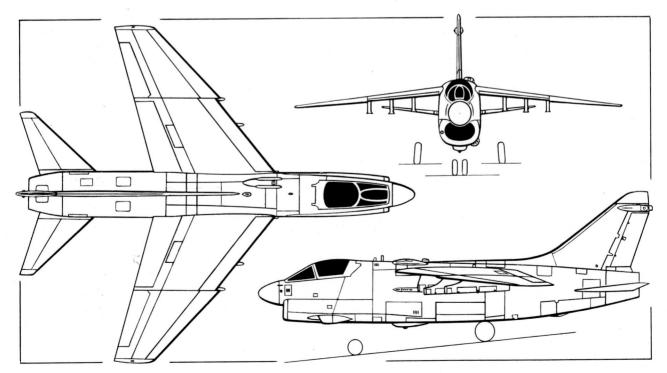

(Above) To conform to USAF standard practice, the Vought A-7D is equipped for "flying boom" in-flight refuelling, with a receptacle in the upper fuselage aft of the cockpit. (Below) The first 16 A-7Ds, however, were equipped for Navy-style probe-and-drogue refuelling, demonstrated in this picture with an A-7B acting as the buddy refuelling tanker.

formation on the more than 100 weapons that may be carried by the A-7D. Taking the necessary data on aircraft speed, heading, attitude, altitude and so on, from on-board equipment, the computer can solve the equation for any one of nine different weapon delivery modes, or for navigation to a pre-selected point. The weapon delivery modes include visual or radar target selection straight ahead or offset, return to an overflown target and gun or rocket attacks. Weapons are released automatically in a diving, level, climbing, or "over-the-shoulder" attack.

Data is fed to the Elliott AN/AVQ-7 head-up display (HUD) to provide essential aircraft performance and attack, navigation or landing guidance on a single display. The information is provided in the form of symbols projected on to a transparent mirror and focused at infinity, so that they remain clearly visible to the pilot while he keeps his eyes focused on the terrain ahead. The computer provides the attack, navigation and landing data while the aircraft performance data comes from the aircraft flight sensors. In navigation situations, the flight director symbol, by its position in relation to the flight path marker, provides azimuth command for resolution of steering errors, displaying both azimuth and elevation steering while in the terrain-following mode.

Selection of one of the computed attack modes causes the HUD to present weapon delivery symbology appropriate for the weapons chosen. The command is always to fly the flight path marker to the bomb fall line. The aiming symbol can be manually slewed or the aircraft can be flown to position the symbol over the target. After "lock-on", visual contact with the target is no longer required, allowing consecutive approaches on the target without redesignation. If visual contact with the target is maintained, aiming refinements can be made by slewing the aiming symbol.

When the target is within maximum computed weapon delivery range, solution cues appear and start moving down toward the flight path marker. A valid solution occurs when the desired solution cue (for loft delivery, level laydown, dive and toss deliveries) intersects the flight path marker. When guns or rockets are selected, an in-range cue appears on top of the aiming diamond. The aiming reticle indicates the impact point. The HUD also has a standby reticle which is similar to a fixed gunsight on other aircraft. This can be used for computed bombing or manual delivery of all weapons.

In attack, terrain-following and landing situations, the pull-up command (flashing five times per second) appears on the HUD to dictate an immediate 4 g pull-up to avoid ground obstacles or blast radius. During the enroute navigation and terrain-following modes, steering commands to the selected destination are displayed in addition to the basic HUD symbology and indicators.

Other avionics equipment includes a Garrett CP-953A/ AJQ air data computer, a Singer-General Precision AN/ ASN-90 (V) inertial measurement set (IMS) and AN/APN-190 (V) Doppler navigator, and Texas Instrument AN/

APQ-126 (V) forward-looking radar for terrain following, terrain avoidance, ground mapping and air-to-ground ranging for weapon release computation. A Computing Devices of Canada ASN-99 projected map display unit (PMD) has also been adopted by the USAF, and is located low down on the starboard side of the instrument panel. This equipment carries 35-mm slides of maps in two scales covering any area over which the A-7D may be required to fly, and is linked with the IMS to provide a continuous indication of aircraft position. Provision is made for fitting LORAN and laser rangefinder.

The communications/navigation fit comprises VHF and UHF radio, a UHF direction finder, IFF transponder, TACAN, ILS, radar altimeter, radar beacon and Juliet 28 secure voice coder. A radar homing and warning system is fitted and a Westinghouse QRC-335A ECM system can be pod-mounted on the outboard wing pylons to supplement the internal ECM capability.

Among the smaller changes specified by the Air Force were those associated with the landing gear. The A-7D has larger tyres and more powerful brakes, with an anti-skid braking system, and the catapult launch strut on the nose-wheel of the Navy aircraft is deleted. An externally obvious difference is the use of standard camouflage finish on the A-7Ds, with a pattern of tan, green and dark green on the upper surfaces and grey on the undersides, whereas Navy aircraft are finished light gull grey (matt) on the upper surfaces and insignia white (gloss) on the undersides and all control surfaces. A Douglas Escapac zero-zero rocket ejector seat is fitted in the A-7D, as in the Navy aircraft.

Most of the new features introduced by the USAF in the A-7D were adopted by the Navy for its A-7E version, now in production. This has, like the A-7D, an M 61 gun, bombing computer, head-up display, improved hydraulic system and anti-skid brakes. The Navy also decided to adopt the Allison Spey engine and introduced this after 67 A-7Es had been delivered with the 12,200 lb st (5 534 kg p) TF30-P-408 engine (similar to the engine used in the A-7B). The A-7E was able to benefit from the further development work on the American-built Spey variants, being fitted with the TF41-A-2 version rated at 15,000 lb st (6 804 kg p), and this permitted an increase in gross weight to 42,000 lb (19 050 kg) — the highest weight at which a single-engined aircraft has been operated from US Navy aircraft carriers.

Operational introduction

When the Corsair II was taken into operation for the first time over Vietnam late in 1967, a USAF contingent, including three pilots, was attached to Navy squadron VA-147 to obtain first-hand experience with the type in the "Coronet Stallion" evaluation. Four officers and 21 airmen sailed with the USS *Ranger*, the unit being headed by Major (now Lt-Col) Charles McLaren, who had had operational experience flying F-105s over North Vietnam before being attached to the Tactical Fighter Weapons Center at Nellis AFB, California. During an 18-month period, the three USAF pilots amassed almost 1,200 hours on the A-7A, each averaging 65 combat missions and 125 carrier landings.

Major McLaren later said of the A-7A that it was "a very rugged, stable and long-legged aircraft. Performance was not impressive with the small engine but fuel economy was astounding. We rarely flew with a full internal fuel load (10,300 lb; 4 672 kg). Six thousand pounds of fuel would normally give us 1¾–2 hr flying time; the full load gave us a 3-hr-plus capability, with a 2,100-mm (3 894-km) range. Add two external tanks and we could make Hawaii from the West Coast.

"Ordnance loads were tailored for the specific missions — low drag configurations for Alpha strikes and anything we could hang on the aircraft for the 'down south' work. Our

A great variety of underwing loads can be carried by the A-7D, up to a maximum of 15,000 lb (6 800 kg). Illustrated in this column are some typical loads, comprising, from top to bottom, fifteen M117A general purpose 750-lb (340-kg) bombs; eight such bombs and four CBU-46/A bomblet dispensers; four 500-lb (227-kg) general purpose bombs and two SUU-42/A parachute flare dispensers; two Sidewinder AAMs and ten 1,000-lb (454-kg) bombs in the configuration evaluated by the Swiss Air Force; and four AGM-65A Maverick TV-guided missiles, two 300 US-gal (1 136-l) drop tanks and two QRC-335A ECM pods.

standard Alpha strike configuration was four MK 84 bombs and in other missions we carried 12 MK 82 high drag bombs as a standard load. In any of these configurations we could go out 350 nm (650 km), stay on station 30 minutes, and return to the ship with a one-hour fuel reserve. We decided that external tanks were more trouble than they were worth because of reconfiguration problems, and with three hours available on internal fuel, who needed them?

"Our maintenance people thought the A-7A one of the most easily maintained aircraft that they had encountered. We had 14 aircraft and kept 12 airborne most of the time while flying an average of 30–36 sorties per day. The A-7A contract guarantee of 11·5 Direct Maintenance Manhours Per Flight Hour (DMMH) was met during our operations, and we feel that the A-7D guarantee of 9·5 DMMH will also be met . . . in spite of the increased system complexity."

VA-147, under the command of Cdr J C Hill, USN, flew its first combat strikes with the A-7A on 4 December 1967, hitting bridges and highway targets around Vinh. Missions were flown in direct support of the besieged US Marine positions around Khe Sanh and many attacks were mounted on anti-aircraft gun and SAM emplacements around Hanoi, Haiphong and other major North Vietnamese cities. Over North Vietnam, the Corsair IIs usually stayed between 3,500 ft and 15,000 ft (1 067–4 572 m), since small arms fire was troublesome below this altitude band, and above it the aircraft were vulnerable to Soviet missiles. As one pilot reported: "Coming off the ground, the Soviet missiles are slow, and if you pick them up it's easy to dodge them because they can't turn as tightly as the plane. Higher up, they reach Mach 2 speed and they're harder to evade."

The *Ranger* with VA-147 on board was directed to the Sea of Japan at the end of January during the *Pueblo* crisis, and spent two months in operation on the Korean east coast, providing the Corsair II with further experience in different weather conditions. Additional Corsair units began to arrive in the Vietnam area in the spring of 1968, with VA-82 (the "Marauders") and VA-86 (the "Sidewinders") going into operation from the USS *America* on 31 May and VA-27 and VA-97 starting to operate from USS *Constellation* in July. These units continued to fly a wide variety of missions, including "Rolling Thunder" operations against North Vietnam and "Steel Tiger" reconnaisance flights over Laos, for which the A-7s could carry four cameras in the fuselage. Most of the effort involved attacks with conventional weapons — bombs, rockets and air-to-surface missiles — on targets of opportunity; other missions included gun spotting along the Vietnamese coastline, armed route reconnaissance, RESCAP rescue patrols, flack suppression and close air support. Following declaration of cease-fire in November 1968, operations continued against targets south of the demilitarised zone. The A-7B, first flown on 6 February 1968, entered combat about the same time.

The original Corsair unit, VA-147, was also the first to use the A-7E on operations, flying from the USS *America*, together with VA-146, against Vietnamese targets on 9 June 1970. With additional squadrons from the USS *Kitty Hawk* and USS *Ranger*, A-7Es had flown over 9,100 operational sorties by March 1971, with only two aircraft lost in combat. By May 1971, the Navy had 27 operational squadrons of Corsair IIs, in addition to two A-7E training units — VA-122 at NAS Lemoore and VA-174 at NAS Cecil Field — and an A-7A/B training unit, VA-125 at NAS Lemoore. Introduction of the A-7Es released A-7As and A-7Bs for use by Naval Air Reserve squadrons, in which they began to replace A-4s early in 1971. The first Reserve squadron to fly A-7As was VA-303 at Alameda, California.

In order to make an early start on flight development of the new nav/attack system in the A-7D, the USAF agreed to accept the first two aircraft with TF30-P-6 engines, and the

first of these prototypes flew on 6 April 1968. It had been intended to fit P-8 engines in these aircraft, but problems with the second stage turbine blade shroud had been encountered with this power plant model when introduced in the A-7B two months before the A-7D's first flight. The first flight with a TF41-engined A-7D was made on 26 September 1968, preceding the first A-7E by two months.

As already noted, the first 16 production A-7Ds were non-standard in respect of flight refuelling provision and in other ways, including lack of the projected map display. Deliveries of these aircraft to the USAF, however, starting on 23 December 1968, allowed a start to be made with various essential pre-operational test phases.

Test and evaluation teams were formed by the USAF around the nucleus of personnel who had obtained experience of the A-7 in "Coronet Stallion", with one unit at Edwards AFB for Category II performance, stability and control tests, and another at Luke AFB for Category III performance tests. The latter unit, a detachment of the Nellis AFB-based 57th Fighter Weapons Wing, was formed in April 1969 under the command of Major Charles McClaren, and in July 1970 received the first four definitive (TAC 1 Mod 1) A-7Ds.

Also at Luke AFB is the 58th Tactical Fighter Wing, with responsibility for instructor pilot training, and a tactical fighter training squadron for the training of all future USAF A-7D pilots. Climatic tests, using two of the pre-standard aircraft, were conducted at Howard AFB, Panama Canal Zone (tropical phase); Eielson AFB, Alaska (Arctic phase); El Centro, California (hot, dry, dusty phase), and Wright Patterson AFB, Ohio (rainstorm phase). Weapons compatibility tests, under the code-name "Seek Eagle" were the responsibility of the Armament Development and Test Center at Eglin AFB.

Earmarked as the first of three tactical fighter wings to fly the A-7D was the 354th TFW, a component of Tactical Air Command serving until 1970 in South Korea. This unit returned to Myrtle AFB, South Carolina, in order to re-equip, and by January 1971 two of its three squadrons, the 511th and 355th Tactical Fighter Squadrons, had become operational on the A-7D; they were followed by the 4456th TFS, while training became the responsibility of the 310th TF Training Squadron. By that same date, about 100 aircraft in USAF configuration had been completed by Vought Aeronautics Division of LTV in Dallas, out of the total planned USAF procurement of 387 A-7Ds. Production of 512 A-7Es is planned, of which about 300 have been delivered.

Mission capabilities

Operating with Tactical Air Command, the A-7D will be employed primarily in the VFR close air support rôle under the Tactical Air Control System. The aircraft also possesses a limited interdiction and all-weather weapons delivery capability, and although it is not planned to be used in an anti-radiation or nuclear weapons rôle, most A-7Ds are being completed with the necessary nuclear weapons control wiring installed. The provision for Sidewinder AAMs to be carried on the fuselage stations gives the A-7D its limited air-to-air capability.

The overseas deployment capability of the A-7D was demonstrated during the Category II testing at Edwards AFB, when two aircraft flew a non-stop, unrefuelled mission of 3,052 st miles (4 912 km). The flight actually terminated at Homestead AFB, Florida after a dog-leg over the Gulf of Mexico, but it represented a California–Hawaii ferry flight, the longest deployment likely to be required of the A-7D. The two aircraft landed with 2,500 lb (1 134 kg) of fuel remaining. Previously, in 1967, two Navy A-7As had

continued on page 163

(Above) The boom pilot's view of an A-7D during heavy-load refuelling trials, with 12,500 lb (5 670 kg) of bombs on the wing pylons. (Below) During weapon trials at Eglin AFB, this A-7D carried 20 Snakeye bombs each weighing more than 300 lb (136 kg). Some of the bombs are striped for photo-assessment purposes, and a strike assessment camera is carried in the small fairing just ahead of the tail hook.

THE CALAMITOUS 'COBRA

Few aeroplanes have had a less auspicious beginning to their operational careers than the Bell Airacobra, which received its combat initiation at the hands of the RAF in October 1941. Dogged by armament problems, low serviceability and an unreliable compass, the Airacobra flew only four missions before it was taken off operations, subsequent events preventing its use by the RAF in the ground-attack rôle for which it proved well suited in other combat theatres. This article describes, for the first time, the background to Britain's order for the unconventional Bell fighter, the disappointments over its performance, the struggle to bring it up to an acceptable operational standard and the eventual fate of the 675 Airacobras built under British contracts.

As soon as the wraps of secrecy were removed from the Bell XP-39 Airacobra in February 1939, the new fighter was subjected to a blaze of publicity which glowed far beyond the shores of the American continent. With its 37-mm gun in the nose — the largest weapon ever mounted by an American single-engined fighter — a tricycle undercarriage, an engine buried in the fuselage amidships and a claimed top speed in the 400 mph (644 km/h) class, the Airacobra appeared to *deserve* all the publicity that it enjoyed. Unfortunately for its subsequent reputation, however, its makers had made no distinction between the capabilities of a lightly-loaded, highly-polished prototype and the likely performance of a fully-equipped operational version ready to go to war. By the time the Airacobra had reached the production stage, not only had its weight increased by a staggering 30 per cent, but its performance had been emasculated by the removal of the all-important engine turbo-supercharger. Consequently, an interceptor of high promise was translated into a mediocre fighter suitable for little other than ground attack duties where enemy air opposition was negligible. Furthermore, many snags had to be worked out of airframe and systems before the Airacobra was fit to operate even in *that* limited rôle!

In 1940, however, few voices were raised to question the accuracy of the Bell company's claims for its fighter, and those that were received scant attention. Among those with high hopes for the Airacobra were the members of the British Direct Purchase Commission in the USA, which added the type to its shopping list of aeroplanes during 1940. Subsequent events were to prove that decision to be calamitous, for of the 675 aircraft ordered (and built) for Britain, only four were destined ever to fly on operations with the RAF. They flew, between them, four missions across the English Channel, and twice fired their guns in anger, surely representing a ratio of achievement to expectation unequalled by any other combat type acquired by the British armed forces during World War II.

RAF experience with the Bell "back-to-front" fighter was not entirely typical of that of other users, it is true, and the Soviet Air Forces eventually used the Airacobra with considerable success and in large numbers in the ground-strafing rôle. There can be no doubt that the RAF paid a high price for being the first to try to use the Airacobra operationally, at a time when most US combat aircraft had still to benefit from the essential feed-back of data obtained under real war conditions. In this particular respect, the deficiencies in the Airacobra were no exception.

Like its American contemporaries, which also suffered their full share of snags and teething troubles when first exposed to actual combat, the Bell Airacobra had been designed by a team of engineers whose ideas, however ingenious, were not matched by practical experience of the needs of the fighter pilot when the chips were down. The result was an aeroplane that in the words of the then Assistant Chief of the Air Staff (AVM R S Sorley), was "pleasant to fly and very easy to take-off and land". The pilots of the only RAF squadron to use it "formed a very favourable opinion of its possibility as a day fighter", but it had to be taken off operations because of difficulties with the compass. As AVM Sorley put it in his report, with typical RAF understatement, "low flying offensive operations against ground targets over the other side demand a reliable compass. Without it pilots get lost".

Many other problems and deficiencies were discovered in the brief period that the Airacobra served with the RAF, and remedial action was taken, although this was not always easy because of a chronic shortage of spares. However, the compass problem was less tractable, since it derived from the original location of the Kollsman unit being too close to the guns in the aircraft's nose; consequently, when the guns were fired, compass deviations of anything from 7° to 165° were recorded. The optimum solution would have been installation of a distant reading compass, but before this action could be initiated, a policy decision was taken to divert the bulk of the British Airacobras to the Soviet Union, which through Lend-Lease allocation was eventually to receive over half of the 9,584 Airacobras built.

Development of the Airacobra had begun at the Bell company's Buffalo, NY, works in mid-1936 under chief engineer Robert J Woods, and formal submission of the project to the US Army Air Corps was made in May 1937. Woods had decided to design his fighter round a buried engine installation, locating a turbo-supercharged Allison V-1710 in the fuselage *behind* the cockpit, with a long shaft running forward under the cockpit to turn the propeller. This arrangement left ample room in the nose for armament, which was to comprise a 25-mm or 37-mm gun firing through the propeller hub, and two 0·50-in (12,7-mm) machine guns. Other advantages included an aerodynamically better fuselage, and space to retract the stalky nose-wheel leg, allowing the Bell fighter to be one of the first single-engined combat types to have a tricycle undercarriage. The Bell submission indicated a top speed of 400 mph (644 km/h) at 20,000 ft (6 096 m) with a gross weight of only 5,550 lb (2 517 kg) — figures which were to prove overly optimistic.

A prototype of the Bell design was ordered as the XP-39 in October 1937, and the first flight was made on 6 April 1938. The weight turned out to be about 6,200 lb (2 812 kg) but a top speed of 390 mph (628 km/h) was recorded, and this performance no doubt encouraged both the Army Air Corps and the British Direct Purchase Commission to place contracts. A service test quantity of 13 Bell Model 12s was ordered by the Air Corps on 27 April 1939, followed by a production contract for 80 (provisionally designated P-45s) on 10 August 1939; Britain ordered the Model 14 version on 13 April 1940 and another USAAC order placed on 13 September 1940 was for 344 Model 15s. The Bell company, having previously produced only a baker's dozen of the unorthodox Airacudas for the Air Corps, was in business in a big way.

Interestingly enough, in view of subsequent events, Bell's claims for the Airacobra had been the subject of a virulent attack in the authoritative British journal *The Aeroplane* on 1 March 1940. The unidentified author of this article dismissed as "absurd" the various figures which had been published by Bell, these figures including a maximum speed of more than 400 mph (644 km/h), a cruising speed of approximately 325 mph (523 km/h), an operating altitude of more than 36,000 ft (10 973 m), a cruising range over 1,000 miles (1 610 km), a fully loaded weight of about 6,000 lb (2 720 kg), and a wing loading of 28·3 lb per sq ft (138 kg/m²), indicating a wing area of 212 sq ft (19,7 m).

"The weight", this article continued, "is not possible with the big motor, the shaft drive, the tricycle undercarriage, the armament specified and the range and wing loading claimed".

The manufacturer "must have discovered some wonderful new law of nature if they can build an aeroplane which weighs no more than a Hawker Hurricane yet has more horse-power, radiators in the wings, about 300 lb (136 kg) more fuel, a cannon, a tricycle undercarriage, a long extension shaft and a constant-speed airscrew". Putting his finger squarely upon the nub of the problem, the author remarked: "The idea seems to be to get a spectacular top speed for advertising purposes at the expense of everything else. That is all right in America where they only use fighters for advertising purposes anyway. When they are exported for War they have to be modified to make them lethal."

Categorising the Airacobra as a "thoroughly ill-conceived aeroplane considered as a fighter, taken in general and in detail", the article concluded with the hope that Britain would not buy it: "As a serious fighter the Bell is all wrong. We trust the British Purchasing Commission in the USA will not be hoodwinked into placing an order."

Whether or not the Commission was "hoodwinked", the order w*as* placed, as noted above. Had the plea in *The Aeroplane* been heeded, a great deal of wasted effort might have been avoided, but in those days of early 1940, with the whole of Europe succumbing to German might, and the RAF about to confront the *Luftwaffe* on anything but equal terms, the overriding need was for fighters and more fighters. American types were available quickly and, in the circumstances, it was hardly surprising that large-scale British contracts were placed for almost anything with wings. In the case of the Airacobra, no British pilot flew an example of the Bell fighter until eight months after the contract was placed, the first evaluation by a British pilot taking place between 31 December 1940 and 4 January 1941, by Christopher Clarkson.

By the time the first of the service-test Bell YP-39s appeared in September 1940, the prototype had been modified into the XP-39B, with a number of changes which included deletion of the turbo-supercharger. In this form, with a V-1710-37 engine, the aircraft reached its best speed at 13,300 ft (4 054 m) whereas the turbo-supercharged version was rated for best performance at 20,000 ft (6 096 m). This decision to use the low-altitude engine without supercharger was a contributory factor in the subsequent operational disappointments to be experienced with the Airacobra.

(Facing page and below) Illustrations of an early British Airacobra (believed to be AH579, the tenth aircraft) prior to delivery. The 20-mm nose cannon is not fitted, and the rather unusual arrangement of the camouflage scheme is noteworthy.

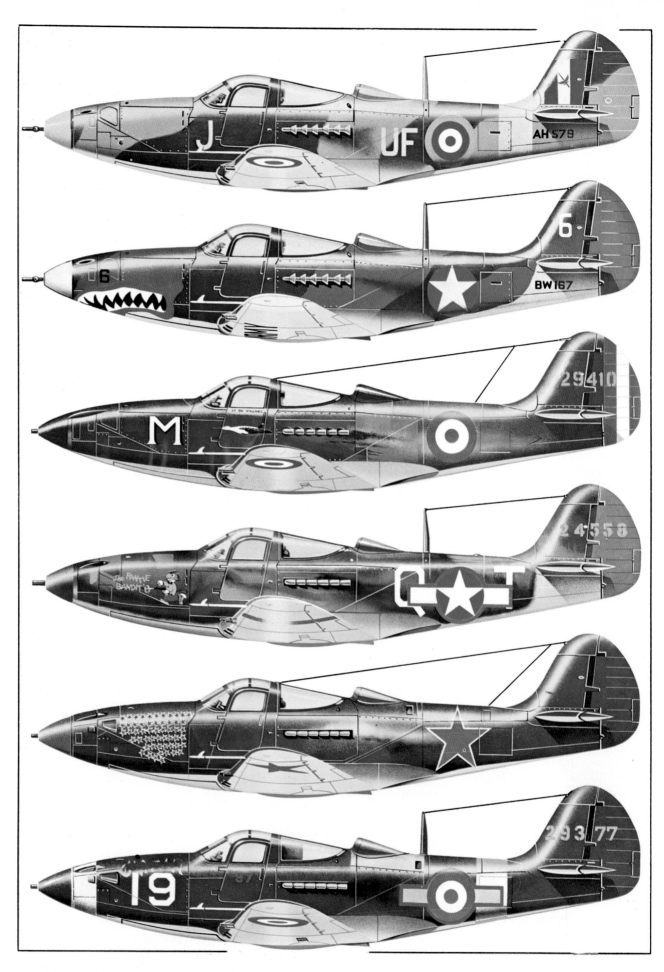

Another factor having a deleterious effect upon performance was the steady growth in both tare and gross weights. The armament was increased by adding 0·30-in (7,7-mm) guns to the 0·50s already located in the nose; first, two of the smaller calibre guns were mounted with the others in the nose (in the P-39C) and then four 0·30s were mounted in the wings (in the P-39D) leaving two 0·50s in the nose together with a 37-mm gun, the 25-mm weapon originally proposed having proved unavailable. Other items adding to weight were bullet-proof tanks, cockpit armour and an armour-glass windscreen.

Deliveries begin

Only the first 20 production aircraft were completed as P-39Cs, these being followed by 429 P-39Ds. The RAF model, which had at first been named the Caribou but for which the American name of Airacobra was adopted in July 1941, was virtually identical with the P-39D with the exception that a Hispano 20-mm cannon replaced the American Armaments Corp 37-mm weapon, with 60 rounds instead of only 30. The 0·30s were provided with 1,000 rounds per gun and the 0·50s with 270 rounds per gun, giving the Airacobra at least three times the weight of armament carried by the Spitfire VB. Bell began test-flying the Model 14 Airacobra I in April 1941, the first few aircraft sporting a somewhat unorthodox version of the approved RAF camouflage scheme. Also tested on AH571, the second British aircraft, was a revised rudder of more angular shape and less area, but although this aircraft was delivered to Britain in this form, the modification was not adopted. A very small dorsal fin did become a standard feature of the RAF aircraft, and was also a distinguishing feature of the P-39D and subsequent versions.

Deliveries of the Airacobra to Britain had to be made by sea, since the air delivery route via Newfoundland, Greenland and Iceland, which was destined to be used by thousands of American warplanes flying to the war in Europe, called for longer stages than the Airacobra could fly. The sea journey was highly hazardous, but the first aircraft, handed over at the Bell factory in May, arrived in England in July, and at a steady rate thereafter. By a margin of a few days, however, the first Airacobra to reach Britain, and subsequently to fly in RAF markings, was not one of the British Direct Purchase contract machines, but a P-39C assigned under Lend-Lease arrangements.

President Roosevelt had signed the historic Lend-Lease Act on 11 March 1941, clearing the way for large quantities of war supplies to be transferred to the Allies with no more then a "gentleman's agreement" that they would in due course be returned or some form of payment would be made. Within three months, a total of 9,756 aircraft of assorted types had been requisitioned by Britain under the terms of the Lend-Lease Act, and among the early requisitions were three P-39Cs — being intended for "war tests"— and a batch of 150 Airacobra IAs (the "A" suffix to mark numbers was consistently used to distinguish between Lend-Lease and Direct Purchase types which were otherwise similar). The Airacobra IAs did not, in the event, materialise as British aircraft, but the three P-39Cs (serials DS173-175) were delivered and used for test-flying and evaluation.

(Opposite page) Airacobras in the markings of five air forces that used the Bell fighter during World War II. Illustrated from top to bottom are: Airacobra I AH579 of No 601 Squadron, RAF, October 1941; P-400 BW167 of the 67th Fighter Squadron (35th Fighter Group) USAAF, in New Caledonia, 1942; P-39N 42-9410 of GC II/6 "Travail" 3eme Escadrille, Forces Aeriennes Francaises Libres, 1943; P-39L 42-4558 of the 93rd Fighter Squadron (81st Fighter Group), Tunisia, 1943; P-39Q used by Major Alexander Pokryshkin (Sov AF) on the Russian Front southern sector, 1943-44; P-39N 42-9377 of the 4 Stormo, Italian Co-Belligerent Air Force in Yugoslavia, 1944.

The first of these P-39Cs arrived, crated, at RAF Colerne on 3 July 1941, followed by the other two the next day. Original plans for the aircraft to be assembled by Scottish Aviation at Prestwick had been switched at the last moment, and the task was assigned instead to a team of BOAC engineers working at Colerne. At the same time, responsibility for airframe repairs was assigned to AST at Hamble (backed up by Reid & Sigrist Ltd at Desford), while the Burtonwood Repair Depot (backed if need be by Sunbeam Talbot Ltd) was to undertake repairs of the Allison V-1710-E4 engine. Bell assigned three field engineers to support the British Airacobra programme, and the USAAC provided two officers and five NCOs. These were responsible for assembly of the first aircraft at Colerne, where it made its initial flight on 6 July 1941, the pilot being 1st Lt Melvin F McNickle, USAAC.

The first P-39C was assigned to the Aeroplane & Armament Experimental Establishment for handling and gun firing tests, being flown to Boscombe Down on 9 July; another of these aircraft went to the Air Fighting Development Unit at RAF Duxford a little later. At this stage, the Airacobra received the first blow to what, up to that point in time, had been an excellent reputation, albeit one based largely on the publicity material of the Bell Aircraft Corporation. In view of the weight growth of the aeroplane, and the deletion of the turbo-supercharger, some degradation of performance had been inevitable. Nevertheless, there seems to have been universal surprise and dismay amongst the RAF personnel concerned with the Airacobra programme when the A & AEE trials revealed a top speed fully 33 mph (53 km/h) lower than claimed.

In a report to his superiors on this aspect of the trials, dated 19 August, Major George Price, the senior officer of the USAAC team supporting the Airacobra in Britain, said: "The loss of top speed from an alleged 392 to 359 mph (631 to 578 km/h) cannot be accounted for and will have to be investigated in the USA." It was later admitted by Bell that an aeroplane weighing over 2,000 lb (910 kg) less than the British Airacobra, and highly polished, was used for the publicity figures on which the Airacobra's reputation partly rested. This was not unnaturally the cause of much discomfort for the Army Air Corps personnel in Britain, and Lt McNickle made a strong recommendation that all performance figures of American aircraft should be made by the USAAC and not by the manufacturer. "Everyone will agree," his report stated tartly, "that an airplane will perform better without the military load, but it is with this load that the airplanes at the front must operate. The American aircraft manufacturers *must* be made aware of this."

However, the criticism was not all one-sided and the American team, both servicemen and civilians, found frequent cause to be irked at the lack of initiative and haste displayed by the British in assembling the Airacobras and preparing them for operational use. Several times, representations were made through the American Embassy to the Ministry of Aircraft Production to put pressure on the BOAC team at Colerne: but the MAP itself did not rate very highly in the estimation of Major Price, who regarded it as "grossly top heavy and over-organised: thousands of people in thousands of little offices, no one doing too much work and no one taking any responsibility."

Operational deficiencies

The first British-purchase Airacobras began arriving at Colerne before the end of July, and four were flying by mid-August, with another 10 in various stages of assembly. While the A & AEE and the AFDU went on with the task of assessing the characteristics of the Airacobra and recommending essential modifications, a start was made on checking out pilots for the first (and, as it transpired, only)

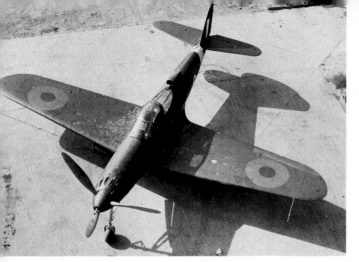

(Above and below) Two views of the P-39C DS173, the first Airacobra to arrive in Britain. This was one of three P-39Cs supplied under Lend-Lease arrangements; the four machine guns in the nose and lack of wing guns distinguish this variant from the British Airacobras.

squadron. Selected for what was destined to prove the somewhat dubious honour of being the first Fighter Command squadron to equip with the American fighter was No 601 "County of London" Squadron of the Royal Auxiliary Air Force. Under the command of Sqn Ldr E J Gracie, DFC, the squadron took its Hurricanes to Matlaske in Norfolk to re-equip with the new fighter. The first two Airacobras, unmodified in every respect, were delivered to the squadron on 7 August 1941, and during the next four days Lt McNickle checked out 19 pilots on the type which was, he reported, "received by the squadron with great enthusiasm and the more they fly it the better they like it".

Two more aircraft were delivered in August but one of these crashed (at Mildenhall) almost immediately. Another seven were delivered during September, and the squadron then moved to RAF Duxford, spending an intensive few weeks working-up on the type and introducing modifications. Some 25 modifications were made at this stage, although not all were classed as operationally essential. Sqn Ldr Gracie, in fact, listed six items which he considered essential to make the aircraft operational, the others being a question of improving operational efficiency and pilot's comfort. The six essential items were:—

   a Introduction of master valve to allow oxygen to be turned on from the cockpit.
   b Modification of the gun sight "to enable the pilot to see out of the front of the aircraft".
   c Changes to the wing ammunition tanks for the 0·30-in guns, which became distorted and caused misfeeds.
   d Modifications to simplify operation of the cockpit harness release.
   e Removal of the IFF (Identification, Friend or Foe) set from behind the pilot, where it obstructed aft view.
   f Introduction of a throttle control quadrant friction damper.

By the second half of September, all these modifications were in hand, but the squadron ground crews were having to spend approximately three days on each Airacobra received from Colerne to bring it to a satisfactory standard. Lack of sufficient engineers at the Airacobra Assembly Unit con-

tributed to the difficulties, but the fundamental problem was that the aircraft were being crated straight off the Bell production line without being assembled and test flown. As a result, the Airacobras reaching Duxford during this period had such faults as guns only partly de-greased, harmonisation and synchronisation faults, incorrect wiring causing R/T trouble, compasses unswung, and instruments at fault. The spares situation was critical; spares had been sent from the States but "had not yet been located in this country", according to Sqn Ldr Gracie. There was also concern about keeping the 0·50-in guns serviceable as the barrels had a life of only 2,000 rounds each, and there appeared to be no spares in Britain!

A tactical assessment

The Air Fighting Development Unit, meanwhile, had received a British Airacobra I from Boscombe Down on 30 July, and completed its report on Tactical and Armament Trials on 22 September. This report found that the aircraft was "pleasant to fly and very easy to take-off and land. The controls are well balanced and although heavier than those of the Spitfire at normal speeds, do not increase appreciably in weight at high speed as in the Spitfire. It is difficult to hold the aircraft in a dive at high speed unless it is trimmed nose-heavy. During a turn the Airacobra gives ample warning of a high speed stall by severe vibration of the whole airframe".

Handling in formation and formation attacks was found to be good, although deceleration was poor because of the Airacobra's aerodynamic cleanliness. Take-offs and landings in close formation were not considered practicable since, although it was possible to check a swing, there was considerable difficulty in bringing the aircraft back to its original path.

With its internal fuel capacity of 100 Imp gal (455 1) the Airacobra had an approximate endurance of 1 hr 20 min at maximum continuous cruising speed at 6,000 ft (1 829 m), 1 hr 5 min at 12,000 ft (3 658 m) and 1 hr 35 min at 20,000 ft (6 096 m). The true airspeeds at these three altitudes were 287 mph, 327 mph and 308 mph (462, 526 and 496 km/h) respectively. Under most economical cruise conditions, the endurance increased to about 3 hr 20 min, the relevant speeds being 183 mph (295 km/h) at 6,000 ft (1 829 m), 217 mph (349 km/h) at 12,000 ft (3 658 m) and 215 mph (346 km/h) at 20,000 ft (6 096 m).

Time to climb to 20,000 ft (6 096 m) was about 15 min under maximum continuous climb conditions and the operational ceiling was considered to be about 24,000 ft (7 315 m), although there was a marked decrease in performance above 20,000 ft (6 096 m). At the Airacobra's rated altitude of about 13,000 ft (3 962 m), it was approximately 18 mph (29 km/h) faster than the Spitfire VB. However, the speed fell off rapidly above that height (the planned turbo-supercharger having been discarded), and the two types were almost exactly matched at 15,000 ft (4 572 m). At 20,000 ft (6 096 m) the Spitfire VB was about 35 mph (56 km/h) faster, and at 24,000 ft (7 315 m), 55 mph (89 km/h) faster. The take-off performance was regarded as something of a limitation for operation from the smaller fighter fields, the ground run distance of 2,250 ft (655 m) comparing with 1,470 ft (448 m) for the Hurricane II and 1,590 ft (485 m) for the Spitfire V.

To assess the Airacobra's manoeuvrability in dog-fighting, the AFDU compared it with a Spitfire VB and a Messerschmitt Bf 109E, and reported as follows: "The Airacobra and the Bf 109 carried out dog-fighting at 6,000 ft (1 829 m) and 15,000 ft (4 572 m), the latter aircraft having a height advantage of 1,000 ft (305 m) in each case. The Bf 109, using the normal German fighter tactics of diving and zooming, could only get in a fleeting shot.

"The Bf 109 cannot compete with the Airacobra in a turn and even if the Bf 109 is behind the Airacobra at the start, the latter should be able to shake him off and get in a burst before two complete turns have been carried out.

"The Bf 109 then tried diving on the Airacobra from above and continuing the dive down to ground level after a very short burst of fire. It was found, however, that the Airacobra could follow and catch up on the Bf 109 in a dive of over 4,000 ft (1 220 m).

"The Airacobra then carried out a similar trial with the Spitfire V, and it was found that although the Airacobra has a superiority of speed up to 15,000 ft (4 572 m), it was out-climbed and just out-turned by the Spitfire.

"These combats show that when fighting the Bf 109E below 20,000 ft (6 096 m) the Airacobra is superior on the same level and in a dive. Unless it had a height advantage the Airacobra could not normally compete with an aircraft similar to the Spitfire V. If on the same level or below, at heights up to about 15,000 ft (4 572 m), the Airacobra would have to rely on its superior level and diving speeds and its ability to take negative 'G' without the engine cutting. Above this height, the Airacobra loses its advantage in level speed. Both the Spitfire VB and the Bf 109E have a superior climb to that of the Airacobra."

The Airacobra was considered to be very suitable for low flying because of the excellent view and controllability, and was fully manoeuvrable at speeds above 160 mph (257 km/h). It was comparatively easy for instrument flying, and could be trimmed "hands and feet off". The aircraft was not difficult to fly at night but the exhaust flames, invisible to the pilot, were visible to another aircraft flying three miles astern. An even bigger deterrent to night operations was that the flash from the 0·50s in the nose was blinding, and could cause the pilot to lose not only the target but also his night vision. Another serious consequence of firing the nose guns was carbon-monoxide contamination in the cockpit, and this could reach a lethal level.

Apart from its disappointment over the Airacobra's performance, the RAF found the greatest problems to be those associated with the armament. All the guns were inaccessible, and maintenance was troublesome. Difficulties included the time taken to remove the recoil mechanism of the 0·50s, problems with loading and unloading the 0·30s because the cockpit charging handles were too stiff to operate, and alignment of the ammunition tanks for the 0·30s. Other problems associated with the armament, already mentioned, were the large compass deviation, CO contamination and flash making night operation undesirable.

In air firing trials conducted by No 601 Squadron at Sutton Bridge with seven aircraft, ten stoppages of the 0·30-in guns occurred (ie, 36 per cent), and three stoppages of the 0·50-in guns (22 per cent). The 20-mm cannon behaved satisfactorily in these tests.

Into action — and out again

With four of its Airacobras modified up to operational status by the end of September, No 601 received sanction to take its aircraft on operations. For this purpose, the four aircraft were detached to RAF Manston, and from there, two aircraft took off on 9 October, and flew across the Channel on a *rhubarb* (the RAF code-name for small-scale raids by fighter or fighter-bombers against targets of opportunity). An enemy trawler was shot-up in the vicinity of Gravelines. Two Airacobras visited the same area next day, without result. On 11 October, two aircraft flew to Gravelines and Calais, attacking some enemy barges, and then three Airacobras flew to Ostend, but no targets were found. The aircraft all returned from Manston to Duxford the same evening.

After these four missions in three days, totalling nine sorties by four aircraft, the Airacobra was taken off operations because of the difficulties with the compass. The British propaganda machine had somewhat belatedly gone

into action, however, and could not be stopped; six days *after* the last operation, on 17 October, No 601 Squadron received the press at Duxford to display their new aircraft, and the next day the press duly carried illustrations of the Bell Airacobra, "now operational with Fighter Command pilots", who, it was reported "have been highly impressed by the performance of this aircraft on active service"(!). An interesting sidelight at the time of this press visit was that the squadron's aircraft were seen to feature an unofficial deviation from the approved form of squadron markings, the "County of London" unit's famous red winged sword being painted in the white band of the fin flash. The squadron CO had also singled out for himself the appropriately-numbered Airacobra AH601, and had the squadron symbol repeated on the forward fuselage in place of the individual letter.

Subject to the introduction of flame dampers for the exhaust pipes, flash elimination for the 0·50-in guns, introduction of a distant reading compass, and improvements in the re-arming procedure, the RAF concluded that the Airacobra would make an excellent day fighter at altitudes below 20,000 ft (6 096 m), and was well suited to the attack of ground targets. Its operation in four rôles was recommended: close escort to bombers at medium altitudes; attack on bombers; low attack of shipping and ground targets, and "cat's eye" night fighter.

However, before these plans could be implemented, a major policy directive effectively ended further operation of the Airacobra by the RAF. The decision was made to divert the bulk of the British contract for Airacobras to Russia, as a part of British aid offered following Germany's attack on the Soviet Union on 22 June 1941. On the face of it, at least, the decision appeared to be in line with Soviet needs, since the Airacobra's low-altitude capabilities would be useful in attacks on German ground targets and in combating low-flying German raiders. It may be that this decision was also seen as an opportunity for the RAF to disengage from a re-equipment programme that was proving an embarrassment.

By the time this decision was taken, production of the British-contract Airacobras had reached four a day at Bell's Buffalo plant. The initial contract for 170 aircraft (AH570 to AH739) had been completed before the end of September, and all but six of these aircraft were actually shipped to

(Above) The fourth British Airacobra, AH573, was used briefly for handling and performance trials at the A & AEE, Boscombe Down, but crashed before the tests were complete. Unlike the earlier examples, this aircraft had divided stubs on each of the six exhaust pipes.

(Below) The second Airacobra I, AH571, showing the modified rudder and a fairing round the exhaust stubs.

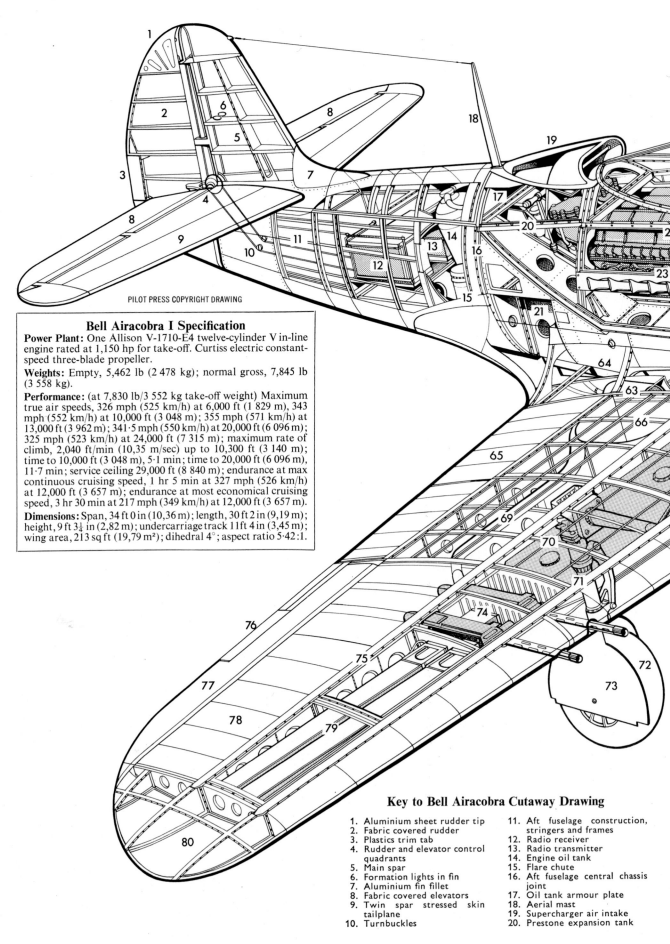

PILOT PRESS COPYRIGHT DRAWING

Bell Airacobra I Specification

Power Plant: One Allison V-1710-E4 twelve-cylinder V in-line engine rated at 1,150 hp for take-off. Curtiss electric constant-speed three-blade propeller.

Weights: Empty, 5,462 lb (2 478 kg); normal gross, 7,845 lb (3 558 kg).

Performance: (at 7,830 lb/3 552 kg take-off weight) Maximum true air speeds, 326 mph (525 km/h) at 6,000 ft (1 829 m), 343 mph (552 km/h) at 10,000 ft (3 048 m); 355 mph (571 km/h) at 13,000 ft (3 962 m); 341·5 mph (550 km/h) at 20,000 ft (6 096 m); 325 mph (523 km/h) at 24,000 ft (7 315 m); maximum rate of climb, 2,040 ft/min (10,35 m/sec) up to 10,300 ft (3 140 m); time to 10,000 ft (3 048 m), 5·1 min; time to 20,000 ft (6 096 m), 11·7 min; service ceiling 29,000 ft (8 840 m); endurance at max continuous cruising speed, 1 hr 5 min at 327 mph (526 km/h) at 12,000 ft (3 657 m); endurance at most economical cruising speed, 3 hr 30 min at 217 mph (349 km/h) at 12,000 ft (3 657 m).

Dimensions: Span, 34 ft 0 in (10,36 m); length, 30 ft 2 in (9,19 m); height, 9 ft 3¼ in (2,82 m); undercarriage track 11 ft 4 in (3,45 m); wing area, 213 sq ft (19,79 m²); dihedral 4°; aspect ratio 5·42:1.

Key to Bell Airacobra Cutaway Drawing

1. Aluminium sheet rudder tip
2. Fabric covered rudder
3. Plastics trim tab
4. Rudder and elevator control quadrants
5. Main spar
6. Formation lights in fin
7. Aluminium fin fillet
8. Fabric covered elevators
9. Twin spar stressed skin tailplane
10. Turnbuckles
11. Aft fuselage construction, stringers and frames
12. Radio receiver
13. Radio transmitter
14. Engine oil tank
15. Flare chute
16. Aft fuselage central chassis joint
17. Oil tank armour plate
18. Aerial mast
19. Supercharger air intake
20. Prestone expansion tank

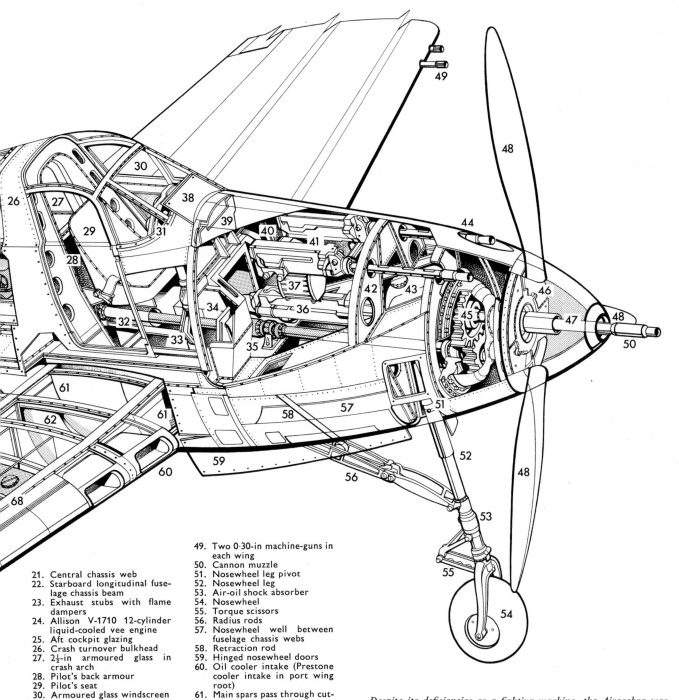

49

48

44

46

48

47

50

51

52

53

55

54

30

26 27 38

29 31 39

28 40

41

37 42 43

36 45

34

32 35

33

61

62 61

59 60

68

56

57

58

21. Central chassis web
22. Starboard longitudinal fuselage chassis beam
23. Exhaust stubs with flame dampers
24. Allison V-1710 12-cylinder liquid-cooled vee engine
25. Aft cockpit glazing
26. Crash turnover bulkhead
27. 2½-in armoured glass in crash arch
28. Pilot's back armour
29. Pilot's seat
30. Armoured glass windscreen
31. Instrument panel
32. Drive shaft sleeve
33. Control column yoke
34. Rudder pedal (port pedal not shown)
35. Drive shaft centre bearing
36. 20 mm Hispano Suiza cannon
37. Cannon ammunition magazine, 30 rounds
38. 7-mm steel plate armour overlapping windscreen
39. Forward armour plate
40. Nose machine-gun magazine boxes forward of cockpit bulkhead
41. Two 0·50-in machine-guns
42. Bulkhead former
43. Propeller reduction gear oil tank
44. Machine-gun port
45. Propeller reduction gear
46. Reduction gear steel armour plate
47. Cannon barrel within hollow prop shaft
48. Three-blade Curtiss electric constant speed propeller

49. Two 0·30-in machine-guns in each wing
50. Cannon muzzle
51. Nosewheel leg pivot
52. Nosewheel leg
53. Air-oil shock absorber
54. Nosewheel
55. Torque scissors
56. Radius rods
57. Nosewheel well between fuselage chassis webs
58. Retraction rod
59. Hinged nosewheel doors
60. Oil cooler intake (Prestone cooler intake in port wing root)
61. Main spars pass through cutaway in fuselage chassis
62. Intake trunking
63. Cylindrical oil radiators in aft wing sections
64. Prestone radiator under fuselage
65. Electrically operated dural sheet flaps
66. Mainwheel well
67. Self-sealing integral fuel tanks
68. Front spar
69. Wing machine-gun heater duct
70. Undercarriage retraction drive
71. Main undercarriage leg
72. Starboard mainwheel
73. Mainwheel door
74. Machine-gun bays in wing
75. Rear spar
76. Laminated plastic trim tabs
77. Fabric covered aileron
78. Wing skinning
79. Machine-gun ammunition boxes
80. Upswept wingtip

Despite its deficiencies as a fighting machine, the Airacobra was pleasant to fly and was regarded by most British pilots as an advanced type with its tricycle undercarriage and other "modern" features.

PAGE 141

The thirteen Airacobra Is of No 601 Squadron were displayed to the press at Duxford in October 1941 — after the aircraft had been withdrawn from operations because of problems with the compass, armament and generally low serviceability. Flame damping exhausts had been fitted at this stage.

Britain. A number remained in their crates after arrival, however, and were shipped on to the Soviet Union. Between 80 and 100 aircraft are believed to have been assembled and flown in Britain by the end of 1941, and these were gathered at Maintenance Units for final modification action before being re-crated and shipped to Russia during 1942. In all Russia received 212 of the British Airacobras and 49 more were lost at sea en route.

No 601 Squadron took its 13 Airacobras to Acaster Malbis in January 1942, and there relinquished them in March, in favour of Spitfires, after losing two more in fatal crashes. One Airacobra was fitted with an arrester hook and used for deck landing trials at the RAE, Farnborough, where it survived the war and was still visible in 1948 — certainly the last of its type in Britain.

Not only the Soviet Union needed aircraft urgently at this time, for the USA also found itself launched into all-out war with the Japanese onslaught on Pearl Harbour on 7 December 1941. One consequence of this attack was that quantities of aircraft and other war supplies in production in the USA for export were promptly requisitioned—among them nearly 200 British direct-purchase Airacobras. Although similar to the P-39D, they were not identical, and they were therefore known as P-400s (other requisitioned British types were similarly given non-standard designations, such as P-322 for Lightnings and L-37 for Venturas). The P-400 designation had, in fact, been associated with the British Airacobras for contractual purposes as early as August 1941.

The initial allocation of the P-400s requisitioned at the factory was to squadrons despatched to the Southwest Pacific area, the first unit to become operational with the type apparently being the 67th Fighter Squadron of the 347th Fighter Group. The personnel of this squadron arrived at an airfield at Tontouta, 35 miles (56 km) from Noumea, in New Caledonia, on 15 March 1942, followed a week later by 47 crated Airacobras, all but two of which were repossessed British machines. The problems that confronted ground crews and pilots assigned to operate these aircraft made the earlier difficulties encountered at Colerne and Duxford pale into insignificance, for they arrived with no instruction manuals, no assembly tools, no spare parts and only two pilots with previous experience on the P-39!

Nevertheless, the first P-400 was flying at Tontouta only six days after it arrived, and 41 had been assembled within 29 days, the other six being cannibalised for spare parts. After a period of training, this unit moved to Guadalcanal in August 1942, and used its Airacobra for ground attack duties, using 500-lb (227-kg) bombs. Improvisation remained the order of the day throughout the operational career of the P-400s in the Southwest Pacific. It was therefore not long before hybrids began to appear, the 67th FS being responsible for fitting a P-39D wing to a P-400, and a little later the 68th FS producing a P-400 fuselage with one P-39D wing, one P-39K wing and an Allison V-1710-63 engine. Other units flying requisitioned British Airacobras included the squadrons of the 35th Fighter Group (Nos 39, 40 and 41 Squadrons), and of the 8th Fighter Group. The latter had established Group Headquarters in Brisbane, Australia, in March

Below, over 250 of the British Airacobras, including some of those that had served with No 601 Squadron, were shipped on to Russia from the UK, and eventually gave satisfactory service in the ground attack rôle.

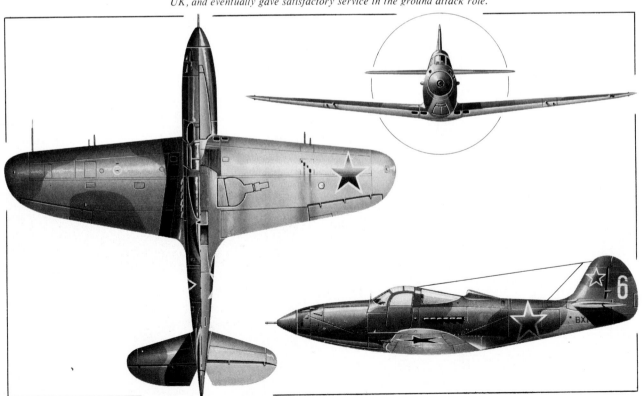

1942, and its P-39s, believed to include some of the ex-British contract aircraft, went into action in defence of Port Moresby, New Guinea, on 30 April. Between July 1942 and November 1943, a total of 22 Airacobras operated with the RAAF in Australia, these comprising P-400s and P-39Fs.

The P-400s also saw a little further use nearer to Britain when 179 of the aircraft sent to the UK were re-acquired by the USAAF. After working up on the type in the UK, the six squadrons of the 81st and 35th Fighter Groups flew P-400s to North Africa in December 1942 and January 1943 to join the Twelfth Air Force. These aircraft had been in crates at the Burtonwood depot since early 1942, and were assembled there by USAAF personnel. The assigned rôle of this Fighter Group was ground attack, for which the P-400s were quite well suited, but the aircraft operating in the Southwest Pacific were called upon to serve as interceptors, and in this rôle they proved to be no match for the Mit-subishi A6M (Zero) fighters in air-to-air combat, while they were unable, because of difficulties with the oxygen supply, even to reach the Mitsubishi G4M ("Betty") bombers raiding from above 25,000 ft (7 620 m).

Although later Airacobra variants produced for the USAAF and, on Lend-Lease contracts, for the Soviet Union and the Free French Air Force, had various improvements, none ever lived up to the promise of the original Bell proto-type. The Bell fighter's ability to absorb battle damage and its usefulness in low-altitude strafing operations earned it a measure of acclaim, especially among Soviet pilots, but this was outweighed, in the final balance, by its less-than-satisfactory performance at higher altitude. When the official history of the USAAF in World War II came to be written, the Airacobra was described as "especially disappointing"; and so far as British experience with the type was concerned, *that* assessment was all too accurate. □

Viewed from the Cockpit

by H A Taylor

ONE OF the minor curiosities of aircraft made between the wars was the way in which specific national design characteristics developed. Taking an extreme case, the air-craft produced by two or three European countries, notably France, had a throttle-lever operation which was the reverse of that normal elsewhere — you pulled the lever back for more power. American aeroplanes of the 1930s developed a series of features which were quite unlike those of aircraft made in Britain — pedal-operated brakes, for instance, inertia starters and the wide use of electric actuators. Their engines were without the automatic boost and (often) mixture controls which saved the pilots so much trouble (and occasional red faces) when handling British engines. American aeroplanes even *smelt* quite differently because of the use of a particular kind of anti-corrosion finish.

So, when US military aircraft started to appear in quantity during 1940–1 the ferry and test pilots — even those who had flown the North American Harvard trainer at the start of their brief conversion course at the RAF's Central Flying School — had to do some fundamental re-thinking of cockpit and other drills. These aeroplanes, with their mass of electrickery and aura of technical complexity, were like something from another world — and none more so than the hub-cannon-armed Bell Airacobra fighter.

Towards the end of 1941 a dozen or so of these quaint devices were delivered to No 48 Maintenance Unit, RAF Hawarden, Chester, for temporary storage and preparation, and the unit's chief test pilot and I set about the business of finding out how they worked on the ground and in the air. Nothing could have been in greater contrast to the Hurricanes, Lysanders, Wellingtons, Ansons and Herefords which, with their typically untidy British layouts, then formed the bulk of the unit's aircraft population — but we had previously held and flown some odds and ends like Curtiss P-40 Tomahawks and Douglas (Northrop) A-17A Nomads, so we were not too bewildered and baffled by the serried rows of switches and strange gauges of the Airacobra. I had even had some experience with nosewheel aeroplanes, so I volunteered to make the first flight.

The Airacobra was nothing if not a highly claustrophobic affair. You entered it through a narrow car-type door on the starboard side, trying hard not to put your left foot through any delicate and vital bit of equipment as you eased it round the control column. When the door (complete with roll-down window) was slammed shut it was like sitting in an extremely complicated and wildly instrumented two-door single-seat saloon car. Initially you were very conscious of that vee-twelve Allison engine sitting in the small of your back, but you soon forgot it was there as you looked through the curved and steeply sloping bluish-coloured screen over the forward cowling — which might just as well have had an engine inside it. There was a doubt-lessly apocryphal story going around about the engine fitter, new to the type, who topped up the reduction gear oil-tank with a glycol coolant mixture.

On releasing the push-pull handbrake lever (in the middle of the dashboard), treadling the pedal brakes to pump them up and opening the throttle, I made the first displeasing discovery: the propeller shaft between my outstretched legs whipped about with noisy frenzy under varying power loads. Thereafter we learned to taxi under fairly high and constant power, controlling the speed on the brakes and by periodi-cally snapping the throttle shut. After rolling a few yards to get the nosewheel straight, the take-off was normal enough, and it was nice to have nothing to do when raising the undercarriage but flick up a three-position switch and centralise it in the "off" position when the pictorial indi-cator showed that all three wheels were up. A nearer and more or less identical switch-unit operated the flaps, which could be left in any intermediate position (shown below on the same indicator) by centralising the switch.

All of this seemed very neat and rational — but we were to have a little trouble, of our own making, with that too-easy undercarriage operation. The emergency system was, by contrast, a very solid blacksmith affair of handcrank (below on the right) and bicycle chain, with a two-position selector. On one early flight the chief test pilot taxied out and took-off without knowing that the selector had been set, for some maintenance reason, and left in the manual posi-tion. He tripped the little switch and nothing happened — except a scream from the undercarriage-up warning horn when, in due course, he closed the throttle. He landed and

continued on page 166

Airscrew versus Jetpipe

A FEW MINUTES before settling in front of the typewriter to tap out this month's commentary on the modelling scene, the pulsating roar of a quartet of piston engines sent your columnist scurrying to the window, binoculars at the ready, to watch, not without some nostalgia, an aged DC-6A freighter climb away sedately in a northerly direction. The drone of its Pratt & Whitney radials had barely faded and we were once again facing the typewriter when the reverberations of the mighty turbines of a passing airliner — mayhap a Boeing 707 or a McDonnell Douglas DC-8 — set a coffee cup vibrating in its saucer. Had it not been for the departed DC-6A it is unlikely that the noise of the passing turbine-driven aircraft would have impinged on the conscious, being — in this office at least — so commonplace as to pass unnoticed.

This event, unimportant in itself, gave cause for reflection. Had the DC-6A exercised some magnetic attraction because of its pounding pistons which recalled a near-bygone era when air transportation possessed some indefinable romance which evaporated with the widespread acceptance of the gas turbine as the means of propelling airliners? Was it, as so many members of the modelling fraternity insist, that aircraft lost much of their character when the airscrew made way for the jet orifice? Or was it simply that the large piston-engined aircraft is becoming the *rara avis* in the skies over our major airports, and the noise that it emits has become discordant in this day and age of the whining turbine, and is thus the more likely to attract attention?

Whatever the reason, there can be no doubt that, to a substantial number of modellers, an aircraft is not a worthy subject for modelling unless it *has* one or more airscrews. We have never been able to appreciate this attitude though we can sympathise with it and would defend to the death any modeller's right to it. We are perhaps fortunate that we find fascination in a good model of *any* aircraft, be it a representation of the Wright Brothers' conglomeration of stick, string and canvas, or the latest in service combat aircraft, such as Hawker Siddeley's unique Harrier, illustrated on the facing page, but there is much evidence to suggest that, among serious modellers, the *specialist* is beginning to preponderate; the modeller confining his activities to, say, the fighter aircraft of WW II, the piston-engined transport, or current jet combat aircraft.

There is something to be said for such specialisation, for concentration on a specific category of model should, logically, result in a very high standard being achieved, but we cannot but help feel that such modellers miss much of the pleasure that may be derived from variety, and to your columnist, variety is, as the old adage says, the spice of (modelling) life — today the Harrier, tomorrow the Sopwith Camel, and next week . . . who knows . . . the MiG-23?

This month's colour subject

As a genuine service aeroplane, the Hawker Siddeley Harrier is unique today and bids fair to remain so well into the future. It is generally accepted that the Soviet Union still has some way to go before it can add to the V-VS hardware inventory a combat aircraft capable of emulating the Harrier's V/STOL capabilities while carrying a worthwhile ordnance load. The adoption of this British warplane by the US Marine Corps is tribute enough to the remarkable potentialities of the Harrier, and this strike and reconnaissance fighter is illustrated (opposite) in both RAF and USMC markings. It is worthwhile noting, incidentally, that while both services employ the same camouflage pattern, that of the RAF is applied in *matt* polyurethane while that of the USMC is gloss!

While there are rumours in the grapevine of something big and spectacular in the form of Harrier kits being on one manufacturer's stocks, at the time of writing only two kits of this aircraft are available; that from Frog at 47p which is a re-release of a Hasegawa original, and an offering from Airfix at 24p. Both are excellent specimens of the kit manufacturer's art and both are generally if not completely accurate. The Frog Harrier .is moulded in high-quality white plastic which aids painting considerably, and the fit of the component parts of both kits is good, although the Airfix kit is definitely the easier of the two to assemble as the exhaust fairings are moulded into the fuselage halves rather than as separate parts, and its wings are in halves instead of having inserts, although this produces somewhat thicker trailing edges.

The forward fuselage contours of the Frog Harrier score over those of its Airfix competitor, the nose and canopy of the latter being too short, but the awkward sections around the engine exhausts are handled better by Airfix which also gains points by reason of the ingenious extension which allows all four exhausts to pivot simultaneously whereas those of the Frog kit only swivel individually. The surface detail of the two kits compares closely in quality, although the Frog offering enjoys a slight edge by reason of its recessed control surface hinge lines and beautifully-detailed exhausts. In so far as decals are concerned, it can only be said that Frog now enjoys the reputation of offering the finest decal sheets in the buisness, and the decals accompanying the Harrier kit do full justice to this reputation. The best that may be said of Airfix's decal sheet is that, as far as it goes, it is satisfactory. Both kits make up into the Harrier GR Mk 1 and both are to 1/72nd scale. The Airfix kit, at little more than half the (UK) price of its competitor, is excellent value, but this is not to suggest that the Frog kit is over-priced as it, too, offers good value — keenly competitive pricing is, however, becoming increasingly the name of the modelling game!

A couple of choppers

Revell has now released a 1/32nd scale kit of the Hughes OH-6A Cayuse light observation helicopter with alternative parts to permit its completion in civil Model 500 form. This model is accurate, easy to assemble (though care must be taken with the transparencies), and neatly detailed, forming a worthy companion for the company's larger kit of the Bell UH-1D Iroquois. From one end of the rotorcraft size scale to the other, the VEB of the German Democratic Republic is now offering a kit of the gargantuan Mil Mi-10K crane-type helicopter to 1/100th scale. The standards attained by this East German concern now compare favourably with those of many western manufacturers, and the Mi-10K is neatly moulded in good-quality material and assembles easily. It includes adequate transparencies, though shrinkage has endowed those for the cabin windows with an odd lens-like effect.

The decal sheet is of good quality, but the decals themselves, owing to their considerable thickness, call for extremely careful handling. The sheet provides the markings of SSSR-29115 which was exhibited two years ago at the *Salon de l'Aéronautique*. These are given in two shades of blue but to the best of our recollection were, in fact, in two shades of *grey*. We have been unable to ascertain the price of this kit and, in any case, it is unobtainable in the West through the normal channels. However, it is not too difficult to arrange an exchange of kits with modellers in Eastern Europe, and the Mi-6 is worth taking a little trouble to obtain.

A pair of trijets

The dawn of the airbus age has naturally produced its quota of airbus kits, and the latest additions to their ranks are the Lockheed L-1011 TriStar from Airfix and the McDonnell Douglas DC-10 from

— continued on page 168

(Opposite) Top side view, first Hawker Siddeley AV-8A (Harrier GR Mk 50), BuAer 158384, of the US Marine Corps; USMC badge on fin shown enlarged, upper right; also location of USMC markings indicated on upper and under plan views — camouflage pattern is same as RAF aircraft. The two plan views and bottom side view show Harrier GR Mk 1 of No 4 Squadron, RAF, based at Wildenrath, Germany; squadron badge on nose shown enlarged, lower left. Scrap views, upper left and lower right, show forward and rear fuselage of Harrier T Mk 2 XW175. RAF finishes are matt, USMC finishes are high gloss.

WF
158384
MARINES

XV
783

WF

MARINES

XW175

E

XV783

IN PRINT

"Dora Kurfurst und rote 13, Vol IV'
by Karl Ries jr
Verlag Dieter Hoffman, Mainz,
West Germany, DM 23·50
192 pp, 6½ in by 9½ in
THIS volume, published a short time ago, concludes the well-known Karl Ries series of pictorial volumes covering Luftwaffe aircraft of the period 1933-45. Like the three earlier volumes, this book comprises a selection of photographs, together with captions, but no text. The pictures fall into no particular pattern, being grouped according to the primary rôle of the aircraft depicted — trainers, fighters, bombers, etc — but include a number of considerable interest to students of German aircraft development as well as modellers and others with an eye for the unusual in unit markings.

The final chapter in this volume is devoted to aircraft armament, and in contrast to the other volumes, this one includes a few illustrations of models of projects that were not completed. Also included is a comprehensive index to the contents of all four volumes.

"Battle of Britain"
by Bruce Robertson
J W Caler Publications Corp, California,
$3·95
52 pp, 8½ in by 11 in, illustrated
THIS new addition to the Caler Illustrated Series is of primary interest as a collection of photographs. The narrative account with which the book opens is restricted to about 6,000 words and consequently provides only a broad-brush picture of the Battle, which has been the subject of many hundreds of thousands of words in other publications.

The illustrations, drawn from a wide variety of sources in Britain, Germany and elsewhere, are devoted principally to operational views of the aircraft of both sides, with a fair sprinkling of wrecks and a few personalities. Somewhat surprisingly, the centre spread of the book is given over to a full colour painting of a Fiat CR 42 — hardly one of the most significant of the aircraft engaged in the Battle. Attractively drawn and produced side view profiles of the Bf 190E-4 and the Spitfire I are presented as loose inserts in the book and are suitable for framing.

"Gloster Aircraft since 1917"
by Derek N James
Putnam & Co Ltd £5·50
446 pp, 5½ in by 8½ in, illustrated
A SIGNIFICANT addition to Putnam's series of British aircraft company histories is made with this volume. The Gloster company has a unique place in British aviation history as the builder of the nation's first jet aircraft and the RAF's first jet fighter. The Meteor and the Javelin (Gloster's last aeroplane) culminated a long line of fighters for which the company name became famous, but as this volume shows, design activities embraced the whole range of aircraft types.

Although handicapped by a lack of preserved records and the fact that the

An illustration from Volume IV of "Dora Kurfurst und rote 13" showing a Junkers Ju 90 prototype taking off at Obertraubling in June 1941 with a Messerschmitt Me 321 glider in tow. The latter has four underwing rocket packs to assist the take-off.

company ceased to exist ten years ago, Derek James has done full justice to Gloster's 44 years of achievement in this volume. Following the established Putnam formula, he provides a brief historical review; a detailed history of each aircraft type produced; and a series of appendices dealing with production lists, unbuilt projects and other data.

Mr James has assembled many interesting and little-known photographs and writes authoritatively, particularly of the later years during which he was himself closely involved in the events he describes. Of special interest, having previously gone virtually unrecorded, are the project studies illustrated in the appendix, including a series of commercial aircraft projected between 1958 and 1961 when the Gloster design team was engaged in a futile attempt to turn the company away from the production of swords and towards plough-shares.

"Rocket Fighter"
by William Green
Ballantine Books, Inc, New York, $1·00
160 pp, 5⅜ in by 5¼ in, illustrated
THIS NEW ADDITION to the "weapon book" series of Ballantine's Illustrated History of World War II tells the story of the rise and demise of a concept — the rocket-propelled point defence interceptor. The narrative spans almost two decades; from the spectacular if insignificant attempts of the early 'thirties to propel aircraft by means of batteries of powder rockets to the early 'fifties and the proposals for rocket-driven fighters tendered in the UK to meet the demands of specification F.124. The author traces the career of the rocket fighter in wartime Germany, in the Soviet Union, and in Japan; covers in detail the abortive attempt in the USA to produce such a warplane, and presents much intriguing and hitherto unrecorded factual matter. The book is profusely illustrated with many interesting photographs, high-quality cutaway drawings and general arrangement drawings of a comparable standard, though many of these suffer, more is the pity, from the publisher's method of presentation. Many of the exotic interceptors intended for this

highly volatile means of propulsion but which, for one or another reason, never *rocketed* into the sky, such as Northrop's XP-79, the Arado E-381, and the Short P.D.7, are included in the pages of this interesting paperback.

"Civil Aircraft Markings"
by J W R Taylor
Ian Allan, Shepperton, England, 25p
152 pp, 5 in by 7¼ in, illustrated
THE 1971 edition of a "spotters" classic. Contains the complete British and Irish civil aircraft registers corrected to the end of 1970, plus details of the fleets of the operators whose aircraft serve British airports regularly.

"Leaflet Operations in the Second World War"
by James M Erdmann
Dept of History, University of Denver,
Colorado 80210, USA
399 pp, 11 in by 8½ in, illustrated
A PRIVATELY PUBLISHED account of the "psychological warfare" waged by the Allies against the Axis in Europe, in which 6,500 million propaganda leaflets were dropped by the RAF and the USAAF. The author writes from personal experience and covers a subject which has been largely ignored until now.

"German Aircraft of the First World War"
by Peter Gray & Owen Thetford
Putnam & Co Ltd, London, £5·25
600 pp, 5½ in by 9 in, illustrated
THE second edition of this title, containing a good proportion of new illustrations, plus textual revisions in the light of additional research.

"The Red Falcons"
by Robert Jackson
Clifton Books, Brighton, England, £2·25
236 pp, 5½ in by 8½ in, illustrated
SUBTITLED "the Soviet Air Force in action, 1919-1969", this work appears to be based on a variety of published sources, mostly outside the Soviet Union. Provides a useful summary of the actions in which the Soviet Air Forces have been engaged but few details of the aircraft which were used.

JUNGLE FLIERS

OF THE RUBBER STATE

MALAYSIA, a federation of 13 states which together are responsible for producing 60 per cent of the entire world's rubber supply, not to mention 20 per cent of all the tin and substantial quantities of iron ore, palm oil, coconut oil and copra, enjoys a strategic importance out of all proportion to her size. Located in the South China Sea area with Indonesia, Thailand and the Philippines as her neighbours, Malaysia has had a turbulent history, and if the last few years have seen a lessening of the tension between herself and Indonesia, the imminent withdrawal of the British military presence from east of Suez may serve as the catalyst for a new round of violence. If so, the youthful but increasingly proficient Royal Malaysian Air Force (*Tentera Udara Diraja Malaysia*) will certainly have a vital rôle to play.

As a cohesive force, the RMAF has existed for barely more than 13 years, but the Malayan association with military aviation goes back much further. The states that now make up Malaysia have been part of the British sphere of influence from 1786 — and, since achieving independence in 1957, the Federation has remained a loyal member of the British Commonwealth. As a Crown Colony before World War II, Malaya forged close links with the Royal Air Force; a Malayan Auxiliary Air Force was formed, providing many volunteers for active service during the war, and two RAF bomber squadrons, Nos 97 and 214, were named after the Straits Settlements and Federated Malay States, respectively, to mark the support given by the colony.

A more active phase of participation in RAF activities came soon after the end of the war, when the Malayan peninsular became the target for attack by Communist terrorists. For most of the war, the territory had been occupied by Japan; after its liberation, British forces remained in Malaya and were soon in action against the guerrillas who, having resisted the Japanese occupation with considerable effect, were unwilling to accept renewed subjection to the British Crown. To help in putting down the guerrillas, a task which was to last until 1960, the Malayan Auxiliary Air Force was resuscitated in 1947, to provide an emergency reserve, and was able to make use of seconded RAF equipment, including Tiger Moths and Chipmunks. In the same year, the RAF Regiment (Malaya) was formed and provided an opportunity for Malay nationals to enlist and thus give direct service in the fight against the Communists. At this time, the pre-war Federation of Malay States had been reinstated, with 11 member states.

While the Royal Air Force remained wholly responsible for all offensive actions in Malaya, following the declaration of a State of Emergency in May 1948, a new indigenous support organisation was formed in 1950 as the Royal Air Force (Malaya), to recruit and train as tradesmen men of all local races, for service at RAF airfields in the Federation and in Singapore. The Malayan population is made up of several different races, with about half Malays, one-third Chinese and the remainder Indian and others, but those of Chinese origin have tended to outnumber other races in the Malayan air force organisation, largely because of superior educational qualifications.

While the Malayan emergency dragged on through the 'fifties, the RAF continued its "Operation Firedog" against the terrorists with a succession of aircraft, starting with Spitfires and including such unlikely types as Harvards, Sunderlands and Lincolns as well as Austers, Beaufighters, Brigands, Hornets, Vampires, Venoms, a variety of transport

The Royal Malaysian Air Force relies heavily upon STOL aircraft and helicopters to fulfil its mission of supporting the ground forces at remote jungle sites. Alouette III helicopters (below right) have taken over this task from the Scottish Aviation Pioneers (below left) used originally. For ground attack and light strike duties, the RMAF uses Canadair CL-41Gs (top).

A: Sud Alouette III of No 3 Squadron at Kuala Lumpur; B: CAC Sabre 32 of No 11 Squadron; C: Canadair CL-41G Tebuan of No 9 Squadron at Kuantan; D: DHC-4 Caribou of No 5 Squadron; E: Handley Page Herald 401 of No 4 Squadron; F: Hunting Provost 51 of the Flying Training School at Alor Star; G: Sikorsky S-61A-4 Nuri as used by No 10 Squadron at Kuantan.

aircraft types, and the all-important helicopters. On the political front, in the same period, progress was made toward achieving independence, an event marked by ceremonies on 31 August 1957 — *Merdeka* Day — although this did little to placate the terrorists who had been resisting British rule.

Birth of an Air Force

Under the terms of an Anglo-Malayan defence agreement drawn up at the time that the independent Federation of Malaya was founded, provision was made for the formation of an independent air force while Britain remained responsible for the continued defence of Malaya. The rôle of what was to be known as the Royal Malayan Air Force was to be primarily that of providing air support for the ground security forces and it was to rely heavily upon the RAF for help in its formative years.

Inauguration day for the RMAF (*Tentera Udara di Raja Persekutuan*) was set for 1 June 1958, a birth taking place, appropriately enough, nine months after Malayan independence, and an RAF officer, Air Commodore (later Air Vice Marshal) A V R Johnstone, DFC, was appointed its first Air Officer Commanding. For a year prior to independence, Air Cdre Johnstone had been acting as Air Adviser to Tunku Abdul Rahman Putra Al-Haj, the Malayan Prime Minister, and he had indicated that the aircraft chosen to equip the RMAF should be simple to operate, multi-purpose in rôle, and able to operate from jungle strips. He listed eight rôles that aircraft of the RMAF might be required to perform: short-range freight and personnel transport; VIP transport; troop carrier; air ambulance; voice broadcasting; light bomber; coastal patrol and photography/reconnaissance.

In the light of RAF experience in the area, the Scottish Aviation Pioneer and Twin Pioneer appeared to be well suited to fulfil these rôles, and four of each type were ordered as initial equipment for the RMAF, lack of funds precluding the simultaneous purchase of helicopters as recommended by the AOC. First of the RMAF's aircraft to arrive was one of the Twin Pioneers, ferried out from Prestwick, Scotland, to Malaya, and flown on the last stage of the flight by Air Cdre Johnstone himself, arriving at Kuala Lumpur on 17 April 1958. Kuala Lumpur, in the western "wing" of Malaya, was at that time a main RAF base, and considerable RAF assistance was extended to the new air force at this time, for by the end of April 1958 the strength of the RMAF itself comprised only one aircraft, three officers, three NCOs and no other ranks!

Delivery of the remaining Scottish Aviation Twin Pioneers and the four Pioneers followed after the official founding of the RMAF, and, forming the equipment of No 1 Squadron, these aircraft were soon in action in the primary rôles of supporting the ground forces and providing communication and liaison flights. Part of the RMAF responsibility was to support the Police Field Force which manned the forts along the Thai/Malay frontier, this task including the ferrying of supplies and air-lifting personnel to and from the forts, which were accessible only to helicopters or STOL aircraft like the Pioneers.

Pilot training initially proved something of a problem for the RMAF, and most recruits were sent on training courses in the UK and elsewhere. The situation was alleviated somewhat by the provision by the RAF of six Chipmunks during 1958, however, and these were used to establish a Flying Training Squadron at Kuala Lumpur in December of that year. With its three types of aircraft, the RMAF now set about creating a firm foundation upon which the planned future expansion could be based. In the event, expansion was to come at a much greater rate than had been anticipated, and the two years or so spent in consolidation were to prove well used.

By 1960, the activities of the Communist terrorists had declined to the point where the State of Emergency could be declared at an end — on 31 July — and the RMAF appeared to be set for a period of steady if undramatic growth. As a first step in this process, the base at Kuala Lumpur was taken over from the RAF on 25 October 1960, and in 1961 the first phase of the planned expansion was put in hand with an order for the first six of an eventual total of 18 Hunting Provost* piston-engined trainers to supplement and, in due course, replace the Chipmunks. The Provosts entered service at Kuala Lumpur in November 1961. Also during 1961, the RMAF took delivery of four more Pioneers and 10 Twin Pioneer Series IIIs; subsequently the four original "Twinpins" were updated from Series I to Series III standard with more powerful Leonides engines.

To provide for communications and VIP flights in somewhat greater comfort and at higher speeds than possible with the Twin Pioneers. two Cessna 310s were acquired in the initial period of RMAF operation, but these were replaced in September 1961 when three Doves were delivered from the UK, one being flown out by the RMAF's new Commander, Group Captain J N Stacey. One of this trio of aircraft was a Dove 8, furnished as a five-seat executive transport; the other two were Dove 7s with eight seats each and provision for the installation of air survey cameras. The Doves, later supplemented in the communications and survey rôles by two Devons provided by the RNZAF, were significant in being the first aircraft to be operated by the RMAF possessing sufficient range to make communication flights outside Malaya. The transport capability was further expanded with the addition of two Herons, both the Doves and Herons being operated by No 2 Squadron, which also took the single-engined Pioneers on strength after the additional Twin Pioneers had been delivered to No 1 Squadron.

The first helicopters

The next significant stage in the history of the RMAF came in 1963. Helicopters had always been regarded as being of vital importance for supply and liaison missions in the Malayan jungle, the RAF's experience in this respect naturally rubbing off on RMAF personnel. The RAF, in fact, had formed its first-ever operational helicopter unit, No 194 Squadron, on Dragonflies in Singapore in February 1953, and this squadron had subsequently seen service in Malaya during most of "Operation Firedog", together with other units flying Whirlwinds and the Belvedere, the latter being known affectionately to the Malayans as the "Flying Longhouse". As its initial helicopter equipment, the RMAF chose the Sud Alouette III, and deliveries of the first of several batches (air-lifted to Malaya in Carvair freighters) began in August 1963, bringing a considerable expansion in the capabilities of the RMAF, and permitting the formation of No 3 Squadron. Training remained a problem, however, and pilots for the Alouettes were sent to a civilian school in France for helicopter conversion courses.

On 16 September 1963, Malaya was joined by Singapore, Sarawak and Sabah (the former territory of North Borneo) to form the Federation of Malaysia. This led to a considerable expansion of the territory to be covered by the RMAF, since Sabah was some 500 miles (805 km) to the east of the Malayan peninsula across the South China Sea, on the northern border of Indonesia. Marking the change in the Federation, the RMAF became the Royal Malaysian Air Force (*Tentera Udara Diraja Malaysia*), and the yellow star in its insignia was modified to embody 14 points, having previously comprised 11 — one for each state in the Federa-

---

The Provost being out of production, the RMAF orders were filled by ex-RAF aircraft and — included in the first batch of six — the former Hunting demonstration aircraft, G-AMZN.

tion. Singapore was destined to remain within this Federation for only two years, but no further change in the insignia was made after Singapore's withdrawal in August 1965.

The creation of the Federation of Malaysia provoked a new crisis in relations with Indonesia, which had inspired the Brunei Revolt in 1962 and had thereby succeeded in keeping Brunei out of the Federation. Propaganda against Malaysia built up into direct military action with attacks across the borders with Sarawak and Sabah, and armed landings in Singapore and Malaya. While the re-equipment programme for the RMAF was speeded up, Malaysia invoked the defence treaty with the UK to bring the RAF into action once again, and for what was to prove to be a three-year "confrontation" with Indonesia.

Transport capability

In March 1963, the RMAF had ordered four Handley Page Herald 401s, and the first of these medium range transports with air-drop capability arrived in Malaysia in November of the same year as the nucleus of the equipment of No 4 Squadron. Delivery of the batch of four was completed by February 1964, and four more were ordered for delivery between September 1964 and November 1965. A further addition to the transport fleet was made in 1965 with the gift of four Caribou by the Canadian government.

The acceleration in the introduction of Alouettes and Heralds into RMAF service, resulting from confrontation, put an even greater strain on the training facilities available in Malaysia. This applied not only to the training of aircrew, but to all technical levels of ground crew as well. Pilots continued to be trained on a Provost/Twin Pioneer sequence at Kuala Lumpur or under contract in other countries — including the UK, Canada, New Zealand, India and the USA. Some technicians received training at the RAF ground schools at Halton and Locking, and, until August 1963, others were trained on an *ad hoc* basis by the Technical Wing at Kuala Lumpur. A School of Technical Training was opened in August 1963, however, at Kinrara, near Kuala Lumpur, and this now has the ability to train 320 airmen a year in airframes, engines and electronics. Replacing the Flying Training Squadron at Kuala Lumpur, a Flying Training School was formed at Alor Star, at the northern

extremity of western Malaysia, in July 1964, and additional batches of Provosts were purchased from the UK for its equipment. These aircraft, like the first six, were refurbished RAF Provosts, but they were brought up to armed standard with a 0·303-in machine gun in each wing root and wing pylons for rockets or light bombs, thus providing the RMAF its first aircraft capable of offensive operation.

The delivery of new aircraft from 1963 onwards allowed the RMAF to expand territorially into new bases. In December 1963, a flight of Twin Pioneers was detached to Labuan, an island off the west coast of Sabah, where it served alongside an RAF detachment, and in July 1964 a unit of Alouette IIIs was detached to Tawau, in eastern Sabah, also to serve alongside the RAF. Alouette pilots also began training in the operation of the helicopter, with floats replacing the wheel undercarriage, from temporary landing decks on RMAF frigates.

A specification for an attack aircraft with a bigger punch than the Provost was drawn up by the RMAF in 1964, and after an evaluation of the various armed trainers and light strike aircraft on the market, the decision was taken to order the Canadair CL-41G. This was the uprated, strengthened and armed version of the RCAF's CL-41 Tutor basic jet trainer, and the Malaysian order for 20 was destined to be the only export sale of this type. Deliveries began in 1967, the RMAF bestowing the name *Tebuan* (an indigenous species of wasp) on its new type, for which a new base was constructed at Kuantan, on the east coast of the Malaysian peninsula.

A fighter force

During 1968, as the full implications of the planned British withdrawal from Malaysia were studied and with the Philippines offering provocative action against Malaysian shipping off the coast of Sabah, plans began to mature for the creation of a full-scale fighter component within the RMAF. A number of offers were made by aircraft-manufacturing nations, among which the UK suggested an interim force of Hunters to be replaced in 1971 by a squadron of Harriers; Canada offered a package that included CF-5 strike fighters, Twin Otters and CL-215 amphibians; France offered 16 Mirage 5s, and Sweden put together a package of Saab 105X strike/trainers and Draken fighters. An evaluation mission from the RMAF toured the production centres early in 1969, but on 15 April a temporary solution presented itself when the Australian Government announced the gift to Malaysia of ten ex-RAF Commonwealth Sabre 32s, plus a simulator and support equipment. This gift was made under the terms of the Australian Defence Aid Programme for Malaysia.

Before deliveries of the Sabres to Malaysia began, the RMAF had grown in personnel strength to some 3,000

The RMAF's first aeroplane, Scottish Aviation Twin Pioneer FM1001 (above), was named Raja Wali *and was followed by 13 more of the same type. They have now been replaced in the transport rôle by Handley Page Heralds (below) and DHC Caribou.*

AIRCRAFT THAT HAVE SERVED WITH THE RMAF

	Rôle	Year of Intro.	Quantity
Scottish Avn Twin Pioneer 1	Tpt	1958	4
Scottish Avn Pioneer 2	U	1958	8
De Havilland Chipmunk	T	1958	6
Cessna 310	C	1959	2
Hunting Provost T.51	T	1961	18
Scottish Avn Twin Pioneer 3	Tpt	1961	10
HS Dove 7	C/AS	1961	2
HS Dove 8	C	1961	1
HS Devon	C/AS	1962	2
HS Heron	C	1962	2
Sud Alouette III	H	1963	25
HP Herald 401	Tpt	1963	8
DHC Caribou	Tpt	1965	14
Canadair CL-41G	S/T	1967	20
Sikorsky S-61A-4	H	1967	16
CAC Sabre 32	F	1969	10
HS 125 Srs 400	C	1970	1
Scottish Avn Bulldog	T	1971	15

Rôles: C, Communications and staff transport; C/AS, Communications and air survey; F, Fighter; H, Helicopter; S/T, Strike and trainer; T, Trainer; Tpt, Transport; U, Utility.

(Above) Largest helicopters operating in the south-east Asia area are the 16 Sikorsky S-61A-4s purchased by the RMAF.

For communications and VIP transport duties, the RMAF has purchased a series of aircraft from Hawker Siddeley, comprising the Dove and Heron (above) and an HS 125 (below).

officers and men, and its 100-odd aircraft were dispersed in ten squadrons. Four of these units were based at Kuala Lumpur, comprising No 1 Squadron with Caribou (nine more of which had been purchased in 1969 and a tenth having been bought in 1971 to supplement the original gift of four); No 2 with Doves, Devons, Herons and Twin Pioneers; No 3 with Alouette IIIs, and No 4 with Heralds. The second Alouette squadron was No 5, based at Labuan, together with the second Caribou unit, No 8 Squadron.

The CL-41G Tebuans at Kuantan now form two squadrons, No 9 for training and No 6 for light strike duties. Pilots are posted to No 9 Squadron after basic training, either on Jet Provosts in the UK or on the Provosts still soldiering on at the Flying Training School at Alor Star. New equipment for the FTS is now a high priority and the purchase of 15 Scottish Aviation Bulldogs was confirmed shortly before this issue closed for press. A Hawker Siddeley HS 125 was delivered in 1970 to replace the VIP Dove. Nos 7 and 10 Squadrons fly Sikorsky S-61A-4s, the first of which were delivered in 1967, against an initial order for 10 helicopters of this type, this order having since been supplemented by an order for a further six, with deliveries commencing this month (August). Able to carry 28 troops or lift a maximum slung load of 8,000 lb (3 630 kg), the S-61A-4, which is known to the RMAF as the *Nuri* (Yellow Bird), is the largest helicopter in the region, and has proved of great value in helping the RMAF fulfil its close support rôle in counter-insurgency operations. One squadron of S-61As, No 7, is based at Kuching in east Malaysia (Sabah), while No 10 is at Kuantan. The latter squadron received acclaim during the recent flooding around Pahang when four of its *Nuris* helped to rescue trapped villagers, and ferried over 20,000 lb (9 070 kg) of supplies to the stricken area during the course of a week.

Sabre delivery
Delivery of the 10 Sabres, fully overhauled by Commonwealth Aircraft in Australia, was made on 1 October 1969 at the RAAF base at Butterworth on the Malaysian west coast, and they were accompanied by a total of 104 RAAF personnel, including four pilots, to assist in training RMAF ground and air crews. This activity was based at Butterworth, where the RAAF had had its own Sabre squadrons based continuously from 1958 to 1969, and the first five RMAF pilots completed their Sabre conversion course in April 1970. By the end of the year, 11 pilots had been trained, providing the flying personnel for No 11 Squadron which is now operational as the newest unit of the RMAF.

The Sabres have an operational life of at least six years, but for the RMAF they are clearly no more than an interim step towards achieving the goal of acquiring a combat aircraft that can be integrated with the air defence system which is being established jointly with Singapore, the chosen type being the Dassault Mirage 5.

In the 13 years of its existence, the *Tentera Udara Diraja Malaysia* has come a long way and achieved a great deal, as demonstrated by its operations of large helicopters and relatively advanced jet fighters. Reliance on seconded personnel from the RAF, RAAF and RNZAF is diminishing, and Malaysians are acquitting themselves well in what is still to them the relatively strange environment of flying. A former Chief of Air Staff of the RMAF has said that "although Malaysians of all races tend to be slow in learning to fly, they amply repay persistence and extra tuition in training, and they become thoroughly sound and very able pilots with experience. Few seem to have high natural ability to fly but, given the will to learn, they do very well indeed once they *have* learned. The final product after a tour in a squadron compares favourably with pilots anywhere".

The RMAF today is "young in years but mature in performance" — a cornerstone in the fight waged by Malaysia against infiltration and aggression of the kind that appears to be endemic to the area. It might well adopt as its motto, from the characteristics of the rubber that Malaysia produces in abundance, "Tough and Resilient". □

The "Halfway-House" Fokker

At the end of each working day in the summer of 1935, Doctor Schatzki, the chief design engineer of the NV Nederlandsche Vliegtuigenfabriek Fokker, and the members of his small team from the Papaverweg factory, alongside the Johan van Hasselt Canal in Amsterdam-Noord, must surely have found much to discuss over their glasses of *Oude Genever* in the *Tolhuis* while awaiting the ferry that would carry them to their homes across Amsterdam Harbour.

The mid 'thirties were witnessing radical transformation of the international fighter scene; a metamorphosis of which few were aware other than those directly concerned with the design and development of such warplanes. Dr Schatzki and his colleagues *were* concerned, for, in the Papaverweg factory, construction had just begun on a new single-seat fighter; the first fighter bearing the world-famous Fokker appellation to be evolved under the ægis of Dr Schatzki, who had assumed design engineering responsibility from Ir F H Hentzen during March of the previous year.

What was being referred to widely as the *nouvelle vogue* in fighter design had become an intriguing topic for discussion; the single-seat fighter marrying an all-metal stressed-skin monocoque structure to a low-wing cantilever monoplane configuration and embellishing this union with a retractable undercarriage and a cockpit canopy. Increasing international tensions were adding to the restrictions already imposed by commercial considerations on the free interchange of information, but as in most professions, a grapevine existed in the aircraft design business; one fighter development team, if unable to obtain *detailed* information concerning the lines along which its contemporaries abroad were working, could at least glean rudimentary facts with which to build up a broad picture of activity in its particular field of interest.

How detailed a picture of the international fighter design scene *was* available to Dr Schatzki and the Fokker team during that summer of 1935 must remain a matter for conjecture, but awareness of only a small proportion of the work being undertaken abroad on fighters conforming to the *nouvelle vogue* must have given rise to speculation during those early-evening discussions in the *Tolhuis* of the possibility that their new fighter was already outmoded in concept. By comparison with fighters being evolved abroad, the new Fokker fighter, which had been allocated the designation D XXI, was very much a halfway-house — a compromise between the long-accepted and the newly-fashionable. Its concessions to modernity were a cantilever low wing and an enclosed cockpit which were combined with Fokker's time-proven constructional recipe — a welded steel-tube fuselage and wooden wings — and a fixed undercarriage.

The D XXI was not, by any stretch of the imagination, an inspired design. It was a sturdy, relatively simple warplane, as was demanded by the specification to which it was designed. This specification had been formulated by the Air Division, or *Luchtvaartafdeling,* of the Royal Netherlands Indies Army, the *Koninklijk Nederlands Indisch Leger* (KNIL), and was unambitious in so far as its performance requirements were concerned. Emphasis was placed on ease of maintenance in the field and simplicity of operation, and had the KNIL been aware of the lines along which fighter development was proceeding in Japan, it is likely that the service would have paid more attention to level speed and climb rate when formulating its fighter requirement.

Dr-Ir E Schatzki was undoubtedly aware of the trend towards the all-metal stressed-skin monocoque fighter, but the extensive jigging and tooling demanded by the new manufacturing techniques employed for such warplanes were, at that time, beyond Fokker's capability, and in justification of the retention of the traditional Fokker-type structure for the D XXI, it was believed that fighters conforming to the

WARBIRDS

nouvelle vogue would be incapable of operating under relatively primitive conditions. Furthermore, analysis of the potential export market suggested that the smaller air arms, many of which were already Fokker customers, would be loath, at that point in time, to purchase the highly-sophisticated fighters that were under development in Britain, France, Germany and elsewhere.

Thus, while the D XXI was designed around the KNIL requirement, Dr Schatzki and his team endeavoured to evolve a basic design capable of taking a wide variety of power plants and armament combinations, and which could be offered with either fixed (as specified by the KNIL) or retractable main undercarriage members. The wing was built up on two wooden box spars with plywood ribs and bakelite-plywood skinning. Where necessary diagonally-glued strips of plywood were used, the overlapping sections being bevelled to obtain a smooth finish. The metal-framed, fabric-covered ailerons featured inset ground-adjustable tabs, and hydraulically-operated split flaps occupied the entire wing trailing edge between the ailerons with which they were interconnected.

The fuselage was a welded chrome-molybdenum steel-tube structure, the forward portion of which to the wing trailing edge was covered by detachable Elekton panels, as was also the fuselage upper decking, the remainder being fabric covered. The cockpit was enclosed by a jettisonable Plexiglass canopy, the port half hinging to provide access, and immediately aft of the canopy the structure was braced to provide a turnover pylon. Alternative fixed and retractable undercarriages were designed for the D XXI, these comprising cantilever legs with oleo-pneumatic struts, the former being enclosed by streamlined Elekton fairings and the latter being operated hydraulically and swivelling laterally into wing wells. Compressed-air brakes which could be operated differentially were provided, and the fixed tailwheel was coupled to the rudder but could be disengaged and re-engaged by the pilot at will.

The D XXI was so designed that it could be powered by virtually any liquid-cooled or air-cooled engine in the 600–1,100 hp category, and when fitted with the former provision was made for a 59·4 Imp gal (270 l) capacity fuel tank immediately aft of the engine with a 28·6 Imp gal (130 l) tank in the wing centre section, offset to port, while installation of the latter was accompanied by an increase in the capacity of the main tank to 64 Imp gal (290 l) and a reduction in the capacity of the auxiliary tank to 24 Imp gal (110 l). Armament called for by the KNIL specification comprised one synchronised 7,92-mm FN-Browning M-36 machine gun with 500 rounds offset to starboard in the fuselage, and two similar weapons each with 300 rounds mounted in the wings outboard of the airscrew disc, but a variety of alternative armament arrangements were envisaged, including both 7,92-mm FN-Browning machine guns and 20-mm Madsen, Oerlikon or Hispano cannon.

An ace of still-birth

The KNIL had specified the installation of a 645 hp Bristol Mercury VI-S nine-cylinder radial air-cooled engine in the prototype, which was finally completed in February 1936, and transported to Welschap airfield near Eindhoven for flight testing. A few weeks later, on the morning of 27 March, with Fokker's Czechoslovak test pilot, Emil Meinecke, at the controls, the D XXI was flown for the first time. Meinecke was elated by the delightful handling characteristics evinced by the little fighter, and was in no doubt that the D XXI met and, from several aspects, exceeded all the demands of the KNIL requirement. Unfortunately, the future of Fokker's fighter was already in doubt, and there seemed every possibility that this promising new warplane would be still-born.

While the Fokker team had been preparing the D XXI for flight testing at Welschap, the KNIL had announced a radical change in its defence thinking. It had concluded that single-seat fighters distributed in penny packages throughout the vast territory of the East Indies could offer no serious defence, and that the only realistic form of defence was attack. The need, therefore, was not for fighters but for bombers. The KNIL had contracted for only one prototype of the D XXI, and by the time flight testing commenced the results were of little more than academic interest to the KNIL *Luchtvaartafdeling,* while the possibility of a production order from the home-based service appeared remote.

On 17 March, 10 days before the D XXI's maiden flight, Dr H Colijn, the Minister for Colonial Affairs, had informed the Defence Ministry of the new defence policy formulated for the East Indies, stressing the fact that no funds would be available in the immediate future for the procurement of single-seat fighters on behalf of the KNIL, and suggesting that the Defence Ministry might give consideration to the possibility of taking over the development contract on behalf of the home-based *Luchtvaartafdeling* (LVA). On 28 March, the Commander of the LVA, Col P W Best, whose opinion had been requested by the Defence Minister, reported as follows:

"According to my Plan III concerning the number of first-line aircraft required, we need 18 ultra-modern single-seat fighters for air defence. Although we do not have reliable information about the number and performance of modern fighters in service with neighbouring nations, I think it reasonable to suppose that speeds of the order of 450 km/h (280 mph) may be expected. In any case, I consider that the speed difference between our present Fokker

(Above) The prototype Fokker D XXI, and (below) D XXIs of the 5e Jachtvliegtuigafdeling in the national insignia used prior to October 1939. Note lack of rod-mounting for ring-and-bead sight.

(Above) One of the 10 D XXI fighters manufactured under licence in Denmark seen prior to the mounting of 20-mm Madsen cannon under the wings, and (below) the 20-mm Madsen cannon standardised on Danish D XXI fighters.

D XVII (370 km/h — 230 mph), which dates from 1933, and the D XXI (411 km/h — 255 mph) is insufficient for the latter to be viewed as an ultra-modern fighter. In my opinion, we must try for a speed such as that anticipated for the new Koolhoven fighter* now under development (520 km/h — 323 mph). We will have to establish such a performance requirement if we want to maintain our policy that our fighters are at least 50 km/h (31 mph) faster than potential enemy bomber aircraft, and some of these we know to be capable of reaching 400 km/h (248 mph) or more.

"Another objection to the D XXI is that the prototype has a fixed undercarriage, and the cost of a D XXI prototype with a retractable undercarriage but without engine will be as high as Fl 76,350, meaning that the total cost of the prototype will be barely less than Fl 100,000. Series production aircraft are likely to cost little less, and comparing this cost with that of a *luchtkruiser* (a heavy fighter in the German *Zerstörer* category) procurement of the latter is highly preferable. To sum up, I would advise you *not* to accept the D XXI as all available funds will be needed for the urgent procurement of training aircraft and the greatest possible number of aircraft in the *luchtkruiser* category."

Despite this rejection on the part of the LVA Commander, on 24 November 1936 the Minister of Defence ordered an intensive flight test programme to be undertaken by the LVA with the D XXI prototype at Soesterberg. This evaluation was performed between 25 November and 10 January 1937, and the results were summarised in a letter of 12 January to the Minister of Defence from the Director

of the *Luchtvaartbedrijf* (Air Materials Department) as follows:

"There are no technical objections against the procurement of a number of Fokker D XXI aircraft on behalf of the LVA. The evaluation that we have just conducted has not been entirely comprehensive, however, as less than six weeks were available in which to perform all tests, and there has not been time to make an intensive study, for instance, of the fuel system. Nevertheless, it can be confirmed that the construction of the D XXI is fully in accordance with the high quality of earlier Fokker aircraft delivered to the LVA. It should be pointed our that the use of chrome-molybdenum in the airframe will dictate some re-training of technical pesonnel, and in times of war this alloy may prove difficult to obtain. The following modifications and changes should be made:

(1) Cockpit instruments and equipment should be relocated to conform to International Standards.
(2) To facilitate an engine change it is suggested that a similar type of mounting to that employed by the Fokker T V bomber should be adopted.
(3) Rubber vibration-damping should be incorporated in the engine mount.
(4) The surface finish of both wings and fuselage should be improved.
(5) The fabric skinning of the leading edges of the tailplane and vertical fin should be strengthened to prevent distortion in flight.
(6) The rudder-actuating cable attachment points should be redesigned to reduce wear.
(7) The single rudder-actuating cable should be replaced by a double cable in accordance with LVA specifications.
(8) The aileron actuating levers should be improved and strengthened.
(9) The elasticity of the tailwheel shock-absorbing leg should be increased.

In the meantime, the Fokker sales team had been making a determined effort to obtain export orders for the D XXI. The Fokker catalogue distributed at the *Internationella Luftfartsutställningen* held in Stockholm in May 1936 offered the D XXI with either fixed or retractable undercarriage,

---

(Opposite page) A Fokker D XXI of the 2e Jachtvliegtuigafdeling at Schiphol, Spring 1940. Upper surfaces finished in irregular pattern of green, beige and olive, and engine cowling (apart from cowling ring) and under surface finished in dark reddish brown. Orange inverted triangles outlined in black on fuselage sides and upper and lower wing surface, orange rudder outlined in black, and white fuselage numerals. (Drawing reproduced by courtesy of BPC Publishing Limited.)

This, the Koolhoven F.K.55, was a radical aircraft in which the engine was mounted immediately aft of the pilot's cabin, driving contra-props via an extension shaft, but was to prove an abysmal failure.

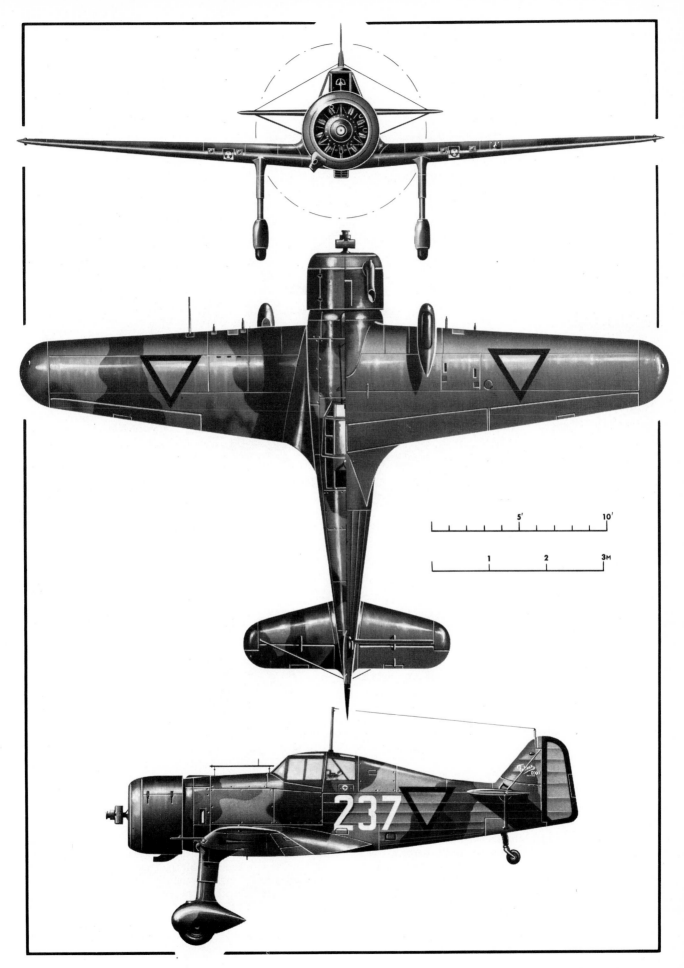

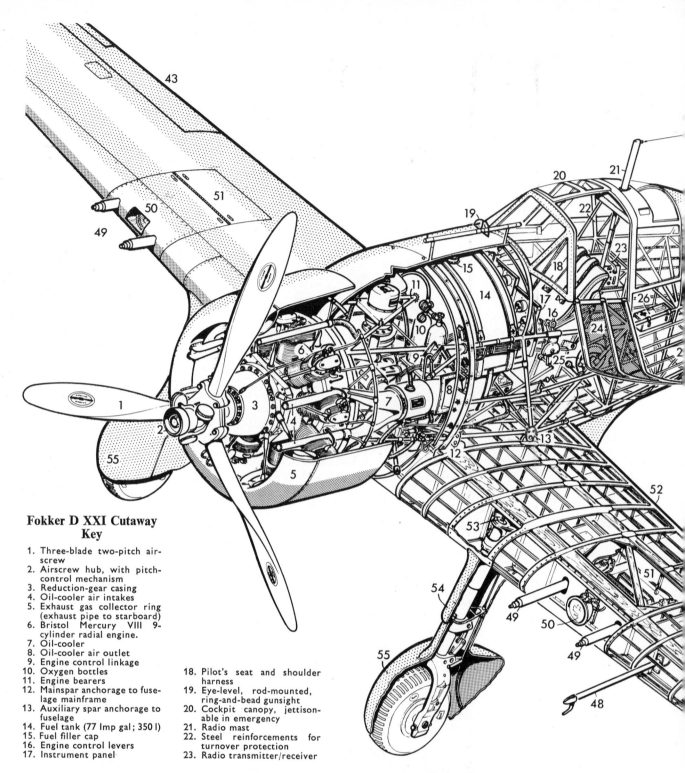

**Fokker D XXI Cutaway
Key**

1. Three-blade two-pitch airscrew
2. Airscrew hub, with pitch-
   control mechanism
3. Reduction-gear casing
4. Oil-cooler air intakes
5. Exhaust gas collector ring
   (exhaust pipe to starboard)
6. Bristol Mercury VIII 9-
   cylinder radial engine.
7. Oil-cooler
8. Oil-cooler air outlet
9. Engine control linkage
10. Oxygen bottles
11. Engine bearers
12. Mainspar anchorage to fuse-
    lage mainframe
13. Auxiliary spar anchorage to
    fuselage
14. Fuel tank (77 Imp gal; 350 l)
15. Fuel filler cap
16. Engine control levers
17. Instrument panel

18. Pilot's seat and shoulder
    harness
19. Eye-level, rod-mounted,
    ring-and-bead gunsight
20. Cockpit canopy, jettison-
    able in emergency
21. Radio mast
22. Steel reinforcements for
    turnover protection
23. Radio transmitter/receiver

and with the 1,100 hp Hispano-Suiza 14 Ha 14-cylinder air-
cooled radial, the 925 hp Hispano-Suiza 12 Ycrs 12-cylinder
liquid-cooled engine, or the 830 hp Bristol Mercury VII
nine-cylinder radial. The fact that the D XXI demanded no
sophisticated jigs and tools for its manufacture now stood it
in good stead, for the Finnish government, seeking a new
fighter that could be licence-manufactured by the State
Aircraft Factory, or *Valtion Lentokonetehdas,* was attracted
by the relative simplicity of the Fokker fighter, and, on 18
November 1936, placed an order for seven D XXIs powered
by the Mercury VII, and took an option on a manufacturing
licence for a further 14 aircraft. This option was turned into a
firm contraction 7 May 1937, and supplemented some six

weeks later, on 15 June, with a contract for the licence
manufacture of a further 21 aircraft.

At the time the Finnish government placed its initial
order for the D XXI, Fokker calculated that the Mercury
VII-powered model would attain a maximum speed of 286
mph (460 km/h) with a fixed undercarriage and 295 mph
(475 km/h) with a retractable undercarriage at rated altitude,
but owing to the higher weights of the latter — empty and
loaded weights with a retractable undercarriage being
3,119 lb (1 415 kg) and 4,277 lb (1 940 kg) as compared
with 2,976 lb (1 350 kg) and 4,134 lb (1 875 kg) when
fitted with fixed gear — climb rate suffered marginally, and
the Finnish air arm, *Ilmavoimat,* opted for the simpler

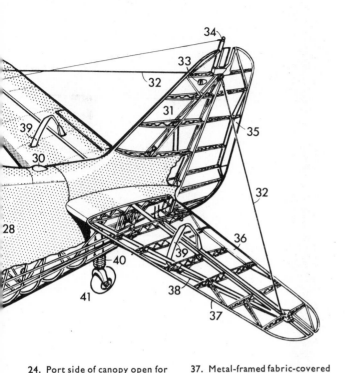

Fokker D XXI Specification

Power Plant: One Bristol Mercury VIII nine-cylinder radial air-cooled engine rated at 730 hp at 2,650 rpm for take-off, with international rating of 825 hp at 13,000 ft (3 962 m) and maximum output (for five min) of 840 hp at 14,000 ft (4 267 m).
Armament: Four 7,9-mm FN-Browning M-36 machine guns with 300 rpg.
Weights: Empty equipped, 3,197 lb (1 450 kg); normal loaded, 4,519 lb (2 050 kg); wing loading, 25·9 lb/sq ft (126,5 kg/m²).
Performance: Maximum speed, 286 mph (460 km/h) at 16,730 ft (5 100 m); cruise (66% power), 240 mph (386 km/h), (55% power), 228 mph (368 km/h); range, 578 mls (930 km); climb to 3,280 ft (1 000 m), 1·45 min, to 9,840 ft (3 000 m), 4·05 min, to 16,400 ft (5 000 m), 6·6 min, to 22,965 ft (7 000 m), 10·1 min; service ceiling 33,135 ft (10 100 m).
Dimensions: Span, 36 ft 1 in (11,00 m); length, 26 ft 10¾ in (8,20 m); height, 9 ft 8 in (2,95 m); wing area, 174·375 sq ft (16,2 m²).

24. Port side of canopy open for pilot access
25. Elevator trimwheel
26. First-aid stowage
27. Recessed foothold
28. Welded chrome-molybdenum fuselage structure, fabric-covered aft of cockpit
29. Detachable dural top decking panels
30. Signal lamp
31. Metal-framed fabric-covered fin and rudder
32. Fin/tailplane bracing wires
33. Tail navigation light
34. Aerial attachment to fin
35. Ground-adjusted rudder trim tab
36. Flight-adjusted elevator trim tabs
37. Metal-framed fabric-covered tailplane and elevator
38. Bracing strut beneath tailplane
39. Elevator mass balances
40. Compressed rubber shock-absorber
41. Steerable tailwheel with optional link to rudder.
42. Wingtip navigation light
43. Metal-framed fabric-covered ailerons
44. Ground-adjusted aileron trim tabs
45. Wing structure of plywood ribs and bakelite/plywood skin
46. Wooden box auxiliary spar
47. Wooden box mainspar
48. Pitot head
49. Barrels of four 7,9-mm FN/Browning M-36 machine guns
50. Landing lamps
51. Machine gun and ammunition bay
52. Split landing flap
53. Undercarriage anchorage to main spar.
54. Fairing on oleo leg.
55. Wheel spats.

fixed-gear version. The Finns specified an armament of two fuselage-mounted and two wing-mounted 7,9-mm FN-Browning M-36 machine guns, and supplied PZL-built Mercury VII engines for the seven aircraft ordered from the parent company, these driving three-blade two-pitch metal airscrews. Most of the modifications and changes recommended by the LVA were incorporated in the Finnish machines, and deliveries to Finland began during the course of the summer of 1937.

By this time, the prospects of the D XXI had radically improved, the *Tweede Kamer* having voted funds for the limited expansion of the LVA, and an order having been placed for 36 D XXIs essentially similar to those purchased by Finland apart from having British-manufactured Mercury VIII engines and the entire quartet of FN-Browning M-36 machine guns mounted in the wings. The Spanish Republican government had contracted for a single pattern D XXI and a manufacturing licence to produce the fighter at Carmoli, and the Danish government was negotiating a contract for two D XXIs and a licence to manufacture the type in the Army Aviation Troops' Workshops, the *Hærens Flyver-troppers Værksteder.*

The first D XXI for the LVA (Nr 212) was completed and flown during the early summer of 1938, and was tested on behalf of the LVA Committee by Dr Ir van der Maas whose report on the characteristics of the fighter was extremely favourable. Speed proved to be slightly higher than that specified at all altitudes, but some problems were encountered with the airscrew pitch setting. With the airscrew at coarse pitch, climb rate to specified ceiling proved excellent, but above this altitude both engine revs and oil and cylinder temperatures increased rapidly. With the airscrew in fine pitch, climb rate was exactly in accordance with specification up to the normal service ceiling but subsequently fell away rapidly. It was decided, therefore, to stick to the standard procedure of taking-off in coarse pitch and immediately switching to fine pitch for the climb, and, meanwhile, to investigate the effects of reducing the coarse-pitch angle of the airscrew.

During the third week of July 1938, the first production D XXI was demonstrated to the LVA at Schiphol by Emil Meinecke, and was officially accepted by the service, brief familiarisation flights being conducted by Capt van Gemeren and Lt de Zwaan before, on 22 July, it was flown to Soesterberg. In the meantime, Dr Schatzki had left Fokker to join NV Koolhoven Vliegtuigen, for which concern, utilising experience gained with the D XXI, he had initiated the design of the F.K.58 single-seat fighter. Early in August, the second and third LVA production aircraft (Nrs 213 and 214) were tested and accepted by the service, the fourth and fifth aircraft (Nrs 215 and 216) following later in the month, all five aircraft being pronounced ready for operational service at Soesterberg. The contract called for the 36th and

(Above left) Twin Wasp Junior-powered D XXIs with decoratively-painted undercarriage fairings in Finnish service in 1942, and (above right) the sole example of the D XXI to be completed with retractable main undercarriage members.

last D XXI for the LVA (Nr 247) to be ready for delivery by 3 March 1939, although, in the event, this aircraft was retained by Fokker for test and development tasks, being finally accepted by the LVA on 8 September.

The D XXI goes to war

Interspersed on the assembly line with LVA machines had been two D XXIs for the Danish Army's air component, the *Hærens Flyvertropper*. These differed from LVA fighters primarily in armament, comprising two fuselage-mounted 7,9-mm machine guns and two 20-mm Madsen cannon, which, manufactured by the Dansk Industri Syndikat in Copenhagen, were mounted beneath the wings outboard of the undercarriage. The two pattern aircraft (J-41 and J-42) reached Denmark on 29 April 1938, these being Mercury VIII-powered but one having a two-blade two-pitch airscrew and the other having a three-blade two-pitch airscrew. The manufacture of a further 10 D XXIs began at the *Værksteder* in 1939, three being completed by the end of the year and issued to 2.*Eskadrille* at Værløse. The remaining seven D XXIs had been completed by the *Værksteder* by 9 April 1940 when the *Luftwaffe* made its surprise dawn attack on Værløse, although one had still to be flight tested. On that date, 2.*Eskadrille* had eight D XXIs on strength, but the unit was not fully operational, and none of the D XXIs had an opportunity to join combat, one being destroyed and four being damaged by the Bf 110s which repeatedly strafed the airfield. Three of the D XXIs were subsequently re-built, and, together with those that had survived the attack on Værløse, were placed in storage at Kløvermarken until 29 August 1943 when they were seized by the *Wehrmacht,* their ultimate fate being unknown.

Work on the licence manufacture of the D XXI at Carmoli in Spain had come to a sudden halt in mid-1938 when the rapid advance of Nationalist forces towards the factory had necessitated its destruction, together with the semi-completed airframes on the assembly line, but in Finland the State Aircraft Factory had flown its first D XXI fighter in October 1938, and by 30 November 1939, when Soviet forces invaded Finland, had completed all 35 of the D XXIs called for by the two licence manufacturing contracts. These were powered by the Tampella-built Mercury VII (which was identical to the Mercury VIII installed in Dutch and Danish D XXIs apart from the reduction gear ratio) and, bearing the Finnish serials FR-83 to FR-117, were on the strength of HLeLv 24. The Finnish-flown fighters thus gained the distinction of being the first D XXIs actually to fire their guns in anger, the first "kill" being recorded on 1 December 1939 when Lt Eino Luukkanen, flying D XXI FR-104, destroyed a Tupolev SB near Jääski.

During the so-called "Winter War", which terminated in

an Armistice on 13 March 1940, the D XXI bore much of the brunt of the defensive fighting in Finnish skies, and *Ilmavoimat* had good cause to be grateful for the structural simplicity and ease of maintenance offered by the Fokker monoplane, being forced to operate the D XXIs in small detached units from the most primitive of bases under conditions that would have grounded more sophisticated fighters within days of hostilities commencing. The D XXI was popular with its pilots whose skill more than compensated for the inferiority of certain aspects of performance which it suffered by comparison with most of the Soviet fighters by which it was opposed.

The D XXI provided a stable gun platform and its handling characteristics were, in general, extremely good, although it did not suffer novices gladly, and several aircraft were lost by the Finns in training accidents, these usually occurring during the landing approach when, if the angle of attack was too high, the tail surfaces were partly blanketed and control was inadequate. At the end of the "Winter War", the first-line inventory of *Ilmavoimat* included 29 D XXIs of which 22 were serviceable, and it is a remarkable fact that, despite having been responsible for the bulk of the 200 Soviet aircraft claimed by the Finns as destroyed in aerial combat, only 12 D XXIs had been lost, and half of these had been casualties of accidents, one, for example, having crashed in thick fog and another having been destroyed when a companion aircraft landed on top of it during a blinding snowstorm.

Less than two months after the Soviet-Finnish Armistice it was the turn of the LVA D XXIs to be blooded in action. When the *Wehrmacht* invaded the Netherlands on 10 May 1940, three *Jachtvliegtuigafdelingen* (each roughly equivalent to an RAF squadron) were operating the D XXI: 1e JaVA based at De Kooy with 11 serviceable aircraft; 2e JaVA based at Schiphol with 10 serviceable aircraft, and 5e JaVA based at Ypenburg with eight serviceable aircraft. Each JaVA was divided into three flights, or *patrouilles,* only the flight leaders' D XXIs being fitted with full R/T.

During the first morning of the invasion all three JaVA suffered heavy casualties, only one D XXI of 5e JaVA surviving the first hours of the *Blitzkrieg* during which the unit accounted for three Ju 52/3m transports, a Messerschmitt Bf 109E fighter, a Bf 110 fighter and an He 111 bomber. In the confused aerial warfare over Dutch territory, the D XXIs gave a good account of themselves despite the air superiority established by the *Luftwaffe* from the outset. For example, eight D XXIs of 1e JaVA engaged a similar number of Bf 109Es in a dogfight in the vicinity of De Kooy at an altitude of about 1,200 ft (365 m), destroying four of the *Luftwaffe* fighters without loss, although virtually all the D XXIs participating in the mêlée were damaged. With the JaVA bases under repeated attack, on the afternoon of 10 May it was decided to evacuate the bases of De Kooy, Schiphol and Ypenburg, and concentrate all surviving D XXIs in a combined JaVA at the auxiliary airfield of Buiksloot, north of Amsterdam, where, on the morning of the second day of the invasion, 12 D XXIs were gathered, these comprising five from 1e JaVA, six from 2e JaVA, and one from 5e JaVA. The combined JaVA operated continuously from Buiksloot for four days until the capitulation at 19.00 hours on 14 May, flying escort missions, offensive patrols, and intercept and ground strafing sorties, and throughout the *Luftwaffe* failed to locate the field from which the unit operated. Throughout this period serviceable strength varied between five and nine aircraft, and nine D XXIs were available for operational use at the time of the capitulation.

Although the D XXI had reached the end of its career in the service of its country of origin, it was to continue in Finnish service for some time to come. What is more, it was to be reinstated in production in 1941 by the Finnish State Aircraft Factory which, during 1940, concentrated on the repair and overhaul of aircraft that had survived the "Winter War". HLeLv 24 relinquished its D XXIs in favour of Brewster 239s, the former being issued to a new fighter unit, HLeLv 32, and some forming a mixed unit with Hurricanes as HLeLv 30. With the decision to resume manufacture of the D XXI the problem of a suitable engine presented itself, all output of Mercury engines from the Tampella Machine Works being required for Blenheims. Fortuitously, 80 Pratt & Whitney R-1535 Twin Wasp Junior SB4-C and -G 14-cylinder two-row radials of 825 hp had been acquired from the USA in 1940, and it was decided to adapt the D XXI to take this power plant.

With the installation of the Twin Wasp Junior engine it was no longer possible to accommodate two of the FN-Browning machine guns in the fuselage, and these were therefore transferred to the wings, and the greater weight of the US engine necessitated a substantial increase in the size of the vertical tail surfaces. The transparent panelling to the rear of the cockpit was extended to improve aft vision, and the telescopic gunsight was replaced by a reflector sight. Weighing 3,380 lb (1 533 kg) empty and 4,820 lb (2 186 kg) in normal loaded condition, the Twin Wasp Junior-powered D XXI was not to enjoy the same popularity with its pilots as had its Mercury-engined predecessor. Obviously underpowered, it was slower and some-

what sluggish by comparison with the earlier model, attaining a maximum speed of 272 mph (438 km/h) at 9,000 ft (2 740 m), an altitude of 3,280 ft (1 000 m) being reached in 1·4 min, and 9,840 ft (3 000 m) demanding 4·5 min.

Fifty Twin Wasp Junior-engined D XXIs were ordered from the State Aircraft Factory, all of which were completed during the course of 1941, and when, on 25 June 1941, what was to be known to the Finns as the "Continuation War" started, HLeLv 32 had 19 fighters of this type on strength, together with 17 Mercury-engined D XXIs, but from a relatively early stage in the new conflict both Mercury- and Twin Wasp Junior-powered D XXIs were relegated to the tactical reconnaissance rôle, initially performing this task with TLeLv 14 and, later, TLeLv 12. Several Twin Wasp Junior D XXIs had the glazed panelling aft of the cockpit still further extended, and one example was experimentally fitted with retractable main undercarriage members, these retracting inwards into wells in enlarged wing-root fairings. Somewhat surprisingly, a further five D XXIs were assembled by the State Aircraft Factory from spares as late as 1944.

The D XXI had been overtaken by developments in the international fighter scene before it flew in prototype form, and it was obsolescent by the time that the first shots of WW II were fired. Nor was it favoured by chance for it was invariably operated under conditions of enemy air superiority. Nevertheless, it acquitted itself well as it wrote *finis* to the story of Fokker fighters which had spanned a quarter-century. □

(Below) A late-production Twin Wasp Junior-powered D XXI (FR-150) with glazed panelling extended along the rear decking to improve view for the tactical reconnaissance rôle, and (above) a similarly-modified D XXI (FR-129) with ski undercarriage.

One afternoon in late July of 1944, a Japanese anti-aircraft shell exploding in Manchurian skies initiated a chain of events that was to cost the North American tax-payer billions of dollars. Its immediate effect was to provide the Soviet Union with an example of the Boeing B-29 Super-fortress, around a Soviet copy of which was to be created a powerful strategic air arm. Its longer term effect was to provide a basis for the development, by genealogical processes stretching over a decade, of the only Soviet bomber capable of performing strikes across the polar regions, and the sole factor justifying the fantastic cost of building the immense chains of radar installations and interceptor bases intended to shield the North American continent from manned aircraft attack. Part One of this fascinating story appeared in the July issue of AIR ENTHUSIAST.

THE BILLION DOLLAR BOMBER

PART TWO

THE AVIATION Day parade over Tushino, Moscow, on 3 August 1947 must have provided Chief Marshal of Aviation Golovanov, Commander-in-Chief of the Soviet Long-range Air Force, the *Dal'naya aviatsiya*, with something of a traumatic experience. Golovanov was piloting the third pre-production example of the new Soviet strategic bomber which had yet to complete its flight test phase. Flying with him and forming a loose vic were two of the most experienced of Soviet test pilots, Nikolai Rybko to port at the controls of the first pre-production bomber, the *yedinitsa,* and Mark Gallai to starboard in the second pre-production bomber, the *dvoika.* The practice flights for the parade had passed without untoward incident, but, on the day, the timing of the individual items on the programme slipped and Golovanov suddenly realised that the main formation was flying on an opposite course to the trio of bombers that he was leading — and at the same altitude! He immediately put the nose of his aircraft down and, with Rybko and Gallai clinging to his tail, succeeded in avoiding

a spectacular collision and swept across Tushino at low altitude.

If Golovanov's experience was traumatic it was probably hardly more so than that of the US air and military attachés who witnessed the fly-past of the bombers from the ground, for they were identical in every respect visible to the eye, naked or otherwise, to the Boeing B-29 Superfortress! Since late 1946 there had been rumours that the Soviet aircraft industry was endeavouring to produce a copy of the B-29s that had been sequestered after landing in Soviet territory during the final phases of WW II, but even when it became known that agents acting on behalf of the Soviet government had attempted to purchase in the USA tyres, wheels, and brake assemblies for the B-29, the rumours had been given little credence. The consensus of opinion was that the magnitude of the task of breaking down and copying so large and sophisticated an aeroplane was far beyond the capabilities of Soviet technology, even if, by means of espionage, the Soviet aircraft industry had acquired rather

(At head of page) The sole prototype of the Tu-70, the commercial transport derivative of the Tu-4 strategic bomber, and (below) a sectioned view of the proposed production version of the Tu-75 military transport with two remotely-controlled gun barbettes, tail armament, and ventral loading ramp.

more knowledge of manufacturing techniques employed in producing the B-29 than could be gleaned from the three intact examples of the bomber that had fallen into Russian hands.

If further evidence was wanted that the Soviet aircraft industry *was* indeed building copies of the B-29 it was forthcoming when Fedor Opad'chy flew what was obviously a transport derivative of the bomber across Tushino. These aircraft, which, according to the commentator, "displayed the achievements of Soviet aircraft construction and design", may have seen birth on the drawing boards of Edward C Wells and his team at Boeing more than seven years earlier, but they were not to be taken lightly. It was calculated that, for one-way "suicide" missions, they would give the *Dal'naya aviatsiya* a reach of some 3,000 miles (4 830 km), and thus, from bases on the Siberian side of the Bering Strait, could carry a useful warload as far south as San Francisco. From a seized base in Iceland they would be capable of covering the industrial areas of New England, New York, Pennsylvania and Ohio, and from bases in Greenland they could reach as far as New Orleans and Denver. Thus, not only had Soviet technology advanced far more rapidly than had been foreseen in the West; the Soviet Union was at last in possession of a weapon with which it could attack North America!

A crash programme

Throughout WW II the Soviet air arm, the V-VS RKKA, had remained primarily a tactical force. The Soviet aircraft industry had undertaken much pioneering work in the late 'twenties and early 'thirties with long-range strategic bombers, and as early as 1936 heavy bombers had been concentrated in a strategic force known as the *aviatsiya osobennovo naznacheniya* (Special Purpose Aviation), or AON. In 1940, this force was re-styled *Aviatsiya dal'novo deistviya* (ADD), or Long-range Aviation, and two years later became a separate force under the command of General A Ye Golovanov, but equipped primarily with twin-engined Il-4 medium bombers, it could not be considered as a true strategic bombing force. The Pe-8 heavy bomber under development at the beginning of WW II had proved an abysmal failure, only 79 aircraft of this type being built, but the exigencies of the times had dictated the expenditure of all development effort on tactical aircraft. Thus, it can be imagined that the ADD witnessed the operational début of the B-29 Superfortress with envy.

By mid-1944, when the B-29 first appeared on operations, it was obvious to the Soviet High Command as it endeavoured to look further into the future than the end of the so-called "Great Patriotic War" that the ADD should be given *true* strategic bombing capability at the earliest opportunity, but it was equally obvious that even if work on an aircraft comparable with the B-29 began immediately, it would be

the early 'fifties before it could possibly attain operational status with the ADD. The solution to this problem literally fell into the Soviet lap on 29 July 1944 with the arrival in Soviet territory of Capt Howard R Jarrell's B-29 Superfortress which had been damaged by a Japanese anti-aircraft shell over Anshan. By copying every detail of the B-29 it seemed likely that the Soviet aircraft industry could overcome the immense technological problems involved in the development of so advanced an aircraft in a fraction of the time that would be demanded if it was to start from scratch.

The Soviet aircraft industry was not entirely devoid of knowledge of the B-29 and the radically new features that it embodied — Soviet intelligence agents were as active in ferreting out the secrets of the Soviet Union's allies as they were those of its enemies — and the arrival of an intact specimen of this advanced warplane set in motion a crash programme already formulated to manufacture a copy of the B-29, a programme to which Iosip Stalin himself allocated the highest priority. How the Soviet Union would have acquired examples of the B-29 for use as pattern aircraft had such not fortuitously been presented must remain a subject for conjecture. It suffices to say that the three examples acquired in August and November 1944 were adequate for the task set the Soviet aircraft industry.

Overall responsibility for copying the B-29 airframe in every detail was assigned to the highly-experienced bomber design bureau of Andrei N Tupolev, while the engine bureau of Arkadii B Schvetsov was awarded the task of duplicating the Wright Cyclone R-3350 turbo-supercharged 18-cylinder air-cooled radial. A wide area of Soviet industry was involved in the programme from the outset, together with numerous scientific research institutes, as many unfamiliar technological processes were involved, embracing not only new production techniques and materials, but highly sophisticated electrics, avionics, and pressurisation and remote-control armament systems. The actual task of breaking down into sub-assemblies and components two of the USAAF B-29s that had landed on Soviet territory — the third being retained intact for flight evaluation — had commenced before the end of 1944, and during the first quarter of 1945, while the B-29 was still engaged in bombing the heart out of Japan's industry, work on the Tu-4, as the Soviet copy of the B-29 was designated, actually began.

(Above right and below) The prototype Tu-75 military transport was flown for the first time in 1950. This lacked the defensive armament and ventral loading ramp proposed for the production derivative which failed to materialise.

Owing to the magnitude of the test programme and the urgency attached to its completion, a factory on the Volga was given the task of building no fewer than 20 test and evaluation aircraft, and, almost simultaneously, two factories behind the Urals began tooling for the assembly of the series production Tu-4. In the late summer of 1946, the NII V-VS (*Nauchno-Issledovatelsky Institut V-VS*), the Scientific and Research Institute of the Air Forces, was informed that the first Tu-4 was ready to be collected. One of the Institute's senior and most experienced test pilots, Nikolai Stepanovich Rybko, was sent to collect the aircraft and ferry it to the NII V-VS test centre, where, a few weeks later, it was joined by the second Tu-4 which had been ferried from the factory airfield near the Volga by another senior NII V-VS test pilot, Mark Gallai.

Rybko and Gallai were primarily responsible for the initial handling trials of the Tu-4, the first three pre-production aircraft being assigned to these and to performance testing with subsequent aircraft being allocated to checking out the various systems. At the time the Soviet Union acquired its B-29s this advanced and complex warplane was still suffering a variety of teething troubles, and it was hardly to be expected that the Soviet copy of Boeing's bomber would prove trouble free. Early in the test programme, a considerable amount of difficulty was experienced with the electrically-actuated undercarriage, and on one occasion an NII V-VS pilot, Rafael Ivanovich Kaprelyan, found it impossible to lock down the port undercarriage leg. Despite Kaprelyan's order to bale out, the test crew insisted on remaining with the aircraft. As the Tu-4 touched down on its starboard main gear, Kaprelyan cut both engines on that side and gave full power to the port engines, and then, as speed dropped, he cut the port engines and actuated the emergency fuel cocks. The aircraft came to a standstill, suffering only some superficial damage to the port airscrews and wingtip, and Andrei Tupolev subsequently presented the pilot with a photograph of the Tu-4 inscribed: "To R I Kaprelyan as a reminder of our work together and of the first landing of this aircraft on one main leg — General Designer A Tupolev."

Another problem frequently encountered during the early test phase of the Tu-4 was runaway airscrews. Gallai had one which refused to feather until the feathering mechanism had been actuated 40 times, and another test pilot, V P Marunov, had a similar experience, but his runaway airscrew parted company with its engine and narrowly missed hitting the aircraft. Test pilot Vasilchenko experienced a fire in the air, nine of the crew members baling out, and the pilot and his observer, Filizov, remained aboard and succeeded in bringing the aircraft in to a wheels-up landing. The Tu-4 was a total write-off but at least it was possible to ascertain the cause of the fire.

One early problem that was hardly destined to popularise the Tu-4 with its test crews was the distortion of vision provided by the extensively-glazed fuselage nose. Mark Gallai was subsequently to refer to this as being akin to a fairground "Hall of Mirrors", and many months were to elapse before this problem was finally overcome. Shortly after the public début of the Tu-4 in the Aviation Day parade over Tushino on 3 August 1947, the initial long-range trials of the bomber were started, Gallai flying the second Tu-4 on a non-stop flight from the test centre near Moscow to the Crimea and back, and on the same day another Tu-4 left the test centre to fly to the new Tu-4 production complex behind the Urals.

Gallai's Tu-4 was not fitted with de-icing equipment, and heavy ice began to build up on the windscreen, the wing leading edges, and on the airscrews, forcing the aircraft to descend to a lower altitude. The outward flight was otherwise uneventful, but shortly after starting the return leg the aircraft suddenly yawed. It was quickly discovered that the cooling gills of No 4 engine were malfunctioning owing to an electrical fault, and were fully open. Gallai and his crew were still struggling to rectify the fault when the "bomb doors open" light began to glow. The bomb doors refused to answer to the control, and the flight engineer was even lowered into the bomb-bay at the end of a rope in an attempt to ascertain the cause of the malfunction, but no effort was of avail, and Gallai was forced to return at low altitude with No 4 engine cut and its airscrew feathered and with bomb doors open. At the test centre work began immediately on checking out the bomb door actuating system, and the cause of the malfunction was quickly discovered. A radar technician had been flying with Gallai as a supernumary crew member to check the functioning of the *Cobalt* radar during the flight, and had accidentally tripped the emergency bomb door switch which overrode that of the pilot!

After this flight, the second Tu-4 was transferred to the *Dal'naya aviatsiya* (as the ADD had meanwhile been redesignated), and Gallai went on to test the twentieth and last of the pre-production Tu-4s which reached the test centre in the late autumn of 1947. Many teething troubles with both the Tu-4 systems and Shvetsov's copy of the R-3350 engine, the ASh-73TK, remained to be resolved, however, one of the most persistent being the formation of air locks in the oil pressure lines, and Gallai's first four landings in "020" were made on three engines as a result, but State Acceptance Trials had been completed by the beginning of 1948 when the first production Tu-4s were cleared for delivery to the *Dal'naya aviatsiya*.

Transport from bomber

In 1945, *Aeroflot* had formulated a requirement for a 50–60 passenger long-range transport, and Andrei Tupolev's

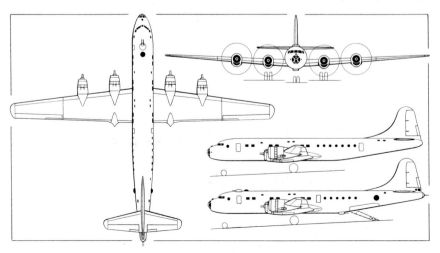

This general arrangement drawing depicts the proposed production version of the Tu-75 military transport, the upper side view illustrating the Tu-70 commercial transport for comparison purposes.

design bureau conceived the idea of marrying the wings, power plants, tail surfaces and undercarriage of the Tu-4 bomber to an entirely new fuselage to meet this requirement. Allocated the design bureau designation Tu-70, this transport was flown for the first time on 27 November 1946 by Fedor F Opad'chy. The circular-section fuselage of the Tu-70 was substantially larger than that of the Tu-4, and was mounted higher on the wing. The greater girth of the fuselage resulted in an increase in the distance between the centrelines of the outboard engine nacelles from 28 ft 2⅞ in (8,67 m) to 30 ft 1¼ in (9,48 m), overall wing span being increased from 141 ft 2¾ in (43,05 m) to 145 ft 2⅛ in (44,25 m), the latter being accounted for by the larger fuselage and marginal extension of the wingtips which raised wing area from 1,736 sq ft (161,28 m²) to 1,787·89 sq ft (166,1 m²). The new fuselage increased overall length from 99 ft 0¼ in (30,18 m) to 116 ft 1¼ in (35,4 m), and maximum loaded weight was 105,822 lb (48 000 kg).

Powered by four ASh-73TK turbo-supercharged engines each offering 2,200 hp for take-off and possessing an emergency rating of 2,400 hp, the Tu-70 accommodated 48 passengers and a crew of seven, and, like the bomber from which it was derived, was displayed over Tushino on 3 August 1947. Shortly after this event, Opad'chy was flying the sole Tu-70 prototype at an altitude of 13,780 ft (4 200 m) when the turbo-supercharger system malfunctioned, and all four engines were "overblown" and wrecked. Displaying considerable skill, Opad'chy succeeded in bringing the stricken aircraft down for a dead-stick landing in deep snow. The prototype suffered negligible damage and soon resumed its flight test programme.

The Tu-70 evaluation revealed a maximum speed of 353 mph (568 km/h) at 29,530 ft (9 000 m) and a range with a full complement of passengers of 3,045 miles (4 900 km), but by the time this transport had completed its test programme, *Aeroflot* had undertaken a reappraisal of its immediate requirements, and had concluded that the forecasts of traffic volume over its longer routes made immediately after WW II had been overly optimistic, and thus, although the Tu-70 fully met the *Aeroflot* specification, no orders were forthcoming.

However, Tupolev was loath to waste all the work undertaken on development of the Tu-70, and therefore, in 1948, when a requirement was formulated for a long-range military freighter for use by the *desantno-transportniya aviatsiya,* the aviation of the airborne forces, the basic design was re-worked to meet the demands of this requirement which called for the ability to carry 22,046 lb (10 000 kg) of freight, including light vehicles, and suitability for paradrop operations. Designated Tu-75, the military transport derivative of the Tu-70 was flown for the first time in 1950 by V P Marunov, and featured fore and aft ventral loading hatches which could be opened in flight for paradrops. Maximum loaded weight was 124,782 lb (56 600 kg), and it was proposed that defensive armament should be provided in the form of a remotely-controlled forward upper gun barbette mounting twin 23-mm NR-23 cannon, a similar ventral barbette forward of the aft loading hatch, and a third pair of NR-23s in the extreme tail. The prototype Tu-75 was not fitted with defensive armament which it was proposed should be introduced on the production model, the latter also differing in having the fore and aft hatches replaced by an hydraulically-operated ventral loading ramp. The Tu-75 was demonstrated over Tushino in July 1951, 33 parachutists and several freight containers being dropped, but this was to be the swan-song of the transports derived directly from the Soviet copy of Boeing's B-29 Superfortress, for further development of the Tu-75 had already been abandoned. ☐

(To be continued next month.)

A-7D ———————————————— *from page 132*

flown non-stop from the US East Coast to Paris; carrying four external tanks each, they took 7½ hrs for the journey and had fuel remaining for another 1½ hours. Actual fuel consumption of the Spey per hour is greater than that of the TF30, so that the endurance of the A-7D and A-7E is rather less than that of the earlier Navy versions.

On a typical close support mission, the A-7D would take off at a weight of 37,456 lb (16 990 kg) of which 6,650 lb (3 016 kg) would be accounted for by eight M 117 bombs on two pylons. In this configuration, the maximum speed at 5,000 ft (1 524 m) would be 544 knots (1 007 km/h), and 594 knots (1 101 km/h) after the bombs had been dropped. At a distance of 400 nm (740 km) from base, over 30 minutes loiter time over the target would be available. The legs to and from the target would be flown at optimum cruise altitude between 30,000 ft and 42,000 ft (9 144–12 800 m) at 460–470 knots (853–871 km/h), with the loiter at 241 knots (446 km/h) at 5,000 ft (1 524 m). At this altitude and a weight of 32,675 lb (14 821 kg) the A-7D can turn in a radius of about 4,500 ft (1 372 m), pulling 2·5 g in an equilibrium turn at 350 knots (649 km/h). Clean and at 25,000 lb (11 340 kg) with no external stores, the A-7D will turn in 4,600 ft (1 402 m) at 5 g and 500 knots (927 km/h).

The latest announced variant of the Corsair II is the A-7G, this being the designation of the version offered to Switzerland in its Venom-replacement evaluation (see AIR ENTHUSIAST June 1971) and named as winner in the first phase. Basically similar to the A-7D, the A-7G would have a more powerful version of the Allison TF41 Spey and a different avionics fit to meet the Swiss requirement. The precise standard of avionics and nav/attack system fitted has an important bearing on the price of the Corsair II, reported to be about $3m (£1·25m) for the A-7E.

With production of about 1,300 examples of the A-7 for the USN and USAF now assured, the Corsair II has become one of the most successful combat aeroplanes initiated during the 'sixties. The integrated weapon system fitted to the A-7D and A-7E makes these versions, in particular, among the most potent of close support aeroplanes available today, and the marriage of these systems to a relatively unsophisticated airframe with a pedigree going back to 1955, when the XF8U-1 Crusader made its first flight, has produced an aeroplane with an exceptionally good maintenance record. Taken into combat after one of the shortest development cycles of any warplane in the last two decades, the A-7 has proved capable of delivering a lethal punch whilst taking a great deal of punishment, and the accuracy of the nav/attack system in the latest versions is such that weapons can be released at a sufficient range from the target to avoid small arms fire, as shown by the low combat loss rate for the A-7E to date. As well as adding significantly to the striking force of USAF's Tactical Air Command, the A-7 seems destined to find other markets outside the US which should keep it in production for several years to come. ☐

Vought A-7D Specification

Power Plant: One Allison TF41-A-1 (Rolls-Royce RB.168-62 Spey) rated at 14,250 lb (6 465 kg) static thrust. No afterburner.

Weights: Empty 19,490 lb (8 840 kg); internal fuel, 9,750 lb (4 423 kg); external load, 15,000 lb (6 804 kg); gross weight, 42,000 lb (19 050 kg).

Performance: Maximum speed (clean) 607 kt (1 123 km/h); tactical radius with internal fuel, 3,600 lb (1 623 kg) external weapon load, 700 miles (1 127 km); ferry range over 4,000 miles (6 437 km); take-off ground run with combat load, 5,800 ft (1 768 m).

Dimensions: Span, 38 ft 8½ in (11,79 m); length, 46 ft 1½ in (14,05 m); height, 16 ft (4,88 m); undercarriage track, 9 ft 6 in (2,89 m); wing area, 375 sq ft (34,84 m²); dihedral, −5°; aspect ratio, 4·0:1; sweepback at quarter chord, 35°.

FIGHTER A TO Z

(Above and below) The Airco D.H.5 in standard production form.

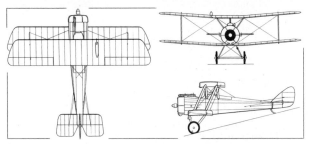

AIRCO D.H.5   GREAT BRITAIN

Characterised by the pronounced negative stagger of its mainplanes, resulting from an attempt on the part of Geoffrey de Havilland to combine the performance of the tractor biplane with the cockpit visibility of a pusher aircraft, the D.H.5 was flown late in 1916 and entered service in May 1917. Powered by a 110 hp Le Rhône 9J rotary, the D.H.5 carried a single 0·303-in (7,7-mm) Vickers machine gun. Some 550 were built but the D.H.5 enjoyed only limited success and had been withdrawn from operations by the end of January 1918. Max speed, 102 mph (164 km/h) at 10,000 ft (3 050 m), 89 mph (143 km/h at 15,000 ft (4 570 m). Time to 10,000 ft (3 050 m), 12 min 25 sec. Empty weight, 1,010 lb (458 kg). Loaded weight, 1,492 lb (677 kg). Span, 25 ft 8 in (7,82 m). Length, 22 ft 0 in (6,71 m). Wing area, 212.1 sq ft (19,7 m²).

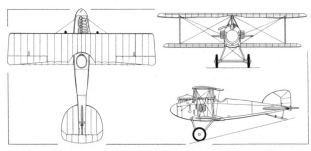

(Below) Albatros D I prototype and (above) the standard production D I.

ALBATROS D I   GERMANY

Designed by Thelen, Schubert and Gnädig, the D I was the first fighter to be developed by the Albatros-Werke, and was introduced in August 1916. Powered by a 150 hp Benz Bz III or 160 hp Daimler D III six-cylinder water-cooled engine, the D I employed conventional fabric-covered wooden wings but the wooden fuselage was unusual in being a semi-monocoque. Armament comprised two synchronised 7,92-mm Spandau machine guns, and between 50 and 60 D Is were built. Max speed, 109 mph (175 km/h). Time to 3,280 ft (1 000m), 6 min. Endurance, 1·5 hr. Empty weight, 1,484 lb (673 kg). Loaded weight, 1,980 lb (898 kg). Span 27 ft 10⅔ in (8,50 m). Length, 24 ft 3⅛ in (7,40 m). Wing area, 246·49 sq ft (22,9 m²).

Albatros D II (Serial D.497/16).

ALBATROS D II   GERMANY

In an attempt to improve the forward and upward field of vision provided the pilot of the D I, Albatros evolved the D II which supplanted the earlier fighter in production in the autumn of 1916. The primary difference was a reduction in the gap between the wings, the upper wing being brought closer to the fuselage. The standard D II had the 160 hp Daimler D III engine, and licence production was undertaken by LVG as the LVG D I. Production of the D II was also undertaken by Oeffag with a 185 hp Austro-Daimler engine. The twin-Spandau armament was retained. Max speed, 109 mph (175 km/h). Time to 3,280 ft (1 000 m), 5 min 30 sec. Empty weight, 1,404 lb (637 kg). Loaded weight 1,958 lb (888 kg). Span, 27 ft 10⅔ in (8,50 m). Length, 24 ft 3⅛ in (7,40 m). Wing area 263·72 sq ft (24,50 m²).

ALBATROS W 4   GERMANY

The W 4 fighter floatplane, which entered service in September 1916, employed what was basically a D I fuselage married to larger wings and tail surfaces, and a 160 hp Daimler D III engine. Armament comprised either one or two synchronised Spandaus, and floats of various types were fitted. Approximately 120 W 4 float fighters had been de-delivered when production ceased late in 1917. Max speed, 99 mph (160 km/h). Climb to 3,280 ft (1 000 m), 6 min 30 sec. Empty weight, 1,742 lb (790 kg). Loaded weight, 2,359 lb (1 070 kg.). Span, 31 ft 2 in (9,50 m). Length, 26 ft 1¼ in (8,26 m). Wing area, 340·14 sq ft (31,60 m²).

ALBATROS D III   GERMANY

Flown late in 1916, the D III entered service early in the following year, and, by comparison with the D II, featured longer-span wings with redesigned tips, the lower wing being of reduced chord, revised control surfaces, new interplane bracing, relocated radiator and a Daimler D IIIa engine boosted to 180 hp by means of an increase in compres-

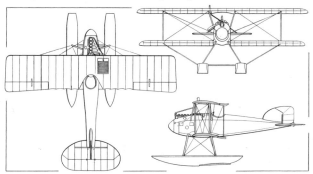

(Above and below) The Albatros W 4. The photograph depicts W 4 serial 1486 of the fourth production batch with wing-mounted radiator (as shown by general arrangement drawing) which replaced the earlier box-like fuselage-mounted radiators.

sion ratio. The standard pair of synchronised Spandau guns was retained, and much of the airframe comprised standard D II assemblies. D IIIs licence-built by OAW and Oeffag were fitted with Austro-Daimler engines of 185, 200 and 225 hp successively. Max speed, 102 mph (165 km/h). Time to 3,280 ft (1 000 m), 3 min 45 sec. Empty weight, 1,457 lb (661 kg). Loaded weight, 1,953 lb (886 kg). Span, 29 ft 8 in (9,04 m). Length, 24 ft 0⅝ in (7,33 m). Wing area, 220·66 sq ft (20,50 m²).

ALBATROS D IV   GERMANY
Evolved in parallel with the D III, the D IV was fitted with an experimental, specially-geared Daimler D IV engine of 220 hp. By comparison with the D III, the D IV embodied major fuselage redesign, but difficulties experienced with the power plant resulted in only one prototype being built and flown, and this, tested late in 1916, did not mount the

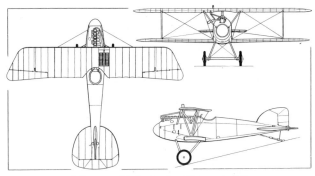

(Above and below) The Albatros D III.

planned twin-Spandau armament. Max speed, 102 mph (165 km/h). Endurance, 2 hr 10 min. Span, 29 ft 8 in (9,04 m). Length, 24 ft 0⅝ in (7,33 m). Wing area, 220·66 sq ft (20,50 m²).

ALBATROS D V   GERMANY
Introduced mid-summer 1917, the D V introduced a fuselage of more truly elliptical section than featured by the D III, and a more curved rudder. The high-compression Daimler D IIIa engine with oversize cylinders offered 180 hp, and armament consisted of two Spandaus. The D Va differed solely in reverting to the arrangement of aileron control cables used by the D III, and both models were manufactured in large numbers. Max speed, 115 mph (185 km/h). Time to 3,280 ft (1 000 m), 4 min 20 sec. Empty weight, 1,514 lb (687 kg). Loaded weight, 2,066 lb (937 kg). Span, 29 ft 8 in (9,04 m). Length, 24 ft 0⅝ in (7,33 m). Wing area, 228·19 sq ft (21,20 m²).

The sole prototype of the Albatros D IV.

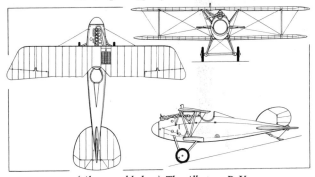

(Above and below) The Albatros D V.

ALBATROS D VI   GERMANY
The D VI twin-boom single-seat pusher biplane powered by a 180 hp Daimler D IIIa engine was built during 1917, and was flown for the first time in February 1918. The undercarriage was damaged during the initial landing and had still to be repaired in May when further work on the D VI was suspended due to the higher priority allocated to other projects, and the engine was removed for another application. Armament comprised a fixed forward-firing 20-mm Becker cannon and a 7,92-mm Spandau machine gun. Empty weight, 1,406 lb (638 kg). Loaded weight, 1,940 lb (880 kg). Span, 32 ft 1⅞ in (9,80 m). Length, 25 ft 5⅛ in (7,75 m). No photograph is known to exist.

taxied in to where I was waiting on the apron. The horn was still blaring and as soon as he opened the Airacobra's starboard door I could see what was amiss. The legs were, indeed, unlocked, but were held safely in the "down" position by the irreversible manual gear. So, while everyone was scurrying around looking for lifting jacks, I just wound the crank a quarter of a turn or so until the locks were home and the horn stopped yelling.

After the initial familiarisation flights we got down to more serious test outings, and discovered that the Airacobra's performance was unremarkable at anything but relatively low levels, and that it had, presumably because of an aft centre-of-gravity position, a nasty habit of tightening-up in steep turns and demanded firm handling when recovering from high-speed descents. Quite heavy and carefully-gauged forward pressure on the control column was needed in tight turns and, unless forward trim had been applied, in pull-outs. For us, there was nothing particularly unusual about this: our four-cannon Hurricane IICs behaved in much the same way when carrying long-range tanks. The forward pressure on the stick was quite instinctively applied — you just had to make sure that you maintained it, or you were liable to jettison your tanks, if not your wings.

We, and the chief test pilot in particular, had some displeasing sessions with the Airacobra when Group Headquarters told us that trouble had been experienced with reserve-fuel-feed failure and that this should be investigated. The reserve supply was below a stack-pipe and on more than one occasion fuel had not reached the engine after this had been selected "open". After examination and re-assembly, we had to fly on the main supply until the engine cut before switching to reserve. The relatively silent windmilling period was probably no more than five seconds, but when the chief test pilot made the first test it probably seemed more like five minutes.

The Airacobra, and its stalky nosewheel leg in particular, struck us as being a bit on the precious side, and not likely to be capable of taking hard knocks. How wrong we were.

On 5 December 1941 — two days, as it happened, before Pearl Harbour — the Ministry of Aircraft Production arranged a show at Speke Airport, Liverpool, of most of

This illustration shows the cockpit of the XP-39E, an experimental version of the Airacobra, but the layout of the instruments is similar to that of the British Airacobras.

the latest American aircraft bought by the Purchasing Commission or obtained under the new Lend-Lease arrangements. Among the more familiar Douglas Bostons, Lockheed Venturas, Curtiss Kittyhawks and so on there was one of the first batch of North American Mustang Is. An example of the still newish Airacobra was not available at Speke and Maintenance Command's No 41 Group (our masters) was asked to find one. The nearest — and possibly the only immediately available ones — were ours, and an urgent signal was sent to No 48 MU to prepare one and to fly it over to Speke first thing on the morning of the show. Looking forward to an easy day, and possibly to a better than average official luncheon, I volunteered to take AH660 on a test flight, to land it at Speke and to hang about until everybody had had a chance to peer around this extraordinary concept of a fighting aeroplane. Only after reporting to Air Traffic Control (called the "Watch Office" in those days) was I told that a demonstration flight was required before adjourning for luncheon. In due course the pilot of the Mustang and I were instructed to taxi out.

By then the weather seemed to me to have become pretty borderline for any such aid-less outing. The forecasters had said that a front would be going through from noon onwards. There would be drizzle, but the cloudbase would, they said, not be unreasonably low — and who was I to question the reliability of this prophecy? As soon as the Airacobra was airborne, with the undercarriage up and the boost and revolutions brought back to climb power, we were popping in and out of cloud. What the forecast had not said was that below the main base there would be seven-tenths broken cloud at about 300 feet. I made the best I could of a 360-degree turn — appalled as much by the lack of forward visibility through the Airacobra's rain-streaked screen as by the actual weather conditions — dived past the spectators and other aircraft on the apron and made as tight a turn as I dared so as to stay over the narrowing neck of the Mersey Estuary while having a look around.

There was obviously little future in staying airborne or in trying to get back into Speke, and the weather seemed to be sitting on the ground towards the south and Hawarden. Then I remembered Hooton Park aerodrome on the southwest bank of the Mersey. This should, I thought, be lying directly ahead with the boundary just about to show up through the murk. I closed the throttle, pushed the constant-speed lever forward and flicked the undercarriage and flap switches. As the boundary went by underneath I kept the Airacobra floating with a little power until the undercarriage had locked and then pushed it into the ground. We were still travelling pretty fast, with the braked wheels sliding on the wet grass, when we ran into what appeared to be a fog-bound builder's yard. I had forgotten that a runway was under construction at Hooton.

They cheerfully dug and manhandled the Airacobra out of the mud and debris and parked it on a nearby piece of greensward. The stalky nosewheel leg had saved the propeller and nothing appeared to have been damaged, but I decided that it would be sensible to fly it back to Hawarden with the undercarriage down and to organise a ground retraction test. In due course the weather cleared, leaving a cheerful post-frontal clarity and no more than a few broken cumulus clouds. There were no unusual vibrations as I ran the engine up and checked the operation of the constant-speed propeller, so I taxied out to the boundary, chose the longest into-wind run I could see on what remained of the grass field and took off. There was a little farewell committee on the apron, so I held the Airacobra down and — without thinking of anything but a wish to give an impression of efficiency — absentmindedly flicked the undercarriage switch. The legs went up — and came down again for the landing, ten minutes later, at Hawarden. □

PLANE FACTS

This column is intended to provide an information service to readers requiring information that is not found in readily-available reference books. We would ask readers to confine their requests to ONE aviation subject or aircraft and we will answer as many of these as space permits. The editors reserve the right to use their discretion in selecting for reply in this column those requests which in their opinion will be of the widest general interest and which cannot be easily dealt with by a visit to the nearest local library.

Readers are asked to note that the editors cannot reply by post to requests for information, nor supply copies of illustrations appearing in the column. Until a flow of readers' requests has been established, items of special interest have been drawn from our files for inclusion on this page. As sufficient requests for information are received, *Plane Facts* will be expanded to two or more pages each month.

A "TSR-2" from Bristol

WHEN the British Air Staff drew up its General Operational Requirement (GOR) 339 for a Canberra replacement in 1957, eight airframe companies were invited to study the requirement and comment upon its feasibility. The chain of events then put in progress led eventually to the production, testing — and cancellation — of the TSR-2, one of the best-known aeroplanes of the 'sixties. Little has ever been published, however, about the other design submissions to GOR 339. One of the most interesting of these was the Bristol Type 204, a general arrangement drawing of which is published below for the first time anywhere.

Although Bristol was to become, in 1961, part of BAC and therefore indirectly associated with the TSR-2, the company was still independent at the time the Type 204 submission was made early in 1958. To achieve the required gust alleviation properties for low-level sorties at high subsonic speed, whilst retaining the stipulated take-off and landing performance, the Bristol design featured a version of the delta wing known as "gothic" (for its resemblance to a Gothic arch in plan view) plus a foreplane. The latter, also of gothic planform, was pylon-mounted under the nose and was in two portions, with 10° upwards movement on the front portion and 40° downward movement on the rear section. Like the noseplanes now adopted for the Dassault Milan, the foreplane on the Type 204 helped to reduce the take-off and landing performance, which would otherwise have been excessive because of the use of a delta wing.

Type 204 had a crew of two in tandem, both pilot and navigator being provided with an excellent field of view. The navigation compartment, complete with seat, equipment and instruments, was separately suspended from the structure to provide additional damping of gusts during low-level flight.

Two 22,520 lb st (9 215 kgp) Bristol Olympus B Ol 22 turbojets with simplified reheat were mounted side-by-side in the rear fuselage, taking air through a "letter-box" intake above the wing. The primary weapon was to be a tactical megaton bomb (TMB) and weapon stowage was in an internal bay, with provision for bombs or rockets to be swung down clear of the fuselage before release. Wing pylons were provided for about 2,000 lb (900 kg) of rockets, rocket pods, or 200-gal (909-l) drop tanks. Pro-

(Above and below) The Tsibin glider.

vision was made for a flight refuelling probe and equipment included forwards and sideways looking radar and an APU.

The Type 204 had a max take-off weight of 78,000 lb (35 380 kg) and max landing weight of 50,000 lb (22 680 kg), with total internal fuel weight of 31,270 lb (14 184 kg). On this tankage, the radius of action was 875 nm (1 620 km), increasing to 1,000 nm (1 850 km) with drop tanks. Max speeds, as required by the specification, were Mach 0·95 at low level and Mach 2·0 at 35,000 ft (10 670 m). For a 1,000 nm (1 850 km) radius of action, the Type 204 required a 2,000 ft (610 m) take-off strip. Principal dimensions included: span, 32 ft 0 in (9,75 m); length, 79 ft 6 in (24,23 m); height, 20 ft 9 in (6,32 m); gross wing area, 820 sq ft (76,18 m²); aspect ratio, 1·25; foreplane span, 8 ft 5 in (2,57 m); foreplane area, 50 sq ft (4,65 m²).

Soviet forward-sweep research

Although it is widely known that the Soviet Union displayed interest in the Junkers Ju 287 bomber featuring forward wing sweep, completing and testing the Ju 287 V2, it is not generally known that, during the late 'forties, a number of manned gliders embodying forward wing sweep were built and tested in the Soviet Union, these being launched in the air from "mother" aircraft. One of these, designed by V P Tsibin, is illustrated by the photographs on this page, which are published in the West for the first time.

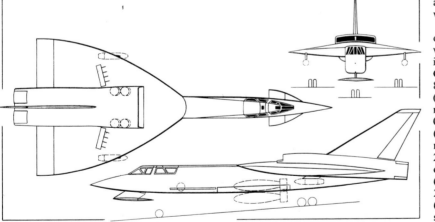

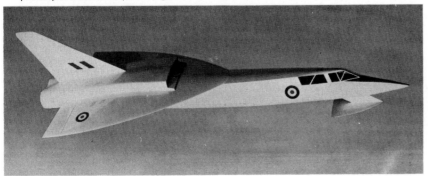

(Below) A model, and (above) general arrangement drawing of the Bristol Type 204.

MODEL ENTHUSIAST

From page 144

Aurora, both to 1/144th scale. The TriStar is neatly moulded in white polystyrene of excellent quality, possesses fine and clear surface detail, and enjoys very good component fit. In contrast to some Airfix kits, this model is *not* marred by a poor-quality decal sheet, and this, providing the insignia of Air Canada with registrations and fleet numbers for no fewer than six aircraft, is extremely well printed. We have never been enamoured with the all-drawing type instructional sheet, although this one is admittedly adequate. Unfortunately, it includes a colour scheme depicting the entire undersurfaces of the aircraft as light grey, whereas, in fact, only the centre third of the wings and tailplanes are so painted, the other two-thirds being highly-polished natural metal. Spanning 1 ft 1 in (33 cm), Airfix's TriStar is, in our opinion, excellent value for money.

Sad to say, Aurora's DC-10 does not, as a kit, compare with Airfix's TriStar, apart from its inclusion of a first class decal sheet which offers the colourful livery of American Airlines. It is cleanly moulded in silver plastic, and the component parts fit together well, but there is no surface detail whatsoever if the control surface hinge lines are discounted. The sole transparency provided is that enclosing the flight deck, and the undercarriage is, by today's standards, crude in the extreme. At least Aurora's DC-10 provides a relatively accurate representation of the full-scale original, and is not over-priced.

A mighty Hurricane

Revell would seem to have made 1/32nd scale peculiarly its own, and has steadily developed its range of WW II fighters produced to this scale. The quality of individual models varies, naturally enough, but the standard of kits manufactured by Revell to this scale has risen steadily, and the company's most recent release, the Hurricane I, is among the best in the entire series. As with the P-38 Lightning included in this 1/32nd scale range and reviewed last month, the fit of the component parts is almost unbelievably good, and, with the exception of the undercarriage, the entire model may be assembled without cement and, what is more, holds together! A remarkable feat of manufacturing precision.

Revell's Hurricane represents its subject very accurately, except for minor and easily-corrected errors in the shapes of the undercarriage cover plates and the windscreen. Surface detail is of a very high standard, while the sliding canopy and detachable cowling reveal plenty of cockpit and engine detail. The decal sheet, which provides the markings of one of the aircraft flown by Sqdn Ldr Stanford Tuck, is of good quality, but Revell gives no instructions regarding the application of the serial numbers. This is less obvious than may seem at first sight, as the numbers were partly overpainted by the Sky fuselage band. The split planview in the instruction sheet is useless, both as regards the camouflage pattern and the arrangement of the mark-ings. When *will* kit manufacturers appreciate the fact that *full* plan drawings of both upper and lower surfaces, and both port and starboard views, are necessary if the markings of a camouflaged aeroplane are to be depicted properly?

Minor shortcomings apart, Revell's Hurricane is a superb model and deserving of every success, and the boxtop painting — by an unidentified artist — is in full accord with the high standards of the box's contents.

A Nipponese whopper

The Japanese Nitto Company would seem now to be concentrating on models of current airliners to larger scales than the widely-used 1/144th. Nitto's efforts hitherto have, however, been marred by various faults which have not excluded inaccuracy of outline, but, happily, a new high standard has been achieved with its kit of the McDonnell Douglas DC-8 Super 62, the company's latest product to reach us. Described as being 1/100th scale but *actually* 1/102nd, it is accurate; its component parts fit together very well indeed, and the surface detail, if somewhat sparse, is very neat, using incised panel lines. The mouldings are very clean, and particularly those of the smaller parts which are beautiful specimens of the toolmaker's art, while the wheels even have separate tyres.

The decal sheet accompanying the kit is superb. Absolutely accurate, highly detailed, and well printed, it provides the markings of one of JAL's aircraft. Nevertheless, the kit falls short of perfection in that the paucity of surface detail is very noticeable on a model as large as this. Furthermore, the major components, pressed in white plastic, have been sprayed silver in those areas which, on the full-scale aircraft, are left in natural metal finish. Pre-finishing of this nature is never satisfactory as it inevitably suffers damage during assembly, necessitating re-painting. Nitto would have done better to have moulded their DC-8-62 in really good silver plastic such as is used by Monogram, leaving the modeller to apply the white finish to the top decking of the fuselage. Available from VHF Supplies Limited in the UK at £2·98, the Nitto kit is far from cheap, but it *is* one of the best airliner kits yet issued, and certainly the best of those portraying the DC-8. □

W R MATTHEWS

RECENTLY ISSUED KITS			
Company	Type	Scale	Price
Nitto	DC-8 Srs 62	1/100	£2·98
Fujimi	Fw 190A	1/50	—
Fujimi	Alouette III	1/50	67p
VEB	Yak-40	1/100	—
VEB	Tu-144	1/100	—
KP	Avia 534	1/72	—

AIRSCENE

From page 120

Hawker Siddeley Argosy: US-owned, Luxembourg-based Nittler Air Transport has one Srs 100, ex-Universal. □ UK-based Saggitair has acquired four, delivered in June, for charter operations.

Hawker Siddeley Comet: Dan-Air has now acquired both of Kuwait Airways' two Srs 4Cs.

Ilyushin Il-62: United Arab Airlines plans to wet-lease two to supplement its Boeing 707 fleet. □ KLM leases one from Aeroflot for its trans-Siberia service (with Dutch cabin crews).

Lockheed Electra: One of two ex-Falconair (no longer operating) is now flying with Sterling's Swedish subsidiary. The other has returned to the US.

Yakovlev Yak-40: Three have been purchased by Bakhtar Afghan Airlines, with initial delivery made already.

MILITARY CONTRACTS

Cessna 182 Skylane: The Venezuelan air arm, the *Fuerzas Aereas Venezolanas,* has purchased 12 Cessna 182 Skylanes for personnel transportation, training and forward air control missions. The aircraft were delivered during July.

Scottish Aviation Bulldog: The delivery of 15 Scottish Aviation Bulldog 100 basic trainers to the Royal Malaysian Air Force, the *Tentera Udara Diraja Malaysia,* is scheduled to commence in October with completion by July 1972. With spares and services the order is worth more than £320,000.

Lockheed L-100-20: The Kuwaiti government has purchased two Lockheed L-100-20 (stretched commercial version of the C-130 Hercules) which are being operated by the Kuwait Air Force in military markings. Delivery was effected earlier this year.

Boeing 737: The USAF has ordered 19 examples of a modified advanced navigational training version of the Boeing 737 under a fixed-price contract valued at $17·98m (£7·49m), and has also taken options on 10 additional aircraft.

Beech T-42A: The Beech Aircraft Corporation delivered five T-42As (military version of the B55 Baron) to the US Army in June for transfer under MAP to the Turkish government for use as multi-engine transition and instrument trainers.

Hughes Model 500: Governmental approval has been given Argentina's naval air component, *Aviación Naval,* for the purchase of six Hughes 500 helicopters. The Model 500 is essentially similar to the OH-6A Cayuse, 12 examples of which were supplied last year to the *Fuerza Aérea Argentina,* together with two Model 500s.

Hawker Siddeley HS.748: Two examples sold to the Royal Australian Navy in June bring total orders for the HS.748 to 250.

Beech U-21F: Five of these military versions of the King Air A100 have been sold to the US Army for communications duties.

AIR
Enthusiast

Volume 1 Number 4 September 1971

CONTENTS

AIRSCENE

MILITARY AFFAIRS

ARGENTINA
It is anticipated that the 12 Dassault **Mirage IIIEs** and two **Mirage IIIBs** currently on order for the *Fuerza Aérea Argentina* (FAéA) will be included on the strength of the *VII Brigada Aérea* at Moron Air Base, Buenos Aires, the *III Grupo Casa-Bombardero*, which was disbanded with the retirement of the last of its Gloster Meteor F 4s in January, possibly being re-formed to operate the new equipment. The only combat unit on the strength of the *I Brigada Aérea*, the I Counter-Insurgency Group which was formed late in 1967 with FMA-assembled B45 Mentors, has now been disbanded and its task assigned to the *I Grupo Exploración y Ataque*, which, with a dozen Hughes 500M (OH-6A) and nine Bell UH-1H Iroquois helicopters, now forms part of the *VII Brigada Aérea*.

AUSTRALIA
Senior Australian defence officials are evincing confidence that the government will confirm its order for **Convair (General Dynamics) F-111C** strike fighters late this year, but the 24 aircraft are unlikely to be available until 1973 owing to the Australian wish that some USAF F-111As should first go through the IRAN (Inspect and Repair as Necessary) programme. The Australian mission that visited the USA earlier this year reported favourably on the progress of the test programme, and a 12–15 year operational life is estimated for the F-111C which is now expected to commence operations from Amberley, Queensland, during 1974. The Australian Air Minister has stated that as of 6 April $US228·288m (£95·12m) had been paid for the F-111Cs and that the final cost of the programme is subject to agreement still to be reached between the US and Australian governments.

HMAS *Sydney* was scheduled to return to Australia from San Diego during August with the eight A-4Gs and two TA-4Gs comprising the second batch of **McDonnell Douglas Skyhawks** for the RAN. The cost of the 10 Skyhawks has been officially quoted as $A12m (£5·58m).

A team from Hawker Siddeley held discussions with the Department of Air and Supply in July with a view to interesting the Australian government in the **HS 1182** as a potential successor to the RAAF's Aermacchi M.B.326H trainer, deliveries of which — 86 for the RAAF and 10 for the RAN — are expected to be completed during the first half of 1972. It is anticipated that the RAAF will require a replacement for the M.B.326H in the late 'seventies.

BRAZIL
The *Fôrça Aérea Brasileira* (FAB) anticipates taking delivery of its first Neiva T-25 (IPD-6201) **Universal** basic trainers during October, almost exactly one year behind the schedule announced at the time of the placing of an order for 150 examples for the FAB. The Universal is to replace the T-6 Texan in the FAB inventory, but no decision has yet been taken concerning development of the projected PT6A-powered Turbo-Universal which is being offered to the FAB for the counter-insurgency rôle.

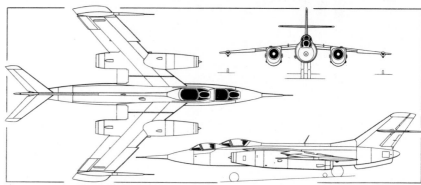

Not previously illustrated in the West, the Yak-28U, codenamed "Maestro", is a trainer version of the tactical attack version of the Yak-28 (Brewer) and not a derivative of the Yak-28P (Firebar) as has been stated by some sources. It has the original short fuselage and two Tumansky RD-11 turbojets.

The designation **TF-26 Xavante** has been given by the FAB to the Aermacchi M.B.326GB which is being partly manufactured in Brazil and assembled by EMBRAER. Deliveries of the Xavante to the FAB will commence next year at a rate of two per month against an order for 112 aircraft. The Xavante is to be used both for basic jet training and for the light strike rôle.

CANADA
A decision concerning the future of the substantial number of **Canadair CF-5s** currently surplus to the Canadian Armed Forces' requirements was believed imminent at the time of closing for press. The last of 115 CF-5s (comprising 89 single-seat CF-5As and 26 two-seat CF-5Ds) was recently completed by Canadair at a total programme cost of $C315m (£127m), but 44 of these are in storage at Trenton and North Bay. The 54 currently operating with the CAF are distributed between the one CF-5-equipped combat unit, No 433 Sqdn at CFB Bagotville, and No 434 Operational Training Squadron at Cold Lake, Alberta. The two-seat CF-5Ds present no problem as these can be employed as replacements for the CAF's ageing T-33A-N Silver Stars in the advanced training rôle.

CONGO (KINSHASA)
It has been reported from France that the seven Aérospatiale **SA-330 Puma** helicopters delivered to the *Force Aérienne Congolaise* (FAéC) by 2 June (see this column July issue) represented only an initial batch against total Congolese orders for 30 helicopters of this type. Among the first seven for FAéC is one fitted with a nine-seat VIP layout for presidential use.

It is now known that two and not *three* Lockheed **C-130H Hercules** transports have been delivered to the FAéC. Both aircraft participated in a special demonstration before President Mobutu and senior Congolese officials on 26 June during which 108 paratroops were dropped from the Hercules.

DENMARK
From 1 July the air component of the Danish Army has been known as the *Hærens Flyvetjeneste* (Army Flying Service). Operating from Vandel, near Vejle, Jutland, the *Hærens Flyvetjeneste* now has only two Piper L-18s remaining of the 16 supplied to the Danish Army in 1957, and is

currently in process of taking the **Hughes 500M** helicopter into its inventory. Twelve Hughes 500Ms have been ordered, the first of these having been assembled at Værløse and flown on 28 April.

The 22 **Starfighters** surplus to Canadian Armed Forces requirements and purchased by the *Flyvevåben* (see this column August issue) are now known to comprise 15 single-seat CF-104s and seven two-seat CF-104Ds. Tenders for the modification of these aircraft, currently in storage at Prestwick, have been made by a number of companies, including Scottish Aviation, Avio Diepen and MBB, and it is anticipated that it will be several months before these aircraft are taken into the inventory of the *Flyvevåben*. The additional Starfighters will not be used to re-equip an F-100D Super Sabre squadron, as was suggested, but will supplement the aircraft of the two Starfighter squadrons, ESK 723 and ESK 726, operating from Aalborg. The *Flyvevåben* originally received 25 F-104Gs and four TF-104Gs during 1964-5, and only two of these have been lost in accidents.

FRANCE
The *Aviation Légère de l'Armée de Terre* (ALAT) plans to acquire a total of 166 Aérospatiale **SA 341 Gazelle** helicopters by the end of 1977, and it is anticipated that orders for these will be placed this year. Approximately 100 Gazelles will be in service with the ALAT by 1975, and some of these will be equipped to perform the attack rôle, supplementing the armed Alouette IIIs currently being operated. All 130 SA 330 Puma helicopters scheduled to serve with the ALAT will have been delivered before the end of 1972.

FEDERAL GERMANY
By the beginning of July the *Luftwaffe* had accepted delivery of 11 of the 18 North American **OV-10B Bronco** target tugs currently on order. Twelve of these are to be fitted with pylon-mounted auxiliary turbojets.

Delivery of the 175 **McDonnell Douglas F-4E(F) Phantom** air-superiority fighters, for which *Bundestag* Defence Committee approval was given on 24 June, is scheduled to commence in January 1974 with completion by the end of June 1976. The largest single weapons contract ever placed by the Federal German government — the price quoted by McDonnell Douglas being

DM 4,037m (£458·3m) — the F-4E(F) order still falls short of the *Luftwaffe* requirement by some 45 aircraft, and the number of aircraft ordered may be reduced if Germany insists on a heavy spares stockpile. The F-4E(F) single-seat air superiority fighter will introduce several aerodynamic refinements, including leading-edge slats for increased manœuvrability, and substantial weight saving will result from a reduction in avionics by comparison with the current F-4E and the inclusion of solid-state radar. McDonnell Douglas claims a maximum speed in excess of Mach 2·0, the ability to climb to 35,000 ft (10 670 m) in 1 min 3 sec, a combat ceiling approaching 60,000 ft (18 290 m), and a radius of more than 800 mls (1 290 km) in the intercept rôle with full external fuel, M-61A1 20-mm rotary cannon and AAMs.

INDIA
The Indian Defence Minister, Jagjivan Ram, recently stated that consideration is being given to the production of a **military freighter** version of the HS 748. HAL is understood to be engaged in the conversion of an HS 748 airframe to freighter configuration, the primary modification apparently being the introduction of wide freight doors, and the evaluation aircraft is expected to commence flight trials shortly.

On 9 July two **Westland Sea King** helicopters for the Indian Navy arrived at RNAS Culdrose, Cornwall, to equip the Indian Naval Training Flight. Twelve officers and 31 ratings of the Indian Navy are currently at Culdrose to receive training in operational anti-submarine warfare techniques. Two other Sea Kings for the Indian Navy were airlifted to India in March, and are currently based at INS *Garuda* (Cochin) where they will shortly be joined by the remaining two Sea Kings of the six currently ordered. The six Sea Kings will eventually equip a new Indian Navy unit, No 330 Squadron, which will operate from INS *Garuda* and from the carrier *Vikrant*. An Indian Navy Helicopter Training School is currently being formed at INS *Garuda*. This will provide initial training on four Hughes 300 helicopters followed by advanced instruction on Alouette IIIs.

ISRAEL
There is no substance in widely-circulated reports that the Air Section of the Israel Defence Forces, *Heyl Ha'Avir,* is the primary customer for the **IAI-101 Arava** light STOL general utility transport on which 82 options are claimed. It is understood that no order for the Arava has been placed on behalf of the *Heyl Ha'Avir,* and that no order will be placed until a military prototype has been evaluated. Suggestions that the Arava will replace the C-47 have no foundation in fact as the *Heyl Ha'Avir* does not currently possess a requirement for a C-47 replacement, and anticipates that it will continue to operate the elderly Douglas transport for a number of years. The military Arava 201 currently projected by Israel Aircraft Industries will have a swing tail and bench seats for up to 20 paratroops, anticipated unit cost being £185,410 ($445,000).

Although the unacceptably high heat level of the exit duct of the General Electric J79 installed in a **Mirage III** under the *Project Salvo* programme has been overcome by redesign of the duct, there are currently no plans to re-engine further *Heyl Ha'Avir* Mirages with the J79 as no spares or overhaul difficulties are being experienced with the standard SNECMA Atar engine.

The already-substantial transport component of the *Heyl Ha'Avir* is being supplemented by an unspecified number of **Lockheed C-130 Hercules**. At the present time the transport inventory of the *Heyl Ha'Avir* includes a dozen modified Boeing 377 Stratocruisers and KC-97 Stratotankers, about 30 Nord Noratlases, and a substantial number of Douglas C-47s.

Hughes Aircraft has sold some $10m (£4·16m) worth of ALQ-71/ALQ-72 jamming pods and related electronic **countermeasures equipment** to Israel to date, and has completed a study for a new computer-based air defence system. Israel is shortly expected to receive 30 Northrop MQM-74 Chukar turbojet-driven drones which, used by the US Navy as targets, are being supplied for an unspecified non-target application.

JAPAN
The Defence Agency has issued the basic specification for the successor to the Kawasaki P-2J maritime patrol aircraft with the Maritime Self-Defence Force. Designated **PX-L**, the aircraft will have a low-positioned swept wing and will be powered by four 9,000 lb st (4 082 kgp) General Electric TF34-GE-2 turbofans. Gross weight will be of the order of 120,000 lb (54 430 kg) which will include a 14,000-lb (6 350-kg) weapons load and a crew of 10. The MSDF requirement specifies the ability to reach a patrol area 520 mls (840 km) from base within one hour with seven hours endurance on station at 253 mph (407 kmh), and a maximum cruise of 575 mph (925 km/h). The PX-L programme will commence in 1972, and two static test and two flight test prototypes will be built, the first flying prototype commencing its test programme in 1977. Kawasaki Heavy Industries will be the prime contractor, and the first PX-L squadron is expected to be formed in 1980.

The first two McDonnell Douglas-built **F-4EJ Phantom** pattern aircraft reached Japan on 23 July, and these are to be followed by 11 more F-4EJs in broken-down form for assembly by Mitsubishi's Nagoya plant. The first F-4EJ built by the parent company was flown on 14 January, and the first flight of a Mitsubishi-assembled F-4EJ is scheduled to take place in May 1972. A further 91 F-4EJ Phantoms have been ordered for the Maritime Self-Defence Force with the procurement of an additional 54 programmed, and it is anticipated that the Defence Agency will finalise an order for an indeterminate number of RF-4EJ Phantoms next year, 18 being included in the 4th five-year programme.

MALAYSIA
Some confusion has arisen over future **fighter procurement** for the Royal Malaysian

Air Force (*Tentera Udara Diraja Malaysia*). Tenku Ahmad Rithaudden, the Deputy Minister of Defence, speaking in Kuala Lumpur on 8 June, stated categorically that 10–12 Mirage Vs would be purchased from Avions Marcel Dassault but that delivery dates had yet to be finalised. Confirmation of this statement followed from another government spokesman, who added that, in view of the short lifespan of the Australian-donated Sabre squadron, Malaysia had to look ahead, and that the Mirage V represented the least expensive and most adaptable type of sophisticated fighter in terms of RMAF requirements. Subsequently, a statement issued by the Office of the Malaysian High Commissioner in London said that no definite decision regarding the Mirage had been taken! It has since been announced that the RMAF is to establish a *second* Sabre Squadron. According to a statement made by the Australian Foreign Minister during a recent visit to Kuala Lumpur, the second squadron will be formed with seven ex-RAAF Sabres shortly to be supplied.

SOUTH AFRICA
It is now reported that the 40 **Aerfer-Aermacchi AM.3C** battlefield surveillance and forward air control aircraft now being built are to be supplied to the South African Army Air Corps (rumoured Ethiopian interest in the AM.3C was reported in this column August issue). The Army Air Corps, which has one unit, No 41 Sqdn, manned by regular personnel and another, No 42 Sqdn, manned by Active Citizen Force personnel, is primarily equipped with Cessna 185s, but some expansion of this component has been reported, and in addition to new fixed-wing equipment it is believed that a number of helicopters are being acquired for use by the AAC.

Asked to comment on reports that a **further 36 Mirages** (in addition to those now to be manufactured under licence — see page 173) have been ordered from Avions Marcel Dassault for use by the SAAF, the chairman of the South African Armaments Development and Production Corporation, Prof H J Samuels, said that "this was purely a defence matter", but Defence Minister P W Botha refuted this report on his return to South Africa from a recent visit to France, saying: "I made no purchases . . . South Africa is *not* militarising on a large scale!"

SWITZERLAND
The Swiss Defence Ministry has established a requirement for up to 100 **air-superiority fighters**, and preliminary briefings have been given to potential suppliers. The requirement is unlikely to be announced officially before the final choice of a close-support fighter has been decided and this is unlikely before September of next year.

Delivered earlier this year to the Fuerza Aerea Colombiana was this Fokker F.28 Mk 1000 which now operates from Bogota as a presidential transport.

Despite the fact that the McDonnell Douglas A-4M Skyhawk was selected as runner-up to the Vought A-7G in the recent second evaluation, it is understood that the Dassault Milan S is once again running second to the A-7G, and may still emerge as winning contender despite considerable political resistance to the purchase of another aircraft from Avions Marcel Dassault.

TANZANIA
The Tanzanian Defence Force Air Wing is to receive some 24 **MiG-17** fighters from China to equip two squadrons, and the first 50 Air Wing personnel are currently undergoing flying and technical training on the MiG-17 in China. The Chinese government has agreed to train 250 Tanzanians as pilots and technicians, and all candidates are being rigorously examined in Tanzania by a Chinese mission as to their suitability for training.

TURKEY
It was recently revealed in Ankara that the Turkish Navy is to receive 12 **Grumman S-2E Trackers** from March 1972 with which to form an ASW squadron. Two S-2A Trackers were delivered to Turkey on 15 July to enable the Turkish Navy to initiate training at Karamürsel, near Gölcük, and three Agusta-Bell AB 205s are shortly to be delivered to the newly-created naval air arm.

The Turkish Air Forces, *Türk Hava Kuvvetleri* (THK), are in process of receiving 16 **Transall C.160D** transports under the Federal German military aid programme. The first two C.160Ds reached Turkey late in June and a new unit is being formed to operate them. These transports are surplus to *Luftwaffe* requirements.

The vertical lift capability of the THK is being increased and modernised by the delivery of **Bell UH-1D Iroquois** helicopters. The dozen or so aged Sikorsky UH-19s that have hitherto comprised the sole helicopter force of the THK are now in process of phasing out, and the first batch of 10 UH-1Ds has been taken on strength.

UNITED KINGDOM
The first RAF **SA 330 Puma** helicopter squadron is scheduled to become fully operational this autumn with No 38 Group Air Support Command. This unit, No 33 Squadron, was officially re-formed on 14 June, and will attain full strength this month (September).

By the end of this year most Argosy C Mk 1s will have been withdrawn from the transport rôle in the RAF, and a number of aircraft of this type are currently receiving internal modifications to enable them to perform the flight check rôle in which they will be known as **Argosy E Mk 1s.**

USA
The US Navy's first mine **countermeasures helicopter squadron,** HM-12, was formally commissioned at Norfolk NAS, Virginia, in April. HM-12, which was previously known as Detachment 53 of Helicopter Support Squadron 6 (HC-6), is currently operating USMC Sikorsky CH-53A and CH-53A helicopters in an interim mine-sweeping configuration but will eventually re-equip with the specialised mine countermeasures RH-53D which will incorporate an improved rotor blade and uprated engines which will permit an increase in maximum gross weight to 50,000 lb (22 680 kg).

In July the USAF retired the **North American F-100 Super Sabre** from combat service, the last combat unit to employ the Super Sabre having been the 35th Tactical Fighter Wing with four assigned tactical fighter squadrons which operated from Phan Rang, Vietnam.

The USAF has rejected a proposal that the projected Convair (General Dynamics) **FB-111G** should be adopted in place of the North American Rockwell B-1. The FB-111G was a projected derivative of the FB-111A with a lengthened fuselage and other aerodynamic changes, which, it was claimed, would be capable of carrying four SRAM (Short-range Attack Missile) weapons over a distance of 4,100 miles (6 600 km) at high altitude and 2,200 miles (3 540 km) at low altitude. The unit cost of the FB-111G was calculated at $9m (£3·75m) more than the $16m (£6·66m) unit price of the basic F-111.

AIRCRAFT AND INDUSTRY

ARGENTINA
After a practical evaluation of the two prototypes of the **FMA IA 58 Pucará** two-seat counter-insurgency aircraft (one fitted with 904 ehp Garrett AiResearch TPE-331-U-303 turboprops and the other with 1,022 ehp Turboméca Astazou XVIG turboprops) the Astazou has been selected as the power plant for the production Pucará of which an initial batch of 50 examples is expected to follow a pre-series of five aircraft for service evaluation. An order for 130 Astazou XVIG engines is being negotiated, and the total *Fuerza Aérea Argentina* requirement for the Pucará reportedly amounts to 80 aircraft.

CANADA
A small twin-turboprop executive aircraft has been projected by de Havilland of Canada as the DHC-8. It would seat five passengers and with two Pratt & Whitney PT-6 engines would sell for less than $500,000 (£208,300).

CZECHOSLOVAKIA
The roll-out of the first production **Aero L 39** has reportedly been delayed by some three months. Difficulties have been encountered with the zero-zero rocket ejection system, canopy trajectory problems having been encountered during high-speed ejection tests, and it now seems possible that the Ivchenko AI-25 turbojets for the L 39 will have to be acquired from the Soviet Union rather than be licence-built by Motorlet which lacks the production capacity to meet anticipated demands.

The Aero Group is engaged in the development of a new light helicopter, the **HC-4,** powered by a single M 601 turboshaft engine in the 700 eshp category. The HC-4 is expected to become a joint effort with the Polish WSK-Swidnik organisation.

FEDERAL GERMANY
First flight of the **VFW-Fokker 614G-1** was made at Bremen on 14 July. On a second flight on 19 July, the aircraft was ferried to Lemwerder where it will undergo resonance testing before making any further flights. The prototype G-2 is expected to fly before the end of this year and G-3 a few months later. The VFW-Fokker 614 is the first aircraft to fly with Rolls/Royce-SNECMA M-45H turbofans, one of which is mounted above each wing.

FRANCE
Two new prototypes of the **Aérospatiale SN 600 Corvette** are under construction with the first to fly at the end of 1972. They will have JT15D-4 engines and, compared with the prototype that crashed earlier this year, a longer fuselage and redesigned internal layout. Marketing in the US is expected to benefit from a recently concluded agreement between Aérospatiale and Piper.

INTERNATIONAL
All 16 airlines holding options on 74 **BAC/Aérospatiale Concordes** have now renewed their options for a further period while final details of performance guarantees are worked out for contractual purposes. United Air Lines was the last to renew, extending its option on six for six months early in July. Flying of both prototypes continued at a high rate during July (beyond the point reached in our record of Concorde test flying elsewhere in this issue), and a further series of handling flights was made by airline pilots on 002 during the month. On one of these flights, with Capt Scott Flower, director of research and development for Pan American in the left hand seat and accompanied by Brian Trubshaw, BAC chief test pilot, the aircraft was subjected to stresses beyond the normal design limit, and was grounded for a few days while a full investigation of the test recordings was made. The incident occurred during an evaluation of stick force per *g*, and the design value of 2·5*g* was exceeded for about 1½ seconds, peaking at 3·1*g*. No damage was suffered by the airframe and flying was to resume during August following some routine servicing.

ITALY
Rinaldo Piaggio has released preliminary details of a project for a small STOL transport in a similar category to the LET 410. Designated **Piaggio RP 169,** it is a high-wing monoplane with a T-tail, powered by Lycoming T53 L 7015 turboprops and seating 22–26 passengers. Gross weight would be 17,483 lb (7 930 kg), cruising speed 280 mph (450 km/h) and take-off and landing distances, 1,312 ft (400 m) and 1,378 ft (420 m) respectively.

JAPAN
The first **Mitsubishi XT-2** tandem two-seat advanced trainer prototype was flown for the first time on 20 July at Nagoya. The aircraft was in the air for a total of 38 minutes during which it attained a speed of 530 mph (853 km/h) and an altitude of 20,000 ft (6 100 m). The take-off weight for the first flight was 19,620 lb (8 900 kg) and the take-off run was 810 yards (740 m). A second XT-2 is scheduled to join the test programme shortly, and both prototypes will be delivered to the Air Self-Defence Force before the end of this year.

The first prototype of the **Mitsubishi MU-2K,** the latest version of this light business and executive transport, has attained a speed of 373 mph (600 km/h) in level flight during its test programme, and total orders for all versions of the MU-2 have now attained 222, and of these 204 had been delivered by the end of June.

The first production **Shin Meiwa PS-1** maritime patrol flying boat is scheduled to be rolled out before the end of the year and to commence its flight test programme in February. Five are to be delivered to the Maritime Self-Defence Force by August 1973. During the 4th Five-year Defence Programme (1972-6) the MSDF will receive 16 PS-1s and three examples of a search and rescue version, the PS-1S, and the replacement of the current T64-GE-10 turboprops by T64-P4C units of 3,310 hp is currently being studied.

NEW ZEALAND
A four-seat derivative of the Airtourer is to be built by Aero Engine Services Ltd. To be named the **Aircruiser,** it is expected to fly next Spring.

SOUTH AFRICA
The South African Armaments Development and Production Corporation (Armscor) has signed an agreement with Avions Marcel Dassault for the **licence manufacture** of both the Mirage IIIE and the Mirage F1. No official comment has been forthcoming on the quantities of aircraft involved, but licence manufacture would obviously be uneconomic for quantities of less than 50 of each aircraft type involved. The licence arrangement with Armscor is part of a broader technical and industrial co-operative agreement authorised by the two governments, and includes Dassault assistance to Atlas Aircraft — a subsidiary of Armscor — in developing a modern aircraft production capacity. It is difficult to overrate the significance of the South African decision to manufacture advanced combat aircraft, yet world reaction has been mild compared with the outcry that greeted the British decision earlier this year to release seven Wasp ASW helicopters to the South African Navy, and what little protest has been made has been referred to by Foreign Minister Maurice Schumann as a "campaign of lies". In a radio interview, M Schumann stated that France had refused to sell arms to South Africa which could be used for purposes of internal repression but could not refuse to sell weapons for South Africa's external defence. Referring specifically to the recently-concluded licence agreement between Armscor and Avions Marcel Dassault, he justified this by saying "as far as the Mirages are concerned, they cannot be used for anti-guerrilla or repressive operations!"

SPAIN
The Construcciones Aeronauticas SA (CASA) plant at Seville is now approaching completion of its manufacturing programme of 70 **Northrop SF-5s** for the *Ejército del Aire Español.* All 34 two-seat SF-5Bs (CE.9) have been delivered, and the last of the 36 single-seat SF-5As (C.9) is now in final assembly.

SWEDEN
The Swedish government has approved plans for continued development work on the **JA 37** version of the Saab Viggen optimised for the intercept rôle, and has authorised SwKr 113m (£9·15m) for the purpose. The JA 37 will, unlike the AJ 37 version of the Viggen, have Mach 2·0 capability, this being achieved by the introduction of variable-geometry intakes and an uprated Volvo Flygmotor RM 8 turbofan. Armament will include a built-in 30-mm cannon. Saab is currently quoting a flyaway cost of SwKr 18m-19m (£1·45m-£1·53m) for an export version of the AJ 37 which has an operating cost of about SwKr 2,300 (£185) per hour.

UNITED KINGDOM
Westland Helicopters Limited is currently offering a high-capacity army support helicopter using the proven dynamic components of the Sea King. Named the **Commando,** the proposed helicopter will be capable of all-weather operation and will have a disposable load equal to 50 per cent of its all-up weight. In the troop transport rôle it will carry 30 troops.

Type Approval of the Rolls-Royce Bristol **Pegasus Pg 11** has been obtained at a rating of 21,500 lb st (9 750 kgp). This variant will be fitted in future production AV-8A Harriers for the USMC and it is expected that Pegasus 101s in RAF Harrier GR Mk 1s and T Mk 2s will be uprated to the same standard as Pegasus 103s. The designations Harrier GR Mk 3 and T Mk 4 have been reserved for RAF aircraft with this engine standard.

USA
The first **Convair (General Dynamics) F-111F** for the USAF was scheduled to commence its flight test programme in August. Seventy F-111F tactical strike fighters are on order for the USAF, this version differing from the F-111E in having a simplified avionics system and the TF30-P-9 turbofans replaced by TF30-P-100 series engines of increased thrust.

The prototype **Piper PE-1 Enforcer** (see this column August issue) crashed and was totally destroyed at Patrick AFB on 12 July, the pilot ejecting safely, but a second prototype has since been completed and flown. The Enforcer is a newly-tooled counter-insurgency aircraft based on the design of the P-51 Mustang and powered by a 2,535 hp Lycoming T55-L-9 turboprop.

CIVIL AFFAIRS

INTERNATIONAL
A Swearingen Merlin III owned by Battenfeld GmbH and flown by J H Blumschein and Fritz Kohlgruber won the 5,851 mile (9 412 km) **London-Victoria Air Race** in July. Competing with 56 other aircraft, the Merlin III was outright winner and placed first in two of the individual stage lengths. The pilots also won the Canadian Prime Minister's Trophy for best overall performance, the Premier of British Columbia Trophy for best performance between Ottawa and Victoria, and other awards. The Merlin III made stops at Prestwick, Keflavik, Frobisher Bay, Goose Bay, Quebec, Ottawa, Winnipeg, Regina and Calgary.

MALAYSIA
The division of **Malaysia-Singapore Airlines** into two separate carriers for Malaysia and Singapore is expected to take effect at the end of 1972. Thereafter, the Singapore airline (which may be called Metropolitan

Singapore Airlines in order to retain the MSA initials) will concentrate on building up its international operations, while the new Malaysia Airlines Berhad will devote its energies to regional and domestic operations. The Singapore airline is expected to retain the Boeing 707 fleet (at present numbering five) plus most of the Boeing 737s and F.27 Mk 500s. Malaysia Airlines will keep the three B–N Islanders and probably order more, and has already ordered six new Advanced Boeing 737s. Qantas has been retained by the new Malaysia Airlines to supply technical and managerial assistance.

UNITED ARAB REPUBLIC
United Arab Airlines is now using one of its recently-acquired Ilyushin Il-62s on the Cairo-London route in place of Boeing 707s. The **UAA fleet** is being built up with the acquisition of two Il-62s, three Tupolev 154s (starting in October), four Tu-134s (later this year) and two Antonov An-12s (next year).

UNITED KINGDOM
Britain's new "second force" airline Caledonian/BUA will start services between **London and Paris** on 1 November, with a frequency of four a day using One-Eleven 500s (three a day with One Eleven 200s on Saturdays and Sundays). The services will operate via Gatwick and Le Bourget and a city-centre to city-centre time of 2½ hours is scheduled. From the same date, BEA will cease to serve Le Bourget, concentrating its service from Heathrow to Orly.

Air Holdings (Sales) Ltd has acquired appropriate licences allowing it to operate **Vickers Vanguards** on ad hoc charters. The Vanguards were accepted as trade-ins from Air Canada in the deal to sell Lockheed L-1011s to the Canadian airlines. One Vanguard converted to all-cargo configuration is at present available in Air Holdings markings.

USA
Following simultaneous delivery of fully-certificated **McDonnell Douglas DC-10** Series 10s to United Airlines and American Airlines on 29 July, regular passenger service with the type was due to begin early in August. According to plans announced up to the time this issue went to press, American Airlines would start service on its Los Angeles-Chicago route on 17 August (in place of a Boeing 747). United had planned to start West Coast-to-Washington

Awaiting delivery recently were these four Yak-40s all with the new thrust reverser on the centre engine. Nearest the camera are two for Bakhtar Afghan Airlines.

operations on 1 September, but moved the date forward to 16 August in order to beat American. Certification of the DC-10, first flown on 29 August 1970, has come about three months ahead of the preliminary schedule, and after some 1,500 hrs of flight development. First DC-10 Series 20 with JT9D engines is No 28 on the line and due to fly in March 1972.

First use of a **de Havilland Caribou** in the fire fighting rôle results from the purchase of a DH demonstrator by Intermountain Aviation of Arizona. It is to be used to paradrop "smoke jumpers" and supplies in support of US Forest Service fire-suppression operations in the states of Arizona, New Mexico, Idaho and Montana.

CIVIL CONTRACTS AND SALES

Aérospatiale Caravelle: Euralair in France will acquire two Caravelle VI-Rs early in 1972 to replace two F.27s. □ Swissair sold one to Catair. □ Transavia sold a Srs III to Royal Air Maroc.

Aérospatiale N 262 Frégate: A previously unannounced order, placed earlier this year, is for one aircraft for radio calibration in Kenya, Uganda and Tanzania. Order was placed by the East African Community and the Frégate will be based in Kenya.

Boeing 707: BEA Airtours contracted to buy seven 707-436s (Conway engines) from BOAC, for delivery between January 1972 and Spring of 1973. □ The two additional 707s recently acquired by Malaysia-Singapore Airlines are ex-Braniff, not new from Boeing. □ Donaldson has confirmed its intention of buying a third -320 from Pan American, and placed an option on a fourth. □ Air Commerz of Hamburg has added two 707-138Bs to its fleet, ex-Standard Airways. □ THY is reported to have taken delivery of its third 707. □ Varig has purchased one -385C from American. □ Latest 707 acquired by Caledonian/BUA is a -323C from American, on lease for six months.

Boeing 720: Boeing has confirmed the previously-reported sales of ex-Eastern aircraft to Trans European (one), Danish charter operator Conair (five) and German charter company Calair (five).

Boeing 727: Icelandair acquired its second -100C, through Grant Aviation Leasing Corp. □ Avianca has acquired one, previously operated by Executive Jet.

Boeing 737: Malaysia Airlines Berhad, which will begin operations in January 1973 following the splitting up of Malaysia Singapore Airlines, has ordered six Advanced 737-200s, for delivery in July-September 1972. □ Boeing has confirmed the order, reported in this column last month, from Saudi Arabian Airlines for five Advanced 737-200s. □ Boeing also confirms the sale of four 737s to Southwest Airlines, the new Texan intra-state airline. □ Dominicana has leased one for one year from Aer Lingus.

Boeing 747: Qantas has ordered one more -237B for delivery next September. Four others are on order, with first delivery made in July.

Business jets: One Jetstar Dash 8 was sold to Grupe Cresi in Mexico.

Canadair CL-44: Argentine non-scheduled operator Rio Platense SA acquired one from Airlift earlier this year. □ Aerotransportes Entre Rios, also of Argentina, will acquire its second CL-44 shortly.

Fairchild FH-227: Nordair of Canada bought one, for delivery in late summer, in Cargonaut configuration in support of USAF and ITT personnel manning the DEW line.

Fokker F.27: Maersk Air in Denmark ordered its fourth Mk.500, for delivery in November; company also has two Mk 600s. □ Turavia in Italy ordered one Mk 600 for July delivery. □ Indian Airlines acquired one Mk 400 previously operated on lease by Botswana Airways.

Fokker F.28: Netherlands Govt ordered one for delivery next January to replace an F.27 as a Royal and government executive aircraft. □ Garuda Indonesian Airways ordered two Mk 1000. □ Two aircraft are reported to be reserved for Aeroco-op Peru.

Hawker Siddeley Argosy: The four Saggittair aircraft (this column last month) are ex-Universal, Srs 102s. □ Another Universal Argosy is in use in Alaska on lease by Jet International.

Hawker Siddeley HS 748: An order for two from Merpati Nusantara Airlines has now been confirmed. □ Botswana Airways uses one, leased from SAA. □ Philippine Air Lines added two to its fleet, one ex-Austrian Airlines and one ex-Bahamas Airways. □ Two others, ex-Bahamas, went to Mandala.

McDonnell Douglas DC-8: Finnair acquired its third Sixty Series by obtaining a -62CF from United. □ A new Finnish charter company, Spear Air, plans to acquire two, second-hand, in 1972.

McDonnell Douglas DC-9: Purdue Aviation has disposed of its three DC-9s — two to Hughes Air West and one to Inex-Adria.

Short Skyvan: A new sale is to Helicopter Utilities in Sydney, to be used in contract to International Nickel. □ Pan-Alaska converted its lease on one to an outright purchase.

Vickers Viscount: Air International, a new UK operator, has one Viscount 702 from Field Aircraft Services. □ Three ex-Falconair Viscounts are now operated by Malmo Aero. □ Arkia in Israel has added one Type 814, ex-Lufthansa, to its fleet. □ Restructuring of BEA into separate operating divisions means that four Viscounts now operate in Scottish Airways markings and four others in Channel Islands Airways markings.

Tupolev Tu-134A: Aviogenex of Yugoslavia now operates three new aircraft, apart from the one written off in May, presumably in place of three Tu-134s previously operated.

MILITARY CONTRACTS

Northrop F-5: Saudi Arabia is reported to have placed an order with the Northrop Corporation for a batch of 27 F-5 fighter-bombers.

Kawasaki-Vertol KV-107: The Swedish government has signed a contract with Kawasaki for the supply of seven KV-107 ASW helicopters for 1973-4 delivery.

Bell 206 JetRanger: The Jamaica Defence Force recently took delivery of a Bell 206 JetRanger which will be used primarily for the transportation of Jamaica's Prime Minister but will also be available for search and rescue missions.

Short Skyvan 3M: The Royal Nepalese Army has ordered a second Short Skyvan 3M, the first having been ordered in December 1970. □ The Sultan of Oman's Air Force has placed an order for an additional two Skyvan 3Ms to supplement the six previously delivered.

AESL Airtourer: Four AESL Airtourer primary trainers are to be supplied to the Singapore Air Defence Command.

Fairchild Hiller FH-227B: Two Fairchild Hiller FH-227B long-body Friendships with cargo doors have been ordered by the *Força Aerea Uruguaya* (FAU). The FAU has been operating two Fokker F.27 Friendship 100 transports since last year.

Yakovlev Yak-40: The Yugoslav Air Force has purchased an unspecified number (believed to be two) Yak-40 transports.

Bell AH-1G HueyCobra: A contract for 70 additional AH-1G HueyCobras has been awarded Bell Helicopter by the US Army, contract value being $16,690,688 (£6,954,450). The additional order, which calls for deliveries between August 1972 and January 1973, brings to more than 1,100 the number of HueyCobras so far ordered, more than 900 of these having been delivered.

Vought A-7: The US Defense Department announced on 1 July the award of a $108,977,536 (£45,407,300) contract to LTV Aerospace for 118 A-7D and A-7E (88 A-7Ds for the USAF and 30 A-7Es for the USN) aircraft for delivery during the 1972 calendar year.

Fairchild C-119K: Three Fairchild C-119K transports were delivered to the Imperial Ethiopian Air Force during June and a further two aircraft were scheduled to follow in August. Converted from C-119Gs to the jet-augmented C-119K configuration, these five aircraft join five C-119Ks delivered to Ethiopia late last year.

IAI Commodore Jet 1123: Two Commodore Jet 1123s were delivered to the Uganda Air Force as presidential and VIP transports in July. Both aircraft were originally IAI demonstrators.

Britten-Norman BN-2A Islander: Three BN-2A Islanders have been ordered by the Iraqi Air Force for photographic survey tasks and will be delivered later this year.

Hawker Hunter: Eight refurbished Hawker Hunters are to be supplied by Hawker Siddeley to the Lebanese Air Force.

Lockheed C-130E Hercules: The *Fuerza Aérea Argentina* is to receive two additional C-130E Hercules transports this year. These will join three C-130Es already in FAéA service.

Fairchild Hiller Porter: The Argentine Navy has ordered three Porters. □ Two Porters were delivered in June by Fairchild Hiller to the Ecuadorian Army.

BAC 167 Strikemaster: A second batch of six Strikemaster Mk 83s has been delivered to the Kuwait Air Force.

LANCER

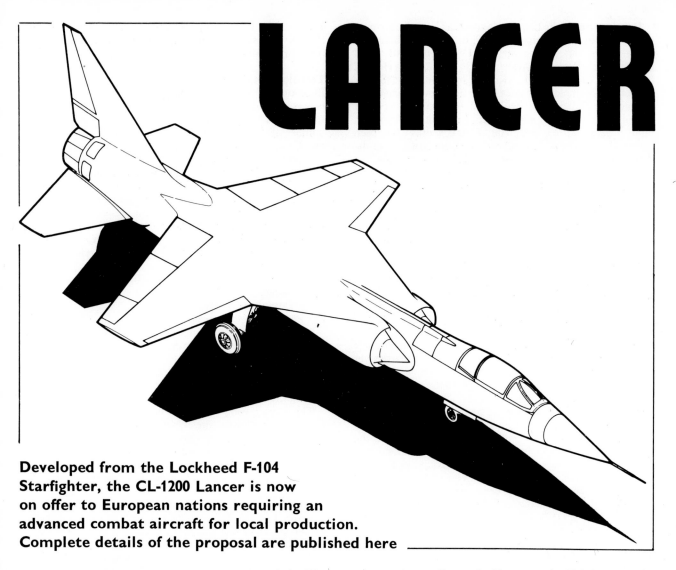

**Developed from the Lockheed F-104
Starfighter, the CL-1200 Lancer is now
on offer to European nations requiring an
advanced combat aircraft for local production.
Complete details of the proposal are published here**

A COMPREHENSIVE market survey conducted by the Lockheed-California Company indicates a possible world-wide market for over 7,500 advanced-design, reasonably-priced fighters during the next decade. This forecast is based upon a study of existing fighter forces and the assumption that they will be updated on a one-for-one basis. It *excludes* the USA, Britain, France and Sweden, since these nations are expected to meet their own requirements with domestic designs.

The size of the potential fighter demand explains the current intense competition, particularly among US manufacturers, to launch a new aircraft capable of capturing at least some of this lucrative market. Even a 10 per cent share, representing, say, 750 aeroplanes by 1980, would be a worthwhile programme, but it is by no means impossible that the right aeroplane could capture as much as 30-40 per cent of the market. There is, of course, no single specification which would meet the needs — apparent or real — of all the various nations making up the total fighter market, but the requirement can be broadly defined as being for an air-superiority fighter to counter the threat of air space domination by large numbers of high-performance enemy (Soviet-manufactured) fighters based only a short distance away.

To counter this threat, the new fighter must be able to react quickly to any airborne intrusion by enemy aircraft, and must have superior combat performance and duration. Unit cost is also of great significance, however, for this will determine the number of aircraft that any given nation can afford, regardless of the numbers actually needed to counter

a given threat. For a significant portion of the potential market, the opportunity to share in production of the new fighter is also important.

The current range of air-superiority fighters was discussed in some detail in AIR ENTHUSIAST Vol 1 No 1 ("The Fight for the Skies", page 5 *et seq*), but so far as the Lockheed-defined market is concerned, the major contenders now appear to include the Dassault Mirage F1; the Northrop P.530 Cobra; the McDonnell Douglas F-4E(F) and Lockheed's own proposal, the CL-1200-2 Lancer. Of these four types, the Mirage F1 is flying and is in production for one customer; the F-4E(F) is a major modification of the Phantom II (over 4,000 built) and has been ordered by one customer; the Lancer is a major modification of the Starfighter (nearly 2,500 built) and not as yet ordered, and the Cobra is a wholly new design also yet to be ordered.

In proposing the Lancer, Lockheed has sought to derive maximum benefit from its Starfighter experience, not only in the design of the new aircraft through commonality of structure and systems wherever possible, but also by making use of past investment in tooling, factory facilities and equipment, and of experience in organising consortium-production. About half of total F-104 production (1,212 aircraft) was undertaken by a consortium of four European nations, and two of these, Germany and Italy, still have the type in production. Moreover, nine European countries are at present operating a combined total of about 1,170 F-104s, and six other nations are also flying the Starfighter elsewhere in the world.

Development of the Lancer, if a go-ahead is received, would be handled by the "Skunk Works", or, to give this unique organisation its proper name, Lockheed's Advanced Development Projects (ADP) unit in California. Headed since June 1943 by C L "Kelly" Johnson, the "Skunk Works" has complete facilities for design, development and production, and operates within an environment of strict security and minimum administrative restraints, permitting rapid progress at low cost. The "Skunk Works" has been responsible for the design and prototype construction of the P-80 Shooting Star, the XF-90, the JetStar, the F-104, the U-2 and the YF-12/SR-71. It has also handled all work on the Lancer to date, including configuration studies and wind tunnel tests.

Minimum cost

By using as much as possible of the F-104 and a production-type engine, coupled with the inherently low costs of the "Skunk Works" operation, Lockheed estimate that the cost of developing the Lancer is only $70·5m (£34m). Assuming that the entire production was handled by a European consortium similar to that set up for the F-104, and making use of existing tooling wherever possible, the unit cost of the Lancer is calculated to be $2·7m (£1·12m) for a run of 500 and $2·4m (£1m) for production of twice this quantity. These prices allow for Lockheed to recover development costs based on a fixed price incentive contract, but do not include the cost of five F-104Gs which would provide the basis for constructing Lancer prototypes and would be supplied, it is assumed, by the consortium.

Lockheed has conducted an intensive investigation into the initial support costs and the operating costs of the Lancer spread over 10 years of operation. Initial support costs, comprising such items as spares, ground equipment, manuals, maintenance training, flight training and mobile training sets, are assumed to benefit from prior operation of the F-104, and come out at $330m (£137m) for 500 aeroplanes and $180m (£75m) for 1,000.

The 10-year operating costs assume aircraft utilisation of 20 hours per month, with aircraft disposed in 12-aircraft squadrons and 15 pilots per squadron. For a 500-aircraft force (with 420 operational) the total 10-year cost is $2,085m (£870m); for 1,000 aircraft (with 840 operational) it is $4,063m (£1,690m). Adding the first cost, initial support and operating costs together, the total Lancer programme costs over 10 years work out at $4,025m (£1,675m) for 500 aircraft and $7,193m (£2,398m) for 1,000 aircraft. Lockheed claims that these figures represent total savings, compared with similarly-calculated F-4E(F) costs, of $1,386m (£578m)

and $1,955m (£815m) respectively. Compared with Mirage F1 costs, the savings work out at $76m (£31.2m) and $207m (£86·2m) respectively.

Technical details

Evolution of the Lancer design is part of a continuous process aimed at getting maximum benefit from the basic Lockheed F-104 configuration. This process has led to a number of design proposals for improved F-104s, of which the CL-1200-2 is the latest. The research, development, test and evaluation phase of the latest project began in November 1969, since which time the company has conducted (at its own expense) extensive configuration studies, wind tunnel tests and preliminary design phases.

The major changes to the F-104 basic design to produce the Lancer were made with the primary goal of improving manoeuvrability and eliminating pitch-up. They include:

a. Enlarging the wing area by 53 per cent and relocating it in a high position.
b. Enlarging the tailplane and relocating it low on the fuselage.
c. Enlarging the fin area.
d. Providing a 46 per cent increase in internal fuel capacity, by lengthening the fuselage and including wings tanks.
e. Improving the air intake design.
f. Improving the high lift devices on the wings.
g. Providing additional stores positions on the wings.
h. Substituting a Pratt & Whitney TF30-P-100 engine for the General Electric J79 used previously.

The Lancer wing comprises outer panels which are virtually identical with the wings of the F-104. New inner portions are fitted, increasing the span and area, and delta-shaped surfaces extend from the air inlet to the leading edge of the new wing portion, the effect of these extensions being to improve the fighter's airfield performance and lower the trim drag in high speed manoeuvres. Trailing edge flaps in the new wing portions double the available flap area to improve the field performance, and very low landing speeds are claimed for the Lancer without resort to boundary-layer control.

Automatic operation of the wing leading and trailing edge flaps is provided during aerial combat — if the appropriate mode is selected by the pilot — as a function of the load factor, speed and altitude. Also to improve manoeuvrability, the maximum deflection angle of the variable incidence tailplane is increased from 17 deg to 25 deg.

Choice of the TF30-P-100 engine (as used in the F-111F) gives 60 per cent increase in thrust at maximum power for a weight penalty of only 190 lb (86 kg) compared with the

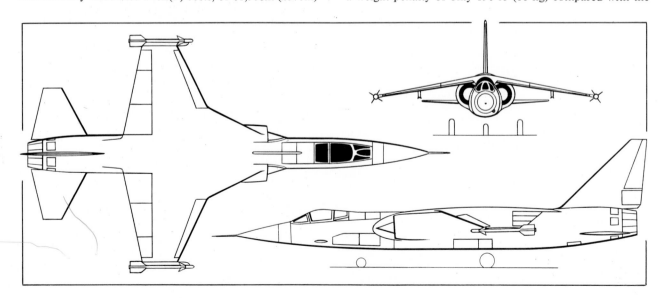

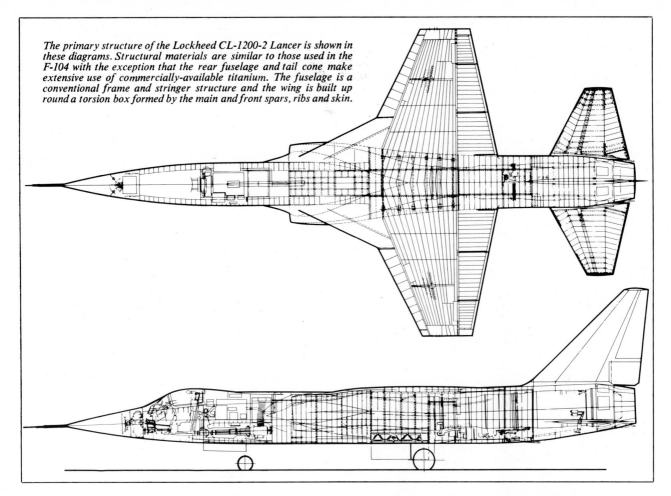

The primary structure of the Lockheed CL-1200-2 Lancer is shown in these diagrams. Structural materials are similar to those used in the F-104 with the exception that the rear fuselage and tail cone make extensive use of commercially-available titanium. The fuselage is a conventional frame and stringer structure and the wing is built up round a torsion box formed by the main and front spars, ribs and skin.

J79-GE-11A in the F-104G. The diameter of the TF30 is slightly greater, requiring an increase in the cross section of the Lancer's rear fuselage, and the intakes are both larger and stronger to cope with the increased air mass flow. The intakes incorporate translating shock cones — with about 4 in (10 cm) movement — in place of the F-104's fixed cones, and embody boundary layer diverter plates. Operation of the shock cones to the optimum position for any given Mach number is by electro-mechanical actuators through a control unit receiving inputs from the air data computer.

Structurally, the Lancer airframe is similar to the F-104. The fuselage is a semi-monocoque structure and the forward section (identified as segment 231 in the F-104 breakdown) is identical with that of the F-104G with the exception that the windshield is of F-104S-type to withstand the increased aerodynamic heating due to the higher Mach number at which the Lancer flies. The commonality of this fuselage section means that any proven Starfighter front fuselage can be used on the Lancer, including the two-seat TF-104G section, the F-104S nose with Sparrow missile electronics system for all-weather interception operations, and reconnaissance sections with radar, infra-red or photographic equipment.

Aft of the forebody retained from the F-104, a 30-in (76,2-cm) long barrel section is added to provide additional volume in the fuselage for fuel, if required. The landing gear and support structure are retained from the F-104, but the structural load path of the fuselage centre section is simplified, as the fuselage no longer has to carry the wing bending loads. As a result, the fuselage structure is simplified, with improved fatigue life. Structural materials in the forward and centre fuselage sections are similar to those in the F-104G and S, but the rear fuselage materials are radically different, in that extensive use is made of commercially-available titanium alloy, 6A1-4V, for the frames, major

longerons and tail cone around the engine jet pipe.

The wing design uses a multi-beam box structure with machine-tapered skins. Although the outer panels are the same as the F-104G wings, the complete structure is one continuous element, and structural provision is made for a supplementary fuel tank between the two main spars. If all the additional fuel and weapon load options are taken up, the gross weight of the Lancer may go as high as 35,000 lb (15,875 kg) compared with the normal maximum take-off weight of 28,800 lb (13,065 kg), which is the same as the Starfighter's. For the higher gross weight, an optional strengthened landing gear is available, at a weight penalty of 180 lb (82 kg).

Flight control surfaces are conventional and similar to

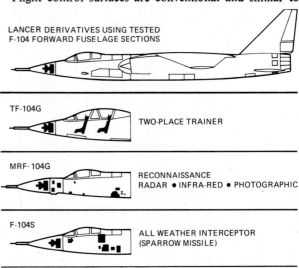

LANCER DERIVATIVES USING TESTED
F-104 FORWARD FUSELAGE SECTIONS

TF-104G — TWO-PLACE TRAINER

MRF-104G — RECONNAISSANCE
RADAR • INFRA-RED • PHOTOGRAPHIC

F-104S — ALL WEATHER INTERCEPTOR
(SPARROW MISSILE)

those of the F-104G, apart from the introduction of clam-shell type speed brakes in the base of the fin, beneath the rudder. These take the place of the speed brakes in the rear fuselage of the F-104. No provision is made for automatic pitch control, as used in the F-104, since wind tunnel tests have shown that the CL-1200-2 remains completely controllable at extreme angles of attack. No artificial stall warning devices are needed.

The electrical system is essentially unaltered from the F-104G, other than using brushless-type generators and solid-state voltage control devices. The hydraulic system differs only in the size of the actuator components. Avionics fit depends on the requirements of the user, the basic F-104G package being offered as a reference point. Alternative off-the-shelf equipment is available to provide a more advanced capability at lower weight.

Fuel tanks in the centre fuselage contain a total of 833 Imp gal (3 787 l) and are the same as those in the F-104G. This internal capacity can be increased to 1,090 Imp gal (4 955 l) by using the optional extra fuselage tank and wing tanks. In addition the wing is stressed to carry four external fuel tanks, two of 142 Imp gal (646 l) each at the tips and two of 162 Imp gal (736 l) on outboard pylons. A further 100 Imp gal (454 l) can be carried in tanks replacing the built-in gun armament when the Lancer is used for extended-range or ferry missions.

Weapons and performance

The Lancer retains, as built-in armament, the 20-mm General Electric M-61 multi-barrel cannon, located in the lower front fuselage to port. This gun can be used for air-to-air or air-to-ground firing and the Lancer carries 725 rounds for it, sufficient for just over 7 seconds firing at the gun's maximum rate of 6,000 rpm. A 30-mm DEFA gun can be fitted in the fuselage, with 400 rounds, as an alternative.

External pylon stations can be provided under the fuselage (one), under each wing (three) and at the wing-tips, providing for a wide variety of external stores, either air-to-air or air-to-ground. For the air-to-air interceptor rôle, AIM-9D Sidewinders can be carried on all eight stations plus two on the fuselage pylon, and for short-range air-to-ground missions, up to 12,000 lb (5 445 kg) of external ordnance can be lifted on the six wing pylons. The actual combination of stores that can be carried depends both on the weights of individual items and clearances for loading, carriage and release, but some idea of the possibilities is given in the accompanying diagram.

Lancer in the form described here is a Mach 2·4 aeroplane, and the maximum placard speed is 800 knots (1 483 km/h), equivalent to Mach 1·21 at sea level, compared with 750 knots (1 390 km/h) or Mach 1·13 for the F-104G. Increased thrust-to-weight ratio and wing area bestow significant improvements upon take-off and landing performance. For example, in a typical intercept configuration, the take-off distance of the Lancer is only 52 per cent of the distance required by the F-104G, being 1,450 ft (442 m) to lift-off at 156 knots (289 km/h). Similarly, the landing run, clean with 1,000 lb (454 kg) of fuel remaining, is 2,060 ft (628 m) with touchdown at 118 knots (219 km/h) compared with the 2,450 ft (747 m) required by the F-104G after touching down at 156 knots. A drag parachute is used for landing, and an arresting hook is fitted as standard; if the optional 35,000 lb (15 875 kg) gross weight version is adopted, a 16·5-ft (5,0-m) diameter ribbon type drag 'chute is fitted.

Weapons capability of the Lancer is shown here. Four of the wing pylons, at stations 75 and 148.5 each side, are optional and apply only when the aircraft is configured for the ground attack rôle.

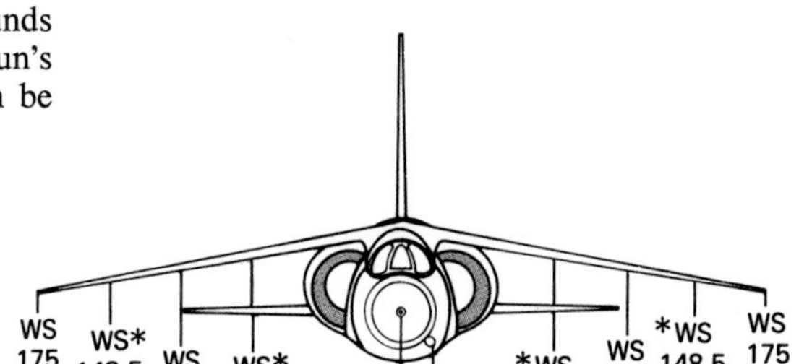

TYPE		STORE OR WEAPON	WT EACH (LB)	WS 175 1500	WS* 148.5 1000	WS 118 2000	WS* 75 2500	2000	*WS 75 2500	*WS 118 2000	*WS 148.5 1000	WS 175 1500	TYPICAL LOADING
		EJECTOR RACK (TER/MER)			T	T/M	T/M		T/M	T/M	T		
INT GUNS		20 mm M61A1, 725 RNDS (STD FEED)	800					●					1
		30 mm DEFA (400 RNDS) *	950					●					1
MISSILES	A-AIR	AIM-9B OR 9D SIDEWINDER	164	¤	¤	¤	¤	¤ ¤	¤	¤	¤	¤	6
		AIM-7N SPARROW *	498		¤		¤		¤		¤		4
	A-G	AS-30 NORD AVIATION *	1150			¤			¤				2
ROCKETS AIR-GND		LAU-3/A POD 2.75 FFAR (19)	415		●●	●●●			●●●	●●			6
		LAU-10/A POD 5.0 ZUNI (4)	533		●	●●			●●	●			6
BOMBS		MARK 82 LDGP/RETARDED	530/560		◇	✺✺	✺	◇	✺	✺✺	◇		7
		M117A1 GP/117D RETARDED	823/857		◇	✺✺	✺✺	◇	✺	✺	◇		5
		MARK 83 LDGP	985		◇	◇◇	◇◇	◇	◇◇	◇◇	◇		5
		MARK 84 LDGP	1970		¤	¤	¤		¤	¤			3
		SPECIAL WEAPONS			¤		¤		¤				3
		M129E1 (LEAFLETS)	750	¤	¤¤		¤		¤¤	¤			5
		CBU-24 B/B (BOMBLETS)	820		◇	◇◇		◇		◇◇	◇		4
		BLU-1/B	750		◇	◇◇◇	◇	◇	◇	◇◇◇	◇		5
		BLU-32/B	610		◇	✺✺	✺	◇	✺	✺✺	◇		7
		SUU-20/A (PRACTICE)	457	⊛	⊛				⊛	⊛			2
		SUU-25/A FLARES (8)	340	●●	●●				●●	●●			6
FUEL TANKS		170 GAL (TIP) *	1335	○	·							○	2
		195 GAL (PYLON) *	1445		○					○			2

∗ OPTIONAL INSTALLATION

Rate of climb of the Lancer at sea level is greater than 60,000 ft/min (305 m/sec), reflecting the aircraft's high SEP (specific excess power). In the intercept rôle, the Lancer could, it is calculated, be expected to make an interception at 59,900 ft (18 257 m) nine minutes after brake release, and the mission radius in this case would be 138 naut miles (256 km), allowing for the use of the afterburner from take-off and all the time until combat, with two minutes at Mach 2·0 before the return to base. For an air superiority mission, providing cover for ground attack aircraft, the Lancer has a mission radius of 398 naut miles (738 km/h).

Ground attack is a secondary rôle for the Lancer, for which, as already noted, it can carry up to 12,000 lb (5 445 kg) of bombs or other weapons. On a typical mission of this kind, with four 1,000 lb (454 kg) Mk-83 bombs and a full load of gun ammunition, it would have a mission radius of 366 naut miles (678 km) flying at Mach 0·7 at sea level and allowance for five minutes at military rating over the target. At a distance of 200 naut miles (370 km) from base, it could loiter for one hour.

The configuration options for the Lancer have already been noted, including improved avionics systems, additional wing pylons, extra fuel, etc. Other possibilities are a Nord AS-30 missile installation; linkless feed ammunition for the 20-mm gun; Sparrow missile installation; and self-contained starter. A longer-term possibility is the development of a Mach 3·0 version of Lancer, using titanium skin structure, new inlets and an advanced engine such as the Pratt & Whitney F401-PW-400 under development for the Grumman F-14B.

Programme

In order to launch the Lancer, Lockheed requires a firm commitment from one or more nations which would ensure recovery of its development costs of $70·5m (£34m). The proposed programme provides for the construction of four CL-1200-2 prototypes plus a specimen for static structural testing. To produce these in minimum time and at lowest possible cost, five F-104G airframes would be used, or at least five sets of F-104G assemblies and components that are common to the CL-1200-2.

First flight of the first prototype would take place 12 months after go-ahead, and the last of the four within 18 months. Category I tests, covering handling and performance, would be conducted by Lockheed ADP, with the customer or consortium of customers conducting the Category II evaluation to obtain a fully qualified air-superiority fighter 24 months after go-ahead.

In a separate proposal, Lockheed has suggested that the USAF should support the construction of two prototypes of the CL-1200-2 to obtain, at low cost, high-performance research aircraft which would be capable of testing various advanced-technology high-performance engines and items of equipment. The proposal is being studied, and the designation X-27 has been assigned to the Lancer by the USAF for this purpose. Should construction of these prototypes go ahead, at least a portion of the development costs would then be covered, making the Lancer proposal look even more attractive to the potential customers in Europe and elsewhere.

With the Lancer, Lockheed is pursuing the philosophy of taking a good product and making it better. The widespread use of the F-104G, especially within NATO, makes the project of immediate interest and the opportunity for European and other nations to participate in the development and production of a modern high-performance warplane whilst at the same time making maximum use of past investments in tooling and plant facilities enhances that interest.

In addition, a number of potential markets outside the European aerospace manufacturing community can be identified and therefore a manufacturing consortium would stand a good chance of earning useful revenues from exports. For example, Lockheed has noted 20 countries that now operate jet fighters — in many cases of obsolescent types — and that in the past have spent or shown a willingness to spend their own resources to acquire jet fighters. These countries include Norway, Denmark, Italy, Greece, Turkey, Spain, Switzerland, Portugal, Saudi Arabia, Iran, Lebanon, Israel, Pakistan, Jordan, New Zealand, Australia, India, South Africa, Malaysia and Indonesia.

Modern aerodynamic design and provisions for future growth by incorporating more modern avionics equipment and more advanced engines as they become available, give the Lancer good potential for long operational life as a front-line fighter. This, coupled with economy both in first cost and continuation costs, provides the Lancer with excellent military capability within budget constraints. Certainly none of the Lancer's competitors can be launched at such low cost and this factor alone must give the project a good chance of going ahead. With several significant national fighter selections nearing completion, the first Lancer could well be in the air before the end of 1972. □

F-104G/LANCER COMPARISON

	F-104G	CL-1200-2
Span, ft & ins (m)	22 0 (6,71)	29 2 (8,89)
Length, ft & ins (m)	54 9 (16,69)	57 3 (17,45)
Height, ft & ins (m)	13 6 (4,11)	17 2 (5,23)
Wing area, sq ft (m²)	196.1 (18,21)	300 (27,87)
Powerplant	J79-GE-11A	TF30-P-100
Sea level static thrust, lb (kg)	15,800 (7 167)	25,000+(11 340+)
Zero fuel weight, lb (kg)	14,962 (6 787)	17,885 (8 112)
Clean take-off weight, lb (kg)	20,786 (9 428)	24,385 (11 061)
Max take-off weight, lb (kg)	28,800 (13 064)	*35,000 (15 875)

*Alternative high gross weight version.

J79-GE-11A/TF30-P-100 COMPARISON

	J79-GE-11A	TF30-P-100
Weight, lb (kg)	3,570 (1 619)	3,760 (1 705)
Length, in (m)	208 (5,28)	241.7 (6,13)
Diameter, in (m)	38.6 (0,98)	49.8 (1,26)
Turbine inlet temp, °F	1,690	2,300
Max a/b thrust, sea level static, lb (kg)	15,800 (7 167)	25,000+(11 340+)
SFC, lb/hr/lb (kg/hr/kg)	1.97 (1,97)	—
Military thrust, sea level static, lb (kg)	10,000 (4 536)	15,000+(6 804+)
SFC, lb/hr/lb (kg/hr/kg)	0·84 (0,84)	—
Cruise SFC at M=0·9, 35,000 ft (10 668 m)	1·06 (1,06)	—
Airflow, lb (kg)/sec	162 (73,5)	260 (118)
Max thrust/weight ratio	4·44	6·66+

F-104G/LANCER WEIGHT COMPARISON

	F-104G lb	F-104G (kg)	CL-1200-2 lb	CL-1200-2 (kg)
Wing (including ailerons and flaps)	1,183·1	(537)	2,420	(1 097)
Tail (including stabiliser and rudder)	614·8	(279)	880	(400)
Fuselage	2,906·7	(1 318)	3,135	(1 422)
Landing Gear	947·2	(430)	950	(431)
Flight Controls (including boosters)	753·8	(342)	860	(390)
Propulsion (including engine and controls)	4,678·9	(2 122)	5,905	(2 678)
Instruments/Navigation	234·2	(106)	105	(476)
Hydraulic System	250·7	(114)	265	(120)
Electrical System	447·1	(203)	400	(181)
Electronics	782·8	(355)	780	(354)
Armament	351·6	(159)	370	(168)
Furnishings/Equipment/Auxiliaries	396·2	(180)	400	(181)
Air Conditioning	224·9	(102)	225	(102)
Weight Empty	13,772	(6 247)	16,640	(7 548)
Crew	240	(109)	240	(109)
Operating items	43	(19,5)	43	(19,5)
Trapped fuel, oil	120	(54)	120	(54)
Gun/Ammo (725 Rounds)	787	(357)	787	(357)
Zero fuel weight	14,962	(6 787)	17,885 *18,175	(8 112) (8 244)
Internal Fuel	5,824	(2 642)	6,500 *8,500	(2 948) (3 855)
Clean take-off weight	20,786	(9 428)	24,385 *26,675	(11 061) (12 099)
Payload (Ordnance, External Fuel)	8,014	(3 635)	4,415 *8,325	(2 000) (3 776)
Max take-off gross weight	28,800	(13 064)	28,800 *35,000	(13 064) (15 875)

*High gross weight version.

Conflict over the Carpathians

Six months before World War II, the armed forces of Hungary and Slovakia met in a short, sharp engagement as the former Czechoslovak province sought to defend its frontiers. In the course of two days, the newly-created Slovak Air Force suffered severe losses, as related in this detailed account of the action by Juraj Rajninec and James V Sanders.

ON 23 March 1939, Hungarian troops crossed the eastern frontier of Slovakia, bringing to a climax a bitter, six-month, border dispute with their northern neighbour. This undeclared war between two little states in eastern Europe, while little more than a footnote in the chronicle of the tragic journey of the world into the holocaust of World War II, is not without interest. Not only does the study of this conflict provide insight into a turbulent period of history, but it is of

Avia B-534 fighters (aircraft of the 1st production series illustrated above and 2nd series below) of the newly-created Slovakian Air Force provided the principal aerial opposition to the Hungarian invasion of Carpatho-Ukraine. The Slovakian B-534s at the head of the page carry the markings adopted subsequent to the Slovakian-Hungarian hostilities.

particular interest to the student of aviation because of the dominant rôle played by the air arms of the combatant forces.

To provide a better understanding of the air battles that ensued from Hungary's warlike gesture of March 1939, it is necessary to return to 1 October 1938. The Munich Agreement, which in Neville Chamberlain's words was to assure "peace in our time", had just been signed. In accordance with that Agreement, German troops began the occupation of the regions of Czechoslovakia known as the Sudetenland. Abandoned by the Western Democracies, the Czechoslovak government's only hope was that part of the Munich Agreement which promised that Germany, Italy, France and England would guarantee the new borders once the Czechoslovak government reached agreement with its Polish and Hungarian minorities. However, Czechoslovakia's neighbours were placing a most liberal interpretation upon what these "agreements" were to be.

On the same day that German troops marched into the Sudetenland, the Polish government demanded 400 square miles of territory, including the cities of Moravska-Ostrava and Teschen. This region had been in dispute since 1920 when the Paris Peace Conference divided the old Duchy of Teschen between Czechoslovakia and Poland. The Prague government had to accede to the Polish demands, and the territory was handed over on 1 November. Hungary demanded the return of the provinces of Ruthenia and Slovakia, demands so excessive that Prague could not possibly acquiesce. On 8 October, representatives from Czechoslovakia and Hungary began negotiating the Hungarian demands.

Border incidents
Meanwhile destructive forces were at work at home. The German-backed nationalistic elements in both Slovakia and Ruthenia pushed for a greater degree of self-determination, and before October was out, both provinces had been granted autonomy by the harassed Prague government. The Republic of Czechoslovakia had become Czecho-Slovakia, hyphenated in both name and fact.

It was not long before the Hungarian government sought

to test the will of the Czechoslovak people to resist further territorial loss. On the morning of 5 October, an armed band of 500 men of the Hungarian organisation *Nyilas Kereszt* (Arrow Crosses) attacked the railway station at Borzava and killed a railway man. The invaders confidently pitched a tent camp in nearby woods but Czechoslovak troops reacted swiftly. They surrounded the Hungarian camp and with the help of three Letov S-328 bombers of the Czechoslovak Air Force (*Ceskoslovenské Letectvo*) from the airfield at Uzhorod, forced its surrender, the Hungarians losing 80 men dead and 400 captured.

Realising that this was only the opening move on the part of the Hungarians, the Prague government began to reinforce the air element at strategic Uzhorod with a further three S-328 reconnaissance bombers, three Marcel Bloch MB 200 bombers and twelve Avia B-534 fighters. Later in October, a Hungarian sabotage group succeeded in damaging the airfield at Uzhorod.

Unable to reach any kind of an agreement on new boundaries, Czecho-Slovakia and Hungary submitted the problem to arbitration by Italy and Germany. Meeting in Vienna, the German and Italian Foreign Ministers, Ribbentrop and Ciano, announced their decision on 2 November. Hungary received 4,200 square miles of territory and 1,060,000 people from the border regions of Slovakia and Ruthenia. Included in this so-called Vienna Award were the important cities of Kosice and Uzhorod.

Evacuation of the Ruthenian provincial government from their ceded capital at Uzhorod to the mountain village of Chust was aided by the Czecho-Slovak army. On 4 November, an Avia-built Fokker F IX (OK-AFG) of the Czecho-Slovak Airlines transported a load of government documents from Uzhorod to Prague, and on the same day a CSA Savoia-Marchetti SM 73 (OK-BAD) took three kilograms of radium from the hospital of Uzhorod to safety. All military aircraft were transferred from Uzhorod to the airfield at Spišská Nová Ves in the eastern part of Slovakia.

Conditions in the remaining part of Ruthenia now became extremely difficult for the central government. The loss of Uzhorod had effectively severed all rail transport into Ruthenia. The provincial government, encouraged by the German Foreign Office, continued to make trouble for Prague. The central government was still responsible for the defence of Ruthenia, and Czecho-Slovak troops were repeatedly in action against irregular marauders from both Hungary and Poland.

On 1 January 1939, the autonomous Ruthenian régime adopted a decree changing the name of the province to Carpatho-Ukraine. In this action they were strongly encouraged by the German Foreign Office which was planning to use this government as a rallying point for anti-Soviet Ukrainian sentiments.

March madness

The German Army had been disagreeably surprised by the strength of the Czech defences in the Sudetenland, and was worried that, were the Czechs to remain strong, they would be a threat to the German southern flank when the time came for operations against Poland. In addition, Germany reasoned that the forthcoming campaign would be much easier if Poland could be attacked from a third side by German troops from Slovakia. Therefore, it was decided that an independent Slovakia under German control would be much better than to have Slovakia ceded to Hungary. The Carpatho-Ukraine government proving too incompetent to be of much help against Russia, Germany decided that this province could be given to Hungary as partial compensation for not regaining Slovakia.

By mid-March, the Germans were ready for the final step in the dismemberment of the Czechoslovak Republic. On 14 March, both Slovakia and Carpatho-Ukraine, acting on cue from Berlin, declared their complete independence from the central government. On 15 March 1939, Hitler declared that the unrest in Czecho-Slovakia was a threat to German security, and sent his troops into Prague. The eastern provinces of Czecho-Slovakia were proclaimed to be the German protectorate of Bohemia and Moravia and the Republic of Czechoslovakia had ceased to exist.

Two days earlier, on 13 March, Hungary had learned that the Germans would not object to a Hungarian take-over of Carpatho-Ukraine. Therefore, the Carpatho-Ukrainian declaration of independence on 14 March was taken as the cue for a Hungarian invasion. Carpatho-Ukrainian irregulars, without support from either Prague or their friends in Berlin, were quickly routed, and on 16 March, Hungary formally annexed the territory. Hungary, however, was still not satisfied with her border with Slovakia. On 23 March, Hungarian troops began an advance from the River Uh into Slovakia.

For this action against Slovakia, the Royal Hungarian Air Force (*Magyar Királyi Légierö*) concentrated six-plane elements from each of three bomber squadrons at the Debrecen airfield. These squadrons were 3/3 *Sárga vihar* (Yellow Tempest), 3/4 *Vörös sárkány* (Red Dragon), and 3/5 *Hüvelyk Matyi* (Tom Thumb), and were equipped with German-built Junkers Ju 86K-2s powered by Manfred Weiss-built Gnôme-Rhône 14K Mistral-Major engines and armed with Hungarian Gebauer machine guns. The fighter

Fiat CR 32 fighters of two Royal Hungarian Air Force squadrons participated in the fighting between Slovakian and Hungarian forces, displaying an ascendancy over the opposing Avia B-534s.

force, comprising the 1/1 *Ricsi* and 1/2 *Ludas Matyi* squadrons from Börgönd, Veszprém, was deployed to Uzhorod and Miskolc, and was equipped with Italian-built Fiat CR 32s. Reconnaissance flights were made by Heinkel He 170s of the 1st Independent Long-range Reconnaissance Group from Kecskemét.

The Slovak Air Force (*Slovenské Vzdušné Zbrane*) had had only nine days to organise at the time of the Hungarian attack. After the declaration of Slovak independence on 14 March, all airmen of Czech nationality were obliged to leave the territory of the new Slovak State. The equipment of the Czecho-Slovak 3rd Air Regiment which had been stationed in Slovak territory now became the equipment of the new Slovak Air Force, which for the time being retained the same national insignia on its aircraft. This equipment consisted of Avia B-534 fighters and Letov S-328 reconnaissance bombers, the force being stationed at Spišská Nová Ves and at Sebastova near Presov.

On the morning of the Hungarian attack, three SAF S-328s from Spišská Nová Ves carried out a reconnaissance of the front lines and returned with information concerning the disposition of the advancing Hungarian columns. Shortly after noon, three Letov S-328s took off with orders to attack the Hungarian troops. The crews of these aircraft were: pilot Flt Lt Wagner, observer Flt Lt Simko; pilot Flt Lt Svento, observer Flt Sgt Szalatnay; and pilot Flt Lt Slodicka, observer Flt Cpl Kotran. After being joined by a fighter escort of three Avia B-534s, this group ran into heavy Hungarian anti-aircraft fire over the village of Sobrance. The lead S-328 was hit but was able to maintain formation. One of the B-534 escorts, also hit, was less fortunate. Its pilot, Flt Cpl Devan, attempted to land his badly damaged aircraft but crashed and was killed. The remaining SAF aircraft returned to base.

At 1500 hours, another group of three B-534s, led by Flt Off Hergott, left Spišská Nová Ves for the front lines. The

A Ju 86K-2 (actually belonging to the 3/1 "Istennyila" bomber squadron) and He 170As of the Royal Hungarian Air Force. The latter, belonging to the 1st Independent Long-range Reconnaisance Group, flew sorties during the Slovakian-Hungarian fighting.

wingman was Flt Lt Svetlik and the third machine was piloted by Flt Cpl Danihel. Near the village of Ulic, the group discovered a column of Hungarian tanks and trucks, and although immediately subjected to heavy ground fire, they went in for a low level attack. On the first pass, Flt Lt Svetlik's plane was hit and began to burn. He made a successful landing but, like Flt Cpl Devan a few hours earlier, perished in the burning wreckage of his aircraft. The remaining two B-534s attacked the Hungarian column and returned to base with all ammunition having been expended.

On this first day of combat, the SAF had lost two B-534s and their pilots to ground fire. The enemy air force had not been encountered, but the second day was to be different.

Avias versus Fiats

At half past five in the morning of the second day, six B-534s (piloted by Flt Off Hergott, Flt Lts Palenicek and Prhacek, and Flt Cpls Martis, Danihel and Zachar) took-off from Spišská Nová Ves. At almost the same time, nine Hungarian CR 32s of the *Ricsi* fighter sqdn were taking-off from the airfield at Uzhorod, led by Flt Lt Aladar Negro. The two formations met in the air near the village of Sobrance, and the ensuing battle did not go well for the Slovak Air Force.

Flt Lt Palenicek was shot down and killed. Flt Lt Prhacek was badly wounded, made a forced landing and then perished when the bombs on his aircraft exploded. After the B-534 piloted by Flt Cpl Zachar was hit by gunfire, he lost his orientation and crash-landed in Hungarian occupied territory, where he was taken prisoner*. At this point in the air battle the Hungarian fighters left the area, and the three remaining B-534s attacked a column of Hungarian tanks and trucks with bombs and machine guns. They were successful in demolishing a number of trucks before their ammunition was exhausted and they were forced to return to their base.

A few hours later, three S-328s took off from Spišská Nová Ves to bomb Hungarian tanks near the villages of Ladomirov and Sobrance. During a low level attack, one of the S-328s was hit by tank gunfire; the pilot, Flt Lt Ondris, was badly wounded, but his observer, Flt Lt Slodicka, was able to fly the damaged plane by reaching over the unconscious pilot. At this time, the S-328s were intercepted by nine CR 32s, but with the almost simultaneous appearance of three B-534s, the Hungarians turned away without giving combat.

The last air battle between the SAF and the HAF took place that same afternoon. At 1500 hours, three S-328s took off from Spišská Nová Ves on a reconnaissance mission. They were escorted by three B-534s piloted by Flt Off Hergott, Flt Sgt Hanovec and Flt Cpl Danihel. The Hungarian ground warning service reported the approach of these Slovak aircraft to headquarters at Uzhorod, and nine CR 32s, under the command of Flt Lt Bela Csemke, were sent to intercept.

The ensuing engagement is best described in the words of Flt Sgt Hanovec. "Our group of B-534s, flying at an altitude of 3,100 feet, is led by Flt Off Hergott. Our task is the escort of the three S-328s which are now flying some miles in front. It is half past three. I look down and the village of Pavlovce is under us. The fire of Hungarian anti-aircraft guns comes up to meet us. I attack with four bombs and with the gunfire of my four machine guns. The anti-aircraft defences are silent. I climb and look around . . . Fiats!

"Nine Hungarian fighters are attacking our S-328s. The aircraft of Flt Cpl Pazicky is hit, and I watch as his observer, Flt Lt Svento, parachutes from the burning plane. One of the Fiats approaches him and fires . . . fires . . . and kills my defenceless friend . . . it is dreadful! The other Avias are

---

This Avia B-534 was later repaired by the Hungarian Air Force, with which it served until 1943 with the serial number G.190, and it then survived another two years in civil use as HA-VAB.

Letov S-328s of the Slovakian Air Force flew bombing and reconaissance missions, operating primarily from Spišská Nová Ves in the Carpathian Mountains.

already fighting with a large number of Fiats. I look for the remaining S-328s. I see that they are flying back home and I am glad that they have escaped. At this time I am attacked by two Fiats. I climb quickly and escape into a cloud.

"When I emerge from the cloud, I see that the two Fiats are still waiting. I attack, but simultaneously I am attacked by four other Fiats. I turn my B-534 around and fly right towards one of my attackers. The Fiat is approaching . . . his silhouette increases quickly in my gunsight . . . 400 yards . . . 300 yards . . . I open fire . . . 250 yards . . . A collision seems imminent . . . at the last second the Hungarian pilot dives sharply with a stream of smoke pouring from his engine. Shot down! I see another Fiat go down burning, shot down by Danihel. My joy lasts but a moment, for another Fiat pumps a long burst into the engine of my Avia. I am hit and I must go down. Fortunately, I make a successful forced landing. On the ground, I meet Flt Off Hergott, who was shot down a few minutes before. Sadly we watch the remaining B-534, piloted by Cpl Danihel, as it turns towards home."

The air combat of the second day had finished with a very unfavourable score for the SAF. Five Avias and one Letov had been shot down with the loss of four crew members. In exchange, the Slovaks claimed two Fiats.*

The last attack
The Hungarians were not yet finished, however. From the captured Slovak pilot, they learned that the main SAF base was at Spišská Nová Ves. At 1300 hours on 24 March, the eighteen Ju 86 bombers of the three bomber squadrons, led by Col Elemer Kovacs, took off in two groups of nine from the airfield at Debrecen with orders to bomb the airfield at Spišská Nová Ves. An escort of nine CR 32s of the *Ludas Matyi* squadron, led by Capt Istvan Timar, were to meet the bombers over Miskolc. The pilots of the fighters were: Flt Offs Odon Turmezei, Joszef Juhaskiss, Adam Major and Antal Laszlo, and Flt Lts Zoltan Kiss, Laszlo Gyenes, Laszlo Pottondy and Antal Bankhidi.

As the bombers were approaching the Hungarian airfield at Miskolc, the bombardier of the Ju 86 navigated by Flt Cpl Tula, thinking that he was already over the target, released his bombs over his own airfield. The remaining Ju 86s and their escort continued, but not far from the target, the lead bombers became lost and had to turn back without bombing. The remaining nine bombers began the attack on Spišská Nová Ves at 1645 hours. They attacked in three waves and destroyed or damaged ten S-328s and B-534s and a single B-71 bomber (Avia-built SB-2). The damage would have been greater had many bombs not failed to explode; the airfield was very soft from recent rain. During the raid, Flt Lt Slodicka attempted to get his S-328 into the air, but it was hit by a bomb and his observer was killed. One B-534 did get airborne but was unable to intercept the attackers.

The Hungarian bombers returned directly to Debrecen while the fighters went on to attack the SAF airfield at Sebastova. They cast their bombs on to an empty airfield. The day before, all nine B-534 fighters normally stationed here had been sent to reinforce the airfield at Spišská Nová Ves.

Meanwhile, Hungarian ground forces had come to a halt after advancing some ten to twenty miles into Slovakia. The Hungarian government was well aware that on 23 March, Germany and Slovakia had signed an agreement in which the Germans promised to guarantee the Slovak borders. They apparently knew that the Germans would not feel obliged to worry about any territorial changes which took place before the agreement became effective on that day. On 31 March, a Hungarian-Slovak commission legalised the Hungarian occupation by granting 400 square miles (1 036 km²) on the west bank of the River Uh to Hungary.

The dismemberment of the Czechoslovak Republic was complete. Soon, the attention of the world would turn to other areas and Czechoslovakia would disappear under the darkness of Nazi terror. However, Czechoslovak airmen would make their way to the West where they would write stories of bravery in the skies above France and England — something which politics had prevented them from doing in defence of their own homeland except for two courageous, if largely ineffective, days of fighting in the air above the Carpathian Mountains. □

Hungarian sources give a different account of this engagement, claiming nine Slovak B-534s shot down without Hungarian loss. This figure may include the B-534s destroyed on the ground later on 24 March.

Concorde

The Evolution of an SST

NOTHING about the Concorde is commonplace. The most expensive aeroplane ever built in Europe, it has become the subject of bitter attacks by an organised, voluntarily-financed opposition group. It has attracted more publicity, good and bad, than any aerospace project other than the moon-shots — yet many believe the prospects of Concorde ever operating airline services are as remote as a passenger shuttle to the moon. Certain to be sold at a loss, it offers airlines the prospect of making a higher return on investment than possible with all-subsonic operation of a comparable network. It cruises ahead of its own sound waves yet paradoxically the noise it makes imposes severe restraints on its operation. But, white hope or white elephant, Concorde is coming, ready to enter service some 20 years and £900m after the first tentative pencil was put to paper to sketch a supersonic airliner.

Although the Concorde is in every sense a joint Anglo-French project and neither BAC nor Aérospatiale would claim that it is "their" design, it is impossible to describe the evolution of this aircraft without placing emphasis upon the work done in Britain in the late 'fifties and early 'sixties. In the earlier stages of this activity, virtually the whole of the aircraft industry was engaged in some way; gradually, as ideas were crystallised, the design studies of the Bristol Aircraft Limited became the focus of attention. By 1962, the Bristol studies and those under way in France by Sud Aviation, had individually reached almost identical conclusions about the size and shape of a possible supersonic transport, so that the merging of these two designs into the Concorde was relatively simple as well as being highly logical. Since that time, the identity of the individual inputs has been submerged in the joint effort.

Prelude

Serious studies of the possibilities of a supersonic airliner began in Britain in 1955, and are unlikely to have reached a more advanced stage any earlier in other countries. Although supersonic travel as a concept had been frequently discussed prior to that date, aerodynamic "state-of-the-art" was such that economic operation of an SST appeared beyond reach. During 1955, however, improvements in the lift/drag ratios that could be achieved at supersonic speeds began to indicate that operating costs of an SST could be brought down to a commercially feasible level.

To investigate the possibilities that were opening up, the government set up, through the then Ministry of Supply, a Supersonic Transport Aircraft Committee (STAC), the purpose of which was "to initiate and monitor a co-operative programme of aimed research designed to pave the way for a possible first generation of supersonic transport aircraft". This committee of 24, including representatives of nine airframe companies (Avro, Armstrong Whitworth, Bristol, de Havilland, English Electric, Fairey, Handley Page, Short Bros and Vickers) and four engine companies (Armstrong Siddeley, Bristol, de Havilland and Rolls-Royce), held its first meeting on 5 November 1956. No display of fireworks ensued, but for the next two years the committee directed a research programme that cost over £700,000, led to the production of 400 written papers and proved the feasibility of the SST concept.

Throughout the period of the STAC-directed research, the airframe companies conducted studies of a variety of different configurations, for two categories of SST. One was to be a medium range type (about 1,500 mls — 2 414 km) cruising at about Mach 1·3 and the other would be a longer range aeroplane (3,450 mls — 5 552 km) cruising at Mach 1·8–2·0, which was regarded as the limit for aluminium alloy structures. For the slower aeroplane, much interest was shown in the M-wing planform, this being essentially a variation of the classical swept-back wing, while the studies for the higher speed tended to be based on a delta wing.

In this period, evolution of the delta planform from a simple "triangle" to the slender delta eventually adopted for the Concorde represented something of a breakthrough in aerodynamics. The simple delta had unsatisfactory characteristics at low speed, but this difficulty was overcome by using curved "streamwise" tips, and this in turn led to the Gothic wing planform when the entire leading edge was given a convex curve. Further refinement produced the ogee wing, in which the leading edge curve had a point of inflection.

Another problem affecting the wing shape was that of trimming. It is a characteristic of the slender delta wing that the aerodynamic centre of lift moves aft as speed increases.

Consequently, the aircraft has to be re-trimmed after take-off, and since any aerodynamic means of achieving this (such as the use of trailing edge flaps) would increase drag, the idea of pumping fuel backwards or forwards between appropriately located fuel tanks was suggested — and this is now a fundamental feature of Concorde operation.

Some enthusiasm was shown during the early stages of the STAC investigation for so-called integrated designs — that is to say, using a thick-section wing in which the fuselage diameter could be wholly or partially submerged. The disadvantages of this arrangement, including lack of conventional windows for the passenger cabin, were eventually found to outweigh the advantages.

As an extension of the wind tunnel research into slender wings, in which over 300 models were tested, the Ministry of Supply subsequently sponsored the production of two research aircraft. One (Specification X.197) was for low-speed research and appeared as the Handley Page HP.115; the other (Specification ER.193D) was for high and low speed investigation, this being the Bristol (BAC) 221 conversion of the Fairey FD.2. Both these aeroplanes played a significant rôle in defining the aerodynamics on which the Concorde was eventually to be based.

The STAC, under the chairmanship of Morien (now Sir Morien) Morgan, made its formal report to the Ministry of Supply on 9 March 1959. The report summarised the work that had been done, illustrated three aircraft configurations that were typical of the assessment studies undertaken, and recommended that "serious detailed design work" should be started on two supersonic aircraft, the 100-seat medium range type and 150-seat long range variant. Broad specifications were given for each of these types, the committee suggesting that the target date for airline service of the long-range type should be 1971–2, and that the shorter-range aircraft might be available two or three years earlier. Noting that development costs were "difficult to estimate" the committee nevertheless indicated a figure of £95m as the highest likely cost to build six prototypes and complete certification.

Variations on a theme

One of the three aircraft configurations shown in the STAC report — selected from many dozens of paper studies — was identified as the Bristol Type 198 and showed an eight-engined slender-delta with a high-mounted ogee wing and a foreplane. This was not, in fact, the first layout to bear the Type 198 label, nor was it the last, for this type number was used by the Bristol company to cover the range of layouts studied for STAC and subsequent feasibility studies until 1961.

The earliest Type 198 drawings, dated February 1957 and reproduced in these pages for the first time anywhere, show

(Below) One of the earliest project studies in the Bristol 198 series was this medium-range SST with a modified M-wing, investigated as part of the work done for the Supersonic Transport Aircraft Committee during 1957.

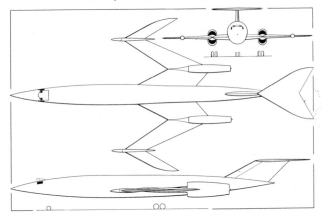

a modified M-wing layout, with vertical pairs of engines on the wing and a slender, streamlined fuselage. This was typical of the kind of study made for STAC during 1957, but the Bristol design team led by Dr A E Russell (the company's representative on the STAC) spent relatively little time investigating the medium range Mach 1·3 SST, and the Type 198 of October 1958 (which was the version illustrated in the STAC report) was a much more detailed study of a potential North Atlantic Mach 2 design. By this time, Bristol was clearly aware of the merits of the slender delta-wing, both the ogee and Gothic type being studied in parallel; the company was, however, unimpressed by the arguments for an integrated design, preferring at this stage a "unified wing-body" arrangement.

The Type 198 of October 1958 was a 314,000 lb (142 430 kg) aeroplane with a span of 80 ft (24,38 m) and eight engines of 15,400 lb st (6 985 kgp) each. It could seat 120 passengers and the maximum payload of 26,400 lb (11 975 kg) could be carried for a distance of 3,450 miles (5 552 km). Use of a blown flap on the foreplane (with 10 per cent of the wing area) was required to provide a nose-up moment when the trailing edge flaps/ailerons were selected down for landing; the foreplane flap was also to be used, unblown, for pitch control in cruising flight. The use of retractable foreplanes was also considered at this time — more than 10 years before Dassault adopted a similar arrangement on the Milan (see AIR ENTHUSIAST, June 1971).

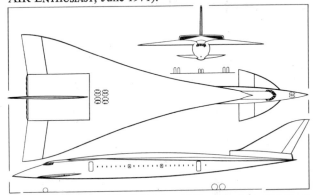

(Above) The Type 198 of October 1958, with eight engines, a high-mounted slender delta wing and a two-piece foreplane of Gothic planform. (Below) Another Type 198 configuration studied in 1959, with six engines, a Gothic wing and no foreplane.

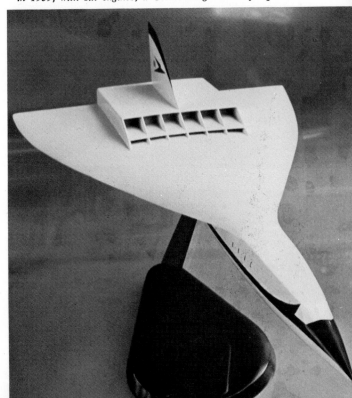

While the STAC report was receiving official consideration, the Ministry of Supply placed further contracts to allow specific work on some aspects of the studies to continue. One such contract, with Bristol Aircraft, was to compare the structural design of a light alloy aircraft for Mach 1·8 with that of a Mach 3·0 steel aircraft (the Bristol 213). The official view during 1959 hardened against a Mach 1·3 medium range aircraft, and in September, Hawker Siddeley and Bristol were asked to undertake a joint feasibility study to establish the most suitable type of long-range SST, having regard to economics, time-scale and costs. Essentially, this study was a comparison between the Avro 735, which had an integrated wing/fuselage design, and the Bristol 198 with its discrete wing and fuselage.

By January 1960, the Bristol 198 had changed considerably, as a result of these studies. The company had become steadily more convinced that a form of ogee wing offered the best prospects for economic operation and reasonable low speed handling, but by this time had made the Type 198 a *low* wing aeroplane, with no foreplane and with six Olympus 591 engines grouped three-a-side in nacelles under each wing. This was the first version of the Bristol 198 in which the ultimate lines of the Concorde could be seen.

The outcome of the feasibility study was a clear victory for this layout and in October 1960 the British Aircraft Corporation (which had by then merged the aviation interests of Bristol, Vickers and English Electric) received a design study contract valued at £350,000 for work to continue.

The submission made under this contract in August 1961 showed the Type 198 to be substantially unchanged from the form it had taken in January 1960. It was to be powered by six Bristol Siddeley Olympus 593/3 engines of 26,700 lb st (12 110 kgp) each, and would carry 136 passengers at 33-in (84-cm) seat pitch. The gross weight was 385,000 lb (174 633 kg), payload 33,000 lb (14 970 kg), and the range with this payload, 3,750 miles (6 035 km). Wingspan was 84 ft (25,60 m) and gross area, 5,000 sq ft (464,5 m²).

Details of the Mach 3 Bristol 213 were submitted to the Ministry together with the Type 198 but on grounds of cost and timescale the steel and titanium aircraft was less attractive than the Mach 2·2 type. However, during 1961 doubts also began to grow about the sonic boom problem

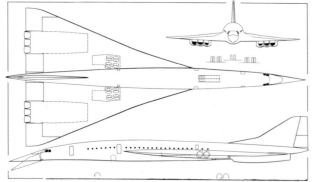

(Above and below) The final configuration of the Bristol Type 198 featured a low-mounted narrow-delta wing and six engines in two underwing groups.

that might be encountered with an aeroplane as large and heavy as the Type 198, and an alternative project was prepared in parallel as the Type 223.

Retaining the same overall layout as the 198, the Type 223 was scaled to carry about 110 passengers between London and New York at Mach 2·2. A major saving in weight was achieved by switching to a four-engined layout (still with Olympus 593/3s), the weight being only 260,000 lb (117 935 kg) with a payload of 23,000 lb (10 432 kg). The span was 70 ft 4 in (21,44 m). By the end of 1961, the Type 198 had been abandoned and all effort was concentrated on the Type 223.

Counterpoint

The possibility of Anglo-French collaboration on an SST was first mooted officially early in 1960, following a visit to France by Duncan Sandys, the then Minister of Aviation. French design activity had, in fact, been under way since the beginning of 1957, the government having issued an outline requirement to industry based on an Air France specification for a medium-range SST carrying 60–80 passengers for about 2,000 miles (3 210 km). Nord, Sud and Dassault made individual studies but early in 1960, Sud and Dassault combined their resources to make a common submission later in the year and Nord dropped out of the competition. Evolution of the Sud design passed through similar phases to that of the British 198. At one stage a canard arrangement was considered, with a plain delta-wing, but this was dropped and the wing leading edges were extended forwards down the fuselage sides instead. A three-engined arrangement was also studied.

At the Paris Air Show in May 1961, Sud unexpectedly displayed a model of its design, called at that stage the Super Caravelle, and although the secrecy that still surrounded the British work at the time prevented most observers from making the comparison, this model showed a dramatic similarity to the British Type 223. Just how close the two designs were is shown by the three-view drawings published here; both types had the same four Olympus engines, similarly located; a similar ogee wing and virtually identical dimensions!

The point was not lost on government officials of the two countries and the advantages of collaboration were again noted when Peter Thorneycroft, as Minister of Aviation, met Robert Buron, the French Transport Minister, on 2 October 1961. The immediate outcome of this meeting was that the respective ministries wrote to BAC and Sud Aviation on 9 October calling upon them to embark upon a collaborative effort. Responding to this request, senior executives of the two companies met in London on 13 October 1961 to take the first steps to combine the projects into a single entity.

The largest point of divergence at this time was in the design range of the aircraft. Although some work had continued at Bristol on a medium-range SST, there was little British enthusiasm for this version; equally, Sud had studied long-range aircraft but had optimised its design for the shorter ranges. As an initial basis for collaboration, it was agreed to proceed with a common design in medium and long-range versions, and this proposal was incorporated in an agreement exchanged between BAC and Sud-Aviation on 25 October 1962. Government approval and sanction for the provisions of this agreement followed on 29 November when an appropriate protocol between the two governments was signed in London by G de Courcel (French Ambassador to the Court of St James), Julian Amery and Peter Thomas. The Concorde was born.

Subsequent evolution of the design up to the prototype standard is well documented, and need only be summarised here. Work proceeded on the two versions until May 1964, when a new specification was drawn up for the long-range aircraft based on uprated Olympus engines, and the medium-haul project was then dropped. The 32,430 lb (14 700 kg) rating of the new Olympus engines made it possible to enlarge the design overall, from a span of 78 ft (23,78 m) to 83 ft 10 in (25,50 m) and from a gross weight of 262,350 lb (119 000 kg) to 326,000 lb (148,000 kg). The basic objective of the specification, to carry a payload of at least 20,000 lb (9 072 kg) between Paris and New York with 85% regularity against the "worst" headwinds, remained unchanged.

In this form, construction of the prototypes proceeded, for flight testing to begin during 1969. Concorde 001 and 002 have now explored the flight envelope in sufficient detail for peformance guarantees to be drawn up, and they are continuing to build up flight hours at Mach 2. A complete record of all Concorde flights made up to press-time is included with this feature.

Mounting tempo

During the time that the prototypes were under construction, further changes have been progressively introduced in the specification of the production model. These changes are being made in two stages, some being incorporated in the first pre-production Concorde, 01, with others introduced on the second pre-production model, 02, to make it aerodynamically representative of the production type.

Concorde 01 was rolled out at Filton on 31 March 1971 and after undergoing resonance testing it is now being completed in preparation for a first flight before the end of September. The principal external change, compared with the prototypes, is that the fuselage is lengthened by 8 ft 6 in (2,59 m) and a revised nose visor is fitted. The fuselage stretch, which is in the form of a constant-diameter section forward of the wing, is combined with an internal change to extend the usable length of the cabin by moving the rear bulkhead farther aft and allows for 128 passengers to be carried in a one-class arrangement at a seat pitch of 34 inches (86,4 cm). The rear freight hold now occupies the full width of the fuselage behind the pressure bulkhead, and there is a new freight loading door in the starboard side, aft of the wing trailing edge. An additional passenger door is provided, over the wing leading edge, but in most cases, and especially if the aircraft is operated in a one-class layout, only the forward door is likely to be used. The size of the cabin windows has been slightly reduced to meet FAA requirements.

Changes in the visor arrangement spring from the views of customer airline pilots, who were unhappy with the severely restricted view at supersonic speeds with the visor up on the prototypes. In place of the two-piece visor and periscope, 01

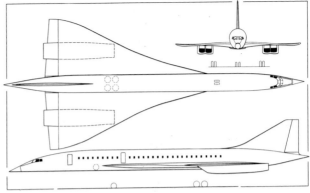

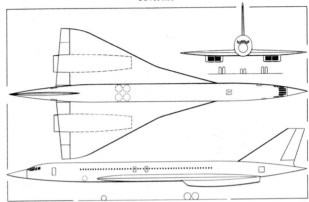

(Above) The Bristol 223 in 1961 and (below) the Sud Super Caravelle of about the same date. The striking similarity of these two independent designs, which were merged to become Concorde, is obvious.

has a one-piece transparent visor and there is a small "step" for the windscreen even when the visor is up, so that forward-vision remains possible.

A small change in the geometry of the wing-tips has been introduced to give an improvement in lift, this being the result of continuing wind tunnel development. Rolls-Royce Olympus 593-4 engines of 36,800 lb st (16 690 kgp) are to be fitted on Concorde 01, compared with the 34,730 lb st (15 753 kgp) Olympus 593-3Bs at present flying in 001 and 002. At a later stage, an Olympus 593-6, rated at 38,050 lb st (17 260 kgp) and representative of the initial production stage, will replace one of the -4 engines.

Among the internal changes made on Concorde 01 (and subsequent) are an increase in the capacity of the air conditioning system, with four instead of three generation systems, and an increase in the capacity of the electric

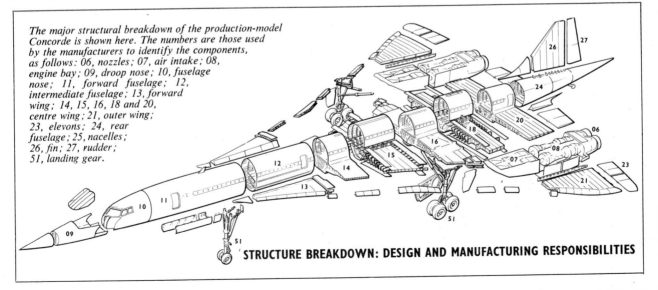

The major structural breakdown of the production-model Concorde is shown here. The numbers are those used by the manufacturers to identify the components, as follows: 06, nozzles; 07, air intake; 08, engine bay; 09, droop nose; 10, fuselage nose; 11, forward fuselage; 12, intermediate fuselage; 13, forward wing; 14, 15, 16, 18 and 20, centre wing; 21, outer wing; 23, elevons; 24, rear fuselage; 25, nacelles; 26, fin; 27, rudder; 51, landing gear.

STRUCTURE BREAKDOWN: DESIGN AND MANUFACTURING RESPONSIBILITIES

system, from 40 Kva to 60 Kva, plus a hydraulically-driven 30 Kva generator as emergency standby. Fittings for air starting are now on each engine in place of the two GTS units which, on the prototype, are used to start the engines and provide an emergency power source. Changes have been made in the arrangement of the hydraulic systems to provide a better division of the three systems between the four engine-driven pumps.

Centre cells have been added for the No 9 and No 10 fuel tanks, and the rear tank has been enlarged, giving Concorde 01 an increased fuel capacity. Larger main wheels are fitted and there is some structural beefing-up of the undercarriage because of the greater weights involved. Maximum droop angle of the movable nose section is now 17½° compared with 12½° to which the prototypes have been limited in practice.

Concorde 02, which is now expected to make its first flight in mid-1972, will have the new features of 01 noted above, plus additional changes making it externally similar to the planned production standard. The most obvious change will be the new fuselage tail cone, extending well aft of the rudder and adding a further 11 ft 4 in (3,45 m) to the overall length. This has been adopted primarily to reduce the drag, although it does also allow a further increase in fuel tankage, and also provides a convenient stowage for the monofuel emergency power unit (MEPU) which takes the place of the two GTSs used on the prototypes for emergency hydraulic and electric power.

A further wing change is introduced on 02, which has

CONCORDE TEST FLIGHTS, MARCH 1969 TO SEPTEMBER 1970

DATE	FLT No	hr	min	REMARKS
2- 3-69	1/1		42	First flight 001, Toulouse. 10,000 ft, 250 kt
8- 3-69	1/2	1	17	U/c retracted. 15,000 ft, 300 kt
13- 3-69	1/3	1	12	Systems tests and cruise performance
17- 3-69	1/4	1	20	No 1 engine cut and relit. 20,000 ft, 330 kt
21- 3-69	1/5	1	33	Engines cut and relit. 350 kt
28- 3-69	1/6	1	38	26,000 ft, Mach 0·75
30- 3-69	1/7	1	25	Performance measurement and handling
30- 3-69	1/8	1	16	Afterburning switched on in flight
2- 4-69	1/9	1	41	30,000 ft, 350 kt, Mach 0·80
9- 4-69	2/1		42	First flight 002, Filton-Fairford. 8,000 ft, 280 kt
16- 4-69	2/2	1	12	9,000 ft
28- 4-69	2/3	1	04	15,000 ft, 350 kt. Systems testing
8- 5-69	2/4	1	22	Flying qualities investigation. 25,000 ft, 350 kt
13- 5-69	2/5	1	30	Engine relights; simulated 2-engine approach
23- 5-69	1/10	1	39	Nose operation to 17.5 deg.
25- 5-69	1/11	1	57	Visor functioned
27- 5-69	1/12		40	Rehearsal for Paris Air Show
28- 5-69	1/13	1	39	Check flight for CDV approval
29- 5-69	1/14	1	32	Toulouse to Le Bourget, Paris
1- 6-69	1/15		41	Demonstration at Paris Air Show
1- 6-69	2/6	2	10	Systems testing. 30,000 ft, Mach 0·80
3- 6-69	2/7	2	02	Assessment by ARB pilot
4- 6-69	2/8	2	00	Mechanical control mode in approach
5- 6-69	1/16		31	Demonstration at Paris Air Show
7- 6-69	1/17		31	Demonstration at Paris Air Show
7- 6-69	2/9	2	15	Demonstration at Paris Air Show, from Fairford
8- 6-69	1/18		31	Demonstration at Paris Air Show
8- 6-69	2/10	2	00	Demonstration at Paris Air Show, from Fairford
9- 6-69	1/19	1	10	Demonstration at Paris Air Show
10- 6-69	1/20	1	25	Le Bourget to Toulouse
10- 6-69	2/11	1	35	Visor up and down
13- 6-69	1/21	2	10	First flutter testing flight
14- 6-69	2/12	2	05	Buckingham Palace fly-over
19- 6-69	1/22	1	55	Afterburner lit at 36,000 ft, Mach 0·80
20- 6-69	1/23	2	04	Flutter tests
22- 6-69	1/24	2	03	Flutter tests and Toulouse Air Show
25- 6-69	1/25	1	59	Flutter tests. 38,000 ft
1- 7-69	1/26	1	48	Flutter tests. 380 kt
4- 7-69	1/27	1	55	Flutter tests. Mach 0·85
9- 7-69	1/28	2	02	Flutter tests. Mach 0·90
10- 7-69	1/29	2	00	40,000 ft, Mach 0·90
12- 7-69	1/30	1	46	Flutter tests. Mach 0·8, 410 kt
16- 7-69	1/31	1	57	Flutter tests. Mach 0·94, 330 kt
18- 7-69	1/32	1	49	Flutter tests. Airfield noise measurements
18- 7-69	2/13	1	45	"Barn-door" auxiliary air intakes fitted.
18- 7-69	1/33	1	23	Continuation of 1/32
19- 7-69	2/14	1	30	Tests of new intakes
22- 7-69	1/34	1	58	Flutter tests
22- 7-69	2/15	1	50	Handling at aft CG
25- 7-69	1/35	2	06	Flutter, aft CG tests. Mach 0·95
25- 7-69	2/16	2	07	Continuation of 2/14 and 2/15
26- 7-69	2/17	1	44	Engine handling
28- 7-69	1/36	2	12	Air data calibrations, various configurations
30- 7-69	2/18	2	00	Handling at aft CG
30- 7-69	2/19	1	40	PE runs, various configurations
31- 7-69	1/37	2	03	Flutter and handling tests
31- 7-69	2/20	1	57	Handling and performance at aft CG
1- 8-69	1/38	2	00	Handling from Mach 0·90 to 0·94
2- 8-69	2/21	1	45	Handling at aft CG
6- 8-69	1/39	1	57	Handling by CEV crew
6- 8-69	2/22	2	05	PE determination at high altitude
7- 8-69	2/23	1	46	Take-off and approach noise measurement
7- 8-69	2/24	2	01	Performance and handling
21- 9-69	1/40	2	08	"Barn-door" intakes fitted. Mach 0·95
23- 9-69	1/41	2	19	Handling to Mach 0·97
24- 9-69	1/42	2	18	Flutter tests, rear fuel tank full
27- 9-69	1/43	2	14	New intake door checks
28- 9-69	1/44	2	12	Flutter tests and turn manoeuvres to Mach 0·97
1-10-69	1/45	1	51	First supersonic, 9 min at 36,000 ft, Mach 1·05
3-10-69	1/46	2	03	Flutter and afterburner tests. 43,000 ft
7-10-69	1/47	1	52	Tests at Mach 1·05/380 kt
8-10-69	1/48	2	13	Continuation of 1/47
9-10-69	1/49	2	07	Handling without artificial feel
13-10-69	1/50	2	02	Flutter tests, Mach 1·05
14-10-69	1/51	1	43	Mach 1·15/350 kt/43,000 ft for 40 minutes
21-10-69	1/52	1	53	No 2 engine replaced. Tests at Mach 1·25/360 kt
23-10-69	1/53	2	00	Flutter and handling to Mach 1·15
24-10-69	1/54	1	00	Flight aborted due to recorder failure
25-10-69	1/55	2	11	Preparation for airline pilots' handling flights
27-10-69	1/56	2	16	Flutter and handling to Mach 1·25
6-11-69	1/57	1	58	Nos 3 and 4 engines replaced. Mach 1·3/380 kt
8-11-69	1/58	2	05	PAA evaluation flight
8-11-69	1/59	1	59	BOAC evaluation flight
10-11-69	1/60	1	54	TWA evaluation flight
10-11-69	1/61	1	51	Air France evaluation flight
12-11-69	1/62	1	54	Flying qualities. First night landing
14-11-69	1/63	2	26	Structural and handling tests
17-11-69	1/64	2	06	Reheat relights at Mach 1·15/35,000 ft
20-11-69	1/65	2	08	TO weight up to 309,000 lb. Descent on two engines
21-11-69	1/66	1	58	Mach 1·40/410 kt
24-11-69	1/67	1	53	Structural and engine tests
26-11-69	1/68	1	54	Mach 1·35/470 kt
28-11-69	1/69	2	12	Structural and performance tests
29-11-69	1/70	1	34	Navigation flight at Mach 1·35
4-12-69	1/71	1	50	First SS engine relight. Mach 1·46/370 kt
10-12-69	1/72	1	35	Performance and de-icing tests
11-12-69	1/73	1	45	Sonic boom measurements
12-12-69	1/74	1	38	Structural and engine tests
13-12-69	1/75	2	10	Mach 1·50/470 kt, 47,000 ft
17-12-69	1/76	1	45	Mach 1·50/500 kt
18-12-69	1/77	2	04	Engine shut-down at Mach 1·5
8- 1-70	1/78	2	25	Reverse thrust used in flight, for rapid descent
10- 1-70	1/79	2	08	Further tests with reverse thrust in flight
12- 1-70	1/80	2	11	Air data calibration
14- 1-70	1/81	2	37	Further tests with reverse thrust in flight
15- 1-70	1/82	2	24	Performance tests
17- 1-70	1/83	2	26	Performance, handling, reverse thrust
20- 1-70	1/84	2	29	Engine cut at VR
21- 1-70	1/85	2	02	Evaluation by CEV crew
22- 1-70	1/86	2	41	Continuation of 1/85
23- 1-70	1/87	2	11	Evaluation by ARB crew
27- 1-70	1/88	2	09	Continuously supersonic for 66 min
29- 1-70	1/89	1	45	Simulated engine cut at VR
29- 1-70	1/90	2	30	Performance tests
31- 1-70	1/91	2	23	Pitch and yaw tests at ss speed
21- 3-70	2/25	1	20	Variable geometry air intakes fitted. Olympus 593-2B
25- 3-70	2/26	2	02	First supersonic by 002
2- 4-70	2/27	2	16	Systems testing
7- 4-70	2/28	2	28	Handling and flutter testing
9- 4-70	2/29	2	07	Continuation of 2/28
10- 4-70	2/30	2	06	Continuation of 2/29
12- 8-70	2/31	1	46	Olympus 593-3B fitted
14- 8-70	2/32	2	03	General handling
17- 8-70	2/33	2	07	Performance and flutter
22- 8-70	2/34	2	00	TO weight 315,040 lb. Intake checks. Flutter and performance.
25- 8-70	2/35	2	06	TO weight 319,450 lb. Intake checks. Flutter
26- 8-70	2/36	1	23	Aborted through instrument defect. Fuel jettisoned
29- 8-70	2/37	2	16	Performance and flutter tests.
1- 9-70	2/38	2	25	TO weight 321,435 lb. Mach 1·68, 43,500 ft
5- 9-70	2/39	1	59	TO weight 322,980 lb. Performance and handling

modified outer leading edges resulting in a small increase in area. Twin tail wheels (retractable) will be fitted in place of the tail skid. Olympus 593-6 engines are expected to be installed from the start, with the new SNECMA-developed TRA (thrust-reverser aft) nozzles. The shape of the engine nacelles will be changed in detail to give more accessibility round the engines and also to improve the aerodynamics. Considerable development of the flight-deck layout has also occurred since the prototypes were designed and 02 will conform to the full airline standard as now planned.

Production Concordes — the first of which is scheduled to fly at the end of 1972 — will be outwardly similar to 02 as described above. Structurally they will differ in that greater use is made of machined panels for the fuselage skins, and

This model of Concorde, displayed at the Paris Air Show, clearly illustrates the revised arrangement of transparencies in the droop nose of the production version.

CONCORDE TEST FLIGHTS, SEPTEMBER 1970 TO JUNE 1971

DATE	FLT No	BLOCK TIME hr min	REMARKS	DATE	FLT No	BLOCK TIME hr min	REMARKS
7- 9-70	2/40	1 16	Farnborough Air Show fly-over	19-12-70	2/66	2 45	Performance, take-off and landing tests, PE flights
8- 9-70	2/41	1 40	Farnborough Air Show fly-over	22-12-70	2/67	2 53	Subsonic performance and handling
11- 9-70	2/42	2 08	Farnborough Air Show fly-over	23-12-70	1/119	2 42	No 2 engine replaced
12- 9-70	2/43	1 23	Farnborough Air Show fly-over	23-12-70	2/68	1 15	Flight curtailed, pressurisation system malfunction
13- 9-70	2/44	46	Farnborough Air Show fly-over; Heathrow diversion	23-12-70	1/120	3 05	Supersonic performance. Night flight.
14- 9-70	2/45	1 20	Heathrow take-off. Return to Fairford	1- 1-71	2/69	2 31	Performance, PEs at Aberporth. 57,700 ft
18- 9-70	1/92	2 00	Variable geometry intakes and Olympus 593-3B	2- 1-71	2/70	2 01	Performance and structural test
24- 9-70	1/93	2 11	Aircraft and engine handling. SS for 75 min	5- 1-71	2/71	3 13	Subsonic performance
28- 9-70	2/46	2 02	No 1 engine changed after washer disintegrated	9- 1-71	2/72	2 05	Supersonic performance
30- 9-70	2/47	2 05	Engine slam tests	11- 1-71	2/73	2 42	Performance levels
2-10-70	1/94	2 24	Performance measurement	23- 1-71	1/121	2 33	Nacelle sealing for drag reduction
2-10-70	2/48	2 14	TO weight 328,490 lb. Mach 1·78, 44,000 ft	26- 1-71	1/122	2 32	Intake ramp drive failure in No 4 engine
7-10-70	2/49	2 19	Performance measurement. TO weight 332,456 lb	3- 4-71	2/74	2 15	Modified intake ramp drive mechanism
8-10-70	1/95	2 29	Handling and intake tests at Mach 1·7	6- 4-71	2/75	2 30	Engine checks
8-10-70	2/50	2 40	Pitot-static calibration and performance tests	8- 4-71	2/76	1 27	Aborted due to faulty instrument indication
10-10-70	2/51	2 35	TO and landing performance, handling tests. Mach 1·83	14- 4-71	2/77	1 10	Aborted due to hydraulic system fault indication
12-10-70	2/52	2 22	Mach 1·8. 45,000 ft	15- 4-71	1/123	3 14	Modified intake ramp drive mechanism
16-10-70	1/96	2 34	Sustained Mach 1·8 for 1 hr. 50,000 ft	17- 4-71	1/124	3 24	Longest flight to date
19-10-70	1/97	2 23	Handling and performance	19- 4-71	2/78	2 12	Engine checks and performance
21-10-70	1/98	2 31	Mach 1·86. 52,000 ft	19- 4-71	1/125	58	Pitch auto-stabiliser optimisation
23-10-70	1/99	2 17	Speed calibration and level performance	21- 4-71	1/126	1 08	Auto-stabiliser optimisation tests
26-10-70	2/53	2 01	Performance, engine and flutter tests	22- 4-71	1/127	1 01	As 1/126
28-10-70	2/54	2 30	Mach 1·89, 46,650 ft	22- 4-71	2/79	2 10	Performance and engine checks
29-10-70	1/100	2 38	Performance and handling to Mach 1·90, 48,000 ft	24- 4-71	2/80	2 12	Surge test on new ramp
2-11-70	1/101	2 29	Handling and engine tests, to Mach 1·93	26- 4-71	1/128	2 39	Noise measurements, take-off and approach
4-11-70	2/55	1 22	Mach 2 attempt curtailed, No 2 engine overheating	26- 4-71	1/129	2 44	As 1/128
4-11-70	1/102	2 16	First Mach 2, sustained for 53 min, 50,200 ft	27- 4-71	1/130	2 26	Air intake and engine tests
6-11-70	1/103	2 38	Mach 2 test, including No 4 engine cut	28- 4-71	2/81	2 32	Engine surge tests
9-11-70	2/56	1 14	No 4 engine shut down after oil loss	29- 4-71	1/131	2 24	Performance tests
9-11-70	1/104	2 43	TO weight 336,200 lb. Two engines cut at Mach 2	30- 4-71	2/82	2 29	Deliberate engine surges
12-11-70	2/57	2 07	First Mach 2 by 002, sustained for 42 min	3- 5-71	1/132	1 37	Checks for Flight 1/134
				4- 5-71	2/83	2 24	As 2/82
12-11-70	1/105	2 26	Structural resonance tests. Mach 2·05/530 kt	6- 5-71	1/133	1 18	Ferry, Toulouse to Le Bourget
19-11-70	2/58	2 02	Performance and engine handling	7- 5-71	1/134	1 37	Demonstration, with President Pompidou
26-11-70	1/106	2 57	Performance, handling, engine tests. 53,000 ft	7- 5-71	2/84	1 32	Ferry, Fairford to Toulouse
27-11-70	1/107	2 59	Longest flight to date. Performance tests	7- 5-71	2/85	45	Aborted due u/c door failure
30-11-70	1/108	2 54	Simulated all-engine failure at Mach 1·95	8- 5-71	2/86	1 20	Ferry, Toulouse to Fairford
1-12-70	2/59	2 26	Engine handling tests	11- 5-71	1/135	2 36	Performance measurements
2-12-70	1/109	2 57	Performance and simulated emergency descent	13- 5-71	1/136	2 43	Performance and handling
2-12-70	2/60	3 04	Longest flight to this date, all subsonic	14- 5-71	1/137	2 33	Performance, two automatic landings
4-12-70	1/110	2 51	Performance and structural tests. 56,000 ft	15- 5-71	1/138	2 27	Performance and Paris Show rehearsal
4-12-70	2/61	1 55	Fuel jettisoned after cabin instruments overheated	17- 5-71	1/139	1 34	Structural tests
4-12-70	2/62	2 28	Performance and engine tests	22- 5-71	2/87	2 54	New No 1 engine with Mk 602 LP compressor
7-12-70	2/63	2 41	Performance and engine tests	23- 5-71	1/140	2 40	Performance and handling
8-12-70	1/111	3 02	Subsonic flight to Bretigny	23- 5-71	1/141	25	Take-off with anti-smoke additive
8-12-70	1/112	2 38	Supersonic performance, return to Toulouse	24- 5-71	2/88	2 25	Sundry measurements
10-12-70	1/113	2 46	Performance and air-to-air photography	25- 5-71	2/89	1 43	Sundry tasks; abandoned due fire warning defect
12-12-70	1/114	53	Subsonic flight to Bretigny	25- 5-71	1/142	2 34	First intercontinental, Toulouse-Dakar
12-12-70	1/115	2 41	Performance and return to Toulouse	26- 5-71	1/143	2 52	Return from Dakar, 2 hr 7 mins SS
12-12-70	2/64	2 18	No 2 engine replaced after bevel gear disintegrated	26- 5-71*	2/90	3 23	Radio polar diagram; engine handling
15-12-70	2/65	2 30	Drag measurements, performance	27- 5-71	1/144	22	Paris Air Show demonstration
16-12-70	1/116	2 58	Performance, compressor blade strain-gauging	27- 5-71	2/91	2 35	Engine performance and sundry tests
17-12-70	1/117	3 02	As 1/116	27- 5-71	2/92	1 09	Aborted due engine oil overfill warning
18-12-70	1/118	2 44	Performance testing	1- 6-71	1/145	2 10	Supersonic press demonstration flight
				1- 6-71	1/146	2 03	As 1/145
				3- 6-71	1/147	1 58	Supersonic demonstration flight
				3- 6-71	1/148	49	Flight test after a/b malfunction
				4- 6-71	1/149	1 57	Supersonic demonstration flight
				5- 6-71	1/150	30	Paris Air Show demonstration
				6- 6-71	1/151	26	Paris Air Show demonstration
				7- 6-71	1/152	1 15	Ferry, Le Bourget to Toulouse
				8- 6-71	2/93	2 13	Engine handling; 3-engined landing
				10- 6-71	2/94	2 07	Engine surge tests. Nosewheel steering failure

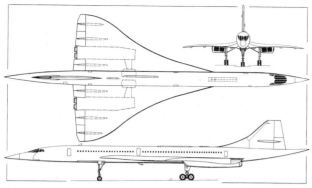

The production configuration of the Concorde is shown here, with the lengthened front and rear fuselage, TRA nozzles and small changes in the wing tips and leading edges.

there will be some changes in materials used, particularly in the area of the intakes.

Engines for the production Concordes will be Olympus 602s in the first instance, this being the version of the engine with annular instead of cannular combustion chambers. Rated thrust will be 38,050 lb st (17 260 kgp) but this figure will subsequently be increased to 38,400 lb st (17 420 kgp) and the uprated engine will then be known as the Olympus 612. Flight experience with the production standard engine will be obtained with an Olympus 602P installed in Concorde 01 during 1972. The ultimate engine rating envisaged at present is 39,940 lb st (18 116 kgp) from the Olympus 621, this version being timed for introduction from the 41st aircraft onwards.

Flying septet

Between now and the end of April 1974, seven Concordes (001, 002, 01, 02 and the first three production aircraft) are scheduled to complete over 3,800 hours of flying to obtain a full certificate of airworthiness. The programme for these aircraft is as follows:

- 001: August 1971 to December 1972. 560 hours divided between performance and system testing.
- 002: August 1971 to December 1972. 525 hours, as 001.
- 01: October 1971 to January 1974. 160 hours development flying and envelope clearance; 25 hours high incidence tests; 100 hours powerplant; 270 hours handling, systems and bad weather certification with Olympus 602; 100 hours as engine test-bed.
- 02: December 1972 to November 1973. 175 hours development flying and envelope clearance; 15

hours high incidence tests; 50 hours handling and development; 140 hours development and handling with auto-pilot; 40 hours system certification.
- 1: December 1972 to November 1973. 125 hours development flying and envelope clearance; 205 hours handling and autopilot certification.
- 2: March 1973 to October 1973. 30 hours performance; 220 hours performance, powerplant and systems certification.
- 3: July 1973 to April 1974. 1,000 hours crew training and route proving.

Seven more production Concordes are now in course of manufacture, in addition to the seven listed above and two test specimens. In addition, the two governments have authorised the manufacturers to order long-dated material and components for another six (Nos 11–16). The production schedule provides for aircraft to be available for airline service from April 1974 onwards, immediately following certification. Options are held by 16 airlines on 74 Concorde delivery positions and at the time of writing it seemed likely that Air France would become the first of these airlines to confirm an order contractually — although this point will probably not be reached until later this year. Such a step would be a logical extension of the support already extended to Concorde in France, culminating in the Mach 2 flight in 01 made by President Pompidou on 7 May.

Until one airline places a firm order for Concordes, the option-holders have the right to drop their options without incurring a financial penalty. Once an order has been placed, however, subsequent cancellation of options will result in forfeiture of the deposits paid to secure the delivery position. Finalisation of the initial Concorde contracts — BOAC and Pan American being among the likely early signatories, in addition to Air France — has awaited preparation of guaranteed performance figures, which in turn was possible only after 001 and 002 had explored the complete flight envelope up to Mach 2. This has now been done.

If Concorde's future remains uncertain, it is not on technical grounds. The programme completed to date has shown that Britain and France can work together in a field of advanced technology, and that the aircraft industries of the two countries can design and build a large supersonic transport capable of meeting its design objectives. Concorde has been described by one respected American observer as "the strongest challenge to American domination of the international transport market in history". Its future now lies largely with the airlines, whose task it will be to evolve suitable operating techniques to allow Concorde to achieve its full potential. □   **FGS**

Rolled out at BAC Filton in March, the Concorde 01, G-AXDN, is to pre-production standard, with a lengthened forward fuselage, revised visor and other significant internal changes, but without some of the modifications planned for the production aircraft.

FIGHTERS IN THE RAF

Part I – Biplane Era: The Flight from Reality

SINCE the Royal Air Force was formed on 1 April 1918, the rôle of its fighter aircraft has broadened considerably, and taken on far greater importance than could have been foreseen at that time. Although employed to some extent in both direct and indirect support for ground forces, and to intercept Zeppelin and bomber raids, the fighter scouts of the Kaiser's war were charged primarily with gaining air superiority above the battleground. The main aim was then to ensure continuity of aerial reconnaissance and fire direction, while denying these facilities to the enemy, and to safeguard our troops from strafing. In contrast, the principal rôles of RAF fighters today are to ensure protection from the genocide of the nuclear bomber, and to provide a tactical deterrent against the rush of Soviet armour across the plains of Northern Germany.

The growing responsibilities of the fighter have been matched by dramatic improvements in performance and operational efficiency, improvements which have been achieved only by surmounting a whole series of development obstacles.

Expansion of the flight boundaries gave rise to the problems of flutter, metal fatigue, compressibility, high altitude operation, and of producing materials and fabricating structures which maintain their strength at high temperatures. Obtaining thrust at higher speeds and altitudes involved the development of superchargers, variable-pitch multi-blade propellors, high octane fuels, and more recently the gas turbine, with the inherent surge problem of the axial-flow compressor.

While these powerplant changes were in progress, the biplane was being superseded by the monoplane, and then wings were thinned to decrease form drag and delay compressibility effects. Wings were next swept back to avoid the difficulty of constructing even thinner surfaces, and the problem of pitch-up was born. This in turn was overcome by various leading edge "fixes" and by raising the wing above the tailplane, despite the high wing position incurring a loss of fin effectiveness.

As speeds increased further, fuselages were made more slender and combined with smaller wings, which resulted in inertial cross coupling. In spite of these and other tribulations, the supersonic fighter has survived, although few of

(Top of page) Introduced into service during the closing stages of World War I, the Sopwith Snipe remained as the RAF's first major post-war fighter, serving with a total of 20 squadrons. (Right) The first significant fighter to enter service after the end of the war was the Gloster Grebe, shown here in the markings of No 56 Squadron. The Grebe was the first of a series of Gloster fighters which were in service with the RAF almost without interruption from 1923 to 1967.

**by Roy Braybrook,
BSc, CEng, AFRAeS**

the current types would be flyable without autostabilisation.

In parallel with this pioneering work on airframes and engines, equipment and operational techniques have advanced to ensure that these vastly more expensive machines can be employed effectively and relatively safely, if necessary in the darkest night and the foulest weather. The incessant technical changes tend to obscure the fact that many fundamental principles of air combat have, however, come down unchanged over more than half a century of fighting.

The game's the same

If you had asked the Red Baron how he was fixed for specific excess power in his Fokker Dr I, he would have been as much at a loss as the generation of pilots who followed in their CR 42s, P-40s, Me 262s, etc. However, cutting through the advanced technology jargon, the pilots of both wars knew as well as any today the need for healthy acceleration and climb rate, and the value of having some travel left on the throttle quadrant to keep the speed up in a hard turn.

By the same token, Max Immelmann never heard the term "energy height", which is used nowadays to combine an aircraft's energy components due to height and velocity. However, he was well aware that height gave him an advantage over his opponent, since it could readily be traded for speed or position. He went on to develop what became known as the Immelmann Turn, ie, a steep climb and a virtual stall turn between diving attacks. This left a low performance target little chance to escape, and enabled Immelmann to maintain a superior energy height, rather than wasting it in a high speed level turn, which would also have taken him further from his prey. In its pure form this manœuvre did not survive long, but a series of diving attacks linked by climbs and fast wing-overs may still be used against docile targets.

A dive out of the sun has traditionally held the prospect of achieving surprise, and thus increasing the chance of destroying the target and escaping unscathed. This is still practised when time permits, although the higher speeds and fuel flow rates of jet aircraft naturally reduce this possibility.

Nonetheless, a formation of fighters cruising at altitude would generally be stepped up down-sun, so that each element has a clear view of the most probable and dangerous direction of attack. This also makes the task of the attacker more difficult, in that his closest target (ie, the up-sun element) is supported by other elements which have superior energy height. The formation would probably be led from one of the up-sun sections, since it is in this direction that most attention would be concentrated by the rest of the pilots.

The tacticians of the first World War also laid down rules regarding the use of cloud cover. In essence, clouds could be useful in stalking a target or in escaping from an attacker, but aircraft should never cruise just below cloud, since they could too easily be jumped. Nor should they cruise immediately above cloud, since both they and the shadows they cast then showed up clearly against a bright white background The principal difference today is that thick clouds have become even more significant in providing escape from infra-red homing missiles.

The need for a good field of view was also appreciated at the start, and is now undergoing a revival of importance, after a long decline in the quest for speed. The weak point of biplane scouts was upward view in the forward hemisphere, despite the cut-outs in the upper centre section. As a result a line astern formation would be stepped up toward the rear, whereas anyone who has seen the Red Arrows knows that this stagger is now reversed.

Follow the leader

Pilots had discovered long before the RAF was formed that two pairs of eyes can search the sky far more effectively than one, and therefore that it was safer and more productive to operate in pairs, even when information could only be passed by waggling wings and making gestures. In pairs operation the German Air Service led the way, and this has been true of many other aspects of fighter tactics in both wars.

The German ace Boelcke is generally credited with having pioneered the technique of operating an element (*Rotte*) of two aircraft in combat, with the leader carrying out the attack while his No 2 guarded his tail. Despite the inequality of this system, it was found to give a far better kill ratio than operating fighters singly, and the two-aircraft element is now universally accepted, even in Communist air forces.

By increasing the number of aircraft in a patrol to a section of four (*Schwarm*), the sky could be scanned even better, although for higher numbers the improved chance of a sighting was found to be sometimes offset by the ease with which the formation could itself be seen. Nonetheless, bigger formations were built up from individual squadrons (*Jagdstaffeln*) or whole wings (*Jagdgeschwadern*) for escort missions and fighter sweeps, and these could be attacked only by the Allies assembling groups of comparable size.

Before going on to discuss how formations developed, it may be worthwhile to review the fundamental considerations that were taken into account. The formation had to be arranged so that the sky could be searched in all directions, especially astern and up-sun, while keeping half an eye on the leader. The various elements should also be disposed so that they could quickly come to the assistance of other elements, yet not so close that there was risk of collision. All the elements had to be equally protected. The layout of the formation had to be such that it could be maintained by inexperienced pilots in both straight flight and turns, without accidents or large throttle movements. A power margin was often not available, due to the need to fly close to the aircraft's ceiling.

Line astern formation was the easiest to fly, but had little military value, as it failed to provide mutual cross-cover and made the flak task too simple. Echelon flying was difficult and relatively dangerous, and gave uneven coverage, but it facilitated hand signalling and was later found useful to group aircraft for a peel-off.

The two most practical aircraft arrangements used in World War I were line abreast and vee formations, and it might be said that what followed in later years was based on a combination of these ideas.

Line abreast was flown at a spacing of about five wingspans, although (as for other formations) this was tightened up considerably for cloud flying. Unfortunately, cross-cover became progressively uneven as the size of formation increased, and it was also more difficult to make turns. However, for a pair of aircraft line abreast was (and remains) the optimum arrangement, with either pilot looking towards his opposite number and scanning a 180-deg sector from dead astern to dead ahead.

A vee arrangement reduced the overall size of a large formation, thereby making it easier for one element to support another, and for the group to make turns together. Simple vee formations of up to six aircraft were found to be manageable, but for larger numbers the tendency was to have separate vee-sections of three to five aircraft, and to set up the sections in an overall vee arrangement, or a series of vees at different levels.

This, then, was the way in which the RAF flew to battle in 1918, in wings of up to 36 aircraft, the pilots crouched in cockpits like undersized dustbins, blazing away at the Bosche with unreliable belt-fed Vickers guns or Lewis with 97-round drums. Parachutes were left to the cowardly-

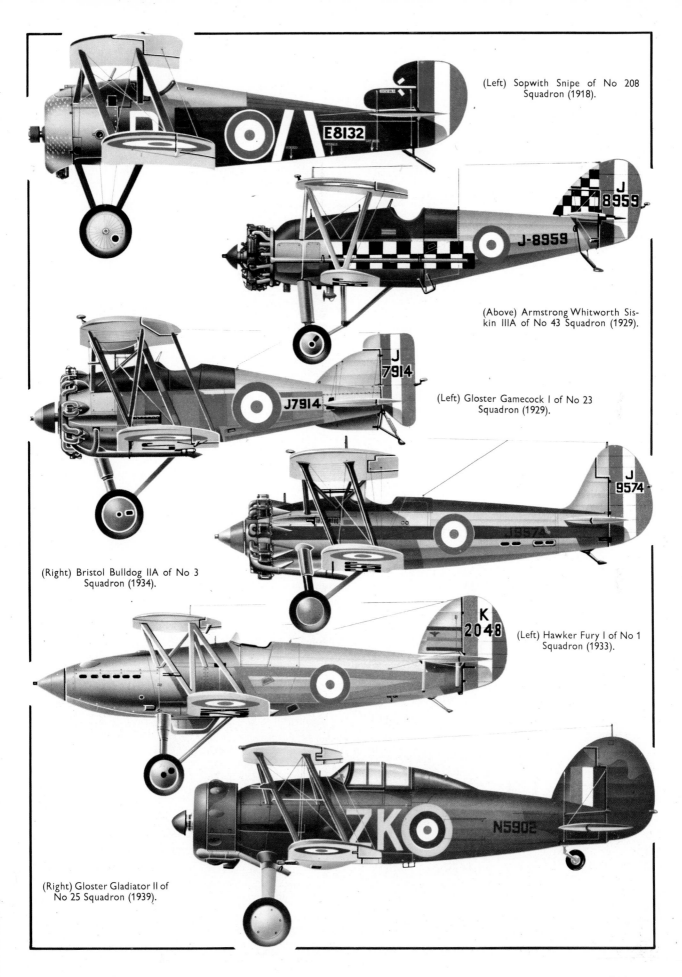

(Left) Sopwith Snipe of No 208 Squadron (1918).

(Above) Armstrong Whitworth Siskin IIIA of No 43 Squadron (1929).

(Left) Gloster Gamecock I of No 23 Squadron (1929).

(Right) Bristol Bulldog IIA of No 3 Squadron (1934).

(Left) Hawker Fury I of No 1 Squadron (1933).

(Right) Gloster Gladiator II of No 25 Squadron (1939).

The Armstrong Whitworth Siskin IIIA replaced the Grebe in many squadrons, serving with a total of 11 RAF units. With its uncowled radial engine and twin 0·303-in machine gun armament it was entirely typical of the 'twenties.

bully Huns, the same caricatures who were to start bombing cities and invent guided weapons in the next lot. In retrospect, one can only regret that the war was not to be won by hurling such epithets, for then the enemy's hash might have been settled with a couple of issues of *Punch*!

"Boom" takes the offensive

Technical leadership had ebbed and flowed with each new type of aircraft and armament. Fighter aces arose, and were praised (probably out of all proportion to their real contribution to the war effort) by a press tired of the unprintable horror of trench warfare. The German Air Service had led in fighter tactics, but Britain was to lead in air strategy. Better than most of his generation, Trenchard (the RAF's first Chief of Air Staff) appreciated that the third dimension made any static system of air defence unprofitable, and that his fighting scouts should therefore be used aggressively. In contrast, his opponents generally preferred their own side of the lines, and thus failed to exploit their tactical and technical advantages.

By the end of the war, Sopwith Aviation had achieved a pre-eminent position among British manufacturers of single-seat scouts. Their Snipe, which was to see the RAF through the first six post-war years, had entered service in September 1918, and in three months of combat had proved a good match for Germany's finest fighter, the Fokker D VII. The Snipe's empty weight was 1,312 lb (595 kg), and it was powered by a 230 hp Bentley Rotary BR 2 engine. It had a gross weight of 2,020 lb (916 kg), a top speed of 121 mph (194 km/h) at 10,000 ft (3 050 m), and a service ceiling of 19,500 ft (5 950 m).

The early 1920s were a barren period in fighter procurement, as taxpayers recovered from "the war to end all wars". When defence planning ultimately began again, it was done against the background of a rolling "Ten Year Rule", ie, no foreseeable threat to the country's security within that time. Since some potential "enemy" was necessary for even the simplest planning, France was put forward as being the only military power within aerial striking distance. Half-a-century later this may seem a fatuous choice, but about that time

Gloster's successor for the Grebe, evolved to an official specification in the mid-twenties, was the Gamecock, illustrated here in the markings of No 56 Squadron. Only seven squadrons were equipped with Gamecocks.

The elegant Hawker Fury was one of the large family of biplanes originated on Sydney Camm's drawing boards in the inter-war periods. Seven RAF squadrons flew Furies, including No 25, illustrated performing a "tied-together" display.

American defence planners took the idea of a British attack from West Indian bases more seriously than any threat from Japan!

When procurement funds finally became available in 1924, the principal consideration in their use seems to have been to build up the aircraft industry. Five new types of fighter were ordered into production during the 1920s, but the average in-service life proved to be less than six years.

All were radial-engined biplanes, with power growing steadily from the 400 hp Armstrong Siddeley Jaguar IV in the Gloster Grebe of 1924 to the 490 hp Bristol Jupiter VIIF in the 1929 Bristol Bulldog. Empty weight increased rather more rapidly, from 1,720 lb (780 kg) for the Grebe to 2,412 lb (1 100 kg) for the Bulldog. In spite of this, maximum speeds went up from 152 mph (243 km/h) to 174 mph (278 km/h), while service ceilings were raised from 23,000 ft (7 000 m) to 27,000 ft (8 230 m).

All British fighters of that period were armed with two Vickers guns, synchronised to fire through the airscrew. Some of the later ones had provision to carry four 20-lb (9-kg) bombs, presumably as a result of the RAF's policing rôle in Iraq, which began in 1922. The Gloster Gamecock followed two years behind the Grebe, and appears to have represented only a minor improvement. The Armstrong Whitworth Siskin III entered service in 1924 relatively underpowered, but came back into the ball game as the IIIA in 1927, using a 425 hp "blown" Jaguar IV.

Following the voluntary liquidation of Sopwith Aviation, a new company had appeared at Kingston in 1920 under the name of H G Hawker Engineering. Their first product, the Hawker Woodcock, was a night fighter with a wing loading much lower than its contemporaries. Its mediocre performance gave little indication that this company was to rise — under the design leadership of the late Sir Sydney Camm — to dominate the British fighter business.

Zero feed-back

Taken in isolation, the overall trends in fighter power and performance may seem to indicate good steady progress, being the sort of figures that are trotted out in most company reports to persuade the shareholders to "let the present directors get away with it for another year". In reality, both Air Staff and fighter constructors appear to have been living in a dream world, paying scant attention to the Schneider Trophy seaplane racers, which by the end of the decade were capable of over twice the speed of the RAF's operational fighters.

This divergence was already clear in 1925, when the Supermarine S.4 monoplane with a 700 hp supercharged Napier Lion in-line engine set a world speed record of 226.8 mph (362,9 km/h). After further development the Lion VIIB for the S.5 reached 900 hp in 1927, ie, *twice* the power of the supercharged Jaguar IV in the Siskin IIIA then entering service. By the time of the next race in 1929, the steady divergence of the pace-setters from standard production fighters had become a run-away. The Lion engine then gave 1,400 hp, while a new Rolls-Royce design was pulling the S.6 through the air with no less than 1,900 horses, ie *four*

(Above) The RAF's first Hawker fighter was the Woodcock; two squadrons were equipped, Nos 3 and 17 (illustrated). (Below) The Bristol Bulldog IIA equipped six squadrons, mostly in succession to the Gamecock, but remained firmly cast in the mould of the British fighters of the 'twenties.

Last of the RAF's biplane fighters, the Gloster Gladiator was flown, at various times, by no fewer than 36 squadrons, but by the time it entered service it had already been outmoded by the new monoplanes that were under development.

times the output of the Jupiter VIIF in the RAF's latest fighter!

In 1931 the Schneider Trophy was won outright for Britain by the 2,300 hp Rolls-Royce 'R' engine and the S.6B. Boosted to 2,600 hp, in the same year this combination set a new world speed record of 407.5 mph (652 km/h). For comparison, the Fury biplane was introduced also in 1931, and gave a 19% speed improvement over the Bulldog, but this still only amounted to a best performance of 207 mph (331 km/h)!

Some readers may feel that the accusation of ignoring technical advances might equally well be levelled at later British Operational Requirements (OR) staffs and fighter designers who happily turned out subsonic products long after "Chuck" Yeager had gone through the barrier in the Bell X-1. However, America's supersonic research programme was based purely on aerodynamic and structural test vehicles, artificially launched into the upper reaches of the atmosphere, where their rocket motors could propel them to high Mach numbers.

The Schneider Trophy racers were much more closely related to practical production machines. It is true that they did make use of short-life engines (TBO was about 1 hr) to achieve incredible power, and that it took service engines ten years to catch up. Also that the use of seaplanes had been specified to avoid the restrictions of operating from the poor airfields then available. However, these aircraft did take off under their own power, and they achieved far higher speeds than contemporary fighters despite the penalty of carrying two barges under their wings!

To give some idea of the enormous wing areas then accepted as normal for fighters, the average wing loading for in-service types at the turn of the decade was approximately one-quarter of that for the S.6B! It seems incredible that no-one had the initiative to develop trailing edge flaps (which had been used during World War I, notably by Fairey), and thus open the way for smaller wings and higher in-flight speeds. Stranger still is the fact that Gloster, having produced aircraft for most of these contests, and set a world speed record of 336 mph (538 km/h) with the Gloster VI seaplane in 1929, went right on manufacturing biplane fighters, and only surpassed this speed with production aircraft some 20 years later in the form of the Meteor and Javelin!

The ostrich syndrome
It can be argued that the potential "enemies" of the 1920s, in their Farmans and Lioré-et-Oliviers, could not even reach 125 mph (200 km/h), yet it must have been clear that foreign bomber designers would ultimately be inspired by the technical advances then in progress. The Heinkel He 70 of 1932 reached 222 mph (355 km/h), and even Britain's own 156 mph (250 km/h) Fairey Fox and 184 mph (294 km/h) Hawker Hart bombers outperformed all contemporary RAF fighters on entering service, in 1926 and 1930 respectively.

Aside from lagging in performance, the principal British fighter manufacturers appear to have been too busy with production contracts to experiment with forms of construction that might have paved the way for major advances in the 1930s. Junkers had pioneered metal construction with the iron and steel J1 in 1915, and with the duralumin J4 in 1917. In Britain this work was followed up with the all-metal Short Streak in 1920, and two prototypes of their Springbok two-seat biplane fighter in 1923.

In the same year Handley Page had constructed the HP 21, a wooden stressed-skin fighter for the US Navy, and Avro tested a similar method in the Type 566 Avenger of 1925, progressing to the all-metal Type 584 Avocet in the following year. Vickers used the Wibault type of metal structure in the Type 121, which also flew in 1926, and in the next year built the Vireo monoplane fighter as a private venture on similar structural lines.

continued on page 208

The Gloster Gauntlet was one of nine types of fighter biplane ordered into production for the RAF between 1918 and 1938, each representing only a small advance in performance over its predecessor. Illustrated are the Gauntlets of No 46 Squadron; another 21 squadrons flew the type.

FLYING THE "FLYING SUITCASE"

THE 1939-45 war historian, writing with the advantages of hindsight, may well consider the Handley Page Hampden to have been an out-dated and somewhat inadequate bomber, but most pilots —at least those who flew it non-operationally, as I did, on test work or ferrying — will remember it as a near-perfect, if mildly eccentric, flying machine. It handled in tight turns and other off-beat manoeuvres like a fighter; the single-seat cockpit — cramped and cluttered though it was — provided a view from wing-tip to wing-tip, and it was as viceless as made no matter. The few who disliked the Hampden did so either because they didn't trust that slender semi-monocoque rear fuselage, or because they never managed the trick of handling it on hold-off and landing; a firm and early check, helped by trimmer action, was necessary if it was not to behave like a well-mannered kangaroo after touchdown.

The Hampden was the first medium-heavy bomber — and the first military twin apart from the Blenheim on which we'd been "converted" — that I was allowed to fly when, in November 1939, as a very inexperienced ferry pilot, I took one from Radlett to No 9 Maintenance Unit at Cosford, near Wolverhampton. It weaved about a bit, overcorrected with rudder pressures and unnecessary throttle differential, while the tail was being lifted on take-off, and made an unplanned, but reasonably soft, wheeler when landing on the small and mist-shrouded grass field at the destination — but all had gone reasonably well and this no doubt partly explains the soft spot I afterwards retained for this strange bomber on ferrying flights and later test-flying outings.

The soft spot was to be further developed a few weeks later when a Hereford — that smooth but ear-splitting Napier Dagger-engined variant of the Hampden — had to be delivered from the Short & Harland Queen's Island strip at Belfast to the MU at Aston Down in the Cotswolds. I was not to know it, but the airspeed indicator had either stuck or was reading 30 mph (48 km/h) high on final approach and the Hereford settled-in, semi-stalled, from 50 ft (15 m) or so touched gently on a level keel just outside the marked boundary, bounced over a slit trench and ran no more than 100 yd (91 m).

Delivery trips with Herefords from Belfast were to become something of a milk run for the ferry pilots. Later, as a test pilot in the RAF, I was to learn just how troublesome the engines of these aeroplanes could be — but more about that

later. Not many people know, or remember, incidentally, that another differently-engined variant of the Hampden was planned and a prototype built and flown as the Mk II. This had Wright Cyclone engines, but unlike the Pratt & Whitney-powered Wellington IV of much the same era, it did not go into production.

In company with most British military aeroplanes of its period, the Hampden had a uniquely individualistic cockpit layout, with devices which were not to be found in any other aircraft type. The constant-speed propeller control levers, for instance, were interconnected by a removable linkage which, by turning a knurled nut, provided vernier adjustment of the relative positions of the levers for accurate synchronisation of the engine speeds. This was fine in theory, but propellers never held their constant-speed settings for longer than a few tens of seconds, so one could be constantly making adjustments to the vernier setting.

In those days hydraulic power had usually to be selected on or off, as required, by lever or plunger. The Hampden had a device called a "power bolt" below the right-hand side window of the cockpit — looking for all the world like a window-locking or jettisoning control. More than once, ferry pilots who had not done their homework properly — or had forgotten it — failed to move this bolt (like that on a door, or rifle), found that there were no results on selecting

Viewed from the Cockpit

by H A Taylor

The cockpit of the Hampden was "cramped and cluttered", says the author of this article, but was well positioned to afford the pilot a view from wing-tip to wing-tip. This illustration clearly shows how the control column (with gun-firing button below the three-quarter wheel) obscured the directional gyro on the standard blind-flying panel, as well as one of the fuel gauges and the compass, on the floor.

undercarriage down with the lever in the four-position undercarriage/flap gate below on the right, and resorted to the use of the emergency system. This action — requiring the pulling of a mysterious length of red chain from the bowels of the aircraft — afterwards involved the ground crews in unacceptably long and tedious labours to get the system back to normal.

For us, a minor discomfort — which may well have been a major difficulty for operational pilots — was the fact that the control-column and its three-quarter wheel for the ailerons effectively hid the directional gyro at the bottom of the standard RAF instrument-flying panel. The column carried, below the wheel, a button for the fixed forward-firing Browning gun. This was, no doubt, a relic of the original day-bomber specification; heaven knows, dashing though the Hampden's behaviour was, what the pilot was expected to attack with this gun.

The pleasing behaviour led us to treat the Hampden with, perhaps, rather too little respect in, for instance, tight S-turns on the approach and even in sideslips — until we learned that it was possible to stall the rudders so that they locked over. That back end twitched about a good deal, but I only once saw a Hampden actually lose its tail. That was after an exceptionally heavy landing by a "new boy", who somehow managed to stall it completely at about 100 ft (30 m) on the approach. I often wondered how they managed to get the control pulleys and other equipment into that narrow tailboom. It was only later that I learned that, as with the Mosquito, the fuselage was made in two halves like the pod of a pea and mated after most of the equipment and fittings had been installed.

At the time of the heavy landing incident I was on attachment to No 23 MU, Aldergrove, Northern Ireland, where the unit had been suffering from a shortage of test pilots qualified for twin-engined aircraft and had built-up a backlog of Hampdens and Whitleys due to be tested and/or ferried to and from pleasant satellites with names like Ballywalter and Murlogh. That was a rare trouble-free interlude in wartime existence, with real shell-egg instead of egg-powder breakfasts, and evenings out in a city that was hardly touched by the war. Almost the only discomforts were the nightly rude awakenings as Coastal Command Liberators took off over one's hut from 03·00 hours or so.

A day or two after returning to the harder world of my own unit — then No 29 MU, High Ercall, near Wellington, Shropshire — I learned just what a beating the Hampden's tail could take and survive. Luckily, as it happened, the day was a dreary one of solid cloud at 2,000 ft (610 m), with visibility reduced to a couple of miles or less by drizzle and Midlands smoke, so flying was restricted to routine test-checks. I was cruising gently along in Hampden AD851, making a few notes on my knee-pad, when there was a sudden jolt and muted crash followed by violent shudderings through the airframe. Almost as soon as the rumpus started I'd pulled the throttles closed and slid the canopy back so that I could have a good look round. There was no need to look far; the starboard Pegasus engine had shed its main cowlings, laying bare the works and leaving only the exhaust ring and now-useless cooling gills. As the speed dropped the shudderings died down, though the rudder pedals continued to twitch. I guessed that something had been damaged at the tail end and that the more violent shudderings might have been caused by the tangled airflow behind the un-cowled engine. Very gingerly, and keeping my eye on the starboard cylinder-head temperature gauge, I opened the throttles and headed for home.

After some low-speed checks, power-on and power-off, to make sure that the stalling speed had not been unduly increased and that lowered flaps did not adversely affect the elevator control, the Hampden was brought in to a fast, powered, wheeled-on landing. An examination on the apron showed that the cowlings had cut gashes in the skin of the centre-section and starboard fin, and had chopped off the mass-balance of the starboard rudder. Only when a local farmer brought in the cowlings did we discover the cause of the failure. They had been fitted with the toggle fasteners held only by the single rivets used by the manufacturers to keep the halves temporarily together pending necessary adjustments and re-riveting.

The handleability of the Hampden and its excellent view for the pilot were adequately demonstrated during at least one early delivery flight, again from Radlett to Cosford. Very low cloud and rain from the north-west had moved in much more quickly than had been forecast, so that eventually the Hampden was trapped, with low cloud on hills from the Cotswolds in the south to the Lickeys in the north. Booming along at 200 ft (60 m) above the Severn after an unsuccessful attempt to get through to Cosford, I spotted the little grass aerodrome north-east of Worcester, where Battle Trainers were then being assembled, made a tight circuit while under-carriage and flaps were coming down, and motored in as low as I dared over the boundary. The field wasn't really big enough for a Hampden, but there was no option and we stayed inside with the help of a judiciously applied ground-loop. Two days later, on a bright spring morning, the tail of the Hampden was pushed into the boundary hedge and I released the brakes only when the engines were giving full power — to float off the ground with hundreds of yards to spare.

We had had very little trouble with Herefords on delivery flights from Belfast — though I arrived one day to find that a fellow ferry-pilot had parked one back on the strip that morning with its starboard Dagger spewing connecting-rods, gear-wheels and other odds and ends. The ground crews at Queen's Island knew their business and the vagaries of those 24-cylinder H-type engines which overheated on the ground and were over-cooled in the air. The troubles began for me when they were being re-tested and prepared for delivery at No 48 MU, Hawarden, Chester — which, in 1941, was a

continued on page 218

(Above) The Hereford was an unsuccessful Hampden variant with Napier Dagger 24-cylinder "H"-type engines. (Below) A Hampden TB Mk I torpedo-bomber (with enlarged bomb-bay) in the markings of No 489 Squadron, a New Zealand unit serving with the RAF.

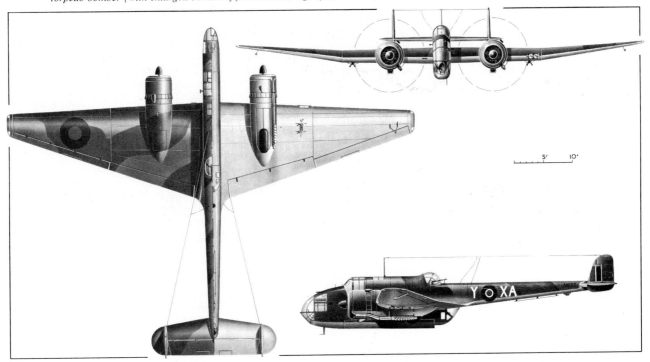

(Top to bottom) F-4D Phantom of the 306th Fighter Squadron, Iranian Imperial Air Force; RF-4B Phantom (153099) of the US Marine Corps Reconnaissance Squadron VMCJ-2; F-4E Phantom of the Israel Defence Forces/Air Section; RF-4E Phantom (97448) of Aufklärungsgeschwader 51 "Immelmann" of the Federal German Luftwaffe; F-4E Phantom of the 34th Tactical Fighter Squadron, 388th Tactical Fighter Wing, USAF; Phantom FGR 2 (XT 901) of No 17 Squadron, RAF.

Loading the competition dice

PERHAPS the most invidious task in the entire modelling spectrum is that of the judges of a modelling contest. The standards set by the competing models are, today, invariably very high indeed — a model conforming with competition standards of a decade ago would probably have difficulty in meeting the minimum standards now demanded — and when faced with a profusion of superb models, the unfortunate adjudicator is likely, in sheer desperation, to fall back on personal preference. Complicating the issue is the fact that when appraising the merits of two models of equal standard the judges almost always award first place to the larger model of the two. Unfair this undoubtedly is, particularly so as the finishing of a small model calls for more expertise than does a larger model, but it is an inescapable fact that the larger the model the more attraction it draws.

Some contest organisers have, of course, attempted to overcome this problem by dividing the entries into classes or categories, but even this apparently simple solution presents its quota of problems. For example, does one divide the contenders on the basis of physical size or on that of scale? The latter is the division usually adopted, primarily on account of the difficulty posed by the selection of suitable dimensions for a comparison based on size. The wing span of a model might *seem* to be the obvious dimension to be taken for comparison purposes, but one need mention only two aircraft types from the same manufacturer, Lockheed's F-104 and U-2, as an indication of the anomalous situation that would result from any such classification. But categorisation by scale is hardly less problematic. Not only is there the unavoidable difficulty that models of such widely differing size, such as the B-52 Stratofortress and the Gnat, have to be combined in one class; there is the basic problem of deciding where exactly the dividing lines between the classes should fall. Astonishing though the fact may seem to many modellers, it is possible to make models of at least one aircraft type from kits in scales from 1/27th to 1/114th, not to mention a considerable number of additional scales outside this range!

Thus, if scale provides the yardstick for classification some compromise is inevitable if the number of classes is not to be impossibly large. In fact, most model kits readily available would seem to fall into one of five groups centred approximately on 1/30th, 1/50th, 1/75th, 1/100th and 1/150th scales. If these are taken as the bases it is usually not *too* difficult to determine the parameters of the group. For example, Aurora's 1/110th scale C-141 StarLifter seems to fit naturally into the 1/100th scale group, while Nichimo's Focke-Wulf Fw 190 to 1/65th scale quite obviously comes within the 1/75th rather than the 1/50th group.

Happily, the situation is gradually resolving itself as the vast majority of kits nowadays make up into models of one or other of the scales considered to be standard, thus easing the tasks of the competition organisers if not those of the judges. In so far as the competition modeller is concerned, if he is not averse to exercising a little gamesmanship, we would recommend that he selects for his entry what would seem to be the least popular scale class and then enter the largest possible model in this group. Other things being equal, he will have loaded the dice in his favour!

Flying enterprise

We metaphorically doff our hat to Humbrol Limited for its enterprise in participating with two other big Hull manufacturing companies in a week's high-speed sales tour of Europe for which Hawker Siddeley HS 125 G-AYOK was used. In seven days from 28 June, Humbrol's Deputy Managing Director, John Gooding, and Sales Director Dick Bond flew some 16,000 miles (25 750 km) and visited 18 cities, the flight plan ranging from Helsinki to Lisbon and from Athens to Oslo. Humbrol — the name is almost a household word in virtually every country in which the art of modelling is pursued — has a record of consistent growth in the export field with its high-quality modelling enamels, and the seven-day sortie in G-AYOK, which enabled John Gooding and Dick Bond to confer with many of Humbrol's European agents in a short space of time, undoubtedly did much to consolidate the company's export efforts which, following record orders at the 1971 Nuremburg Toy Fair, have included an international marketing convention in Sweden in May, and exhibitions of Humbrol products aboard the motor yacht *Saluja II* in Portugal, Gibraltar, Spain and Malta.

This month's colour

For any piston-engined aircraft model protagonist who bases his preference on the allegation that the modern jet "lacks character", the McDonnell Douglas F-4 Phantom II which provides the subject for this month's colour page is perhaps the one aircraft that could give him cause to think and possibly tempt him into broadening his interest, for few combat aircraft of today possess *more* character than this welter weight from Missouri. No southern belle by any criterion, the Phantom appears to bulge in all the wrong places, refuting the old adage that a plane must look right to be right, but what she lacks in beauty the Phantom certainly compensates for with character. Having been around for nigh on 14 years, the Phantom is no débutante, and today serves with air arms from Australia to Germany and from Israel to South Korea in a variety of versions. In view of its production longevity and its widespread use, it is to be expected that a large number of model kits of the Phantom should have reached the stockists' shelves and the modellers' workbenches.

For those with a penchant for fairly large models there is the now rather elderly kit from Aurora to 1/48th scale which represents the F-4A quite adequately, and to similar scale is a superb RF-4E from Fujimi which is probably the best of all the Phantom kits. Kits to 1/72nd scale have been issued by Hasegawa depicting the F-4B, the F-4J and the F-4K, the two last-mentioned being also included in Frog's list with different — and rather better — decals, and by Airfix and Revell, while a really excellent Phantom kit to 1/100th scale comes from Tamiya. This list is by no means exhaustive, although all the kits mentioned are of good quality and represent their subject well.

A Czechoslovak biplane

The fighter biplanes of the 'thirties exercise a considerable fascination for many modellers, and there can be few more attractive little aeroplanes of this category and period than the Czechoslovak Avia B 534, a contemporary of the Hawker Fury. No model kits of Czechoslovak aircraft appear in the lists of western manufacturers, and we therefore particularly welcome the efforts of the Czechoslovak company, Kovozavody Prostejov, to bring the best of these to the attention of modellers, these efforts being doubly welcome in view of the quality of the kits. This manufacturer's first kit, the L 29 Delfin, was considered almost universally as an astonishingly good effort, and the second offering from the Czechoslovak company, the Avia B 534 to 1/72nd scale, *almost* reaches the standards attained by western manufacturers of long experience.

From *most* aspects the B 534 *does* attain the highest western standards, but it suffers a good deal of quite thick flash and the fit of the component parts leaves much to be desired. The correction of these faults has to be undertaken with a great deal of care as the plastic used for the kit is very soft, and it is all too easy when preparing for assembly to cut away too much material and ruin the model. However, if the necessary care is taken, the result is an accurate model with first class surface detail. The decal sheet, which provides markings for a pre-15 March 1939 Czechoslovak Air Force fighter and for a wartime Slovak Air Force machine, is excellent, and the cockpit transparency is notably thin and clear. In short, the application of a little expertise will result in a first class model of an extremely interesting aeroplane, and it is well worth going to the trouble of arranging an exchange with a Czechoslovak modeller in order to obtain an example of this kit.

A contrasting pair from Fujimi

Fujimi's latest offerings to 1/50th scale could hardly represent two more diverse types — the Focke-Wulf Fw 190A and the Aérospatiale Alouette III. Fujimi is known to modellers the world over for the quality of its kits, and the Fw 190A and Alouette III are fully in keeping with earlier standards. Their parts fit together with extreme precision, while the standard of surface detail achieved on these two models is almost unbelievable.

The Focke-Wulf kit includes extra parts which enable it to be assembled as any one of nine versions of this famous fighter. The basic sub-types are the Fw 190A-6, A-7, A-8 and A-9, and the excellent decal sheet provides, in addition, full markings for an Fw 190A-7/R3, an A-8/R1 and an A-8/R3, but no colour scheme drawings are provided by the instruction sheet. The kit itself is generally accurate in outline, although the wingtips are rather too square. The component parts are moulded in white plastic and fit together very well indeed — even the least experienced of modellers should find no difficulty in assembling this kit. Full cockpit detail is provided, although its effect is rather spoiled by a thick transparency, and it may be assembled in motorised form (the motor is *not* included in the kit). At 300 Yen in Japan — and available in the UK from VHF Supplies — this kit is far from expensive, and, despite minor criticisms, is certainly the best Fw 190 kit on the market.

Fujimi's Alouette III is not a new kit, but it has recently been re-released in the very attractive markings of the Tokyo Metropolitan Fire Department. Two colours of plastic are used, a brilliant orange-red and black, and the kit assembles so well that very little painting is necessary if sufficient care is taken during the assembly process. Surface and interior detail are to the usual high Fujimi standards, and this kit can be highly recommended as it makes up into an attractive and simple model of one of the world's most widely-used helicopters. At 200 Yen in Japan (or 67p from VHF Supplies) it offers good value for money.

SST from VEB
Singularly few Soviet aircraft of any generation are represented by plastic kits, more is the pity, but thanks to the efforts of the VEB of the German Democratic Republic a fairly wide range of kits of contemporary Soviet commercial aircraft, mostly to 1/100th scale, can be obtained by exchange, and the latest addition to the VEB range is the Tupolev Tu-144. This kit is neatly pressed in good-quality white plastic, its component parts fit together well, and the model in which it results appears to be generally accurate in outline. There is plenty of surface detail, but this, despite being of the straight-line type, is decidedly on the heavy side. This fault is somewhat surprising in view of the steady improvement in this respect that was being shown by the products of this East German concern, and is shared by the VEB's recent Yak-40 to which we refer later. A drooping nose is provided but its operation is inaccurate; the visor and windscreen have been omitted, and the nose outline remains strictly supersonic, endowing the model with a decidedly odd appearance when the nose is drooped. The modeller is thus faced with the choice of discarding the undercarriage and mounting the model on a stand — a pity as the component parts of

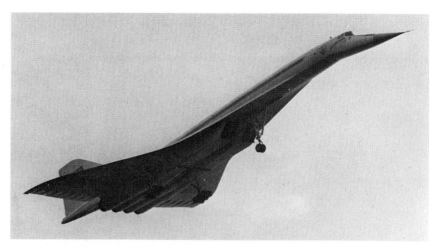

The first Tupolev Tu-144, SSSR-68001, as demonstrated at the Paris Air Show in June.

the undercarriage are beautifully moulded — or undertaking a good deal of difficult conversion work in order to add an accurate windscreen and visor.

The decals are of good quality but extremely thin, and therefore demand careful handling. Fortunately, several spares are provided so that if one decal is broken during application the model is not necessarily spoiled. Naturally enough, the markings are for the first Tu-144, SSSR-68001, and are accurate in outline, although the blue would seem to be somewhat dark in shade. Though not faultless by any stretch of the imagination, this is basically a good kit and worth some effort to obtain, especially so as it provides an excellent companion for Heller's Concorde to the same scale.

Another product of the VEB that is worth acquiring is the recently issued kit of the Yak-40, also to 1/100th scale. Accurate, neat and easy to assemble, this kit suffers heavy-handed surface detail, and although the decals are well printed, they are, like those of the Tu-144, exceedingly thin and, in consequence, call for the utmost care in application. This kit, too, is moulded in good-quality white plastic, and since most Yak-40s are finished primarily in white, little painting should be necessary provided that care is taken in assembly.

Decals and accessories
The latest issue of decals received from Max Abt of Paris consist of two sheets, both representing to 1/72nd scale markings of WW II *Luftwaffe* aircraft of considerable interest. The first of these, No 115, offers the markings of an all-black Bf 110E of 1./NJG 3 operating in the Mediterranean theatre during the summer of 1941, while the second, No 119, provides the markings of a Ju 52/3m employed in the air ambulance rôle during the summer of 1940. A model completed with the markings provided by either of these sheets, and particularly the latter, will definitely catch the eye at any exhibition, and these products are to be strongly recommended.

Humbrol is, of course, world-famous for its paints, but it may not as yet be generally known that this concern also manufactures a wide range of other modelling accessories, several of which we have been able to test recently. The first of these accessories is a

really excellent miniature spray gun powered by an aerosol canister which will spray one pint (0,57 l) of paint or other liquid. The liquid to be sprayed is housed in a glass jar which screws into the gun. As this jar is provided with a lid it may be detached before its contents are exhausted and replaced by another jar containing a different liquid. The gun is simple to operate and produces an excellent finish — unlike aerosol cans it does not emit blobs. While surprisingly small areas may be sprayed, it should be remembered that this device is a spray gun and *not* an airbrush, and is designed to be used over fairly *large* areas. It is unquestionably the best spray gun of its kind available, and at its very low price of £1·25 it represents outstanding value. Incidentally, a larger aerosol canister capable of spraying two pints (1,14 l) is available at 75p.

Another excellent Humbrol product is plastic wood. This is finer and easier to apply than plastic wood of other brands, dries quickly and does not shrink excessively. It is available in both tubes and tins at 9p and 15p respectively, but for modelling purposes we recommend the former. Humbrol is also marketing a new universal adhesive. Unlike many similar adhesives, it will stick polystyrene, though it offers no real advantage over normal polystyrene cement for this purpose. It is, however, exceedingly useful for cementing materials such as wood to plastic, and, of course, has innumerable domestic uses. It is priced at 10p per tube.

Strongest of all adhesives are the apoxy resins, and Humbrol (under the name of Britfix) now markets this also. It is unlikely to find much use in modelling, except on occasions when an extra strong adhesive is called for, when, provided that the instructions are followed closely, it will perform very efficiently indeed. The price of a two-tube (resin and hardener) pack is 30p, and our only criticism of this and the other Humbrol products mentioned is the difficulty that may be experienced in obtaining them as most dealers, in the UK at least, do not seem to stock them. We hope that Humbrol will make as great a marketing effort in the UK as it is on the European continent to ensure the success that these excellent products deserve. □

W R MATTHEWS

NATTER

The "Throwaway" Fighter

As the spring of 1944 translated to summer, the lengthening of the days was viewed with trepidation by the *Oberkommando der Luftwaffe*. During the hours of daylight the skies over Germany were now rarely free from the aircraft of the US 8th and 15th Air Forces. The battle for supremacy in German airspace had already been lost by the *Luftwaffe*. The *Jagdflieger* were overstretched and hard-pressed; many of their units had been decimated in the holocaust above German industry, and only inclement weather now afforded relief. Orthodox means of countering the escalating daylight offensive had failed abysmally, and such was the situation that serious consideration had to be given to the *unorthodox*.

Ingenious schemes for countering the Allied bomber offensive were legion; few survived more than a cursory examination of their practicability. One of the most radical proposals that *was* destined to complete the transition from drawing board study to actual hardware, however, called for a *throwaway* fighter — a simple ramp-launched, rocket-propelled interceptor employing an airframe built of low-grade, non-essential materials and which would serve for a single operational sortie, after which the pilot would bale out and the rocket motor descend to the ground by parachute for retrieval and re-use. The originator of this scheme, one Erich Bachem, thinking perhaps of the lengthening summer days ahead and possibly motivated by a line from Shakespear's *Julius Caesar*, "It is a bright day that brings forth the Adder," dubbed his proposed *throwaway* fighter the *Natter* (Adder).

The *Technische Amt* (Technical Office) of the State Air Ministry, the *Reichsluftfahrtministerium*, had issued a requirement for a small, inexpensive rocket-propelled target-defence interceptor suitable for mass production, and Dipl-Ing Bachem promptly submitted his *Natter* proposal as a contender, but equally promptly the pundits of the

(At head of page) Oberleutnant Siebert climbing into a pre-series Natter for the first manned test, and (below) pre-series Nattern under construction at Waldsee.

Technischen Amt turned the scheme down. Despite the exigencies of the times, the official view was that the *Natter* proposal had been uninvited and had, in any case, been submitted for consideration through abnormal channels, Bachem having enlisted the support of *Generalleutnant* Adolf Galland who had himself passed the project to the RLM with his recommendation that it be developed.

This unorthodox procedure had incensed the *Technische Amt* which considered such an approach to be a supreme example of meddling on the part of the *General der Jagdflieger,* although in defence of the *Technischen Amt* it could be claimed that the *Natter* proposal did *not* meet the demands of the official target-defence interceptor requirement which did not envisage a *throwaway* fighter as propounded by Bachem.

A number of design studies had been submitted to meet the rocket-propelled interceptor requirement, differing only in degree of exoticism, but none was quite as radical as the *Natter*. The RLM possessed its hard core of traditionalists which had sufficient difficulty in accepting the rocket motor as a viable means of propelling a fighter without taxing its imagination still further by coupling such a prime mover with a disposable airframe; a sort of halfway house between interceptor and missile.

But Bachem was not to accept defeat so easily. He was convinced of the feasibility of the *Natter* concept which he had progressively refined over a number of years while Technical Director of the Gerhard Fieseler Werke, and with the rejection of his proposal by the RLM he immediately sought and was granted an interview with Heinrich Himmler. The imagination of the SS Chief was immediately captured by the project, and Himmler vouchsafed his full support. Within 24 hours, Bachem was informed by the *Technischen Amt* that it had reconsidered its rejection of the *Natter* proposal which would now receive the highest development priority!

Erich Bachem had meanwhile acquired a small factory at Waldsee in the Schwarzwald, some 25 miles (40 km) from Lake Constance, and here he was joined by Dipl-Ing H Bethbeder, formerly a Technical Director of the Dornier-Werke, and Ing Grassow from the Walter-Werke at Kiel. By August 1944, when the *Natter* came within the orbit of the Emergency Fighter Programme, or *Jägernotprogramm*, work had begun in earnest. By this time, the RLM had completed its evaluation of the competing proposals for a rocket-powered target-defence interceptor that had been submitted through *orthodox* channels, and had selected Heinkel's *Projekt* 1077 as the winning contender, the runner-up being Junkers' EF 127, these being referred to under their respective cover-names of *Julia* and *Dolly*. The RLM did not welcome the addition of the *Natter* to this programme, but the interest expressed by Himmler provided adequate insurance against any attempt to terminate the project, and the official designation 'Ba 349' was allocated to Bachem's 'throwaway fighter'.

In its definitive form the *Natter* differed in some respects from Dipl-Ing Bachem's original proposals. Bachem had envisaged an initial attack on the intruding bomber formation with a battery of rocket missiles, the pilot of the *Natter* subsequently using the remaining kinetic energy to climb above the bombers and, after selecting a second target, perform a ramming attack. The pilot was expected to eject himself from the cockpit of the *Natter* immediately before impact, activation of the ejection seat triggering explosive bolts which would detach the portion of the fuselage housing the rocket motor, an automatically-deployed parachute lowering this to the ground. The provision of an ejection seat tended to introduce complexity into a design that was intended to provide the essence of simplicity. Furthermore, the dimensions of the cockpit proposed for the *Natter* were insufficient to permit the installation of an effective ejection seat. In consequence, the decision was taken to dispense with the ramming attack, the pilot jettisoning the fuselage nose complete with windscreen after discharging the battery of rockets and clearing the target formation, simultaneously deploying a parachute to decelerate the remainder of the aircraft and throw him forward and clear.

Various forms of armament were considered during this initial development stage, including the so-called *Trommelgerät,* or drum-apparatus, which was a semi-automatic cylindrical magazine for 40 30-mm shells, a *Rohrbatterie* (tube-battery) of 49 30-mm SG 119 rocket shells, and a *Bienenwabe* (honeycomb) arrangement of hexagonal tubes for 73-mm Hs 217 *Föhn* (Storm) missiles or quadrangular tubes for 55-mm R4M missiles. The *Bienenwabe* for the larger missile comprised 24 tubes and that for the smaller missile had 33 tubes, and these honeycomb installations were finally adopted. Detail design proceeded in parallel with testing in the DVL (*Deutsche Versuchsanstalt für Luftfahrt* — German Aviation Experimental Establishment) wind tunnel at Braunschweig, and during tunnel testing speeds in excess of Mach 0·95 were simulated without adverse stability or compressibility effects.

Structural simplicity

The entire airframe of the *Natter* was of wood, metal being used only for control push-rods, hinges, and load-supporting attachment points. The fuselage was a semi-monocoque with a laminated skin, and the wing possessed a single laminated wooden spar which was continuous from wingtip to wingtip, and passed between the fuselage fuel tanks. No movable surfaces were carried by the wing, rolling control being obtained by differential operation of the elevons forming part of the horizontal tail surfaces. The tail assembly was of asymmetrical cruciform design in that the tailplane was mounted above the fuselage and the vertical surfaces extended below the fuselage. The tailplane, being large by comparison with the wing, contributed an important proportion of the total lift, and both wing and tailplane were of rectangular planform without dihedral, taper or sweep. The wing utilised a symmetrical aerofoil, thickness to chord being 12 per cent and maximum thickness being located at 50 per cent chord.

A jettisonable plastic fairing enclosed the forward end of the *Bienenwabe* missile housing prior to firing, and considerable importance was attached to the protection of the pilot, the forward cockpit bulkhead being constructed of armour

(Above) The Natter M 1 on the towing trolley for mounting beneath an He 111 bomber for air-launching trials, and (below) the Natter M 3 fitted with a rudimentary fixed undercarriage.

A pre-series Natter mounted on a transport trailer and found by the Allies at Waldsee. One of the Schmidding take-off booster rockets may be seen in the foreground.

plate and cut away at the base in order that the pilot's feet could reach the rudder pedals which were positioned one on each side of the *Bienenwabe*; sandwich-type armour was provided on each side of the pilot's seat, and aft protection was provided by a rear bulkhead of armour which divided the cockpit from the fuel bay. Instrumentation and equipment were spartan, and a simple ring sight projecting from the nose ahead of the cockpit was the pilot's sole aid in aiming the rocket armament.

Immediately aft of the rear bulkhead were the two fuel tanks, that above the wing spar housing 95·7 Imp gal (435 l) of *T-Stoff* (80 per cent hydrogen peroxide plus oxyquinoline as a stabiliser), and the tank below the spar accommodating 41·8 Imp gal (190 l) of *C-Stoff* (a 30 per cent hydrazine hydrate solution in methanol and water). Behind the tank bay was the HWK 509A-1 rocketmotor which weighed only 220 lb (100 kg) complete, and comprised two principal assemblies, the forward assembly consisting of the turbine housing, the worm-type fuel pumps geared to the turbine shaft, the control box, a pressure-reducing valve, and the electric starter motor, and the aft assembly being formed by the combustion chamber unit which was connected to the forward assembly by means of a cylindrical tube through which pipes carried fuel to the individual jets. This motor offered a maximum thrust of 3,307 lb (1 500 kg).

Modus operandi

The *Natter* was to be launched from a near-vertical 80-ft (24,50-m) ramp, the wingtips and the tip of the lower fin being reinforced to run in three guide rails. The ramp was pivoted at its base in order that the *Natter* could be "loaded" in the horizontal position, and as the thrust-to-weight ratio was marginally short of 1:1 and therefore insufficient for vertical take-off, it was proposed to augment the thrust of the HWK 509A-1 rocket motor with four 1,102 lb (500 kg) thrust Schmidding 533 solid-fuel booster rockets. Attached to the rear fuselage, these rockets had a firing time of 10 seconds, after which they were to be jettisoned, and it was calculated that the initial acceleration would not exceed

2·2 *g*. However, as a safeguard against the possibility of the pilot blacking out, the elevons were pre-set for the required flight path while the *Natter* was still on the ramp, and a three-axis autopilot ground-controlled by radio link assuming guidance of the interceptor at an altitude of 560–600 ft (170–180 m), at which point the booster rockets were to be jettisoned.

At a range of 1–2 miles (1,5–3,0 km) from the bomber formation the pilot was to override the autopilot, jettison the nose of the *Bienenwabe* to expose the battery of rocket missiles, close with the bombers and fire the entire complement of rockets in one salvo. It was calculated that the rockets would scatter to cover an area that would be occupied by a four-engined bomber at a distance of 600 yds (550 m). The pilot would then turn away from the bomber formation and, at a safe distance, actuate the bale-out sequence. This entailed releasing his seat harness, uncoupling the control column, and releasing the safety catches and mechanical connections holding the nose section. The nose would then fall away from the aircraft, complete with windscreen, instrument panel, forward bulkhead and rudder pedals, simultaneously releasing a parachute housed in the rear fuselage. The deceleration resulting from deployment of the chute was expected to throw the pilot forward and clear of the aircraft, and he would then descend by parachute in the normal way. As the sole purpose of the pilot was to direct the aircraft during the final phase of the attack, the scheme offered the possibility of employing personnel without any training other than that which could be provided on a rudimentary ground rig.

A measure of the impetus placed behind the *Natter* development programme may be gauged from the fact that the first of an initial series of 50 test and evaluation models was completed at Waldsee during October 1944. It was proposed that all 50 pre-series *Nattern* should be used for gliding trials and for testing the Schmidding boosters, and the first example, referred to by the factory as the M 1, was tested in gliding flight during the course of November. Having no form of undercarriage, the M 1 was mounted on a rudimentary three-wheeled trailer so that it could be towed into position for mounting beneath an He 111 bomber adapted to serve as the "mother" aircraft. Ballasted to a weight of 3,748 lb (1 700 kg), the M 1 was carried to an altitude of some 18,000 ft (5 500 m) and then released over the Heuberg Army Test Range.

During this first flight the pilot of the *Natter* dived the tiny warplane to attain a speed of 425 mph (685 km/h), and, after completing the scheduled test programme, baled out. His subsequent report that the stability was excellent, and that the controls remained light and effective from 124 mph (220 km/h) up to the maximum speed attained, led to the idea of trying to land the unpowered *Natter* after each gliding test instead of abandoning the aircraft. Therefore, the third airframe completed, the M 3, was fitted with a fixed tricycle undercarriage and a drogue chute which was to be deployed at touchdown, which was expected to be of the order of 140–150 mph (225–240 km/h).

An attempt to land the *Natter* had still to be made when, on 18 December 1944, an empty airframe was fitted with the four Schmidding booster rockets, mounted on the launching ramp, and the rockets ignited. This test was a complete failure, the *Natter* failing to leave the ramp as a result of the rockets burning through the release cables. A second attempt was made four days later, and on this occasion the aircraft left the ramp as planned and disappeared into the cloud base at an altitude of 2,460 ft (750 m). Ten more successful launchings of unmanned *Nattern* followed, although it was ascertained that climbing speed attained by the time the booster rockets had burned out and been jettisoned — calculated to be closely comparable with that likely to be reached by a fully tanked-up manned *Natter* benefiting from the additional thrust of the HWK 509A-1 rocket — was insufficient to provide full control surface effectiveness.

To remedy this defect the vertical tail surfaces were substantially enlarged, the upper rudder portion being virtually doubled in chord, and the ventral portion of the fixed surface being reduced in depth and elongated. The chord of the elevons was also increased, these changes being introduced on M 16 and all subsequent *Nattern* built. At the same time it was proposed that small water-cooled control vanes be fitted in the rocket exhaust orifice. Although it was anticipated that such vanes would have a life of only some 30 seconds, it was believed that this period would be

Key to Bachem Ba

1   Rudder
2   Rudder post construction
3   Tail fin construction
4   Tailplane construction
5   Elevon
6   Exhaust orifice

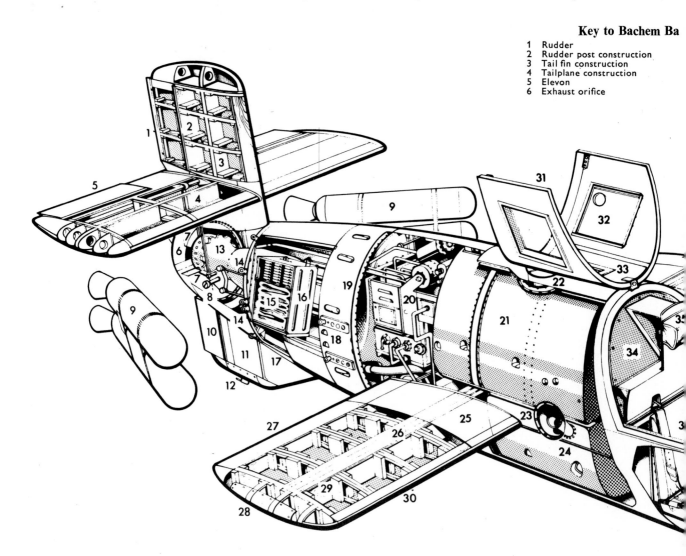

sufficient to enable the *Natter* to remain under full control until a speed could be attained at which the control surfaces became fully effective.

The original plan to use all 50 pre-series aircraft for gliding trials and pilotless launchings had by now been abandoned because of the time factor, and Heinrich Himmler's loss of interest in the programme had resulted in something of the drive behind the *Natter* scheme departing. Indeed, on 22 December 1944, the day on which the first successful launching of a pilotless *Natter* had taken place, the Chief Development Commission (*Entwicklungs Hauptkommission*) for aircraft met in Berlin to consider the various experimental programmes being conducted, and recommended that all preparations for the series production of the *Natter* should be discontinued at once. The Commission suggested that every effort should be made to expedite development of the Me 263, and decided that neither Heinkel's *Julia* nor Bachem's *Natter* held sufficient promise to warrant continued development, the potential of the Me 263 and the mixed-power version of the Me 262 being such as to render all other target-defence interceptors superfluous. It was concluded, therefore, that work

on *Julia* should be discontinued forthwith because of what was referred to as a "totally inadequate" endurance; that work on Junkers' *Dolly* should be suspended, resumption of its development depending on results achieved with the Me 263 and the mixed-power Me 262, and that development of the *Natter*, while opposed on both technical and tactical grounds, should be completed owing to the advanced stage that it had attained, but that all preparations for its series production should be stopped.

The pronouncements of the Commission were no more than recommendations, however, and while they were acted upon in the case of *Julia* and *Dolly*, they were never implemented in so far as the *Natter* was concerned. However, Bachem's creation was facing problems totally unrelated to the disapproval of the pundits in Berlin. The Patin PDS three-axis autopilot was proving unreliable and difficult to synchronise, and the behaviour of the Schmidding booster rockets was discovered to be erratic, burning time and thrust varying, and several units exploding while under test at Waldsee. Furthermore, the promised deliveries of HWK 509A-1 rocket motors by the Walter-Werke had failed to materialise. Indeed, the first HWK 509 was destined not to reach the Bachem-Werke until February 1945.

On the other hand, the airframe had proved virtually trouble-free, demanding only 250 manhours to produce, and being built for the most part by semi-skilled and unskilled labour in small wood-working shops in and around the Schwarzwald. With the arrival of the first Walter motor, preparations were hurriedly completed for the first launching of a complete *Natter*, this taking place on 25 February with a dummy pilot seated in the cockpit. The launch was a complete success, and at a predetermined altitude the nose section and power plant section broke away, and both dummy pilot and Walter motor descended safely by parachute. The RLM, impressed by the results of this test, now performed a complete volte-face, studied disinterest in the *Natter* giving place to a demand that piloted trials of the fully-powered aircraft should begin immediately!

Erich Bachem voiced his opinion that such tests were premature, and in this he was supported by Professor Ruff of the DVL, but a sense of desperation now reigned in the RLM, all objections were overruled, and on 28 February, *Oberleutnant* Lothar Siebert, who had volunteered to test the *Natter*, clambered into a fully tanked-up machine. The *Natter* left the launching ramp successfully, the booster rockets fell away from the accelerating aircraft on schedule and all

20A Natter) cutaway drawing

7	Water-cooled control vanes	22	T-Stoff filler cap
8	Control rod linkage	23	C-Stoff filler cap
9	Jettisonable rocket clusters	24	C-Stoff tank (41·8 Imp gal;
10	Ventral rudder		190 l capacity)
11	Ventral fin	25	Wing skinning
12	Launch-rail strengthening	26	Laminated mainspar
13	Combustion chamber	27	Solid rear spar/trailing edge
14	Rocket attachment eyes	28	Wingtip launch-rail
15	Recovery parachute		strengthening
16	Spring-operated container	29	Wooden wing construction
17	Parachute exit hatch	30	Solid forward spar/leading
18	Forward rocket attachment		edge
	points	31	Hinged cockpit canopy
19	Fuselage break point	32	Side glazing
20	Walter HWK 509A-1 rocket	33	Roof glazing
	motor housing	34	Back armour
21	T-Stoff tank (95·7 Imp gal;	35	Headrest
	435 l capacity)	36	Seat padding
		37	Seat pan and harness
		38	Control column
		39	Instrument panel
		40	Armoured windscreen
		41	Recovery parachute cable
		42	Rudder pedal
		43	Missile control and fusing
			box
		44	Armoured bulkhead
		45	Ring sight
		46	Honeycomb (Bienenwabe)
		47	Hs 217 Föhn 73-mm missiles
		48	Jettisonable plexiglas nose
			cone

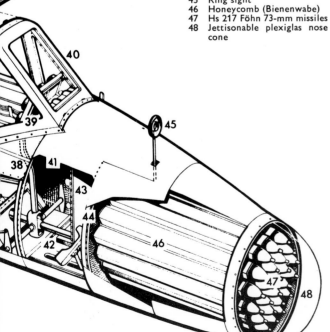

Bachem Ba 349B-1 Specification

Power Plant: One Walter HWK 509C-1 bi-fuel rocket motor possessing a maximum thrust of 4,410 lb (2 000 kg) of which 3,750 lb (1 700 kg) provided by main chamber and 660 lb (300 kg) by auxiliary chamber, plus (for take-off) four 1,102 lb (500 kg) or two 2,205 lb (1 000 kg) Schmidding solid-fuel rockets.

Armament: Twenty-four 73-mm Henschel Hs 217 *Föhn* or 33 55-mm R4M rocket missiles, or (proposed) two 30-mm MK 108 cannon with 30 rpg.

Weights: Take-off (including booster rockets), 4,920 lb (2 232 kg), (with boosters jettisoned), 3,900 lb (1 770 kg), (with fuel expended), 1,940 lb (880 kg).

Performance: Maximum speed, 620 mph (998 km/h) at 16,400 ft (5 000 m); initial climb rate, 37,400 ft/min (190 m/sec); range after climb, 36 mls (58 km) at 9,840 ft (3 000 m), 34 mls (55 km) at 19,685 ft (6 000 m), 26 mls (42 km) at 29,530 ft (9 000 m), 24 mls (39 km) at 32,810 ft (10 000 m); endurance, 4·36 min at 9,840 ft (3 000 m), 4·13 min at 19,685 ft (6 000 m), 3·15 min at 29,530 ft (9 000 m).

Dimensions: Span, 13 ft 1½ in (4,00 m); length, 19 ft 9 in (6,02 m); height (fin base to fin tip), 7 ft 4½ in (2,25 m); wing area, 50·59 sq ft (4,70 m²).

seemed well when, at about 1,640 ft (500 m), the cockpit canopy flew off and the *Natter* flicked on to its back. Inverted, the aircraft continued to climb, but at a shallower angle, and at some 4,920 ft (1 500 m) it performed a half-loop and then plummeted straight into the ground, exploding on impact. Intensive investigation failed to produce an entirely satisfactory explanation for the accident, but it was assumed that the canopy of the *Natter* had been insecurely fastened prior to launching, and that Siebert had blacked out.

Several pilots volunteered to take Siebert's place, and the test programme continued, three successful manned launchings being performed in rapid succession. The RLM now decided that the *Natter* had displayed an acceptable standard of reliability to warrant operational evaluation, and preparations were made to set up 10 fully-armed pre-series aircraft at Kirchheim, near Stuttgart.

Meanwhile, Bachem and Bethbeder, dissatisfied with the powered endurance of the *Natter,* had decided to adapt the airframe to take the HWK 509C motor with auxiliary cruising chamber. This necessitated some redesign of the aft fuselage to accommodate the vertically-disposed rocket pipes, and, for aerodynamic reasons, the lower contours of the fuselage were marginally deepened. No attempt was made to increase fuel capacity, but for cg reasons the attachment

points for the booster rockets were moved aft, and provision was made to supplant the four 1,102-lb (500-kg) Schmidding booster rockets with two 2,205-lb (1 000-kg) units. These modifications resulted in an increase in launching weight of only 127 lb (57,5 kg), and flying weight was virtually unchanged, but powered endurance at 9,840 ft (3 000 m) and 497 mph (800 km/h) was increased from 2·23 to 4·36 minutes. This development, designated BP 20B by the Bachem-Werke and Ba 349B by the RLM, met with the full approval of the *Technischen Amt,* and it was decided that the initial production version, the BP 20A or Ba 349A, would be discarded in favour of the modified model, and that the 51st *Natter* off the line (ie, the first production example) should be completed to Ba 349B standards.

In the event, only three experimental Ba 349B interceptors were to be completed, these being rebuilt pre-series airframes, and only one of these was destined to be flight tested. Similarly, the pre-series *Nattern* set up on ramps at Kirchheim had no opportunity to use their venom, as, with the approach of Allied armour, they were destroyed to prevent them falling into enemy hands. At that time, a total of 36 *Nattern* had been completed at Waldsee, and of these 25 had been flown either as gliders or in powered form, although only seven had been tested with pilots at the controls. □

FIGHTERS ——————————— *from page 196*

Most significant of all, Supermarine had used wooden monocoque construction for the S.4 of 1925, but switched to stressed-skin metal structures for their later racers. In terms of primary structure, fighters had changed from wood in the 1926 Gamecock to metal in the 1927 Siskin IIIA, but for ten more years the RAF was to persist with fabric coverings.

By 1931 the message was perfectly clear: high performance stressed-skin metal monoplanes could be produced around de-rated versions of the Rolls-Royce liquid-cooled racing engines, using leaded fuels to permit high compression ratios without detonation. This high octane petrol had been first tested at Wright Field in 1922, and had subsequently been used by the US Navy from 1926, also in the Schneider Trophy races from 1927, and by the US Army Air Corps from 1928. In spite of this, it was not until 1933 that the first British Air Ministry specification was written for a leaded fuel!

As noted before, it also remained for the fighter designer to make use of trailing edge flaps, so that smaller wings could be used, combining low drag with the high roll rate otherwise only possible in a biplane. In addition, Schneider Trophy racers had shown the need for variable-pitch propellers, since their fixed coarse pitch (optimised for maximum level speed) gave poor take-off, and yet in a fighter would not have been coarse enough for high speed dives. Variable-pitch airscrews were in fact used (along with semi-retracting main gear) in the 188 mph (300 km/h) Boeing YB-9 monoplane bomber of 1931.

Finally, designers had yet to develop low-drag radiators which avoided the vulnerability of the surface-cooling systems used for both water and oil in the Supermarine racers. There can be little doubt that all of these things were technically feasible in 1931; what was lacking was the motivation.

Phoenix from the ashes

Germany had been forbidden to re-create an air force by the 1919 Treaty of Versailles. However, pilot training had continued secretly, firstly at Lipezk in the Soviet Union and later in Italy. In March 1931 the existence of the new Luftwaffe was revealed, and two years later Hitler came to power, intent on re-uniting the German-speaking peoples in a single state, irrespective of existing frontiers.

Although the Nazi menace may not have been taken very seriously in Britain in the early 1930s, a series of plans was formulated to build up the strength of the RAF, and some thought was given to improved fighters.

However, the defence threat was now assessed as formations of bombers from Northern Germany, and thus operating beyond the radius of escorts. For this reason, combined with the doubling of fighter speeds since the end of the previous war, it was assumed that dogfighting was outdated, and that the fighter pilot's trade consisted solely of attacking in tight vics of three. In the 1930s, RAF fighter tactics thus became a matter of pure dogma, every eventuality being covered by one of six standard attacks, laid down in the Fighter Manual and practised assiduously.

Although single-seat biplane fighters lacked the necessary performance advantage over the new bombers, work on this category continued. Indeed, the even slower two-seat fighter enjoyed a revival in the form of the 182 mph (291 km/h) Hawker Demon, which entered service in 1933. The single-seat Gloster Gauntlet of 1935 represented some improvement over the Fury, having a top speed of 230 mph (368 km/h) and a service ceiling of 33,500 ft (10 220 m). However, it retained an open cockpit and the traditional twin-Vickers armament.

With increasing speeds, the problem of firing a lethal burst in the shortening time available was causing concern. The unreliability of the Vickers Mk III gun necessitated its breech mechanism being within reach of the pilot, and hence total rate of fire was limited by both space considerations and the synchronising rear. This led to the more reliable Colt being licensed in Britain as the Browning gun, which had no such restrictions on its installation.

Four of these guns were fitted in the RAF's ultimate biplane fighter, the Gloster Gladiator, which also had the distinction of introducing the closed cockpit. By 1937, when this type was issued to the squadrons, the power of production engines had at last begun to increase, and the Gladiator used an 840 hp Bristol Mercury IX radial to achieve 253 mph (405 km/h).

However, in that same year the Bf 109B-1 entered service, and the time for rehashing the Sopwith Pup and for flying pretty formations at Hendon was over. Germany was on the march again. □

(To be continued)

Emil versus the Luftwaffe

Thirty years ago, in April 1941, the anomalous situation arose in which the Messerschmitt Bf 109E-equipped Jagdgruppen of the Luftwaffe committed to the assault on Yugoslavia found themselves opposed by fighters of the same type as those they were flying. Some 18 months earlier, the Yugoslav Royal Air Force had begun to add the 'Emil' to its inventory, and now Bf 109E was pitted against Bf 109E! The hitherto untold story of this famous German fighter in Yugoslav insignia is related here by Zoran Jerin

"ATTACK all 109s with yellow noses!" This rather curious order was issued to all Yugoslav fighter pilots and anti-aircraft batteries within hours of the launching of Operation *Marita,* as the assault on Greece and Yugoslavia was known to the *Wehrmacht.* The Yugoslav High Command had a cogent reason for issuing such an order; numerically the most important fighter in Yugoslav service was the Messerschmitt Bf 109E, and, apart from national insignia which could not be identified at any great distance, the only discernible difference between the Bf 109E fighters of the opposing sides was provided by the yellow paint that had been applied to the noses of the *Luftwaffe* aircraft.

The situation was, at very least, confusing. Little effort had been placed behind aircraft recognition training in the Yugoslav armed forces, and few could differentiate between a Bf 109E and any other single-engined fighter, with or without a yellow nose. In the chaotic conditions prevailing during those April days of 1941, it was common practice to fire first and ascertain the identity of the target afterwards, and in such a situation tragic mistakes were unavoidable. But if the use of the same fighter by both sides was confusing for the Yugoslavs it must have been equally so for the *Luftwaffe.*

The first Messerschmitt Bf 109E fighters had reached the Yugoslav Royal Air Force, or *Jugoslovensko kraljevsko ratno vazduhoplovstvo* (JKRV), in the early autumn of 1939, some 20 months after the supply of such aircraft had been officially requested. Negotiations for the purchase of Messerschmitts had begun in January 1938 when Yugoslav Premier Stojadinović had visited Germany with the primary purpose of acquiring modern German weapons for the Yugoslavian armed forces. The military attaché at the Royal Yugoslav Embassy in Berlin, Colonel Vauhnik, had been highly impressed by the capabilities of the Bf 109, and had expressed his anxiety that this outstanding fighter should be purchased for the JKRV, and when he and Premier Stojadinović met *Reichsmarschall* Hermann Göring to discuss Yugoslavia's weapons requirements, several dozen Bf 109s were a priority item on the list.

Göring exuded cordiality towards his guests, but when the matter of supplying the JKRV with Bf 109 fighters was reached, the *Reichsmarschall* interrupted Stojadinović brusquely: *"Meine liebe Herren!",* Göring said, "There are no weapons on our secrets list that I would not recommend be released to the glorious Serbian Army which I so deeply admired during the last war, but as your sincere friend, I would dissuade you from the Bf 109. There is too much with which your brave pilots are unfamiliar in this new aircraft. They should convert by degrees to such fast warplanes or they will suffer many casualties, particularly during landings, and will develop a fear of really advanced fighters. I can recommend to you with all my heart the excellent Do 17 bomber . . . ." Stojadinović was of the opinion that Göring was being perfectly sincere, but Colonel Vauhnik persisted in his demand for Bf 109s, and it was on the supply of these fighters and anti-tank artillery that discussions very nearly broke down. But the iron, chrome and copper ore that Yugoslavia was offering in payment for German weapons was of vital importance to German industry, and in the end Göring relented, and agreed to fulfil all Yugoslavia's requirements.

Nevertheless, some 15 months were to elapse before contracts for the supply of Bf 109s were to be finally ratified, the *Reichsluftfahrtministerium* constantly raising minor issues to delay signature. Eventually, on 5 April 1939, the initial contract for 50 Bf 109Es and 25 spare DB 601A engines was signed, and eleven weeks later, on 23 June, a supplementary contract was signed for a further batch of 50 Bf 109E fighters.

The 6th Fighter Regiment (*6. Vazduhoplovni lovački puk*) had been designated the recipient of the Bf 109 fighters, and early in the autumn of 1939, the first three Messerschmitts, ferried from Augsburg, arrived over Zemun where a large group of senior JKRV officers, the German air attaché and his staff, and a number of representatives of the Messerschmitt AG, were gathered to witness the delivery. The high-pitched whine of the Daimler-Benz engine and the nose-up landing attitude of the Bf 109E were novelties to the JKRV personnel, and the first two fighters touched-down smoothly but the third, to the astonishment of the spectators, demonstrated a perfect belly landing! A plausible answer for this accident was given by the ferry pilot, and a totally different explanation was given by the representatives of the Messerschmitt AG, who pointed out that, as the fighter had suffered comparatively little damage, this untoward accident had, in fact, served to demonstrate to JKRV pilots that the Bf 109E could be landed on its belly with impunity!

The fighters delivered to the JKRV were of the Bf 109E-3 sub-type, and these began to arrive in small groups flown by Yugoslav pilots, but on one occasion a group that had taken-off from Frankfurt landed on a Rumanian airfield as a result of poor navigation. In the event, of the 100 Bf 109Es ordered for the JKRV, only 73 fighters were delivered, and owing to totally inadequate spares backing, many of these were to spend their brief lives unserviceable, sometimes being grounded as a result of non-availability of as simple an item as a spare tyre.

Conversion problems

The JKRV soon had cause to wonder if Hermann Göring had not been sincere in his attempt to dissuade the Yugoslavs from acquiring the Messerschmitt fighter. The 6th Fighter Regiment was soon suffering a considerable number of landing accidents. As one former pilot of this Regiment recently recalled: "The usual picture of Zemun airfield with two or three nosed-over Furies scattered around changed after the arrival of the Messerschmitt monoplanes. Now there were invariably two or three Bf 109Es flat on their bellies scattered around instead!" The same pilot is still, nevertheless, full of praise for the general handling qualities of the Bf 109E, which he regards as having been better than those of the Hurricane*, and the combination of slots and flaps was, he considers, particularly to be admired, although take-offs and landings could be nightmares for inexperienced pilots, and the narrow-track undercarriage and tricky variable-pitch VDM airscrew contributed to high attrition.

The major problem was that no suitable transition aircraft

(Above) A Bf 190E-3 with its ferry pilot shortly after arrival at Zemun in the autumn of 1939, and (left) a Bf 109G-12 two-seater after suffering a "slight accident" in postwar service with the Yugoslav air arm.

between the simple and forgiving Fury and the sophisticated Bf 109E was available to the JKRV. An attempt was made to employ the Bf 108B Taifun for this task, but it sometimes proved necessary for former Fury pilots to convert to the docile Hurricane for some 10 hours flying, and follow this with about 20 hours on the Bf 108B before finally converting to the Bf 109E! Various factors, including spares shortages, strictly limited the number of hours flown with the Bf 109E. The logbook of the previously-mentioned pilot included only 50 flying hours on the Bf 109E at the time of the German attack, and this was well above average number of hours flown on the Messerschmitt by its JKRV pilots. Few Yugoslav pilots could be considered to have gained complete mastery over the Bf 109E by the time they were committed to battle; hardly any had experience of flying the fighter at night or in bad weather, and the pilot who tested his oxygen supply for the first time on the day that he fired his first rounds of ammunition in anger — and, incidentally, scored his first "kill", an He 111 — was, if not typical, certainly not the most inexperienced of the JKRV's Bf 109E pilots.

In January 1941, the Bf 109E-3s — of which 64 were included in the order of battle at that time — were all assigned to the 6th Fighter Regiment whose strength also included eight indigenous IK-2s and nine IK-Zs, providing the backbone of the air defences of Belgrade, but during the following two months the JKRV was reorganised and the aircraft complement of each Regiment was reduced. Some Bf 109Es were handed over to the 2nd Regiment and all the IK-2s were transferred to the 4th Regiment, both of these regiments including Hurricanes in their strengths.

During this period, Yugoslavia was subjected to intense diplomatic pressure and threats of force from Germany, and on 25 March Yugoslavia's Prince Regent Paul had reluctantly agreed to join the Axis alliance. Two days later, on 27 March, patriotic anti-German elements in the Yugoslav Army staged a *coup d'état,* overthrew Prince Paul and rejected the Axis alliance. Hitler, infuriated by these events, immediately initiated preparations for an assault on Yugoslavia which commenced on 6 April with an air attack on Belgrade.

The order of battle of the JKRV Bf 109E fighters on that day was as follows: The 6th Fighter Regiment responsible for the defence of Belgrade comprised two Fighter Groups, one of which, the 32nd Fighter Group (*32. Lovačka grupa*), had two squadrons with a total of 22 serviceable Bf 109Es based at the unsuitable but well-camouflaged Prnjavor airfield, near Krušedol, some 30 miles (48 km) from Belgrade, and the other, the 51st Fighter Group, also had two squadrons, one with 10 serviceable Bf 109Es and the other with six IK-Zs, both based at Zemun, the principal Belgrade airfield. The 2nd Fighter Regiment, which had been assigned the task of defending the most important industrial targets, consisted of the 31st Fighter Group with 11 Bf 109Es based at Sušičko polje, near Kragujevac, and the 52nd Fighter Group with 18 Hurricanes based at Knić airfield. There were also three Bf 109Es based at Mostar as part of the Independent Squadron of the Fighter Training Unit (*Samostalna eskadrilja lovaćke škole*) which also had two Hurricanes and two elderly Avia BH-33E fighter biplanes. Thus, only 46 Bf 109Es were immediately available at the time of the German onslaught.

The first JKRV fighters to actually oppose the *Luftwaffe* on 6 April were in fact the IK-Zs of the 51st Fighter Group's 161.*eskadrilja* which took-off from Zemun in defence of Belgrade. Within minutes they were joined by the Bf 109Es of the 32nd Fighter Group from Prnjavor and the

The JKRV had received a dozen Hurricane Is from December 1938, and subsequently ordered a further dozen of which deliveries commenced in February 1940, licence manufacture being initiated by the Zmaj and Rogozarski concerns, but only 40 were on strength at the time of the German assault.

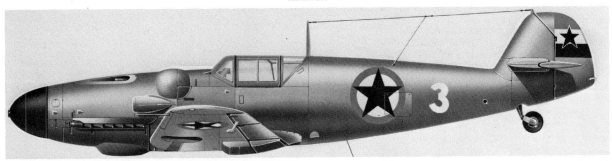

(Above) A Bf 109E-3 of the 6th Fighter Regiment, 1940. Upper surfaces finished black-green overall with pale blue under surfaces. Black identification letter and numerals. (Below) A Bf 109G-6 in postwar service. Pale grey finish overall with red spinner and white numeral.

51st Fighter Group's 102.*eskadrilja,* the Belgrade defenders having been warned on the previous evening of the imminence of the German onslaught. The IK-Zs and Bf 109Es attacked the first wave of the *Luftwaffe* in twos and threes, and succeeded in destroying some 10 of the intruders, but by the end of the day the 6th Regiment had lost 15 of its fighters — two IK-Zs and 13 Bf 109Es — which represented more than 40 per cent of its serviceable strength.

The 2nd Regiment, although based only some 50 miles (80 km) from Belgrade, was forbidden to join the mêlée over the capital as its task was the defence of the Serbian industrial towns. Nevertheless, two Bf 109Es from the regiment's 31st Group took-off from Sušičko polje against orders, and attacked the *Luftwaffe* formations over Belgrade. They succeeded in destroying two Ju 87B dive bombers but one of the JKRV Bf 109Es was shot down in the process, this being, in fact, the only 2nd Regiment Messerschmitt to be lost in the air during the conflict. Another pair of Bf 109Es from the 31st Group destroyed a lone Hs 126 engaged in tactical reconnaissance over eastern Serbia.

Ammunition presented a major problem from an early stage in the fighting as no incendiary bullets for the MG 17 machine guns of the Bf 109Es had been delivered from Germany, and therefore all ammunition from the MG 17s of *Luftwaffe* aircraft brought down was carefully salvaged and delivered to the fighter squadrons.

On 7 April, the pilots of the 6th Regiment were determined to avenge their heavy casualties of the previous day, and to engage the enemy in force, but they had already discovered that they were not opposed solely by the *Luftwaffe!* The order to fire on all 109s with *yellow noses* had had little effect on the understandably-nervous Yugoslav anti-aircraft batteries which had suffered their share of strafing from the *Luftwaffe,* and had soon adopted the practice of firing at anything with wings that came within range. Thus, the fighters of the 6th Regiment were finding it necessary to take evasive action to avoid their own anti-aircraft artillery. Furthermore, aircraft recognition was not their strong point and the greatest care had to be taken to ensure that their potential victims did, in fact, belong to the *Luftwaffe.* During the previous evening, one young pilot of the 32nd Group had shot down a JKRV Blenheim in the belief that it was a Ju 88!

Precious moments had been lost as a result of hesitation to attack Bf 109s in case they belonged to their own air arm.

Although fewer *Luftwaffe* aircraft appeared over Belgrade during 7 April, the 6th Regiment was again engaged in heavy fighting, losing a further 12 Bf 109Es and another IK-Z, while the 2nd Regiment confined its activities to the skies over East Serbia, chasing a few *Luftwaffe* reconnaissance aircraft without apparent results. The following day was relatively uneventful in the air owing to inclement weather, and during the evening the surviving serviceable fighters of the 6th Regiment — four Bf 109Es of the 32nd Group and three Bf 109Es and three IK-Zs of the 51st Group — were transferred to the Veliki Radinci airfield, near Ruma. On 9 April all aircraft were grounded owing to rain and sleet, but on the following day, on which Zagreb fell to the advancing *Wehrmacht,* the survivors took-off several times to engage the *Luftwaffe* in minor skirmishes.

By this time, the Yugoslav High Command had been paralysed and communications were in chaos, and in the belief that Paul von Kleist's Panzer Corps would arrive at its base at any moment, on 10 April the 2nd Regiment burned its fighters to prevent them falling into the hands of the enemy. A similar fate overtook the fighters of the 6th Regiment on the following day as a German armoured thrust approached the Veliki Radinci airfield, and the brief chapter relating to JKRV service in the long history of the Messerschmitt Bf 109 fighter had come to a close. However, the burning of the last serviceable Bf 109E at Veliki Radinci was not destined to be the end of the story of the Bf 109 in Yugoslav colours, for, with the liberation of Yugoslavia and the creation of a new Yugoslav air arm, the *Jugoslovensko ratno vazduhoplovstvo* (JRV), Messerschmitt's fighter once more appeared in the inventory, though in later form than that which had fought the *Luftwaffe* in Yugoslav skies. As the *Wehrmacht* retreated, many Bf 109Gs of both the *Luftwaffe* and the Croatian Air Force were abandoned on Yugoslav airfields owing to lack of fuel or spares, and those that could be made airworthy were incoporrated in the JRV, these later being supplemented by between 50 and 60 Bf 109G-2s, G-6s, G-10s, and two-seat G-12s which had served with the Royal Bulgarian Air Force and which were "presented" to Yugoslavia as war reparations. □

VETERAN & VINTAGE

THE GREAT DANES

SEVERAL excellent collections of historic aircraft exist in Denmark and some of the exhibits in these collections are of special merit. In Copenhagen, the Royal Danish Arsenal Museum is largely devoted to Army weapons and exhibits, but three aircraft are also on show, together with a German Fi 103 flying-bomb and a series of models to depict Danish aviation. The full-size aircraft exhibits include a Hawker Dancock, the only surviving example of this land-based Naval fighter that went into operation with the Danish Naval Air Service (*Marine Flyvevaesenets*) in 1926. In addition to three Dancocks acquired from the Hawker company in Britain, 12 examples were built in Denmark at the Naval Dockyard (*Orlogsvaerftet*) and it is one of these, number 158, that is exhibited.

Another British type built at the Naval Dockyard was the Avro 504N, and one of the four Danish-built aircraft, serial 110, is also on show. The third aircraft exhibit is a Berg and Storm monoplane, type BS 3, similar to the first aeroplane acquired by the Danish Army in 1912. This monoplane was designed by Danish engineers Olaf Berg and Louis Storm, and was powered by a 30 hp three-cylinder engine.

One of three surviving examples of the General Aircraft Monospar is among the exhibits at the *Veteransmuseum* at Egeskov Castle, Kvaerndrup, on the island of Fuen. This museum is devoted to all forms of

(Above) The General Aircraft GAL-25 Monospar OY-DAZ displayed at Egeskov Castle was in use as an ambulance until 1963. (Below) The Berg and Storm BS 3 at the Royal Danish Arsenal Museum in Copenhagen is similar to the Danish Army's first aeroplane, acquired in 1912.

A prized exhibit in the Copenhagen museum of the Royal Danish Arsenal is the Hawker Dancock, only surviving example of this 1927 biplane with an Armstrong-Siddeley Jaguar IV radial engine. No 158 is one of the 12 Dancocks built at the Danish Naval Dockyard.

(Above) The Lockheed 12A displayed at the Veteransmuseum at Egeskov Castle was last in use as OY-AOV and was operated on contract as a target-tug for the Danish Navy. It had been built in 1941 for the Dutch East Indies Airways, subsequently being pressed into military service in the area, and has been restored in Dutch military markings.

(Right) Two views of the Saab B 17A that was presented to the museum at Ekeskov by the Swedish Air Force in May 1970. It was previously operated in Sweden as a target-tug, registered SE-BWC, and was restored by Saab in Danish Air Force colours, recalling the operation of B 17Cs by a Danish squadron in exile in Sweden in 1944.

(Bottom right) Among the exhibits in the museum at Helsingoer is this Donnet-Leveque flying-boat, with a completely new hull but the original (reconditioned) wings built at the Naval Shipyard in Copenhagen in 1914.

Photographs by Howard Levy

ground transportation plus a collection of aircraft. The Monospar ST-25, OY-DAZ, flew as an ambulance in Denmark throughout the German occupation and until 1963. Other exhibits at this site include a Danish-built KZ-II, OY-DOU; a Lockheed 12A that served successively in the Dutch East Indies, Norway, Sweden and Denmark, and has been restored in its original Dutch military markings; a Canadian-built AT-16 Harvard in Danish Air Force markings, a Saab B 17A, ex-Swedish but now in Danish markings; a Saab J 29 in Swedish colours; a Canadian-built Fairchild PT-26 in Norwegian Air Force markings; a de Havilland Tiger Moth and a couple of gliders.

The third significant Danish museum is at Helsingoer and this is very largely devoted to the work of the Danish pioneer J C H Ellehammer. Apart from Ellehammer's 1906 monoplane in which he is claimed to have made the first flight in Europe, in September 1906, this museum contains his 1909 Standard Type and 1911 helicopter. Among the non-Ellehammer exhibits here are a Svendson Glenten, the Danish-built version of the Henry Farman which was the first aircraft acquired by the Danish Navy, in 1911; and a Donnet-Leveque flying-boat, including the original wings and 80 hp Gnome engine of the *Maagen* 2, one of two built for the Navy in Denmark, with a new hull constructed in 1966. ☐

One afternoon in late July of 1944, a Japanese anti-aircraft shell exploding in Manchurian skies initiated a chain of events that was to cost the North American tax-payer billions of dollars. Its immediate effect was to provide the Soviet Union with an example of the Boeing B-29 Super-fortress, around a Soviet copy of which was to be created a powerful strategic air arm. Its longer term effect was to provide a basis for the development, by genealogical processes stretching over a decade, of the only Soviet bomber capable of performing strikes across the polar regions, and the sole factor justifying the fantastic cost of building the immense chains of radar installations and interceptor bases intended to shield the North American continent from manned aircraft attack. Previous parts of this story appeared in the July and August issues of AIR ENTHUSIAST.

THE BILLION DOLLAR BOMBER

PART THREE

THE FOURTH anniversary of the arrival in the Soviet Union on 29 July 1944 of the first AAF Boeing B-29 Super-fortress saw much activity at several of the principal bases of the Soviet strategic bombing arm, the *Dal'naya aviatsiya*. The State Acceptance Trials of the Soviet copy of the B-29 had been completed satisfactorily several months earlier, and now a number of regiments, or *polki*, of the *Dal'naya aviatsiya* were simultaneously engaged in working up to operational status on this new warplane which endowed the *Voenno-vozdushniye Sily* (V-VS), the Air Forces of the USSR, with true strategic capability for the first time in its history.

The duplication of this large and sophisticated aircraft by Soviet industry was a technical achievement of no small moment; it was a feat that many western experts had declared impossible and one for which General Constructor Andrei N Tupolev had deservedly received the highest award that Iosip Stalin could bestow. For obvious reasons it was not an achievement that could be publicised to boost Soviet techno-logical prestige abroad, but the innumerable problems that the task of copying the American warplane had posed and the successful solution of these problems had provided the Soviet aircraft industry with the foundations upon which it

was to build a capability second to none; a capability of which the west was to be provided with ample evidence over the next decade.

During the summer of 1948, as the *Dal'naya aviatsiya* struggled to overcome the difficulties of phasing into service the Tu-4, as the Soviet copy of the Boeing bomber had been officially designated, the B-29 Superfortress was still the mainstay of the USAF Strategic Air Command which had yet to take on strength the first fully operational Convair B-36Bs. The threat that the Tu-4 posed the industrial centres of North America was undeniable. Assuming that year-round operations could be mounted from airfields at latitudes as far north as 70°, the Tu-4 possessed the necessary range to strike the principal centres of Chicago, Los Angeles and New York with a worthwhile load on one-way "suicide" missions. The interception of the Tu-4 did not, in itself, present any serious difficulties to the USAF Air Defense Command, but if the *Dal'naya aviatsiya* launched a suffici-ently large number of Tu-4s to penetrate the North American defensive periphery simultaneously from different points and at different altitudes, the chances were that enough Soviet bombers would evade interception to wreak serious damage.

(At head of page) The Tu-80 was a refined derivative of the Tu-4 with "wet" wing and lighter structure, and the drawing below shows the interior arrangement of the proposed production version of the Tu-80. It will be noted that the forward upper gun barbette was to have been semi-retractable.

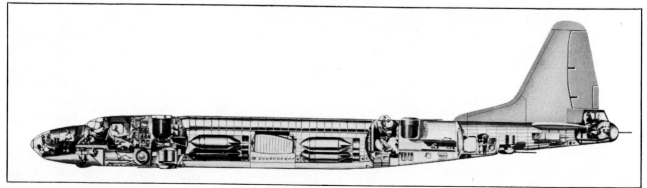

However, US intelligence agencies believed that it would be at least mid-1949 before the V-VS could conceivably have sufficient combat-ready Tu-4s with which to launch such an attack.

The US assessment of the Tu-4 delivery situation was not far wide of the mark as the production programme was bedevilled by delays in the delivery of equipment and by teething troubles that had not revealed themselves during the initial test and evaluation programme. Series production Tu-4s suffered continuously from malfunctions in the remotely-controlled gun barbette system and in the pressurisation system, and the reliability of the turbo-superchargers of the ASh-73TK engines left much to be desired. Quality control at the series production factories was tightened up, and by early 1949 the more serious problems had been resolved. The first full DA *Divisiya* to be equipped with the Tu-4 and comprising three *polki* each with three 10-aircraft *eskadrilii* had attained full operational status by mid-1949, and some 300 Tu-4s had entered the DA inventory by the end of the year. In addition, a small number of Tu-4s equipped for the long-range maritime patrol task had been delivered to the Naval Air Force, the *Aviatsiya Voenno-morskovo Flota* (AV-MF).

Powered by four ASh-73TK engines affording 2,200 hp for take-off, a war emergency rating of 2,400 hp and a maximum continuous rating of 2,000 hp at 25,590 ft (7 800 m), the Tu-4 allegedly achieved a maximum speed of 261 mph (420 km/h) at sea level and 354 mph (570 km/h) at 32,808 ft (10 000 m), such speeds presumably being attained at weights somewhat lower than the 135,584 lb (61 500 kg) quoted as the maximum take-off figure, this including 5,873 Imp gal (26 700 l) of fuel in the wing tanks. The range of the Tu-4 with an 11,023-lb (5 000-kg) bomb load at long-range cruise power at 9,845 ft (3 000 m) was 3,107 miles (5 000 km), average cruising speed being 224 mph (360 km/h), and with a 6,614-lb (3 000-kg) bomb load and weapons-bay auxiliary fuel tanks the range could be extended to 4,100 miles (6 600 km). At maximum continuous power at 25,590 ft (7 800 m) and an average cruising speed of 310 mph (500 km/h), range with an 11,023-lb (5 000-kg) bomb load was 1,926 miles (3 100 km).

Variations on the theme

While the *Dal'naya aviatsiya* had been struggling to bring the Tu-4 to operational status, Andrei Tupolev's design bureau had been engaged in refining the basic design of the bomber in an attempt to meet, at least in part, the demand for a greater operational range, and the result had been the Tu-80, the first prototype of which had begun its flight test programme early in 1949. The Tupolev team retained the basic Tu-4 structure but had increased the fineness ratio of the fuselage by lengthening the nose, the overall length being increased from 99 ft $0\frac{1}{4}$ in (30,18 m) to 112 ft $7\frac{1}{8}$ in (34,32 m), simultaneously introducing a stepped windscreen and enlarged vertical tail surfaces. In order to reduce drag the nacelles of the ASh-73TK engines were redesigned, and a "wet" wing was fitted, this increasing fuel capacity by some 15 per cent. Although embodying a revised structure, the wing retained the contours and aerofoil section of the Tu-4, and overall span was marginally increased from 141 ft $2\frac{3}{4}$ in (43,05 m) to 142 ft $6\frac{2}{3}$ in (43,45 m).

It was intended that the Tu-80 should carry similar offensive loads to those of the Tu-4 and feature a similar defensive system of remotely-controlled barbettes mounting

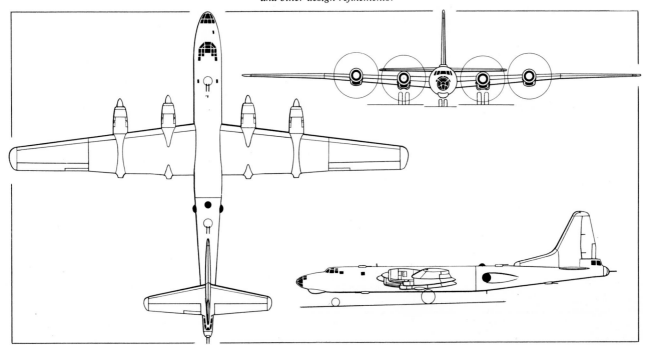

The Tu-80 (above and below) was essentially similar to the Tu-4 apart from the lengthened fuselage nose, enlarged vertical tail surfaces and other design refinements.

a total of 10 23-mm cannon or 12,7-mm machine guns. However, the gun barbettes were not fitted to either of the two Tu-80 prototypes as the requirement to which this Tu-4 derivative had been evolved had already been overtaken by a new specification aimed at providing the *Dal'naya aviatsiya* with a bomber comparable in capability with the Convair B-36 of the USAF Strategic Air Command. Thus, the Tu-80 progressed no further than the prototype evaluation stage, a fate shared by another derivative of the Tu-4 evolved to meet the same requirement, the DVB-202, which, featuring a shoulder-mounted wing, had been developed by a team led by Vladimir Myasishchev.

To meet this new specification, the Tupolev design bureau was, by mid-1949, immersed in the development of what was destined to be the largest aircraft built in the Soviet Union to that time, and also the last Soviet warplane to be powered by piston engines. Several engine design bureaux were engaged in the development of powerful turboprops and turbojets but Soviet-US relations were steadily deteriorating and time was of the essence — the *Dal'naya aviatsiya* had to be given true intercontinental atomic bombing capability with the least possible delay. It was manifestly obvious that several years must elapse before the new power plants could conceivably attain acceptable stages of reliability and economy for their use in an intercontinental bomber, but both the bureau of Arkadii D Schvetsov and that of V A Dobrynin were engaged in the development of large piston engines in the 4,000-plus hp class as a back-up programme, and therefore Tupolev's intercontinental bomber was designed from the outset to be powered by one or other of these.

It appears likely that the specification demanded the ability to carry an 11,023-lb (5 000-kg) bomb load over a distance of 4,350 mls (7 000 km) and then return to base without refuelling, and Tupolev's answer to this requirement was, to all intents and purposes, a substantially scaled-up Tu-80 which, in turn, had been merely a refined Tu-4. The proven aerodynamics of the earlier warplane were evidently adopted as a means of reducing the time factor in the development cycle of the new bomber to the absolute minimum, and the success of the Tupolev bureau in cutting the elapsed time from issue of specification to prototype flight test may be gauged from the fact that barely more than

two years were required by the Soviet team whereas more than *five* years had been needed by Convair to bring the closely-comparable B-36 to a similar stage.

During the course of 1950 bench running of both the Shvetsov ASh-2TK and the competitive Dobrynin VD-4K was initiated, both being 24-cylinder units in the 4,000 hp category. In the event, the latter was selected for installation in Tupolev's intercontinental bomber, the first prototype of which began its flight test programme as the Tu-85 early in 1951. A 24-cylinder turbo-supercharged engine, the VD-4K offered 4,300 hp for take-off and possessed a maximum continuous rating of 3,800 hp, specific consumption for cruise being 0·38 lb (175 gr)/hp/hr.

Weight consciousness was emphasised throughout the design of the Tu-85, although no attempt was made to employ in the structure such exotic materials as magnesium which saw extensive use in the Tupolev bomber's US counterpart, the B-36. The wing, which spanned 183 ft 6¼ in (55,938 m) and possessed a gross area of 2,947·56 sq ft (274,36 m²), made considerable use of thick-walled load-carrying panels, the number of ribs being reduced to a minimum, and housed tankage for up to 13,470 Imp gal (61 234 l) of fuel. The fuselage, which had an overall length of 128 ft 11¼ in (39,30 m), was a metal semi-monocoque structure built up of a series of circumferential bulkheads and frames, extruded longerons and stringers and butt-jointed stressed metal skin. Comprising five sections, it embodied three pressurised compartments, one forward and one aft of the weapons bays and one in the extreme tail. A crawl tunnel some 46 ft (14 m) in length over the weapons bays interconnected the two forward compartments but the tail compartment was isolated. These compartments housed a total of 16 crew members, and defensive armament followed the pattern set by the Tu-4, comprising four remotely-controlled barbettes each with two 23-mm cannon, two above and two below the fuselage, and a further pair of cannon in the tail position. Up to 44,092 lb (20 000 kg) of bombs could be accommodated by two weapons bays, one of some 14·3 ft (4,35 m) length forward of the wing carry-through structure, and one of about 21·45 ft (6,55 m) length immediately aft.

The Tu-85 had a maximum take-off weight of 235,892 lb (107 000 kg), this including 97,000 lb (44 000 kg) of fuel and

The Tu-85 (above and below) remains dimensionally the largest aircraft ever built in the Soviet Union, and was the last combat aircraft of Soviet design to be powered by piston engines.

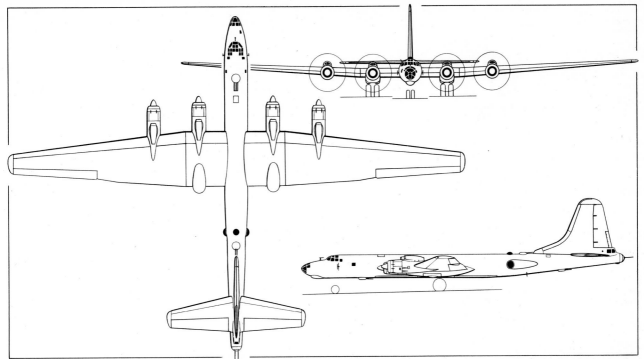

The general arrangement drawing above and photograph below of the Tu-85 reveal clearly its B-29 ancestry. Experience gained in the development of this immense aircraft was used in the design of the turboprop-powered Tu-95 which was to enter V-VS service as the Tu-20.

oil and an 11,023-lb (5 000-kg) bomb load, and flight testing revealed that this offensive load could be carried over a range of 7,456 miles (12 000 km) at an average cruise of 339 mph (545 km/h) at 32,810 ft (10 000 m). The normal range of the Tu-85 was considered to be 5,500 mls (8 850 km), and for shorter-range missions a maximum weapons load of 44,092 lb (20 000 kg) could be carried. Cruising speed at 32,810 ft (10 000 m) ranged from 310 mph (500 km/h) to 367 mph (590 km/h) according to weight, and the maximum attainable speed at this altitude (presumably in lightly loaded condition) was 404 mph (650 km/h), but there can be little doubt that the Tu-85's capabilities fell somewhat short of the requirements of the *Dal'naya aviatsiya* in respect of range, a factor motivating against series production of this type as a successor to the Tu-4 in the ranks of the strategic bombing force. Furthermore, the decision of the USAF to order the B-52 Stratofortress into production announced in February 1951, shortly before the Tu-85 commenced its flight test programme, must have underlined the approaching obsolescence of the piston-engined strategic bomber. Thus, the decision was taken to abandon the further development of the Tu-85 and concentrate on more advanced, turbine-powered strategic bombers capable of rivalling the payload-range characteristics of the B-52.

Vladimir Myasishchev's bureau was promptly assigned the task of designing a turbojet-powered strategic bomber which was to emerge several years later as the M-4, while Andrei Tupolev's bureau was instructed to evolve in parallel a turboprop-driven strategic bomber for which suitable power plants were being developed by Nikolai D Kuznetsov's bureau. At this time — mid-1951 — the Tupolev team was well advanced in the construction of prototypes of the

Tu-88 twin-jet medium bomber which was to enter service in the mid 'fifties as the Tu-16. The Tu-88 embodied much of the fuselage structural design of the Tu-80 of 1949, and a similar approach was made by the Tupolev team to the design of the new turboprop-powered strategic bomber, this being allocated the design bureau designation of Tu-95 and embodying an essentially similar fuselage to that of the Tu-85. The Tu-95, which was to join the *Dal'naya aviatsiya* in the mid 'fifties as the Tu-20 and provide the backbone of this strategic force for nearly a decade, was thus to be directly related to the Tu-4 and, in consequence, the B-29 Superfortress and the Boeing drawing boards.

The decision to proceed no further than prototype evaluation with the Tu-85 meant that the *Dal'naya aviatsiya* would have to soldier on into the second half of the 'fifties with the rapidly ageing Tu-4, and proposals were made by the Tupolev bureaux to re-engine the bomber with turboprops in order to boost its performance. At the time the decision was taken to discontinue further work on the Tu-85, the development bureaux of both Nikolai D Kuznetsov and Aleksandr G Ivchenko were bench-running turboprops in the 4,000–4,500 eshp category, the former bureau being concerned with the TV-4 (TV = *Turbovintovy*) and the latter with the TV-02. The Tupolev bureau prepared studies of the Tu-4 powered by four Kuznetsov TV-4 turboprops each driving six-blade contra-props and by four Ivchenko TV-02s driving four-blade airscrews, but both power plants were at early stages in their development and the time factor in bringing these turboprops to acceptable stages of reliability and economy motivated against the adoption of the re-engining scheme.

In addition to operational service with the DA and the AV-

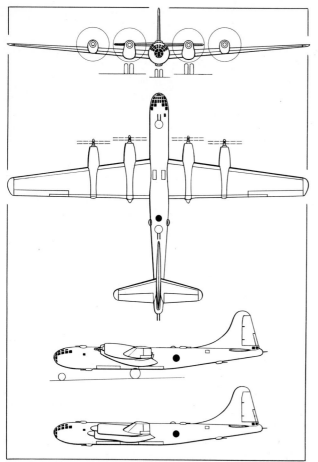

The general arrangement drawing above depicts the Tu-4 in the form proposed with TV-4 turboprops, the lower sideview illustrating the alternative version with TV-02s.

MF, the Tu-4 participated in numerous development programmes associated with the equipment and features of its intended successors, the M-4 and Tu-20. Fedor F Opad'chy, who had been responsible for the flight testing of the Tu-70 and had subsequently transferred from the Tupolev to the Myasishchev bureau, performed a number of trials with one such Tu-4 modified to have an undercarriage of similar configuration to that of the M-4 jet bomber then under development (ie, tandem main bogies and outriggers) and fitted with 19,840 lb (9 000 kg) thrust take-off rockets to simulate the anticipated take-off characteristics of the Myasishchev-designed bomber.

Other Tu-4s were used for flight refuelling trials, and as the type was progressively withdrawn from operational service with the *Dal'naya aviatsiya* during the second half of the 'fifties it was introduced into service with the air transport force, the *voenno-transportnaya aviatsiya,* to supplement the short-range Li-2s and Il-14s. For the transport rôle the Tu-4 was fitted with loaded containers beneath the wings. These, understandably enough, seriously impaired the performance of the aircraft, and with the availability of the Antonov An-12, the Tu-4 was withdrawn from the transport task. By the time the 'sixties were born, the Tu-4 had also virtually disappeared from the inventory of the shore-based maritime patrol force which had provided this Soviet copy of the B-29 Superfortress with its last operational rôle, at least, in Russian service. Sufficient Tu-4s had been supplied to the Sino-Communists to provide at least a token bombing force, and the type was reported as being still in service in China as recently as 1968 — thirty years after the B-29 had begun its journey across the Boeing drawing boards.

(To be concluded next month)

HAMPDEN————————————————*from page 199*

quagmire with very limited hard paving. When taxying over the heavy ground the Daggers over-heated and often could not be started again. Most, if not all, of our Herefords were being prepared at that time so that they could be flown away for re-engining with Pegasus.

In the end — ashamed of sending for ferry pilots to collect aircraft which might or might not be serviceable on their arrival — we decided that the best plan was to fly them on test and, if all went reasonably well, to carry straight on with the delivery flight. On one of these trips — to Tollerton, Nottingham, with a very troublesome Hereford that had finally been made to go — I got myself well and truly lost in bad visibility north-east of Birmingham while avoiding the cloud-covered southern Pennines which were on the direct track to Tollerton. For the better part of half-an-hour I'd recognised nothing down below and none of the many railway lines matched those on my rain-sodden map. Then I saw an aerodrome with a totally unfamiliar runway layout and decided I'd better put down even if the engines could never be started again.

Temporarily avoiding the embarrassment of admitting the truth of the situation, I strolled, with my engine-fitter passenger (who had come for the ride), to the watch office to report my arrival and proposed destination. My luck was in; the watch office had a wall map with a compass rose and bearing string. As usual the map was worn away inside the rose, but I could see roughly where we were. While chatting about the prospects of an early weather clearance I realised that the place must be RAF Lichfield and relaxed in the comforting knowledge. In due course the weather lifted; both Daggers started without trouble and ran like noisy sewing machines during the 20-minute flight to Tollerton.□

Three variants of the Hampden shown here are (top to bottom) the prototype K4240 with original "bird-cage" front-end and early dorsal gun emplacement; Hampden TB Mk I torpedo-bomber and Hampden II with Cyclone engines, distinguished by the short-chord cowlings.

PLANE FACTS

High-flying Junkers

I shall be glad if you will include in 'Plane Facts' a general arrangement drawing and some details of the Junkers Ju 49 research plane of the early 'thirties.

Christian Emrich, 8051 Neufahrn, Germany

The Junkers Ju 49ba was a specialised high-altitude research aircraft, development of which was initiated in 1928 in collaboration with Dr-Ing Asmus Hansen of the *Deutsche Versuchsanstalt für Luftfahrt* (German Aviation Experimental Establishment). Initially registered D-2688 and later D-UBAZ, the Ju 49ba embodied a noteworthy innovation by comparison with previous high-altitude aircraft, a pressure cabin accommodating two crew members. The pressure cabin, constructed of *Elektron* alloy, was inserted in the airframe as a capsule, small portholes providing forward and side vision for the pilot and observer. Power was provided by a Junkers L 88a 12-cylinder liquid-cooled engine rated at 800 hp for take-off and driving a large-diameter four-bladed airscrew, but various problems delayed the commencement of flight testing until 2 October 1931. The L 88a engine proved extremely troublesome, necessitating the abrupt termination of most test flights, but in September 1933 an altitude in excess of 29,530 ft (9 000 m) was attained. Two years later, in 1935, the Ju 49ba reached 42,650 ft (13 000 m).

The Ju 49ba. had empty and loaded weights of 7,937 lb (3 600 kg) and 9,370 lb (4 250 kg) respectively, and overall dimensions included a span of 92 ft 8½ in (28,26 m), a length of 56 ft 5⅛ in (17,20 m), and a height (tail down) of 15 ft 7 in (4,75 m).

The Junkers Ju 49ba high-altitude research aircraft is seen above in its original form with vertical exhaust stack. The general arrangement drawing below illustrates the definitive exhaust arrangement.

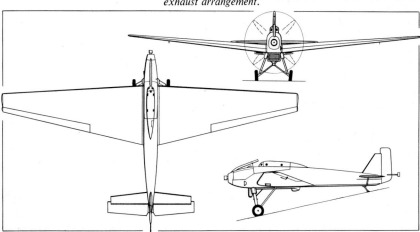

Argentine Arrow

Could you please give details in your 'Plane Facts' column of the Pulqui I fighter developed in Argentina by Emile Dewoitine immediately after World War II?

Terence Ansell, Watford, Herts

Work on the I.Ae.27 Pulqui (Arrow) single-seat fighter was initiated at the Instituto Aerotecnico, Cordoba, by Emile Dewoitine in 1946. Of all-metal construction, the Pulqui employed an I.Ae.242-1 high-speed laminar-flow aerofoil and a Rolls-Royce RB.37 Derwent 5 centrifugal-flow turbojet rated at 4,000 lb (1 814 kg) for take-off. Fuel

capacity was 264 Imp gal (1 200 1) and armament comprised four 20-mm Hispano cannon. The sole prototype was flown for the first time on 9 August 1947, but performance proved to be substantially below that calculated, maximum speed, for example, being 447 mph (720 km/h) compared with the 528 mph (850 km/h) estimated, and development was abandoned in favour of the more advanced I.Ae.33 Pulqui II designed by Kurt Tank. The performance of the I.Ae.27 Pulqui included a cruising speed of 373 mph (600 km/h) at 32,810 ft (10 000 m), an initial climb rate of 4,920 ft/min (25 m/sec), a service ceiling of 45,930 ft (14 000 m), an endurance of 1 hr 30 min, and a range of 559 miles (900 km).

Empty and loaded weights were 5,198 lb (2 358 kg) and 7,937 lb (3 600 kg) respectively, and overall dimensions were: span, 36 ft 11 in (11,25 m); length, 31 ft 9½ in (9,69 m); height, 11 ft 1½ in (3,39 m); wing area, 212·05 sq ft (19,70 m²).

Mikoyan-Gurevich wooden twin

A feature on the aircraft produced by the Mikoyan-Gurevich team which appeared in a recent issue of a magazine sent to me from Holland included a drawing of an unusual twin-engined aircraft apparently in the same category as the Westland Whirlwind. The drawing was simply captioned "DIS", and no details of the aircraft were included in the accompanying text. I would welcome some details of this aircraft in your 'Plane Facts' column, with, if at all possible, a general arrangement drawing and a photograph.

Pierre M Lemoign, 63 Clermont Ferrand, France

The DIS (*Dvukhmotorny Istrebitel Soprovozhdenya* — Twin-engined Escort Fighter) designed by Artem Mikoyan and Mikhail Gurevich was a single-seat twin-engined aircraft of wooden construction initially flown in 1941. Although designed primarily for the escort rôle, the DIS was also suitable for the ground-support and torpedo-bombing rôles, and was powered by two 1,400 hp Mikulin AM-37 12-cylinder Vee liquid-cooled engines. Armament comprised one 37-mm cannon mounted in the abbrevi-

The I.Ae.27 Pulqui (below) designed by Emile Dewoitine proved to possess a disappointing performance, and development was abandoned in favour of the more advanced I.Ae.33 Pulqui II.

ated forward fuselage together with two 7,62-mm ShKAS machine guns, and a further four ShKAS guns in the wing centre section. Loaded weight was 17,770 lb (8 060 kg), and performance included a maximum speed of 379 mph (610 km/h), a range of 1,417 miles (2 280 km), the ability to climb to 16,400 ft (5 000 m) in 5 min 30 sec, and a ceiling of 35,760 ft (10 900 m).

Development of the AM-37-engined DIS was discontinued but a revised version powered by 1,700 hp Shvetsov ASh-82F air-cooled radials was flown in 1942. Apart from the power plants, the second version of the DIS differed from the initial model in having a slightly lengthened fuselage nose and revised armament. The quartet of ShKAS machine guns in the wings was retained but fuselage-mounted armament comprised twin 23-mm VYa cannon and two 12,7-mm BS machine guns. Performance of re-engined DIS included a maximum speed of 375 mph (604 km/h) and a range of 1,553 miles· (2 500 km). An altitude of 16,400 ft (5 000 m) was attained in 6 min !8 sec, and ceiling was 32,150 ft (9 800 m). Although plans existed to manufacture the DIS in series as the MiG-5 no production of this twin-engined fighter was, in fact, undertaken.

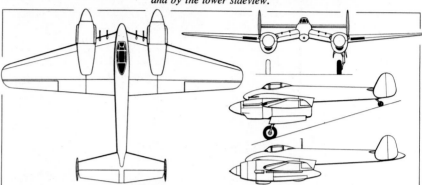

Designed by Artem Mikoyan and Mikhail Gurevich, the DIS single-seat fighter was designed primarily for the escort rôle. The initial version with AM-37 engines is illustrated by the general arrangement drawing, and the definitive version with ASh-82 radials is illustrated above and by the lower sideview.

Ryan rarity

Will you please publish a picture of the YPT-25 if such exists? I have seen nothing concerning this Ryan trainer in print, and Fahey's US Army Aircraft 1908-1946 does not include a photo of this plane.

Charles Istasse, B-6000 Charleroi, Belgium

The Ryan Model ST-4, or YPT-25, was produced in 1942 in answer to a request made by the US Army to the Ryan Aero-

Five examples of the Ryan Model ST-4, or YPT-25 (above and below) were built but the shortages of strategic materials that this trainer was designed to overcome did not materialise, and series production was not therefore undertaken.

nautical Company for the rapid development of a new primary training monoplane which would make no demands on supplies of strategic materials. All previous Ryan trainers had employed metal construction, but the YPT-25 was built almost entirely of plastic-bonded plywood, and, apart from the engine cowling, its airframe used no aluminium alloys, forgings, castings or extrusions. By the time Ryan had conducted

the initial development work and had built a service test quantity of five YPT-25s (42-8703 to -8707) the anticipated shortage of aluminium had not materialised, and therefore no production orders were placed for Ryan's wooden trainer.

A tandem two-seat monoplane powered by a Lycoming O-435-1 six-cylinder horizontally-opposed engine rated at 185 hp for take-off and possessing a normal maximum rating of 175 hp, the YPT-25 possessed a maximum speed of 149 mph (240 km/h) and a cruising speed (at 75% power) of 134 mph (216 km/h). Initial climb rate was 1,590 ft/min (8,07 m/sec), an altitude of 6,560 ft (2 000 m) being attained in five minutes. Service ceiling was 20,300 ft (6 187 m), and cruising range at 110 mph (177 km/h) was 378 miles (608 km). Loaded weight was 1,800 lb (816,5 kg), and overall dimensions included a span of 32 ft 10½ in (10,02 m), a length of 24 ft 3¼ in (7,397 m), a height of 6 ft 7¾ in (2,025 m), and a wing area of 161·2 sq ft (14,97 m2).

Rohrbach over Spain?

I wonder if 'Plane Facts' can supply the answer to a question that has puzzled me for some time? In a book concerning the Spanish Civil War — I regret that I cannot recall the title — the author referred to a "Rohrbach aircraft" which he says was flown by the Nationalists. I seem to remember him stating that it was a twin-engined aircraft. I shall be greatly obliged if you can clear up this mystery.

M J Steward, Beckenham, Kent

We have carefully checked our records regarding aircraft operated by the combatants during the Spanish Civil War, but we have been unable to unearth any reference to the use of an aircraft produced by the Rohrbach Metall-Flugzeugbau GmbH by either side. We can only assume that the author of the book incorrectly identified the aircraft to which he referred. □

FIGHTER A TO Z

ALBATROS Dr I GERMANY

The Dr I was essentially a D V fuselage, tail surfaces, and undercarriage married to three sets of wings, and flown in the summer of 1917 for comparison trials with the standard biplane. Powered by a Daimler D IIIa engine, the Dr I proved to offer no advantage over the D V, and progressed no further than prototype evaluation. Armament consisted of the usual pair of Spandau machine guns. Span, 28 ft 6½ in (8,70 m). Length, 24 ft 0⅝ in (7,33 m).

ALBATROS D VII GERMANY

Flown in August 1917, the D VII was powered by a 195 hp Benz Bz IIIb eight-cylinder water-cooled vee engine. Strut-linked ailerons were carried by all wings, and armament comprised two Spandau machine guns. The characteristics of the D VII offered an insufficient advance to warrant development further than prototype status. Max speed, 127 mph (204 km/h). Time to 6,560 ft (2 000 m), 7 min. Endurance, 2 hr. Empty weight, 1,389 lb (630 kg). Loaded weight, 1,951 lb (885 kg). Span, 30 ft 6⅞ in (9,32 m). Length, 21 ft 8¼ in (6,61 m).

(Above) The Albatros Dr I which was essentially a triplane version of the D V.

(Below) The Benz Bz IIIb-powered Albatros D VII.

ALBATROS D IX GERMANY

Unlike previous Albatros fighters, the D IX featured a slab-sided, flat-bottomed fuselage. The wings were similar to those of the D VII, as were also the tail surfaces, power was provided by a 180 hp Daimler D IIIa engine, and armament consisted of twin Spandaus. The sole prototype appeared early in 1918, but performance proved disappointing and development was discontinued. Max speed, 96 mph (155 km/h). Time to 3,280 ft (1 000 m), 4 min. Endurance, 1·5 hr. Empty weight, 1,492 lb (677 kg). Loaded weight, 1,977 lb (897 kg). Span 34 ft 1½ in (10,40 m). Length 21 ft 9⅞ in (6,65 m).

The sole example of the Albatros D IX.

The Albatros D X flown in the second D-type Competition.

ALBATROS D X GERMANY

Developed in parallel with the D IX and possessing a similar slab-sided fuselage, the D X was powered by a 195 hp Benz Bz IIIbo eight-cylinder water-cooled engine, and participated in the second D-type Contest held at Adlershof in June 1918. Armament comprised the usual pair of Spandau machine guns. Development progressed no further than prototype trials. Max speed, 106 mph (170 km/h). Endurance, 1·5 hr. Empty weight, 1,466 lb (666 kg). Loaded weight, 1,995 lb (905 kg). Span, 32 ft 3⅛ in (9,84 m). Length 20 ft 3⅛ in (6,18 m).

ALBATROS Dr II GERMANY

The Dr II was, in effect, a triplane variant of the D X biplane with a similar 195 hp Benz Bz IIIbo engine. Ailerons were fitted to all wings, and the parallel-chord, heavily-staggered wings were braced by broad I-struts. Armament consisted

The Albatros Dr II was built for comparison purposes with the D X.

of two Spandau machine guns, and the sole prototype of the Dr II was flown in the spring of 1918. Empty weight, 1,490 lb (676 kg). Loaded weight, 2,039 lb (915 kg). Span, 32 ft 9⅔ in (10,00 m). Length, 20 ft 3⅓ in (6,18 m). Wing area, 286·32 sq ft (26,6 m²).

ALBATROS D XI   GERMANY
Flown for the first time in February 1918, the D XI departed from the traditional Albatros formula in numerous respects. Like its predecessors, it was of wooden construction with fabric-covered wings and plywood-covered fuselage, and the unequal-span staggered wings had inclined aerofoil-section I-struts braced from their bases by pairs of diagonal struts which eliminated the need for wire bracing. For the

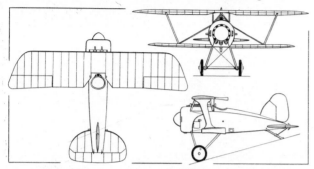

(Above and below) The second prototype Albatros D XI.

first time in an Albatros fighter a rotary engine was employed, this being a 160 hp Siemens-Halske Sh III, and the unusually large airscrew necessitated an exceptionally tall undercarriage. Armament comprised the usual twin Spandau machine guns, and two prototypes were built, the first having a four-blade airscrew and balanced parallel-chord ailerons, and the second having a two-blade airscrew and unbalanced ailerons of inverse taper. Max speed, 118 mph (190 km/h). Time to 6,560 ft (2 000 m), 4 min 40 sec. Endurance, 1·5 hr. Empty weight, 1,089 lb (494 kg). Loaded weight, 1,519–1,594 lb (689–723 kg). Span, 26 ft 3 in (8,00 m). Length, 18 ft 3½ in (5,58 m). Wing area, 199·13 sq ft (18,5 m²).

ALBATROS D XII   GERMANY
The last Albatros fighter of World War I actually completed and flown, the D XII featured the slab-sided plywood-covered fuselage introduced by the D X, and the first of two prototypes was flown in March 1918 with a 180 hp Daimler D IIIa engine. The second prototype, fitted with a Bohme undercarriage embodying compressed-air shock absorbers, and unbalanced ailerons of inverse taper in place of the balanced parallel-chord ailerons of the first prototype, followed in April 1918, and was later fitted with a 185 hp BMW IIIa engine for participation in the third D-type contest of October 1918. Max speed (D IIIa), 112 mph (180 km/h). Endurance, 1 hr. Empty weight, 1,279 lb (580 kg). Loaded weight, 1,676 lb (760 kg). Span, 26 ft 10¾ in (8,20 m). Length, 18 ft 11½ in (5,78 m). Wing area, 213·55 sq ft (19,84 m²).

The Albatros D XII with Bohme undercarriage.

ALBATROS L 77v   GERMANY
The L 77v two-seat fighter and reconnaissance aircraft was developed in the late 'twenties by the Albatros-Flugzeug-werke GmbH from the L 76 Aeolus tandem two-seat trainer manufactured under licence by Focke-Wulf and Heinkel for the *Deutsche Verkehrsfliegerschule* (DVS). The L 77v was powered by a 600 hp BMW VI 12-cylinder liquid-cooled engine and carried an armament of one fixed forward-firing 7,9-mm machine gun and a similar weapon on a free mounting in the rear cockpit. A small number of L 77v fighters was built by Ernst Heinkel for use at the clandestine German military aviation school at Lipezk in the Soviet Union. Max speed, 137 mph (220 km/h) at 4,920 ft (1 500 m). Span, 41 ft 10⅓ in (12,76 m).

ALBATROS L 84   GERMANY
The L 84 tandem two-seat fighter was flown for the first time in 1931 and was powered by a 660 hp water-cooled 12-cylinder BMW VIu engine driving a four-bladed fixed-pitch airscrew. The first prototype was destroyed during flight testing in 1932, but in February of the following year a second prototype was completed by the Focke-Wulf Flugzeugbau AG which had, in the meantime, amalgamated with the Albatros-Flugzeugwerke. No further examples were built and development was abandoned.

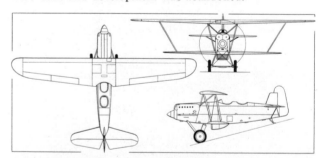

(Above) The Albatros L 77v tandem two-seat fighter, a few examples of which were built clandestinely.

(Below) One of the two prototypes of the Albatros L 84 two-seat fighter.

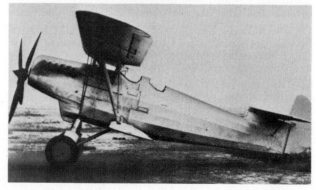

IN PRINT

"British Gliders and Sailplanes, 1922–1970"
by Norman Ellison
Adams & Charles Black, London, £3·25
296 pp, 9 in by 7 in, illustrated

ALTHOUGH gliding as a pastime is pursued with great enthusiasm by its devotees, there appears to be little available literature devoted to the aircraft themselves. Norman Ellison and his publishers have thus done the movement a great service by producing this excellent record of all the British gliders and sailplanes built to date.

The main part of the book comprises descriptions, arranged in alphabetical order, of every known machine, plus three-view line drawings of the most important. This section will thus be of special interest to the gliding fraternity. For those with a more specific interest in individual gliders, an appendix lists details of each machine built, including information on construction number, registration or BGA number and other notes.

The lack of photographs (other than a few in the opening chapters giving an outline history of gliding and glider manufacture in Britain) will be regretted, but the volume as a whole is an important historical work of permanent reference value.

"The Story of 609 Squadron"
by Frank H Ziegler
Macdonald & Co (Publishers) London, £3·25
352 pp, 5½ in by 8½ in, illustrated

AMONG the many histories of RAF squadrons already published, this new offering stands out on two grounds: it is an excellently written account, and it is concerned with a squadron of more than average interest. No 609 was formed in February 1936 as the West Riding Squadron of the Royal Auxiliary Air Force and it was disbanded, with the remainder of the R Aux AF, in February 1957.

During the war, it saw service over France and then fought throughout the Battle of Britain with great distinction. It was the first Spitfire squadron officially credited with

(Above and below) The first glider described in "British Gliders and Sailplanes" (noted on this page) is this 1922 single-seater designed by G A Handasyde, F P Raynham and Sydney Camm and built by Air Navigation Co for Handasyde Aircraft Co.

one hundred victories, and later scored a second century on Typhoons. Many well-known fighter pilots commanded, or served with, the squadron.

Frank Ziegler, author of this account, was 609's Intelligence Officer for three years, giving him a unique insight into the squadron's activities and the personalities of those who served in it. He has supplemented his own diaries with research among the surviving members of the squadron to present an intimate account of the way in which they flew, fought and — too often — died together. These men — some of Dowding's "Few" — helped to change the course of history and to read this book is to gain a rare glimpse of the qualities that helped to make their achievement possible.

This is the first in what the publishers describe as a major series of squadron

histories (the stories of Nos 1, 74 and 27 are in preparation). If the future stories are as well told as this one, the RAF will be well served indeed — FGS.

"Civil Airliner Recognition"
by J W R Taylor
Ian Allan Ltd, Shepperton, Surrey, 25p
72 pp, 4¾ in by 7¼ in, illustrated

THE emphasis in this recognition volume (1971 edition) is upon the aircraft seen in the skies above the UK. Section 1 provides a photograph, three-view silhouette, specification and an outline of development and service for each of 50 types of airliner operating regularly in Europe or about to enter service. Another 22 types of third-level airliner and business twins receive similar treatment in Section 2, but without the silhouettes.

Finnish Air Force Brewster B-239 BW-387 in the markings of the Lynx Squadron of HLeLv.24 — one of the illustrations in "Brewster B-239 ja Humu", the first in a significant new series of volumes on Finnish Air Force history, noted overleaf.

"Flying Boats and Seaplanes Since 1910"
by Kenneth Munson
Blandford Press, London, £1·15
164 pp, 4½ in by 7 in, illustrated
LATEST addition to the Pocket Encyclopaedia of World Aircraft series, following the format of 12 previous titles by the same author. For readers unfamiliar with these books, the treatment comprises a series of colour tone drawings (covering 58 types) followed by descriptive text relating to each of these drawings. There are no photographs. The colour drawings show a side view and a split plan view (half top, half underside), with no head-on view. The accuracy of the colours depicted can be criticised in detail but must be viewed in relation to the cost of the book, which by today's standards is reasonable.

"The Power to Fly"
by L J K Setright
George Allen & Unwin, Hemel Hempstead £3·75
224 pp, 6 in by 9¼ in, illustrated
FOR much of the history of powered flight, the availability of suitable power plants has been the pacing factor in achieving higher performance with aircraft. Surprisingly, therefore, little has been written about the history of aero-engines, although there is a large and indeed increasing literature devoted to the history of aeroplanes.

"The Power to Fly" goes some way to make up the deficiency, although it also leaves much still to be written. The author has set out to give a broad view of the evolution of piston aero-engines from the first unit to fly in 1903 to the engines of 3,500 hp and more which represented the ultimate in aircraft power plants when the turbojet took over. The story is told in terms understandable by the layman and with proper regard for the personalities involved. There are some irritations in the text, such as the spelling of de Havilland's famous unit as the Gypsy (it was always Gipsy) and the appendices somewhat arbitrarily include complete tabulations of Fiat and Mercedes/ Benz/Daimler-Benz aero engines, but nothing comparable for US or British manufacturers.

"Hot Air Ballooning"
by Christine Turnbull
Speed and Sport Publications Ltd, London, £1·50
128 pp, 8¾ in by 5¾ in, illustrated
YOUTHFULNESS is clearly no impediment to the pursuit of this fast growing sport: at 21 the authoress had seven years' ballooning experience behind her. She writes pleasantly and enthusiastically of the sport.

"Brewster B-239 ja Humu"
"Dornier Do 17Z/Junkers Ju 88A-4"
by Kalevi Keskinen
Kalevi Keskinen, Helsinki, Finland
100/104 pp, 7 in by 9¾ in, illustrated
THESE two volumes are the first in a promised historical series devoted to the aircraft of the Finnish Air Force. Each volume, with card covers, includes a large number of photographs of the aircraft and crews in Finnish operation, plus a series of colour profile drawings showing selected aircraft in their Finnish Air Force markings.

In addition to describing the operational careers of these aircraft, the author lists, by serial number, each individual example and traces its life history in Finland. The text is in Finnish but a brief English summary is available on request.

Copies of these volumes can be ordered direct from the author/publisher at Postilo-kevo 13117, 00130 Helsinki 13, Finland, at a price of £2 each, inclusive of postal and other charges.

"Militaire Straalvliegtuigen"
by Hugo Hooftman
La Rivière Voorhoeve NV, Holland, FR 16·90
220 pp, 6 in by 9 in, illustrated
LATEST volume from this Dutch author, the work describes and illustrates (photographs only) the range of major military aircraft currently in production or in large scale service. Fighters, bombers and transports are included and the photographs are well-chosen for maximum interest.

"Flight Directory of British Aviation"
1971 Edition
Kelly's Directories, London, £2·75
518 pp, 5½ in by 8½ in
LITTLE introduction is needed for this hardy annual which is indispensable as a quick guide to names and addresses of companies and organisations involved with aviation in Britain. The style, format and contents remain substantially unchanged from the original "Aeroplane" Directory, and the information is up-dated to late 1970.

"Business Jets International"
Edited by P G Dunnington
An 'Air Britain' Publication, 90p to non-members
94 pp, 7 in by 9¼ in, illustrated
A NEW title in the series of 'Air Britain' publications devoted to aircraft registrations, this handy volume provides production lists for ten types of business jet. For each aircraft built, the book lists the construction number, external registration and. owner, plus details of changes of ownership where appropriate. Available from Air-Britain Sales Dept, Stone Cottage, Great Sampford, Saffron Walden, Essex.

"Open Cockpit"
by John Nesbitt-Dufort
Speed and Sport Publications Ltd, London 90p
77 pp, 8½ in by 5½ in, illustrated
A SERIES of pilot's assessments of 12 pre-war aircraft in service with the RAF, first published in "Pilot and Light-aeroplane".

"Scramble"
by John Nesbitt-Dufort
Speed and Sport Publications Ltd, London £1·00
125 pp, 8½ in by 5½ in, illustrated
MORE than a dozen aircraft of World War II vintage flown by the author are described, most of the chapters having previously appeared in "Pilot and Light-aeroplane".

"2nd Tactical Air Force"
by Christopher F Shores
Osprey Publications Ltd, Reading, £4·00
298 pp, 7 in by 9¾ in, illustrated
LATE in 1943, Second Tactical Air Force, RAF, came into being as a component of the Allied Expeditionary Air Force, in preparation for the assault on enemy-occupied Europe. Thereafter, the fighter and light bomber squadrons of "2nd TAF" played an increasingly important rôle in the operations leading up to, and following upon, the Invasion. Full justice to the story is done in this finely-produced volume, which combines excellent narrative accounts of the operations from November 1943 to the end of the war, with concise data showing the disposition of squadrons and other useful information.

In addition to a large number of half-tone illustrations, the book includes eight pages of colour profiles of 2nd TAF aircraft. These give a useful indication of the colours and markings used, but are not up to the standard of the remainder of the book.

"Flypast"
A history of aviation in sound
Flypast Productions, London, £2·75
SOUND recordings are playing an increasingly important part in the preservation of history, and several organisations — the Royal Aeronautical Society and the American Aviation Historical Society among them — have taken active steps to record and preserve the recollections of aviation pioneers. Whilst such recordings may be available to serious researchers, however, they are not on general release. The 12-in LP record "Flypast", recently released in London, is therefore doubly welcome, as one of the very few aviation records available to the public, and as a genuine contribution to history.

The publishers have gathered together a unique collection of about 30 aircraft sounds, and have supplemented these tracks with recordings of the voices of a dozen pioneers, to produce 45 minutes of fascinating — and for many, nostalgic — listening. Among the voices are those of the Wright Brothers' mechanic Charles Taylor; Donald Douglas; Glenn Martin; Amy Johnson; Igor Sikorsky; Sir Barnes Wallis and Winston Churchill (speaking his famous tribute to "the Few"). The aircraft include several of World War I vintage, and such oddities as the Me 323 and the V-1 from World War II, as well as the more significant British and German combat types.

Of historic interest are the R/T conversations during the first launch of the X-15, first flight of TSR-2 and first supersonic flight of Concorde. The juxtaposition of some tracks is odd, and the overall order of items appears to be meaningless, but these shortcomings, and the necessarily poor quality of a few of the recordings, are outweighed by the overall interest and value of the record.

The publishers of the record also offer, as a special package to UK customers only, a copy of "Kitty Hawk to Concorde", published in 1969 as "Jane's 100 Significant Aircraft" to mark the 60th anniversary of "Jane's All the World's Aircraft". This volume, by J W R Taylor and H F King, contains material relevant to many of the tracks, and the record sleeve is appropriately cross-referenced. The combined package costs £5·00 from Flypast Productions, 30 Old Bond Street, London W1X 3AD; the record alone from the same address costs £3·00 including post and packing. FGS ☐

Volume 1 Number 5 October 1971

CONTENTS

AIRSCENE

MILITARY AFFAIRS

ARGENTINA
The I and II *Escuadrones de Ataque* of Argentina's *Aviación Naval,* which have been operating F9F-2 Panthers and T-28 Fennecs respectively, are currently in the initial stages of converting to the McDonnell Douglas **A-4B Skyhawk,** 20 refurbished examples of which have been acquired for the *Aviación Naval* from the USA. Both *Escuadrones* deploy detachments aboard the ARA *25 de Mayo.* The *Aviación Naval* is to receive four Sikorsky S-61D-4 helicopters early in 1972 and these will be operated in the SAR and ASW rôles by the I *Escuadron de Helicopteros* currently equipped with Sikorsky SH-34s and UH-19s.

BELGIUM
The *Force Aérienne Belge* (FAéB) retired its last Douglas DC-4 at the beginning of May. The aircraft, which had been part of the inventory of the 15ème *Wing de Transport* at Melsbroek, Brussels, had accumulated 23,800 hours of flying. The current inventory of the 15ème *Wing* includes four DC-6As and DC-6Cs.

BRAZIL
Contrary to earlier reports published elsewhere, the *Fôrça Aérea Brasileira* (FAB) has not as yet selected the **Breguet Atlantic** to replace the Lockheed P-2E Neptunes currently operated by the 7° *Grupo de Aviação* from Salvador, Bahia. Two aircraft are competing for the FAB order, the Atlantic which is reportedly being offered at a unit cost of £2·7m and the Lockheed P-3C Neptune at £2·96m, and the selected type will be operated from two new maritime bases being constructed for the FAB at Florianopolis, Santa Catarina, and São Luis, Maranhao State. An initial batch of eight aircraft is required for 1973-4 delivery.

FEDERAL GERMANY
The reorganisation of the transport support element of the *Luftwaffe,* the *Transportgruppe,* is now well advanced, current plans calling for one *Lufttransportgeschwader* (LTG) in North Germany and one in South Germany, both equipped with the **Transall C.160D.** LTG 63 at Hohn, Rendsburg, has already completed conversion to the C.160 from the Noratlas, and has some 50 aircraft on strength, and LTG 61 at Landsberg is in process of conversion and will eventually have a similar number of Transalls. VFW-Fokker, which had delivered 36 C.160s by the beginning of this year, is scheduled to deliver the last transport of this type in August 1972. Production of the Transall C.160 in France is also approaching completion, 35 having been delivered by the beginning of this year. As recorded in this column last month, 16 of the C.160Ds surplus to the requirements of the *Luftwaffe* are currently in process of delivery to Turkey. Proposals to use surplus N.2501D Noratlas transports to equip reserve units are not being acted upon owing to personnel shortages, and many of the Noratlases withdrawn from the *Lufttransportgeschwader* have now been disposed of to Greece and Israel.

The *Marineflieger der Bundeswehr* has been negotiating with Sikorsky for the conversion of six of its 23 SH-34J helicopters to **S-58T** standard (by the installation of UACL Twin-Pac turbines) for interim use pending commencement of deliveries of the 22 Westland Sea King Mk 41 helicopters currently on order for the ASW rôle.

GUATEMALA
Unconfirmed reports suggest that the *Fuerza Aérea Guatemalteca* (FAéG) is increasing its **counter-insurgency** (COIN) capability with US assistance, and that recent deliveries have included some half-dozen Cessna A-37s and a small number of Bell UH-1 Iroquois armed helicopters. The FAéG comprises two combat squadrons, a small transport component and several communications and liaison flights, plus a flying and technical training school. Until recently, the combat element consisted of the *Escuadron de CazaBombardeo* with several Cavalier-remanufactured F-51D Mustangs and four AT-33 armed trainers, and the *Escuadron de Bombardeo* operating a small number of B-26B Invaders.

SOUTH AFRICA
It has been reported that the South African Air Force has ordered an additional batch of **Piaggio P.166s** to supplement the nine currently included in the SAAF inventory. The P.166, known as the Albatross in the SAAF, is operated by No 27 Squadron.

SYRIA
According to press reports in the USA which have been confirmed by the US State Department as "essentially correct", the Syrian Air Force, *Al Quwwat al-Jawwiya al Arabia as-Suriya,* received from the Soviet Union between the beginning of April and the end of June a total of 21 MiG-21s, five Sukhoi Su-7s, 22 troop-carrying helicopters (presumably Mil Mi-8s) and nine other aircraft of unspecified type. The first-line combat aircraft inventory of the Syrian Air Force is reportedly some 30 per cent larger than that prior to the 1967 conflict, but training attrition is alleged to be extremely high, comparing closely with that of the Air Force of the UAR which is said to be four times as high as the lowest Western standard. It may be assumed that the recent deliveries of aircraft from the Soviet Union are to make good unit attrition.

THAILAND
The 16 **North American Rockwell OV-10C Bronco** forward air control aircraft purchased by the Royal Thai Air Force (see pp 231–238 of this issue) were formally inducted into service in a ceremony at Don Muang, Bangkok, during July. The aircraft had been "cocooned" and delivered by sea from California to Cam Ranh Air Base, Vietnam, where they were prepared for flight and then flown to Bangkok. After the induction ceremony, attended by the Thai Prime Minister, the C-in-C of the RTAF, Air Chief Marshal Boon-Chu Chandarubeksa, and other dignitaries, the OV-10Cs were transferred to Koke Krathiem in Lopburi for service with the 2nd Wing, which provides the RTAF's principal counter-insurgency component.

UNITED KINGDOM
The RAF's famous **No 74 Tiger Squadron** has disbanded at Tengah at the conclusion of a four-year spell of service in Singapore. Its BAC Lightning F Mk 6s have been transferred to No 56 Squadron in Cyprus, which had previously been equipped with Lightning F Mk 3s. The ferry flight was made by Lightnings non-stop in about 13 hours, with seven in-flight refuellings by Victor tankers.

A **Slingsby T 61A** powered glider (XW983) is undergoing evaluation by the Air Training Corps as possible future equipment for its 28 gliding schools and two gliding centres. The T 61A is a license-built version of the Scheibe SF-25B Falke, powered by a 45 hp Volkswagen engine. The aircraft was tested at the A & AEE Boscombe Down for six weeks and is now at the No 2 Gliding Centre, RAF Spitalgate, Lincs, for a six-month period. The ATC at present has 160 two-seat Kirby Cadet Mk 3s and Sedberghs, as well as a few single-seat Swallows and Prefects. These are winch-launched, and the 63 twin-drum winches available to the ATC are due for replacement by 1975. The T 61A, classed as a self-launched glider (SLG), would allow greater flexibility in the training syllabus and could be used from smaller fields.

USA
Kaman Aerospace Corp delivered the first two **Kaman SH-2D** helicopters to the USN on 2 August for the start of BIS (Board of Inspection Survey) acceptance trials at Patuxent River. The aircraft are the first of 10 HH-2Ds scheduled for conversion to interim LAMPS configuration (Light Airborne Multi-Purpose System), with all ten to be delivered by November. They will serve aboard nine DLG and one DLGN class ships. On 11 August, Kaman announced it had received a second Navy contract for the conversion of 10 more Sea-Sprites to SH-2D configuration, and it is expected that all H-2 series helicopters in the USN inventory will eventually be so converted. The conversion comprises airframe modification for the installation of a search radar, sonobuoy rack, magnetic anomaly detector, torpedo mounts, smoke launchers, associated antennae and control systems and displays, including a new console in the aft cabin.

US Navy will equip two of its Fleet Air Reconnaissance squadrons, VQ-1 and VQ-2, with the **Lockheed EP-3E Orion.** Delivery of 12 of these specially-equipped aircraft began in July and will be completed by July 1972. Ten of the new Orions are being converted by Lockheed from P-3As; the other two have been operating as EP-3Bs with VQ-1 since 1969 and will be updated to EP-3E standard. For the electronic warfare rôle, the EP-3Es are fitted with solid state signal intelligence equipment; they have large "canoe" radar fairings above and below the fuselage as well as a ventral radome, but the Orion's usual MAD stinger fairing is not carried. Another new USN variant is the **WP-3A Orion,** equipped for weather reconnaissance and now replacing WC-121 Constellations in US Navy squadron VW-4. The VW-1 weather squadron, also flying WC-121Ns for the past ten years, has recently been disestablished and its functions absorbed by VQ-1.

US Navy helicopter squadron HC-2, which is based at NAS Lakehurst, NJ, and provides detachments of helicopters aboard aircraft carriers for plane-guard duties, recently started to convert from Kaman

HH-2Ds to the **Sikorsky SH-3G Sea King.** More than 20 examples of this new variant are scheduled for delivery to HC-2, including six equipped with Minigun pods for use on search and rescue missions in combat conditions. Delivery of the new aircraft (which are converted and updated SH-3As) will allow HC-2 to operate in two additional rôles — combat search and rescue and anti-submarine warfare.

VENEZUELA

The Venezuelan government is reported to have approached the Canadian government with a view to discussing the possible purchase of 20 **CF-5s** surplus to the requirements of the Canadian Armed Forces for the *Fuerzas Aéreas Venezolanas* (FAV). It is not known what price the Canadian government is putting on these aircraft, but the original unit cost is believed to be of the order of £530,000. Venezuelan interest in the CF-5 does not necessarily mean any loss of interest in the Mirage, the acquisition of which was reportedly being discussed between the Venezuelan government and the Dassault-Breguet group earlier this year,

as the bulk of the FAV fighter equipment is now obsolescent, and unless the fighter component is to be substantially reduced, the FAV has a requirement for some 50-60 new aircraft. At the peak of its strength the FAV fighter component comprised five fighter squadrons, No 34 *Escuadron* with Venom FB 4s, No 35 and No 36 *Escuadrones* with Vampire FB 5s, and No 37 and No 38 *Escuadrones* with F-86F-30 Sabres, all of which had been acquired between 1950 and 1956. Fifty-one F-86K Sabres were obtained from Federal Germany in the late 'sixties but it is not known how many of these were placed in service with the FAV as, in 1969 at the time of the "Football War", four F-86Ks still in their shipping crates were sold by the Venezuelan government to Honduras for £625,000, these aircraft reportedly still existing in their crates in a hangar at Tegucigalpa.

YUGOSLAVIA

The Yugoslavian air arm, *Jugoslovensko Ratno Vazduhoplovstvo* (JRV), is reportedly still evincing interest in the possibility of purchasing the **Northrop F-5** (see this

column June), although the Soviet Union is believed to have agreed to sell the JRV about a dozen late-model MiG-21 fighters to supplement the 62 MiG-21F and -21FL interceptors currently in the JRV inventory. Other Soviet equipment supplied to the JRV in recent years allegedly includes 18 Mil Mi-4 helicopters and 13 Ilyushin Il-14 transports, plus one Il-18 and two Yak-40s. In addition to the indigenous Galeb trainer and Jastreb light strike aircraft, the cosmopolitan inventory of the JRV includes a Caravelle, a small number of Alouette IIIs, Whirlwinds, Lockheed T-33As and RT-33As, Douglas DC-6Bs, Canadair Sabre 4s, and North American F-86D Sabres.

AIRCRAFT AND INDUSTRY

AUSTRALIA

The **Project N2** twin-turboprop light STOL transport prototype made its first flight at the Government Aircraft Factory airfield at Avalon on 23 July. The initial launching funds of £1·46m for the N2 were authorised early last year, covering two flight prototypes and a static test airframe, and it is anticipated that at least 50 of the proposed military utility version, the N22, will be produced for the Australian armed forces. An enlarged version of the basic design, the N24, is to be offered on the civil market, this having a somewhat larger payload than the 3,300 lb (1 497 kg) of the military model. The N22 will make provision for the installation of ejection seats for two crew members, self-sealing fuel tanks, and hard points for bombs, rockets or gun pods.

CANADA

Saunders Aircraft Corp of Manitoba, Canada, has plans to develop a follow-on to its ST-27 feederliner designated the **Saunders ST-30.** Whereas the ST-27 is a conversion of the DH Heron with lengthened fuselage and two PT6A-27 turboprops, the ST-30 would be jig-built as a new aeroplane, although retaining much Heron technology. Engines would be uprated PT6A-34s.

FEDERAL GERMANY

VFW-Fokker is reported to have stopped development of the **VC-400** tilt-wing V/STOL aircraft, on which work had been proceeding for several years. The R & D programme had been supported financially by the Federal German government but the tilt-wing formula did not find favour in the Thalau Committee evaluation of various V/STOL projects for possible military or civil use.

Production of the **Dornier Do 28** has been concluded, with delivery of the 120th and last example recently. The customer for the last aircraft was the French Customs authority, which had previously acquired three for coastal protection and patrol duties. Production of the Skyservant, for which orders now total nearly 200, is continuing.

FRANCE

The first production Aérospatiale **SA 341 Gazelle** five-seat light utility helicopter was flown for the first time at Marignane on 6 August. The first SA 341 from Westland's assembly line at Yeovil is scheduled to fly next month (November). Production plans for the SA 341 are based on the supply of over 400 helicopters of this type to the French and British governments over the next five years.

Recent additions to the inventory of the Turkish Air Forces, the Türk Hava Kuvvetleri (THK), have included 10 Bell UH-1D Iroquois helicopters as recorded in this column last month, one of these being illustrated above; sufficient Northrop RF-5A photographic aircraft to equip one squadron of the THK's 3rd Tactical Air Force, one of these being illustrated immediately below, and three ex-THY Vickers Viscount 794s (bottom) which are being used exclusively as VIP transports.

The **Dassault Mercure 01** prototype completed its initial phase of flight testing on 22 July and was then grounded for the August vacation period. Ground resonance testing was begun during September and the prototype will then be fitted with definitive JT8D-15 turbofans before the second phase of flight testing begins.

Aérospatiale has published total sales figures for its helicopters up to the end of June, as follows: **SE 3130 and SA 318C Alouette II,** 1,238 for 104 customers in 44 countries (762 export); **SA 315 Lama,** 27 for six customers in four countries (24 export); **SA 316/319 Alouette III,** 945 for 99 customers in 60 countries (795 export); **SA 341 Gazelle,** 110 for two customers in two countries (60 export); **SA 330 Puma,** 230 for 11 customers in 10 countries (99 export); **SA 321 Super Frelon,** 57 for eight customers in six countries (37 export).

The final stage in the merger between Breguet and Dassault has recently been reached with the formation of a new single company known as **Avions Marcel Dassault – Breguet Aviation.** The activities of the two former companies are being fully integrated under the direction of B C Vallieres as president–chief executive.

Thrust reversers for the General Electric CJ610 turbojets on the Dassault Falcon 20 have been developed in the US by Aeronca (previously responsible for the reversers on the DC-8, DC-9 and DC-10). The CJ610 reversers are of the cascade type and add 70 lb (31,5 kg) to the weight of each nacelle; reverse thrust equal to 40 per cent of normal engine thrust is achieved.

ITALY
The Agusta **A 109C Hirundo** (Swallow) eight-seat twin-turbine helicopter was flown for the first time on 4 August at Agusta's Cascina Costa plant. The Hirundo is powered by two Allison 250-C20 turbines, and three more prototypes are being built for the flight test and certification programme, with a further example for ground-running trials. The first production deliveries of the Hirundo are scheduled to commence late next year.

The second prototype of the **Fiat G.222** transport made its successful first flight on 22 July at Caselle Airport, Turin. Although no order for the G.222 on behalf of the *Aeronautica Militare* had been placed at the time of closing for press, the flight programme with the first prototype at the Pratica di Mare airfield has reportedly proved extremely successful, and there is now little doubt that funding will shortly be provided for the production of the G.222 against an *Aeronautica Militare* requirement for some 40–50 transports.

NETHERLANDS
Fokker-VFW is developing wing leading edge slots for the **Fokker F.28.** The modification will allow operation of the F.28 Mk 2000 at higher weights for a given take-off distance, or will improve the field performance at given weights.

UNITED KINGDOM
First flight of the **BAC/Breguet Jaguar B-08** was made at Warton on 30 August. This is the last of the eight prototypes and the only one to British Jaguar T Mk 2 two-seat trainer standard. Prototypes S-06 and S-07 represent the RAF's Jaguar GR Mk 1 single-seat strike aircraft.

First flight of the **Lockspeiser LDA 1** was made at Wisley on 24 August. This unusual aircraft, which has been under development for 14 years, has a canard layout, with a short foreplane mounted low on the box-section fuselage in the area of the cockpit, and the rear wing mounted high at the end of the fuselage. The aircraft has a central fin and two outrigged fins and rudders and is powered by an 85 hp Continental engine driving a pusher propeller. The initials in the designation stand for Land Development Aircraft and the prototype (G-AVOR) is intended as a 7/10ths scale representation of the planned production version, which is expected to have a wide variety of utility rôles.

USA
US Navy production orders for the **Grumman F-14B** variable-geometry air-superiority fighter are being postponed, primarily because of the delay of between six and 12 months in the development of the Pratt & Whitney F100-PW-100 advanced-technology engine. The US Navy had originally planned to purchase 20 F-14B Tomcat fighters in Fiscal 1972 plus 28 F-14A Tomcats powered by the Pratt & Whitney TF30-P-412, and now the US Navy proposes to exercise its Fiscal 1972 production option for the full 48 Tomcats but all of these will now be F-14As. Some of these may eventually be reconfigured as F-14Bs when the advanced-technology engines become available. Grumman contends that rising costs and higher-than-anticipated inflation levels have rendered the existing US Navy contract for the Tomcat unrealistic, the unit price of the aircraft having risen by at least $1m (£416,000) to around $11·5m (£4·79m).

The Department of Defense has concluded a restructured contract with Lockheed to cover completion of the **AH-56A Cheyenne** armed helicopter development programme by December 1973. Terms of the original contract have been in abeyance since the Army cancelled its production order for the Cheyenne in May 1969. The new contract covers the work originally specified plus two new items — development of an advanced mechanical control system, and integration of an improved tube-launched optically-tracked wire-guided TOW anti-tank missile with a night vision system for the co-pilot/gunner.

North American Rockwell has proposed a three-engined version of the **RA-5C Vigilante** to meet USAF requirements for an air defense interceptor. The proposal is in line with Department of Defense pressure to utilise existing airframes and technology in the development of new projects. The proposed USAF version of the Vigilante would have a third engine installed in the bomb bay, with the jet pipe on the centre line above the two existing engines. Six air-to-air missiles would be carried in semi-recessed installations under the fuselage. Production of the RA-5C for the US Navy ended last November but all tooling for the aircraft is still intact at the Columbus factory.

Robertson Aircraft Corp has obtained FAA approval for its STOL conversion of the **Twin Comanche R/S,** the first twin-engined type to receive this treatment, which involves modification of the wing, control system and tail unit. At a cost of $4,450, the conversion gives the Twin Comanche improved take-off run and landing characteristics, better single-engined performance, payload and range and improved low-speed control.

US certification for the **Britten-Norman Trislander** was obtained on 4 August. Orders and options for the new three-engined transport (see pp 239–244) now total 25.

First flight of the fourth **Lockheed L-1011 TriStar** was to be made early in September as this issue went to press, with the fifth aircraft to join the programme later this year. By August, the first three TriStars were about a quarter of the way through their flight test programme with about 420 hours flown. Final certification and first deliveries are now expected to be achieved in April 1972.

Reflecting the current decline in air transport growth, **Boeing** expects to deliver only 48 new aircraft in the second half of 1971, after delivering 98 in the first half of the year. The totals delivered (and for delivery) are, **Model 707,** 7 (3); **Model 727,** 29 (9); **Model 737,** 16 (13); **Model 747,** 46 (23). Long-term market forecasts by Boeing indicate that production of the Model 727 will eventually exceed 1,000. With 873 ordered to date, the 727 is already the world's most successful jetliner, the McDonnell Douglas DC-9 being second with 677 orders.

FAA certification has been awarded to the **Interceptor 400,** a high-performance, pressurised four-seat light aircraft powered by a 400 shp Garrett AiResearch TPE 331-1-101 turboprop engine. First flown in July 1969, the Interceptor 400 cruises at 275 mph (443 km/h) for 1,000 miles (1 610 km). Deliveries have begun, with the first ten aircraft scheduled for delivery by mid-1972, when the production rate will reach two a month.

The USAF has been evaluating a special version of the Beech Debonair in its Pave Eagle programme to supplement Lockheed EC-121Rs in use as electronic data relay platforms in SE Asia. With increased engine power and larger wing span than the commercial model, these aircraft are designated QU-22B and can be operated as drones, carrying one occupant to monitor the sensors.

Boeing has given some details of its studies leading toward a new generation of jet transports. Provisionally called the **Boeing 767** (see illustration), the new project is for a very long aeroplane to be used on "thin" routes — ie, those where traffic loads are important but not sufficient to support operation of the Boeing 747 or other large transports. The 767 would cruise at Mach 0·98, or about 100 mph (161 km/h) faster than the Boeing 707, and would have operating costs some 15 per cent lower than the current generation of jet transports. The configuration under study at present has an area-ruled fuselage with a narrow waist, a highly swept wing based on the latest supercritical design philosophy, two engines beneath the wing and two on the rear fuselage. It would seat 200 with six abreast seating and two aisles. These studies are company-funded but Boeing also has a $1m contract from NASA (together with Lockheed and Convair) to study the application of supercritical air flow and other advanced technology features to long-range transports.

Acquisition of the assets of **Swearingen Aircraft** by Piper Aircraft Corp has been approved by the board of the latter company, subject to certain final points. Under the terms of the proposal, Piper would form a new, wholly-owned subsidiary to assume the present aircraft manufacturing, sales and service operations of Swearingen; Piper would also assume responsibility for all sales of the Metro airliner, and international sales of the Merlin, as well as complementing the present Garrett AiResearch domestic Merlin sales programme.

Reported to be close to completion but delayed because of the recent decline in the business aircraft market is the **Swearingen Merlin 5**, a light executive twin with turbofan engines. Aircraft would have a greater range than its near competitors, which comprise the Cessna Citation, Dassault Falcon 10 and Learjet 26.

Learjet plans to introduce **thrust reversers** on the General Electric CJ610 engines on all Learjet aircraft at the end of 1971. The reversers, to be manufactured by Dee Howard Company of San Antonio, Texas, will be available for retrofit on existing Learjets. Dee Howard is undertaking prototype development and flight testing.

More details are now available of the **Dominion Aircraft Skytrader** (AIR ENTHUSIAST August p 119). It is a high-wing twin-engined aeroplane of simple construction and orthodox appearance, available on floats or with a tricycle undercarriage. Powered by 400 hp Lycoming IO-720-B1A engines, it will accommodate two pilots and 12 passengers, with a useful load of 3,527 lb (1 600 kg) as a landplane and 2,866 lb (1 300 kg) as a seaplane. Gross weight is 8,488 lb (3 850 kg). The Skytrader is expected to have a cruising speed of 177 mph (285 km/h) at 9,840 ft (3 000 m) at 75 per cent power, with a range of 930 miles (1 500 km). Take-off distance to 50 ft (15,2 m) is given as 885 ft (270 m) and landing distance as 720 ft (220 m). The prototype, due to fly in March 1972, is being built in Renton, Washington, and will be assembled and flown in Vancouver. Projected price of the landplane is $130,000 (£54,170): the floatplane will cost $28,000 (£11,650) more.

Kaman Aerospace Corps won a $7·5m (£3·1m) contract from the USAF to develop a prototype **Airborne Weather Reconnaissance System.** The new equipment is to be

An impression of the Boeing 767 (described alongside), the subject of current studies by the Boeing Company into long-range transports with high subsonic cruising speeds and supercritical wing design.

installed in a WC-130 Hercules for prototype tests at the end of 1972 and if put into production will be installed in 22 WC-130B/E aircraft of the Air Weather Service.

The first of 50 modified **North American T-28D-10** light strike and close-support aircraft to be rebuilt from T-28A trainer airframes by the St Augustine facility of the Fairchild Aircraft Service Division under a $3·5m (£1,458,000) contract placed by the USAF in November was rolled out early in May. The programme calls for the complete disassembly of the aircraft, the installation of wing weapon pylons, a gunsight system, new cockpit panels, switches, wiring, and an updated avionics package.

The latest version of the Sikorsky SH-3 Sea King to be announced is the **SH-3H** which features new lightweight sonar and three additional ASW sensors — active and passive sonar buoys, magnetic anomaly detection equipment and radar — plus an anti-ship-missile-defence system. The SH-3H is to be powered by T58-GE-10 turboshafts and will have a maximum take-off weight of 21,000 lb (9 525 kg), a prototype being scheduled to commence testing next year. The initial US Navy contract calls for the conversion of 11 SH-3G Sea Kings to SH-3H configuration, but it is planned to convert all existing SH-3As and SH-3Gs. A new US Navy squadron, HS-15, is to be commissioned to operate the first SH-3Hs.

Two new versions of the North American Rockwell Sabreliner executive jet have been introduced. The **Sabre 75** is an enlarged version of the Sabreliner 60, offering

increased headroom in the cabin. Powered by 3,300 lb st (1 500 kgp) Pratt & Whitney JT12A-8 turbofans, the Sabre 75 has a gross weight of 21,200 lb (9 615 kg), and a range of 1,700 miles (2 720 km). At the lower end of the range, the **Sabre Commander** has the same engines and dimensions as the Sabreliner 40; it seats nine including two pilots and is offered at slightly less than $1 million (£416,650).

CIVIL AFFAIRS

CAMEROUN
Following its withdrawal from the Air Afrique consortium, the Cameroun has set up its own airline with the name **Cameroun Airlines** (or Cam-Air). Air France has a 30 per cent interest and is now giving the new company technical help in selecting equipment.

NORWAY
After only a few months of operation, using an ex-Eastern Air Lines Boeing 720, the Norwegian charter company **Transpolar** is reported to have gone out of business.

INTERNATIONAL
The first **transatlantic flight by Concorde** was made on 6 September when prototype 001 arrived in Rio de Janeiro, Brazil, on what was described as part of its flight development and sales promotion programme. The itinerary provided for departure from Toulouse on 4 September and arrival in Rio de Janeiro on the 6th, the eve of Brazil's Independence day, with intermediate stops at Isla do Sal and Cayenne.

Beech Aircraft Corporation has introduced the King Air A100, powered by two 680 shp PT6A-28 turboprops. Compared with the earlier King Air, the A100 has more fuel, increased gross weight of 11,500 lb (6 805 kg) and other improvements.

Among the current operators of Vickers Vanguards is Thor Cargo of Iceland, which has these two, ex-Air Canada, acquired from Air Holdings.

On 8 September, 001 was to visit Sao Paulo for the opening of the France 1971 exhibition, and demonstration flights for airlines and government officials were to be made from Rio and Buenos Aires before the return to Toulouse on 17 September. Both the Concorde and the Tupolev Tu-144 have been formally invited to participate in the Transpo-72 exhibition at Dulles Airport, Washington. The exhibition, from 21 May to 4 June, will cover all transportation technology.

HONG KONG

Orient Pearl Airways is a new operator formed in Hong Kong to undertake freight charters. The initial equipment is a single Douglas DC-6.

UNITED KINGDOM

BEA Helicopters Ltd, by arrangement with the government, is to assign three of its Sikorsky S-61N helicopters for **air-sea rescue duties** off the North East coast of Scotland. The aircraft are being fitted with specially-developed winches and crews will receive training and participate in occasional exercises.

Britain's "second force" airline, formed by a merger between Caledonian Airways and British United, has now changed its name to **British Caledonian Airways.**

USA

First revenue service by **McDonnell Douglas DC-10** was flown between Los Angeles and Chicago on 5 August by American Airlines, just one week after simultaneous delivery of the first two DC-10s to American and United. American had previously announced a 17 August inaugural but brought the date forward to beat United Airlines' first service, flown on 16 August. Full FAA certification and a production certificate were acquired on 29 July.

The FAA has initiated a feasibility study of a close-in, **offshore airport** to serve metropolitan New York City. Possible sites are Long Island Sound or the Atlantic Ocean south of Long Island. Technical, social and economic factors will be studied under contract to the FAA by a division of Litton Industries.

Piedmont Airlines has decided to equip its entire fleet with McDonnell Douglas **EROS II collision avoidance systems,** becoming the first airline to do so. Similar equipment has been in use in USAF and USN F-4s flying in the St Louis area since 1965. Certification testing of Eros II in Piedmont Boeing 737s will begin in 1972 with delivery of production equipment scheduled for the first quarter of 1973.

A McDonnell Douglas DC-9 Srs 10 opera-ted by the FAA has been certificated for operation with a nitrogen **fuel tank inerting system.** It is the first approval of the system of protection against accidental ignition of fuel vapour in the tank, achieved by diluting air in the fuel tank and vent lines with nitrogen so that the oxygen concentration is always below the level to support combustion. About 270 lb (122 kg) of liquid nitrogen would be needed for a typical day-long airline operation. FAA is now evaluating the system for reliability and operating costs.

CIVIL CONTRACTS AND SALES

Aérospatiale Caravelle: Disposal of Swissair fleet includes two to China Airlines (not from United, as previously reported); three to Transavia and one to Catair in France.

Aérospatiale N.262: Two of the Nord 262s of Danish Cimber Air have been re-registered to the German associate, Cimber Air GmbH in Kiel. □ Société de Travail Aerien (STA) in Algiers has acquired at least five Nord 262s from Allegheny in the USA.

Antonov An-24: Two have entered service with Brazzaville-based Lina Congo.

BAC One-Eleven: Merpati Nusantara has leased one Series 400 ex-American Airlines for three months, pending delivery of one from American following modifications.

Beech 99: Golden West Airlines, currently negotiating a take-over of bankrupt Los Angeles Airways, has acquired two.

Boeing 707: TWA has sold one -131 to Club International of Seattle. □ Israel Aircraft Industries has bought 13 -131s from TWA for overhaul and resale.

Boeing 727: Air Mali has leased one from World Airways for a period of one year. □ Icelandair's second aircraft (reported last month) is ex-American Flyers.

Britten-Norman Islander: First Brazilian customer is Taxi Aero Cesar Aguiar of Rio de Janeiro. Delivery of two was made in August.

Convair CV-990 Coronado: Belgian International Air Services has purchased two Coronados for delivery in the near future.

DHC Twin Otter: Bakhtar Afghan Airlines has ordered two to add to two already in service.

Fokker F.28 Fellowship: Garuda Indonesian Airways has ordered a third for December delivery. The first was handed over on 19 August and the second in September.

Hawker Siddeley Comet: Channel Airways has acquired one Srs 4, originally in the BOAC fleet, from Mexicana.

Hawker Siddeley Trident: Northeast Airlines (BEA associate) will acquire three 1Cs from BEA later this year. □ Contract for six Trident 2Es for China was confirmed on 24 August, with deliveries to be completed by 1973.

Lockheed JetStar: Recent sales include one to the Mexican government and one to Cerveceria Moctezuma, also in Mexico.

McDonnell Douglas DC-8: Air New Zealand has acquired two -52s from United Air Lines. □ Air Congo recently took delivery of a second -63F. □ Two -63s ex-American Flyers are now operated by Trans International. □ Trans International has leased from American Airlines three -61Fs acquired by American through the merger with Trans Caribbean. □ Universal has acquired one -61 from Trans International.

NAMC YS-11: Indonesian oil company Pertamina has ordered two, for 1972 delivery.

Sikorsky S-58T: Sikorsky Aircraft has begun delivery of five of these turbine-engined conversions to Helicopter Utilities of Sydney.

Vickers Vanguard: Invicta International is now operating two, and a third, with cargo door, flies in the markings of Invicta Air Cargo.

Vickers Viscount: Alitalia sold one Type 785 to Anesa Andes Airlines in Ecuador.

MILITARY CONTRACTS

Bell OH-58A Kiowa: It was announced on 29 July that the US Army has placed a contract with Bell Helicopter valued at $26,183,948 (£10,910,000) for a further 400 OH-58A Kiowa light observation helicopters for delivery between November 1972 and July 1973, this order bringing to 2,200 the total of Kiowas ordered by the US Army since 1968.

Britten-Norman Islander: Three Islanders have been ordered by the Mexican Presidency, to be operated on behalf of the President by the Mexican Air Force, and the first two were handed over on 18 August in Mexico City. The aircraft are fitted with Rajay superchargers on the standard 260 hp Lycoming engines (see pp 239–244 of this issue) and are fitted with executive interiors convertible to either commuter or cargo configuration.

Kaman SH-2D: Kaman Aerospace Corp received a $1,770,000 contract from US Navy with authorisation for the second batch of 10 SH-2D interim LAMPS conversions of HH-2D SeaSprite helicopters (see item in "Military Affairs" section).

Scottish Aviation Bulldog: The Swedish government has taken up options on 20 Bulldogs, bringing total orders to 78. The additional aircraft are reported to be for use by the Swedish Army.

Vickers Viscount: One Type 839 has been acquired in Britain for operation by the Omani Air Force in Muscat and Oman, presumably as a VIP transport.

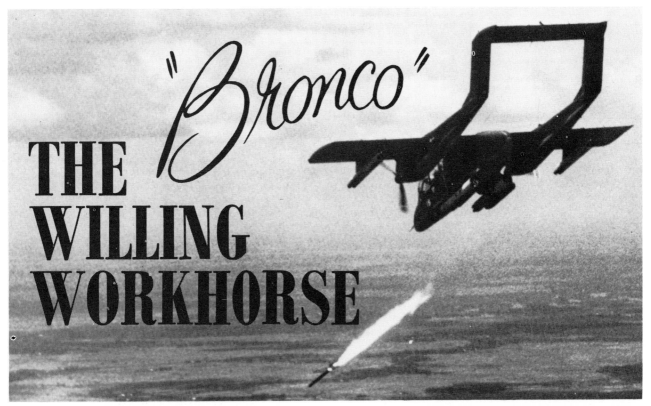

THE "Bronco" WILLING WORKHORSE

COMBAT OPERATION of the North American Rockwell OV-10A Bronco began in Vietnam a little over three years ago. Since that time, the Bronco has amply proved the validity of the original, somewhat unusual design concept for a light, armed reconnaissance aircraft and has given noteworthy service with three of the American armed forces — the Air Force, the Navy and the Marines. Service experience with the type has shown that it is ruggedly simple, able to absorb considerable damage from small arms fire and requiring a very low ratio of maintenance man-hours to flight hours, even in the far-from-optimum conditions at South Vietnam operational bases.

Success, nevertheless, has not been achieved by the Bronco without a measure of difficulty. Changes in the specification during the period that the prototypes were being built called for a major redesign before production could begin, and flight testing threw up its own share of problems. Although the OV-10A as delivered to the USAF and USMC eventually met all major contractual performance guarantees, including maximum level flight speed, take-off and landing distances, ferry range, loiter time, empty weight and single-engined ceiling, the time taken from initial go-ahead to initial deliveries was three-and-a-half years — excessive for what set out to be an essentially simple aeroplane. The political objective of combining several requirements into a single specification played its part in this delay, which was especially unfortunate in view of the urgent need that developed for aircraft of this category in Vietnam.

Once deliveries *did* begin, however, there was no holding the Bronco. Training crews and working-up squadrons to operational readiness occupied only a few days over four months from the initial Service deliveries, and the first operational sortie was flown by a Marine Corps OV-10A less than three *hours* after the aircraft arrived in South Vietnam. Within a year of arriving in Vietnam, the OV-10As completed more than 100,000 combat flight hours — a figure that can be put in perspective by considering that the USAF took four years to complete 100,000 hrs flying on the Convair F-111 with a greater number of the latter aircraft in service.

Production of the OV-10A for the US Services ended in April 1969, but orders were subsequently obtained from Germany and Thailand for a total of 34 more aircraft and the line was re-opened. Deliveries to these two countries continued into 1971. In addition, while the Bronco remains in service in Vietnam, primarily for forward air control (FAC) and close air support, it is currently the subject of several development programmes aimed at increasing its capabilities in the rôles in which it has already proved so useful. Details of some of these variants and the export models appear later in this feature.

Early days

Although the USAF is now the biggest single user of the OV-10A, the Marine Corps played the major rôle in fostering interest in the concept of a counter-insurgency (COIN) aircraft, leading to development and production of the OV-10 being undertaken in the first place. The Marine Corps, with its special responsibilities for supporting ground action at close range, participated in the study of COIN-type aircraft during the 'fifties, and in 1961 two Marine Corps officers made a detailed proposal for a "Light, Light Marine Attack Airplane" called the L2VMA. With a wing span of only 20 ft (6,1 m) and, in its initial version, an empty weight of 1,680 lb (762 kg) this simple aeroplane was intended to "live" in the battlefield environment at the battalion or regimental level, and was to be available for operation in an armed rôle as well as for observation.

During 1962, while studies of this concept continued within the Marine Corps, the office of the Director, Defense Research and Engineering in the Department of Defense, began to take an interest in the work. This interest was occasioned by the need for a new small, simple and inexpensive aeroplane that could be supplied through the Military Assistance Programme to overseas countries as a replacement for North American T-6s and T-28s operated in the COIN rôle. With the US Navy acting as the development agency, a steering group was set up in 1963 to help formulate an appropriate specification for an aircraft meeting the Marine Corps and DDRE requirements.

Representatives of all three services and of the DoD served on this steering committee, so that the specification, completed by the end of 1963, effectively had tri-service approval.

At the end of 1963, the aircraft industry was formally invited to make design proposals to the specification, which was now identified as the LARA (Light Armed Reconnaissance Aircraft). Bids were to be submitted within three months. The specification required the LARA to have a crew of two for a close air support mission carrying 2,400 lb (1 088 kg) of ordnance with a one-hour loiter time at a radius of 50 naut miles (92,6 km), and to be able to fly a three-and-a-half-hour visual reconnaissance mission when carrying no ordnance.

In alternative rôles, it was to be able to carry and air drop up to 2,000 lb (907 kg) of cargo, or six paratroopers, to be capable of carrying stretcher-wounded and able to operate as a float-equipped amphibian. Severe rough field capability was written into the requirement, and other points were an airframe stressed for $8g$; internal armament of four 7,62-mm machine guns with 2,000 rounds; ability to carry 20-mm gun pods and Sidewinder AAMs and the capability to perform a 50-degree dive from 25,000 ft (7 620 m) entered at maximum speed. Required to be twin-engined, with turboprops, the LARA was expected to be able to survive an engine failure at the critical point during a maximum performance take-off or landing.

Proposals were made in response to the Navy specification by 11 companies and of these the projects by Beech, Douglas, General Dynamics/Convair, Helio, Lockheed, Martin and North American were considered to be good enough for detailed evaluation. This process, conducted by the Navy, led to selection of the North American NA-300 design by the Navy steering committee in August 1964, with formal ratification by the DoD and the contract go-ahead following on 15 October.

In some of the early phases of the study leading to the LARA specification, the unit cost of the proposed aeroplane was, with singular optimism, put at $100,000. US Navy

(Above) A USAF OV-10A with armament of four machine guns only and (below) a US Marine Corps aircraft with rocket pods on the stub wings.

studies more reasonably indicated a minimum unit cost of $300,000, but changes in the specification made while the prototypes were under construction eventually pushed the unit price up to almost double the latter figure for OV-10As equipped to full USAF operational standard.

North American chose a twin-boom layout (in common with Martin and Convair) for its LARA submission, with the fuselage pod suspended from the parallel-chord wing — an arrangement providing a good field of view for the pilot and observer in tandem, and ease of access to fuselage-mounted equipment during maintenance periods on the ground. Conventional aluminium monocoque construction was adopted in order to avoid development delays, and emphasis was placed upon design features that would simplify maintenance and servicing.

One of the few innovations in the OV-10 was the design of the spoilers, used to supplement aileron power for lateral control. Since the OV-10 was required to have virtually STOL performance, the minimum control speed had to be low. However, little span remained available on the wing for ailerons after allowing for the large, double-slotted flaps that were needed to achieve STOL in the first place. Spoilers were therefore essential, and they had to be manually operated in order to avoid the complication of hydraulic boosters in the control circuit. The spoiler design adopted for the OV-10 is believed to be unique, comprising, in each wing half, four flat quarter-circular plates that rotate up through slits in the upper skin. They are geared into the aileron operating circuit in such a way that deflection of the spoilers on the appropriate wing begins with 5 degrees of up-aileron movement and progresses to a maximum of 89 degrees "up". The effect is to give the Bronco an outstanding rate of roll with very light stick forces.

Another notable feature of the OV-10 design is the undercarriage, each main unit of which is a double-telescoping oleo design to absorb very high rates of descent and to permit operation from extremely rough fields. Sink rates of 1,200 ft/min (6,1 m/sec) had to be absorbed to meet the specification, and normal operating technique is to put the OV-10 on the ground with a descent rate of about 800 ft/min (4,05 m/sec).

The OV-10A was designed around either two 650 shp Pratt & Whitney (United Aircraft of Canada) T74 turboprop engines or 660 shp Garrett AiResearch T76s, with the latter type selected by North American at the time the initial DoD contract was placed. The engines are "handed" so that the propellers turn in opposite directions and cancel out torque. Aiding the short landing performance, the propellers are reversible in pitch, with scissor switches incorporated in the oleo legs to prevent selection of reverse pitch until the aircraft is firmly on the ground.

In its initial form, the OV-10A had a wing span of 30 ft 3 in (9,1 m), and ailerons with a span of only 3 ft 6 in (1,07 m) at each tip. The designed empty weight was 5,267 lb (2 390 kg) and the gross weight 10,550 lb (4 795 kg). Stub wings under the fuselage, in which the fixed armament of four 7,62-mm machine guns was mounted, were flat, like the mainplane, with neither dihedral nor anhedral. Engines in the prototypes were designated YT76-G-6 or G-8 according to the direction of rotation.

Seven prototype YOV-10As were ordered on the original development contract and the first of these aircraft made its first flight at Port Columbus, Ohio, on 16 July 1965, with North American Columbus Division's chief test pilot Ed Gillespie at the controls. Further aircraft followed on 30 November, 31 December and 31 January. The No 5 YOV-10A flew on 15 August 1966 with new Hoerner-type wingtips that increased the span to 34 ft (10,36 m). The seventh and last of the prototypes did not fly until 7 October 1966, the DoD having decided that this YOV-10A should be fitted with Pratt & Whitney T74 engines for comparative trials and to provide information for an alternative production model in case early difficulties with the T76 could not be overcome. Two more complete airframes, less systems, were built on the original contract, for static and fatigue tests.

Flight testing of the prototypes was in the hands of the manufacturers for most of 1966, in the standard Category 1 phase. During this period, the OV-10A underwent the Navy Preliminary Evaluation (NPE) and was also evaluated by an All Service Evaluation Group (ASEG) — the first of its kind — in which the USAF, Marine Corps and Army were represented. Special emphasis was placed upon rough field testing, for which purpose an artificial undulating runway at NATC Patuxent River was used. Category 2 operational testing by the customer began after the YOV-10A No 7 had been accepted by the Navy on 27 March 1967 and was delivered to Patuxent River, followed by No 6 a week later. Category 3 testing, a six-month investigation of the aircraft's performance under simulated operational conditions, began in January 1968 at the Special Air Warfare Center, Eglin AFB, using two production model OV-10s.

Into production

Production of the Bronco for the USAF and USMC had been authorised in October 1966. By that date, however, a number of changes in the specification had been requested, leading to a considerable redesign. Among the changes were the addition of some 300 lb (136 kg) of armour protection to reduce vulnerability to small arms fire, self-sealing protection for the entire internal fuel system, added communications equipment and other items. The effect of these changes was to increase the empty weight of the aircraft to the point where the performance requirements could no longer be met.

Consequently, North American had to embark on a major modification of the design, including a 10 ft (3,05 m) increase in wing span, with the tail booms moved further apart, while Garrett AiResearch had to increase the power output of the T76 without any increase in dimensions. In its production configuration, the OV-10A had an empty weight some 30 per cent higher than that of the prototypes at 6,893 lb (3 127 kg), the span was 40 ft (12,19 m), aileron span 7 ft (2,13 m) on each wing, the max gross weight 14,444 lb, and the engines, rated at 715 shp each, were designated T76-G-10 or G-12. Another external change was that the stub wings were angled sharply downwards, and small dorsal fins were fitted on each boom. Flight testing of the production configuration, using a modified prototype, began in March 1967, with a second modified prototype flying in July.

The OV-10A is provided with seven external store stations — one on the fuselage centre line with a 1,200 lb (544 kg) capability, two on each stub wing with a 600 lb (272 kg) capacity each, and provision for a 500 lb (227 kg) load to be carried under each wing if required. The fully-equipped OV-10A, with maximum internal fuel load of 1,638 lb (743 kg), crew of two and 2,000 rounds for the four internal machine guns, has a margin of some 4,350 lb (1,973 kg) for external loads. External fuel — a 150-US gal (568-l) or 230-US gal (870-l) tank — can be carried on the fuselage centre line only, and this station can also carry a 20-mm gun pod. The four stub wing stations are wired for rocket pods, and these plus the fuselage station can accommodate mini-gun pods or bombs. AIM-9D Sidewinder AAMS or rocket pods can be carried under the wings.

Weapon aiming is by means of a general purpose, illuminated gun/bomb-sight in the front cockpit, for fixed reticle delivery of bombs and rockets and for directing air-to-ground gunfire. The sight incorporates an elevation adjustment to permit depressed reticle bombing and air-to-ground rocketry.

To meet a German requirement for target tugs, North American Rockwell developed this jet-boosted version of the Bronco, with a J85 engine pod-mounted above the fuselage. The intake (as shown below) can be covered when the engine is not in use.

Navigation and communication equipment varies between the USAF and USMC versions of the OV-10A, which are otherwise virtually identical apart from the external finish. The Marines aircraft have only compass, TACAN and UHF/ADF for navigation, whereas the USAF aircraft have, in addition, LF-ADF, VOR and ILS glide slope. Both types have UHF-AM, VHF-FM and HF-SSB for communications, the USAF having VHF-AM also. Both carry IFF/SIF, and the USAF type has a radar beacon also.

The initial production contracts covered a total of 185 OV-10As, of which 109 were for the USAF and 76 for the USMC; subsequent orders brought the total for these two services to 157 and 114 respectively. First flight of the first production OV-10A was made at Port Columbus on 6 August 1967 by the company's senior engineering test pilot, Don McCracken, this being one of the USMC aircraft.

Production of the 271 OV-10As for the US services ended in April 1969, at which time North American was in the final stages of negotiating a contract with the West German government for the sale of OV-10s to be used as target-tugs. These negotiations were conducted through the foreign military sales programme of the US Department of Defense and were concluded successfully in May, the formal contract being placed with North American by the US Navy at the end of June.

The German order was for 18 aircraft, of which six were to be OV-10Bs and the remainder OV-10B(Z)s, the latter differing in having a General Electric J85 turbojet pod-mounted above the fuselage to increase the maximum speed. North American flew the first OV-10B on 3 April 1970 and began deliveries in June.

The first flight of an OV-10B(Z) with the jet pod fitted was made at Columbus on 3 September 1970 by the director of flight operations, Dick Wenzell, although the J85 was not started on this flight. In subsequent flights, the configuration was fully evaluated and the performance verified, this work being completed by February 1971. With 715 shp T76-G-410 and G-411 turboprops plus a 2,950 lb st (1 338 kgp) J85-GE-4 turbojet, the OV-10B(Z) has a maximum TAS at 10,000 ft (3 050 m) of 341 knots (632 km/h) compared with 244 knots (452 km/h) for the unboosted OV-10B. Comparative take-off

Various colour schemes have been adopted for OV-10s, including those illustrated on this page. The US Navy aircraft (above), borrowed from the USMC, make use of an overall field green scheme, with gull grey undersurfaces and a white top wing surface, to aid visibility from the air when the aircraft is operating low down in the forward air controller rôle. The US Marine Corps scheme (left) is similar but the upper wing surface carries a white panel bordered by olive, and there is also a white panel on the top tailplane surface. The Royal Thai Air Force has borrowed USAF camouflage patterns, using brown and two shades of green for its OV-10C (below) with gull grey undersurfaces. USAF OV-10As in Vietnam, however, are gull grey overall with white on the upper surface of the wing.

runs, at a gross weight of 12,000 lb (5 443 kg), sea level ISA, are 550 ft (168 m) with lift-off at 79 knots (146 km/h) and 1,130 ft (344 m) at 84 knots (156 km/h), while the comparative initial rates of climb are 6,800 ft/min (345 m/sec) and 2,300 ft/min (117 m/sec).

Provision is made for the OV-10B and OV-10B(Z) to carry a 230-US gallon (870-l) external fuel tank on the fuselage centre line, adding to the internal fuel capacity of 252 US gal (954 l). Both pilot and observer have zero-zero ejection seats, as in the OV-10A, but the stub wings are not fitted. Only one OV-10B(Z) has been completed by North American, for flight testing; the other 11 are having the jet pods fitted in Germany by Rhein-Flugzeugbau.

Final production version of the Bronco is the OV-10C, now in service with the Royal Thai Air Force which ordered 16 at the end of 1969. The OV-10C is virtually identical to the OV-10A, although it lacks the provision for underwing rockets and missiles, which are optional on the latter, and has a KB-18 strike camera fitted. The engines in the OV-10C, rated at 715 shp each, are designated T76-G-412 and G-413. First flight was made at Port Columbus on 9 December 1970.

Operational use

Delivery of the first two OV-10As was made on 23 February 1968 in an unusual if not unique simultaneous hand-over to the two users, Air Force and Marine Corps. The USAF had by this time specified the rôle of the OV-10A as Forward Air Control, replacing or supplementing the Cessna O-1F and O-1G Bird Dogs in Vietnam, while the USMC classified its OV-10As as light armed reconnaissance aircraft with helicopter escort and a variety of other missions specified.

With training of crews the first priority, the two OV-10As departed Port Columbus International Airport in different directions, the gull grey-and-white USAF Bronco heading south for Eglin AFB (Hurlburt Field), Florida, and the olive green-and-white USMC aircraft going west to Camp Pendleton, Calif. At Eglin, the 4410th Combat Crew Training Wing had responsibility for getting the OV-10A into service, while at Pendleton, VMO-5 had been designated as the Bronco training unit. The latter responsibility subsequently passed to HML-267 at the same base, while the 4409th CCT Squadron at Hurlburt provides training facilities for the USAF on the OV-10A.

Little time was lost in getting the Bronco into service, with the Marine Corps gaining a short headstart on the USAF. The first operational unit was VMO-2, attached to the First Marine Air Wing at the Marble Mountain air facility south of Da Nang. Six OV-10As were ferried by aircraft carrier from San Diego, California, to Cubi Point in the Philippines, from where they were flown to South Vietnam by VMO-2 pilots on 6 July 1968. The first combat sorties were flown less than three hours after arrival at Marble Mountain.

Six OV-10As (crated and airlifted in Douglas C-133s) arrived at Bien Hoa, in South Vietnam, on 31 July and were attached to the 19th Tactical Air Support Squadron, 504th TAS Group, for combat evaluation. Subsequently, this unit, plus the 20th and 23rd TAS Squadrons, were fully equipped with OV-10As for service in Vietnam.

A second Marine Corps unit, VMO-6, equipped with the Bronco in Vietnam late in October 1968 and started operations on 1 November. Standard establishment for these units was 18 aircraft, deliveries of all but the first six being aboard aircraft carriers, for which purpose the aircraft had to be "Cocooned". The Marine Corps also used the OV-10A in the US to equip VMO-1 at MCAS New River, and two Reserve squadrons, VMO-4 at Grosse Ile, Michigan, and VMO-8 at Los Alamitos, California. Establishment for the Reserve units is nine aircraft each.

Although the US Navy acted as programme manager and procurement agency for the OV-10As on behalf of the Marines and Air Force, it showed little specific interest in the type until operational results from Vietnam became available. These were so dramatic that the Navy then acquired 18 OV-10As from the USMC and used them to equip VA(L)-4, operating riverine patrols in the Mekong Delta. This unit was commissioned on the Bronco on 3 January 1969, moving to Vietnam later in the year; four of the Navy OV-10As remained in the US in the hands of VS-41 at NAS North Island, which was responsible for training Navy pilots on the type.

In Vietnam, all three services flying the Bronco found that it was possible to achieve very high utilisations. For example, the first six aircraft serving with VMO-2 completed 500 combat hours in 250 missions within 3½ weeks of entering service. Operational availability figures averaged between 80 per cent and 90 per cent, and utilisation of more than 100 hours a month was consistently achieved by many aircraft. This was in face of climatic conditions described by the Marine Corps as being "as adverse as could be found anywhere in the world". Temperatures ranged from 45 deg F to 115 deg F, with 30–35 knot (56–65 km/h) winds blowing fine sand for as much as six hours of the day. Periods of as much as two months went by, in the rainy season, when the sun was never seen and the moisture content was "unreal". The ease with which the Bronco could be serviced and maintained in these conditions was a great vindication of the care taken in the design — at about 3·3 man hours per flight hour, the figure was at least 10 per cent below specification.

Pilots of the Bronco are unanimous in their praise of the aircraft as a flying machine, and its ability to absorb punishment and still get back home has endeared it further to its crews. The armour protection, self-sealing fuel tanks, duplicated control circuits and single-engine capability all contribute to the aircraft's high survivability —, and in the few instances when OV-10As have had to be abandoned, the North American-developed ejection seats have performed well.

The Marine Corps used its OV-10As with armament from the outset in Vietnam, using them for close support with 2·75-in and 5-in (6,99 and 12,7 cm) rockets or 7,62-mm mini-gun pods on the external points to supplement the built-in guns. Operating in the Forward Air Controller rôle, the aircraft carries MK-45 flares externally, and hand smoke grenades internally, dropped through the message drop door to mark targets for tactical strike aircraft. The Navy unit in Vietnam, known as the "Black Ponies", made maximum use of the Bronco's load carrying ability, usually operating with a 20-mm gun pod on the centre line, three 7,62-mm mini-gun pods and eight 5-in (12,7-cm) rockets on the stub-wings and a pod of ten 2·75-in (6,99-cm) rockets on each outer wing station. In a year, the Black Ponies flew over 4,000 missions in support of the river patrol boats in the Mekong Delta, providing 24 hr scramble readiness.

The USAF initially intended to operate its aircraft for FAC duties without any armament. However, six OV-10As from the 19th TASS were armed with rockets for a 9-week evaluation in April–June 1969, and the results of this trial, code-named *Misty-Bronco*, showed that the Bronco could respond to battlefield requests for close air support in an average time of 5 minutes, compared with 50 minutes for other tactical aircraft. Consequently, the commander of the 7th Air Force ordered that all USAF OV-10As should operate in the armed rôle forthwith, using both the fixed machine guns and external rockets.

The fuselage of the OV-10A provides 75 cu ft (2,12 m³) of space for cargo, and this can be increased to 110 cu ft (3,11 m³) if the ejection seat, instruments and primary flight

The US Marine Corps is testing two YOV-10Ds as Night Observation/Gunship Systems, with ventral turret and infra-red sensors in the extended nose. This picture shows the flying mock-up of the new model, based on a YOV-10A.

and engine controls are removed from the rear cockpit. Access to the cargo bay is through the hinged door forming the rear fairing of the fuselage; this door can be completely removed when the OV-10A is being used to drop paratroops. Typical loads comprise 3,200 lb (1 452 kg) of cargo, five fully-equipped paratroops or two stretcher patients with an attendant. Paradrops have been demonstrated by the Marine Corps in Vietnam, in non-combat operations.

In another demonstration of the OV-10A's capability, landings and take-offs have been made from the decks of aircraft carriers, without the use of arrester gear or catapults. The original LARA specification had required that the aircraft should be capable of operating from LPH-4-class carriers, and the demonstrations were made to prove that the OV-10A met the requirement. On 26 October 1968, two full-stop landings and two touch-and-go landings were made on the 720-ft (220-m) long angled deck of the USS *John F Kennedy,* take-offs from the deck being made into a 25-knot (46-km/h) headwind. A longer series of tests in August 1969 was made aboard the smaller USS *Boxer,* with 24 take-offs and landings at weights ranging from 9,000 lb to 12,000 lb (4 082–5 443 kg). Landings were made to a full stop in under 300 ft (91,4 m) using only the aircraft's brakes and reversing propellers, and the take-off roll averaged less than 700 ft (213 m).

A number of alternative armament and equipment installations has been proposed or tested on the Bronco to extend its capabilities in Vietnam. One early trial was the installation of a 105-mm recoilless rifle slung under the fuselage, with ammunition carried in bins in the cargo compartment and a special device to pick the shells from the bin and load them into the chamber of the gun. Installation of side-firing and rearwards-firing cannon was also studied.

More recently, North American's Columbus Division converted two USMC OV-10As to YOV-10D configuration as Night Observation/Gunship Systems (NOGS). The distinguishing features of the YOV-10D are a Hughes forward-looking infra-red (FLIR) sensor in the nose, and a General Electric XM-197 three-barrel 20-mm cannon in a ventral turret under the rear fuselage. First flight of the YOV-10D configuration was made on 9 June 1970, the original No 2 YOV-10A having been converted to an aerodynamic test vehicle with mock-up turret. The two OV-10Ds were accepted at the end of 1970 when they were ferried to the Naval Weapons Center, China Lake, Calif, for installation of operational equipment and weapon firing trials.

Object of the YOV-10D conversion is to improve the night operational capability of the Bronco. A similar programme is under way for the USAF under the code name *Pave Nail,* although in this case no additional armament is fitted. The *Pave Nail* aircraft, intended to operate as night forward air control and strike designation aircraft, are fitted with a combination Laser rangefinder and target illuminator made by Martin-Orlando and a Varo Inc stabilised night periscope sight. Once a ground target has been located with the help of the night sight, it can be ranged accurately by the laser. Loran equipment is then used to determine the precise position of the OV-10A and the relevant position of the target, and thus provide an off-set vector for accompanying strike aircraft. Alternatively, the OV-10A can illuminate the target with the laser for the benefit of a laser-seeking bomb, which can home on the energy reflected from the target. LTV Electrosystems is responsible for the *Pave Nail* conversion programme and at present has contracts to convert 13 OV-10As at a cost of $12 million. □ FGS

Viewed from the Cockpit

by Ed Gillespie
Chief Test Pilot
North American, Columbus

REGARDLESS of rumours or appearances, this aeroplane does not fly quite as easily as a "double-breasted Cub" and it does demand more of the pilot than merely reading the handbook. Throughout the Bronco's development period, we tried to optimise the flying qualities for the operational missions that the aeroplane was intended to perform. This meant short, low-speed take-offs and landings, lengthy cruise flight periods, tight and fast manœuvrability for forward air control and armed reconnaissance missions, and a stable but responsive weapons platform for delivering ordnance on target.

The cockpit layout is functional and displays Air Force, Army and USN commonality. Engine instruments are on the right side of the instrument panel and the UHF transceiver controls on the left. The "basic six" flight instruments are grouped in the centre. The cockpit reflects the best efforts of tri-service pilots and mock-up board members to integrate all of the requirements and practices of the individual services into one layout acceptable to all—in itself, no easy trick.

Although the OV-10 is of simple and proven construction, it behoves "Bronco Busters" to acquaint themselves thoroughly with the operating characteristics of solid shaft turboprop engines, the flying qualities and performance, particularly in the areas of take-off and landing, and single-engine control and flyaway speed characteristics.

Preflight, start and taxi

A walkaround preflight inspection can be quickly conducted. Primarily, checks should cover the conditions of canopy and access doors for security, tyres and brakes for wear, intakes and exhaust for plugs removed and cleanliness, wings for fuel leaks, cargo compartment for loose gear, hydraulic reservoir level, and angle-of-attack and pitot-static covers off. Entrance into the cockpit is awkward, but once in, the pilot's sitting position is comfortable and easily adjusted. After strapping in properly and arming the seat, starting is simple. Either battery or external power starts are available. With the engine master switch on, condition levers at FUEL SHUT-OFF, and power control levers at the FLIGHT IDLE position, only advancement of the condition levers to NORMAL FLIGHT and monitoring of engine temperature is required after the start switch is actuated.

Following the start, it is necessary to reverse each prop to ensure that prop locks have been retracted prior to taxi. The Bronco taxies moderately fast on the level with power control levers at about the FLIGHT IDLE position. Reverse thrust is available for braking and manœuvring through very sharp turns. Nose wheel steering is also available, but is generally used only when manœuvring across unprepared terrain. Taxi-ing is usually performed with the condition levers in the NORMAL FLIGHT position for low noise levels.

Take-off and climb

Prior to moving into the take-off position on the runway, double check that all canopies are locked (both fore and aft edges) and all trims at zero. Condition levers should be in the "take-off and land" mode (TO/LAND) and throttles advanced to the limiting engine temperature or limiting torque depending on pre-take-off calculations. The aircraft can be maintained statically with max rated thrust applied by holding brakes. Engine temperature is slow to rise and the Bronco will probably be allowed to roll before the engine temperature stabilises.

Rudder effectiveness is available almost immediately upon brake release. Direct cross winds of up to 20 kts (37 km/h) can be tolerated although lateral effectiveness is low during the initial take-off roll. In this case, the aeroplane can be allowed to assume a list, and directional control is maintained by utilising rudder until airborne. Good lateral effectiveness is immediately available upon main gear lift-off.

Normal take-offs on adequate runways are performed with zero flap as this enables the Bronco to attain a safe single-engine climb (flyaway) speed more quickly. For short field take-offs, half flap is used, and the aeroplane is rotated briskly once the minimum single-engine speed is obtained. The undercarriage is retracted by depressing the gear handle latch and firmly punching the handle upward. The gear then comes up in about 10 seconds and locks home with a firm "klunk".

Reconnaissance cruise is typically conducted at altitudes from 500 to 1,000 ft (152 to 305 m) with power settings of 90 per cent to 92 per cent rpm which provides approximately 180 kts (334 km/h) IAS. A typical cross country cruise is made at 5,000 to 8,000 ft (1 524 to 2 438 m) with power settings of 89 per cent to 91 per cent providing a true airspeed of approximately 200 kts (371 km/h). Long-range (ferry) flights are performed at the oxygen altitudes at lower power settings and true airspeeds of approximately 170 kts (315 km/h). In all cases, the engines are in the "normal" mode to provide best miles per pound of fuel and the lowest noise levels.

Trimmability is outstanding and the Bronco can be maintained in wings level flight utilising the rudders for directional and roll control while the pilot is studying the necessary navigation paraphernalia. The view from the cockpit is outstanding under any flight condition.

Manœuvrability

The Bronco has a load factor range from +8·0 to −3·0 g within the allowable speed envelope and the flight characteristics displayed in this wide range during manœuvring

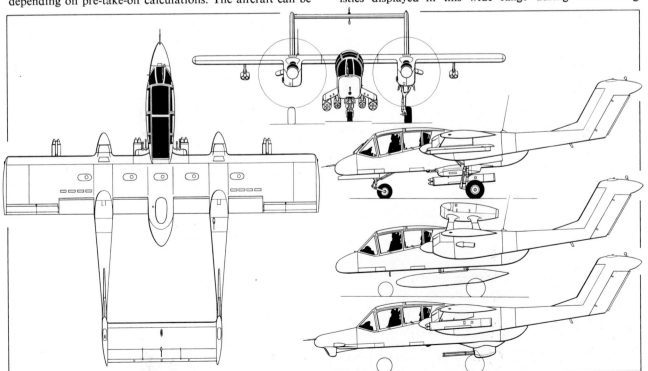

flight with a normal cg show the aircraft to its best advantage. Relatively light control forces and high response are exhibited during manœuvring at normal manœuvring speeds of 160 to 350 kts (297 to 649 km/h) IAS. The light forces enable the aircraft to be turned continuously without causing excessive pilot fatigue. Control displacement requirements are relatively low, again reducing the workload of the pilot. Lateral response is good at the normal approach speed of 75-100 kts (139-185 km/h), and is very good above 250 kts (463 km/h). Adverse yaw is present in the landing configuration, but is easily corrected by the very effective rudders. Moderate pitch excursions are noticeable with large lateral deflections during high-speed rolls, but are not noticeable during normal corrective control movements. The yaw damper effectively prevents small amplitude directional snaking during tracking runs in turbulence. It is recommended that the damper be turned off for other flight conditions. The turning radius is remarkably small and a 360-degree turn could be completed within the confines of the average college football stadium, although this is not recommended on Homecoming Day!

The Bronco displays no unusual characteristics during high-speed flight, although steep dive angles are required to obtain speeds in excess of 350 kts (649 km/h) IAS. Control forces, laterally and longitudinally, remain light at high speed. Rudder forces increase, but are still moderate. With no external stores, airframe buffet is negligible. Engine torques increase significantly with a decrease in altitude and/or an increase in airspeed; consequently, in high-speed flight, especially in cold weather, attention to engine limits is absolutely required, or the engine will be unacceptably over-torqued.

Stalls and spins

Normal stall characteristics of the Bronco are about the only similarity it has to a Cub. An artificial stall warning is provided by a rudder pedal shaker 3-10 kts (5,5-18,5 km/h) above stall depending on power setting (the higher the thrust, the lower the stall warning). In addition, in cruise and landing configuration, stalls are characterised by natural airframe shake 5 to 7 kts (9,3-13 km/h) before stall occurs with full aft stick. Stall breaks are not finite and the wings can be maintained level, fully stalled, by judicious use of rudders; however, a high sink rate develops.

Take-off stalls occur with less stall warning, but in an abnormally steep attitude. This stall is characterised by an increase in sink rate, moderate directional wandering and, if continued, the nose will slice either left or right until recovery is made. In all cases, recovery is immediate with relaxation of aft stick and application of aileron and rudder to control any roll or yaw tendencies. Because of the lack of a finite stall break, actual stall speeds are nebulous, but at nominal gross weight of 9,500 lb (4 310 kg) minimum speeds are about 55 kts (102 km/h) IAS in the take-off configuration, 64 kts (119 km/h) IAS in the landing configuration,

North American Rockwell OV-10A Specification

Power Plant: Two Garrett-AiResearch T76-G-10 (left) and -12 (right) fixed-shaft turboprop engines, each rated at 715 shp. Fully reversible, fully-feathering three-bladed 8 ft 6 in (2,59 m) diameter aluminium propellers.
Armament: Four 7,62-mm M60 C machine guns in sponsons. One centreline fuselage station, four sponson stations and two underwing stations for guns, rockets, bombs, missiles, flares, etc. Total external stores capacity 4,600 lb (2 086 kg), subject to aircraft structural and gross weight limitations.
Weights: Empty, 7,190 lb (3 260 kg); normal loaded, 12,500 lb (5 670 kg); maximum take-off weight, 14,444 lb (6 550 kg).
Performance: Maximum speed, 244 knots (452 km/h) at 10,000 ft (3 048 m); limiting dive speed, Mach 0·7=456 knots (845 km/h) EAS at 4,000 ft (1 219 m); cruising speed, 195 knots (361 km/h) at 18,000 ft (5 486 m); loiter speed, 110 knots (204 km/h) at 5,000 ft (1 524 m). Take-off distance to 50 ft (15,2 m), ISA, sea level, 20 deg flap, 1,800 ft (548 m) at 12,500 lb (5 670 kg); landing distance from 50 ft (15,2 m), ISA, sea level, 40 deg flap, 1,050 ft (320 m) at 9,000 lb (4 082 kg). Rate of climb at sea level, clean, 9,500 lb (4 105 kg) weight, 2,800 ft/min (14,2 m/sec) on two engines, 850 ft/min (4,31 m/sec) on one engine. Typical close support mission radius, 165 naut mls (306 km) with full internal fuel, 2,800 lb (1 270 kg) ordnance load and 1-hr loiter at 5,000 ft (1 524 m); typical FAC mission radius with full internal fuel, 150-US gal (568 l) external tank and two 5-in rocket pods, 145 naut miles (269 km) with 3-hr loiter; ferry range 1,240 naut mls (2 300 km).
Dimensions: Span, 40 ft (12,19 m); overall length, excluding nose probe, 39 ft 9 in (12,12 m), with nose probe, 41 ft 7 in (12,67 m); overall height, 15 ft 1 in (4,62 m); wheelbase, 11 ft 7½ in (3,54 m); wheeltrack, 14 ft 10 in (4,52 m); tailplan span, 14 ft 7 in (4,45 m); gross wing area, 291 sq ft (27,03 m²).

60 kts (111 km/h) IAS in the power approach configuration, and 70 kts (130 km/h) IAS in the cruise configuration.

Because of the high manœuvrability and the operational requirement for pilots (particularly FAC) to maintain their attention outside of the cockpit, it is reasonable to expect that accelerated stalls will be encountered more often in the Bronco than in other types. Fortunately, accelerated stalls are easily recognised and stall and recovery characteristics are excellent. As g is increased and/or airspeed is decreased during turning flight or symmetrical pull-ups, moderate to heavy airframe and rudder pedal shake occurs 5 to 8 kts (9,3-15 km/h) prior to stall. Stall occurs with a moderate but firm pitch-down and a mild tendency to roll either left or right, which can easily be stopped with lateral control. Recovery is effected by relaxation of aft stick and the aeroplane is immediately under full control.

Because of mild stall characteristics and strong rudder effectiveness, the Bronco resists spinning in any configuration. However, if desired, it can be made to spin to the left or right and can also be recovered very quickly. Spins are characterised by pitch oscillations during the first turns from about 60-80 degrees, then becoming steadier at the third turn. After the third turn, the spin usually slows of its own accord for one-half turn, then speeds up and repeats the cycle if pro spin controls are maintained full. Recovery can be effected within one-fourth turn by application of opposite rudder and neutral stick. The only difference between normal 1·0 g and accelerated entry spins is that the aeroplane will perform a couple of snap rolls in the direction of applied rudder following an accelerated entry prior to entering the normal upright spin.

Single-engine characteristics

Intentional shutdown or loss of an engine in the normal flight envelope is easily handled. The wind-milling propeller automatically seeks a no-thrust blade position which varies

—continued on page 260

FROM
ISLANDER TO TRISLANDER

DESIGNING replacements for successful civil transport aeroplanes is a field of endeavour in which many have tried but few have succeeded. Because certain aeroplanes, over the years, have succeeded beyond their designer's wildest dreams — the Douglas DC-3 is, of course, the prime example — many companies have been attracted by the possibility of producing a replacement type which, in theory, would clean up the market and sell in equally impressive and profitable numbers.

All too often, this "replacement market" has proved to be a chimera and the attempt to capture it has brought ruin rather than riches. The pages of aviation history covering the past 30 years are littered with projects and prototypes for "Dakota replacements", "Rapide replacements", or similar proposals. Few aeroplanes projected from the outset in one or other of these replacement categories have ever reached production, and of those that have, most have sold in quantities far smaller than the sanguine hopes of their creators.

One of the very few exceptions is the Britten-Norman Islander, which, by the definition of its original specification, was firmly placed in the category of a Rapide replacement and which is now selling and operating in just that category. Very few Islander operators, it is true, have actually bought the aircraft to replace Rapides, and this fact underlines one of the basic misconceptions that has led to the downfall of many projects. The true replacement for aeroplanes like the Dakota, the Rapide, the Dove and so on is not one that is sold to the same operator on a one-to-one basis: for a hundred different reasons, those operators who bought the original aeroplanes 20 or 30 years ago, now have a hundred different sets of requirements, and no one aeroplane can hope to meet more than a small segment of *that* market. The successful replacement aeroplane is the one that can be sold to the new operators who are today in the same stage of evolution as the purchasers of the older types in the decade after the end of World War II. In the case of the Islander, this boils down to the air taxi/third level airline market, not only in the "grown-up" environment of Europe and North

America but also in the very rough, tough and unsophisticated conditions found in the less-developed areas of the world.

With the 300th Islander delivered in August and production continuing at a healthy dozen a month, there can be no doubt that the simple little twin-engined eight-seater from the Isle of Wight has found its niche in the transport world. It is interesting to recall that the crystallising factor in getting the Islander designed and launched was the need of one specific operator who was unable to find on the market a twin with "more than six seats" having the required combination of short take-off, single-engined performance, ability to operate with minimum ground support in rugged conditions, and low cost. The operator in question was Cameroon Air Transport, in which the Britten-Norman company, through its subsidiary Crop Culture (Aerial) Ltd, had a 25 per cent interest.

While a larger or more experienced manufacturer might have been tempted to conduct an extensive market survey that most probably would have served to blur the image of the required aeroplane, John Britten and Desmond Norman set about designing an aeroplane to meet CAT's needs, feeling instinctively that there was a large market for such a type. This was in late 1963. The two founders of the Britten-Norman enterprise had been working together since 1951, largely in agricultural aviation, building up a fleet of 70 ag-planes for operation throughout the world. Their sole experience of aircraft design was the far from successful BN-1F ultra-light single-seater, but it was perhaps a happy augury for the prospective Rapide replacement that the two originators were both ex-students of the de Havilland Technical School.

If one analyses the reasons for the success of the Rapide in the years immediately before and after the war, it is clear that the combination of relatively large capacity with "lightplane" engineering and operating concepts was the dominant factor. Britten-Norman realised that the expanding third-level and commuter airline market of the 'sixties needed a similar approach, in direct contrast to the emphasis placed

(Above) The prototype Islander in its original guise, and (below) after conversion to take Lycoming engines in modified nacelles and an increase in wing span.

(Above) The first supercharged Islander with Rolls-Royce Continental engines, and (below) the current "IO-Islander" with uprated Lycoming engines. The example illustrated also has wingtip fuel tanks.

by most manufacturers of light twins upon higher performance with relatively small payloads. Thus the BN-2 project took shape as a high-wing, lightly loaded monoplane that could carry 10 people on two 210 hp engines.

Reckoning that potential operators of the BN-2 were unlikely to fly stage distances of more than 100 miles (161 km) or so, the designers accepted a modest cruising speed, and could thus afford to use a large wing, with simple single-slotted flaps and ailerons, to achieve the kind of STOL performance that faster-cruising contemporaries could match only with expensive and complicated high-lift devices. To obtain maximum capacity in the smallest possible fuselage with minimum frontal area, the fuselage width was reduced to that of a double seat, with no aisle, access to each pair of seats being gained through individual doors in the fuselage side. A non-retractable tricycle undercarriage was adopted in keeping with the emphasis upon maintenance and simplicity of operation.

Design of the BN-2 was completed in the first half of 1964, at which time the company employed about 100 persons at its small factory at Bembridge on the Isle of Wight. To supplement its own resources, Britten-Norman contracted with F G Miles for the services of a group of designers.

First metal was cut for a prototype of the BN-2 in September 1964, the same month in which news of the project was first made public, with a model on display at the SBAC show at Farnborough. The brochure available at that time defined an aircraft with two 210 hp R-R Continental IO-360 engines, useful load of 1,970 lb (894 kg), gross weight of 4,750 lb (2 155 kg) and wing span of 45 ft (13,7 m).

Working hard and fast, Britten-Norman completed a prototype to this specification in less than 10 months from the start of construction, the first flight being made at Bembridge on 13 June 1965. This prototype, G-ATCT, with a special category C of A, flew to Le Bourget four days later to participate in the Paris Air Show. Following its return to the UK, the serious work of flight testing began, with 187 flights totalling more than 100 hours made in the next three months.

Two things soon became obvious. One was that interest in the BN-2 was running at a high level and the sales potential was considerable. The other, less satisfying, was that there were problems with the prototype that would require some fairly major modifications to be made before the aircraft — by this time named the Islander — could enter production. While handling was satisfactory, performance figures indicated that the aircraft was high on drag, with cruise speed and single-engine climb below specification. The eventual solution adopted was to extend the wing span by 4 ft (1,22 m) to 49 ft (14,9 m), refine the rear fairing of the engine nacelles and fit 260 hp Lycoming O-540-E engines. All these changes were incorporated on G-ATCT, which flew in the revised form on 17 December 1965. The increased empty weight was more than offset by the higher permitted gross weight of 5,200 lb (2 360 kg) at this stage, and eventual certification at 5,700 lb (2 585 kg) gave a further bonus of 500 lb (226 kg) in the useful load.

Production begins
A second, fully modified prototype flew on 20 August 1966, and this aircraft, G-ATWU, appeared with the first Islander at the 1966 SBAC Display. Up to this point, Islander development had been wholly-financed by the still-small Britten-Norman company. To finance the cost of launching production and building a new factory on the airfield at Bembridge, government aid was sought and obtained during 1966, with £190,000 in launching aid repayable from a levy on sales and £550,000 in interest-bearing loans to cover expansion of the production line.

Islander number 3 (G-AVCN) was the first to be assembled in the new factory, flying for the first time on 24 April 1967 in time to be shown at the Paris Air Show that year. It appeared in the colours of Glosair, destined to be the first operator to take delivery of an Islander, on 13 August. The full C of A had been granted by the ARB on 10 August, and in a modest ceremony on 15 August, Islander number 4 was handed over to Loganair, a third level service in the Orkney and Shetland Isles, that was entirely typical of the kind of operations for which the aircraft had been designed.

Overseas markets had always been a primary objective of the Britten-Norman enterprise, and this was stressed by the allocation of Islanders 5 and 7 as US demonstrators, being flown across to America in September and October 1967 respectively. FAA certification for the BN-2 was obtained on 19 December, and deliveries of customer aeroplanes to the US began in January 1968.

The demand for Islanders snowballed, and Britten-Norman soon found that its own facilities could not match the required delivery schedules. Quick delivery was essential, however, if the potential business was to be retained; customers eager to cash in on the booming third-level airline business were not prepared to wait a year or more for their aeroplanes — they would go shopping elsewhere instead.

BRITTEN-NORMAN ISLANDER VARIANTS

DESIGNATION	WING TIP TANKS	LE DROOP	FLAP DROOP	SPEED PAK	ENGINES		BROCHURE EMPTY WEIGHT		MAX. TAKE-OFF WEIGHT						SINGLE ENGINE CEILING*	
									BRITISH VFR & IFR		US		AUSTRALIAN			
											VFR	IFR	VFR	IFR		
					hp	Type	lb	(kg)	lb (kg)		lb (kg)	lb (kg)	lb (kg)	lb (kg)	ft	(m)
BN.2A	—	—	—	×	260	O-540	3,583	(1 625)	6,300 (2 858)		6,000 (2 722)	6,000 (2 722)	6,000 (2 722)	6,000 (2 722)	5,600	(1 707)
BN.2A-1	×	—	—	×	260	O-540	3,660	(1 660)	6,300 (2 858)		6,300 (2 858)	6,300 (2 858)	6,300 (2 858)	6,170 (2 799)	6,600	(2 012)
BN.2A-2	—	×	×	×	300	IO-540	3,723	(1 689)	6,300 (2 858)		6,300 (2 858)	6,300 (2 858)	6,300 (2 858)	6,300 (2 858)	7,800	(2 377)
BN.2A-3	×	×	×	×	300	IO-540	3,800	(1 724)	6,300 (2 858)		6,300 (2 858)	6,300 (2 858)	6,300 (2 858)	6,300 (2 858)	8,800	(2 682)
BN.2A-4							PROJECTED HIGHER GROSS WEIGHT VERSION									
BN.2A-5							PROJECTED HIGHER GROSS WEIGHT VERSION									
BN.2A-6	—	×	—	×	260	O-540	3,588	(1 627)	6,300 (2 858)		6,200 (2 812)	6,170 (2 799)	6,300 (2 858)	6,000 (2 722)	5,600	(1 707)
BN.2A-7	×	×	—	×	260	O-540	3,665	(1 662)	6,300 (2 858)		6,300 (2 858)	6,300 (2 858)	6,300 (2 858)	6,170 (2 799)	6,600	(2 012)
BN.2A-8	—	×	×	×	260	O-540	3,588	(1 627)	6,300 (2 858)		6,200 (2 812)	6,170 (2 799)	6,300 (2 858)	6,000 (2 722)	5,600	(1 707)
BN.2A-9	×	×	×	×	260	O-540	3,665	(1 662)	6,300 (2 858)		6,300 (2 858)	6,300 (2 858)	6,300 (2 858)	6,170 (2 799)	6,600	(2 012)
BN.2A-10	—	×	×	×	270	TIO-540	3,740	(1 696)	6,300 (2 858)		6,300 (2 858)	6,300 (2 858)	6,300 (2 858)	6,300 (2 858)	12,500	(3 810)
BN.2A-11	×	×	×	×	270	TIO-540	3,817	(1 731)	6,300 (2 858)		6,300 (2 858)	6,300 (2 858)	6,300 (2 858)	6,300 (2 858)	15,000	(4 572)

* At 6,300 lb (2 858 kg) gross weight.

Consequently, the company decided in 1968 to sub-contract the entire airframe production to British Hovercraft Corporation (a subsidiary of Westland Aircraft Ltd), conveniently located in the former Saunders-Roe aircraft factory at Cowes, also on the Isle of Wight. An initial contract was placed with BHC for 236 airframes, with another 134 contracted subsequently, and Islander number 22 was the last of wholly Britten-Norman construction. The Bembridge factory remains responsible for final assembly and flight test, but in another move to spread the load, aircraft from No 47 onwards have been flown as bare unpainted shells from Bembridge to Cambridge, for fitting out and painting by Marshalls.

Another aid to Islander production arose somewhat fortuitously in 1967, when the Rumanian airline TAROM began shopping for jet equipment in the west. With the BAC One-Eleven high on the airline's short list, Rumania indicated that a favourable decision would be reached only if she could obtain, as part of a package deal, help in re-establishing a local aircraft industry. BAC put the proposal to Britten-Norman and the outcome was an agreement between the latter company and IRMA, the state-owned factory in Bucharest, whereby an Islander production line was set up with full Britten-Norman assistance, and a contract was placed with IRMA for 215 Islander airframes.

To help the new production group up the learning curve, Britten-Norman supplied complete sets of parts for the first few Islanders to be assembled in Rumania. The first of these (the 85th British Islander) was flown in Bucharest on 4 August 1969 and subsequently delivered to the UK. With a gradual swing to complete Rumanian production, a steady flow has been built up and 50 Islanders had been completed and flown by July this year. They are identical with the UK models and are marketed through the Britten-Norman organisation in the normal way.

Product improvement

The Islander has been subjected to a process of continuous improvement in the course of the four years since deliveries began. Many of these refinements are relatively minor in nature and have been fed into the production line as features of the basic specification; others represented more clearly-defined steps in the evolution of the BN-2, while a further series of modifications are available as customer options.

The first significant improvement was to recertificate the Islander at a gross weight of 6,000 lb (2 722 kg) early in 1968. This produced the BN-2A, introduced on the production line at position 25; most BN-2s were subsequently modified to the same standard. A second step was reached in June 1969 with effect from aircraft number 70, when the BN-2A series 2 introduced a separate baggage compartment door (previously access was through the door serving the rear row of seats) and other small refinements.

The current basic BN-2A is certificated at a gross weight

of 6,300 lb (2 857 kg), this further stage having been reached in 1970 with the introduction of the so-called Speedpak nacelle and other changes. The Speedpak modification comprises a new shape to the rear nacelle plus an increase in the chord of the main leg fairings. For US certification, another modification is required, comprising a "drooped" or cambered wing leading edge between the fuselage and

The four stages in the life of the second prototype Islander are shown in this sequence of pictures. (From top to bottom) As first flown; with stretched fuselage; converted to Trislander prototype, and with interim top fin.

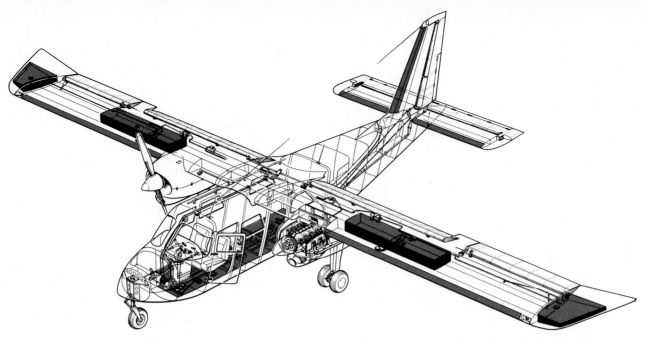

nacelle. This reduces the stalling speed and allows the Islander to comply with the FAR single-engined climb requirement at the higher gross weight. A modification permitting a significant increase in maximum zero fuel weight is to provide a fixed droop of 6 deg on the flap, and this is also now standard on production aircraft.

Other standard new features which, like those mentioned above, can be purchased as "customising kits" for incorporation on earlier Islanders, include such things as Cleveland brakes and wheels; exhaust pipe stays; pilot's rams-horn handwheel; improved seat webbing and end trim; additional cabin air extractor; improved roof interior trim; separate baggage compartment door; engine cowl fastener improvements and modular engine instrument cluster. Other kits are for features that are customer options on new aircraft such as castoring nosewheel; Whelan HR strobe anti-collision beacons; increased baggage bay area, and windshield de-icing panel.

An option that makes an immediately noticeable difference to the appearance of the Islander is the installation of

wing-tip fuel tanks. Although the Islander was conceived with ultra-short range operations in mind, the progressive increases in useful load/gross weight, and the inevitable demands of some operators for more range, led the company to develop a long-range option offering an extra 50 Imp gal (227 l) of fuel (an increase of some 44 per cent on the basic internal fuel capacity of 112 Imp gal; 509 l). The method by which this increase is obtained is typical of Britten-Norman's inventiveness — the extra tanks are carried in new wing tip fairings that replace the standard Hoerner-type tips and increase the span to 53 ft (16,15 m). The effect of this modification is to increase the range to 800 naut miles (1 482 km) with 800 lb (363 kg) payload; the standard aircraft flies only 540 naut miles (1 000 km) with this payload, but has superior payload over shorter ranges.

In an effort to improve the "hot and high" performance of the Islander, the number 9 airframe (G-AVUB) was modified in 1968 to have turbo-supercharged engines, after first serving as the test-bed for BN-2A certification. Installation of two 300 hp RR-Continental TSIO-520 engines was made by Miles Aviation and Transport at Ford, and the aircraft flew as the BN-2S on 6 September 1968, just in time to appear at that year's Farnborough display. This installation, however, did not prove entirely satisfactory and no production ensued. Instead, a variant with 300 hp Lycoming IO-540K engines was developed, the first installation being in airframe 145 (G-AYBI), first flown on 30 April 1970 and displayed at the 1970 SBAC display. The "IO-Islander", as it is called around Bembridge, has already proved of special interest in Australasia, which is Britten-Norman's second most important sales area after the Americas, and deliveries of this version began in November 1970. Australian certification of the basic Islander had been obtained on 15 March 1968; as with the FAA in America, the Australian DCA had slightly different requirements from the British ARB, with the result that different operating weights apply to the same aircraft according to whose rules apply locally — as shown in the accompanying table of variants.

The latest engine variation is a new supercharged version, the "TIO-Islander", flown for the first time on 30 April 1971 — and thus somewhat predictably maintaining the company's tradition of introducing something new at each Paris (and Farnborough) show. The TIO installation comprises two 270 hp Lycoming TIO-540-H engines and is expected to be of particular interest to operators using high altitude airfields in South and Central America, Asia and elsewhere.

Britten-Norman BN-2A Islander Specification

Power Plant: Two Lycoming O-540-E4C5 flat-six air-cooled engines, each rated at 260 hp. Hartzell constant speed fully-feathering two-bladed propellers, 6 ft 8 in (2,04 m) diameter.

Weights: Empty weight, 3,588 lb (1 727 kg); max zero fuel weight, 6,000 lb (2 722 kg); gross weight, 6,300 lb (2 857 kg).

Performance at 6,300 lb (2 857 kg): Max speed at sea level 170 mph (273 km/h); cruising speeds, 75 per cent power at 7,000 ft (2 140 m), 160 mph (267 km/h), at 67 per cent at 9,000 ft (2 750 m), 158 mph (253 km/h), at 59 per cent at 13,000 ft (3 960 m), 154 mph (248 km/h); stalling speed, flaps down, 42-47 mph IAS (67-76 km/h). Take-off distance to 50 ft (15,2 m), 1,090 ft (333 m); landing distance from 50 ft (15,2 m), 960 ft (293 m). Rate of climb at sea level, two engines, 1,050 ft/min (5,3 m/sec), one engine, 190 ft/min (0,96 m/sec). Service ceiling, 14,600 ft (4 450 m); single-engine ceiling, 5,600 ft (1 710 m). Range (standard fuel), 717 st mls (1 154 km) at 160 mph (267 km/h), 870 st mls (1 400 km) at 154 mph (248 km/h); range (with tip tanks), 1,040 st mls (1 674 km) at 160 mph (267 km/h), 1,263 st mls (2 035 km) at 154 mph (248 km/h).

Dimensions: Span, 49 ft 0 in (14,92 m); span with tip tanks, 53 ft 0 in (16,15 m); length, 35 ft 7¾ in (10,9 m); height, 12 ft 7 in (3,77 m); wheelbase, 13 ft 1¼ in (4,0 m); wheeltrack, 11 ft 10 in (3,6 m); wing area 325 sq ft (30,2 m²); wing area with tip tanks, 337 sq ft (31,25 m²).

Britten-Norman's principal distributor on the US East Coast, Jonas Aircraft and Arms Inc, has alternatively proposed a supercharged version of the standard Islander with Riley-Rajay turbosuperchargers fitted to the O-540 engines.

Military possibilities
Another of the aircraft "unveiled" by Britten-Norman for the 1971 Paris Air Show was called the Defender, this being the name adopted for military versions of the Islander. In fact, a few Islanders have already been sold for military or police duties, including three to the Abu Dhabi Defence Force and two to the Guyana Defence Force, while three more are on order for the Iraqi Air Force. With virtually no modification, the Islander provides a useful vehicle for light transport, liaison, communications and patrol duties.

The Defender takes the idea one stage further with the provision of four underwing pylons, nose-mounted weather radar and options for internal layouts that provide for stretcher-carrying, para-dropping with a door removed, aerial survey and long-range patrol. Each inboard wing pylon can carry 700 lb (318 kg), and the outboard positions carry 250 lb (113 kg) each. Loads can include mini-gun pods, rocket pods, practice bombs, flares, etc, or a 50-Imp gal (227 l) tank can be carried on each inboard position. Even without these tanks, but with the optional wing tip tanks, the Defender has an endurance of over 10 hours and a range of more than 1,000 miles (1 610 km). The standard nose-mounted radar with a 30-mile (48-km) range permits the Defender to perform search and rescue missions, anti-smuggling patrols, etc.

For use as a demonstrator, Britten-Norman allocated Islander airframe 235 (G-AYTS), and subcontracted design and installation of the pylons to F G Miles at Shoreham. First flight after modification was made on 20 May 1971, and the nose radar was then fitted before the aircraft flew to Paris.

Enter the Trislander
Making up a trio of new Britten-Norman exhibits at this year's Paris Show, the Trislander is the latest expression of the company's highly innovative approach to aircraft development. The appearance of the prototype at the SBAC Display at Farnborough last September on the very day that it made its first flight and only 60 days after the decision to go ahead was yet another demonstration of the company's flair for solving problems in minimum time and in such a manner as to achieve maximum publicity.

The speed with which the Trislander (ie, three-engined Islander) was completed is attributable to the degree of commonality between the two types. The wing of the Trislander is a standard Islander unit with the extended wing tips, and the two Lycoming O-540s are unchanged. The fuselage cross section is the same, the only significant difference being the extra 7 ft 6 in (2,29 m) section ahead of the wing. The only major new item is the strengthened fin structure, carrying the third engine, also an O-540. The tailplane is similar but of increased span. Some 75 per cent of the basic Islander is retained; the modification, however, increases maximum seating to 18, gives a 50 per cent increase in engine power and is offered for less than £100,000, including avionics.

To prove the concept, Britten-Norman decided, in mid-1970, to modify the original second prototype Islander, G-ATWU. This aircraft had previously been the subject of an earlier experimental "stretch", when the fuselage was lengthened by 30 in (0,76 cm). In this form it flew on 14 July 1968, but little development was undertaken as the market was not then ready for a larger Islander. Further study of the requirement led to definition Trislander based on a "more than two, less than four" engine argument that had

PAGE 243

Britten-Norman Trislander Specification
Power Plant: Three Lycoming O-540-E4C5 flat six air-cooled engines, each rated at 260 hp. Hartzell constant-speed fully-feathering two-bladed propellers, 6 ft 8 in (2,04 m) diameter.
Weights: Empty weight (equipped, less avionics) 5,638 lb (2 557 kg); max zero fuel weight, 9,050 lb (4 105 kg); gross weight, 9,350 lb (4 240 kg).
Performance at 9,350 lb (4 240 kg); Max speed at sea level, 187 mph (301 km/h); cruising speeds, 75 per cent power at 6,500 ft (1 981 m), 180 mph (290 km/h), 67 per cent at 9,000 ft (2 750 m), 175 mph (282 km/h), 59 per cent power at 13,000 ft (3 960 m), 170 mph (274 km/h). Take-off distance to 50 ft (15,2 m), 1,800 ft (549 m); landing distance from 50 ft (15,2 m), 1,490 ft (454 m). Rate of climb at sea level, three engines, 1,120 ft/min (5,7 m/sec), two engines, 400 ft/min (2,03 m/sec). Service ceiling, 14,500 ft (4 420 m); engine-out ceiling, 10,000 ft (3 048 m). Ranges, 160 st mls (257 km) with max payload at 170 mph (274 km/h), VFR reserves; 700 st mls (1 127 km) with 2,400 lb (1 089 kg) payload at 175 mph (282 km/h), IFR reserves.
Dimensions: Span, 53 ft 0 in (16,15 m); length, 43 ft 9 in (13,3 m); height, 14 ft 6 in (4,42 m); wing area, 337 sq ft (31,25 m²).

been applied by Hawker Siddeley and Boeing in producing the Trident and Boeing 727. This philosophy, with the "engine in the tail" solution, had not, however, previously been applied to a piston-engined aircraft.

In its new guise, G-ATWU flew at Bembridge early in the morning of 11 September 1970, and after a couple more flights during the day, it was flown to Farnborough for static display at the SBAC Display the same evening. At this stage, the third engine was mounted on top of the fin. A lack of directional stability was apparent right from the start of flight trials, and soon after the return from Farnborough, some necessary extra fin area was added above the engine nacelle. Thus modified, G-ATWU attained its fourth and final configuration; after completing an initial series of flight tests, it was grounded to become a structural test specimen, needed in particular to prove the strength of the third engine installation in the landing case, when severe downward loads might be encountered. These tests have now been successfully completed, and G-ATWU has ended its useful life.

Trislander flying continued with a production prototype, G-AYTU, modified during production from Islander number 245 and first flown on 6 March 1971, with the definitive top fin shape, larger in area than that on the prototype. This aircraft is now assigned as a company demonstrator. The first genuine production Trislander is G-AYWI, initially flown on 29 April and displayed at Paris in the colours of Aurigny Air Services, the first customer. Aurigny is already operating a fleet of eight Islanders and has ordered three Trislanders for use on its scheduled services between the Channel Islands and France. Following ARB certification of the Trislander on 14 May, delivery of this particular aircraft was made to Aurigny on 29 June. On the same day, the second production aircraft (G-AYZR) left Bembridge for participation in the London-Victoria Air Race and subsequent delivery to Islander Sales Corporation in Florida for use as a demonstrator.

ISC, one of Britten-Norman's distributors, has four

The production prototype Trislander was the first to feature the definitive top fin as shown here, and was finished in the overall yellow of Aurigny Air Services, the first Trislander customer.

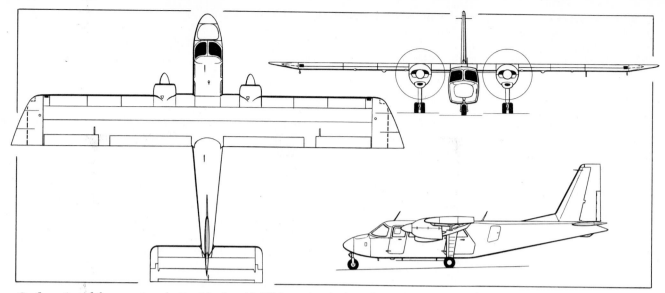

Configuration of the current production Islander with 260 hp engines is shown here. The standard wingtip shape is shown by dotted line. The extended tips contain extra fuel.

Trislanders on order; Jonas has ordered three and International Aircraft Sales, on the US west coast, one. Other orders and options from distributors and customers in other parts of the world bring the Trislander backlog to about 25 at the present time. Production is planned at a rate of two a month for the remainder of 1971.

Islander production, meanwhile, is continuing at about 12 a month, inclusive of the 3-4 a month from Rumania, and deliveries, after something of a hiatus in the first half of 1970, have picked up to a similar level. Annual delivery figures for the Islander have been: 1967, nine; 1968, 30; 1969, 79; 1970, 97 and 1971, 85 to the end of July. Thus, the 300th Islander was delivered in August, with suitable ceremony, to North Cay Airways, the largest Islander operator with 15 in service and five on order.

For the future, several possible further developments of the Islander have been considered at Bembridge. Higher gross weight can probably be certificated for the "IO-Islander" with little difficulty. Versions on skis or amphibious floats have been studied and could be developed with little effort. A more dramatic innovation for agricultural use has also reportedly reached the mock-up stage, and involves locating the pilot in a new cockpit above the wing, for improved visibility. Standard Islanders have already been fitted with crop-spraying gear beneath the wings, including micronair rotary atomisers, with an insecticide tank in the fuselage.

A longer-term programme might well see Britten-Norman establishing their position firmly in the third-level/commuter market with something approaching a Dakota-replacement — a 30-seater with the same simple, low cost design approach already demonstrated with the Islander and Trislander. If this should prove to be the case, then the foundations already laid with the existing types should give this company a better-than-average chance of success where so many others have failed. □

The initial production configuration of the Trislander on which the extended wingtips are standard.

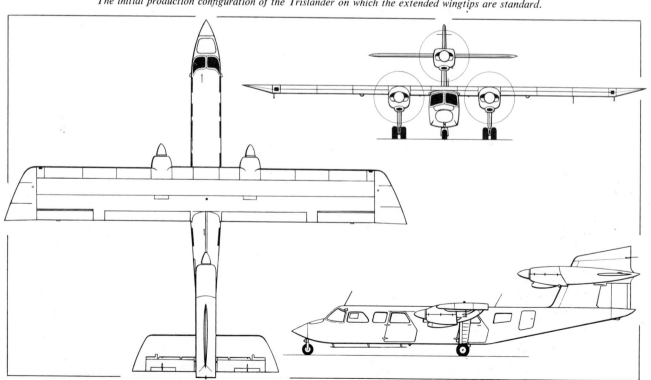

IN PRINT

"Polish Aircraft 1893-1939"
by Jerzy B Cynk
Putnam & Co Ltd, London, £7.50
760 pp, 5½ in by 8⅜ in, illustrated

WITH few exceptions, the aeronautical books published by Putnam and Company are "musts" for any self-respecting reference library, and "Polish Aircraft 1893-1939" cannot be numbered among the exceptions. It is the end product of what must have been a truly gargantuan task of research and compilation, consuming, according to the Introduction, no less than a quarter-century. The subject matter is perhaps somewhat more esoteric than most other titles issued by Putnam, and at first sight the published price seems somewhat high at £7.50 in the UK — despite the virtual doubling of book prices over the past two or three years, they are still substantially lower in the British Isles than virtually anywhere else in the world — but, in this reviewer's opinion, "Polish Aircraft 1893-1939" is worth every newpenny!

The book is divided into four sections — Early Pioneers, Main Aircraft Establishments, Individual and Amateur Aeroplane Designs, and Gliders and Sailplanes — and fascinating reading it makes. The author's style is erudite, and his text is complemented by excellent general arrangement drawings contributed by Waclaw Klepacki, and innumerable intriguing photographs, a large proportion of which have certainly not appeared hitherto in an English-language publication. The author has not been content to include merely the aircraft that were built and flown; he has spiced the repast with many intriguing projects, such as the PWS Ciolkosz twin-boom bomber, the RWD 25 Sokól fighter, and others indicating the advanced lines along which Polish aircraft designers were working at the time of the German assault.

As previously commented, the subject matter of this book is perhaps esoteric for the average aviation enthusiast, and the publisher is to be complimented on its courage, therefore, in including "Polish Aircraft 1893-1939" in its list.

"Camouflage and Markings — RAF Fighter Command, Northern Europe, 1936 to 1945"
by James Goulding and Robert Jones
Ducimus Books, London, £4.50, and Doubleday, New York, $14.95
294 pp, 9⅞ in by 7⅜ in, illustrated

THE NUMBER of books devoted not to the aircraft of a specific period but to their external finishes and markings is growing rapidly. Not so many years ago the *average* member of the modelling fraternity was willing to accept the instructions for markings and finishes that accompanied the plastic aircraft kit of his choice, but not today. Applying the external finish and markings to an aircraft model is an art in itself, and with the growth of expertise has been born a desire on the part of the modeller to display individuality by applying *different* "warpaint" to that suggested by the kit manufacturer, or to at least check the

The line of high-wing fighters designed by Zygmunt Pulawski for PZL occupy a significant portion of the important new book Polish Aircraft 1893-1939 *reviewed on this page. First of the line was the PZL P.1, the second prototype of which is illustrated above.*

authenticity of the schemes offered. Should the modeller's choice be an aircraft of RAF Fighter Command operating in northern Europe between 1936 and 1945, then this first volume of "Camouflage and Markings" from Ducimus comes within the "must buy" category.

But there is a growing interest in aircraft finishes, particularly of the WW II period, among non-modellers. Each change was made for a specific purpose, as is competently revealed by this book, which covers in detail the 18 fighter types that served during 1936-45 with RAF Fighter Command in northern Europe, as well as three (ie, Tornado, Mohawk and Welkin) which did not. Profusely illustrated by photographs of high quality, excellent line drawings revealing the patterns in which camouflage was applied and the markings that the aircraft sported, and a number of pages of colour art, this book hardly makes leisure reading. Its phraseology tends to be stilted, but such is to be expected of a work which is intended primarily for the reference shelves.

"Military Aircraft of the World"
by John W R Taylor & Gordon Swanborough
Ian Allan Ltd, Shepperton, Surrey, £1.50
242 pp, 5½ in by 8½ in, illustrated

THIS new title in the series of Ian Allan aircraft handbooks replaces the former "Warplanes of the World", from which it differs primarily in having more and larger pages. The book comprises two major sections, of which the first deals with first-line aircraft (defined as fighters, bombers, tactical and strategic transports, and anti-submarine aircraft). Each of 141 aircraft types in this section is illustrated by a photograph and a three-view silhouette; there is a data summary, and an outline of development and service testing, with reference to all major versions and quantities ordered by each customer.

Part two of this work comprises photographs and brief data, together with development and service histories, for 156 "second-line combat aircraft, obsolescent types, light transports and trainers". There are also eight pages of colour photographs and a comprehensive index.

A number of the illustrations have suffered by being located so close to the top of the page that portions of the subject have been trimmed off; the quality of reproduction also is poor in places. The book is nevertheless exceedingly good value and provides a comprehensive guide to virtually every type of aircraft in military service in 1971.

"Skyvan"
Compiled and edited by B Tomkins
Airline Publications & Sales, Staines, Middlesex, 50p
24 pp, 6½ in by 9 in, illustrated

SECOND in a series entitled "Airlines and Airliners", this booklet provides a record of Short Skyvan production up to the end of 1970. A production list gives details of 54 airframes, including delivery dates, registration and ownership changes.

The body of the book is taken up with illustrations of the Skyvan in the markings of the majority of operators to have used the type to date, plus notes on markings, and there are eight side views in colour depicting various liveries.

"Messerschmitt — An Aircraft Album"
by J Richard Smith
Ian Allan, Shepperton, £2.25
144 pp, 7 in by 9 in, illustrated

UNIFORM in series with previously-published volumes on Handley Page and Heinkel, this landscape format "Aircraft Album" possesses somewhat more text than its predecessors, but adds nothing to known and published information, although, in fairness, it is presumably not intended to do so, being primarily the vehicle for the publication of a collection of photographs of aircraft designed by Willy Messerschmitt or under the aegis of this distinguished German designer. If this *is* the intention of the publisher, however, then it fails rather miserably, as the half-tone illustrations are, with one or two notable exceptions (eg, the Me 328B on pages 122-3), thoroughly hackneyed, and, what is more, seem to have been badly reproduced from poor copy prints of what were presumably quite respectable originals.

FIGHTERS IN THE RAF

by Roy Braybrook, BSc, CEng, AFRAeS

Part II – Trial by Fire

THE previous instalment of this survey took the story of RAF fighters and their tactics up to the advent in 1937 of the Gloster Gladiator, the first of the closed-cockpit generation and more significantly the last of the biplanes. Despite the nostalgia that surrounds that era, it must have been obvious for at least ten years that OR staffs and designers were flogging a dead horse, painfully coaxing speeds up from the 156 mph (250 km/h) of the 1927 Siskin IIIA to the 253 mph (405 km/h) of the Gladiator, while monoplane record speeds were increasing more than twice as fast, reaching 440.6 mph (705 km/h) by 1934.

The engines used in record-breaking flights were admittedly a far cry from those available for operational service, but this did not excuse the failure to take advantage of the new developments in aviation technology. By the early 1930s monoplane transports and bombers were beginning to appear with virtually the same speed capability as the RAF's biplane fighters. For example, 1932 saw the appearance of prototypes of the 222 mph (335 km/h) Heinkel He 70a and the 208 mph (333 km/h) Martin XB-10. These aircraft had closed cockpits and fully retracting undercarriages, and the latter featured an internal weapon bay. Between 1934 and 1936, Italy flew the Breda Ba 65, Caproni Ca 135, Fiat B.R.20, and Savoia-Marchatti S.M.79, all capable of about 270 mph (432 km/h).

The reluctance to switch to monoplanes may well have been a partial hangover from the earliest days of British military aviation, when a series of structural failures had led the War Office to introduce a ban on their use, lasting from 1912 almost until the start of the Second World War. Aside from avoiding the monoplane's stigma, biplanes had the practical advantages of a vastly more extensive development background, plus the ability to combine structural stiffness, light weight, large wing area for manœuvrability and short field performance, and short span for high roll rate. However, overriding all these considerations was the unacceptable drag of their fixed landing gear, struts and bracing wires. Looking at the performance trend, the biplane fighter had an ultimate potential for only 300 mph (480 km/h) or thereabouts.

Notwithstanding development and procurement for the RAF of four new types of biplane fighter in the 1930s, some thought was fortunately also being given to the monoplane.

In tendering to Specification F.7/30, which was eventually fulfilled by the Gloster Gladiator, Supermarine had produced in 1933 the Type 224 monoplane. This had an enormous thick wing of inverted-gull form and 45.9-ft (14-m) span, a fixed trousered undercarriage, and the steam-cooled 660 hp Rolls-Royce Goshawk which had been tipped as the favoured engine. Its top speed of 230 mph (368 km/h) was only marginally better than that of the similarly-powered Hawker PV3 development of the Fury biplane, yet this Supermarine project was to prove the springboard for what was in some respects the greatest fighter of the Second World War.

Even as the Type 224 was being built, chief designer Reginald Mitchell was planning an improved version which ignored the Air Ministry's demand for low wing loading, and featured a much smaller wing with split flaps, a retractable landing gear, and a closed cockpit. This proposal was then revised around the Rolls-Royce PV12 engine, and became the Type 300 project.

Old pro and brilliant beginner

While the changes were in progress, the threat of the RAF being outclassed by *Luftwaffe* equipment had become apparent, and in 1933 Hawker proposed a monoplane fighter, derived from the Fury airframe and equipped with a Goshawk engine and a spatted undercarriage. With the availability of the Rolls-Royce PV12 this project was modified and formed the basis for Specification F.36/34, which was later amended to include retractable landing gear and an armament of eight Brownings. The resulting Hurricane flew on 6 November 1935, an initial batch of 600 was ordered in mid-1936, and deliveries began before the end of 1937.

On a marginally later timescale, Specification F.37/34

was written around the Supermarine Type 300 project, and the resulting Spitfire left the ground on 5 March 1936. A production order for 310 aircraft was placed in the middle of that year, and deliveries to the service began 24 months later.

The fact that the Hurricane was manufactured so much more quickly and easily proved a decisive factor in the Battle of Britain, and may be attributed to Hawker Aircraft's infinitely greater experience of mass-producing fighters. Supermarine, with a background of seaplanes and racing, had turned out an aircraft that was, in comparison, an aerodynamicist's dream but a production engineer's nightmare, relying on more sophisticated tooling and skilled labour. On the other hand this freedom from the conservatism that comes from decades of continuous work in one field had enabled Supermarine to take a much greater step forward, possibly analogous to Lockheed's post-war entry into jet fighter and high speed helicopter activities.

Comparing the initial production versions, the Hurricane had a gross weight of 6,220 lb (2 820 kg) and a top speed of 320 mph (512 km/h), whereas the Spitfire grossed 5,332 lb (2 450 kg) and reached 355 mph (568 km/h), using the same 1,030 hp Merlin II. The Spitfire's lighter weight and better aerodynamics made it superior to the Hurricane, and the obvious choice for continuing development in the air combat rôle and for high altitude interception. On the other hand, the Hurricane proved more suitable than the Spitfire for later development for ground attack.

However, before going on to discuss how these aircraft were used, consideration must be given to contemporary equipment developments, without which even these high speed monoplanes would have been useless.

First and foremost, Britain's limited fighter force had to be employed efficiently, not squandered in standing patrols which relied on the possibility of locating incoming raids visually. Fighter Command had to know when, where, and in what strength each attack would arrive, and the answer to all these questions lay in what was then known as radio location, and later as radar.

Radio waves had been used as early as 1921 in meteorological experiments to locate and track thunderstorms. This work formed the basis of Watson-Watts' proposals early in 1935 to detect aircraft and distinguish between friend and foe. In that same year preliminary trials were held, and aircraft were detected at up to 40 st miles (64 km) out to sea.

However, this was still only a starting point. In the four years of peace that remained, it was found possible to improve detection range to 100-120 st miles (160-190 km), thus giving the fighters roughly 20 minutes' warning of a raid. There also remained the race against time to build the chains of coastal radar stations, back them with the Observer Corps for inland tracking, and develop the control technique necessary to bring the fighters into contact with the enemy.

The spoken word

Radar would, of course, have been useless without good radio communications between Sector Control and the squadrons in the air. In the First World War air-surface communications had been limited to MF telegraphy. This was used in two-seat reconnaissance aircraft, with 300 ft (90 m) of trailing wire aerial, and an observer tapping away in morse code, which was transmitted on 500-1500 kHz. By the 1930s radio telephony (ie voice communications) was available, using HF of about 6000 kHz, and by 1940 four-channel VHF of 100-150 mHz was entering RAF service. These relatively short wavelengths not only permitted smaller antennae, but also limited range and thus gave much clearer reception by cutting out a great deal of interference from other transmitters.

Having got the target in sight, there was still the task of

(Head of opposite page) The RAF's first fighter monoplane, the Hurricane I, L1599 of No 56 Sqdn being illustrated, and (above), a Hurricane squadron flying in tight vics.

aiming the guns accurately. Since World War I most fighters had been equipped with simple ring-and-bead sights, which gave the pilot no assistance in judging target range, and necessitated accurate head positioning and changes in eye focus. With the Gladiator a new reflector sight was introduced which projected an aiming display focused at infinity, thus alleviating the last two problems.

In addition, the pilot could, by adjusting a switch according to the wingspan of the particular type of target, control the length of a horizontal "range bar" forming part of the display. From this he could judge (in a tail chase) when the target was at the distance for which his guns were harmonised. This technique is known as stadiametric ranging, and in modified form is still in use today.

Finally, the reflector sight displayed a "100 mph (160 km/h) circle", by which the pilot might judge the necessary lead angle according to his estimate of the target's crossing speed component. For all this, every pilot that the writer has talked to has said that, at the time of the Battle of Britain, the average driver could be relied upon to hit the enemy only by firing a long burst from dead astern, and at point-blank range.

Other important improvements included bullet-proof windscreens, armour plate behind the pilot, and variable-pitch airscrews, all of which were introduced prior to the Battle of Britain.

The testing time

It had become clear during the fighting over France that the RAF's textbook formation of a tight vic of three fighters was tactically inferior to the *Luftwaffe's* loose line-abreast *Schwarm* of four which had been developed in Spain. The vic, designed to concentrate firepower on an unescorted and tightly-packed formation of bombers, placed the fighters too close together to manœuvre safely as a unit, or to permit more than one aircraft to fire effectively. Most

Following the Hurricane into RAF service, the Spitfire I with early two-blade airscrew is seen here with No 19 Sqdn in 1938.

A failure in the day fighter rôle, the Boulton Paul Defiant was switched to night interception with indifferent success, a Mk II night fighter of No 151 Sqdn being illustrated above. The Blenheim IF, illustrated (left) in service with No 54 OTU, was the only RAF night fighter for the first 12 months of WW II.

important of all, the concept of operating in flights of three aircraft was at fault, as in combat one wingman would inevitably be separated from the vic and destroyed.

It may also have been significant that the tight vic dated back to open cockpit biplanes with unrestricted rear vision. With the closed canopies of the new monoplanes, rear vision was limited, and this could be offset only by having widely-spaced elements, each protecting the other's tail. Alternatively, a formation could guard against rear attack by having either one aircraft (sometimes called a "swinger") or an entire section weaving at the rear, although once this had sighted the enemy it was quite often shot down.

Starting from the basic vic formation, RAF fighters frequently used a line-astern attack, which was referred to by the Luftwaffe as *die Idiotenreihe* (the row of idiots), presumably because it reduced the time available for firing and simplified the bombers' defence task.

By the end of the Battle of Britain (ie, autumn of 1940) most RAF fighter squadrons had changed on their own initiative to operating as three sections of four aircraft. In essence, one side of the 60 deg vic of the section formation was extended to take the fourth man, at the same distance as the separation between the other three. This "finger four" was still relatively tight, but at least it could break down into two individual fighting elements. Whatever may have been laid down in the pre-war Fighter Manual, what actually happened was one co-ordinated attack, and then it was every element for itself!

Of the fighters used in the Battle of Britain, the Spitfire II with 1,150 hp Merlin XII was probably best, followed by the Bf 109E-3, Spitfire I, Hurricane, and the Bf 110C in that order. The Hurricane is reported to have enjoyed a far higher serviceability than the Spitfire, and to have been a better gun platform and able to take more battle damage. The Bf 109 was less manoeuvrable than either, but was virtually as fast as the Spitfire II, and when necessary could escape by making use of its petrol injection system's unrivalled ability to keep the engine going in negative-*g* manoeuvres.

The Mosquito NF II (DD609 illustrated) with "arrowhead" AI Mk 4 radar which joined the RAF night fighting component in 1942.

The main controversy in RAF fighter tactics at that time was whether enemy raids should be intercepted well before they reached their target, or whether a delay should be accepted until a wing of three, four, or even five squadrons could be built up at a suitable height. Keith Park, who commanded No 11 Group in the front-line of SE England, preferred to attack with a wing as small as two squadrons if this was the only way to intercept before the bombs started falling.

In contrast, Leigh-Mallory, commanding No 12 Group in Central England, claimed that from Duxford he could assemble five squadrons at 20,000 ft (6 000 m) over Hornchurch within 25 minutes and that this concentration paid off with more enemy aircraft destroyed. Bader backed him up by saying "I would rather shoot down fifty of the enemy bombers after they have reached their objective, than ten before they have done so". In the end, the "big wing" concept appears to have prevailed, and large formations were certainly to be required after the tide had turned in 1941, and fighter sweeps were being made over the Continent to bring the enemy to battle.

As RAF fighters spent increasing amounts of time in flak-defended areas, sections opened out into a "fluid four" again following German practice. The two pair elements might then be spaced 200-300 yd (180-270 m) apart in line abreast, with 50-75 yd (45-70 m) between the aircraft of one element. The No 2s flew in the wing positions, so that they could watch the sky behind the other element, while keeping half an eye on their own leaders. Each No 2 was set back, so that in an emergency the leader could break across his nose. This broader spread decreased the risk from flak, improved searching in the rear hemisphere, and put each element in a better position to come to the assistance of the other.

Failure by night

Notwithstanding its brilliant success in the Battle of Britain, the RAF had failed to provide effective cover above Dunkirk, due to the limited range of UK-sited radars and the fact that (even when flying from Manston) its fighters could stay 40 minutes at most over the embarkation area. Drop tanks had been used by *Luftwaffe* fighters since the Spanish Civil War, but had not yet been specified for the RAF. This Dunkirk episode, and the lack of air support during the later withdrawals from Greece and Crete, gave rise to a bitterness on the part of the Army that was to linger through much of the war.

However, these failures were of little account compared with the almost total helplessness of the RAF in defending

the UK centres of population against night attacks during 1940. The lack of any effective countermeasure (despite the night raids of World War I and the demonstration of German bombing in the Guernica affair) must surely mark the early months of the Blitz as the lowest point ever reached by Fighter Command, and must act as a lasting condemnation of those charged with defence planning in the 1930s. From books on that period, one might conclude that the pre-war RAF had little taste for night flying, and expected its enemies to be similarly inclined!

For almost the first twelve months of the war the only available night fighter was the Mk IF conversion of the Bristol Blenheim bomber, with a total of five machine guns and a maximum speed of 260 mph (416 km/h). Airborne interception (AI) radar had been installed in three Blenheims immediately before the war, and came into general use in June 1940, but the combination of short detection range, slow speed, and poor ground control resulted in very few contacts.

In daylight operations the controller need only bring his fighters within 10-12 st miles (16-19 km) of the enemy for a sighting, but at night even halving this separation was inadequate, since the omni-directional AI transmission limited its range to the fighter's height above the ground, typically less than 4 st miles (6,4 km). Even if AI contact were obtained, the operator still had a very difficult task, with his one metre wavelength radar and primitive displays, to bring the fighter within 1,000-3,000 ft (300-900 m) for visual acquisition.

In late 1940, the Boulton Paul Defiant turret fighter, having been found easy meat for Bf 109s, was fitted with AI radar and switched to night fighting. Introduced on roughly the same timescale and likewise capable of just over 300 mph (480 km/h) at a typical interception height of 15,000 ft (4 500 m), the Bristol Beaufighter had an armament of four 20-mm Hispano cannon and six Brownings. Combined with GCI (Ground Controlled Interception) and assistance from searchlights, the Beaufighter enabled the RAF at last to begin taking a significant toll of bombers in 1941. The new technique of GCI was based on small precision radars and controllers using individual radar scopes to direct either one or two fighters, rather than judging relative positions from markers on the operations room table. The greater accuracy of GCI enabled a fighter to be turned directly behind the bomber, and placed within about 3 st miles (4,8 km).

In 1942 a further advance was made with the introduction of the de Havilland Mosquito NF II, which had four Hispano and four Brownings, better manoeuvrability and ceiling than the Beaufighter, and a top speed of 370 mph (592 km/h). This was followed in 1943 by the Mk XII with centrimetric AI Mk VIII, which concentrated its emission in a narrow beam and thus overcame the height limitation on detection range. In later variants this range was increased up to 10 st miles (16 km), and Mosquitoes thus equipped had considerable success in escorting night raids by Bomber Command, using IR for identification of "friendlies".

In the course of these support operations the Mosquito was also used against enemy night fighters taking-off and landing at their home bases, and this gave rise to what became known as *Moskitopanik* among Luftwaffe pilots. It may be felt that, rather than having wasted effort in a succession of futile attempts to develop a useful turret fighter, Air Ministry should have pressed the development of the intruder fighter in the 1930s. Even without AI, this might well have stopped more night-flying Heinkels, Junkers and Dorniers than the pathetic efforts of the line-patrolling Blenheims of 1940.

Day fighter developments

Following the end of massed bomber attacks on the UK,

the RAF turned to the offensive with the low level fighter sweeps ("rhubarbs") and heavily-escorted daylight bomber raids ("circuses") across the Channel throughout 1941 and 1942. The principal fighter used in these efforts to wear down the *Luftwaffe* (and later to reduce pressure on the Eastern Front) was the Spitfire V, powered by the 1,440 hp Merlin 45, and fairly well matched to the Bf 109F then in service.

However, in September 1941, the Messerschmitt began to be supplemented in combat by the Fw 190. This aircraft could reach 408 mph (653 km/h) compared with the 374 mph (598 km/h) of the Spitfire V, and had up to four 20-mm cannon and two 7,9-mm machine guns, whereas even the Mk VB had only two 20-mm and four 7,7-mm. The Fw 190 was also highly manoeuvrable and had a roll rate that exceeded even that of the clipped-wing Spitfires. It was this German aircraft more than any other that was to force the pace of British fighter development throughout the remainder of World War II.

With the introduction of the Fw 190 the loss rate swung heavily in favour of the *Luftwaffe*, but by mid-1942 the RAF had two new types to redress the balance. First of these was the Spitfire IX, which was powered by the 1,660 hp Merlin 61 and (like the Mk V) was produced with various wingtip shapes to suit different operating altitudes. At this stage the gross weight had risen to 7,500 lb (3 400 kg), and maximum speed with the elliptic wing was equal to that of the Focke-Wulf Fw 190.

The Typhoon (a Mk IA R7700 used for armament trials at Boscombe Down being illustrated above) had an inauspicious start to its career and the Tempest V (seen at foot of page in service with No 501 Sqdn) was essentially an aerodynamically-refined development.

The Spitfire IX was also the first fighter to have as standard fit a gyro gunsight (GGS Mk II), enabling lead angles to be predicted automatically from the spin rate of the sight line to the target. Despite initial opposition from some "natural shots", the British GGS gave a substantial improvement in deflection shooting for the average pilot, and came to be accepted by both the USAAF and USN. With various detail refinements the GGS remains in widespread service today, although it is essentially a crude steady-state device, incapable of correcting lead angle for sight line acceleration, or of modifying ballistic drop according to closing rate or bank angle.

The second RAF fighter to counter the Fw 190 (although only at low level) was the Hawker Typhoon. A far heavier aircraft than either the Spitfire or Hurricane — it grossed 11,400 lb (5 200 kg) clean — it nonetheless reached 412 mph (659 km/h) by virtue of a Napier Sabre of 2,200 hp, ie, twice the power of the Merlin II used in the Battle of Britain.

The Typhoon suffered teething troubles with the Sabre engine, and was probably the only Hawker aircraft ever to experience structural failures in flight. It was eventually discovered that failures of the monocoque rear fuselage (the centre fuselage retained a tubular steel primary structure) were being caused by elevator flutter, induced by faulty positioning of the remote mass balance weights.

Despite this inauspicious start, the Typhoon went on to become one of the best ground attack aircraft of the war. In the Dieppe raid of August 1942 it was found that even four 20-mm Hispano were incapable of knocking out well-protected gun emplacements, and this led to development of the 3-in (7,6-cm) rocket, which has only recently been phased out in favour of the more accurate 6,8-cm SNEB. Using eight RP, the Typhoon achieved great success against armour and radar installations, and it could alternatively carry two 1,000-lb (450-kg) bombs, which was double the load of the Spitfire.

Performance development of the latter aircraft nevertheless continued, with the 1,735 hp Griffon-engined Spitfire XII entering service in 1943. In the following year the Spitfire XIV appeared with a five-blade Rotol airscrew absorbing the 2,050 hp of its Griffon 65, to give a top speed of 448 mph (717 km/h). This aircraft was one of the few capable of catching the V-1 flying bomb, and accounted for over 300 of the 1,771 destroyed by the RAF.

A further 638 of these guided missiles were destroyed by the Hawker Tempest V, essentially an aerodynamically-refined Typhoon. Because of compressibility effects encountered in diving the latter (due to its 18 per cent thick wing), the Tempest aerofoil was reduced to 14.5 per cent at the root and 10 per cent at the tip. Maximum speed was 427 mph (683 km/h).

The 2,200 hp Sabre-engined Tempest V was followed by the 2,520 hp Centaurus-engined Mk II, which was just entering service as the war ended. The de Havilland Hornet was likewise designed for long-range operations in the Far East, but was powered by two closely-cowled 2,030 hp Merlins, and reached 472 mph (755 km/h), compared with the 440 mph (704 km/h) of the Tempest II. The Hornet had an empty weight of 12,880 lb (5 860 kg), and grossed up to 20,900 lb (9 550 kg). It was the fastest piston-engined fighter to be used by the RAF, and was employed operationally in the ground attack role during the Malayan Emergency (together with Spitfires and Tempests). It was also the last of these types to remain in service, being phased out in 1955.

This brief outline of the principal RAF piston-engined fighter types has necessarily omitted a number of interesting aspects, such as the development of pressurised high-altitude variants of the Spitfire and Mosquito, and the use of US fighters (notably the P-51, P-47 and P-40) to supplement British products. Neither is space available to discuss the Westland Whirlwind, which was used in limited numbers for daylight interdiction missions, nor the same company's Welkin, which was designed as a high altitude interceptor but never used operationally.

The big picture

Notwithstanding these omissions, the broad picture of RAF fighter development is fairly clear. Pre-war planners had failed lamentably to foresee the need for night interception, the night interdiction of enemy airfields, and for long-range escort fighters. Clear weather daylight interception and short-range air superiority were the only aspects to receive any real attention, and in these aims Britain succeeded only by the narrowest of margins.

The Hurricane was arguably not as good as it should have

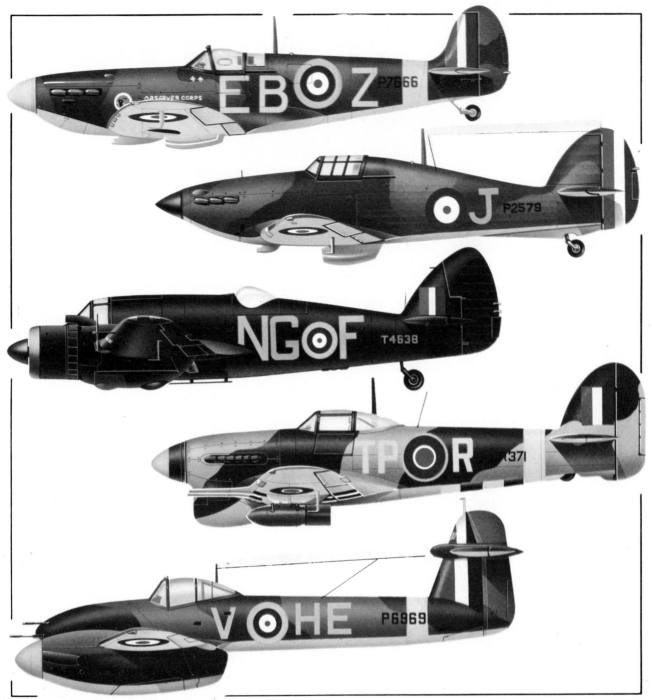

A representative selection of RAF fighters: (from top to bottom) a Spitfire II (P7666) of No 41 Sqdn, Autumn 1940; Hurricane I (P2579) of No 73 Sqdn, France, Winter 1939–40; Beaufighter I (T4638) of No 604 Sqdn, Spring 1941; Typhoon IB (JR371) of No 198 Sqdn, Summer 1944; Whirlwind (P6969) of No 263 Sqdn, Winter 1941–2.

been, but was present in the numbers required to blunt the daylight bomber attacks of 1940. The Spitfire was equally biased in the opposite direction (ie, in favour of performance at the expense of production build-up), and this enabled it to meet the Bf 109 on equal terms and (given 60 per cent more power) even to hold its own against the Fw 190.

The Hawker Typhoon proved disappointing at high altitude, but excelled below 20,000 ft (6 000 m) and provided the close support capability that the British Army had long admired in the Ju 87. In 1943 the Hawker company at last found its form, and produced in the Tempest an aircraft that could not only perform the ground attack mission, but (having released its rockets or bombs) could take on with confidence the best propeller-driven fighters that Germany could produce.

The Fury derivative of the Tempest never reached RAF service, but its speed of approximately 485 mph (776 km/h) was indicative of the ultimate limit for the piston-engined fighter. Just as biplane performance had flattened off below 300 mph (480 km/h) in the 1930s, so monoplane fighters with reciprocating engines appeared to be restricted to below 500 mph (800 km/h). To overcome this limit meant a radically new form of powerplant: the gas turbine on which both Germany and Britain had been working since the 1930s.

The final part of this series will examine how RAF jet fighters have developed, and how tactics changed (and are still changing) to suit the revolution that the turbine engine and guided weapons have brought to air combat operations. □

MIG-3

IT WAS WITH some surprise that, in April 1941, the German air attaché in Moscow, *Oberst* (later *Generalleutnant*) Heinrich Aschenbrenner, received information that a German delegation would be permitted to visit certain Soviet aircraft factories and air force bases. The invitation was surprising in view of the fact that, despite the 20-month-old Soviet-German "Pact of Friendship", tension between the two countries had never been higher; Soviet forces were heavily concentrated along the western frontier of the Soviet Union and German forces were massing along the eastern periphery of Germany. Soviet motives behind the gesture were not immediately obvious, but the invitation was accepted with alacrity, and arrangements were made to ensure that representatives of *5 Abteilung*, the intelligence branch of the *Luftwaffe* in Berlin, would be included in the visiting delegation.

One of the most difficult tasks of *5 Abteilung* had been the securing of reliable intelligence concerning the composition of the Soviet Air Forces, the V-VS RKKA, and the capabilities of the new warplanes with which they were known to be re-equipping. Successive air attachés had failed to uncover anything of real importance; the wealth of reports that had reached the Department from Finland during the Russo-Finnish conflict in the winter of 1939-40 had been of little real value owing to the limitations of the Soviet objectives and the fact that only aircraft assigned to the Leningrad Military District had been deployed, while the clandestine reconnaissance missions flown over the Soviet periphery by the ostensibly *civil* aircraft of the special unit commanded by *Oberstleutnant* Theodor Rowehl had, of necessity, terminated with the signing of the Soviet-German Pact on 23 August 1939.

The photographic reconnoitring had, in fact, been resumed after October 1940 by the *Aufklärungsgruppe Rowehl*, which operated out of Werder and, later, Fritzlar, flying pressurised Junkers Ju 86Ps at extreme altitudes over Soviet territory. The information derived from these photographic sorties was strictly limited, however, and the Soviet government was all too well aware of the fact that German aircraft were performing such missions. Indeed, early in 1941, a Ju 86P

Key to MiG-3 cutaway drawing

1 Airscrew spinner
2 VISch-22E variable-pitch metal airscrew (later supplanted by VISch-61Sch)
3 Muzzle port for asymmetrically-mounted UBS machine gun
4 Muzzle ports for twin ShKAS machine guns
5 Oil cooler air intake
6 Intake trunking
7 Mikulin AM-35A 12-cylinder liquid-cooled engine
8 Ejector exhaust stubs
9 Gun blast tubes
10 Port 7,62-mm ShKAS machine gun
11 Ammunition tank
12 Breech of 12,7-mm UBS machine gun
13 Fuel tank (23 Imp gal; 109 l capacity)
14 Mast for RSI-3 (later RSI-4) receiver
15 One-piece windscreen
16 PBP-1 reflector sight
17 Instrument panel
18 Dural forward-fuselage skinning
19 Rudder pedal
20 Control column and gun trigger
21 Seat
22 Back armour (9-mm)
23 Aft-sliding canopy
24 Aft vision glazing
25 RSI-3 radio compartment
26 Fuselage stringers
27 Fuselage frames
28 Rudder cable
29 Tailplane attachment points
30 Integral fin structure

31 Fabric skinning
32 Trim tab
33 Rear navigation light
34 Fixed tailwheel
35 Control cables
36 Radiator trunk housing
37 Dural wing-root fillet
38 Fuel tank (54 Imp gal; 245 l capacity)
39 Ventral radiator bath
40 Radiator outlet flap
41 Integral stub wing root
42 Inner mainwheel cover plates
43 Fuel tank (33 Imp gal; 147,5 l capacity)

44 Wing root induction air intake
45 Outer wing panel attachment point
46 Mainleg actuating jack
47 Wooden outer wing panel structure
48 Plywood outer panel skinning
49 Port navigation light
50 Aileron
51 Mainwheel oleo leg
52 Brake cooling gills
53 Port mainwheel

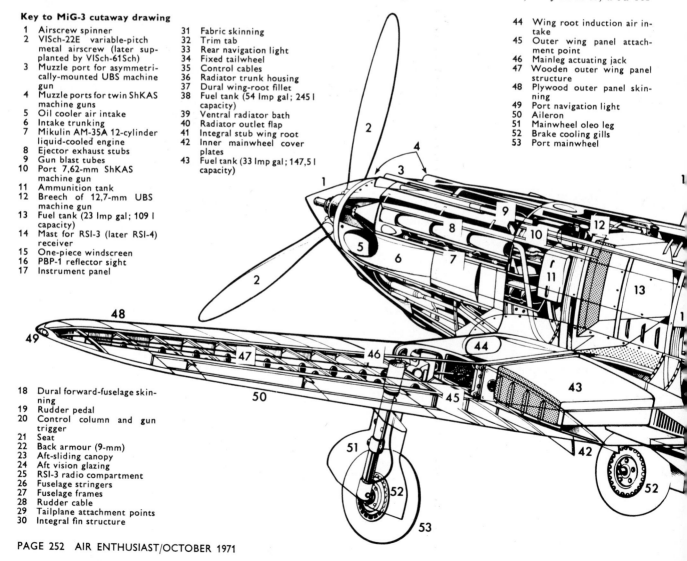

UNDISTINGUISHED FOREBEAR OF A DISTINGUISHED LINE

had been forced down in bad weather near Vinnitsa and its *Luftwaffe* crew captured.

The reasoning behind the invitation was soon to become all too obvious, the visit of the German delegation, led by *Oberst* Aschenbrenner, was to be reminiscent of the game of bluff and counterbluff played by France and Germany during the last year of uneasy peace in Europe, delegations being exchanged and the hosts endeavouring to convince their guests that their respective re-equipment programmes had attained more advanced stages than they in fact had.

The German delegation was taken to several IAP-VO (*Istrebitelnaya Aviatsiya Protivo-vozdushnoi Oborony* — Fighter Aviation of the Anti-aircraft Defence) bases in the vicinity of Moscow at which it saw substantial numbers of a new fighter monoplane, which, its Soviet hosts announced, was the MiG-3 and "already in widespread service with our fighter aviation!" Among the factories shown to the Germans was *Zavod 1* at Vnukovo in which the MiG-3 was being manufactured in quantity. The delegation, although not so naïve as to believe what it saw to be fully representative of

all V-VS units and Soviet aircraft factories, was impressed, and its leader, *Oberst* Aschenbrenner, was soon in no doubt as to the Soviet government's reason for the invitation; a motive summed up succinctly in a comment by a certain Artem Ivanovich Mikoyan, chief engineer of *Zavod 1*, and quoted verbatim in Aschenbrenner's subsequent report to the *Generalstab der Luftwaffe*: "We have now shown you everything that we have and what we can do, and whoever attacks us we will destroy!"

The visit of the German delegation was destined to have exactly the reverse effect to that for which the Soviet government had hoped when issuing the invitation. Heinrich Aschenbrenner was impressed not so much by the young Soviet engineer's unequivocal warning but by the standard of workmanship displayed by the MiG-3 and the volume of production of this fighter already obviously being achieved at *Zavod 1*. While the delegation had been afforded no means of assessing the capabilities of the MiG-3 as a combat aircraft, it could be in no doubt that the performance of this new warplane was markedly in advance of that of the ageing

WARBIRDS

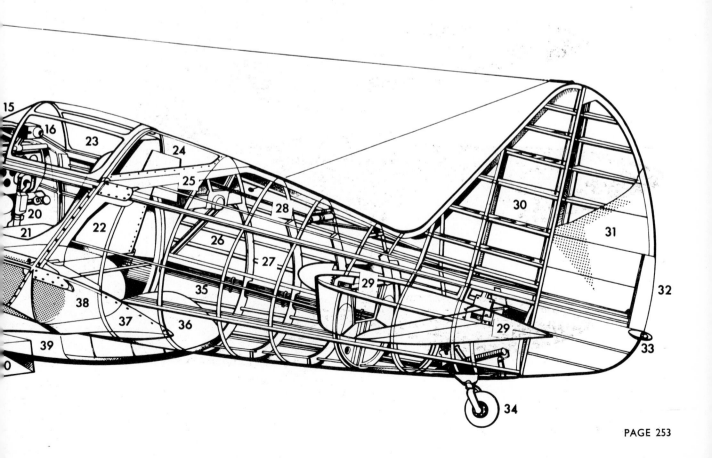

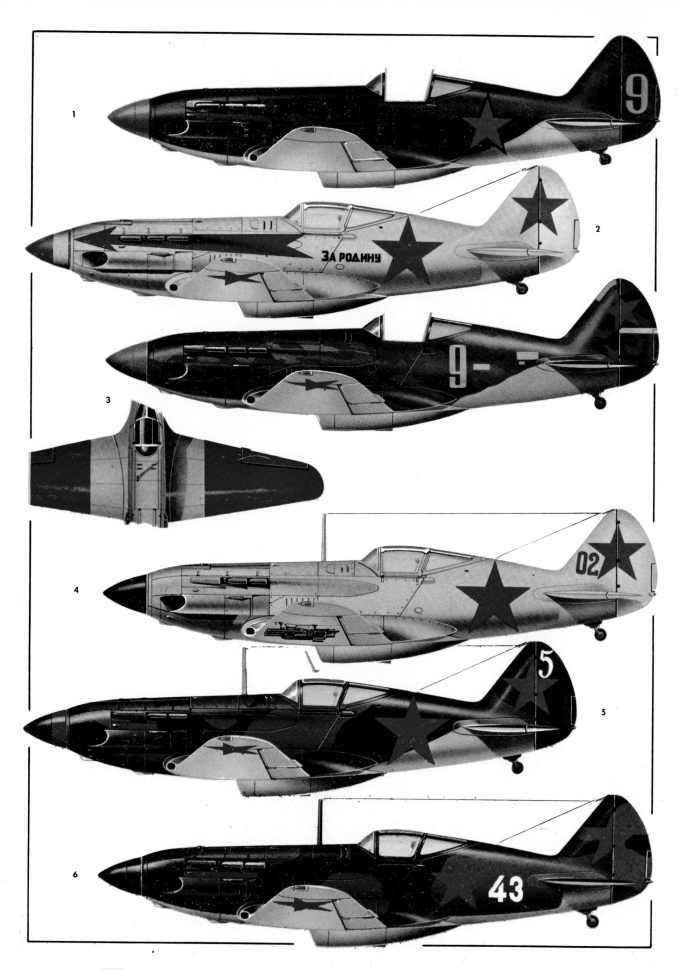

fighters with which it was convinced that the bulk of the V-VS was still equipped.

During 1931, Aschenbrenner had been attached to the 20th Brigade of the V-VS, and, like so many former German Army officers who had undergone clandestine flying training in the Soviet Union — and several of whom now held positions of seniority in the *Generalstab* — he had held an extremely low opinion of Soviet capabilities in the area of workmanship as applied to combat aircraft. His report to the *Generalstab* made immediately upon his return to Berlin from Moscow stressed the remarkable transformation in quality standards that appeared to have taken place in the Soviet aircraft industry. Adolf Hitler reacted to Aschenbrenner's report with the following comment: "Well, you can now see how far they have already progressed. We must begin our attack immediately!"

A high-altitude interceptor

Aschenbrenner was correct in his belief that the MiG-3 signified a major advance in Soviet standards of workmanship*, and *5 Abteilung* was not far out in its assessment that, on the eve of Operation *Barbarossa,* as the assault on the Soviet Union was designated, relatively few fighter *eskadrilii* had achieved full combat status on the MiG-3. They had no means of knowing, however, that this new Soviet fighter that had so impressed the German delegation in April was presenting the V-VS with serious problems, and was, from several aspects, failing to fulfil the sanguine expectations of its creators. Nor could they have foreseen that the acronym "MiG", borne for the first time by this fighter, was to become known from Afghanistan to the Yemen and from Albania to Yugoslavia long after their Third Reich had become no more than so many chapters in history books.

The MiG-3 had stemmed from a V-VS requirement formulated in 1939 for a single-seat interceptor fighter offering maximum performance at altitudes above 18,290 ft (6 000 m), and powered by the supercharged AM-35A engine newly developed at the TsIAM (Central Institute of Aviation Motor Construction) by a team led by Aleksandr Mikulin. At the same time requirements had been issued for new medium-altitude fighters, their preparation having coincided with an important change in the Soviet policy of fulfilling official aircraft specifications.

Whereas previously all important combat aircraft requirements had been passed to either the TsAGI (*Tsen-*

---

*The veteran Soviet test pilot Pyotr M Stefanovsky comments in his memoirs that the MiG-3 possessed a somewhat better finish than the contemporary Yak-1 and LaGG-3.

---

(1) MiG-3 of zveno leader of detached eskadril *from IAP assigned to the IA-PVO of the Leningrad Army Region and shot down by Finns over the Karelian Isthmus, September 1942. Note that the aircraft was flown without cockpit canopy or radio mast. (2) MiG-3 of the 34 IAP operating from Vnukovo (Western Sector, Moscow Corps Command IA-PVO) in the winter of 1941-2. Legend beneath the cockpit reads "For the Fatherland". (3) MiG-3 of an unidentified reconnaissance* eskadril *operating over the Central Sector of the Eastern Front and shot down by the Luftwaffe during the spring of 1942. Note unusual camouflage finish applied at unit level, and lack of cockpit canopy and radio mast. (4) MiG-3 of 12 IAP attached to Moscow Army Region IA-PVO during winter of 1942-3. This unit began operations in the summer of 1941 equipped with the Yak-1 but converted to the MiG-3 during the following year when it was awarded the "Guards" title. The plan detail (applicable to this aircraft) shows high-visibility outer panels of wings. These were intended to ease location of a fighter forced down over snow-covered terrain. (5 & 6) MiG-3s of an unidentified IAP in service during the summer of 1942. Note temperate camouflage scheme.*

tralny Aerogidrodinamichesky Institut — Central Aero and Hydrodynamic Institute), whose Department of Experimental Aircraft Design was headed by Andrei N Tupolev, or the TsKB (*Tsentralny Konstruktorskoye Byuro* — Central Design Bureau) whose experimental aircraft section was supervised by Nikolai N Polikarpov, each of these organisations possessing a number of individual design brigades, the more talented of the brigade leaders had now been given a greater degree of autonomy and had been encouraged to establish their own experimental design bureaux, these being intended to compete one with another in fulfilling official specifications.

This change in policy and desire to create a competitive spirit in the design of combat aircraft in general and fighters in particular had stemmed from a somewhat belated appreciation by the *Narkomavprom* and the *Narkomat Oborony* (the People's Commissariats for the Aviation Industry and for Defence) of the fact that the standard V-VS fighters, all of which had been born on the TsKB drawing boards, had fallen behind world standards and no successors were in prospect. The utmost urgency had been attached to the development of new fighters, and among the many new, semi-autonomous design bureaux hurriedly created to evolve such combat aircraft had been one formed by Artem Mikoyan, a highly-creative but relatively inexperienced young designer, and an older and appreciably more experienced engineer, Mikhail Gurevich. The two friends had decided ideas on the lines along which fighter development should proceed, and had immediately put their ideas into practice in an endeavour to fulfil the V-VS high-altitude fighter requirement. This demand particularly appealed to Mikoyan as it called for an aircraft optimised for speed, this taking precedence over all other capabilities, whereas the parallel medium-altitude fighter requirement dictated compromise, manœuvrability being regarded as important as other aspects of its performance.

The AM-35A 12-cylinder liquid-cooled Vee engine with a single-stage supercharger that had been stipulated for the high-altitude fighter afforded 1,350 hp at 2,050 rpm for take-off, and possessed maximum continuous and cruise ratings of 1,200 hp at 18,290 ft (6 000 m) and 1,150 hp at 21,335 ft (7 000 m) respectively. Unfortunately, it was not ideally suited for fighter installation, being a somewhat large and extremely heavy power plant, its dry weight of 1,830 lb (830 kg) comparing unfavourably with contemporary Merlin and DB 601A engines which had dry weights of 1,376 lb (624 kg) and 1,340 lb (608 kg) respectively. Understandably, power plant weight influenced the design of the fighter to a very marked degree. Structural weight had to be kept to a minimum in order to achieve the desired performance, and the problem posed by the heavy engine was compounded by the limited availability of light alloys which dictated extensive use of wood.

Mikoyan and Gurevich had therefore designed what they considered to be the smallest practicable airframe capable of accommodating the AM-35A engine, a pilot and armament, and the project design had been submitted to the GUAP (*Glavnoye Upravlenie Aviatsonnoi Promishlennosti* — Chief Directorate of the Aviation Industry) under the design bureau designation of I-200 in October 1939. Within two weeks instructions had been received to proceed with the construction of prototypes with all possible haste, work on these having actually commenced in November, detail design proceeding in parallel with prototype construction.

Such was the urgency attached to the provision of modern fighters that early in 1940, although no prototype of the I-200 had flown, instructions had been issued to *Zavod 1* to commence tooling for an initial batch of 100 examples of the new fighter, and by 5 April 1940, when Arkadii N Yekatov had performed the initial flight test of the first

Personnel of the MiG-3-equipped 12 IAP attached to the Moscow Army Region IA-PVO receiving the highly-coveted "Guards" title during the winter of 1941–2. RS-82 missiles may be seen beneath the wings of their MiG-3s. (Opposite page) A MiG-3 of the 34 IAP, winter 1941–2.

I-200 prototype — barely five months from the commencement of construction — work on the assembly line at *Zavod 1* had been making rapid progress.

No novice's aircraft

The I-200 used a welded steel-tube forward fuselage with an integral stub-wing centre section, a wooden monocoque rear fuselage and wooden outer wing panels. The movable control surfaces were metal-framed and fabric-covered, the forward fuselage was covered by light alloy panels, and the rear fuselage and wings had impregnated-plywood skinning, as did also the integral wooden vertical fin and the steel-tube tailplane. All three members of the undercarriage were retracted pneumatically, the main members, which had a track of 9 ft 1¼ in (2,80 m) folding inwards into wells in the wing centre section forward of the fuel tanks. The AM-35A engine drove a VISch-22E three-bladed variable-pitch airscrew, and provision was made for 89 Imp gal (404 l) of fuel distributed between three light alloy tanks (a 23 Imp gal/ 109 l tank ahead of the cockpit and 66 Imp gal /295 l in the wing centre section) for which a measure of bullet proofing was provided by a sheathing of layers of rubber and tough chord fabric. The pilot was seated over the wing trailing edge beneath a sideways-hinging canopy.

On 24 May 1940, Yekatov had attained a speed of 403 mph (648,5 km/h) in level flight at an altitude of 22,640 ft (6 900 m) while flying the first I-200 prototype, and this performance

(Above) The first prototype I-200 with hinged cockpit canopy, and (below) a MiG-3 preparing to take-off during the winter of 1941–2.

had been bettered within two weeks by the second prototype which clocked 404.5 mph (651 km/h) at 22,965 ft (7 000 m), but from the outset of flight testing it had been patently obvious that the I-200 possessed a number of serious shortcomings, and was, at *best*, no novice's aircraft! The haste with which the fighter had been designed and built, and the relatively limited experience of its creators were reflected in its handling characteristics. Smaller than the Spitfire but appreciably heavier, its wing loading without armament or other operational equipment being of the order of 35·8 lb/sq ft (175 kg/m²), the I-200 had revealed extremely poor longitudinal stability, its controls had proved heavy, and it had displayed a tendency to spin at the least provocation. Manœuvrability had been pronounced most unsatisfactory, and both take-off and landing had been found to demand a very high degree of competence on the part of the pilot.

It had been obvious that some of the more fundamental shortcomings — such as the longitudinal stability problem which had resulted from too short a fuselage — were inherent in the design. Engineering test pilots of the NII V-VS (*Nauchno-Issledovatelsky Institut V-VS* — Scientific and Research Institute of the Air Forces) had been responsible for performing the State Acceptance Trials of the I-200, and shortly before these had commenced, the design bureau's test pilot, Arkadii Yekatov, had lost his life when the AM-35A engine of the I-200 that he had been flying failed during a landing approach. The NII V-VS pilot assigned the task of first testing the I-200 had been Andrei G Kotchetkov who had been a fellow pupil of Artem Mikoyan at the Zhukovsky Air Force Engineering Academy.

Another NII V-VS pilot, Pyotr Stefanovsky, who witnessed Kotchetkov's first flight in the I-200 subsequently recorded the event as follows: "We all gathered along the airfield perimeter to watch Kotchetkov perform his first test with the new fighter. He was a first class engineering test pilot but he must have had mixed feelings about his task, following, as it did, almost immediately upon the fatal crash of an experienced comrade in the same type of aircraft. Kotchetkov climbed into the cockpit, started the engine, and shortly afterwards the experimental fighter was roaring across the field. Our eyes followed the aircraft as it gained altitude and Kotchetkov began his first banking turn. Suddenly the engine cut! Our hearts missed a beat. That the new fighter was to claim another victim seemed inevitable as to turn an aircraft such as this through 180 degrees without power and with airscrew windmilling was virtually impossible! Our instructions were clear enough. In the event of an engine failure after take-off the pilot was to level off and then endeavour to land straight ahead. Scores of first-rate pilots had lost their lives in attempts to turn back and land on an airfield with a dead engine, invariably run-

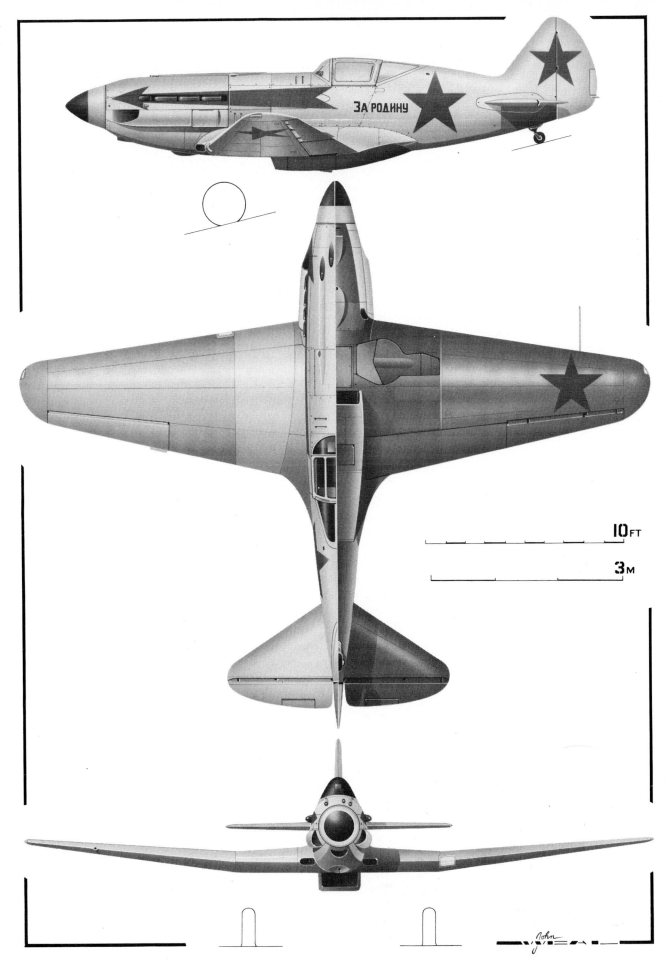

ЗА РОДИНУ

10FT

3M

ning out of height or stalling in. With *this* highly-loaded aeroplane, and in view of what we had been told of its behaviour once power had been lost, Kotchetkov's chances of survival appeared slim indeed. But this outstanding pilot, as on other occasions, pulled off the all-but-impossible feat of completing his turn and bringing the valuable aircraft down in a perfect power-off landing!"

NII V-VS trials with the I-200 had been completed in August 1940. The test pilots, while praising its speed and altitude performance, had been almost unanimous in their disapproval of its general handling characteristics, its manœuvrability and its stability. It had been obvious that only major redesign could eradicate completely the less desirable features of the I-200, but the fighter had already been committed to production, and the first series machines had been following closely behind the last of the I-153 biplanes on the *Zavod 1* assembly line.

The I-200 had not been alone among the new generation of Soviet fighters to suffer problems during the autumn of 1940, as its medium-altitude contemporaries, the I-22 created by S A Lavochkin, V P Gorbunov and M I Gudkov, and the I-26 designed by Aleksandr Yakovlev, had also been beset by difficulties. Both had commenced their prototype trials during the previous March, and both had been committed to production before their characteristics could be thoroughly investigated, but the Russo-Finnish conflict had stressed the inadequacy of existing V-VS fighter equipment, and had served to re-emphasise the urgency already attached to the re-equipment of the combat units, and any means of circumscribing the prototype-to-service process had been considered justified. Thus, despite the seriousness of the shortcomings displayed by the I-200, the Party Central Committee had not considered the risk that it had taken in ordering it into production unwarranted.

The I-200 had undeniably proved itself capable of quite phenomenal speeds at the altitudes at which it was designed to operate, and the chief of the *Narkomavprom,* Aleksei Ivanovich Shakurin, who had been appointed to his post by Stalin in the previous January, with Aleksandr Yakovlev as his deputy for experimental aircraft and research, had adopted the view that the most important difficulties could be ironed out without major redesign and consequent disruption of the assembly line. Two large wind tunnels, T 101 and T 104, had been placed in operation by the TsAGI late in 1939, and it had been decided to perform full-scale wind tunnel tests with a prototype I-200. Simultaneously, intensive flight testing of the remaining prototypes had been undertaken, and Mikoyan and Gurevich had been instructed to expend every effort to improve on certain aspects of the fighter's performance envelope, such as its inadequate range. Meanwhile, production of the initial batch of 100

The remains of a MiG-3 showing evidence of damage from small arms fire on the Central Sector of the Eastern Front in the spring of 1942.

aircraft essentially similar to the prototypes had continued in *Zavod 1.*

By this time, the design bureau designation I-200 had given place to the official designation of MiG-1, and by the end of 1940 a total of 20 MiG-1s (compared to 64 Yak-1s) had been completed and flown. The MiG-1 had retained all the undesirable features of the prototypes, and low-speed characteristics had, in fact, worsened as a result of the addition of armament and a 9-mm armour plate behind the pilot's seat which raised the normal gross weight to 6,770 lb (3 071 kg) and wing loading to 36 lb/sq ft (176 kg/m²). Armament was grouped in the forward fuselage and consisted of two 7,62-mm ShKAS machine guns with 750 rpg and a 12,7-mm UBS machine gun with 300 rounds. The pilot was provided with a simple PBP-1 reflector sight, but instrumentation was spartan to the extreme — for example, no fuel gauge was fitted, the pilot having to attune his ear to the note of the engine to determine the fuel status — and although provision was made for a single-channel RSI-3 receiver this was rarely installed.

The average V-VS pilot's predilection for an open cockpit had been reflected in the MiG-1 by the deletion of the sideways-hinging canopy featured by the prototypes. Apart from a general dislike of a confining hood, however, there had been more cogent reasons for dispensing with the canopy. No provision had been made on the prototypes for the emergency jettisoning of this item of equipment which tended to jam shut, and the Plexiglas used in its manufacture had all the characteristics of bottle glass, seriously impairing vision. In clean, factory-fresh condition, the MiG-1's maximum speed ranged from 298 mph (480 km/h) at sea level to 363 mph (585 km/h) at 16,400 ft (5 000 m) and 375 mph (604 km/h) at 19,685 ft (6 000 m), while 390 mph (628 km/h) was attainable at 23,295 ft (7 100 m). An altitude of 16,400 ft (5 000 m) could be reached in 5·3 minutes, and maximum ceiling was 39,370 ft (12 000 m), but the poorest aspect of the MiG-1's performance envelope was its range, which, with full tanks at long-range cruise, was only 454 mls (730 km). At maximum cruise of 342 mph (550 km/h) with 10 per cent fuel reserve for hold-off and landing, range was reduced to a mere 360 mls (580 km).

Extensive modification

Early in 1941, the first MiG-1s had been distributed to service units for evaluation and familiarisation, the IAP-VO units assigned the task of defending primary targets, and V-VS VMF (*V-VS Voenno-morskovo Flota* — Air Forces of the Navy) fighter units of the Baltic Fleet having priority. Only the most experienced of their pilots had been assigned to the new fighter, and even these had found the MiG-1 a difficult machine to handle. Forward view for taxying was almost non-existent, the fighter was extremely unwieldy at lower speeds and altitudes, controls were heavy at the upper end of the speed range, and extreme caution had to be exercised during aerobatics and landing as the MiG-1 was prone to spinning out of a steep banking turn.

With the MiG-1 the phenomena of the high-speed stall had been encountered by V-VS service pilots for the first time, necessitating careful practice of turns at varying speeds and rates of turn to accustom them to the symptoms of an approaching stall. It was hardly to be expected that these characteristics, coupled with a strictly limited endurance and what was already considered inadequate armament, had endeared the MiG fighter to the V-VS.

While the MiG-1 was being introduced to service units with inauspicious success, the results of the TsAGI wind tunnel testing and trials at various test establishments had been embodied in the basic design. The former had dictated a number of modifications to the air intakes, the supercharger intake being redesigned and cut back; the ventral radiator bath had been enlarged and extended forward, and several degrees of additional dihedral had been applied to the outer wing panels. A cockpit canopy had been re-introduced, this being of entirely new design, using improved Plexiglas and sliding aft; the fuselage decking aft of the cockpit had been cut down at a more acute angle and glazed to improve aft vision; the hinged lower halves of the mainwheel cover plates had been removed from the main covers and transferred to the fuselage, and a 54 Imp gal (245 l) overload tank had been introduced beneath the pilot's seat.

These changes had all been embodied in the 101st I-200 fighter which had left the assembly line late in February 1941, the modified variant being redesignated officially as the MiG-3. The aerodynamic refinements that had been introduced as a result of the TsAGI tunnel tests had marginally improved overall performance, stability had been decidedly improved, and the additional fuel tank had augmented range and endurance to an acceptable degree, the maximum range at long-range cruise at 16,400 ft (5 000 m) being raised to 777 miles (1 250 km), and range at maximum cruise with 10 per cent fuel reserve being extended to 510 miles (820 km).

Weight had become an increasingly critical factor, however, and while the empty weight at 5,950 lb (2 699 kg) had increased by only some 88 lb (40 kg) over that of the MiG-1, take-off weight in clean condition with maximum fuel had risen to 7,385 lb (3 350 kg), and the admittedly inadequate armament had perforce remained unchanged in order to avoid further weight escalation. Still no novice's aircraft, and possessing characteristics demanding a high degree of piloting skill, the MiG-3 was, nevertheless, a major improvement on the MiG-1. Its overall dimensions had remained unchanged, these including a span of 33 ft 9½ in (10,30 m), a length of 26 ft 8⅞ in (8,15 m), a height with tail down of 8 ft 7⅛ in (2,62 m), and a gross wing area of 187·72 sq ft (17,44 m²). Maximum speed in normal loaded condition was 398 mph (640 km/h) at 25,590 ft (7 800 m), and an altitude of 16,400 ft (5 000 m) was attained in 5·7 minutes.

Such had been the impetus placed behind the production programme that the first MiG-3s had begun to reach the V-VS *eskadrilii* during the course of March 1941, and a second factory had completed tooling and was about to supplement the output of *Zavod 1* at the time of the German delegation's visit in the following month when production tempo of the MiG-3 was fast approaching 75 fighters per week! Small wonder that, during the previous month, Artem Mikoyan and Mikhail Gurevich had each received the title of "Stalin Prize Laureate" for their efforts in organising the volume production of the warplane in record time. By 30 June 1941, total MiG-3 production attained the very substantial figure of 1,209 aircraft out of the total of 1,946 "new generation" fighters that left the assembly lines in the first six months of 1941, the balance being made up by 322 LaGG-3s, 335 Yak-1s, and 80 MiG-1s.

Among the first IAPs (*Istrebitel'ny aviatsionnye polki* — Fighter Aviation Regiments) to convert to the MiG-3 were IA-PVO units assigned the task of defending Baku, Leningrad and Moscow, such as the 34 IAP and the 233 IAP assigned to the last-mentioned zone, and whose bases were, incidentally, included in the itinerary of the visiting German delegation. The 34 IAP was commanded by Major Leonid G Rybkin, and its MiG-3s operated from the southern side of Vnukovo, and the similarly-equipped 233 IAP was based at Tushino and commanded by Major Konstantin M Kuzmenko, both units comprising part of the 6th Fighter Aviation Corps when this was formed under Colonel Ivan D Klimov for the defence of Moscow two days before the launching of Operation *Barbarossa*.

The IA-PVO pilots were, for the most part, more experienced and had enjoyed a higher standard of training than other

MiG-3s of the 34 IAP at Vnukovo, Moscow, in the winter of 1941–2.

V-VS fighter pilots, and, in consequence, accepted more readily the idiosyncracies of the MiG-3, but the conversion of the average V-VS fighter unit to the new warplane imposed many problems. Pyotr Stefanovsky relates how, in May 1941, he received instructions to proceed immediately to Kishinev in Moldavia, near the Rumanian border, where an *IA Divisiya* commanded by General Osipenko — the first V-VS Fighter Division to receive a full complement of MiG-3s — was apparently reluctant to relinquish its I-153s and I-16s in favour of the new fighters. Stefanovsky found the *Divisiya* to possess two full sets of fighters, its older Polikarpov types and a similar number of MiG-3s, yet not one pilot had attempted to fly the new fighter, believing it to possess decidedly lethal propensities. The conversion of the *Divisiya* was considered a matter of the utmost urgency, but none of its pilots evinced any enthusiasm to undertake a familiarisation flight in the MiG-3 until Stefanovsky took one of the fighters into the air and demonstrated its capabilities. From that point on the conversion training programme got into its stride, the pilots soon losing their fear of the MiG fighter.

The MiG-3 joins combat

Although several other fighter divisions had converted at least in part to the MiG-3 by the time Operation *Barbarossa* began, they had had little time to familiarise themselves thoroughly with their new mount, and relatively few MiG-3s were encountered in combat by the *Luftwaffe* during the opening phases of the campaign, although Colonel Aleksandr Pokryshkin gained his first "kill" while flying a MiG-3 on the first day of the German assault, Pokryshkin being destined to end the war as the Soviet Union's second ranking "ace". The *Luftwaffe* fighter pilots reported that, in general, the MiG-3 units appeared singularly lacking in aggressiveness. This was partly due to the inexperience and inadequate training of the V-VS pilots, but there was also a more cogent reason: tacit admission that the MiG-3, optimised for the high-altitude rôle, was at a distinct disadvantage in fighter-versus-fighter combat with the Bf 109F below 19,685 ft (6 000 m). Therefore, the MiG-3 pilots generally avoided combat with enemy single-seat fighters, concentrating on the interception of bombers and reconnaissance aircraft.

The V-VS needed fighters for low- and medium-altitude interception and for close-support tasks, these being the rôles assigned to the so-called "frontal fighters" such as the LaGG-3 and Yak-1, but the MiG-3 was available in substantially greater numbers than either medium-altitude contemporary, and despite a manifest unsuitability for any task other than the high-altitude intercept mission, the first six months of its combat career were largely spent fulfilling missions not envisaged by Mikoyan and Gurevich when designing their fighter, nor by the specification that it fulfilled.

As fast as they could be rolled off the assembly lines the MiG-3s were despatched to combat units, the exigencies of the times denying some of these the most elementary conversion training. The inadequacy of the standard three-gun armament led to several attempts in the field to improve armament effectiveness, such as hanging six RS-82 rocket missiles or a pair of 12,7-mm UBS guns beneath the wings. With such supplementary weapons, loaded weight rose as high as 7,705 lb (3 495 kg), and frequently the standard RSI-3 or -4 receiver and even the cockpit canopy were discarded as contributions to the critical weight problem.

The fighter elements of the V-VS VMF of the Northern, Baltic and Black Sea Fleets, largely responsible for the air defence of naval installations and for providing top cover for convoys and naval vessels operating in coastal waters, possessed 763 aircraft when the German invasion of the Soviet Union began, but this fighter inventory included only 72 MiG-1s, MiG-3s and Yak-1s, although other V-VS VMF units were in process of converting to "new generation" fighters. These, too, were to pronounce the MiG-3 disappointing, partly because it was a difficult aircraft to master but primarily because necessity dictated its assignment to rôles for which it was never intended. Thus, as more LaGG-3s and Yak-1s became available, the MiG-3 was progressively withdrawn from the "frontal" fighter *eskadrilii* and re-assigned to IA-PVO units, such as the 12 IAP in the western sector of the Moscow PVO which exchanged its Yak-1s for MiG-3s and so distinguished itself in combat that it was awarded the coveted title of "Guards Regiment" in 1942.

From the early weeks of the conflict, it became painfully obvious that the specialised high-altitude interceptor as epitomised by the MiG-3 was a luxury that the Soviet Union could ill afford at that stage of the war. Virtually all aerial combat taking place at altitudes at which the MiG-3 was decidedly at a disadvantage, and with the transfer of the bulk of the Soviet aircraft industry eastward, coupled with a demand to reduce the number of individual aircraft types to which the industry was committed, the decision was taken to phase out the MiG-3. Thus, before the end of 1941 the MiG-3 reached the end of its production life, a total of 3,322 fighters of this type having been delivered.

The termination of production did not, of course, signify the early demise of the MiG-3 in V-VS service. Numerically it was to remain an important aircraft in the fighter inventory for a further two or three years, being employed for rear area defence until the closing months of World War II, having been progressively withdrawn from the fighter units assigned to the more important PVO regions from early 1943. During this time, a number of modifications were tested with a view to improving specific aspects of the fighter's performance, some of these modifications being applied retrospectively to service MiG-3s, and between 1942 and 1944 maximum speed was raised by 30 mph (50 km/h) at sea level by uprating the engine. One modified version in which take-off weight was raised to 8,818 lb (4 000 kg) by the introduction of additional overload fuel tanks as the MiG-3DD (*Dalny Deistvya* indicating extended range) was tested in the early months of 1944 but not adopted.

From most aspects, the MiG-3 was an undistinguished fighter, but it was the aircraft on which the late Artem Mikoyan's design bureau "cut its teeth", and its turbojet-driven descendants were to achieve the success and fame denied this initial product of a distinguished team.□   WG

BRONCO————————————————*from page 238*

according to the flight speed. In the cruise configuration, it is normal for the Bronco to yaw and roll slightly into the failed engine. The pilot will instinctively apply corrective roll and yaw controls, generally requiring less than half of the total control available. The problem may then be sorted out and the propeller feathered without undue hurry on the pilot's part. The aeroplane can be trimmed up hands-off in single-engine flight.

Loss of an engine in the take-off configuration will result in greater yaw and roll toward the failed engine, but recovery may be made to wings level flight at and above 80 kts (148 km/h) IAS. The aircraft must be cleaned up by retraction of gear and flaps and speed increased above 90 kts (167 km/h) to achieve flyaway. Loss of an engine in the landing approach configuration results in very mild roll and yaw toward the failed engine which are very easily controlled. Power should be added to the operative engine, flaps retracted, and speed increased to at least single-engine flyaway (90 kts; 167 km/h) in case a wave-off is necessary. The Bronco does not have single-engine flyaway capabilities with the gear and flaps down.

Air start procedures are simple, requiring only actuation of the air start switch, replacement of the condition lever to the NORMAL FLIGHT position, and monitoring of the engine temperature. Power control levers should be set between FLIGHT IDLE and half-way to MILITARY. Drag pulses from the starting engine may be expected as rpm increases above 80 per cent.

Approach and landing

Limit for the undercarriage to be down is 158 kts (293 km/h) IAS and operation provides no trim changes. Full flap (40 degrees) limit is 130 kts (241 km/h) IAS and requires about one unit nose-down trim to compensate. Normal field landings can be made at any approach airspeed above single-engine minimum, but 85 to 90 kts (158-167 km/h) IAS is recommended with 20-degree flap selected. A "seat of the pants" instinct should be utilised to determine if power is symmetric during the approach. Due to the rapid thrust response available, small throttle changes will produce considerable thrust variances and if they are not symmetrical, it is quickly apparent to the pilot by sideslip angle or a loss of lift on one side. Therefore, throttle movements during approach should be made smoothly and simultaneously correcting any change in asymmetry by advancing or retarding either or both throttles to maintain balanced flight.

Because of the large amount of lift generated by the propellers, it is recommended that the throttles should not be retarded quickly as the Bronco may yaw unsymmetrically and commence a high sink rate at low approach speeds. Gentle flared landings are possible during landings on normal length runways. Short field approaches should be conducted with full flap at the minimum single-engine approach speed (dependent on gross weight) and as the desired landing spot is reached, the aircraft should be firmly "planted" by symmetric throttle reduction with no attempt to flare. The rugged landing gear enables landings of this type to be made safely though somewhat unnaturally. Reverse thrust can be selected during either normal or short field landings immediately upon main gear touchdown. Adequate directional control is available through use of brakes or rudders during reverse thrust operation. Heavy braking should be avoided until the full weight of the aeroplane is on the landing gear as brake power is high enough to cause the tyres lo skid if landings are made with even light brake pressure applied.

When not being shot at in combat or on GCA final at night in thunderstorms, pilots find the Bronco a fun aeroplane to fly because of its wide speed range, rugged structure, crisp manœuvrability, and outstanding visibility. These characteristics and the airframe component arrangement enable the Bronco to do a multitude of jobs with the workload on the pilot eased due to stable, responsive, and safe flight characteristics. □

FIGHTER A TO Z

ALCOCK A.1   GREAT BRITAIN

Evolved at the RNAS base at Mudros, in the Aegean, by Flt Lt John Alcock during the summer of 1917, the A.1 employed modified components of the Sopwith Triplane (forward fuselage and lower wings), Sopwith Pup (upper wings), and Sopwith Camel (tailplane and elevators) which were married to a rear fuselage and vertical tail surfaces of original design. Powered by a 110 hp Clerget 9Z nine-cylinder rotary and carrying a 0·303-in (7,7-mm) Vickers machine gun, the A.1 (which was also referred to by its designers as the "Sopwith Mouse" in recognition of its part parentage) flew at Mudros in October 1917, but was written off after crashing early in 1918. Approx span, 24 ft 3 in (7,39 m). Approx length, 19 ft 1 in (5,82 m). Approx height, 7 ft 9 in (2,36 m).

The Alcock A.1 (above) employed modified components of Sopwith types. The Alter A.1 (below) was rejected by the Idflieg in 1917.

ALTER A.1   GERMANY

The A.1 single-seat fighter built by the Ludwig Alter-Werke of Darmstadt was designed by Kallweit and Ketterer who undoubtedly found their inspiration in earlier Nieuport designs. Powered by a 110 hp Goebel Goe II rotary engine, the A.1 was initiated in October 1916 and was flown for the first time in February 1917. It was demonstrated for the *Inspektion der Fliegertruppen* but rejected, and no production was undertaken.

AMBROSINI S.S.4   ITALY

The first fighter of canard or tail-first configuration, and one of the first to feature a fully-retractable nosewheel undercarriage, the single-seat S.S.4 was designed by Ing Sergio Stefanutti of the Società Aeronautica Italiana Ing A Ambrosini. Work on the S.S.4 began in 1938, and the initial flight was made on 1 May 1939. All-metal construction was employed, and power was provided by a 960 hp Isotta-Fraschini Asso XI R.C.40 12-cylinder liquid-cooled engine. Proposed armament comprised one 30-mm and two 20-mm cannon, but the prototype was destroyed during 1941 when the engine failed during a landing approach. Max speed, 335 mph (540 km/h) at 16,400 ft (5 000 m). Loaded weight, 5,401 lb (2 450 kg). Span, 40 ft 5 in (12,32 m). Length, 22 ft 1½ in (6,74 m). Height, 8 ft 1¾ in (2,48 m).

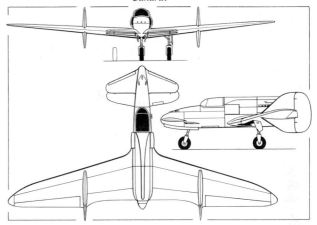

(Above and below) The sole example of the Ambrosini S.S.4 Canard.

AMBROSINI S.A.I.107   ITALY

A lightweight single-seat fighter of wooden construction, the S.A.I.107 was a direct derivative of the S.A.I.7 tandem two-seat trainer, and was flown for the first time early in 1942 with a 540 hp Isotta-Fraschini Gamma air-cooled engine. Official trials of the unarmed prototype at Guidonia revealed an exceptional performance, but development of the S.A.I.107 had already been overtaken by a more powerful derivative of the basic design, the S.A.I.207. Max speed, 348 mph (560 km/h). Loaded weight, 2,200 lb (998 kg). Span 29 ft 6⅛ in (9,00 m). Length, 26 ft 3¾ in (7,00 m). Height, 9 ft 5 in (2,87 m). Wing area, 149·618 sq ft (13,9 m²).

The unarmed S.A.I.107 lightweight single-seat fighter prototype.

AMBROSINI S.A.I.207   ITALY

Evolved from the S.A.I.107 and featuring a similar wooden structure with monocoque fuselage, the S.A.I.207 designed by Ing Sergio Stefanutti was flown in 1942. Power was provided by a 750 hp Isotta-Fraschini Delta R.C.40 12-cylinder air-cooled engine, and armament of the production model consisted of two 20-mm Mauser MG 151 cannon and two 12,7-mm Breda-SAFAT machine guns. An order was placed for 2,000 production aircraft but, in the event, this was superseded by an order for the S.A.I.403 Dardo, and only 13 pre-production S.A.I.207s were completed. Max speed, 398 mph (640 km/h) at 14,765 ft (4 500 m). Cruise, 304 mph (490 km/h). Range, 528 mls (850 km). Time to 19,685 ft (6 000 m), 7 min 34 sec. Empty weight, 3,858 lb

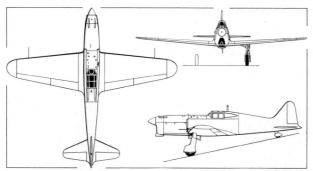

(Above and below) The production version of the S.A.I.207.

(1 750 kg). Loaded weight, 5,324 lb (2 415 kg). Span, 29 ft 6⅓ in (9,00 m). Length, 26 ft 3¾ in (7,00 m). Height, 9 ft 5 in (2,87 m). Wing area, 149·618 sq ft (13,9 m²).

AMBROSINI S.A.I.403 DARDO   ITALY

Basically a refined version of the S.A.I.207, the S.A.I.403 Dardo (Arrow) was completed as a prototype late in 1942, and in January of the following year orders were placed for 3,000 production aircraft, but none had been completed when the Armistice terminated further work. The Dardo was powered by a 750 hp Isotta-Fraschini Delta R.C.21/60 12-cylinder air-cooled engine, and two versions were to have been produced: the Dardo-A light interceptor weighing 5,459 lb (2 471 kg) with two 12,7-mm Breda-SAFAT guns, and the Dardo-B general-purpose fighter with armament augmented by two 15-mm or 20-mm MG 151 cannon. The following figures refer to the latter. Max speed, 403 mph (648 km/h) at 23,620 ft (7 200 m). Cruise, 304 mph (490 km/h). Range, 582 mls (936 km). Time to 19,685 ft (6 000 m), 6 min 40 sec. Empty weight, 4,372 lb (1 983 kg). Loaded weight, 5,820 lb (2 640 kg). Span, 32 ft 1¾ in (9,80 m). Length, 26 ft 10¾ in (8,20 m). Height, 9ft 6 in (2,90 m). Wing area, 155·646 sq ft (14,46 m²).

(Above and below) The S.A.I.403 Dardo lightweight fighter.

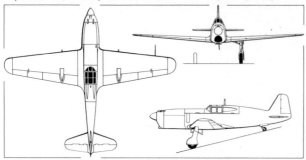

AMIOT-S.E.C.M. 110   FRANCE

Flown for the first time in June 1928, the Type 110 was designed by M Dutartre as a contender in the so-called "Jockey" lightweight interceptor contest in which it competed against nine other types. Two prototypes were built simultaneously, these being of all-metal construction and powered by the 500 hp Hispano-Suiza 12 Mb 12-cylinder liquid-cooled engine, the first prototype having fabric, and the second prototype, metal, wing skinning. Basically a parasol monoplane, the Type 110 was fitted with a jettison-able aerofoil-section fuel tank, inserted in the fuselage aft of the main undercarriage members in the form of stub wings. Armament comprised two Vickers guns. On 1 July 1929 the first prototype was destroyed in an accident, and no further development was undertaken. Max speed, 184 mph (296 km/h). Range, 310 mls (500 km). Time to 13,120 ft (4 000 m), 6 min. Empty weight, 2,469 lb (1 120 kg). Loaded weight, 3,307 lb (1 500 kg). Span, 34 ft 5⅓ in (10,50 m). Length, 21 ft 4 in (6,50 m). Height, 9 ft 2¼ in. Wing area, 226·042 sq ft (21 m²).

The Amiot-S.E.C.M. 110 (the second prototype being shown above).

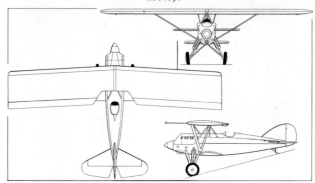

ANF-MUREAUX 114   FRANCE

In 1930, the Ateliers des Mureaux amalgamated with the Ateliers de Construction du Nord de la France (ANF), and the first aircraft to appear after the amalgamation were the Mureaux 110 and 111 designed by André Brunet to participate in the two-seat reconnaissance aircraft programme initiated in 1928. The first of these all-metal parasol monoplanes was flown in April 1931, the initial production derivative being the Mureaux 113. Two early production airframes were completed during the summer of 1933 as two-seat night fighters under the designation Mureaux 114, these differing from the Mureaux 113 primarily in being equipped with searchlights. Powered by a 650 hp Hispano-Suiza 12 Ybrs 12-cylinder liquid-cooled engine, the Mureaux 114 carried an armament of two fixed forward-firing 7,7-mm MAC machine guns in the fuselage and two Lewis guns on a flexible mounting in the rear cockpit. Max speed, 194 mph (312 km/h) at 16,400 ft (5 000 m). Climb to 16,400 ft (5 000 m), 8 min 55 sec. Range, 572 mls (920 km). Empty weight, 3,704 lb (1 680 kg). Loaded weight, 2,644 lb (2 560 kg). Span, 50 ft 6⅓ in (15,40 m). Length, 32 ft 11⅔ in (10,05 m). Height, 12 ft 6 in (3,81 m). Wing area, 375·66 sq ft (34,90 m²).

THE BILLION DOLLAR BOMBER

PART FOUR

One afternoon in late July of 1944, a Japanese anti-aircraft shell exploding in Manchurian skies initiated a chain of events that was to cost the North American tax-payer billions of dollars. Its immediate effect was to provide the Soviet Union with an example of the Boeing B-29 Super-fortress, around a Soviet copy of which was to be created a powerful strategic air arm. Its longer term effect was to provide a basis for the development, by genealogical processes stretching over a decade, of the only Soviet bomber capable of performing strikes across the polar regions, and the sole factor justifying the fantastic cost of building the immense chains of radar installations and interceptor bases intended to shield the North American continent from manned aircraft attack. Previous parts of this story appeared in the July, August and September issues.

IN FEBRUARY 1954, a much respected and usually reliable US aviation journal created something of a sensation in the western world by publishing what were purported to be the first photographs of the "latest Soviet intercontinental weapons"; two swept-wing long-range strategic bombers with four and six turboprops respectively ascribed to the design bureaux of Sergei Ilyushin and Andrei Tupolev. More than 400 of these bombers, the journal claimed, were reported to be based in the northern provinces of the USSR across the polar ice cap from North America, adding that reconnaissance versions had made flights at extreme altitudes over Alaska and the Canadian defence perimeter!

The allegedly genuine photographs were syndicated all over the world, credence being lent them by the respectability of their apparent origin. In fact, the strategic bombers depicted by these remarkable photographs had been conceived somewhat further west than the Soviet drawing boards to which they were attributed; they had been created in West Berlin by an unscrupulous individual who had discovered that he could supplement his income by catering for the western demand for information on military aviation development behind the "Iron Curtain". The expertise of the artist who produced the "photographs" and the gullibility of his customers ensured the success of the confidence trick.

It was not that there was no basis of fact for these illustrations, even if the text that accompanied them was highly imaginative — it was certainly read with considerable amusement in the Soviet Union. German engineers who, since 1950, had been trickling back from Kuybyshev, where, as part of Nikolai D Kuznetsov's development bureau, they had been working on turboprop development, reported that the results of their efforts were being applied to large swept-wing strategic bombers. With this information and a certain amount of imagination, the gentleman in West Berlin encountered no difficulty in producing reasonably credible turboprop-powered strategic bombers, the "photographs" of which, he subsequently claimed, had been smuggled out of the Soviet Union by one of the returning German engineers.

The Soviet strategic bombing arm, the *Dal'naya aviatsiya*, had every reason to wish that the "exposé" in the US aviation journal *had* embodied an element of truth; that it in fact *possessed* bombers having capabilities such as those listed and in the quantities reported. The facts of the matter were *very* different, for, at the time the spurious photographs were seeing widespread publication, the *Dal'naya aviatsiya* was equipped entirely with the decidedly obsolescent Tu-4 which, in view of its Boeing B-29 ancestry, was of pre-WW II concept. The first prototype of the only warplane of Soviet origin to bear the remotest resemblance to the bombers ascribed to the Soviet Union by the US journal had still to commence its flight test programme, and, even accepting the most sanguine estimates, could not reach the strategic bombing *eskadrilii* for at least a further three years.

Almost three years earlier, in 1951, the demise of the piston engine as a viable power plant for high-performance strategic aircraft had been finally accepted in the Soviet Union, and the intended successor to the Tu-4 in the long-range bombing rôle, the Tu-85, had been abandoned and with it all likelihood of endowing the *Dal'naya aviatsiya* with true intercontinental nuclear bombing capability before the late 'fifties. The early phases of the air war over Korea had emphasised the vulnerability to jet interceptors of large piston-engined bombers, stressing the need for a performance far superior to that offered by the Tu-85.

A requirement had already been issued for a turbojet-powered medium bomber with payload-range characteristics adequate for conventional war in the European theatre, and the design bureaux of both Sergei Ilyushin and Andrei Tupolev were actively engaged in fulfilling this demand.

Whereas the Tupolev medium bomber, which bore the design bureau designation Tu-88, was of relatively advanced concept, featuring swept surfaces and two immense AM-3 turbojets which, evolved by Aleksandr Mikulin's bureau, had the quite remarkable rating of 17,637 lb (8 000 kg) and were recessed into the fuselage sides, Ilyushin's offering, which was essentially a back-up programme, was much less ambitious in concept. Designated Il-46 and powered by two 11,020 lb (5 000 kg) AL-5 turbojets developed by Arkhip Lyulka's bureau, the Ilyushin contender for the medium bomber requirement was to all intents and purposes a scaled-up Il-28 light bomber.

Both the Tu-88 and Il-46 entered the flight test phase during the summer of 1952, by which time the Tupolev bureau, together with that of Vladimir Myasishchev, had been assigned the task of developing the complementary heavy bomber potentially capable of destroying North American war capacity by means of strikes across the polar regions.

A unique warplane

There can be little doubt that it was the Myasishchev bomber upon which the primary hopes of the *Dal'naya aviatsiya* were centred, the Tupolev contribution to the programme initially being regarded as a back-up system. The former was a swept-wing bomber in the B-52 Stratofortress category powered by four Mikulin AM-3 turbojets, but the latter, which bore the design bureau designation of Tu-95, was unique in having both airscrews and swept surfaces, a combination then widely regarded, in the West at least, as pointless. Airscrews were believed to be unusable at speeds above Mach 0·75 and swept surfaces were unnecessary below this speed.

The Tu-95 was, in fact, being designed around an extremely powerful turboprop unit under development at Kuybyshev by former Junkers engineers attached to the bureau of Nikolai D Kuznetsov. The specification for this engine, which had been framed in 1950, stipulated a basic sea level output of 12,000 eshp with 8,000 eshp being available at Mach 0·85 at 36,090 ft (11 000 m) with a fuel consumption of 0·35 lb

(161 gr)/hp/hr, the power plant weight (excluding airscrew) being restricted to 5,100 lb (2 330 kg). The requirement to which the Tupolev bureau was working had stipulated the rather ambitious dash speed of Mach 0·85, and a tactical radius of 4,350 miles (7 000 km) at Mach 0·7 long-range cruise at 36,090 ft (11 000 m) carrying an 11,023-lb (5 000-kg) weapons load.

To meet this requirement in the minimum possible time, Tupolev's team retained the structural techniques that had evolved from the Tu-4 and the experimental derivatives of this bomber, and employed what was essentially the fuselage of the Tu-85 to which were married new swept wing and tail surfaces. The fuselage was a circular-section semi-monocoque, embodying, like the Tu-85, three pressurised compartments, those forward of and immediately aft of the weapons bays being interconnected by a crawl tunnel, and the third compartment in the tail being isolated. Defensive armament comprised an aft-positioned remotely-controlled dorsal barbette mounting twin 23-mm cannon, a similar barbette beneath the rear fuselage, and a third pair of 23-mm weapons in the tail. All weapons could be controlled from a sighting station in the tail which featured glazed blisters beneath the tailplane. Alternatively, the dorsal and ventral barbettes could be controlled from a sighting station aft of the flight deck. The overall length at some 155 ft 10 in (47,50 m) was barely less than the overall span of the wing, which was about 159 ft (48,50 m), the wing surfaces being swept 37 degrees inboard and 35 degrees outboard, and being three-spar structures with power-operated ailerons and Fowler-type flaps.

The practice initiated with the Tu-4 of building simultaneously a large batch of test and evaluation aircraft rather than the usual two or three prototypes was maintained with both the Tu-95 and its pure-jet Myasishchev contemporary, and the development of both bombers proceeded on roughly the same time scale. It was planned that both types would participate in the fly-past over Tushino forming part of the 1954 May Day celebrations, but in the event only the first pre-series Myasishchev bomber had commenced its flight test programme and was available to take part in the parade,

The latest variant of the Tu-20 employed for the maritime patrol and reconnaissance task, dubbed "Bear-D" by NATO, features a large belly radome, a new undernose radar scanner, and sensor housings in the form of tailplane-tip bullet fairings.

The "Bear-D" (above) joined earlier conversions of the basic Tu-20 bomber, such as the "Bear-A" (below left), on maritime patrol and reconnaissance and "ferret" missions in 1967. Small sensor housings may be seen under the forward fuselage of the "Bear-A".

flight clearance of the Kuznetsov turboprops having delayed testing of the first pre-series Tu-95.

Flight testing of the Tu-95 began during the late summer of 1954, and during the Aviation Day fly-past over Tushino a year later, on 3 July 1955, a formation of no fewer than seven pre-series aircraft appeared. It had presumably been anticipated that the Myasishchev bomber powered by pure turbojets would possess a range inferior to that of the Tu-95, but that this disadvantage would be offset to some extent by its higher performance which, it could be assumed, would endow it with a greater chance of evading interception. However, there is some evidence to suggest that range fell appreciably below that specified, with the result that planned production of the Myasishchev bomber, officially designated M-4 in *Dal'naya aviatsiya* service and unofficially if somewhat optimistically dubbed the *Molot* (Hammer), was reduced to modest quantities, and its turboprop-driven Tupolev contemporary assumed proportionately greater importance in the re-equipment programme of the Soviet strategic bombing arm.

The public début of the Tupolev bomber over Tushino on 3 July 1955 came as no surprise to western intelligence agencies owing to the information gleaned from the interrogation of German engineers that had worked on the bomber's turboprop in Kuybyshev, and all of whom had, by that time, returned to Germany. Dipl-Ing Ferdinand Brandner, who had led the German team working on this power plant, had stated that it had been designed for a maximum speed of Mach 0·85 at 36,090 ft (11 000 m) at which its four-bladed 18 ft 4½ in (5,60 m) diameter contra-rotating airscrews were to have turned at 750 rpm, the corresponding airscrew tip speed being Mach 1·08. Although these figures undoubtedly proved overly optimistic in practice, it was soon obvious that western experts had been somewhat too *pessimistic* in their calculations of the speed limitations that would be imposed by shock losses at the airscrew tips. The turboprop unit, which had been designated Kuznetsov NK-12, had achieved the specific power within the weight limitation imposed, but the stressing of both power plant and gearbox had been such that they could accept little more than maximum cruise loads; the figure of 12,000 eshp for sea level take-off was only the theoretical output achieved, had it been possible to run the engine at full revs, but the NK-12 was limited in practice to something of the order of 8,000-9,000 eshp. Nevertheless, in so far as the West was concerned, the capabilities of this unique bomber were impressive enough.

From Bull to Bear

During the latter half of 1956, the *Dal'naya aviatsiya* began the task of phasing into its inventory the turboprop-driven Tupolev bomber, which had by this time received the

official designation of Tu-20. The Pentagon-based air intelligence unit responsible for allocating identification names to the progeny of Soviet drawing boards, and which had been dubbed the Tu-4 bomber *Bull*, continued to display its penchant for stockmarket jargon, bestowing the appellation of *Bear* on this B-29 copy's successor.

Although some uncertainty existed in the West of the relative importance attached to Myasishchev's M-4 and Tupolev's Tu-20 by the Soviet strategic bombing force, there was little uncertainty by 1958 as to which constituted the greatest threat to North American industry. Although the dash and cruise speeds of the M-4 were markedly superior to those of the Tu-20, the *Dal'naya aviatsiya* could hardly threaten vital US targets with this type unless it was willing to commit half of its bomber force to tanker operations, and U-2 reconnaissance flights and other sources of intelligence information suggested that the quantities of M-4 bombers available or being manufactured were insufficient to permit such operations. The Tu-20, on the other hand, was apparently being produced in larger numbers, and, with an 11,023-lb (5 000 kg) bomb load, appeared to have a radius of action of some 4,500 miles (7 240 km), reducing to around 4,000 miles (6 440 km) if the bomb load was doubled.

Any lingering doubts in the West as to the Tu-20's potential were dispelled in the spring of 1958 when a version of the bomber with armament and weapons bays removed, and a section of the fuselage aft of the wings pressurised and fitted with windows to form a passenger cabin, flew non-stop Moscow-Irkutsk-Moscow, a distance of 5,280 miles (8 500 km), at an average speed of 503 mph (810 km/h), or Mach 0·76, with a corresponding airscrew tip speed of the order of Mach 0·98. Presumably for the purpose of this remarkable flight the aircraft was designated Tu-114D, a commercial derivative of the bomber with an entirely new

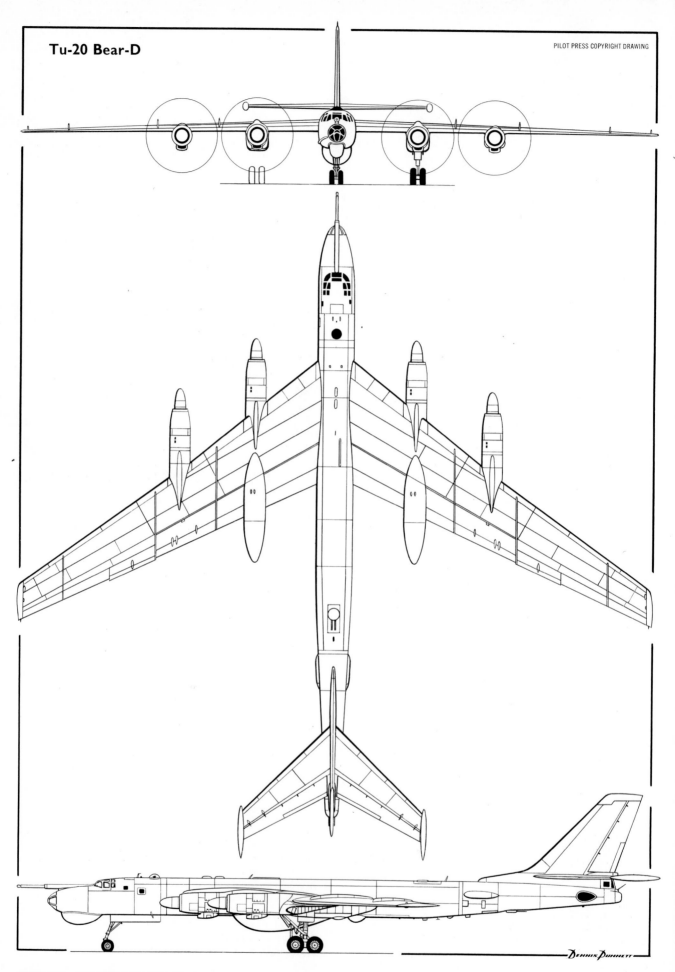

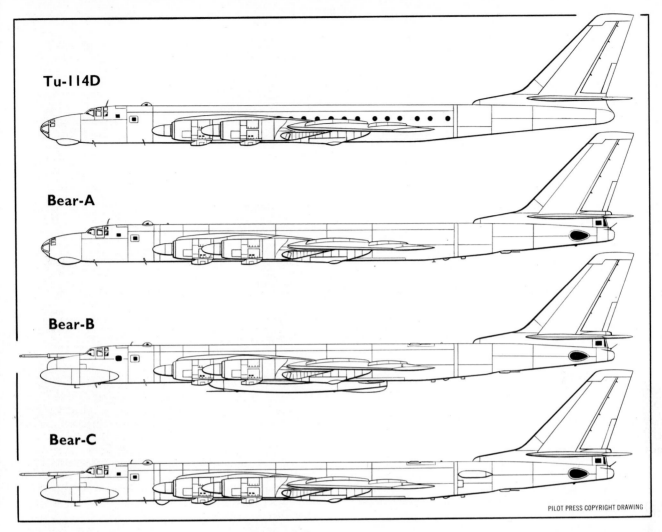

Tu-114D

Bear-A

Bear-B

Bear-C

and substantially enlarged fuselage having flown as the Tu-114 during the previous August. That the USA considered the Tu-20 to pose a serious threat was evidenced by the accelerated construction of interceptor bases and radar systems intended to provide the North American continent with a shield against manned aircraft attack.

No less impressive than the Tu-20 was its *genuine* commercial counterpart, the Tu-114, which, at the time of its appearance, was the world's largest and heaviest airliner. Accommodating up to 220 passengers, the fuselage has an appreciably larger diameter and was some 21 ft (6,40 m) longer, overall length being 177 ft 6 in (54,10 m). The engines, undercarriage and tail assembly were similar to those of the Tu-20, as was also the wing, although chord was increased inboard, which, together with a marginal increase in overall span (to 167 ft 8 in/51,10 m), resulted in a gross wing area of 3,349 sq ft (311,10 m²). Development of the Tu-114 was by no means trouble-free, and the prototype was, in fact, lost during the test programme. On this occassion the Tu-114 was being flown by Tupolev test pilot Aleksei Perelyot. While performing high-speed trials at 22,965 ft (7 000 m) number three engine caught fire. Initially it seemed that the fire had been extinguished but the engine suddenly burst into flames again, rapidly burning through the engine nacelle which fell away from the aircraft. Perelyot ordered the crew to abandon the aircraft, but the chief engineer, Chernov, elected to remain with him and assist in an attempt to land the crippled prototype. All seemed to be going well, and the aircraft was in its final approach when aileron control failed, followed by the control boosters. The aircraft adopted 40 degrees of bank, all control was lost and the

aircraft crashed, both Perelyot and Chernov losing their lives. This accident delayed the introduction of the Tu-114 by *Aeroflot*, which eventually placed the type in service in April 1961 on the route between Moscow and Khabarovsk.

A bevy of Bears
By June 1961, when a variant of the Tu-20 embodying sufficient external changes to warrant the application of a suffix letter to its NATO identification name appeared over Tushino during Aviation Day, the primary rôle of the turboprop-driven Tupolev bomber had been largely taken over by the ballistic missile. This modified version — promptly designated *Bear-B* in the West, the original model thus becoming the *Bear-A* — differed primarily in that the nose configuration had been adapted to accommodate a "duck's bill" radome for high-definition search radar, and in having cutaway weapons bay doors to permit a large turbojet-powered stand-off missile dubbed *Kangaroo* to be carried semi-internally.

Some examples of the *Bear-B* were fitted with a large flight-refuelling probe over the nose, and with the progressive reduction of the intercontinental bombing commitment of the *Dal'naya aviatsiya* and the commensurate phase-out of the Tu-20 from its inventory, the bombers were, after modification to suit them for maritime reconnaissance and patrol, and for "ferret" missions, gradually transferred to the shore-based elements of the *Aviatsiya Voenno-morskovo Flota* (AV-MF), the naval air arm. For its AV-MF tasks, the *Bear-B* had the recess previously occupied by the *Kangaroo* missile faired over, camera ports let into this fairing, and a streamlined blister, presumably housing non-optical sensor

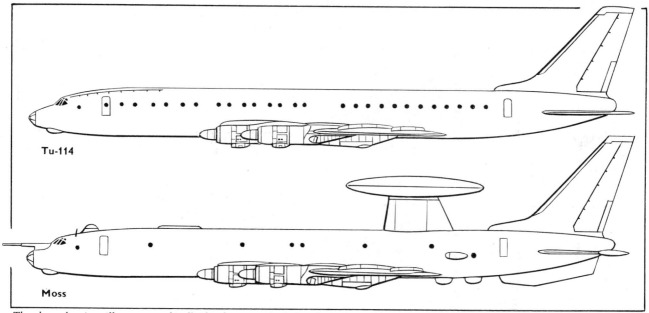

Tu-114

Moss

The above drawings illustrate graphically the changes embodied in the basic Tu-114 commercial transport to produce the airborne warning and control aircraft named "Moss" by NATO. The wing, power plants and undercarriage are apparently unchanged.

equipment, introduced on the starboard side of the rear fuselage.

The *Bear-B* was joined in AV-MF service by the *Bear-C*, which, first identified in September 1964 when it appeared in the vicinity of Allied naval forces engaged in *Exercise Teamwork*, differed from the *-B* primarily in having a series of small ventral radomes, and the streamlined blister that had appeared only on the starboard side of the rear fuselage of the *-B* duplicated on the port side. In mid-1967 yet a further maritime reconnaissance model appeared. This, the *Bear-D* which was first seen by US Coast Guard ice-breakers in the Kara Sea, introduced an undernose radar scanner not un-like that of the Canadair Argus Mk 2, reintroduced the glazed nose cone, which was accompanied by a lengthened refuelling probe, and, for good measure, added a large belly

The photographs above and below depict the "Moss" which locates low-flying intruders and directs missile-armed inter-ceptors.

radome. Sensor housings in the form of tailplane-tip bullet fairings appeared on this variant for the first time.

At much the same time as the *Bear-D* was introduced into the inventory of the AV-MF, western intelligence agencies became aware of yet another derivative of the original Tu-20 in Soviet military service. This, named *Moss* by Nato, was an AWACS (Airborne Warning and Control System) aircraft derived from — and probably actually converted from — the commercial Tu-114, and intended to operate in conjunction with interceptors, locating low-flying intruders and directing missile-armed fighters towards them. By comparison with the Tu-114, the *Moss* has considerably taller vertical tail surfaces, a substantial ventral fin, a flight refuelling probe over the nose, various sensor-housing blisters, including tailplane-tip bullet fairings similar to those of the *Bear-D*, and an immense pylon-mounted "saucer" type early warning scanner housing with a diameter of about 37 ft 6 in (12,00 m). The wing, power plants and undercarriage of the *Moss* are similar to those of the Tu-114, and overall length, including probe, is of the order of 187 ft (57,00 m).

Although the Tu-20 was never employed in anger and the question of how it would have fared had it attempted to penetrate the North American defence screen in earnest is now a matter of academic interest only, the fact that it gave the Soviet Union an economic victory of considerable importance cannot be denied. The very existence of the Tu-20 had necessitated the fantastic expenditure on interceptor bases and warning radar established in North America to counter manned aircraft attack. The cost to the North American tax-payer was many billions of dollars; the cost to the Soviet Union was merely that of developing and building perhaps fewer than 300 examples of the turboprop-driven Tupolev bomber. Against the latter can be offset the tremendous technological prestige value of evolving a turboprop commercial transport capable of seriously rivalling the payload-range characteristics of American turbofan-powered aircraft.

The technology which produced the Tu-20 and Tu-114 sprang directly from the task of copying the Boeing B-29 Superfortress, and it is intriguing to speculate on the course that Soviet strategic aircraft development may have taken had not a trio of B-29s fortuitously arrived in the Soviet Union 26 years ago. ☐ WG

VETERAN & VINTAGE

A surviving Messerschmitt

ONE of Finland's two Messerschmitt Bf 109G-6s, and one of the best-kept examples of this sub-type anywhere, has recently been restored to its wartime finish at Rissala AFB. This example, MT-507, one of 162 *Gustavs* used by the Finnish Air Force between 1944 and 1954, performed the final flight by a Bf 109 in Finland on 13 March 1954, when the pilot was Erkki Heinila. Subsequently, it was put on permanent display at the Utti AFB, but at that stage was incorrectly painted and was fitted with a non-standard tailwheel leg.

During 1970, the other surviving Bf 109G-6, MT-452, was transferred from Santa Hamina to Utti, and MT-507 was taken to Rissala AFB for restoration by personnel of the *Karjalan Lennosto*. In May this year, the engine was run again for the first time since 1954, and the aircraft was displayed in its new, authentic finish at an air display at Rissala on 22-23 May. It is not intended that the aircraft should be flown. (*With acknowledgements to Eino Ritaranta*)

The oldest Canberra?

THE LONGEVITY of the English Electric (now BAC) Canberra is well known, and there is still an active demand for refurbished ex-RAF examples from such countries as Argentina and India. Not so well known, perhaps, is the fact that among the Canberras still in active service with the RAF itself are examples from the first production batch of B Mk 2s, delivered 20 years ago.

One such aircraft is WD948, now in service with No 85 Squadron at Binbrook. Its interesting career includes service in Germany and the Far East, as well as an eight-year period in storage at an MU.

The service life of WD948 began on 5 October 1951, when it was collected from the EEC at Preston and delivered to No 101 Squadron at Binbrook. There it remained until the squadron re-equipped with Canberra B 6s in October 1954. The aircraft was

(Below) Photographed at Binbrook in April, the Canberra B Mk 2 WD948 has survived 20 years of RAF service. (Right) the same aircraft is seen, nearest the camera, when it formed part of No 101 Squadron at Binbrook in 1951.

The Messerschmitt Bf 109G-6 MT507 with engine running at Rissala on 23 May this year, as described in the item alongside.

taken on charge by No 102 Squadron at Gutersloh on 19 October, but returned to the English Electric Company in the UK on 9 August 1955. It stayed with EEC for six months, returning to RAF Gutersloh on 2 February 1956, but again returned to the UK in August that year arriving at 15 MU, Wroughton, on the 9th. After transferring to 5 MU at Kemble on 3 February 1958, WD 948 was soon sent to the Far East Air Force, being delivered to No 45 Squadron at Tengah, Singapore, on 8 April.

After a stay of 18 months in the Far East, WD948 once again returned to the UK on 20 October 1959, this time to 41 Group, and underwent modifications at Marshall's of Cambridge. Returning to Tengah on 2 February 1960, it was again taken on charge by No 45 Squadron on the 25th of

that month. On 7 December 1961 it was transferred briefly to No 75 Squadron (presumably RNZAF), which was also at Tengah, though this posting was shortlived, and the Canberra came back to 15 MU at Wroughton on 2 February 1962. It apparently stayed at this MU, where it was completely refurbished, until 23 March 1970, when it paid its second visit to 5 MU Kemble, before being sent to No 85 Squadron Binbrook on 4 May 1970. It is still on the strength of this Squadron as a B Mk 2, thus being back at the base to which it was originally delivered 20 years ago. Up to leaving Tengah in early 1962, WD948 had flown approximately 1,700 hours and at 28 April 1971 had still totalled only 2,153·45 hours. (*With acknowledgements to. Peter H T Green*)

Sensibility and salvage

To the sensitive modeller there can be few more traumatic experiences than that of seeing the end product of his labours in pieces on the floor. By their very nature models of aircraft are fragile, and few can survive a fall of a few feet unscathed. A careless elbow, a curtain blown by a puff of wind from an open window, or an inquisitive family cat can, in seconds, wreak the most appalling damage on a model that has consumed innumerable hours of patience and care. It is a thoroughly disheartening experience, a fact to which the writer can testify, having recently been faced with the debris of no fewer than seven models that had been knocked from their exhibition shelves by a somewhat obese cat.

Mastering the temptation to prove the old adage concerning curiosity *killing* the cat, and overcoming the urge to consign the wreckage to the trashcan, or, at least, the spares box, your columnist began the task of sifting through the bits and pieces. Near-despair set in after examination of the battered remains of an Airfix Sunderland into which we had crammed carefully-designed gadgetry to permit accurate operation of the bomb doors, racks, etcetera. Fighting down another surge of anti-feline feeling, however, we persisted, and, tackling the least damaged model first, started what at first sight seemed the onerous business of making good the depredations of the domestic quadruped.

Somewhat surprisingly, as the work progressed it did not prove to be unenjoyable. Possibly our expertise had increased in the time that had elapsed since the models had first been assembled for we found that several of them could be improved upon during the course of repairs, and those that demanded a completely new paint job provided an opportunity to apply finishes and markings not available previously. Two full evenings later all seven models were back on their exhibition shelves, and even the Sunderland's bomb gear was fully operational once more. So if there is a moral to be drawn from this experience, apart from the obvious one that cats and models are a mix fraught with hazard, it is that, in the event of a model suffering an accident, *nil desperandum*; it is cheaper to repair the damage than to start afresh with a new kit, and the effort can be highly rewarding.

This month's colour subject

Japan's Mitsubishi Zero-Sen fighter was unique among the warplanes of WW II in that it created a myth — the myth of Japanese invincibility in the air. The allies credited the Zero-Sen with almost mystical powers of manœuvre, fostering this myth of invincibility, and even the Japanese themselves began to believe it omnipotent, and to them the Zero-Sen was everything that the Spitfire was to the British. Time disproved its invincibility, but it was, nevertheless, the first shipboard fighter capable of besting its land-based opponents and remains one of military aviation's true immortals. To this day, the Zero-Sen — or *Reisen* if you prefer it — remains the *only* Japanese aeroplane that will sell really well outside its homeland in model kit form, a fact which has certainly been accepted by model kit manufacturers, virtually all of which have included a Zero-Sen kit in their lists. Indeed, something like 50 kits of this fighter have found their way onto the market, in a fantastic variety of scales and to an equally fantastic variety of standards which range from the superb to the truly atrocious.

It is obviously possible to mention only a small proportion of these kits in this column. The majority hail from Japan, although many of them have been issued by western manufacturers, and, on the grounds of availability, we will therefore concentrate on these. Most of the Zero-Sen kits represent the A6M5, and, of these, the best is undoubtedly the 1/32nd scale offering from Revell. Accurate, finely-detailed and easily assembled — though our review sample was marred by a great deal of flash — it suffers only one real fault: owing to the fact that the cockpit canopy is intended to open its centre portion is much too high. This fault can be rectified but a great deal of careful filing is necessary. Revell's 1/72nd scale A6M5 is also good, although the engine cowling is too narrow and somewhat square. Nevertheless, it is still superior to the elderly Frog and Airfix kits, primarily on the grounds of its excellent surface detail.

Nichimo of Japan offers good 1/70th scale kits of three of the most widely used variants of the Zero-Sen, the A6M2 Model 21, the A6M3 Model 32 and the A6M5 Model 52, and these are fairly easily obtainable in the West, this manufacturer's kits of the Models 21 and 32 being, so far as we can ascertain, the only kits of these variants currently available in anything approaching 1/72nd scale. Nichimo also offers kits of the A6M5 to 1/48th and 1/35th scales, the former being almost identical to the Monogram kit, and the latter — of quite high standard, incidentally — apparently being original. In addition to the Nichimo-Monogram kit, Tamiya offers the A6M5 to 1/48th scale, as well as a slightly modified version of the kit representing the A6M8 Model 64, although this failed to attain service status. There is little to choose between these kits, Tamiya's A6M5 perhaps being slightly more accurate than its competitors, particularly around the gun breech fairing forward of the windscreen, but Monogram scores on ease of assembly, its kit featuring locating pins which are not provided by the Tamiya kit.

In the UK the Tamiya kits are easier to find than those of Monogram, being regularly imported by Richard Kohnstam Limited and sold under the trade name "Riko", but in the USA the situation is reversed. Lindberg issues what, considering its age, is an astonishingly good kit of the A6M2 Model 21 to 1/48th scale, and Aurora offers the same version to the same scale but to a somewhat lower standard of accuracy. No doubt many readers of this column will have their particular favourites among Zero-Sen kits which have not been included, but those that we have mentioned are sufficient to show that any modeller wishing to add this remarkable warplane to his collection should have no difficulty in obtaining a kit to whatever scale he prefers.

A feline fighter from Monogram

As our opening remarks may have suggested, at the time of writing we have little affection for *Felis domesticus*, and cannot but help consider the appellation bestowed on Grumman's latest fighter to be slightly ludicrous, but there is certainly nothing ludicrous about the aircraft itself, for the Grumman F-14 Tomcat promises to make a highly formidable addition to the US Navy's inventory. Names apart, it was with particular pleasure that we received a kit of the latest Grumman 'cat' from Monogram as it has been so long since that distinguished manufacturer added to its range. It was particularly disappointing, therefore, that we discovered Monogram's 1/72nd scale kit of the F-14A to attain rather lower standards than those that we have come to expect from this company.

It is a competent and workmanlike job, accurate in outline, neatly moulded, and finely, if somewhat sparsely, detailed. It features an ingenious and efficient swing-wing mechanism, and despite this is reasonably easy to assemble, which is perhaps hardly surprising in view of the fact that the kit comprises only 33 parts. However, assembly would be a good deal easier if the fit of the component parts was better, numerous unsightly gaps being left at the joints between the top and bottom halves of the wings and fuselage. On the credit side is a superbly clear canopy which opens to disclose amply-detailed cockpits, but where this kit *really* falls down is on the score of the crudity of such items as the air intakes which are offered with thick

(Opposite page) Top to bottom: Mitsubishi A6M2 of the 12th Combined Air Corps, Hankow Region, China, Winter 1940-1; A6M2 aboard carrier Hiryu for attack on Pearl Harbour, December 1941; A6M2 of the 6th Air Corps, Rabaul, New Britain, November 1942; A6M2 of 402nd Sqdn, 341st Air Corps, Clark Field, Philippines, winter 1944; A6M3 of 251st Air Corps, Formosa, November 1942; A6M5 of Genzan Air Corps (Training), Wonsan, North Korea, winter 1944.

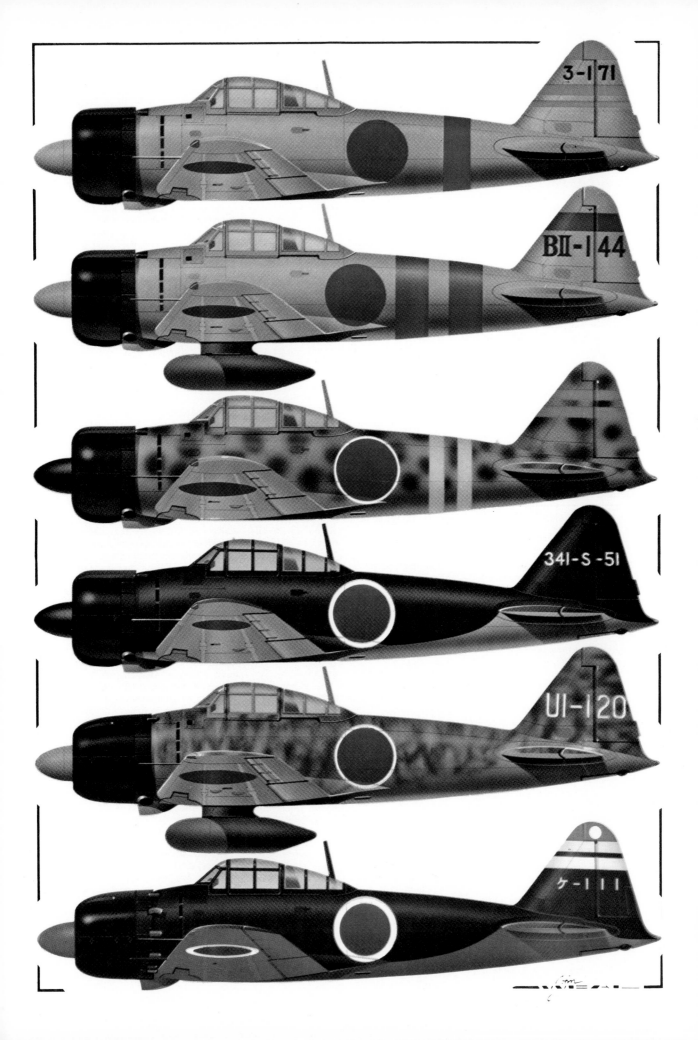

square edges, while the intakes themselves are *solid*, an example of barbarism not perpetrated for something like a decade!

The trailing edges of the wings and tail surfaces are also too thick, a fault that is quite inexcusable in view of the fact that the surfaces are produced from single thicknesses of plastic. The decal sheet is of good quality, although the operational markings provided are, of course, fictitious, and the instruction sheet is good. But let us be fair. Had this kit of the F-14A been issued by *some* manufacturers it would have been acclaimed. As a Monogram product it is a disappointment and we can but hope that with its next kit the company will fully regain its former standards.

Twin and twin-boom from Frog
Having pleaded for years the cause of kits covering that most neglected period of aviation, the half-score or so years immediately following WW II, we have been delighted to see a few of the better-known types of the late 'forties and early 'fifties appearing on the stockists' shelves. Frog has been well to the fore in this "gap-filling" process, and the company's recently-released de Havilland Vampire FB Mk 5 particularly welcome, despite being accompanied by yet another kit of that perennial, the Messerschmitt Bf 110, which, if more glamorous than the Vampire, must be infinitely more familiar to the modelling fraternity.

The Vampire is both accurate and neatly moulded, and suffers virtually no flash. Manufactured in good-quality dark grey plastic, it has fine surface detail and is accompanied by a first class decal sheet providing the markings of an aircraft of No 502 Sqdn RAF and those of an FB Mk 50 of Sweden's *Flygvapen*, all of which should add up to an excellent product. Unfortunately it suffers one serious shortcoming: the fit of the component parts is, frankly, appalling! Literally every major joint has to be filed or filled or both! For a Frog kit this failing is most unusual, yet it is no exaggeration to say that the effort involved in getting the Vampire's joints to fit even reasonably well is such as to spoil the pleasure in assembling what is in all other respects a very good kit, which, at 20 pence, is outstanding value. We can only hope that Frog will rework this kit as soon as possible as, in its present form, it will have an adverse effect on its manufacturer's reputation, particularly among younger modellers.

Frog's companion release, the Messerschmitt Bf 110G, which, like the Vampire, is to 1/72nd scale, is of generally good quality though far from faultless. It is accurate in outline, except for the engines which, in side elevation, present a distinctly odd 'peaked' profile. Fortunately, the correct shape is shown by the box art, and, although by no means easy, the necessary correction can be made by filing. What surface detail there is is fine, but the fuselage itself is devoid of detail, endowing the model with a somewhat naked look. On the other hand the detail of the minor parts, such as the exhausts, the radar aerials, and the undercarriage legs, is very well executed. The fit of the component parts is a good deal

better than that of the Vampire, though care should be taken with the wing root joint as careless assembly will result in insufficient dihedral. The decal sheet provides the markings of a Bf 110G-6 flown by Major Heinz-Wolfgang Schnaufer (one of only two *Luftwaffe* night fighter pilots to be awarded the Oak Leaves with Swords *and* Diamonds) while serving with NJG 4, and a Bf 110G-2 on the strength of NJG 200. At its UK price of 47 pence Frog's Bf 110G represents good value for money and is well worth buying.

A new gleam
"The answer to a modeller's prayer." This description, on the wrap of a new US product named "Bare Metal", is fully justified. "Bare Metal" is a very thin self-adhesive aluminium foil which is supplied in sheets of 14 in by 7 in (35,56 cm by 17,78 cm) with a waxed-paper backing. The piece of foil required is cut to shape complete with backing. The backing is then peeled off and the foil is pressed down on the model with the fingertips, the operation being completed by burnishing with a very soft cloth.

Similar products have appeared on the market previously, but none of them until now has been really practical on account of excessive thickness or stiffness. "Bare Metal" is both thin and pliable, and will take up complex double-curved shapes with ease. Even if it does crease slightly during application, it is so thin that the creases are virtually invisible to the naked eye. Ease of application is its greatest asset, but it is essential that the surface to which it is to adhere should be absolutely free from any grease or dust which will nullify the adhesive qualities of the product.

Our only criticism is that after application "Bare Metal" is rather *too* shiny, but the manufacturer, recognising this, suggests that it be coated with a dulling varnish before removal from the backing sheet. "Bare Metal" may be obtained in the UK from the sole agent, D S Ives (Foil), 104 Longwood Gardens, Clayhall, Ilford, Essex, the price being 47 pence per sheet.

Dora and Ram
Last month we reviewed Fujimi's 1/50th (but ostensibly 1/48th) scale kit of the Focke-Wulf Fw 190A, and this has now been joined by the Fw 190D, or *Dora*, from the same manufacturer. From the viewpoint of accuracy, Fujimi's Fw 190D is generally good, although the nose contours are too rounded and the wingtips too square, while the undercarriage main members are mounted too far aft and the cover plates are incorrectly shaped. Except for the undercarriage positioning, all these errors *can* be corrected, though modifying the nose contours is no simple task. Considered purely as a kit, this Fujimi product is superb; beautifully moulded with precise component part fitting and a wealth of surface detail of a standard that no other kit manufacturer seems to be able to equal.

A tiny electric motor is included with the kit, together with a holder for its battery, the whole forming a unit which may be slipped into the fuselage through the detachable radiator cowling. The decal sheet is truly

magnificent. It is beautifully printed and provides colourful markings for at least five individual aircraft, these including *Werk-nummern* and stencilling. However, illustrations of only three machines are carried by the box, and none of these sports the most interesting markings offered by the decal sheet. Diligent research has so far failed to establish details of those machines that go unillustrated, and this is a tragedy as it means that some superb decals must be wasted. Fujimi's Fw 190D kit is available in the UK from VHF Supplies at £1.20.

Italaerei's 1/72nd scale kit of the Caproni Reggiane Re 2000 has evolved as several different versions of the basic Roberto G Longhi design, and the latest of these to appear is the Re 2002 Ariete II (Ram II) fighter-bomber. Accurate in outline, neatly pressed in good quality dark green plastic, easily assembled and finely detailed, Italaerei's kit makes up into an attractive model. The canopy is very clear, the neatly-printed decal sheet provides *Regia Aeronautica*, *Luftwaffe* and Co-belligerent Air Force markings, and the informative instruction sheet includes good drawings which show the colour schemes applicable to the three services by which the Ariete II was used. This kit is well worth having; unfortunately we do not know its UK price although we believe it to be available from Model Toys of 246 Kingston Road, Portsmouth, Hants.

Hasegawa's big boat
Many are the lost causes of aviation, and numbered among these is often the flying boat, yet waterborne aircraft linger on even if the number of countries still engaged in actively developing and manufacturing such aircraft in amphibious or non-amphibious form has been reduced to two — Japan and the Soviet Union, producing the Shin Meiwa PS-1 and the Beriev Be-12 respectively. While the latter is unlikely ever to be represented in the plastic kit manufacturers' lists, the PS-1 now has that distinction, being offered by Hasegawa to 1/72nd scale.

To any modeller with a knowledge of Hasegawa products it will come as no surprise to learn that this company's PS-1 is one of the finest kits on the market. Moulded in gull grey plastic of excellent quality, its component parts fit together with such precision as to render their assembly remarkably simple despite the fact that there are over 100 of them. The flaps operate, and do so efficiently, while alternative parts are included to permit completion of the model in the projected SS-2 amphibious ASR version of the basic design. The airscrew blades are moulded as separate parts which, if rendering airscrew assembly a little more complicated, makes for a very much higher standard of accuracy.

The outline of the PS-1 model is truly accurate, and the surface detail is beautifully executed. The decal sheet, which provides markings for the two prototypes and the first pre-series aircraft, is excellent, and the instruction sheet, which is in English, incidentally, is easy to follow. The rub, in so far as the UK modeller is concerned, is the price of £3.75 from VHF Supplies, but the quality of this kit is such that it is *still* good value for money. □   W R MATTHEWS

A VIOLENT FINNISH WIND

For the Axis powers a sense of foreboding must have accompanied the spring of 1943. Dramatic events were in the offing and it was unrealistic to imagine that any of them would favour the Axis cause. In the Pacific, Rear Admiral Koyanagi had skilfully evacuated the Japanese forces from Guadalcanal but Admiral William F Halsey was obviously gathering his forces in preparation for a drive northwestward through the Solomons; the Allies had at last gained air superiority over North Africa where the final offensive against the Axis forces was about to be launched, and although Erich von Manstein had miraculously halted the Soviet drive and recaptured Kharkov, few believed that the German front could withstand another Soviet offensive.

At this momentous period in WW II, when the tide had obviously turned in favour of the Allies, few eyes were following the course of events in the extreme north, on the Russo-Finnish front. The winter months had seen only limited and relatively unspectacular fighting between Soviet and Finnish forces, but with the spring thaw had come a rapid increase in aerial activity on the part of the Soviet Air Forces, the *Voenno-vozdushniye Sily* (V-VS). A noticeable improvement in the quality of Soviet pilots assigned to the Russo-Finnish front and in the aircraft that they flew had come about during the closing months of 1942, and the Finnish air arm, *Ilmavoimat*, had been particularly concerned by the début in some numbers of the Lavochkin La-5 fighter. The La-5 had proved itself more than a match for the ageing mélange of fighters with which *Ilmavoimat* was equipped.

The supply of more modern fighters was urgently requested from Germany, and at the beginning of March an initial batch of 16 Messerschmitt Bf 109G-2s had reached Finland, but past experience had taught the Finns the need for a measure of self-sufficiency in the supply of combat aircraft, and early in 1943 the State Aircraft Factory (*Valtion Lentokonetehdas*) at Tampere was instructed to give the highest priority to the design and construction of a modern fighter suitable for both intercept and close-support rôles. The fighter, which was eventually to be named the Pyörremyrsky (Whirlwind), had to place emphasis on the use of indigenous materials coupled with suitability for operations from the small Finnish frontline fields under the most severe climatic conditions.

The State Aircraft Factory was not devoid of experience in the design and development of fighters. In the late 'twenties prototypes of the D.26 Haukka I and D.27 Haukka II had been built, and tooling had just commenced for the series production of the Myrsky (Storm) single-seat fighter powered by a 1,065 hp SFA STWC3-G 14-cylinder air-cooled radial, a Swedish-built copy of the Pratt & Whitney SC3-G Twin Wasp. However, it was patently obvious that the Myrsky would be no match for the La-5 and other modern Soviet fighters that had begun to appear in Finnish skies, and that the development of a more advanced successor was a matter of paramount importance.

The task of designing the Pyörremyrsky was entrusted to Dipl-Ing Torsti R Verkkola, who had graduated from the Technical University of Finland in 1935 and had undertaken post-graduate studies at the Technical University of Berlin. From 1937 he had assisted in the design of several aircraft at the State Aircraft Factory, aiding Dipl-Ing A Ylinen with the Pyry advanced trainer and Dipl-Ing E Wegeluis with the Myrsky fighter, before, at the beginning of 1943, he had been appointed Director of the Design Department.

Many lessons had been learned from the construction and testing of the Myrsky, which, in its initial form, had suffered more than its share of teething troubles. The undercarriage had proved to be too weak, defective bonding had resulted in the plywood skinning peeling off under stress, and fractures of the wing attachment bolts had necessitated the complete redesign and re-stressing of the wing, while inaccurate calculations of drag and weight had resulted in a performance appreciably below that anticipated. Therefore, despite the urgency attached to the development of the Pyörremyrsky, Dipl-Ing Verkkola and his team exercised extreme care in the detail design of the new fighter and with the stress and weight analyses in order to ensure no repetition of the problems encountered with the Myrsky.

The power plant selected for the Pyörremyrsky was the Daimler-Benz DB 605 12-cylinder inverted-vee liquid-cooled engine which, driving a VDM three-bladed metal constant-speed airscrew, offered some commonality with the Messerschmitt Bf 109Gs then entering service with *Ilmavoimat*. Around this engine was designed an extremely clean and compact fighter of mixed construction. The single-spar wooden wing, which had a few degrees of anhedral on the inboard sections resulting in a slight reverse gull effect, had plywood skinning with metal-framed, fabric-covered ailerons and electrically-operated light-alloy flaps. The forward fuselage was a chrome-molybdenum steel-tube structure with detachable metal panels and the aft fuselage was a wooden monocoque of pine and ply. The tail assembly was a ply-covered wooden structure with metal-framed fabric-covered movable surfaces, the fin being built integral with the fuselage.

The relatively spacious cockpit was enclosed by an aft-sliding canopy, and offered excellent visibility in all directions. Aft protection was provided the pilot by a 10-mm armour bulkhead immediately behind the seat, and the radio equipment and most instrumentation were of national origin. Some 101 Imp gal (460 l) of fuel were housed by a self-sealing tank behind the armoured bulkhead, and provision was made for two 33 Imp gal (150 l) drop tanks beneath the wing. Proposed armament comprised an engine-mounted 20-mm MG 151 cannon and two 12,7-mm LKK-42 machine guns mounted in the upper decking of the fuselage nose. For the close-support rôle provision was made for two 441-lb (200-kg) bombs beneath the wing.

Prototype construction was slow, the State Aircraft Factory being preoccupied during the first eight months of 1944 with production of the Myrsky, the adaptation of Morane-Saulnier M.S.406s to *Mörkö-Moraani* standard (with Klimov VK-105P engine and VISh-61P airscrew), and the overhaul and repair of *Ilmavoimien* aircraft. The last-

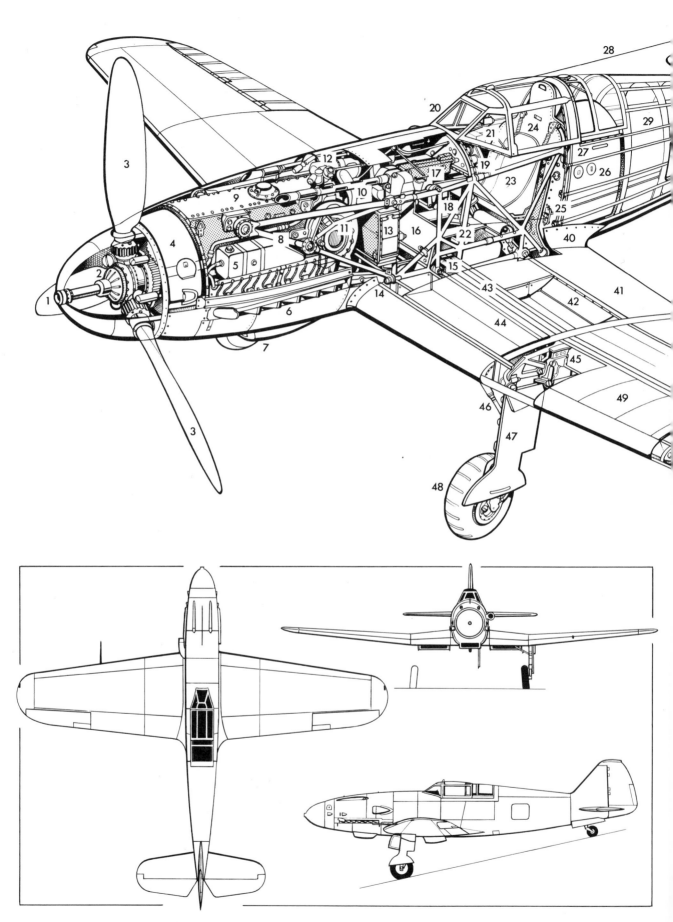

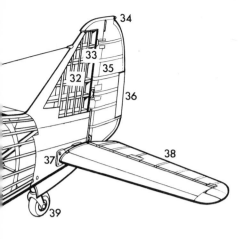

VL Pyörremyrsky Specification

Power Plant: One Daimler-Benz DB 605AC 12-cylinder inverted-vee liquid-cooled engine rated at 1,475 hp for take-off and 1,355 hp at 18,700 ft (5 700 m).

Armament: (proposed but not fitted to the prototype) One engine-mounted 20-mm MG 151 cannon and two 12,7-mm LKK-42 machine guns, plus wing pylons for two 441-lb (200-kg) bombs.

Weights: Empty equipped, 5,774 lb (2 619 kg); normal loaded, 7,297 lb (3 310 kg); wing loading, 35·68 lb/sq ft (174,2 kg/m²).

Performance: (At normal loaded weight) Max speed, 324 mph (522 km/h) at sea level, 385 mph (620 km/h) at 21,000 ft (6 400 m); endurance (internal fuel), 1 hr 30 min, (with auxiliary drop tanks), 2 hr 25 min; climb to 16,400 ft (5 000 m), 4 min 30 sec; service ceiling, 36,910 ft (11 250 m).

Dimensions: Span, 34 ft 0⅜ in (10,38 m); length, 32 ft 3¾ in (9,85 m); height, 12 ft 9¼ in (3,89 m); wing area, 204·514 sq ft (19,00 m²).

mentioned occupation took absolute priority during the violent defensive battles of the summer of 1944 over the Karelian Isthmus when the V-VS enjoyed complete air supremacy. By 4 September 1944, when the Finnish government was compelled to accept Soviet surrender terms, work on the Pyörremyrsky prototype, which had languished for several months, had been brought to a halt.

Shortly after *Ilmavoimien* units returned to their peacetime bases, on 20 January 1945, after operating in support of Finnish forces fighting against the *Wehrmacht* in accordance with the preliminary Armistice terms, work on the Pyörremyrsky was, somewhat surprisingly, resumed by the State Aircraft Factory. A Daimler-Benz DB 605AC engine, which had been removed from a Bf 109G and overhauled, was installed in the prototype airframe, this engine being rated at 1,475 hp for take-off and 1,355 hp at 18,700 ft (5 700 m), and after preliminary taxying trials at Tampere, the fighter was flown for the first time on 21 November 1945 by Capt Esko Halme, the chief test pilot of the State Aircraft Factory.

By comparison with the Bf 109G-6, the Pyörremyrsky proved to possess markedly superior taxying and take-off characteristics, and once in the air was found to have light, responsive controls. Handling was outstanding under all flight conditions; it could outclimb the Bf 109G-6 and was more manœuvrable than the Messerschmitt fighter. The Pyörremyrsky suffered extremely few teething troubles, and those troubles that were encountered were minor in nature. Speed and climb trials were completed by Capt Halme, and four other test pilots had performed familiarisation flights when the decision was taken to terminate the Pyörremyrsky test programme. At this time the prototype had logged 30 hours flying, and its few teething troubles had been largely eradicated, but it was obvious that series production was out of the question. No funds were available for the purchase of new aircraft for *Ilmavoimat,* and, in any case, more than sufficient Bf 109G fighters remained in the inventory to equip the drastically reduced fighter component that was likely to be permitted *Ilmavoimat.* Even had it been possible to manufacture a small series of Pyörremyrsky fighters, redesign for an alternative power plant would have been necessary as there was no likelihood of obtaining a supply of Daimler-Benz engines. Thus, the brief career of the Pyörremyrsky came to an end without this Finnish fighter having any opportunity to show its capabilities. Many years later, Capt Halme recalled the Pyörremyrsky with enthusiasm, commenting: "It was undoubtedly one of the most pleasant and efficient piston-engined aircraft I have ever piloted. If only it had been available in quantity in the summer of 1944 . . .!" □

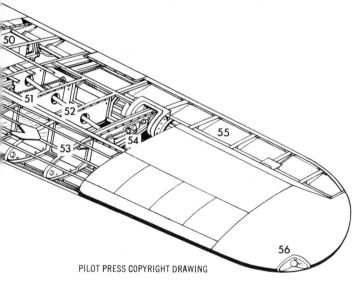

PILOT PRESS COPYRIGHT DRAWING

1 MG 151 cannon muzzle
2 Variable-pitch mechanism
3 VDM airscrew
4 Oil tank
5 Coolant header tank
6 Ejector exhausts
7 Oil cooler intake
8 Engine bearers
9 Daimler-Benz DB 605AC 12-cylinder inverted-vee liquid-cooled engine
10 Port 12,7-mm LKK-42 machine gun
11 Supercharger
12 Machine gun synchroniser
13 Machine for MG 151 cannon
14 Engine bearer/fuselage connection
15 Wing/fuselage bolt
16 Stepped firewall
17 Breech of LKK-42 gun
18 Forward fuselage frame
19 Instrument panel
20 Windscreen
21 Reflector sight
22 Adjustable rudder pedals
23 Pilot's seat
24 Armoured bulkhead
25 Forward/aft fuselage joint
26 Fuel tank
27 Canopy track
28 Aerial
29 Monocoque aft fuselage
30 Access panel
31 Tailwheel actuating lever and rudder control run
32 Integral tail fin
33 Rudder post
34 Rear navigation light
35 Rudder
36 Trim tab
37 Adjustable tailplane
38 Elevator
39 Retractable tailwheel
40 Wing fillet
41 Port flap
42 Underwing cooling radiator
43 Laminated wooden mainspar
44 Retraction gear torsion tube
45 Undercarriage bracket
46 Undercarriage retraction strut
47 Mainwheel door
48 Mainwheel
49 Wing skinning
50 Flap torsion bar
51 Trim tab control wire
52 Aileron control rod
53 Wooden wing construction
54 Aileron control lever
55 Metal-framed aileron
56 Port navigation lamp

PLANE FACTS

A pair from Hiro

Is it possible to cover in your "Plane Facts" column the Hiro G2Y and the Hiro (I believe) H7Y flying boat? I should like to see a photograph, a specification, and a three-view line drawing of each of these.

Charles G James,
San Antonio, Texas 78206

The Hirosho G2H1 (*not* G2Y) was designed by Lieut Cdr J Okamura under the supervision of Cdr Misao Wada at the 11th Naval Air Arsenal (*Dai-Juichi Kaigun Kokusho*) at Hiro as the 7-*Shi* Experimental Attack Bomber. This project, the prototype of which was completed on 29 April 1933, was to exercise considerable influence on the development of Japanese Naval shore-based bombers. Of all-metal construction and of relatively advanced concept for its time, the G2H1 prototype was powered by two supercharged Rolls-Royce Buzzard 12-cylinder liquid-cooled engines each rated at 955 hp at 2,300 rpm at sea level and 825 hp at 2,000 rpm.

The G2H1 was accepted by the Japanese Navy as the Type 95 Attack Bomber, the production model being powered by two Aichi Type 94 liquid-cooled engines each rated at 900 hp at 2,100 rpm and 1,180 hp at 2,300 rpm. A crew of seven was carried, offensive load comprised six 551-lb (250-kg) or four 882-lb (400-kg) bombs, and defensive armament consisted of one 7,7-mm machine gun in an open nose position, a similar weapon in a retractable ventral "dustbin", and twin 7,7-mm guns in an open dorsal position. Performance included a max speed of 152 mph (245 km/h) at 3,280 ft (1 000 m), a long-range cruising speed of 104 mph (167 km/h), the ability to climb to 9,840 ft (3 000 m) in 9 min 30 sec, and normal and maximum ranges of 671 mls (1 080 km) and 967 mls (1 557 km) respectively. Empty and loaded weights were 16,682 lb (7 567 kg) and 24,251 lb (11 000 kg), and overall dimensions included a span of 103 ft 11¼ in (31,68 m), a length of 66 ft 1¼ in (20,15 m), a height of 20 ft 7¼ in (6,28 m), and a wing area of 1,506·95 sq ft (140 m²). The Mitsubishi Jukogyo KK participated in production of the G2H1, but only eight bombers of this type were built.

(Above and below) The Japanese Navy's Type 95 Attack Bomber, the Hirosho G2H1, only eight examples of which were built.

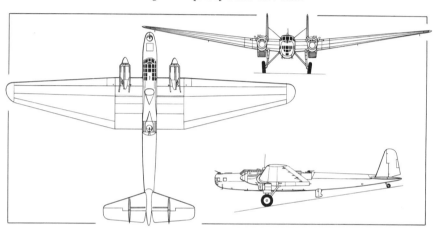

The H7Y1 12-*Shi* Experimental Special Flying Boat was developed in great secrecy by the 1st Naval Air Arsenal (*Dai-Ichi Kaigun Kokusho*). The project was initiated in 1937 with the aim of providing the Japanese Navy with a high performance reconnaissance flying boat with a range sufficient for reconnaissance missions as far as Hawaii from bases in the home islands. Power was provided by four 605 hp Junkers Jumo 205C six-cylinder Diesel engines selected for their low fuel consumption, four crew members were carried, and loaded weight was 39,683 lb (18 000 kg). Emphasis was placed on lightweight construction, but prototype trials revealed a performance substantially below specification and poor stability, and the H7Y1 proved unacceptable, development being abandoned and the prototype scrapped. No photographs or reliable general arrangement drawings of the H7Y1 are known to exist.

Baby Bristol

Will you please publish in "Plane Facts" any information that you may have on the Bristol Babe biplane (1919), and particularly the version powered by the Siddeley Ounce engine. Can you tell me the colour that the Ounce-engined Babe was doped?

W Sneesby, Brayton Selby, Yorks

The Bristol Babe single-seat light sports biplane designed by Frank S Barnwell in 1919 was originally intended for the 60 hp ABC Gadfly five-cylinder air-cooled radial engine, but ABC Motors abandoned aero engine manufacture at this time, and it was proposed, therefore, that one of two Babes put in hand in April 1919 should be temporarily fitted with a 45 hp Viale engine of 1910 vintage for initial trials. The initial flight with the Viale engine was made on 28 November 1919, and, in the meantime, a third Babe had been built specifically as a test-bed for the new 40 hp Siddeley Ounce flat-twin engine. This third aircraft, with an incomplete Ounce installed, was exhibited at the *Salon de l'Aéronautique* held in Paris in December 1919. The Ounce-engined aircraft was known as the Babe Mk II, but, in the event, the Siddeley engine was never

The Bristol Babe Mk I with a 45 hp Viale engine flown on 28 November 1919.

developed to the flight test stage and, in consequence, this aircraft was never registered or flown.

The new lightweight Le Rhône rotary engine of 60 hp was finally selected as a replacement for the ABC Gadfly, and the first two Babes were fitted with this type of power plant and registered as G-EASQ and G-EAQD, the latter having been the original Viale-engined example. With the Le Rhône engine installed the aircraft became the Babe Mk III, and subsequently G-EASQ was fitted with a thick-section cantilever monoplane wing, although it was not flown after this modification. The registration of Babe Mk III G-EAQD lapsed in December 1920, and that of G-EASQ was cancelled in February 1921. The Babe Mk III had an empty weight of 460 lb (209 kg) and a loaded weight of 840 lb (381 kg), and allegedly attained a maximum speed of 107 mph (172 km/h). Overall dimensions were: span, 19 ft 8 in (5,99 m), length, 14 ft 11 in (4,55 m), height, 5 ft 9 in (1,75 m), wing area, 108 sq ft (10,03 m²).

We have been unable to confirm the colours in which the Babes were doped but believe all three were painted red overall. Perhaps one of our readers can assist Mr Sneesby on this point?

Bulgarian recce biplane

I am engaged in researching the history of Bulgarian military aviation, and I shall be grateful if, through your "Plane Facts" column, you can identify for me a single-bay biplane with a radial engine and fixed, spatted undercarriage, which, of Bulgarian design, apparently served in some numbers with the Royal Bulgarian Air Force in the short-range reconnaissance rôle. If you can positively identify the type, I shall be glad if you will publish a photograph of this aircraft together with some details. Can you also publish an illustration of the national insignia used by the Royal Bulgarian Air Force prior to the adoption of the diagonal white cross on a square white field during WW II?

B K Ellis, Croydon, Surrey
CR9 2AR

The aircraft most likely to be that you are endeavouring to identify, Mr Ellis, is the LAZ-3 alias DAR-3a Garvan (Raven) designed by Prof-Ing Svetan Lazarov and built by the State Aircraft Factory (Darjavna Aeroplanna Rabotilnitza) at Bojourishte, Sofia, during 1939-40. The initial model of the LAZ-3 appeared in 1929 with a 400 hp Lorraine-Dietrich 12-cylinder inline liquid-cooled engine, and the basic tandem two-seat single-bay biplane of mixed construction was progressively updated. Embodying some structural strengthening, the LAZ-3 appeared in 1937 with a 480 hp Gnôme-Rhône Jupiter radial, this version being followed by the LAZ-3a or DAR-3a initially powered by a 655 hp Wright Cyclone engine. The definitive version introduced enclosed cockpits, a long-chord cowling, and a 750 hp Alfa-Romeo 126 nine-cylinder radial driving a three-bladed two-pitch airscrew. In this form the aircraft was operated in some numbers for tactical reconnaissance. Unfortunately no specification for the LAZ-3a is available.

The national insignia employed on the wings and fuselage of aircraft of the Royal Bulgarian Air Force immediately prior to the adoption of the black diagonal cross is illustrated by the accompanying photograph. This insignia was in red, green and gold on a circular white field, and was accompanied by (top to bottom) white-green-red horizontal tail striping. The rampant crowned lion (which was sometimes omitted from the wing marking) always faced forward on the fuselage sides and was superimposed on a red disc outlined in gold and, in turn, encircled by a broader green ring, also outlined in gold. The arms of the cross were red and the crossed swords were gold.

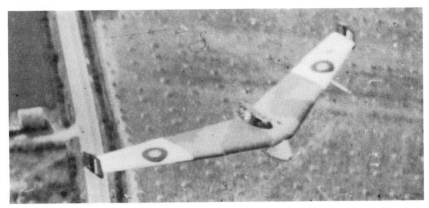

The Baynes "Bat", seen in towed flight above, was a one-third flying scale model of the projected "carrier wing" which it was intended to fit to tanks.

Baynes Bat

Please can you identify the accompanying three-view silhouette which is labelled "Experimental Aeroplane No 209"? I believe that this aircraft was a Baynes project.

R Coles, Brighton,
Sussex BN1 5GH

The silhouette drawing accompanying your letter, Mr Coles, depicts the Baynes Bat, a reduced-scale test vehicle for the "carrier wing" scheme proposed in 1941 by L E Baynes of the Aircraft Section of Alan Muntz & Company of Heston. This scheme envisaged assisting the aerial movement of armoured fighting vehicles by attaching wings directly to the vehicles themselves. The wing was envisaged as a single-spar structure with endplate vertical surfaces and differentially-operated elevons. To investigate the aerodynamic properties and practicability of the project, a one-third scale flying model was built by Slingsby Sailplanes, and was dubbed unofficially the "Bat", the first test flight being undertaken by Flt Lt Robert Kronfeld at Sherburn-in-Elmet, Yorkshire, in July 1943.

The Bat weighed 763 lb (346 kg) empty and 963 lb (437 kg) loaded, and dimensions included a span of 33 ft 4 in (10,16 m), a length of 11 ft 4¼ in (3,46 m), and an area of 160 sq ft (14,86 m²). A small nacelle housing the pilot took the place of the armoured vehicle, and flight trials showed that control

(Above) The insignia of the Royal Bulgarian Air Force and (below) the LAZ-3 alias DAR-3a Garvan two-seat tactical reconnaissance biplane.

in both towed and free flight was satisfactory, and control loads light. Stability was good throughout the speed range, maximum speed attained being 120 mph (193 km/h), and control was good down to the stalling speed of 40 mph (64 km/h). No full-scale carrier wing was built owing to the decision to transport armoured vehicles in large gliders.

Berlin bomber

Can you please include in "Plane Facts" a photograph and details of the Tabor triplane developed during World War I to bomb Berlin from bases in Britain?

Gerry Baylis, Sherman Oaks, California 91403

The Tarrant Tabor, designed by W G Tarrant, built in close collaboration with the RAE and assembled in the balloon shed at Farnborough, was an immense triplane intended to bomb Berlin and built under a wartime contract but not completed until early 1919. The Tabor was originally designed to be powered by four 600 hp Siddeley Tiger engines, but as these did not come up to expectations the prototype Tabor was completed with six 450 hp Napier Lions, four of these being mounted in two tandem pairs between the bottom and middle wings, and the other two being mounted between the middle and top wings. Provision was made for 1,600 Imp gal (7 273 l) of fuel, and it was calculated that, with a 4,480-lb (2 032-kg) bomb load, the Tabor would have a range of 900 miles (1 448 km) at 112 mph (180 km/h) increasing to 1,090 miles (1 754 km) by cruising on four engines at 91 mph (146 km/h). It was believed that an altitude of 5,000 ft (1 524 m) would be attained in 10·5 min and 10,000 ft (3 048 m) in 33·5 min, the service ceiling of 13,000 ft (3 962 m) being reached in just over one hour. Empty and loaded weights were 24,750 lb (11 226 kg) and 44,672 lb (20 263 kg) respectively, and overall dimensions were: span, 131 ft 3 in (40,00 m), length, 73 ft 2 in (22,30 m), height, 37 ft 3 in (11,35 m), wing area, 4,950 sq ft (459,82 m²).

On 26 May 1919, the Tabor was readied for its first flight test, and after some preliminary taxying trials, the pilot, Capt F G

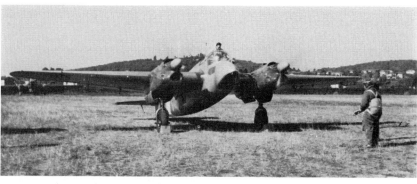

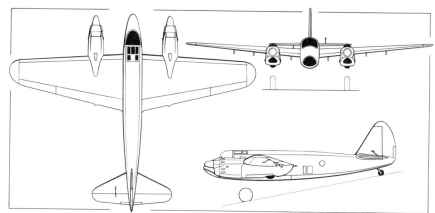

(Above and below) The Savoia-Marchetti S.M.86 single-seat dive bomber in its definitive form with Isotta-Fraschini A.120 R.C.40 air-cooled engines.

Dunn, increased speed, lifted the tail from the ground, and opened up the two upper engines. The Tabor immediately nosed over, slid some yards along the ground until the nose collapsed, and finally came to rest on the leading edges of its wings with the fuselage vertical. Both first and second pilots sustained fatal injuries, and further development of the Tabor was abandoned.

Dive-bombing twin

I have recently seen reference to a dive bomber designated Savoia-Marchetti S.M.86 but have been unable to discover any information concerning this aircraft other than the fact that it had in-line engines. Was this aircraft related in any way to the S.M.85 dive bomber? Can you possibly publish some details of this aircraft and perhaps a general arrangement drawing?

Norman D Allen, Newport, Mon

The Savoia-Marchetti S.M.86 single-seat dive bomber was generally similar in concept to the earlier S.M.85 but aerodynamically a much cleaner design. The first prototype S.M.86 was flown on 8 April 1939 with two Walter Sagitta I-MR 12-cylinder inverted-vee air-cooled engines each rated at 600 hp for take-off with a maximum continuous rating of 550 hp. As the S.M.86 proved to be underpowered with these engines they were replaced at an early phase in the test programme by two Isotta-Fraschini A.120 R.C.40 12-cylinder inverted-vee air-cooled engines each rated at 770 hp for take-off and 710 hp at 13,125 ft (4 000 m).

The S.M.86 was primarily of wooden construction, and carried a single 551-lb (250-kg) or 1,102-lb (500-kg) bomb internally, and a single 20-mm cannon was to have been mounted in each wing root. It emulated the S.M.85 in being extremely stable and possessing excellent diving characteristics, but performance was unspectacular and no production of the S.M.86 was undertaken. The overall dimensions of the S.M.86 included a span of 45 ft 11¼ in (14,00 m), a length of 35 ft 9⅛ in (10,90 m), a height of 10 ft 11⅞ in (3,35 m), and a wing area of 331·53 sq ft (30,80 m²). Empty and loaded weights were 7,401 lb (3 357 kg) and 11,193 lb (5 077 kg), and performance included a maximum speed of 270 mph (435 km/h), a normal cruising speed of 209 mph (336 km/h), and a service ceiling of 20,670 ft (6 300 m).

The Tarrant Tabor long-range heavy bomber after being rolled out at Farnborough in the spring of 1919.

TALKBACK

In Yugoslav skies

WHILE perusing your feature on the Caproni light twins (July issue) I note that you have omitted one small detail. At the beginning of 1941, the Yugoslav Royal Air Force (JKRV) used 36 Ca 310s and negotiations for the purchase of the Ca 311 were well advanced.

Your readers may care to have details of the order of battle of the Hurricanes of the JKRV prior to the German attack to supplement the details of the Messerschmitt Bf 109 in Yugoslav service (September issue). At the beginning of 1941, all Hurricanes except one or two experimental ones in the *Opitna grupa* (Test Group) at Zemun, and a few used for training, were included in the 4.*Vazduhoplovni lovački puk* which had 46 battle-ready fighters of this type. One Group, the 52.*Lovačka grupa*, was subsequently transferred from this Fighter Regiment to the newly-formed 2.*Vazduhoplovni lovački puk*, the 52nd Group having 18 Hurricanes on strength on 6 April 1941. The 4th Fighter Regiment had received some IK-2s, and thus on the first day of hostilities comprised the 14 Hurricanes of the 33.*Lovačka grupa* and the six Hurricanes of one squadron of the 34.*Lovačka grupa*, the other squadron of this Group being equipped with eight IK-2s. The Fighter School at Mostar had two Hurricanes, three Bf 109Es and two Avia BH-33s in its inventory. One Hurricane, one Bf 109 and both Avias were shot down during the brief conflict, but the second school Hurricane was seized by the Italians, this being illustrated by the accompanying photo, still having Yugoslav markings.

Zoran Jerin
61001 Ljubljana

While in transit

HAVING recently obtained the AIR ENTHUSIAST for the first time, I think that fellow readers may find interest in the accompanying photograph of a Cessna 337 of the *Fuerza Aerea Ecuatoriana* taken at Guayaquil in April. Despite being plainly labelled as belonging to the FAE, it carries civil registration letters (HC-CYA), while the TAME DC-6B that can be seen in the background bore both a military registration (FAE 45535) and a civil registration (HC-AVG).

Another interesting item seen recently

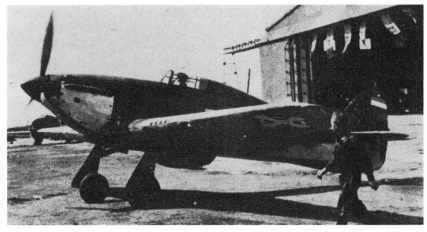

This Hawker Hurricane, one of two at the Yugoslav Royal Air Force Fighter School at Mostar in 1941, was captured by Italian forces when they occupied the base, and in this picture is being operated by Italians, despite the Yugoslav markings.

was a trio of three Tupolev Tu-16s in the insignia of the Libyan Arab Air Force. The aircraft were at the Tripoli airport on 25 June when I arrived and were still there four days later when I left. They were each camouflaged in a different fashion, one being grey-green, another having desert camouflage of sand and brown, and the third being painted in two shades of green in irregular vertical stripes. Flying around was a Bell 206 in Libyan Arab Air Force insignia and the serial U824, and a military Alouette III with desert camouflage but without serial.

Claudio Maranta
00182 Roma, Italy

Airpower over Bangla Desh

IT HAS occurred to me that readers of the AIR ENTHUSIAST may be interested to learn something of the use of airpower over Bangla Desh, or East Pakistan as it is still *officially* known, as this facet of the troubles has received little publicity. During the first seven weeks after the secession of the territory, all PIA Boeing 707s were pressed into service as troop carriers, flying from Karachi to Dacca via Colombo, as the Pakistan Air Force's single C-130 Hercules squadron could not cope with the requirement. It seems certain that Turkish and Iranian C-130s, too, participated in the airlift in the wake of earlier, thinly-disguised air defence exercises involving these CENTO signatories. There is evidence that

the PAF's single F-86 Sabre squadron in East Bengal was reinforced by a MiG-19 squadron, and both saw much action, flying up to 30-40 combat sorties during an average day on strafing, rocketing and napalm missions, while PIA F-27 Friendships and PAF HH-43 Huskies carried troops into the various sectors.

At the time of writing, the PAF had lost three or four aircraft to groundfire or in accidents, and at least three air bases (including those at Lalmonirhat and Shalutikar) had been briefly overrun by Bangla Desh freedom fighters. An unconfirmed report has suggested that two PAF Sabres flown by Bengali pilots defected to the Indian half of Bengal, but what is confirmed is the fact that all Bengali PAF personnel have been "grounded" and placed under virtual house arrest. The Pakistanis are obtaining 18 helicopters (possibly AB 206 JetRangers) fitted with 12,7-mm machine guns for counter-insurgency operations through Iran, and these are presumably intended for use in Bangla Desh. Incidentally, earlier this year, between 24 April and 27 April, a joint Iranian-Pakistani air exercise (IRPAK) was held in the Lahore sector, and it was reported that five of the Northrop F-5s of the IIAF that participated were subsequently evaluated by the PAF to assess their suitability for use in campaigns such as that in East Bengal.

P S Chopra
New Delhi 3

German data

WHILE this may not be the first *Letter to the Editor*, I don't want to be the last to offer my congratulations for a fine magazine, which shows from the outset that it will be a roaring success and will become a "standard" in the industry. Perhaps your readers may be interested in the following comments regarding the AEG fighters ("Fighter A to Z", AIR ENTHUSIAST June 1971).

AEG DI. The Type Test was expected in August-September 1917 after the fuselage had been lengthened 40 cm. Both top and bottom wing possessed only one spar. By

The Cessna 337 Super Skymaster of the Fuerza Aerea Ecuatoriana referred to in the letter from Mr C Maranta above.

29 August 1917 it had not passed its Type Test because it was difficult to fly. Lt Hendrichs, a competent pilot, crashed the day before completion of the Type Test. Even so, a series of 20 aircraft was ordered for front-line evaluation. This order was finally cancelled after the crash of Ltu Höhndorf in an AEG DI on 5 September 1917, with Jasta 14.

AEG DRI This machine was better in climb than the DI but much slower (as was to be expected!)

AEG PE Tested in March 1918. Easy to fly but not particularly fast. Climb, 8,200 ft (2 500 m) in 18 min, 13,123 ft (4 000 m) in 48 min. Fuselage too short giving poor flight characteristics. Tests abandoned in June 1918.

AEG DJ Nr.1 Under construction in April 1918. Completed in May 1918, first flight in July 1918. Still under test September 1918.

AEG DJ Nr.2 A second armoured single-seater was under construction in September 1918. Completion date not known. Engine Bz IIIb.

AEG DJ Nr.3 With 240 hp Maybach Mb IVa engine, under construction May/June. New engine installed in September 1918. No record of flight, but it seems engine was tested in aircraft.

Officially the designation "DJ" was used interchangeably with *D-Flugzeug gepanzester Doppeldecker*. But I do not believe that the designation DJ I was used. Rather DJ was a stenographic shortcut in official documents. The DJ 1 on the fuselage was an AEG appellation.

Peter M Grosz
Berlin

Still room for pilots

IN VIEW of your stated editorial policy of completeness and accuracy in reporting aviation developments, you should advise Roy Braybrook, author of the article "The Fight for the Skies" (AIR ENTHUSIAST June 1971) that he has not done all his homework.

In the concluding paragraph of his article on page 38, under the subheading of "Curtains for Knuckleheads?", the author implies that Remotely Piloted Vehicles (RPVs) could do away with the need for the pilot in the fighter cockpit. Not so!

The RPV is not intended to compete with pilots. It can take on certain missions under very hazardous conditions, and it can, under some circumstances, perform a lead rôle to be followed up by manned aircraft. The RPV is complementary to manned aircraft as well as to the ballistic field. Actually, drones have been flying many kinds of missions for several decades past. During those same decades, the need for piloted fighters has not diminished.

R R Schwanhausser
Vice-president, Aerospace Systems
Teledyne Ryan Aeronautical
San Diego, California

Un chasseur pas unique

MAY I query the statement (July issue) that the Arsenal VB 10 "was unique at the time of its conception and was to remain so". Surely the Kawasaki Ki-64, which was originated in 1939 and flew in December 1943, used essentially the same propulsion system, its 2,350 hp Kawasaki Ha-201 being, in fact, two Ha-40 engines in tandem driving

The only photograph of the Kawasaki Ki-64 known to have survived the war. See letter below from Mr R V Sotheran.

contra-props, and the pilot being positioned between the two power plants.

R V Sotheran
Hartlepool

We metaphorically bow our heads in shame, Mr Sotheran. You are, of course, perfectly correct. Although the VB 10 was conceived (as the VG 20) before the Kawasaki Ki-64, the latter was the first to actually fly, but irrespective of which was first conceived or first flown, the existence of the Ki-64 effectively nullifies any claim to uniqueness on behalf of the VB 10. The Ki-64 was conceived in 1939 by Takeo Doi, but prototype construction was not authorised until October 1940. Its Kawasaki Ha-201 power plant consisted of two Ha-40 12-cylinder liquid-cooled inverted-vee engines which drove contra-rotating three-blade airscrews, the controllable-pitch forward airscrew being driven by the rear engine, and the fixed-pitch rear airscrew was driven by the forward engine. The power plant components were installed fore and aft of the cockpit, and were cooled by steam vapour. The Ki-64 was flown in December 1943, but during the fifth test flight a fire occurred in the rear engine, necessitating a forced landing. Both airframe and power plant were still under repair at the time of Japan's surrender.

Gad, sir, it's not British!

HAVING just purchased the second issue of the AIR ENTHUSIAST, I am more than ever convinced that the title should be qualified by the addition of "non-British"! This makes it all the more surprising that you should have chosen the Harrier to illustrate the front cover. A couple of pages on the Trident, a photograph of the D.H.2, and another of the Sopwith Baby replica, and that is the lot in so far as the *British* enthusiast is concerned. Even if you had managed to include a photograph of the Southampton in Turkish Air Force service it would have made the expenditure of 25p still worthwhile. You may argue that this is the mixture that sells, but I'm certain that you will lose circulation (me for one) unless you are prepared to present a more balanced production.

John C Paton
Airdrie

A more balanced production, Mr Paton? Do you mean that because the AIR ENTHUSIAST is printed in the English language it must necessarily be biased in content matter towards British aviation products or British aviation history and activities? Are you

suggesting that the average British aviation enthusiast is interested only in British items? The interests of the vast majority of air enthusiasts are international, as our post-bag confirms, and therefore the approach of the AIR ENTHUSIAST is international. It is "the mixture that sells", as you so rightly suggest that we might argue. We have to offer such a "mixture" or the AIR ENTHUSIAST would very soon be out of business. The magazine was conceived to appeal to a world-wide readership, and in the Editors' judgement, over-emphasis on British subject matter would lose more circulation abroad than it would gain in the UK.

This is not to say, however, that the balance of material in the first two issues of the AIR ENTHUSIAST represented editorial policies that are being applied over a longer period than two months. There are many factors that dictate the selection of a particular feature for a specific issue, and not all interests can necessarily be satisfied within the 56 pages of any one issue of the magazine. We hope that "British" enthusiasts like Mr Paton, will have enjoyed such items as the first part of a three-part study of "Fighters in the RAF" (to be concluded in the November issue), Tony Taylor's recollections of flying the Hampden in the previous issue, which also included an exclusive coverage of early Bristol design work on supersonic transports, and, in this issue, the story of the Britten-Norman Islander/Trislander family. The AIR ENTHUSIAST is published in the English language in Great Britain, but it is an INTERNATIONAL magazine, for its editorial team believes the great majority of air enthusiasts to have truly international interests.

The wrong digit

IN THE July issue of your magazine you made a mistake in the identification of the Nieuport fighter (page 86). This is not, as you state, a Nieuport 17 but a Nieuport 27. The latter can be identified by the curved fin and rounded fuselage side.

L D Burrows
Warminster

Quite right, Mr Burrows, the photograph does depict a Nieuport 27 of the Turkish air arm and not a Nieuport 17 as stated by us.

Not so hot!

WITH REFERENCE to page 55, I find it rather hard to believe a Milan taking off in 82°C — the pilot would fry before he got airborne surely. I'll believe 82°F.

Jeffrey J Simpson
Pudsey

AIR Enthusiast

Volume 1 Number 6 November 1971

CONTENTS

AIRSCENE

MILITARY AFFAIRS

AUSTRALIA
The last official flight of the **Commonwealth CA-27 Sabre** in RAAF service was marked by a formation fly-past over Melbourne on 30 July, 17 years and 17 days after the initial flight of the Avon-powered Sabre (13 July 1954), and 15 years after the introduction of this fighter into RAAF service. The CAC-built Sabre was operated primarily by No 78 Wing of the RAAF, and 17 refurbished examples have been presented to Malaysia. The remaining survivors will now presumably be scrapped.

Interservice squabbling is threatening the future of the **Project N** twin-turboprop light STOL aircraft which flew for the first time recently (see this column October). Basically, the problem results from the RAAF's reluctance to accept any expansion of the Army's rôle in aviation which would take place if the Army utilises the proposed military production derivative of the Project N, the N22, for the wide variety of missions now under consideration. Originally initiated as a liaison and utility aircraft, Project N has now become a rugged STOL transport capable of accommodating such loads as 14 troops and externally-mounted items such as helicopter blades, and suitable for air strike control reconnaissance, casualty evacuation, supply dropping, and light strike missions, and the RAAF sees no need for these capabilities. At the time of writing the programme was confined to initial prototype funding, but the Government Aircraft Factory hoped that a favourable decision on production would be taken last month (October), and is planning an output of four examples of the military version per month to be attained next year.

BOLIVIA
F-51D Mustangs of the *Fuerza Aérea Boliviano* (FAéB) reportedly played a major rôle in the recent right wing **military "coup"** which overthrew the régime of General Torres. The FAéB supported the rebels led by Colonel Banzer, employing its Mustangs to strafe left wing student groups entrenched in and around La Paz. The 10 Cavalier-remanufactured F-51D Mustangs were supplied to the FAéB in 1968 under what has now proved to be the singularly inappropriate project name of *Peace Condor*.

CANADA
The recent Canadian Armed Forces White Paper indicates that all intercept tasks in eastern Canada are to be fulfilled by the CAF's three squadrons (Nos 409, 416 and 425) of **CF-101 Voodoos**, and that the two CAF squadrons of CIM-10B Super Bomarc surface-to-air missiles (Nos 446 and 447) at CFB North Bay, Ontario, and CFB La Macaza, Quebec, are to be scrapped by September next year. Among other defence changes are the use of **CF-5s** for coastal water photo reconnaissance in place of T-33A Silver Stars, and the assignment of other CF-5s to NATO for use in Norway and Denmark in an emergency. Although the **Argus** maritime patrol aircraft that were to be retired in 1973 appear to have been given a temporary reprieve, a decision will be taken within the next year concerning their replacement or rebuilding. If the decision to rebuild these aircraft is taken this work will be conducted between 1975 and 1977. Contenders as replacements are the Nimrod, the Orion, and a modification of the Boeing 707. About 40 **Trackers** are being retained specifically for patrolling the Gulf of St Lawrence and the Bay of Fundy to guard against any violation of Canada's 12-mile (19,3-km) territorial limit.

FEDERAL GERMANY
Although an agreement has been completed between the Federal German Defence Ministry and the US government under which the *Luftwaffe* will receive 175 single-seat **F-4E(F) Phantoms** for the air superiority rôle, the Federal German Defence Ministry is understood to be initiating an investigation into the cost effectiveness of purchasing standard two-seat F-4Es rather than the single-seat model that is being developed specifically to meet *Luftwaffe* requirements, and has requested recommendations from the US government. The F-4E(F) programme cost, pegged at DM 4,000m (£300m) by *Bundestag* decree, which it is now thought will cover the acquisition of the planned 175 aircraft, will certainly not cover the cost of purchasing a similar number of F-4E Phantoms, therefore reducing the quantity of Phantoms. If a decision is taken in favour of the standard F-4E and the number of aircraft to be delivered to the *Luftwaffe* is not reduced, German funding for the MRCA will be placed in jeopardy as the *Luftwaffe* operational requirement for this aircraft will undoubtedly be reduced. Current plans call for the F-4E(F) to re-equip four *Geschwader*, replacing two F-104G-equipped interceptor *Geschwader*, one F-104G-equipped strike *Geschwader* and two G.91R-equipped light ground attack and reconnaissance *Geschwader* during 1974–6.

INDIA
The Indian government has protested to the Pakistan government over the repeated **intrusion of Indian air space** by aircraft of the Pakistan Air Force. In July, two Pakistan Mirages flew low over Srinagar airfield, and early in September three Pakistan Mirages flew over Nowshera, Jammu, while on 16 September two Pakistan Sabres were intercepted by the IAF over the Khasi-Jaintia hills of Assam and turned back into Pakistan air space.

An inquiry has been ordered into the crash on 26 July of an HF-24 **Marut Mk 1A** near Bangalore, HAL's test pilot, Wing Commander J K Mohlah, losing his life in the crash. During August, the IAF suffered three crashes involving a MiG-21, a Su-7, and an unspecified type rumoured to be "a medium bomber of Soviet origin" with a crew complement of 11 members.

JAPAN
The Defence Agency is currently studying requirements for a successor to the **Grumman S-2A Tracker** serving in the ASW rôle with the Maritime Self-Defence Force, the *Kaijo Jieitai*. Fifty-five Trackers remain in the inventory, and although no decision has yet been taken whether to evolve an ASW aircraft of indigenous design or to purchase an aircraft off-the-shelf from the USA, the latter course is believed to be favoured owing to the comparatively small number of aircraft involved, and the Lockheed S-3A

The MiG-21MF interceptors of the V-VS which paid a goodwill visit to France 6–10 September revealed several features not previously seen, such as the fuselage-mounted armament of twin ultra-short 23-mm cannon. It was previously thought that the conical centrebody in the intake housing the "Spin Scan" search and track radar was fixed, but this is now known to be a three-position centrebody as used by earlier day fighter versions of the MiG-21 (Photo by Roger Demeulle).

therefore appears to be the only possible candidate.

The Air Self-Defence Force (*Koku Jieitai*) and Maritime Self-Defence Force (*Kaijo Jieitai*) are to increase their planned purchases of the **Kawasaki C-1** turbofan-powered transport during the 4th and 5th Five-year Defence Programmes. The former service is increasing its requirement for the transport version from 27 to 38 aircraft, while the latter service intends to increase procurement of a mine-laying version from four to eight. In addition, the *Koku Jieitai* anticipates that it will acquire a stretched-fuselage version of the C-1 transport as well as 15–25 examples of an airborne early warning variant. The decision to develop an AEW version of the C-1 rather than acquire the Grumman Hawkeye was announced by Defence Minister Nakasone on 1 April, and *Yen* 1,700m has been allocated from Fiscal 1972 funds for initial studies. The Defence Agency has stipulated that the unit cost of the AEW C-1 must not exceed the *Yen* 8–9,000m of a comparable aircraft imported from the USA. This will include R&D estimated at about *Yen* 12,000m over a six-year period, two-thirds of which will be devoted to the Japanese phased-array radar system. It is anticipated that the wing of the AEW C-1 will be extensively redesigned to embody integral fuel tankage, and range with a 6·5-ton payload for the AEW mission will be extended to 2,010 miles (3 250 km) and endurance to seven hours, compared with the 1,380 miles (2 220 km) and four hours for the standard transport model with the same payload.

JORDAN
The Royal Jordanian Air Force (*Al Quwwat Aljawwiya Almalakiya Alurduniya*) has now received 18 of the 36 refurbished **Lockheed F-104A Starfighters** under the agreement signed between Jordan and the US government in June 1969. The remaining 18 aircraft are all reportedly in the delivery pipeline, and will presumably reach Jordan via Turkey during the next few months.

LIBYA
It is anticipated that some 15 **Dassault Mirages** will have been delivered to *Al Quwwat Aljawwiya Al Libiyya*, the Libyan Air Force, by the end of this year. The majority of the aircraft so far delivered are tandem two-seaters, and deliveries of the single-seat models are scheduled to accelerate during the first six months of 1972.

PAKISTAN
The government of Saudi Arabia has reportedly placed 10 DHC-2 **Beavers** and two **Alouette IIIs** at the disposal of the Pakistan government, ostensibly for relief operations in East Bengal, but these are apparently being used by the Pakistan Army's Special Service Group, which, trained by the US Army, is patterned on that service's "Green Berets".

The Pakistan Air Force's No 6 Squadron lost one of its C-130 **Hercules** transports at Sylhets' Shalutikor airfield during August as a result of satchel charges placed by Bangla Desh guerillas. No 6 Squadron reportedly includes in its inventory several Turkish and Iranian Hercules transports loaned to Pakistan.

The Pakistan Air Force has admitted that all its **Bengali personnel** (about 1,500) have been "grounded" following an attempt by a Bengali Squadron Leader to fly a Martin B-57 to India after taking-off on a routine

This view of the interior of the cockpit of the MiG-21MF reveals rather more American than European influence (eg, the large vertical panel). The radar scope appears to be relatively low, suggesting the use of a two-position seat (ie, that can be raised for take-off and landing). All the usual flying instruments are on the left, together with a clock (now omitted from most western panels to save space), compass with VOR/ADF needle, and a "vernier ASI". To the right are the engine instruments, fuel flow rate and fuel contents gauges. The central warning panel endows the cockpit with a "modern" look, but the wall switches to starboard are rather passé. The control column handle is rather untidy, with tailplane trim switch and two firing buttons, and the brake handle suggests that steering is performed by differential mainwheel braking rather than nosewheel steering (Photograph by Roger Demeulle).

training flight from Mauripur, north-west of Karachi. The attempt, which took place in mid-August, was allegedly foiled by another PAF officer, Flying Officer Rashid Minhas, who was also flying in the B-57, which crashed 35 miles (56 km) short of the Indian frontier. F/O Minhas lost his life and was posthumously awarded Pakistan's highest award for gallantry.

SOUTH AFRICA
According to a statement made by Defence Minister P W Botha in Pretoria, South Africa will soon be manufacturing an advanced version of the **MB 326 Impala** (presumably the more powerful single-seat MB 326K operational trainer and close-support aircraft), guided missiles and light transport aircraft (believed to be the Aerfer-Aermacchi AM.3C). This statement followed a tour by the Defence Minister of the Armaments Development and Production Corporation (Armscor) facilities on the Witwatersrand and in Pretoria. Mr Botha emphasised that South Africa is now self-sufficient in the production of arms and ammunition for internal defence, production currently including 1,000-lb (453,5-kg) general-purpose bombs, napalm bombs and air-to-ground rockets. Still further expansion of the South African aircraft industry is forecast by an announcement by Prof H J

Samuels, chairman of Armscor, that additional licence agreements for the manufacture of military aircraft will be concluded during the next few months, the next major type to be involved in South African production allegedly being the Aérospatiale SA 319 Alouette III helicopter.

No decision has yet been taken concerning the replacement of the three **HS 125 Mercurius** transports lost on 26 May. One HS 125 Mercurius currently remains in the SAAF inventory.

SOVIET UNION
Some details are now available concerning the six **MiG-21** interceptors of the V-VS that paid a five-day goodwill visit to France in September. The aircraft were of the MiG-21MF (*Fishbed-J*) type powered by an uprated version of the Tumansky R-11 turbojet with a dry thrust of 11,244 lb (5 100 kg) and a reheat thrust of 14,550 lb (6 600 kg), and equipped with a boundary layer blowing system known as SPS (*sduva pogranichnovo sloya*) which reduces the normal landing speed from 193 mph (310 km/h) to 168 mph (270 km/h). Features not previously seen included JATO attachments on either side of the rear fuselage, debris deflectors below the suction relief doors, a rear-view mirror, and the installa-

tion of a pair of ultra-short 23-mm cannon with 100 rpg in the fuselage belly. The installation of these weapons displayed some ingenuity, being extremely narrow to go between the air brakes, and having splayed link chutes to miss the fuselage drop tank.

The MiG-21MF has a take-off weight with four 165-lb (75-kg) K-13 (*Atoll*) AAMs of 18,078 lb (8 200 kg), increasing to 19,731 lb (8 950 kg) with two K-13s and two 110 Imp gal (500 l) drop tanks. With two K-13s and three drop tanks (ie, max loaded condition) take-off weight is 20,723 lb (9 400 kg). Internal fuel capacity is 581 Imp gal (2 640 l) with which range is 683 miles (1 100 km), ferry range with maximum external fuel being 1,118 miles (1 800 km). Maximum speed at low altitude is 808 mph (1 300 km/h), or Mach 1·06, increasing to 1,386 mph (2 230 km/h), or Mach 2·1, above 36,090 ft (11 000 m). Service ceiling is 59,055 ft (18 000 m), and take-off distance at normal loaded weight is 875 yards (800 m), landing distance being 600 yards (550 m). The general finish of the MiG-21MF fighters that visited Rheims was considered poor — access panels fitting badly and a longitudinal lap joint on the rear fuselage — and the plethora of excrescences and stencils sported by these aircraft was noteworthy.

Additional information concerning the appellations bestowed by NATO on variants of the **MiG-21** (see August issue) has recently been made available, and it is now known that the *Fishbed-H* is actually a reconnaissance version of the *Fishbed-J* (MiG-21MF) and *not* a variant of the late-production *Fishbed-D* (MiG-21PFM). The *Fishbed-H* is externally similar to the -J but has a rather obvious buried antenna at mid-fuselage, can be fitted with ECM fairings at the wingtips, and usually carries an external container for a forward-facing or oblique camera, or an ECM pod beneath the fuselage. The *Fishbed-E* is now known to be .similar to the *Fishbed-C* (MiG-21F) but featuring the extended vertical fin similar to that applied to the late-production *Fishbed-D* (MiG-21PFM). The MiG-21PFM was illustrated at the top of page 125, August issue, and was incorrectly labelled as *Fishbed-E* in the comparison drawings on page 122 of the same issue.

The advanced variable-geometry **strategic bomber**, which is known to be undergoing flight testing in the Soviet Union, has now been dubbed *Backfire* by NATO. Attributed to the design bureau of A N Tupolev, the *Backfire,* which was alluded to in March by US Defense Secretary Melvin Laird, is believed likely to attain service status with the V-VS by 1973–4, and according to Pentagon sources has a gross weight of the order of 272,000 lb (123,375 kg) and an unrefuelled combat radius well in excess of 3,000 miles (4 830 km). Maximum speed exceeds Mach 2·0 at altitude, and performance is believed to include low-level supersonic dash capability. Power is allegedly provided by two Kuznetsov turbofans with an afterburning thrust in excess of 40,000 lb (18 150 kg), the fuselage is area ruled, and only the outboard panels of the wing are movable, the fixed centre section being of delta planform.

SWITZERLAND
As was perhaps to be predicted, the long-drawn-out contest to replace the aged Venom as a **ground attack fighter** in the inventory of the *Flugwaffe* has moved from the technical to the political arena, and the Vought A-7G Corsair, selected by the *Flugwaffe* on its technical merits as the

winning contender now seems likely to lose out to the Dassault Milan S, political resistance to the purchase of another product of Avions Marcel Dassault having apparently now evaporated and the Milan S being favoured by the Swiss Federal Council. Earlier, the McDonnell Douglas A-4M Skyhawk was displaced from its position as second choice owing to what President Gnägi — who also happens to be Defence Minister — described as the "politically intolerable" situation of having *two* US aircraft as finalists. Dassault determination to win the contest is evidenced by the company's offer to make a Milan available free of charge for a fly-off contest — whereas the US says that it must charge $165,000 (£66,500) for this phase of the contest — and to place 30 per cent of the value of the programme with Swiss industry. The Federal Council has stated that it will not make its final choice known until December 1972, although there are indications that the decision time will be extended until mid-1973. Dassault, assuming the signing of a contract by 1 January 1973, is guaranteeing the delivery of 91 single-seat and two two-seat Milans between October 1974 and September 1976 at a rate of five aircraft per month.

Deliveries of the batch of 30 refurbished Hawker Hunter Mk 50s for the *Flugwaffe* was initiated during September. Assembly of the Hunters is being undertaken at Emmen.

TANZANIA
Despite Sino-Communist control of the Tanzanian Defence Force Air Wing (see this column September issue), the Tanzanian government has purchased a further eight **DHC-4 Caribou** transports from Canada, these supplementing five Caribou acquired earlier (the first two being donated by the Canadian government). The additional Caribou transports are ex-Canadian Armed Forces aircraft, and the first three reached Canada in June, several of the Tanzanian DHC-3 Otters being returned to Canada at the same time, presumably as trade-ins.

UNITED KINGDOM
The build-up of the **RAF Phantom force** in Germany was completed recently with the activation of the service's fourth squadron. This unit, No 31 Sqdn, is based at RAF Bruggen, and, in common with Nos 2, 14 and 17 Squadrons, flies the Phantom FGR Mk 2. Three of the four squadrons operate in the strike-attack rôle, No 2 Squadron being assigned the reconnaissance task. Home-based Phantom units, all now operational, are Nos 6 and 54 Squadrons assigned to No 38 Group, Air Support Command, with Phantom FGR Mk 2s, and No 43 Squadron in No 11 (Fighter) Group of Strike Command with Phantom FG Mk 1s. A second squadron may eventually be formed in Strike Command operating Phantoms released by the Royal Navy.

USA
Delivery of the 10th Hawker Siddeley **AV-8A Harrier** to US Marine Corps Squadron VMA-513 at MCAS Beaufort was scheduled to be effected during October. This was the last USMC aircraft to be fitted with the Pegasus 10 engine; the remaining two Harriers of the initial batch of 12 are held back in the UK for installation of the Pegasus 11 engine and other equipment, and will be delivered to the USMC next spring. Meanwhile, work is proceeding at Kingston on the second batch which comprises 18 aircraft, and confirmation is awaited of the planned Fiscal 1972 buy of a further 30 for

which long-leadtime items have already been ordered and some preliminary manufacturing is already taking place.

The USAF began the flight testing of the three aircraft selected to participate in the second phase of **Project Pave Coin** at Eglin AFB, Florida, on 10 August, the objective being the selection of an aircraft capable of meeting the specialised tactical requirements for forward air control (FAC) and light strike aircraft (LSA) missions, and is available "off-the-shelf". The three aircraft selected by the Aeronautical Systems Division for testing are American Jet Industries' Super Pinto, the Beech (Aerfer-Aermacchi) AM-3C/E, and the Piper Enforcer. The AM-3C/E is in the FAC category and the Super Pinto and Enforcer are in the LSA category. Flight evaluation of the three aircraft was completed at Eglin towards the end of August, and data evaluation was scheduled to be completed late September.

The **Lockheed S-3A** ASW aircraft was scheduled to be rolled out at Burbank, Calif, on 8 November, and is expected to commence its flight test programme early in January at Palmdale, some two months ahead of schedule. The US Navy will exercise its first production option — for 13 aircraft — by 1 April provided that the first aircraft has flown at least 15 days prior to that date and that a formal laboratory demonstration of the S-3A's integrated avionics system has taken place. Total research and development cost of the S-3A now exceeds $700m, primarily as a result of the reassignment of two production examples of the aircraft to development status. The first General Electric TF34-GE-2 engines for the S-3A were delivered ahead of schedule, and under the current contract Lockheed will build a total of 199 S-3s for the US Navy (including the eight development aircraft), and programme unit cost is approximately $14.7m per aircraft.

AIRCRAFT AND INDUSTRY

ARGENTINA
Production of both the **AESL** Airtourer and **Fletcher FU-24** in Argentina is planned by a newly formed company in Buenos Aires. As a first step, 11 FU-24s and four Airtourers have been purchased from their New Zealand manufacturers, respectively Air Parts (NZ) Ltd and Aero Engine Services Ltd, and final assembly of these has begun in Buenos Aires. Argentine Air Force interest in the Airtourer as a trainer is reported.

BRAZIL
The first **TF-26 Xavante** (Aermacchi MB 326GB) to be assembled from components imported from Italy was flown for the first time at São José, near São Paulo, on 3 September, the initial flight lasting 90 minutes. Deliveries of the Xavante to the *Fôrça Aérea Brasileira* are scheduled to commence next year against orders for 112 aircraft of this type.

CHINA (TAIWAN)
The co-production of **Bell UH-1H Iroquois** helicopters at Taichung, Taiwan, is proceeding on schedule, and the first four of an initial batch of 50 helicopters of this type for the Nationalist Chinese Army had been delivered by the end of July, and a follow-on quantity has now been approved.

FEDERAL GERMANY
The first of the three **VFW-Fokker VAK 191B** VTOL research aircraft was flown for

the first time at Bremen on 10 September with Ludwig Obermeier at the controls. The aircraft took-off vertically and made a 3 min 18 sec circuit of the airfield at well below transition speed. The two additional prototypes of the VAK 191B are virtually ready to join the test programme, and at least one of these may also have flown by the time these words appear. Conceived a decade ago for the low-altitude nuclear-strike rôle, the VAK 191B will now be utilised for a broadly-based research programme.

Dornier, VFW-Fokker and MBB, the three major German aerospace companies, have agreed to join in the development of a VTOL transport aircraft, subject to government financial support. The project would be based on Dornier concepts using separate lift and cruise engines, and an investment of DM 1,000m (£120m) is estimated to be necessary to launch such an aircraft. The VFW-Fokker and MBB VTOL projects have been abandoned, including the VC400 on which some government money had been expended.

FRANCE
Wassmer Aviation is reported to have overcome its recent financial difficulties and is pressing ahead with marketing of the all-plastics **WA-51 Pacific** and **WA-52 Europe,** respectively with 150 hp and 160 hp engines. The WA-51 has obtained ARB certification in the UK and Rollason Aircraft has been appointed UK distributor for Wassmer aircraft.

A two-seat derivative of the four-seat **Avion Robin HR-100** is undergoing flight testing at Dijon. Designated the HR-200, it has a 100 hp Continental engine, but may be offered with a 90 hp engine in production form.

Fitted with Pratt & Whitney JT8D-15 engines, the **Dassault Mercure 01** has resumed its flight trials on 7 September at Istres. Compared with the JT8D-11s fitted for the first series of 20 flights, the -15 engine is rated at 15,500 lb st (7 030 kgp) instead of 15,000 lb (6 800 kg) and is flat rated at temperatures up to 84°F.

The fourth prototype of the **Mirage F1** completed its second series of maintenance and flight qualification tests at Mont-de-Marsan in August. The Mirage F1-04 flew a successful series of radar intercepts and attacks under all weather conditions against drone targets and ground targets. Elements of the first production aircraft are now coming together at Bordeaux, and deliveries to the *Armée de l'Air* are scheduled to commence late in 1972.

INTERNATIONAL
On 9 September the governments of Federal Germany, the United Kingdom and Italy announced the satisfactory outcome of their review of the **MRCA programme.** Detail part production and sub-assembly of the MRCA has now begun, and master fuselage mating jigs are complete. Governmental authorisation of the construction of six prototypes has been given, the first of these being scheduled to commence its flight test programme at Manching late in 1973, and current plans call for three fully-instrumented production aircraft to support the prototype flight test programme. A March 1977 in-service date with the *Luftwaffe* is still planned, and it is currently anticipated that the *Luftwaffe* will receive some 420 Panavia 200s, the RAF taking about 380, and the Italian *Aeronautica Militare* taking the remainder of an intended production run of approximately 900 aircraft.

SWEDEN
Saab-Scania is proceeding with development of the **Saab 105G** advanced attack version of the twin-jet multi-purpose aircraft, and a prototype which will embody the most important features is scheduled to commence its flight test programme in March. It will be powered by two 2,850 lb (1 293 kg) thrust General Electric J85-GE-17B engines, and a significant feature will be a new avionics package including a high-precision navigation and attack system consisting of a Saab BT9R laser ranging bombing and rocket firing computer, a Sperry platform, Decca Doppler radar and map display, an Elliott air data unit and a Ferranti ISIS sight head. Another important feature of the Saab 105G will be the modified wing leading edge which will more than double the aircraft's attainable load factors in the high-speed régime without any increase in drag or deterioration of low-speed characteristics. The air brakes are being enlarged 60 per cent, and maximum flap setting will be increased to provide rapid deceleration in combat, and a much steeper angle of approach for landing on small airfields.

UNITED KINGDOM
First flight of the **Sigma C** sailplane was made at Cranfield on 12 September by H C N "Nick" Goodhart. This advanced high performance sailplane of mixed light alloy and glass-reinforced plastics construction features a full-span flap (incorporating the ailerons) which allows both the area and the camber of the wing to be varied. The Sigma C was built in the BEA workshops at Heathrow following destruction by fire in November 1968 of the first two prototypes during construction by Slingsby Sailplanes.

An extensive evaluation of the **Jetstream 200** has been completed at the A & AEE, Boscombe Down, to assess its suitability for the RAF in a twin-engined advanced flying training rôle. A requirement exists for about 25 aircraft and the Jetstream is on the short list together with a version of the North American Rockwell Hawk Commander.

USA
Piper Aircraft Corp has introduced a new light twin, the **PA-34 Seneca,** with the

Illustrated (right and below) is the Lock-speiser LDA-1, the first flight of which was made on 24 August as reported in our previous issue. Powered by a Continental C-85-12, this aircraft has a span of 29 ft (8,84 m) and gross weight of 1,300 lb (590 kg) and is a 70 per cent scale model of a projected "flying utility truck". The latter would have a 340 hp Lycoming O-540 engine, a span of 44 ft (13,4 m), gross weight of 4,500 lb (2 040 kg) with a 2,000 lb (907 kg) payload and range of 350 naut mls (649 km).

lowest price of any twin-engined type on the US market. Basic retail price is $49,900 (£20,620) which, allowing for inflation, is less in real money terms than the 1954 price of Piper's first twin, the Apache. The Seneca seats up to seven in three seat rows, and is powered by two 200 hp Lycoming IO-360 engines driving opposite-rotating propellers. Empty and gross weights are, respectively, 2,479 lb (1 124 kg) and 4,000 lb (1 814 kg) and overall dimensions are: span, 38 ft 11 in (11,9 m); length 28 ft 6 in (8,7m); height, 9 ft 11 in (3,0 m); wing area, 206.5 sq ft (19,18 m²). The Seneca has a top speed of 196 mph (315 km/h) and cruises at 171-187 mph (275-301 km/h) according to power rating and altitude, with ranges of 860-1,160 miles (1 385-1 866 km) according to speed.

Included in the 1972 range of five Piper Cherokee models is the **Cherokee Arrow II,** a stretched development of the original retractable-undercarriage version of the Cherokee. Compared with the Arrow I, the new model has a fuselage 5 in (12,7 cm) longer, 26 in (0,66 m) more wing span, larger tailplane and a dorsal fin and flaired rudder which is common to all the new Cherokee line. The four-seat Arrow II is powered by a 200 hp Lycoming IO-360 engine. Empty and gross weights are 1,504 lb (682 kg) and 2,650 lb (1 202 kg) respectively; overall dimensions are span, 32 ft (9,75 m), length 24 ft 7 in (7,49 m), height, 8 ft (2,44 m) and wing area, 170 sq ft (15,79 m²). Max speed is 175 mph (282 km/h) and range 900 miles (1 448 km) cruising at 55 per cent power.

Flight testing has been completed at Seattle of the first **Boeing 747B** fitted with Pratt & Whitney JT9D-7 engines. The aircraft is the first for South African Airways and delivery was scheduled before the end of October. Rating of the JT9D-7, a single example of which had previously been flown in the Boeing-owned 747 demonstrator, is 47,000 lb st (21 320 kgp) with water injection; earlier 747Bs have been delivered with JT9D-3As dry-rated at 43,500 lb st (19 731 kgp) and operate at a lower gross weight.

Certification of the **Cessna Citation** to full FAR 25 standards, was confirmed by the FAA in September, with deliveries starting

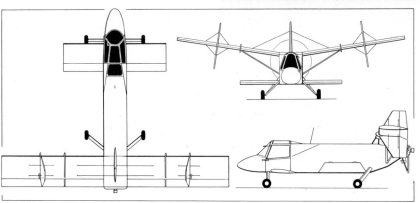

On 28 September the Bell Helicopter Company revealed a new armed helicopter, which, first flown on 10 September, has been named the KingCobra. Powered by a 1,800 shp Pratt & Whitney T400-CP-400 "Twin Pac" power plant, the first prototype KingCobra retains the general configuration of the HueyCobra but has larger 13-ft (3,96-m) stub wings, an enlarged rotor and a tailboom extension. It embodies a multi-sensor fire control system.

before the end of the month. First three aircraft went to American Airlines (for use in its Citation pilot-training service), United Aircraft of Canada (for JT15D-1 engine development) and to Bendix (who supplies the flight director/autopilot).

Following the approval of the US government's loan guarantee for Lockheed by the Senate, the final stages in the lengthy processes were reached on 14 September with the signature of appropriate documents. The government has guaranteed a total of $250 million (£104 million) which has now been advanced by a group of 24 American lending banks to allow Lockheed to complete the **L.1011 TriStar** programme. At the 14 September signature date, Lockheed said it had firm contracts for 98 aircraft from four major airline customers (with 39 more on money-backed option) comprising Air Canada, 10 (plus 9); Delta, 18 (plus 6); Eastern, 37 (plus 13) and TWA, 33 (plus 11). Other customers included in the overall order plus option total of 178 are Air Finance, 2 (plus 1); Air Holdings, 19 (plus 10); Air Jamaica, 1 (plus 1); PSA, 2 (plus 3) and Turner-Haas, 2. Lockheed is now rehiring TriStar production workers laid off early this year and expects to have the aircraft ready for airline service in April, five months behind the original schedule. Rolls-Royce delivered the first three RB.211s with the full rating of 42,000 lb st (18 700 kgp) each early in September and these first flew in TriStar No 2 on 8 September; a second ship-set has since been delivered, bringing the total number of RB.211s sent to Palmdale to 26. Two more TriStars were expected to join the first three in flight test by the end of the year; 500 hours of testing with three aircraft had been amassed by the end of August.

CIVIL AFFAIRS

USA

Latest statistics show that the number of **active pilots** in the US in 1970 reached a new peak at 732,729. This represented a two per cent increase in the figure for 1969. Of the total active pilots (those holding current medical certificates), 41 per cent had private pilot ratings and 25 per cent had commercial ratings. Fewer than five per cent held airline pilot ratings. The total included 6,677 pilots qualified to fly helicopters only, 3,114 qualified only on gliders, and 195,861 student pilots.

NASA has now completed its plans for the development of a **STOL research transport** and, subject to funding, hopes to have two prototypes of a selected design flying by mid-1975. A Proposal Request for an "Experimental STOL Transport Research Airplane" will be issued to industry, with an instruction that primary designs should be based either on the augmentor wing or the externally-blown flap concept; alternative proposals using other means to achieve STOL can also be put forward. At least two designs will be chosen for a six-month design phase starting in January 1972, leading to selection of one design later in 1972 for construction. First flight target for the first of two prototypes is October 1974.

A merger has been agreed in principle between **National Airlines** and **Northwest Airlines** and is now subject to approval by the CAB and the stockholders of the two companies. As National already has transatlantic route rights (Miami–London) and Northwest has trans-Pacific rights (to Tokyo, Manila and Hong Kong), while both airlines have transcontinental US routes, a merger could produce a third major international company to compete with TWA and Pan American. Meanwhile, the two last-named airlines have been conducting studies to see if a merger would be advantageous, although the latest guidelines published by the Department of Transportation suggest that such a merger would be frowned upon officially because the resultant company would be too large. Two other proposed mergers, between American Airlines and Western, and between Delta and Northeast, are expected to receive official approval.

Europe's **busiest airport**, London Heathrow, compares in total annual movements with the 44th busiest in the US (Salt Lake City) according to figures prepared by the Aircraft Owners and Pilots Association. However, only five per cent of Heathrow's movements were in the general aviation category compared with 67 per cent at Salt Lake, and Heathrow has more international movements than any other airport. AOPA figures stress the high level of general aviation (ie, non-airline) activities at US airports, with more than 135,000 aircraft being used.

USSR

Aeroflot began regular route proving flights with the **Tupolev Tu-154** during August, following standard Soviet arrangements for the introduction of new types into passenger services. Freight and mail only were carried on the proving flights, primarily on routes from Moscow to Tbilisi, Mineralniye Vody, Sochi, Simferopol and Khabarovsk.

CIVIL CONTRACTS AND SALES

BAC One-Eleven: British Caledonian has ordered one more Series 500, to make a total fleet of 13 of this model plus eight Series 200s. □ Air Malawi has ordered one Srs 475 for February delivery, replacing one Srs 200 leased from Zambia. □ New German IT operator Interfly is reported to be acquiring two Srs 400s, second-hand.

Boeing 707: Air Cameroun has leased one from Air France, for operations starting in November. □ Britannia Airways has taken a five-year lease on a 707-320C (its second) from Executive Jet; the same aircraft was on lease to British Caledonian until October.

Boeing 727: Western Airlines ordered three -200s for delivery next May and June, and optioned three more. □ All Nippon Airways ordered eight -200s.

Boeing 737: One (with possibly a second to follow) leased from Aer Lingus by Air Cameroun, the new national carrier for Cameroun to take the place of Air Afrique. □ Two -200s have been purchased by Mey-Air, Oslo-based charter operator, with delivery of the first in September and the second in November. □ All Nippon Airways ordered three.

Fokker F.28: Germanair is reported to have taken over an order for four originally placed by Bonn-Air, a new IT operator in process of formation in Germany.

Hawker Siddeley HS 125: Sales total reached 252 with confirmation of contracts for two sold through Beechcraft Hawker in the USA and one sold in the UK.

Helicopters: Asahi Helicopter Company has ordered the first Bell 212 to be delivered in Japan. It will be used to support off-shore rigs. □ Two Hughes 500s have been ordered from Kawasaki in Japan by the Ryukyu government, for November delivery. They will be used for medical duties.

Lockheed Electra: Ansett Airlines of Australia has contracted to have three of its Electras converted to all-cargo layout by Lockheed Aircraft Services.

McDonnell Douglas DC-10: Western Airways ordered four. □ SAS took up an option on two.

Nord Noratlas: At least one N 2502, ex-*Luftwaffe*, is now operated in civil guise by Elbeflug in Germany.

Swearingen Merlin IV: Litton Industries took delivery of one, to become first US operator of the type.

Tupolev Tu-134A: Two have been ordered by Malev, supplementing six Tu-134s in service.

Yakovlev Yak-40: A provisional contract for two, subject to French certification of the type, has been signed by the import-export company Realisations Industrielles et Commerciales.

MILITARY CONTRACTS

Hawker Hunter: A second batch of refurbished Hunters (quantity unspecified) has been sold to Singapore.

Fokker F.27 Friendship: A contract was signed in Lagos on 21 September for the supply of six additional Friendships for the Nigerian government. Two F.27 Mk 600s will be delivered during the next two or three months and four F.27 Mk 400s are scheduled for delivery late next year.

CASA C.212 Aviocar: The Spanish air arm, *Ejército del Aire*, has placed an order for 50 C.212 Aviocar STOL utility transports with deliveries commencing in 1973.

FIGHTERS IN THE RAF

Part III – Decline and Fall?

by Roy M Braybrook
BSc, CEng, AFRAeS

THE ROYAL AIR FORCE ended World War II with a clear lead over all the other Allies' air forces in the field of turbojet fighters. However, within less than five years both the United States and the Soviet Union had leapt ahead in technical development, and (with the temporary exception of V/STOL) they have left this country trailing behind pathetically ever since. Today the "British RAF" is derided abroad as a small force, heavily dependent on American equipment, and with only a few minor actions to distinguish it from the Swiss Flying Corps in terms of operational experience during the past quarter-century.

There have been many reasons for the RAF's decline in fighter equipment. One critical point was that, much in the same way that Italy had produced a wave of military aircraft timed for a war in the 1930s, so Britain made a concentrated effort to develop jet fighters for air superiority in the mid-1940s. In both instances short-term objectives were gained at the expense of long-term effectiveness.

With everything staked on the Gloster Meteor and de Havilland Vampire, Britain chose the quick and easy road to success by adopting straight wings and centrifugal-flow compressors, both of which were to prove blind alleys in development. Metropolitan-Vickers meanwhile carried out the initial development of a family of axial-flow engines having comparable thrust to the R-R Derwent and DH Goblin, but of far less frontal area. However, none of these early axials was chosen for production, and in 1948 this famous engineering company decided it had had enough of the vagaries of aviation.

Coming later to jet fighters, and thus faced with the challenge of leap-frogging Britain's well-established position, America and the Soviet Union took the longer view of gas turbines, and examined German swept wing research data in a more optimistic light. The USAF approved swept wings for the F-86 immediately after the end of the war, virtually three years before the RAF issued its corresponding F.3/48 requirement, which led to the Hawker Hunter. The Soviet Union was also quick to accept the swept wing, and adopted German axial-flow and British centrifugal-flow engines as stop-gap measures, pending second generation axials from their own designers.

The two super powers thus raced down the mainstream of fighter development, while Britain drifted off into the backwaters. Once firmly embedded in the mire, this country threw away the paddle by deciding in 1946 not to risk human lives in an experimental supersonic flight programme. This was one year before America went through the barrier with the Bell X-1. Short of actually bombing our own aircraft factories to rubble, there was little else that this country could have done in the early post-war years to ensure the early demise of the British fighter!

Future historians may conclude that the British were temperamentally incapable of exploiting the open-ended situation created by the advent of the jet fighter. The fundamental problem in aviation development is to choose steps of optimum size, so as to advance at the highest possible rate without falling flat in the process. This is easy enough when the way ahead is restricted by known obstacles

The RAF's first jet fighters were the Gloster Meteor and the de Havilland Vampire. Illustrated below are Meteor F Mk IIIs of No 124 Squadron in 1946 and Vampire F Mk Is of the first Vampire unit, No 247 Squadron, also in 1946. The heading photograph shows a Gloster Javelin FAW Mk 5 of No 151 Squadron.

Last de Havilland fighter to serve with the RAF was the Venom, shown (above) in its day fighter FB Mk 4 version, and (below) in the NF Mk 3 night fighter version, in the markings of No 141 Squadron.

to a series of small progressions, but the sudden limitless advance offered by jet engines and swept wings automatically gave the contest to the United States, a nation reared on taking chances and with the resources to absorb a high failure rate.

The fact that Soviet post-war fighters have generally been somewhat in advance of Britain's is presumably explained by the fact that, although their designers may be equally conservative, for them the urgency of war has never ended.

Export and die

The RAF's first turbine-powered fighter, and the only Allied jet to be used operationally in the 1939–45 war, was the Meteor. The sole squadron (No 616) of Mk 1s began intercepting V-1s in August 1944, and destroyed a total of 13.

At that stage the Meteor was powered by two R-R Wellands of 1,700 lb (770 kg) and was capable of only 385 mph (616 km/h) at sea level, ie, less than the Tempest!

A flight of these aircraft and a squadron of Mk IIIs with 2,000 lb (910 kg) R-R Derwents later saw limited service during the final advance into Germany. Perhaps fortunately, they never encountered the Me 262, which could reach 500 mph (800 km/h) at low level. The Meteor saw active service again six years later, when Mk 8s were used by the Australians in the Korean War, initially as top cover for USAF F-86s, and later in the more plausible rôle of ground attack.

The Meteor Mk 1 grossed only 11,750 lb (5 350 kg) clean, ie, almost exactly the same as the Typhoon and Tempest. Fitted with Derwent 5s of 3,500 lb (1 600 kg), an early Mk IV set a world record of 606 mph (970 km/h) in 1945, and in the following year one with clipped wings reached 616 mph (986 km/h). The ultimate day fighter variant was the Mk 8, which was used by the RAF from 1950 to 1957, and could reach Mach 0·78 on the deck, and Mach 0·81 at the tropopause.

Gloster's day fighter Meteors were paralleled by night fighter variants from Armstrong Whitworth. These aircraft, together with a small number of NF Vampires, replaced the long-suffering Mosquito NF 38 in 1951. Two years later they were supplemented by NF Venoms.

The day fighter Vampire had entered service in 1946, employing a pod-and-booms layout to minimise jetpipe losses. Clean gross was 8,600 lb (3 900 kg), and the engine was a 3,100 lb (1 410 kg) DH Goblin. With detail refinements and slightly more thrust, the Vampire remained in service until 1955.

The Venom was an improved Vampire with a DH Ghost 103 of 4,850 lb (2 220 kg), and tip tanks set on a slightly swept wing of 10 per cent thickness, reduced from the 14 per cent of its progenitor. Although a maximum of 640 mph (1 025 km/h) was slow for its generation (1951–62) it had the useful rate of climb of 9,000 ft/min (46 m/sec) and good manœuvrability at altitude. Like other RAF jet fighters prior to the Hunter, it retained the wartime armament of four 20-mm Hispano.

For the first ten years after the war the main bomber threat was massed daylight formations of Tupolev Tu-4s. RAF fighters therefore practised head-on assaults against B-29s and B-50s, followed by beam attacks, using the techniques developed by Bf 109s and Fw 190s to break up the boxes of B-17s over Germany. At night it was still a matter of vectoring single fighters against individual bombers, turning the fighter behind the target at precisely the right moment and bringing him up in the 6 o'clock

Ultimate version of the Meteor in RAF service was the NF Mk 14 developed by Armstrong Whitworth, illustrated here in the markings of No 85 Squadron.

Three RAF fighter types of the 'fifties are shown together in this photograph, all in the markings of No 66 Squadron. From left to right they are the Hawker Hunter F Mk 4, Canadair Sabre F Mk 4 and Gloster Meteor F Mk 8.

position for visual acquisition. This manœuvre is known by various names, all of which are unprintable.

The big spread
The Korean War provided the first practical experience of combat between jet fighters, and although no British squadrons participated, a number of RAF pilots flew with the USAF and RAAF. Korea had a major influence on RAF day fighter tactics, and a brief discussion of that war is therefore justified at this point.

To some extent the air war was phoney, in that the rule keeping UN aircraft south of the Chinese border prevented air power from becoming a decisive factor. The F-86 was inferior to the MiG-15 in high altitude performance, and US pilots only accepted combat at medium levels, where their superior training, better transonic airframes, and radar ranging gave them a claimed kill ratio of over 10:1.

The normal procedure was for sections of four F-86s to patrol south of the Yalu, watching the MiG-15s take off and climb over China, form up above them in big gaggles, and dive to attack. The emphasis from the Sabre driver's viewpoint was on knowing precisely when to break, and the formations used over Korea were therefore arranged primarily to give rear view well beyond enemy firing range.

USAF pilots flew a modified "fluid four" with 400-500 yd (350-450 m) between pair elements, and 75-100 yd (70-90 m) between the aircraft of one element. The wing men were brought forward into a rough line abreast formation, to make it easier to keep an eye on their leaders while looking directly aft. This change was a direct result of the F-86 hood design, making it possible to scan on both sides of the fin at once.

When Korean War tactics came to be adapted to RAF use, the separation of the elements was doubled to allow for the Hunter's inferior rear view. For the same reason it was possible to stagger the wing men aft in the classic fluid four position, which made an outward break by their leaders far safer.

With the later advent of air-air missiles, this high level battle formation was opened out by a factor of 3 or 4, to give rear vision out to 10 st miles (16 km). Fighters were also flown as high as possible (subject to contrail conditions) to give good warning as the enemy climbed to attack. The two elements could then make an inward break, ie, a fast 180° turn, to present the attackers with a closed-up formation head-on.

Low level battle formations developed from those used in World War II, and have always been subject to wide variations according to the type of terrain, the nature of the opposition, and the ideas of individual squadron com-

manders. Fighters often still operate in fours, but the spacing between elements can be anything from 250 yd (230 m) to 1,500 yd (1 400 m), and the second element may be swept back from the first at any angle from zero to 60°.

Flying low over desert, where the principal risk would be rear gun attack, the section would be opened out in fluid four to give rear vision. At the opposite extreme, flying low over hilly country, where the main problems are to keep the elements together and to avoid creaming the aircraft in the trees, the formation would be closed up almost into a finger four, with even the No 2s looking forward much of the time. At the first sign of rear attack the two elements would then make a split break (a fast 180° outward turn), and hope to sandwich their attackers.

Last day fighter
The Hunter was broadly required to be an improved F-86, with better high altitude performance and a cannon armament that would reduce any bomber to a fine dust. Only a handful of foreign pilots have ever tested in earnest the combined fire power of its four 30-mm Adens against the aircraft targets for which this armament was intended, but they include one character who wiped out five MiG-19s in a fit of pique! Present day thinking is that the Hunter was overgunned, a belief which — had it come in the early 1950s — might have avoided most of the thousand screaming agonies that marked its initial development.

Last minute rethinking on the airbrake delayed the Hunter's service introduction until 1954, when it was overshadowed as an interceptor by the USAF's supersonic F-100, but it later proved an excellent ground attack aircraft, and incidentally one of the best aerobatic fighters of all time. The surprising fact is that the FGA 9 achieved success in ground attack in spite of its poor internal fuel capacity and lack of really good pylon locations, and despite initial official reluctance even to consider it as a Venom replacement. It is some measure of the aircraft's worth that it continues in squadron service, and that the last Hunter will probably not be retired until the 1980s. RAF Hunters have mostly been used for attacks with guns and prehistoric "drainpipe" RPs, and even now their armament has not been developed to the same extent as export models, which variously carry 8-cm rockets, HE bombs, napalm, and Sidewinder.

The ground attack Hunter grosses 18,350 lb (8 350 kg) clean, and its 10,000 lb (4 540 kg) R-R Avon gives a maximum of 0·92 Mach at sea level and 0·94 at the tropopause. Almost identical speeds were achieved by its night fighter contemporary, the Gloster Javelin. Weighing approximately twice as much, the Javelin was fitted with two reheat R-R Sapphires of up to 13,400 lb (6 100 kg) and was in RAF

The shape of the 'seventies, for RAF fighter pilots, is well represented in this picture of a Hawker Siddeley Harrier GR Mk 1 accompanied by a McDonnell Phantom FGR Mk 2.

service from 1956 to 1968. This aircraft was the first fighter employed in Britain's air defence to be equipped with guided missiles, DH Firestreaks being introduced in 1958.

Just as the Venom was an interim type pending availability of the Hunter, so the Javelin had only a four-year lead over its Mach 2 replacement, the BAC Lightning. Going back to the mid-1950s, Air Ministry had issued OR 329 for an all-weather interceptor to carry two Red Deans, and later two Red Hebes, both types being far larger than any air-air missile now in service. The response from industry was a series of proposals for predictably heavy and expensive fighters, ranging from a relatively modest Hawker proposal with a single DH Gyron to an enormous Saunders-Roe delta-wing project, using two Gyrons and four Spectre rocket motors!

The high cost of such fighters, coming at a time when the defence threat was changing from manned bombers to ballistic missiles, led to the abandoning of the OR 329 programme. Instead it was decided to develop an operational interceptor from the English Electric P1, which had been flying since 1954. The big radar-guided missiles were replaced by the infra-red Firestreak and its Red Top development. According to the 1957 Defence White Paper, this was to be the last of the RAF's manned aircraft, after which their various rôles would be taken over by guided weapons.

The resultant Lightning is essentially an aerodynamicists' design, achieving extremely low values of wave drag and lift-dependent drag. The price paid is a vertical arrangement of the two 16,360 lb (7 450 kg) reheat Avons, a heavy wing structure, shortage of internal fuel (as on all previous British interceptors), and a main undercarriage that severely restricts external loads. The payoff comes in rate of climb and in sustained turning performance, in which the Lightning is reckoned to be far superior to both the MiG-21 and Mirage III (*sans* SEPR).

Notwithstanding its unconventional appearance, the Lightning is said to be a real "pilot's aeroplane", handling rather like a large Hunter. The systems complexity of such aircraft is illustrated, however, by the fact that 1st and 2nd line servicing personnel on a squadron of twelve Lightnings amount to 185 men, which is three times the number that serviced twelve Spitfires.

Buy American

US fighters have normally been acquired only to overcome temporary shortages, as in the case of Canadair-built F-86s, which were used from 1953 to 1956 pending large-scale availability of Hunters. However, the McDonnell Phantom II was purchased to last for a whole generation, to replace some Hunter squadrons in the ground attack rôle, to supplement and ultimately replace Lightnings in air defence,

and to provide fighter cover for naval operations following the retirement of RN carriers.

With deliveries commencing in 1968, the R-R Spey-powered F-4M was the first RAF fighter to use inertial navigation, having the same Ferranti system as the Harrier. It also carries Westinghouse AWG-10 pulsed Doppler radar, one of the few AI equipments capable of detecting low-flying aircraft. In addition to its impressive mix of Sparrow III and Sidewinder missiles, the Phantom II can carry an M-61 20-mm gun in a centre-line pod.

The Phantom is undoubtedly an extremely fine interceptor, and the virtual impossibility of designing anything with a significant improvement over its airframe-engine-avionics-missile combination (short of the F-14/F-15 cost ballpark) justifies its purchase as a Lightning replacement. However, whether anglicising the Phantom was better value for money than buying a J79-powered F-4J off the shelf is, in retrospect, very dubious.

The F-4 purchase was one result of the 1965 cancellation of the Hawker P1154 V/STOL fighter. Originally designed to meet a NATO requirement that fizzled out in 1962, this aircraft was then pursued as a national effort to replace both the RAF's Hunter and the RN's Sea Vixen, a notion which may have made more sense financially than it did technically. The P1154 affair failed to arouse the heat of debate that followed TSR-2 cancellation, but some opinions were voiced to the effect that Britain had thrown away the lead in V/STOL, mainly by journalists with no knowledge of its unimpressive BS100 engine.

The other substitute for the P1154 was the Hawker Harrier V/STOL ground attack and reconnaissance aircraft, which entered service in 1969 and is now deployed with RAF squadrons in the UK and Germany. In terms of suitability for world-wide operation from semi-prepared sites, the Harrier makes far more sense than the P1154 would have done, since there was no Western helicopter capable of economical logistic support for the latter, and no realistic way to assess its heightened ground erosion problem, short of forking out the money and building the aircraft. The Harrier restores to fighter operations the flexibility once enjoyed by piston-engined aircraft, to move their airstrips with the ebb and flow of war, to operate from small carriers, make use of minimal strips in island-hopping campaigns, and to get ashore fast on the heels of invasion forces.

The Harrier's maximum clean level speed sounds unimpressive (being little different from that of the Hunter), and conveys nothing of the thrust margin that exists lower down the scale. However, even the ageing Kestrel development aircraft is reported to have "eaten" a T-38 in NASA combat trials. The story is also told of an F-4 driver flying over

Patuxent, who found a Marine Harrier alongside with its pilot gesturing for a race: the former hit both burners and then watched in disbelief as the portly AV-8A accelerated out in front, *climbing*!

Aside from having a large thrust margin at low-level combat speeds, the Harrier can also cruise faster in LO-LO close support missions than nominally supersonic aircraft, including the F-4. The similarly-sized BAC Jaguar, which is scheduled for service next year and has a clean maximum of Mach 1·1 at sea level, has a dry thrust of just over 4 tons, whereas the Harrier has more than 9 tons. To give the other side of the picture, if the Harrier is all poke, the Jaguar is all juice, hence its very commendable 310 nm (570 km) LO-LO radius on internal capacity alone.

Tora! Tora!

Both Harrier and Jaguar have inertial navigation, optically-projected moving maps, and CRT-generated head-up displays, although interchangeability between the two types would make any taxpayer weep. These modern aids are not only giving better navigation and attack accuracies, but are changing flight profiles and the form of the attack itself. Head-up presentation of flight instrument and navigation data makes it possible to cruise much lower with safety. The head-up display of accurately computed target position means that for the first time an aircraft can make a straight run-in at low level.

In previous types a deliberately offset run-in for low attacks was necessitated by inaccurate navigation and poor view directly over the nose. A typical procedure would be to navigate to the right of the target, then pull up, moving into echelon starboard, searching to port and peeling off as the target was sighted. The principal alternative of remaining in finger four gave the pilots a clearer view and would be used when necessary to shorten the attack, but made it more difficult to break to either side after weapon delivery.

A special form of attack was developed for operations such as that in the Radfan, to give the best possible RP accuracy in the absence of flak. The target would be overflown low at 420 kt (775 km/h), followed by an 18-second straight run, 4g pull-up into a half-loop, roll off the top, and 30° dive attack. This set-piece operation naturally went out of fashion with the abrogation of Britain's policing rôle. In the present European scenario, the emphasis is on remaining low and making a first-pass attack, using the flexibility of the weapon aiming computer to allow variations in all the flight path parameters, and thus to cause Soviet fire control systems to predict inaccurately.

In recent years fighters and their tactics have been influenced mainly by operations in Vietnam, although information reaching Britain is incomplete and often out of date, since RAF pilots are not allowed to fly combat missions with either the USAF or RAAF. Considering this in a purely technical light, it *might* be argued that Britain's non-participation has had no serious effect on the development of RAF tactics, since neither the COIN operation in the South or the bombing of the North is strictly relevant to our main threat of a war in Europe.

On the other hand Vietnam has been the first war in which air-air guided missiles and flight refuelling in combat sorties have played any significant part, in which fighters have been controlled routinely from AEW aircraft, and manœuvres have been developed to evade SAMs. Furthermore, as a direct result of this war, US fighter development has leapt ahead in ECM equipment, laser applications for improved weapon delivery, and in protection against small calibre ground fire. Ground attack missions at night have become normal along the Ho Chi Minh trail, whereas the RAF has probably lost its expertise at using fighters in this way.

In some respects the fighter situation is strangely reminis-

cent of that in the 1930s. Transport aircraft are appearing with the same speed capability as the RAF's fastest interceptors. A generation of British pilots has grown up without having had a shot fired at them. Major air forces are amassing experience and developing equipment through conflicts in which the RAF has played no part. New technical advances are in hand (eg short-range, highly manœuvrable air-air missiles) which may revolutionise air warfare.

Supposing that the RAF is to have a new fighter, then what threat is it to counter? In addition to types now in service, American satellite information indicates that there is a variable-geometry *Backfire* Tu-22 replacement under active development. The writer's own far-flung sources (operating more in the saloon-bar orbit) credit the Chinese with a new bomber design (presumably equivalent to *Brewer*), and the Soviet Union with a redesigned Harrier and a ramjet-boosted MiG-25 interceptor.

Aside from the Soviet obsession with high altitude reconnaissance, the emphasis is likely to remain on low level operations, and in any event the Phantom can be modernised periodically to keep abreast of high level intruders. As far as a brand new design is concerned, the more important features are STOL performance (to operate regardless of runway bombing), substantially smaller size and cost than the F-15, and the ability to provide both ground attack and low level tactical air superiority in the period 1980-2000. Several foreign manufacturers already claim to be offering such characteristics in their new projects, but their planning seems to be based on the Soviet Union marking time for the next thirty years!

The major aircraft innovation now waiting to be exploited is the integration of powerplant and airframe for agility in combat. Taking the simplest form of integration, when the first fighter is cleared for thrust reversal at combat speeds, then all other types will be obsolete. Carrying integration one stage further, if engine manufacturers can produce a front fan engine to give high energy jets that are vectorable 180° throughout the flight envelope, then we might at last see a worthwhile advance in fighter design. But if we revert to the philosophy of the early 1930s, that the defence budget cannot run to an advanced technology (in those days: monoplane) fighter, then the RAF may once more be in for a very rude awakening! ☐

Throughout the 'sixties, the English Electric Lightning has provided the backbone of Britain's fighter defences and the latest versions remain in front-line service at home and abroad. Illustrated are Lightning F Mk 1As of No 111 Squadron.

THE FOKKER FOURS

PRODUCTION of airliners and transport aircraft of Fokker design and bearing designations in the F series has recently reached four figures, with the continuing production of the highly successful F.27 Friendship and the new F.28 Fellowship. Although precise figures for production of the early F types are almost impossible to establish, it appears that the F.2 to the F.36* in the pre-war period accounted for just over 500 airframes. The F.27 and F.28 alone account for the second half-century.

Among the Fokker designs making up this total (including those built under licence in other countries) are numerous single, twin- and three-engined types — but only five four-engined transports have been built by the parent company, and a handful more by its American associate. These four-engined Fokkers, all of pre-war design and construction, represented the final expression of the Fokker design philosophy that had placed the company in the fore of commercial transport manufacturers for a decade. Seeking to extrapolate this successful formula into larger sizes, Fokker produced the F.22 and F.36 designs too late to capture a market that was already in process of being revolutionised by the Douglas DC-2 of very much more advanced concept.

Chronologically, the first four-engined Fokker was the F.32 of 1929, a product of the US Fokker Aircraft Corporation. Of conventional Fokker wooden construction and with a cantilever high wing, the F.32 was novel chiefly for its power plant, comprising four 525 hp Pratt & Whitney Wasp or Hornet radials mounted in fore and aft pairs in nacelles slung under each wing, two engines driving tractor propellers and two pushers.

The F.32 was sponsored by Universal Air Lines, which intended to order five, but this plan came to nought after the prototype crashed on 27 November 1929, within weeks of its first flight (on 13 September) following a double engine failure during take-off. The only user of the F.32, therefore, was Western Air Express, which used five on its Los Angeles-San Francisco service for several years. One other was operated as a VIP transport for Anthony Fokker himself during his visits to the US, but following the acquisition of a 40 per cent stake in Fokker Aircraft Corporation by General Motors earlier in 1929, these became increasingly infrequent and in July 1931 he severed his connection with the company completely.

The F.32 is of interest in one other respect. Had Fokker followed its customary designation procedures, this aircraft would probably have become the FXII, in the sequence of "F" numbers used for civil aircraft since 1919, when the FI was designed. However, the American operators, with an eye on publicity, requested Fokker to use a number indicating the number of seats. The F.32 had 32 passenger seats — and its method of designation set a precedent for the European four-engined Fokkers.

KLM influence

The first four-engined transport design by Fokker NV Nederlandsche Vliegtuigenfabriek appears to have been the FXIX, with four 250 hp Gnôme-Rhône Titan engines. This project, dated 1931-2, was not built, but work started at about the same time on the F.36 to meet a KLM requirement for a larger transport that it could use on the route to the Far East. A provisional commitment was made by KLM for six aircraft.

Initially identified as the F.Y, the design took shape during 1933 as a very conventional high-wing monoplane, with a fixed undercarriage and seats for 32 passengers. The crew of four was added to this total to obtain the designation F.36, since F.32 had already been used.

The wing was attached to the fuselage by four bolts, the

* The Fokker company originally used Roman numerals for its aircraft designations, continuing the German practice of World War I. This practice gradually gave way to the use of Arabic numerals, which were in use in Fokker literature by 1936, although many publications continued to use Roman designations after that date. For consistency, Arabic numerals are used throughout this article.

underside of the centre section where it passed through the fuselage being left uncovered between the main spar to permit greater head room. Fuselage structure was of patented Fokker design using welded chrome-molybdenum steel tubes, with fabric covering except for the nose section, which was plywood covered. Tailplane and fin were also steel tube structures with fabric covering; the elevators and rudder were balanced aerodynamically and statically and incorporated trim tabs.

A contemporary Fokker brochure offered the following description of the accommodation in the fuselage; it is reproduced here verbatim as an interesting example of "brochuremanship" 35 years ago!

"COCKPIT — The pilots' cockpit is situated in the nose of the fuselage where the pilots have a clear view on all sides. The sliding windows are rain, snow and draught proof. There is a separate entrance for the crew. Besides being adjustable in a vertical direction the seats may be tilted backwards, a fact which ensures less fatigue on long flights. The wireless operator's seat is to the rear of the pilots on the right, whilst the mechanic sits on the left.

"GALLEY — On the right to the rear of the cockpit is the galley and on the left a compartment for the stewards. The galley is equipped with an electric cooker.

"PASSENGER CABIN — The cabin for the passengers is separated from the cockpit by a sliding door and is divided into 4 equal compartments each containing 8 seats for day flying or 4 sleeping berths for night flying. An aisle runs down the entire length of the cabin. A folding table is affixed to the wall on either side in each compartment of the cabin. The entrance door to the cabin is to the rear of the aftermost compartment.

"The problem of sound proofing, heating and ventilating has been thoroughly studied by Fokker for a number of years and effective provision has been made in accordance with the experience acquired.

"LAVATORY COMPARTMENTS — To the rear of the cabin there are two lavatories and a dressing room."

Baggage compartments were located in the wings: one each side behind the rear spar, accessible from the cabin, and two others between the main spars and between the engine nacelles. Fuel was carried in two tanks between the spars inboard of the inner engines, with a total capacity of 748 Imp gal (3 400 l), and each tank had a motor driven pump plus a manual pump for emergency use.

Registered PH-AJA, the F.36 made its first flight at Schiphol on 23 June 1934, piloted by Fokker's German chief test pilot at that time, Emil Meinecke. It was, and was to

remain, the largest aeroplane ever built in Holland. During its construction, the F.36 was entered for the MacRobertson Trophy Race from Mildenhall to Melbourne, held in October 1934, but was not ready in time. Appropriately, as it happened, the handicap section of the race was won by KLM with its first Douglas DC-2 — for by 1934, the Dutch airline under its dynamic president Dr A Plesman had already realised that aircraft like the F.36 were out of date and that the future lay with the sleek, all-metal monoplanes appearing from the American west coast.

Plans for the purchase of more F.36s were abandoned by KLM but the company did accept the prototype in March 1935, naming it *Arend* (Eagle) for use on routes in Europe. Thereafter, it became a frequent visitor to Croydon on the service from Amsterdam until sold by KLM in 1939. Before delivery to KLM, the F.36 made a demonstration flight to the UK, landing at Croydon on 28 September 1934, accompanied by Anthony Fokker himself. The visit provided a convenient opportunity for discussions between Fokker and the Airspeed Company, which was eager to obtain rights to produce and sell various of the Fokker types. Agreement was eventually reached in January 1935*, covering several Fokker types and the Douglas DC-2 (for which Fokker held the European licence), whereupon Airspeed assigned their type number AS.20 for possible production of the F.36, and AS.16 for the smaller F.22 that was then under construction in Holland.

Under the terms of the agreement, Fokker received an outright fee of £20,000, plus royalties on any Fokker designs or DC-2s built; in the event none were. Incredibly, however, the deal also assigned Fokker one per cent of Airspeed's gross receipts on all production — a provision the company came to regret when production of its own designs, including the Oxford, snowballed in the years immediately prior to the War, while no extra business resulted from the deal with Fokker. Anthony Fokker himself became technical adviser to Airspeed.

(Below) The largest Fokker built in Holland, the F.36 PH-AJA, photographed in its original colours for KLM. (Above right) The same aircraft as G-AFZR in RAF colours at Prestwick circa 1940 (photo by C A Nepean Bishop via A J Jackson). Heading the opposite page is the Fokker F.22 PH-AJR in KLM colours.

(Top) The first F.22, SE-ABA, used by A B Aerotransport, had minor differences from the KLM version. (Above) The last of the four-engined Fokkers, F.22 G-AFZP, at Prestwick in 1946.

This line-up at Prestwick in December 1939 shows the two F.22s and (distinguished by its three-bladed propellers, nearest the camera) the sole F.36.

Work on the F.22 had started in 1934 on the basis of a scaled-down F.36 for KLM's European operations. Similar in all respects to the larger prototype, it had 10 ft (3,05 m) less span, was 7 ft (2,13 m) shorter and as the designation indicated seated 22 passengers — although in this case, the crew was *not* included in this number. The engines were 500 hp Pratt & Whitney Wasp T1D1s.

Four F.22s were built, two being delivered in March 1935 and two in May. One of the first two went to A B Aerotransport in Sweden as SE-ABA *Lappland* and was used by this company (which was one of the founders of SAS after World War II) on a service between Malmö and Amsterdam until June 1936, when it was written-off in an accident at Malmö.

KLM accepted the other three F.22s as PH-AJP *Papagaai* (Parrot), PH-AJQ *Kwikstaart* (Wagtail) and PH-AJR

Roerdomp (Bittern) and used them on various European routes. *Kwikstaart* survived only two months of service, however, being written-off at Schiphol on 14 July 1935. Thus, by 1939, three of the four-engined Fokkers remained in service with KLM — two F.22s and one F.36.

One final attempt was made by Fokker to prolong the life of this type of transport by offering versions of both with retractable undercarriages. A 1936 brochure described the F.36 with 850 hp Wright Cyclone engines and the F.23 with 550 hp Pratt & Whitney Wasp S3H1-Gs or 650 hp Gnôme-Rhône 9 Kfr radials. Apart from the powerplant, retractable landing gear and different weights and performance, these two projects were identical with the F.36 and F.22 respectively. Neither was built.

Another project of about the same period was the F.56, a four-engined mid-wing monoplane with twin fins and rudders

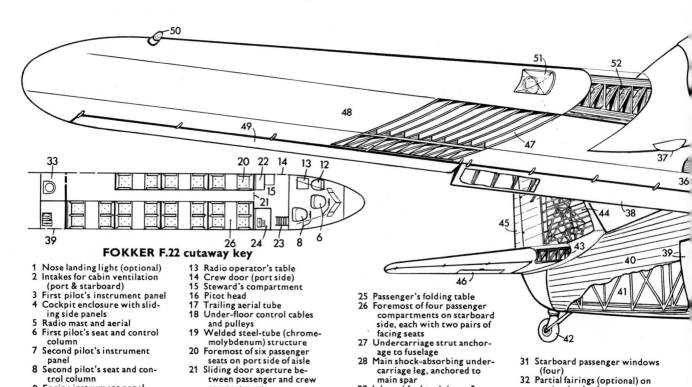

FOKKER F.22 cutaway key

1 Nose landing light (optional)
2 Intakes for cabin ventilation (port & starboard)
3 First pilot's instrument panel
4 Cockpit enclosure with sliding side panels
5 Radio mast and aerial
6 First pilot's seat and control column
7 Second pilot's instrument panel
8 Second pilot's seat and control column
9 Engine instrument panel
10 Compass
11 Control cables and pulleys
12 Radio operator's seat
13 Radio operator's table
14 Crew door (port side)
15 Steward's compartment
16 Pitot head
17 Trailing aerial tube
18 Under-floor control cables and pulleys
19 Welded steel-tube (chrome-molybdenum) structure
20 Foremost of six passenger seats on port side of aisle
21 Sliding door aperture between passenger and crew compartments
22 Steward's store-cupboard
23 Access ladder to flight-deck
24 Forward cargo and baggage bay
25 Passenger's folding table
26 Foremost of four passenger compartments on starboard side, each with two pairs of facing seats
27 Undercarriage strut anchorage to fuselage
28 Main shock-absorbing undercarriage leg, anchored to main spar
29 Inboard fuel tank (port & starboard)
30 Baggage bay in wing (port & starboard, accessible from cabin)
31 Starboard passenger windows (four)
32 Partial fairings (optional) on mainwheels of non-retracting undercarriage
33 Toilet on port side (starboard passenger compartment omitted for clarity)

The Fokker F.22 G-AFZP at Prestwick in 1940, after modification while in service with No 12 EFTS operated by Scottish Aviation. Most noticeable modification is the extension of window area in the cabin. Compasses for use by the student navigators can be seen within the cabin, and the unusual style of fin flash is well shown in this picture.

and of somewhat more modern appearance than those described above, but still relying upon Fokker's classical wooden wing structure and fabric-covered metal fuselage. The number, F.56, again indicated the number of passengers. Other four-engined projects included the 60-seat F.60 with a transatlantic capability and the interesting F.180, with a twin-boom layout and the passenger accommodation buried in the wing centre-section, Burnelli-fashion.

By 1936, the Fokker constructional techniques were clearly outmoded. Realising that it could not embark on a completely new kind of airliner from scratch so soon after the F.36, in which the company invested some 500,000 guilders of its own money, Fokker had concluded an agreement with Douglas in January 1934 giving it rights to build and market the DC-2 in Europe. In the event, no production of the Douglas type took place, however, and 20 years were

to elapse before the next Fokker transport, the F.27, emerged.

Wartime service

By 1939, KLM had no more need of its four-engined Fokkers, and put them up for sale. The first to go was the F.22 *Roerdomp*, which was acquired by British American Air Services Ltd at Heston (a company then being run by Roly Falk, later to achieve fame as an Avro test pilot on the Vulcan). It arrived in the UK in August 1939 and was registered G-AFXR, but did little flying with its new owner.

Then, in September, Scottish Aviation became interested in acquiring the remaining F.22 and the F.36, for use as flying navigation classrooms. This Prestwick-based company was at that time responsible for running No 12 EFTS (Elementary Flying Training School) and was under pressure

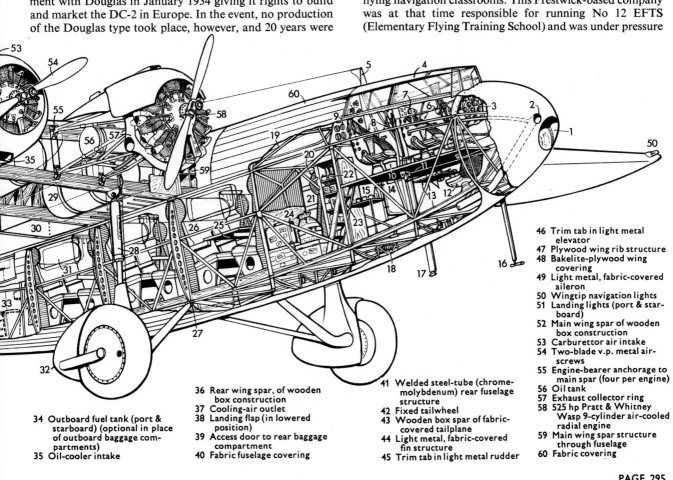

34 Outboard fuel tank (port & starboard) (optional in place of outboard baggage compartments)
35 Oil-cooler intake
36 Rear wing spar, of wooden box construction
37 Cooling-air outlet
38 Landing flap (in lowered position)
39 Access door to rear baggage compartment
40 Fabric fuselage covering
41 Welded steel-tube (chrome-molybdenum) rear fuselage structure
42 Fixed tailwheel
43 Wooden box spar of fabric-covered tailplane
44 Light metal, fabric-covered fin structure
45 Trim tab in light metal rudder
46 Trim tab in light metal elevator
47 Plywood wing rib structure
48 Bakelite-plywood wing covering
49 Light metal, fabric-covered aileron
50 Wingtip navigation lights
51 Landing lights (port & starboard)
52 Main wing spar of wooden box construction
53 Carburettor air intake
54 Two-blade v.p. metal air-screws
55 Engine-bearer anchorage to main spar (four per engine)
56 Oil tank
57 Exhaust collector ring
58 525 hp Pratt & Whitney Wasp 9-cylinder air-cooled radial engine
59 Main wing spar structure through fuselage
60 Fabric covering

		F.22	F.23	F.23	F.36	F.37	F.56
Power plant		P & W T1D1	P & W S3H1-G	G-R 9Kfr	Wright SGR-1820-F2	Wright R-1820-G	Hispano 14 HA
Power, each engine	hp	500	550	650	750	850	1,120
Span	ft in (m)	98 5¼ (30)	98 5¼ (30)	98 5¼ (30)	108 3¼ (33,0)	108 3¼ (33,0)	126 4 (38,5)
Length	ft in (m)	70 7¼ (21,52)	71 10 (21,9)	71 10 (21,9)	78 9 (24,0)	78 9 (24,0)	84 8 (25,8)
Height	ft in (m)	16 1 (4,9)	16 1 (4,9)	16 1 (4,9)	19 8 (6,0)	19 8 (6,0)	21 2 (6,45)
Wing area	sq ft (m²)	1,400 (130)	1,400 (130)	1,400 (130)	1,830 (170)	1,830 (170)	2,153 (200)
Empty weight	lb (kg)	17,857 (8 100)	18,520 (8 400)	18,805 (8 530)	21,825 (9 900)	23,920 (10 850)	33,620 (15 250)
Useful load	lb (kg)	10,803 (4 900)	11,245 (5 100)	12,060 (5,470)	14,551 (6 600)	19,070 (8 650)	15,985 (7 250)
Gross weight	lb (kg)	28,660 (13 000)	29,762 (13,500)	30,865 (14 000)	36,376 (16 500)	42,990 (19 500)	49,605 (22 500)
Max speed	mph (km/h)	177 (285)	189·5 (305)	202 (325)	180 (290)	214 (344)	220·5 (355)
at altitude of	ft (m)	—	—	—	—	8,040 (2 450)	9,265 (2 825)
Cruising speed	mph (km/h)	133·5 (215)	168 (270)	179 (288)	149 (240)	168 (270)	180 (290)
at altitude of	ft (m)	—	13,123 (4 000)	14,763 (4 500)	—	6,560 (2 000)	6,560 (2 000)
Service ceiling	ft (m)	16,076 (4 900)	14,895 (4 550)	17,060 (5 200)	14,435 (4 400)	21,325 (6 500)	23,786 (7 250)
Range	miles (km)	838 (1 350)	(1 760)	(1 580)	838 (1 350)	(1 600)	(1 520)
Number built		4	0	0	1	0	0

to expand as war became imminent. The well-known second-hand aircraft dealer and broker, W S "Bill" Shackleton, was entrusted with the task of negotiating purchase of the Fokkers from KLM, and that they ever reached Britain at all was due largely to his astuteness. On the eve of flying to Schiphol to conclude delivery arrangements, he recalled that at the time of the Munich crisis in 1938, the Dutch government had put a temporary but total ban on all sales of aircraft in Holland. With events in Europe moving swiftly to their inevitable climax in September 1939, Shackleton realised that a similar ban might again be imposed. He therefore took the precaution of having Scottish Aviation's bank, as soon as it opened next day, cable the purchase price of the Fokkers to KLM in advance of his visit. The transaction was consequently concluded a matter of hours before the Dutch government did again "freeze" all such deals, on the very same day that the money was credited to KLM.

Arriving in the UK later in September, the F.22 *Papagaai* became G-AFZP and the F.36 became G-AFZR. In November they were joined at Prestwick by G-AFXR, purchased from British American Air Services. For their operations with No 12 EFTS, the Fokkers were painted in "training" camouflage, green and brown upper surfaces and yellow undersides, with RAF roundels, but retained their civil registrations on the fuselage and beneath the wings. The F.36 was lost on 21 May 1940, when it overran the airport boundary during a take-off attempt, and caught fire.

Both the F.22s were transferred to No 1 Air Observers'

Navigation School in 1940 and formally impressed into the RAF on 15 October 1941, when 'FXR became HM159 and 'FZP became HM160. The former operated during 1942 with No 24 Squadron, bearing the name *Brontosaurus;* and then joined No 1680 Flight at RAF Abbotsinch, where it was renamed *Sylvia Scarlet.* The end came for this Fokker on 3 July 1943, when it caught fire in the air and crashed into Loch Tarbert, killing all 20 occupants.

The last surviving four-engined Fokker, HM160, also served with No 24 Squadron during 1942. On 3 April 1943, it suffered an engine fire while attempting to take off at Prestwick, where it then remained, grounded, for the rest of the war. Scottish Aviation re-acquired the F.22 from the RAF in June 1944, and after complete overhaul it made its first flight again on 18 October 1946. It was painted in the livery of Scottish Airlines but operated on lease to BEA from the end of 1946 to August 1947 on the Prestwick-Belfast route.

Uneconomic, hard to maintain and with marginal performance, the F.22 could hardly be expected to survive once the immediate post-war shortage of aircraft had been overcome. Outmoded before its first flight, this F.22 had nevertheless given useful service in varying and difficult circumstances. It was withdrawn from use in 1947 and eventually broken-up and burnt at Prestwick in July 1952, almost exactly 20 years after its delivery to KLM. Few were there to mourn the passing of an era — the age of the four-engined Fokkers. □

(Below) The Fokker F.22 in the form in which it was delivered to AB Aerotransport.

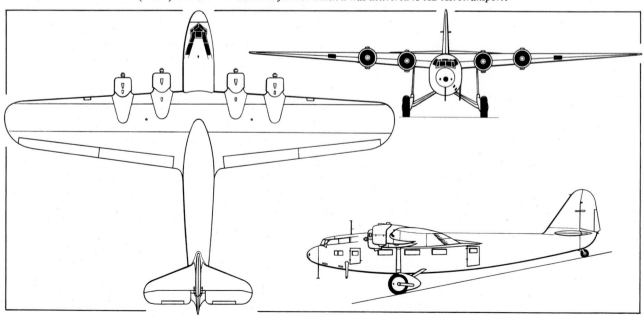

LYULKA

a Soviet pioneer

Gᴇɴᴇʀᴀʟ Constructor Arkhip Lyulka, a self-effacing, modest man with a highly-developed sense of humour and a craggy face ready to break into a broad smile at the least provocation, has for long been honoured in the Soviet Union as one of the most eminent technicians of the aero engine industry. At 62 years of age, Lyulka is today a professor at the Moscow Aviation Institute, a Deputy of the Soviet of Moscow, and an Academician of the Sciences of the USSR, and among the many high honours that have been bestowed upon him are the Stalin Prize, the Order of the Red Flag, and the title of Hero of Soviet Labour. Yet outside the borders of the Soviet Union he is virtually unknown.

It was in 1936 that Arkhip Lyulka, at the age of 28, first began to take an interest in the possibilities of jet propulsion. At that time the holder of a chair at the Kharkov Aviation Institute and a corresponding member of the Academy of Sciences of the USSR, Lyulka checked the theoretical work of Academician B S Stechkine and the Frenchman Maurice Roy, and was convinced of the feasibility of the gas turbine as a means of propelling aircraft. His preliminary calculations met with official scepticism, which was hardly surprising in view of his claim that, with an engine thrust of 1,157 lb (525 kg), he could increase the speed of a fighter from the 280 mph (450 km/h) that was the maximum attainable by warplanes in this category of the period to something of the order of 560 mph (900 km/h)!

Despite this scepticism, Lyulka, aided by two colleagues, I F Kozlov and P S Shevchenko, began work on a prototype gas turbine in the Leningrad works of S M Kirov. Designated VRD-1 and weighing 1,102 lb (500 kg), this power plant had an axial eight-stage compressor, and steel turbine blades. The designed turbine inlet temperature was 700° C (1,292° F) and the calculated thrust was 1,323 lb (600 kg). Work on the VRD-1 was interrupted by the war, the Leningrad factory being evacuated, and as long-term research was now considered to be of little importance, Lyulka had to turn his

In the West the name Arkhip Mikhailovich Lyulka is unfamiliar to all but a few, yet, in the Soviet Union, it is honoured in much the same fashion as is that of Sir Frank Whittle in Britain, for Lyulka is the Soviet father figure in the development of gas turbine propulsion for aircraft. Few western works of reference mention the rôle played by this eminent Soviet engineer in the overall picture of jet propulsion evolution, and Jacques Marmain, who recently interviewed this pioneer of the turbojet, outlines here the career of Arkhip Lyulka, the creator of the first indigenous Soviet aircraft gas turbine.

talents to the development of tank engines. Early in 1942, however, he was sent to a factory near Sverdlovsk where a team under the overall supervision of Professor Viktor F Bolkhovitinov had been working for some time on a rocket-propelled fighter, the BI designed primarily by Aleksander Y Bereznyak. This fighter had been conceived by Aleksei Isaev, who, together with Leonid Duskin, was endeavouring to perfect the 2,425 lb (1 100 kg) thrust D-1A-1100 liquid-fuel rocket motor. Arkhip Lyulka was assigned to the special task force which was endeavouring to overcome the serious problems that were being presented by this highly volatile power plant, but at the end of 1942, at which time the BI was under test at the nearby Koltsova airfield, Lyulka was permitted to return to Leningrad and, with an almost constant background noise of gunfire, resumed work on his turbojet.

At this time, Lyulka was given the task of developing a 2,866 lb (1 300 kg) thrust turbojet in two versions — one with an axial compressor and the other with a centrifugal compressor. After developing and bench-running an experimental 1,543 lb (700 kg) thrust engine, the VRD-2, he concluded that the axial compressor offered the greatest promise,

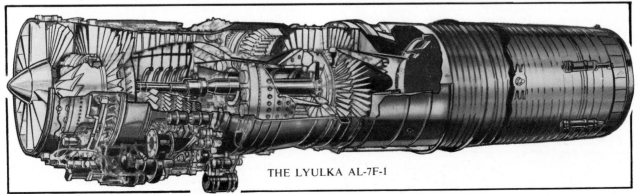

THE LYULKA AL-7F-1

(Above) The first aircraft to fly with turbojets designed by Arkhip Lyulka was the Ilyushin Il-22 flown on 24 July 1947 with four 2,866 lb (1 000 kg) TR-1 turbojets. The Il-22 carried a crew of five and a maximum internal bomb load of 6,614 lb (3 000 kg). With 2,475 Imp gal (11 250 l) of fuel, the Il-22 had a maximum range of 1,160 miles (1 865 km), and performance included a top speed of 446 mph (718 km/h). Overall dimensions included a wing span of 75 ft 5¾ in (23,06 m), a length of 68 ft 11 in. (21,05 m), and a gross wing area of 801·91 sq ft (74,5 m²).

(Above) The Sukhoi Su-11 was another unsuccessful prototype powered by the Lyulka TR-1. A single-seat fighter with a pressurised cockpit, the Su-11 had provision for an armament of two 23-mm and one 37-mm cannon, and, completed late in 1947, weighed 9,910 lb (4 495 kg) empty and 13,838 lb (6 277 kg) loaded. Performance included maximum speeds of 503 mph (810 km/h) at sea level and 528 mph (850 km/h) at 26,250 ft (8 000 m), range being 1,243 miles (2 000 km). Overall dimensions included a span of 38 ft 8½ in (11,80 m), a length of 34 ft 7¼ in (10,55 m) and a gross wing area of 230·35 sq ft (21,4 m²).

(Above) The Lyulka AL-5 at its initial rating of 9,921 lb (4 500 kg) was installed in the Ilyushin Il-30 twin-engined tactical bomber which carried a maximum bomb load of 6,614 lb (3 000 kg). Maximum speed was of the order of 621 mph (1 000 km/h), and a 4,410-lb (2 000-kg) bomb load could be carried over a distance of 2,175 miles (3 500 km). A crew of four was carried and the wings, which had a span of 54 ft 1½ in (16,50 m), were swept 35 deg on the leading edges, overall length being 59 ft 0⅔ in (18,00 m).

Powered by two 11,023 lb (5 000 kg) AL-5s, the Ilyushin Il-46 competed with the Tupolev Tu-16 for production orders. Maximum speed was 576 mph (928 km/h) at 9,840 ft (3 000 m), and up to 13,228 lb (6 000 kg) of bombs could be carried. Maximum range with a 6,614-lb (3 000-kg) bomb load was 3,076 miles (4 950 km), a crew of three was carried, and empty and loaded weights were 54,156 lb (24 565 kg) and 92,594 lb (42 000 kg) respectively, while span and length were 98 ft 5⅛ in (30,00 m) and 88 ft 7⅛ in (27,00 m). The Il-46 commenced its flight test programme on 15 August 1952.

and, with the assistance of I F Kozlov and S P Kouvchyuni-kov, Lyulka began work on the VRD-3, also known as the S-18 and later as the TR-1, which comprised a diffuser cone, an eight-stage axial compressor, an annular combustion chamber, a single-stage turbine and a fixed nozzle. The diffuser cone also served as an oil tank and oil cooler. During the construction of this engine, Lyulka succeeded in reducing the forecast dry weight of 2,205 lb (1 000 kg) to 1,951 lb (885 kg) which resulted in what was, for the period, a very creditable thrust to weight ratio of 1·46, maximum output being 2,866 lb (1 300 kg), thrust and turbine speed being 6,950 rpm.

At the end of 1944 the VRD-3, now known as the TR-1, successfully completed official bench-running trials, and in the following year work began on a small production series of engines, the first application of these being the Ilyushin Il-22 four-jet bomber which was flown for the first time on 24 July 1947. A shoulder-wing monoplane with four TR-1 turbojets mounted in individual underwing pods, the Il-22 made a brief public début in the Aviation Day Parade held over Tushino on 3 August, but the career of the sole proto-type was destined to be short, terminating on 22 September, barely two months after its inaugural flight. The TR-1 also found application in another unsuccessful prototype, Pavel Sukhoi's Su-11 twin-jet fighter, which was flown in 1948, being a re-engined version of the Su-9 that had commenced its test programme in 1946 with two RD-10 (Junkers Jumo 004A) turbojets. Pavel Sukhoi had proposed to employ an uprated version of Lyulka's engine, the TR-1A of 3,307 lb (1 500 kg) thrust, in his Su-10 multi-jet bomber which was to have had the TR-1As superimposed one above the other and slightly staggered in two wing nacelles, but al-though the prototype was virtually complete at the beginning of 1948, the development programme was cancelled and the aircraft was never flown.

A more ambitious engine

During 1946, Arkhip Lyulka and his team had initiated the design of a very much more ambitious turbojet of 9,921 lb (4 500 kg) thrust, the VRD-5 or TR-3 with a seven-stage compressor, an automatic hydraulic regulator and a starter pack. Trials with this engine began in 1950, and it was at this time allocated the designation AL-5, signifying the fact that Arkhip Lyulka had been awarded the title of General Constructor. Production began at an initial rating of 10,140 lb (4 600 kg) thrust, and the AL-5 was selected for installation in the Ilyushin Il-30 twin-engined tactical bomber with swept wing and tail surfaces, and a widely-spaced "bicycle" type undercarriage. During the course of flight trials in 1951, the Il-30 exceeded the then magical figure of 1 000 km/h (621 mph) in level flight, but develop-ment was not pursued beyond the prototype stage. The AL-5

Built in 1954 and flown early in 1955, the Ilyushin Il-54 three-seat light tactical bomber was one of the first aircraft to receive Lyulka's AL-7 of 14,330 lb (6 500 kg) thrust. The wing was swept 55 deg on the leading edge and had a span of 58 ft 4¾ in (17,80 m), overall length being 71 ft 6¼ in (21,80 m). Like the earlier Il-30, it featured a widely-spaced "bicycle" undercarriage with twin-wheel main members and small outrigger wheels, and performance included a maximum speed of 714 mph (1 150 km/h) at low altitude, and a range of 1,490 miles (2 400 km), loaded weight being 63,934 lb (29 000 kg).

of 10,140 lb (4 600 kg) thrust was also to have been installed in the Sukhoi Su-17 which, it was anticipated, would attain Mach 1·08 at 36,090 ft (11 000 m), but work on the prototype of this fighter terminated with the disbanding of the Sukhoi bureau.

The AL-5 was meanwhile uprated to 11,023 lb (5 000 kg) thrust, and in this form powered two experimental fighters, the Lavochkin La-190 and the Yakovlev Yak-1000, both of which flew in 1951. These fighters each had a single AL-5 turbojet and were, for their time, somewhat exotic in concept, a fact which may have motivated against their adoption for series production. The La-190 featured 55 deg of wing sweep and a delta tailplane, while the Yak-1000 had a wing of clipped-delta form, and both had zero-track tricycle undercarriages with small outrigger stabilising wheels. These fighters were both marginally supersonic but their capabilities were evidently eclipsed by those of the MiG-19 which was to gain the distinction of becoming the first genuinely supersonic single-seat fighter to enter the inventory of the V-VS.

Misfortune continued to dog the AL-5 in that each airframe for which it was selected failed to progress further than the prototype stage. The Ilyushin Il-46 medium bomber powered by two AL-5s and flown initially on 15 August 1952 was discarded in favour of its Tupolev competitor, the Tu-16, and several years later, in 1957, when the Tupolev Tu-110 commercial transport made its début with four AL-5s uprated to 12,125 lb (5 500 kg) thrust, it was held to possess no advantages over the Tu-104 from which it had been derived. Meanwhile Arkhip Lyulka had been investigating the prob-

(Above) The Lavochkin La-190 experimental fighter of 1951 received the uprated AL-5 of 11,023 lb (5 000 kg) thrust, flying with this power plant in February of that year. Weighing 20,408 lb (9 257 kg) in loaded condition, the La-190 attained a maximum speed of 750 mph (1 190 km/h) at 16,400 ft (5 000 m) and a range of 715 miles (1 150 km). Armament comprised two 37-mm cannon, and overall dimensions included a span of 32 ft 5¾ in (9,90 m) and a length of 53 ft 7¾ in (16,35 m).

(Below) Contemporary of the La-190 and similarly powered by an AL-5 turbojet, the Yak-1000 transonic fighter featured a clipped narrow-delta wing planform with tapered trailing edges, a tailplane of similar configuration, and a zero-track tricycle undercarriage with outrigger stabilising wheels. A maximum speed of 714 mph (1 150 km/h) was anticipated, but its characteristics were allegedly poor and trials were abandoned at an early stage.

Lyulka's AL-5 turbojet saw commercial application for the first time in 1957 with the début of the Tupolev Tu-110 powered by four uprated AL-5s of 12,125 lb (5 500 kg) thrust. Possessing an all-up weight of 165,347 lb (75 000 kg), the Tu-110 had a high-speed cruise of 559 mph (900 km/h) at 32,810 ft (10 000 m), and a range with one hour's reserves and a 26,455-lb (12 000-kg) payload of 2,050 miles (3 300 km). Overall dimensions included a span of 112 ft 3⅝ in (34,54 m), a length of 126 ft 3¾ in (38,50 m), and a wing area of 1,829·86 sq ft (170 m²). The Tu-110 offered no advantage over the Tu-104 and was not, therefore, placed in production.

The first production application of the afterburning AL-7F engine was in the Sukhoi Su-7 single-seat ground attack fighter, which, first flown in 1956, has been in continuous production for more than a dozen years. Several versions of the Su-7 have been produced in quantity, including the two-seat Su-7UTI illustrated here, and the principal version of Lyulka's engine employed by these is the AL-7F-1 rated at 15,432 lb (7 000 kg) thrust dry and 22,046 lb (10 000 kg) with afterburning. The Su-7MF, which has improved short-field characteristics, has a normal take-off weight of the order of 26,455 lb (12 000 kg) and an overload weight of some 30,865 lb (14 000 kg), and maximum speed without external stores is approximately 720 mph (1 160 km/h) at 1,000 ft (305 m), or Mach 0·95, and 1,056 mph (1 700 km/h) at 39,370 ft (12 000 m).

lems of the supersonic compressor which promised a substantial reduction in both the size and weight of turbojets, the fruits of this work being embodied in the TR-7 which, with an axial compressor and supersonic stages, commenced bench running in 1952 and had reached production in 1954 as the AL-7 with an initial rating of 14,330 lb (6 500 kg) thrust.

One of its first installations was in the Ilyushin Il-54 light tactical bomber which, with two 14,330 lb (6 500 kg) thrust AL-7 turbojets, attained a speed of 714 mph (1 150 km/h) during trials in 1955, this speed representing a Mach number of approximately 0·93 at the altitude flown. While the Il-54 failed to find favour, the fortunes of Arkhip Lyulka's engines were about to see a turn for the better. During the course of 1955, the AL-7 had been fitted with an afterburner affording some 40 per cent increase in thrust, and the afterburning version, the AL-7F, had been selected for installation in the Sukhoi Su-7 single-seat ground attack fighter, the prototype of which was completed in 1956, and in the delta-winged Sukhoi Su-9 all-weather interceptor that had been developed in parallel. Both types were to be manufactured in substantial numbers, powered by progressively uprated versions of Lyulka's engine, the principal model being the AL-7F-1 rated at 15,432 lb (7 000 kg) dry and 22,046 lb (10 000 kg) with full afterburning.

The year 1956 saw the AL-7F flying in the Tupolev Tu-98 and the Lavochkin La-250. The former was a derivative of the Tu-16 medium bomber with two AL-7F afterburning turbojets mounted side-by-side in the aft fuselage, and attained a speed of 769 mph (1 238 km/h) at an altitude of 39,370 ft (12 000 m), or Mach 1·16. The similarly-powered La-250, known unofficially as the *Anaconda,* was a large tailed-delta designed to meet a requirement for an all-weather interceptor and strike-reconnaissance aircraft. The test programme of the La-250 was punctuated by a series of accidents, and development was finally abandoned in 1958. At about this time, a non-afterburning version of the engine, the AL-7PB rated at 14,330 lb (6 500 kg) thrust, was adopted for the Beriev Be-10 maritime patrol and reconnaissance flying boat, which, in 1961, established several impressive class records.

Apart from his large military turbojets, Arkhip Lyulka has produced the small TS-31M turbojet, which, weighing only 51 lb (23 kg) and affording a thrust of 121 lb (55 kg), is used by the Antonov An-13 powered glider, and the TS-20 110 lb (50 kg) thrust APU. Concerning his latest production military turbojet, the AL-9, General Constructor Lyulka is not forthcoming, leaving us to speculate as to the aircraft that it powers, and we would guess that the Sukhoi Su-11 all-weather interceptor is a likely contender. □

The only pure jet flying boat ever to have attained service status, the Beriev Be-10 was powered by two Lyulka AL-7PB turbojets each rated at 14,330 lb (6 500 kg) thrust. The Be-10 was manufactured in limited numbers during the early 'sixties for the maritime patrol and reconnaissance rôle, and is believed to have had empty and gross weights of some 53,000 lb (24 040 kg) and 90,000 lb (40 823 kg) respectively. Operational radius was probably of the order of 1,300 miles (2 100 km) with a nominal weapons load of 4,400 lb (2 000 kg), and dash speed allegedly approached 578 mph (930 km/h).

BEST HIT

Stephen Peltz reports from Eskisehir, Turkey, on the latest AIRSOUTH Fighter Weapons Meet in which Greek, Italian and Turkish teams, plus a combined US Navy-USAF guest team, competed for eight trophies

THE 1971 AIRSOUTH (Allied Forces Southern Europe) Fighter Weapons Meet, "Best Hit", held here at Eskisehir, two-and-a-half hours' drive along a winding road over the Anatolian Plateau from Ankara, between 6 and 17 July, was a noteworthy event from several aspects. This year's meet brought together for the first time after a lapse of 15 years teams from Italy, Turkey and Greece, and whereas in the mid 'fifties when "Best Hit" was an annual event, the participants were the so-called "top guns" of the individual air forces that they represented, the pilots flying in this year's contest were selected from all combat-ready personnel. This has meant that the scores achieved by the participating pilots more closely represented the overall standard. Furthermore, the 1971 meet provided an opportunity to show five different aircraft systems in operation, and, perhaps surprisingly, it was the team flying the oldest of these aircraft, the North American F-100D Super Sabre, that won top honours!

The previous "Best Hit" meets in which all three

Among the teams competing in the 1971 Airsouth Fighter Weapons Meet were (top of the page) the Northrop F-5As of the Hellenic Air Force's 341st Squadron and (below) Vought A-7As from VA-37. The Turkish Air Force Northrop F-5B (above) was used as a 'chase plane during the meet.

A McDonnell F-4E Phantom II from the USAFE's 401st Tactical Fighter Wing, 613th Squadron, in landing configuration at Eskisehir. Phantoms and USN Corsairs formed a composite team competing as guests of the Turkish Air Force.

AIRSOUTH countries competed took place in 1954, in which year the Hellenic Air Force provided the winning team, 1955 when honours went to Turkey, and 1956 when the Italian team gained most points. Last year's meet, held at the Istrana air base, Italy, brought together only Italian and Turkish teams, plus a USAF guest team, but although 114 sorties were flown by the competitors, the meet was ruled incomplete because inclement weather forced cancellation of some of the events needed to select a winner.

Hosted by General Muhson Batur, Commander of the Turkish Air Forces, and Lieut General Fred M Dean, USAF, Commander AIRSOUTH, this year's event was held at No 1 Air Base, the Headquarters of Turkey's No 1 Tactical Air Force. Some three miles (5 km) outside the town of Eskisehir, the centre of the region famous for the manufacture of Meerschaum pipes, the base is about 3,000 feet (915 m) above sea level, was built in the late 'thirties, and is undistinguished, if one discounts the intensity of the heat that it experiences during the day and the pair of storks nesting in a tree alongside the Officers' Club. Housing the School of Technical Training and the principal depot for major overhauls of all Turkish jet aircraft, Eskisehir is the home base of the F-100D-equipped No 111 Squadron and

is shared by No 114 Squadron which flies Republic RF-84F Thunderflashes. The Super Sabres flown by No 111 Squadron, which, in fact, provided the winning team, are, for the most part, ex-USAFE F-100Ds but include a sprinkling of aged F-100Cs brought up to D-model standards, as well as some two-seat F-100Fs which served during the meet in the rôle of 'chase planes.

Competing against the team from Turkey's No 111 Squadron were a team from the Hellenic Air Force's 341st Squadron normally based at Achialos, Volos, as a component of the 110th Wing, and flying Northrop F-5As; the Italian Air Force's 103rd Squadron home-based at Treviso as a component of the 2nd Wing, and flying G.91Rs, and a US team comprising a USAF element of McDonnell Douglas F-4E Phantoms from the 613th Squadron of the 401st Tactical Fighter Wing based at Torrejon, Spain, and a US Navy element of Vought A-7A Corsair IIs from Attack Squadron 37 of Attack Carrier Wing Three from the US Sixth Fleet. The latter, incidentally, has the last three A-7A-equipped squadrons in first line service. The firing range itself, on which old F-84G Thunderjets served as targets, was a few minutes' flying time from Eskisehir for the participants and about 40 minutes by helicopter, and the

Flying North American F-100Ds, No 111 Squadron of the Turkish Air Force took first place in the Fighter Weapons Meet with a substantial lead over the next-best team.

(Above) The Italian Air Force competed with a team of Fiat G 91Rs from the 103rd Squadron, 2nd Wing, usually based at Treviso. One member of this team achieved top score for dive-bombing. (Below) A Vought A-7A from the mixed USN/USAF guest team.

(Below) Taking off at Eskisehir, a Northrop F-5A of the Greek Air Force carries a rocket pod on the fuselage centre line.

One of the USAF McDonnell F-4Es used in "Best Hit". The Phantom II pilots achieved personal successes — one as best individual pilot in the contest, and one as top-scoring pilot in strafing.

entire meet took place within Soviet radar coverage, a fact from which the competing pilots derived a certain amount of amusement.

The teams competed for eight trophies, the COMAIRSOUTH Trophy being awarded to the highest-scoring team, and other awards including the Individual High Score Trophy presented to the highest-scoring pilot of each team, and the Event Trophy awarded to the highest-scoring pilot in each of the strafing, rocketry and bombing events. Range familiarisation flying began on 8 July, and the actual competition flying took place between the 12th and 15th, weather throughout the week providing for the completion of all missions on time, although high winds over the target range on the 14th made it necessary to fit three missions for each of the Italian and Turkish teams into the final day's flying schedule.

The standards achieved by all competing teams in each category were extremely high, but the team from the Turkish No 111 Squadron emerged a clear winner, despite the age of its F-100D Super Sabres, with a total of 596 points. Runner-up in the competition was the combined USAF-US Navy guest team with 538 points. The next best team score was achieved by the Italian Fiat G.91Rs with 464 points, the Hellenic team finishing with 422 points. The individual high-scoring pilots were all members of the guest team with one exception, Capt Omero Cominato of the Italian team obtaining the highest individual score in dive bombing with 34 points. The other individual high-scorers were Capt Harvey Kinsey, USAF, with 35 points in strafing, Lt John Sherm, USN, with 27 points in rocketry, and, as the top-scoring pilot, Maj Roger Jaquith, USAF, with 129 points.

General Dean, in discussing "Best Hit", said: "Since NATO strategy has moved from one of all-out nuclear retaliation to one which includes conventional operations, the emphasis is once again on the ability of each individual pilot to improve his accuracy in the delivery of various weapons. I have long recognised the value that competition plays in increasing the overall ability of a command to accomplish its mission, and I believe that this meet has been a good yardstick to measure the progress we have made in making the pilots of the air forces of NATO's Southern Region combat ready." □

(Above) A Turkish Air Force F-100F used as a 'chase plane during the meet and (below) another view of one of No 103 Squadron's Fiat G 91 Rs.

TALKBACK

MRCA views

May I attempt to correct certain points made concerning MRCA in the article "The Fight for the Ground" which appeared in your otherwise excellent July issue.

1: In meeting tri-service requirements the MRCA is not "falling between three stools" as far as the RAF is concerned. This is proven by the fact that the RAF requirements for the MRCA are more stringent in terms of operational range and performance than they were for the purely UK combat aircraft on which we were working before the tri-service requirements were laid down for the MRCA.

2: MRCA is smaller than TSR.2 because those same RAF requirements make it so.

3: An MRCA unit cost of "between £1·2 and £1·5 million" was reiterated at the Paris Air Show by Panavia's chairman. It is thus within Mr Braybrook's quoted German ceiling of £1·7 million, and is NOT too expensive for the *Luftwaffe*.

4: Air superiority demands a good rate of climb, good acceleration and superior manœuvrability. These are inherent in the high "Specific Excess Power" enjoyed by the MRCA through its new-technology engines and variable-geometry wings. An escort will NOT be required by the MRCA. It will be fully capable of looking after itself.

5: In comparing the MRCA with the Viggen "design-wise" we presume that Mr Braybrook is comparing a canard wing with a variable-geometry wing. There is something to be said for both configurations. Suffice to say we at BAC Military Aircraft Division studied the canard configuration in depth with the aid of wind tunnel programmes and concluded, among other things, that it could not meet the MRCA's STOL requirements. In this connection I just do not understand Mr Braybrook's statement that "the European scenario does not allow the full benefit of the VG wing to be exploited".

We in BAC are not obsessed by variable geometry. It is just a fact of life that swinging the wings is the only feasible and cost effective method of meeting the various requirements this Multi-Rôle Combat Aircraft is designed to fulfil.

Alexander Johnston
Publicity Manager
British Aircraft Corporation
Preston Division
Preston PR4 1AX

Roy Braybrook replies: I believe the principal difference between our statements is that I was considering *fundamentals* whereas Mr Johnston is concerned more with the situation as it has now developed. To clarify my views:

1 and 2: RAF range/payload figures were *originally* as required from TSR.2, but have been scaled down by a large factor as a compromise with German requirements. Had it proved possible to collaborate with a country whose fundamental strike require-

ments are more in line with our own (eg, USA and Australia), then the compromise aircraft would have been much closer to TSR.2, and I believe that the RAF would have been happier. The fact that Britain would have been limited to an even smaller aircraft than MRCA in the event of proceeding on a single-country basis is more a reflection of our sad economic situation than of RAF fundamental needs.

3: Germany's *original* wish was for an aircraft with a unit price of DM 10m, which is more in the Jaguar category than that of MRCA. Germany's views on spending money on military aircraft are further illustrated by the decision to use Alpha Jets in the ground attack rôle, which I think Mr Johnston will agree is interesting, to say the least.

4: I agree that German talk of an escort fighter for MRCA (or any other European strike aircraft) is misguided. However, if an aircraft is to achieve air superiority in the world's toughest combat environment over the period 1980-2000 (which is presumably the Italians' *fundamental* need), then it will require an outstanding thrust/weight ratio. I would not have thought that the SEP for MRCA will be in the same category as that for the F-15, for example, but I would be delighted to be proved wrong.

5: Without access to BAC's wind tunnel tests, I would have thought Viggen's STOL performance was in the same ballpark as MRCA. The outstanding advantages of VG over the canard are in ferry range and HI-LO missions, and thus cannot be exploited in the European scenario. The obvious points against the Viggen are its country of origin and its still unpublished stall/spin characteristics.

Finally, I would assure Mr Johnston that my reservations on MRCA are not symptomatic of an anti-BAC feeling. In my opinion, his company's Super One-Eleven is in a class of its own, and Concorde is Europe's *only* foothold in advanced technology. However, the fact that the only sensible course *now* is to proceed with MRCA (having got so far) does not preclude reflection that the joint requirement must have been an unusually difficult compromise and this was really the essence of the paragraph that Mr Johnston disputes.

This photograph, taken in April 1925, depicts one of the Armstrong Whitworth Siskin Vs ordered by the Rumanian government but subsequently cancelled (see letter from D C Whitworth).

Rumanian Siskin

CONCERNING the very interesting article on Rumanian military aviation (AIR ENTHUSIAST June 1971), I was interested to see the reference to the crash of a Siskin V at Whitley. I have seen recently an original photograph of the remnants of this crash. Captioned "Rumanian Crash", it is dated March 1925. The only surviving part of the structure was the fin and rudder, in obviously Rumanian markings, with the following inscription on the rudder:

TYP 3
No 29
GG ...
GU ... } KGR
GT ...

From personal memory, the pilot was a dashing, oily-haired man whose crash was no surprise to many. The aeroplane was spread over a large area. I hope these facts will be of use to the compiler of the article.

D C Whitworth
Kettering

We believe, Mr Whitworth, that you may have misread the markings on the rudder of the crashed Siskin, and that the inscription actually read "TYP 5" and NOT "TYP 3". The Rumanian order was definitely for Siskin Vs and not Siskin IIIs. We have been unable to ascertain how many Rumanian Siskins were actually built before the order was cancelled, but there were at least 10 and probably more, and the inscription on the fin of the Rumanian Siskin depicted by the accompanying photograph reads:

TYP 5
No 10

There is evidence to suggest that the original Rumanian Siskin order called for a total of 65 aircraft.

Those Caproni "twins"

I FOUND your article about the Caproni light bomber (AIR ENTHUSIAST July 1971) especially interesting since, as mentioned, the Norwegian Army Air Force, the *Hærens Flyvevåben*, ordered a number of Ca 310s. The opportunity to trade *Klipfisk* (Dried Cod) with Italy for Ca 310s and a manufacturing licence was the primary reason for the selection of the Caproni light bomber. Four Ca 310s were purchased but these were not very successful, suffering several minor accidents, and the more powerful Ca 312 was ordered.

The *Hærens Flyvevåben* planned to acquire 15 Ca 312s from Italy and the Army Aircraft Factory, or *Hærens Flyvefabrik*, at Kjeller by 1 July 1940, by which date it was anticipated that a further five would be nearing completion at Kjeller. The Ca 312 was considered only as an interim type, and in 1939 the *Hærens Flyvevåben* was already looking for a successor. By the time Norway was invaded in April 1940, no Ca 312 had been delivered, and the four Ca 310s never saw combat. An interesting corollary to the story of the Caproni "twins" and the *Hærens Flyvevåben* is the fact that, after WW II, the Caproni concern inquired if Norway was *still* interested in acquiring the Ca 312!

I would be interested to learn if any examples of the Ca 310 or Ca 312 remain in existence today.

Bjørn Wallin
1450 Nesoddtangen
Norway

For the sake of completeness, I think that your readers may be interested in the accompanying production list of the light twin-engined aircraft mentioned in the feature "The Caproni that Nearly Joined the RAF" (July issue). This list is the outcome of a most exhaustive research of Count Caproni's archives. There is also some additional information and corrections which may interest you. The Borea was the Ca 306 and not, as some have stated, the Ca 308, which was the designation applied to the Caproni Bergamaschi A.P.1 light attack bombers that were supplied to Paraguay in 1935 in exchange for coffee. The Ca 310 prototype (M.M.401) was eventually converted as a Ca 312 prototype, as was also another Ca 310 (M.M.20837), but there were no production Ca 312s. The aircraft for Norway were of the Ca 312bis type with Ca 313 R.P.B.1-style forward fuselages, and those for Belgium (not mentioned in your feature) had Ca 313G-style forward fuselages. The Ca 313bis was the Ca 313G as initially proposed to the *Regia Aeronautica*.

The Ca 311 *was* the recipient of an export order, as 15 aircraft of this type were ordered for the Yugoslav Royal Air Force.

One of the Ca 310s delivered to the *Hærens Flyvevåben* and referred to in the letter from Mr Bjørn Wallin. The four Ca 310s were to be joined in Norwegian service by the Ca 312bis to which Dr Abate refers in his letter.

In addition to the 12 Ca 310s that you mention, Yugoslavia placed an order for 12 examples of the Ca 310bis and 15 examples of the Ca 311. The Yugoslavian Ca 310bis and Ca 311 differed very little in outward appearance, the only external clues to their identity being the extended window panelling along the fuselage sides of the Ca 311 and, of course, the identification numbers on the fin (Nos 13 to 24 being aircraft of the Ca 310bis type and No 25 onwards being Ca 311s). Incidentally, the last Yugoslav Ca 310bis (No 24) was fitted with Gnôme-Rhône 9K engines and three-bladed airscrews. Of the Ca 311s, five were actually delivered to the Yugoslav Royal Air Force in 1941, the remaining 10 being delivered to the Croatian Air Force in 1942.

Incidentally, the 36 Ca 310s supplied to Hungary were actually delivered during March and April 1939 (the original *schedule* having called for the delivery of a dozen aircraft in each of August, September and October 1938) to the *Regia Aeronautica* which transferred them to Hungary between May and August 1939. The 33 Ca 310s returned in 1940 were in exchange for Ca 135bis Us bombers, and after refurbishing at the Aero-Caproni works in Trento they were passed to the *Regia Aeronautica* in 1940-1.

Dr Rosario Abate
20135 Milano
Italy

High gloss or matt

I find reason to disagree with your statement (August issue) concerning the finish of the RAF's Harriers. Having visited RAF Wittering where No 1 Squadron and No 233 OCU are based, I have been able to study their Harriers at close range. All the Harrier GR Mk 1s are finished in high-gloss polyurethane, and the T Mk 2s are all matt finished. Why this is I do not know. I believe that, as time goes by, the GR Mk 1s may well *appear* to be matt finished, not as a result of respraying but through the "sanding" effect of airflow friction during service usage.   *F A Grant*

Plymouth PL4 8RP, Devon

The first few Harrier GR Mk 1s WERE gloss finished, Mr Grant, but a matt finish supplanted the gloss finish at the same time as white was dropped from the upper wing surface and fuselage side roundels, both changes being made for the same reason, to render the aircraft less conspicuous from above. Aircraft with the new finish, as depicted in our August issue, have been in RAF service since the early summer.

What happened?

I HAVE received the second issue of AIR ENTHUSIAST. What happened to the Raiden fuselage tail in the side elevation of the three-view drawing on page 69?

J McWaters
Potters Bar, Herts

That's exactly what we asked our printers, Mr McWaters. The original drawing was complete.

Airacobras recalled

THE ARTICLE on the RAF Airacobras in the August issue was most interesting. On checking an old diary for 1942, I find that I saw an RAF Airacobra on both 11 June and 20 June in that year, in the Peterborough area. This was some time after No 601 Sqdn relinquished their aircraft.

I also find that there were quite a few Airacobras (presumably USAAF) flying in the Grimsby area in December 1942. I record one on the 20th, three on the 21st, one on the 22nd and several on the 23rd — when they came abruptly to an end. This coincides with your record of six USAAF squadrons moving to North Africa.

Peter H T Green, Irby, Lincs

Type	Customer	Year(s)	Quantity	Remarks
	PRODUCTION LIST OF THE CAPRONI "LIGHT TWINS"			
Ca 306	Ala Littoria	1935	6	—
	R A Command, Libya	1935	2	One fitted with armament
Ca 309	Libyan Government	1936–8	78	Series I and II
	Ala Littoria	1937	1	I-ALAL
	Duke of Aosta	1937	1	—
	Regia Aeronautica	1940–4	165	Series IV to IX
	Paraguay	1938	2	—
Ca 310	Regia Aeronautica	1937–9	161	10 originally for Rumania
	Yugoslavia	1938	12	—
	Norway	1938	4	—
	Peru	1938–9	16	Including two "one-off" machines
	Spain	1939	16	Via Regia Aeronautica
	Hungary	1939	36	Via Regia Aeronautica
Ca 310bis	Yugoslavia	1939	12	—
Ca 311	Regia Aeronautica	1939–42	320	Mostly converted to Ca 311M
	Yugoslavia	1941–2	15	—
Ca 312bis	Norway	1940	15	To R A as transports
	Belgium	1940	24	To R A as torpedo-bombers
Ca 313	France	1940	5	(R.P.B.1) 200 ordered
	Sweden	1940–2	84	(R.P.B.1)
	Regia Aeronautica	1939–41	122	(R.P.B.2)
	Luftwaffe	1942–4	164	(Ca 313G) 137 delivered of 905 ordered
Ca 314A	Regia Aeronautica	1942	73	—
Ca 314B	Regia Aeronautica	1942	80	—
Ca 314C	Regia Aeronautica	1942–3	254	—
	R A = Regia Aeronautica			

THE UBIQUITOUS HAWK

REFLECTING American thoughts on pursuit design in the years immediately preceding World War II, and encumbered by its full share of the *réclame* that inevitably attended US combat aircraft of the period, good, bad or indifferent, the Curtiss Model 75 was fated to have little claim on fame if its extraordinary ubiquity was to be discounted. Yet this rather shapely little fighter was to gain for the Allies the first aerial victories of both European *and* Pacific conflicts. It was to gain the distinction of fighting against all three major Axis powers *and* against the Allies; it was to be purchased by a number of air arms yet operate more widely with air arms that were to obtain it fortuitously. The hard school of war was to reveal the inadequacy of its performance and firepower, and its shortcomings in combat were to be decried, yet its qualities as an aeroplane were to be lauded, and its pilots were to recall with nostalgia its superlative manœuvrability, delightful handling characteristics and beautifully harmonised controls.

The Model 75 was conceived in 1934, a vintage year in fighter annals. Nineteen hundred and thirty-four witnessed the dawn of a new era in fighter development; the beginning of a revolution destined to set a style in design that was to prevail until the introduction of the turbojet — the all-metal stressed-skin monocoque cantilever low-wing monoplane with enclosed single-seat cockpit and retractable undercarriage. Simultaneously and spontaneously in several major aircraft-manufacturing countries, fighters embodying this combination of features were in embryo; what was to be termed the *nouvelle vogue* in fighter fashion was about to be launched.

In the Augsburg drawing office of the Bayerische Flugzeugwerke, Willy Messerschmitt and Walter Rethel were creating the Bf 109, while, in Ernst Heinkel's bureau at Warnemünde, Walter and Siegfried Günter were labouring over the He 112. In France, the Bloch 150, the Loire 250,

the Dewoitine D.513 and the Loire-Nieuport 161 had begun passage across their respective drawing boards, and on the other side of the Channel, an ailing Reginald J Mitchell was working at the Supermarine Aviation Works in Southampton on what was to become the Spitfire, the Vickers drawing offices at Weybridge were creating the Venom, and the Bristol and Gloster companies, too, were working on contributions to the *nouvelle vogue*.

Across the Atlantic, in anticipation of a US Army Air Corps requirement for a Boeing P-26A replacement, John Northrop and his team were developing the Model 3A, while Don Berlin, the Chief Engineer of the Curtiss-Wright Corporation's Airplane Division, was working on the Model 75. All these fighters, German, French, British and American, had one thing in common — their configuration. All had metal stressed-skin monocoque structures, low-positioned cantilever wings, enclosed cockpits and retractable undercarriages. Most were to prove indifferent warplanes and few were to progress further than prototype status, but numbered among those that did achieve quantity production was the Curtiss Model 75.

Addition to the Hawk family

The Curtiss-Wright Corporation's Airplane Division, which had been on top of the fighter business until its biplanes had been elbowed out of the way by the Boeing P-26 monoplane, could have synthesised its years of fighter design experience, translating this into the newly-fashionable monoplane configuration, but Don Berlin was convinced that such an approach would inevitably result in mediocrity; that Curtiss-Wright must produce something *in advance* of the state of the art — a warplane owing nothing to the company's earlier products.

The Curtiss Hawk biplanes had virtually symbolised US pursuit development from the mid 'twenties until the early

WARBIRDS

The Curtiss Hawk 75A-2

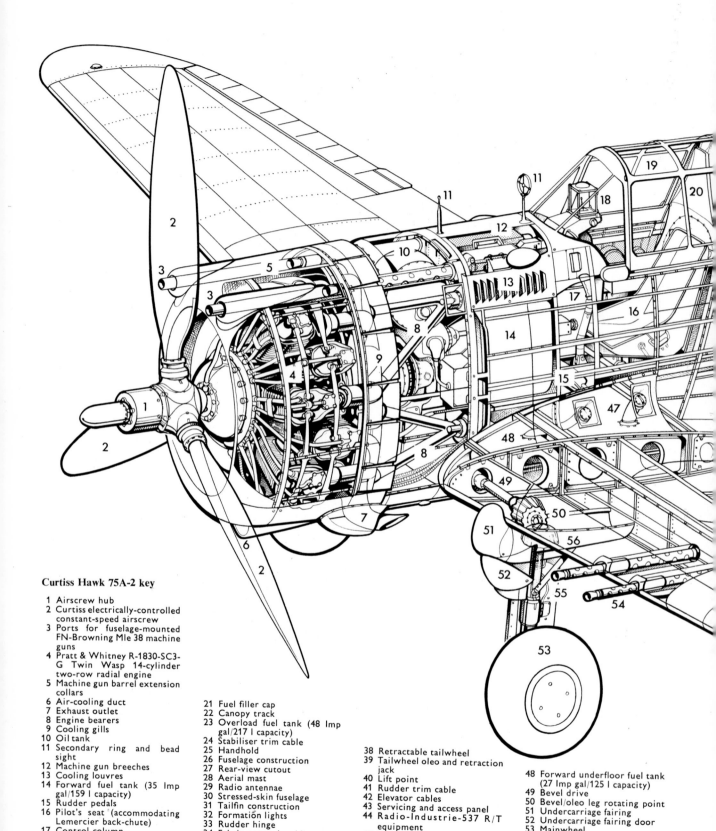

Curtiss Hawk 75A-2 key

1 Airscrew hub
2 Curtiss electrically-controlled constant-speed airscrew
3 Ports for fuselage-mounted FN-Browning Mle 38 machine guns
4 Pratt & Whitney R-1830-SC3-G Twin Wasp 14-cylinder two-row radial engine
5 Machine gun barrel extension collars
6 Air-cooling duct
7 Exhaust outlet
8 Engine bearers
9 Cooling gills
10 Oil tank
11 Secondary ring and bead sight
12 Machine gun breeches
13 Cooling louvres
14 Forward fuel tank (35 Imp gal/159 l capacity)
15 Rudder pedals
16 Pilot's seat (accommodating Lemercier back-chute)
17 Control column
18 Baille-Lemaire gun sight
19 Aft-sliding canopy
20 Pilot's head and back armour

21 Fuel filler cap
22 Canopy track
23 Overload fuel tank (48 Imp gal/217 l capacity)
24 Stabiliser trim cable
25 Handhold
26 Fuselage construction
27 Rear-view cutout
28 Aerial mast
29 Radio antennae
30 Stressed-skin fuselage
31 Tailfin construction
32 Formation lights
33 Rudder hinge
34 Fabric-covered rudder
35 Fabric-covered elevator
36 Tailplane
37 Tailwheel door

38 Retractable tailwheel
39 Tailwheel oleo and retraction jack
40 Lift point
41 Rudder trim cable
42 Elevator cables
43 Servicing and access panel
44 Radio-Industrie-537 R/T equipment
45 Batteries
46 Wing fillet
47 Aft underfloor fuel tank (25 Imp gal/113 l capacity)

48 Forward underfloor fuel tank (27 Imp gal/125 l capacity)
49 Bevel drive
50 Bevel/oleo leg rotating point
51 Undercarriage fairing
52 Undercarriage fairing door
53 Mainwheel
54 Two 7·5-mm FN-Browning Mle 38 machine guns
55 Mainwheel leg
56 Retraction actuator rod

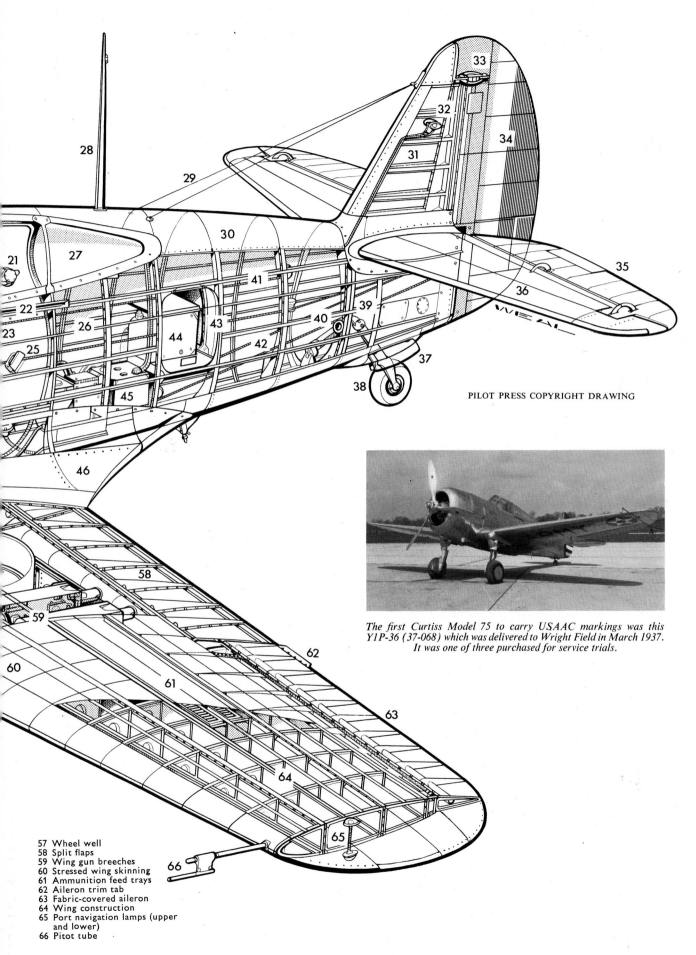

PILOT PRESS COPYRIGHT DRAWING

The first Curtiss Model 75 to carry U.S.A.A.C. markings was this Y1P-36 (37-068) which was delivered to Wright Field in March 1937. It was one of three purchased for service trials.

57 Wheel well
58 Split flaps
59 Wing gun breeches
60 Stressed wing skinning
61 Ammunition feed trays
62 Aileron trim tab
63 Fabric-covered aileron
64 Wing construction
65 Port navigation lamps (upper and lower)
66 Pitot tube

(Above) Curtiss's prototype modified as the Model 75B with Wright XR-1820-39 Cyclone engine, strengthened cockpit canopy and scalloped aft decking for improved rear view. *(Below)* The Model 75 prototype in its original form with a Wright XR-1670-5 engine as submitted at Wright Field in May 1935.

'thirties, and the monoplane that it was now proposed should be added to the Hawk family to recapture the company's position in both home and export markets was to combine the most advanced aerodynamic and structural features; its design was to be uncompromised by tradition or convention. The design team elected to use an aluminium-alloy monocoque fuselage and a multi-spar metal wing with flush-riveted smooth Alclad skinning, hydraulically-actuated split flaps and undercarriage retraction, mainwheels that pivoted about their axes to lie flat in the wing, and an aft-sliding cockpit canopy.

(Above) The Hawk 75R was a company-owned demonstration aircraft with a turbo-supercharger mounted beneath the nose with the intercooler beneath the wing trailing edge.
(Below) The company-owned demonstrator NX22028, previously fitted with a Twin Wasp engine and turbo-supercharger, after being re-engined with a Cyclone.

Designated Model 75, the design was not revolutionary, a fact readily admitted by Don Berlin, but the Curtiss-Wright team believed it sufficiently audacious to enable it to steal a march on other contenders for AAC orders. The team would undoubtedly have been intensely surprised had it been made aware that its contemporaries in several countries had reached identical conclusions as to the form that the next generation of fighters should take.

The power plant selected for the Model 75 was as new as the fighter's concept, being the 14-cylinder two-row Wright XR-1670-5 rated at 900 hp, but the proposed armament — one 0·5-in (12,7-mm) and one 0·3-in (7,62-mm) machine gun synchronised to fire through the airscrew disc — reflected the unimaginative AAC requirement for pursuit firepower in adhering to a standard established shortly after WW I. The first metal was cut on the prototype Model 75 in November 1934, and within a few days over six months, in May 1935, the initial flight test programme had begun at Buffalo, the AAC Material Division having meanwhile called for bids on new single-seat pursuit aircraft to be opened in that month.

The Model 75 was submitted at Wright Field on 27 May, but was the only pursuit ready for AAC evaluation. The Seversky Aircraft Corporation of Farmingdale, New York, had planned to submit the SEV-2XP, a tandem two-seat fighter monoplane featuring a fixed trousered undercarriage and designed by Alexander Kartveli, but due to an accident in transit, the aircraft did not reach Wright Field until 18 June. It was promptly returned to Farmingdale where it was reworked as a single-seater with mainwheels partly retracting into underwing fairings, returning to Wright Field on 15 August as the SEV-1XP. The only other serious competitor to the Curtiss Model 75, the Northrop 3A, had taken-off on its maiden flight on 30 July, headed out over the Pacific and promptly disappeared.

The SEV-1XP was powered by a single-row nine-cylinder Wright R-1820-G5 Cyclone, and both this and the Model 75 were plagued by engine difficulties throughout the Wright Field trials. Nevertheless, it was soon obvious that the AAC favoured the SEV-1XP over the Curtiss contender, and there can be little doubt that a production contract would have been awarded the Seversky fighter had not Curtiss protested that its competitor had been given an unfair advantage by the AAC which had permitted its late arrival. The Material Division therefore agreed to defer a decision until after further competitive evaluation which was to take place in April 1936.

Don Berlin took the opportunity presented by the deferment to replace the thoroughly unsatisfactory XR-1670-5 by a nine-cylinder Wright XR-1820-39 (G5) Cyclone rated at 950 hp at 2,200 rpm for take-off and having a normal maximum output of 850 hp at 2,100 rpm from sea level to 6,000 ft (1 830 m). At the same time, the fuselage aft of the canopy was scalloped to improve rear view and the canopy itself was strengthened. In this form, now known as the Model 75B, weighing in at 4,049 lb (1 837 kg) empty and 5,075 lb (2 302 kg) loaded, the Curtiss fighter returned to Wright Field for the new trials which began on 15 April. By this time, the Northrop 3A design had been purchased by Chance Vought and a further prototype built as the V-141, but the stiffest competition still proved to be that provided by the SEV-1XP.

The new Cyclone engine of the Model 75 proved as unsatisfactory as its predecessor, failing to develop its full rated power and calling for no fewer than four engine changes during the Wright Field trials. The Model 75B proved capable of attaining only 285 mph (459 km/h) compared with the 294 mph (473 km/h) at 10,000 ft (3 050 m) guaranteed by Curtiss-Wright, but the SEV-1XP was also down on performance, reaching only 289 mph (465 km/h) whereas 300 mph (483 km/h) had been guaranteed. Furthermore, while Curtiss-Wright quoted $29,412 per aircraft in lots of 25 and $14,150 in lots of 200, the unit prices quoted by Seversky for similar quantities of aircraft were $34,900 and $15,800. Nevertheless, the SEV-1XP was awarded 812·39 points in the contest while the Model 75B made a rather poor runner-up with a score of only 719·84 points, and a contract for 77 production examples and the designation P-35 passed to the Curtiss-Wright fighter's opponent.

Despite the fact that the Seversky fighter had been selected by the AAC as the winning contender in the contest and, on 16 June 1936, had been the recipient of a production order, only three more weeks had elapsed when, perhaps to hedge its bet, the Material Division gave Curtiss-Wright a contract for three service test examples of its fighter, these being allocated the official designation Y1P-36*. The Division had earlier requested a study of the performance potential of the fighter fitted with the 14-cylinder two-row Pratt & Whitney R-1830 Twin Wasp, and on the basis of

this study, the contract placed with Curtiss-Wright on 7 August 1936 stipulated that the three Y1P-36s (37-068 to -070) should receive the R-1830 engine.

The first Y1P-36 was completed and delivered in March 1937, and successfully passed official trials at Wright Field during the following June. The Y1P-36 was powered by an R-1830-13 Twin Wasp rated at 900 hp at 2,550 rpm at 12,000 ft (3 660 m) and driving an hydraulically-operated constant-speed three-bladed Hamilton Standard airscrew. For take-off the engine had been de-rated from 1,050 to 950 hp, and the aircraft was tested at a gross weight of 5,437 lb (2 466 kg) which included 161 lb (73 kg) of ballast in lieu of armament, empty weight being 4,389 lb (1 991 kg). The aircraft proved to possess a maximum speed of 294·5 mph (474 km/h) at 10,000 ft (3 050 m) with the engine delivering its full rated power. With 660 hp being developed at 2,200 rpm a speed of 256 mph (412 km/h) was clocked at the same altitude, which was reached in 3·44 minutes from take-off. Take-off distance from grass was 450 ft (137 m), a 50-ft (15,2-m) obstacle being cleared within 950 ft (289 m), and initial climb rate was 3,145 ft/min (16 m/sec). Range at normal operating speed with 126 Imp gal (573 l) proved to be 752 miles (1 210 km).

The Wright Field test pilots were highly enthusiastic regarding the fighter's characteristics. Manœuvrability was considered to be of a very high order except for some jerkiness at the start of spins and snap rolls. The effectiveness and operation of all controls throughout the speed range of the fighter were reported as excellent, and stability and ground handling were particularly commended. The only

* *The "1" indicating that the aircraft were being purchased with "F-1" funds rather than from the regular AAC appropriations.*

(Above) The first Y1P-36 after being fitted with two-bladed contra-props, the first such installation on a US aircraft. (Below) A P-36A of the 79th Pursuit Squadron of the 20th Pursuit Group, the first USAAC Group to receive the Curtiss fighter, at Barksdale Field.

criticisms were directed at the location of the undercarriage and flap controls, cabin ventilation, and the curvature of the windscreen which resulted in some distortion of vision during landing.

Lucrative export market

The official performance trials with the Y1P-36s (which subsequently became simply P-36s) were completed on 22 June 1937, and on the strength of these trials the AAC decided to adopt the Curtiss fighter, awarding a $4,113,550 contract for 210 machines on 7 July, this being the largest single order for fighters placed in the USA since 1918. The privately-financed development of the Model 75 had been a calculated gamble which had now begun to pay off.

The Airplane Division of the Curtiss-Wright Corporation had not been allowing the grass to grow beneath its feet, however, while the AAC had been making up its mind whether or not to adopt the Model 75. Late in 1936, with an eye to the export market which had proved so lucrative in the case of the Hawk fighter biplanes, development of a simplified version of the new fighter monoplane intended specifically for export had begun.

Curtiss-Wright was aware that some potential customers whose air arms operated under relatively primitive conditions would look askance at such sophisticated a feature as a retractable undercarriage which promised to afford

Opposite page, top to bottom: P-36C of the USAAF in the finish standardised early 1942; Hawk 75A-3 (CU-562) of Finnish HLeLv 32, Nurmoila, winter 1942-3; Hawk 75A-8 of the Norwegian flying training centre, Island Airport, Toronto, 1941; Hawk 75A-5 of the Nationalist Chinese Air Force, Kunming, 1942; Hawk 75A-7 flown by Colonel Boxman of the 1.Vliegtuigafdeling, KNIL Luchtvaartafdeling, Madioen, December 1941.

maintenance problems. Therefore, an alternative fixed cantilever undercarriage had been designed, the main legs being encased by streamlined light metal fairings, and this had been applied to the original prototype Model 75B which, re-engined with a Wright Cyclone GR-1820-G3 rated at 875 hp at 2,200 rpm for take-off and 840 hp at 2,100 rpm at 8,700 ft (2 650 m), had been registered NR1276 as the demonstrator for what Curtiss-Wright, with its unexcelled expertise in brochure design, now publicised as the Hawk 75. Emphasis in the publicity was placed on ease of maintenance, rough-field capability, and the amenability of the Hawk 75 to different types and combinations of armament, and there was no lack of interest abroad.

A second and definitive demonstration aircraft, NR1277, was completed which differed from its predecessor in several respects, embodying some of the features introduced by the Y1P-36, such as the more deeply scalloped decking immediately aft of the cockpit which improved the pilot's rear view, and revised windscreen arch and canopy framing. Whereas the armament of the first demonstrator comprised a single 0·5-in (12,7-mm) gun with 200 rounds and one 0·3-in (7,62-mm) gun with 600 rounds, both in the upper decking of the forward fuselage, this was supplemented in the second demonstrator by a pair of 0·3-in (7,62-mm) wing guns mounted outside the airscrew arc, and provision was made for the attachment of standard A-3 wing racks for ten 30-lb (13,6-kg) or six 50-lb (22,7-kg) bombs and a Curtiss-designed centreline rack for a single 500-lb (227-kg) bomb.

The Chinese National Government immediately evinced interest in the Hawk 75, purchasing NR1276 at a cost of $36,000, and presenting this to General Claire L Chennault (who had been engaged to reorganise the Chinese air arm) for his personal use. Simultaneously an order was placed for 112 production examples which were to be delivered as major assemblies to the Central Aircraft Manufacturing Company at Loi-Wing. Apart from minor revisions to the undercarriage fairings and adoption of a four 0·3-in (7,62-

(Above) A Hawk 75N of the 5th Wing of the Royal Thai Air Force based at Prachuab Kirikhand in 1941.

(Above) The second Hawk 75 demonstrator (NR1277) demonstrated at Buenos Aires late in 1937 in competition with the Seversky 2-PAL. (Below) The 19th Hawk 75O built in Argentina by the Fábrica Militar and delivered to the Army Aviation Command in 1941.

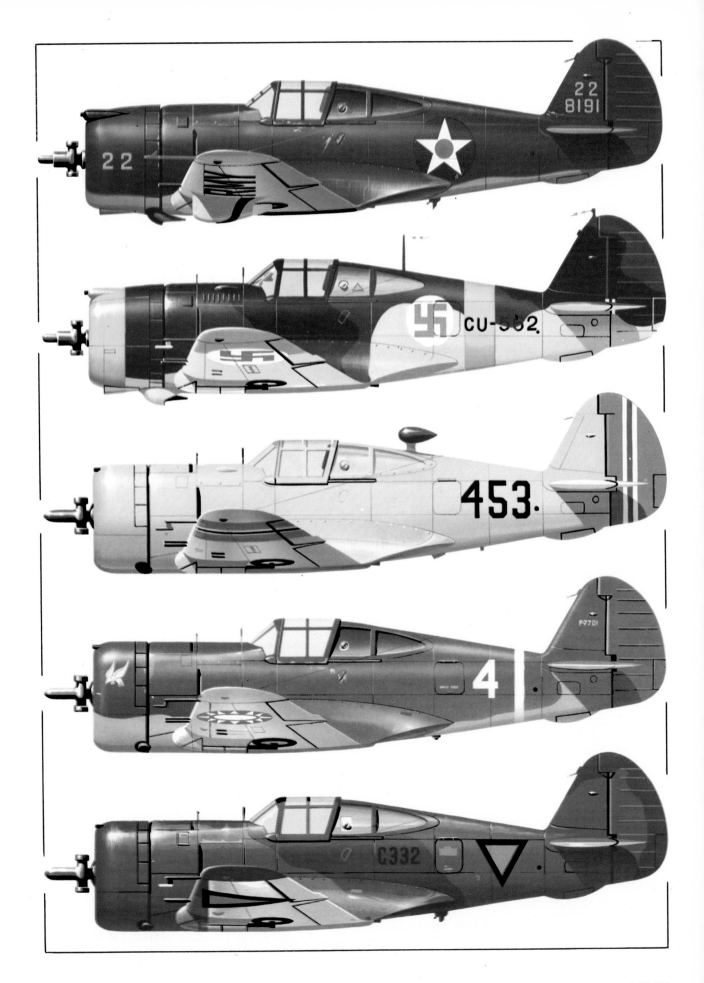

mm) gun armament, the production model for China (retrospectively assigned the designation Hawk 75M by the Curtiss-Wright Corporation) was essentially similar to the second demonstrator. Although flown against the Japanese during 1939-40, the Hawk 75M in Chinese service had strictly limited success, owing to poor serviceability and the indifferent training of both its pilots and ground personnel.

China's lead in purchasing the Hawk 75 was quickly followed by Thailand, the Thai government ordering 25 examples (subsequently allocated the designation Hawk 75N) which differed from those supplied to the Chinese National Government in embodying further redesign of the undercarriage mainwheel fairings and in the armament. The two fuselage-mounted 0·3-in (7,62-mm) weapons were retained, but two 23-mm Danish Madsen cannon were mounted in detachable underwing fairings.

The Thai Hawk 75s were to see some action during the Thai invasion of French Indo-China in January 1941, four Hawk 75s carrying 33-lb (15-kg) bombs accompanying a formation of nine Mitsubishi Ki.21-I bombers attacking the airfield of Nakorn Wat on the 11th of that month. The formation was intercepted by four Morane-Saulnier 406s, and in the subsequent mêlée two Thai Hawk pilots, Warrant Officer Thongkham and Pilot Officer Sangwan Worsab, each claimed the destruction of an MS 406, although these claims were subsequently refuted by the French. Equipping the 5th Wing of the Royal Thai Air Force at Prachuab Kirikhand, the Hawk 75s were to see action again on 7 December 1941, this time against Japanese forces invading Thailand, one-third of the unit's serviceable fighters being lost during the brief but bitter fighting that took place before a ceasefire was agreed.

Deliveries of the Hawk 75N to Thailand had been initiated in 1938 and, in the meantime, Argentina had decided to acquire similar fighters for her Army Aviation Command. After purchasing the second demonstrator (NR1277), the Argentine government placed an order for 30

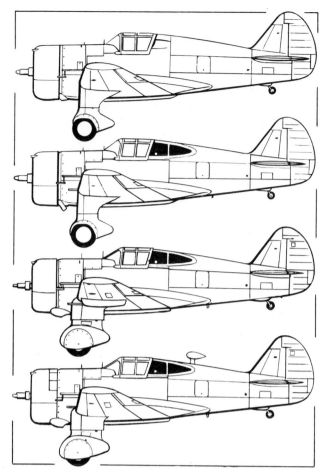

(Top to bottom) Hawk 75 initial prototype; Hawk 75M; Hawk 75N, Hawk 75O.

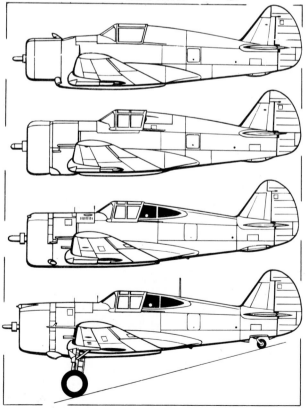

(Top to bottom) Model 75 prototype; Model 75B; Hawk 75A-4; Hawk 75A-3.

of the Curtiss fighters (Hawk 75O) and simultaneously obtained a licence to manufacture the Hawk 75 at the *Fábrica Militar de Aviones*. The Hawk 75O had a similar undercarriage to that of the Thai Hawk 75N, and featured a redesigned engine exhaust system with a semi-circle of electrically-operated gills at the rear of the cowling. Armament consisted of a quartette of 0·3-in (7,62-mm) Madsen M.1935 machine guns, and the first Hawk 75O was completed by the parent company late in November 1938.

Licence manufacture at the *Fábrica Militar* was begun in 1939, and the first FMA-built example was delivered on 16 September 1940, a total of 20 being completed. The Hawk 75O subsequently equipped the *I, II* and *III Regimientos de Caza*, 45 of the 50 aircraft remaining in the active inventory as late as 1945, when they provided the equipment of the *I* and *II Regimientos*, and the Curtiss fighter continued in first-line service well into the 'fifties, the survivors being finally withdrawn from the *II Regimiento* at El Plumerillo, Mendoza, in 1953. For a further year they were flown as advanced trainers at the *Escuela de Aviacion Militar* at Cordoba.

The characteristics of all export versions of the Hawk 75 were basically the same. The Hawk 75O, had empty and loaded weights of 3,975 lb (1 803 kg) and 5,172 lb (2 346 kg) respectively, and an overload weight of 6,418 lb (2 911 kg), and its performance included maximum speeds of 239 mph (384 km/h) at sea level and 280 mph (451 km/h) at 10,700 ft (3 260 m). Normal cruise was 240 mph (386 km/h) at 10,700 ft (3 260 m), initial climb rate was 2,340 ft/min (11,90 m/sec), and an altitude of 23,000 ft (7 010 m) could be attained in 12·52 minutes. Service ceiling was 31,800 ft (9 690 m), and normal and maximum ranges were 547 miles (880 km) and 1,210 miles (1 947 km). □

(To be continued next month)

For the young of all ages

Few reading this column are likely to feel other than does the writer that the modelling of aircraft is a fully adult pastime, yet it invariably comes as something of a surprise when a chance remark by a layman reminds one that this opinion is by no means universal. The suggestion that aircraft modelling is primarily a hobby for children and that to indulge in this form of recreation is to reveal delayed adolescence is by no means rare. The tolerant virtuoso of the polystyrene and enamel will probably mutter something like *"Chacun à son goût"* in answer to such a suggestion—especially if he happens to be French — while the less tolerant may well be taken in charge for intent to injure with a deadly weapon (for example, a flash-trimming knife).

These remarks are engendered by the experience of this columnist at a recent gathering of modelling cognoscenti which included among its guests a member of the Soviet diplomatic corps who was obviously intrigued if perhaps a little bemused by the array of skilfully-finished models. On being asked if he, too, was a member of the modelling fraternity, the Soviet diplomat said that he did not make models himself, adding, somewhat *undiplomatically*, that he found it both surprising and interesting that so many adults should expend so much time and effort on items that he purchased frequently as presents for his children. Although his subsequent comment that the high standard revealed by the models exhibited had given him cause to modify his opinion provided his hosts with some hope that a modelling *aficionado* may yet be infiltrated into the Soviet Embassy, it gave us reason to reflect that plastic aircraft model kits are viewed by their manufacturers primarily as toys, however much care to ensure accuracy they may inject into their products; that perhaps we modelling virtuosi are missing out on some of the pleasure that they have to offer by being overly critical when they fall short of the perfection that we continuously seek.

Are we in fact sometimes too harsh in our criticism of the review kits that arrive on our work bench? Upon further reflection, we concluded that, severe though a critique of a kit may seem at times to the kit's manufacturer, it can, so long as the adverse comments are fully justified, have a salutary effect on kit manufacturing standards with obvious benefit to the modelling fraternity which includes the young of all ages!

Heller and Tamiya via Kohnstam

The firm of Richard Kohnstam Limited, which uses the trade name of "Riko", is not as yet as well known to British modellers as, say, Frog or Airfix, but as one of the largest importers of kits in the UK it performs the thoroughly praiseworthy service of making available kits bearing a number of well-known foreign brands at reasonable prices. Among the most important of these are kits from Heller and Tamiya and, thanks to Richard Kohnstam, a large selection of the products of both companies reached our work bench recently.

The Riko/Heller kits comprised the Curtiss Hawk 75A, the Morane-Saulnier M.S.406, the Dewoitine D-520, the Bloch M.B.151, the Caudron C.714 and the ANF-Mureaux 117, all to 1/72nd scale. Since they have been available in France for some time, some of these kits will be already familiar to our readers, but what is probably not so familiar to UK modellers is their UK price — now only 25 pence. At this they represent outstanding value, as they are, without exception, neatly-moulded, finely-detailed and, in general, accurate. It is a pity that the decals that accompany these kits do not attain a standard comparable with the component parts, being exceedingly glossy and none-too-well printed, and this fault is compounded by the provision of instruction sheets including our particular *bête noire* — split planviews depicting only half of the complex camouflage patterns applied to the upper wing surfaces of these aircraft! We do hope that Heller will not perpetuate this exceedingly annoying practice.

Richard Kohnstam informs us that the entire Heller *musée* range will eventually be available in the UK under the Riko trade name, while for commercial aircraft model enthusiasts, the company will be offering Heller's 1/125th scale Boeing 747 which is to be issued in the not-too-distant future. We eagerly await the arrival of this monster.

The Riko/Tamiya kits that we have received come from the internationally-famous Japanese company's 1/100th scale range, and comprise the MiG-19PM *Farmer-E* interceptor and three helicopters, the Boeing-Vertol Model 107 in both civil and military versions, and the Sikorsky SH-3D Sea King. Until quite recently, 1/100th scale has received scant attention from most modellers, but many are now beginning to appreciate its advantages, the principal of which being that it is literally the *only* scale in which a comprehensive collection can be assembled without the individual models becoming ludicrously small at one end of the range or unmanageably large at the other! Far more kits are available to 1/100th scale than might be immediately apparent, and of these the best are unquestionably those produced in Japan by Tamiya.

The MiG-19 is a little gem, being both accurate and beautifully moulded, and the same applies to the Boeing-Vertol heli-copters, but the SH-3D unfortunately suffers from a lack of surface detail which is a great pity as the full-scale helicopter sports a plethora of prominent rivets and panels. All four of the Riko/Tamiya kits enjoy the virtue of having component parts that fit together so well that assembly is simplicity itself, despite the assembly instructions of all but the MiG-19 being readable only by those with a working knowledge of Japanese. The accompanying decal sheets are, in themselves, works of art, beautifully-printed and semi-matt finished, and each providing three alternative sets of markings for its subject. At 49 pence each, these kits are not particularly cheap in view of their size, but their outstanding quality would seem to justify the price asked.

Also supplied by Richard Kohnstam was Tamiya's 1/50th scale alternative Zero-Sen kit representing the A6M7 variant and *not* the A6M8c as we stated in this column last month. This, too, makes up into an extremely attractive model of this famous Japanese fighter.

Gilding the lily

The Modeldecal range of decal sheets, issued by Modeltoys of 246 Kingston Road, Portsmouth, and designed by Richard Leask Ward (who, incidentally, is responsible for the markings instructions on Frog kit boxes), now has an unrivalled reputation for quality and accuracy, and its latest issue, Sheet No 10 depicting four aircraft of the USAF operating in Vietnam, maintains this in full. To 1/72nd scale, the sheet provides full markings for a McDonnell RF-101C Voodoo, a Republic F-105D Thunderchief, a Douglas EC-47N, and a Douglas A-1H Skyraider, including stencilling and instrument panels, and is accurately printed and matt finished.

A very complete instruction sheet is provided which, in addition to providing full details of the four aircraft, offers a number of useful general notes. Our only criticism of this sheet is that the decals are a trifle thick, but for any modeller with a penchant for aircraft operating in the Vietnam area this product is a must.

This month's colour

Mohawk, P-36 or Hawk 75A; whatever the appellation under which it flew, Curtiss's little radial-engined fighter monoplane is one of those warplanes that, for some indefinable reason having no relation to operational success or association with outstanding exploits, possess a certain élan which the modeller finds irresistibly attractive. Few fighters of its era have borne a wider diversity of national markings nor seen greater ubiquity, as is revealed by the account of the development and career of this *warbird* that commences elsewhere in this issue. Yet, curiously enough, for many years this Curtiss fighter was completely neglected by kit manufacturers who concentrated on later Curtiss progeny which they believed to possess more glamour and, in consequence, greater saleability. Then,

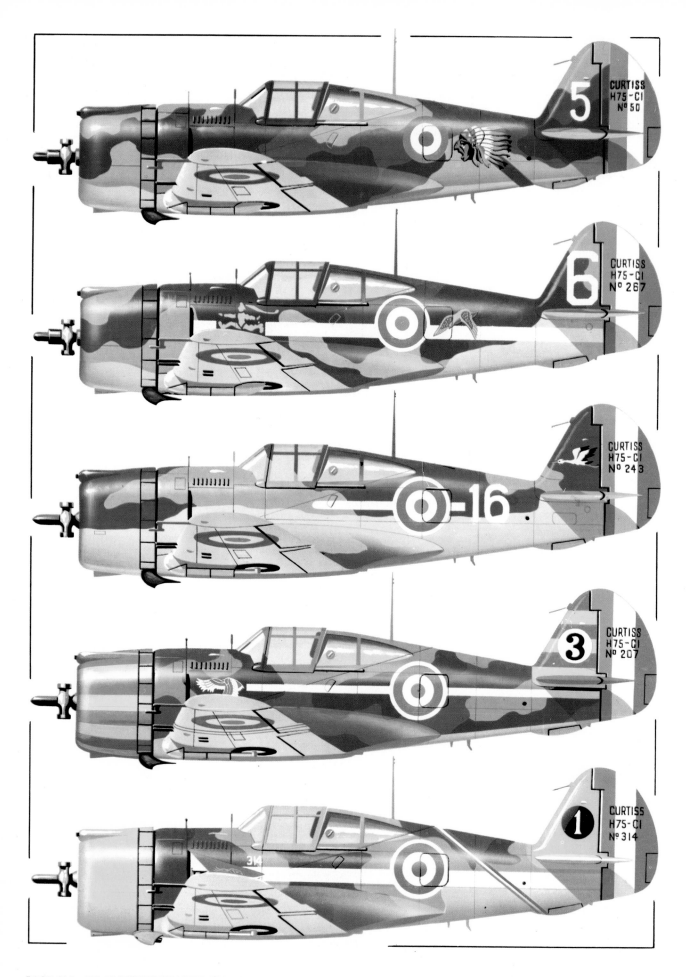

CURTISS
H75-C1
N° 50

CURTISS
H75-C1
N° 267

CURTISS
H75-C1
N° 243

CURTISS
H75-C1
N° 207

CURTISS
H75-C1
N° 314

within the space of barely more than 12 months, no fewer than four kits, all to 1/72nd scale, appeared on the market from Aosima, Heller, Revell and Monogram.

All four kits make up into generally good models; all four suffer their detail inaccuracies, and all represent the Twin Wasp-powered version. The most readily available of these kits in the UK are those offered by Revell and Heller, and either one of these can produce a sound, essentially accurate model of the P-36A or the earlier versions of the Hawk 75A, the largest single operator of which was the *Armée de l'Air,* and it is in the markings of this air arm that we depict Curtiss's little fighter in our colour page this month. For the modeller with a penchant for greater variety, this warplane is illustrated in colour in the markings of other air forces by which it was flown on page 316, but we would point out that conversion of any of these kits of the Twin Wasp-engined fighter to one or other of the Cyclone-powered models is rather more difficult than appears at first sight, and such a conversion should be avoided by all but the most skilled of modellers.

Honourable Spitfire

It is perhaps surprising that, until now, the most widely-used version of the Spitfire, the Mk V, has been poorly represented in kit manufacturers' lists. Fujimi's recent release of a Spitfire Mk VB is therefore particularly welcome. Ostensibly to 1/48th scale but actually slightly less than 1/50th scale — the five per cent difference being very noticeable when this and the genuinely 1/48th scale Monogram kit are placed side by side — the Fujimi Spitfire VB is generally accurate in outline, although the contours of the wingtips, rudder and tailplane are not *quite* right. Fortunately, a little filing rectifies these errors, but the wheels are slightly too large and their cover plates appreciably too small, and their correction presents somewhat more difficulty. On the credit side, Fujimi has made the best attempt that we have so far seen in moulding the Spitfire's curvacious canopy accurately.

In terms of quality this kit is first class. It is equipped for motorisation, although the motor is not actually included in the kit, and the component parts, moulded in high-quality white plastic, fit together with extreme precision. Surface detail attains that superb standard only seen on Fujimi kits, and ample cockpit detail is provided. An excellent decal sheet provides the markings for no fewer than *eight* aircraft,

Among the Heller kits recently made available in the UK by Richard Kohnstam Limited under the trade name of "Riko" is a kit of the fascinating little Caudron Renault CR.714 light interceptor, the first production example of which is seen here at Guyancourt in June 1939.

including complete stencilling for which full instructions are given. Only four of the eight aircraft for which markings are provided are illustrated on the box, but for some inscrutable oriental reason three other Spitfires for which markings are *not* provided by the decal sheet are illustrated by the instruction sheet! All the decals must, incidentally, be cut out with great care as in many cases the film covers more than one design. This particularly applies to the stencilled markings.

This kit was not available in the UK at the time of closing for press, but was expected to reach the market at a price of around £1.00. Its Japanese price is 350 Yen, and at this it represents excellent value for money.

A pair of deltas

Two modern delta-winged combat aircraft added to the stockists' shelves recently are 1/72nd scale kits of the Saab 37 Viggen from Airfix and the McDonnell Douglas A-4 Skyhawk from Frog. It would be nit-picking to suggest that the Viggen kit is anything but superb. It possesses almost all the virtues — accuracy, clean pressings, excellent surface detail, easy assembly, a mass of underwing stores, and so on. Its one shortcoming is provided by its decals which are inaccurately coloured and fuzzily printed, and unworthy in every respect of the fine kit that they accompany. At a UK retail price of 34 pence Airfix's Viggen is quite astonishing value.

The other delta results from Frog's policy of re-issuing kits originally produced by Hasegawa, and is one of the most recent versions of the A-4 Skyhawk, though not *the* most recent. Basically the A-4F model, it can be completed as an A-4H of Israel's *Heyl Ha'Avir* or an A-4K of the RNZAF, markings for both being provided by the high-quality decal sheet, this kit provides a neat, well-detailed model which we believe to be easily assembled. We say *believe* advisedly as our review sample suffered severe distortion of its fuselage halves. Unfortunately, it is not entirely accurate in outline as its overall length of 7·05 inches (17,91 cm) scales out at 42·5 feet (12,95 m) which is the length of the *two*-seater. It should, in fact, have a length of 6·7 inches (17,02 cm) to scale out at the correct length of 40·1 feet (12,22 m). A large selection of underwing stores is included as is also the "avionics hump" which appears aft of the cockpit of the A-4K and is now appearing on

A-4Hs, and the box art illustrates the camouflage applied to both Israeli and New Zealand aircraft completely, that of the Israeli Skyhawk, if accurate, being fascinating. Frog's Skyhawk retails at 47 pence in the UK, and in terms of quality is value for money.

Jaguar trainer

When reviewing Heller's 1/50th scale Jaguar kits in the July issue, we mentioned that we had been unable to obtain a kit of the "E" version. Having now received this kit we can report that, from some aspects, it is the best of the three. Its component parts seem to fit better, without gaps around the nose section, and the enormous canopy, which is extremely well moulded, adds interest to the model. Unfortunately the cockpit interior is spartan indeed, and those assembling this kit would be well advised to add additional detail to obtain a semblance of authenticity.

The decal sheet offers the markings of the Jaguar E prototype and, again, is better than the sheets accompanying the other two kits. For example, the roundels are accurately centred, and there are quite a number of additional markings. However, the sheet is highly glossy, and the markings must be flattened down before application. Furthermore, the individual markings must be cut out separately as, in several cases, one area of film covers more than one design.

Our strongest criticism is, once again, directed at the useless split planview in the instruction sheet which purports to show the camouflage pattern. While open to criticism, this is a good kit by most standards, and can produce a first rate model of an aeroplane that is likely to be of primary importance to the RAF, the *Armée de l'Air* and *Aéronavale* in the years ahead.

RECENTLY ISSUED KITS

Company	Type	Scale	Price
Airfix	Saab Viggen	1/72	34p
Frog	DH Vampire FB 5	1/72	20p
Frog	Bf 110G	1/72	37p
Frog	A-4 Skyhawk	1/72	47p
Fujimi	Fw 190D	1/50	£1·20
Hasegawa	Shin Meiwa PS-1	1/72	£3·75
Heller	Jaguar E	1/50	—
Italaerei	Re 2002	1/72	
Monogram	F-14A Tomcat	1/72	

IN PRINT

"*United States Naval Aviation, 1910-1970*"
edited by Lee M Pearson
US Government Printing Office, Washington, DC, $4·00
440 pages, 7¾ in by 10½ in, illustrated

MANY readers will recall the detailed chronology *United States Naval Aviation, 1910-60* compiled by Lee Pearson and Adrian O Van Wyen twelve years ago. That book was hardly off the presses when Pearson was annotating his own copy and pasting in additions for a new and revised edition. It has now been published and it is excellent. The first edition was 223 pages; the current one is almost twice that size.

The chronology itself runs to 259 pages; it is followed by thirteen appendices. These include a list of the first 250 US Naval Aviators; a list of aviation commands in order of their creation; a list of US aviation ships with the dates of their commissioning and ultimate disposals as of 1968; tables of aircraft on hand; tables of combat aircraft procured with yet another set for helicopters; and there are 28 pages that match BuAer serial numbers with aircraft procured. Another table gives numbers of aviation personnel on active duty from 1920 through to 1965; there is a list of air stations (active and inactive), with information about the men for whom they were named; a list of ships named for Naval aviators; Medal of Honor winners in Naval aviation; a description of aviation ratings; and a chronology which deals with the evolution of air groups in wings.

The text is supplemented by 80 pages of photographs, and each photo is accompanied by its Navy (or other) identification number. This is a practice that deserves to become universal.

For the person interested in US Naval Aviation this book ranks as a "best buy". For all others — well, they simply cannot go wrong for only $4·00. — RKS

"*Encyclopedia of US Military Aircraft*"
by Robert Casari
Volume 1, Aeromarine to Christofferson
64 pp, 6 in by 9 in, illustrated
Volume 2, Curtiss D to G
74 pp, 6 in by 9 in, illustrated
Softbound, $2.50; hardbound, $4.95 each

DESPITE their apparently small compass, these works are a significant contribution to the history of the early aircraft used by the US Army. The scope of Part I, to be completed in an unspecified number of volumes, is the 541 aeroplanes of some 60 different types bought by the Army up to the time the US entered World War I on 6 April 1917. Part II will cover the purchases during the War.

Through extended research, Mr Casari has succeeded in compiling, as the basis for these books, a record of the serial numbers allocated to the Army aircraft purchases; no such record previously existed in any one source. The various types of aircraft are described (in alphabetical order) in considerable detail and the history of each example procured is traced. The illus-

trations comprise half-tone photographs and, in a few cases, general arrangement drawings.

Copies can be ordered direct from the author at 6 Applewood Drive, Chillicothe, Ohio 45601, USA; prices quoted above include postage.

"*Early Military Aircraft of the First World War*"
Volume 1, Landplanes, $1.95
Volume 2, Airships, $2.50
Volume 3, Seaplanes and Motors, $2.50
Flying Enterprises, 3209 Coral Rock Lane, Dallas, Texas 75229, USA
94 pp each volume, 4¾ in by 7 in, illustrated

THESE three volumes together represent a photo-litho reprint of a little-known book entitled "Aircraft of Belligerents in the year 1914". The original was compiled at the US Navy Gas Engine School at Columbia University NY, and was published in November 1918 for "the information and guidance of Aviation Mechanics Schools". Under the general editorship of Lieut Cdr Charles E Lucke, it was itself based on a translation from an unspecified German document.

The contents suffer the limitations that are to be expected in any document dating from the early years, lacking the kind of fine detail that has been applied in retrospective publications dealing with the same period. However, the very fact that these *are* contemporary records gives them a certain value and charm of their own, and the three-view drawings, albeit somewhat crude, cannot easily be located in other published works.

"*The C-5A Scandal*"
by Berkeley Price
Houghton Miffin Co, Boston, $5·95
238 pages 5¾ in by 8½ in

AS ONE may gather from its subtitle, "A $5 Billion Boondoggle by the Military-Industrial Complex", this book is an *exposé*. It details the efforts of the US Congress to unravel the C-5A embroglio after it was discovered that the huge transport would "suddenly" cost two *billion* more dollars than originally calculated. Fortunately, the *exposé* never assumes tones of hysteria, nor does it become vindictive.

The author is more concerned with politics, pressure groups, finances and the Pentagon's deceitful stratagems than with the aircraft *per se*. But there are nevertheless many interesting insights into the conception of this huge and controversial aeroplane, its construction and teething problems. Snippets of how the Air Force chose a Boeing C-5A and the Pentagon's civilian hierarchy awarded the contract to Lockheed make for some fascinating reading. The book also serves to further debunk the myth of Robert S McNamara, whose computerised pedantry held too many persons in thrall during 1961-8.

The book is also instructive in its insights into the extent that American political leadership and its bureaucracy

have sometimes gone to *protect* its aircraft industry; so different from the British experience.

The author is a journalist, but unlike most he thoughtfully includes 12 pages of references at the back of the book for the reader who would pursue the subject further. In sum, this is a very illuminating book as regards current or recent practices in American military aircraft procurement; and it represents a significant piece of history relative to the Lockheed. — RKS

"*The Birth of the Luftwaffe*"
by Hanfried Schliephake
Ian Allan, Shepperton, £3·00
224 pp, 9 in by 5¾ in, illustrated

FROM rise to demise in the short span of a dozen years — two years as a clandestine fledgeling after the assumption of power by the National Socialist Party in 1933, and 10 years from its existence being proclaimed in March 1935 until its total destruction in the final débâcle that engulfed Germany: this is the popular belief of the tenure of life of the *Luftwaffe* of Germany's Third Reich. But the fledgeling had hatched out long before the accession of Adolf Hitler, whose régime merely nourished it and hastened its growth to maturity. Indeed, the *Luftwaffe* existed in embryo within a few years of the end of World War One, a fact which is made clear in "The Birth of the Luftwaffe".

The text in this volume is brief — some 30,000 words occupying 48 pages plus a few pages of appendices — but it will prove of absorbing interest to those who, sated with works of reference on the wartime *Luftwaffe*, are anxious to learn something of the growth of the German air arm in the years prior to WW II. The reader of this book will *not* find that "some old and well-worn myths are crushed" in its pages, as is claimed by the Foreword, and it can in no way be considered a definitive work on the development of military aviation in Germany between 1918 and 1939, the period that "The Birth of the Luftwaffe" encompasses, but it presents for the first time in the English language much fascinating information hitherto unavailable, and on this score alone is to be thoroughly recommended. Readability would undoubtedly have benefited from some competent editing, and we found irksome the numerous and rather naïve over-simplifications (eg, "Heinkel developed the He 111 out of the older He 70") which have perhaps resulted from inadequacies in the translation.

Perhaps the least said about the 144 pages of half-tone illustrations the better. The publishers have exercised little selectivity in their choice, few can be categorised as items for the connoisseur, many display the patterning effect of the coarse block screens used by the newspapers or periodicals from which they have evidently been reproduced, while others seem to be the unfortunate results of photographs taken in a thick mist or through bottle-glass. Another poor feature of this book is the use of singularly uninformative captions. □

FLYING THE "WOODEN WONDER"

So MUCH PRAISE has been lavished on the de Havilland Mosquito that it is difficult to resist a temptation to pull it metaphorically apart — plywood, balsa wood, glue and all. No one will ever question the fact that the Mosquito was one of the few really great war-winning aircraft, or that the story of its conception, design and procurement was one of the most remarkable of the era — but the sound of success has often drowned calmer voices among those who actually flew this beautiful, surprisingly easy, yet sometimes temperamental device.

As an RAF maintenance-unit test pilot, handling many different types week-by-week and even day-by-day, I found the behaviour of the Mosquito to be paradoxical. So long as reasonable care was exercised in a relaxed and confident frame of mind, the take-off and landing presented no difficulties; yet as soon as you started to take too *much* care (when fitting it in, for instance, on a flat-calm day to our 2,700-ft runway) the previously so-amenable Mosquito suddenly became a problem-child — trying to swing on take-off, tending to *g*-stall to a heavy landing and needing too much differential brake during the landing run.

We soon learned to relax and to treat this wooden wonder as we would any other mildly "hot" aeroplane of its era. Delightful in its handling qualities — and particularly so with the stick control and good forward view through the flat bullet-proof screen of the fighter and fighter-bomber versions — the Mosquito's fits of temperament related mainly to some detail characteristics. The ailerons, for instance, were very touchy in adjustments to the servo tabs and, using shims, to the hinge position and shroud gaps. If incorrectly adjusted during re-assembly or after test complaints, the ailerons might ride up dangerously in the dive, or over-balance at lower speeds when the slotted flaps were down. On one occasion our unit had to send for a test pilot and specialist rigger from the manufacturers (Standard Motors, in this case) to deal with a refractory Mk VI fighter-bomber on which our own adjustments had failed to produce an acceptable compromise in lateral-control behaviour.

Then there were the inadequate hydraulics — an apparently off-the-shelf system similar to that in the Airspeed Oxford trainer — which provided slow undercarriage retraction coupled with frequent failures to lock up before the selection lever jumped back into neutral as the pressure increased. We learned on take-off to hold the lever (regardless of possible stress in the pipelines) until all three "greens" came up, and to check the operation, with successive re-cyclings, as they would now be described, at a later stage in the test routine. This unreliable lock-up was a nuisance because the undercarriage, flap and bomb-door levers were neatly grouped in the centre of the dashboard and one's "stick hand" had to be changed, leaving the not-too-reliably-damped throttles unattended on the left.

The Mosquito came into my working life at a time when there was a considerable "flap" among ferry pilots about its capacity to swing to self-destruction on take-off or landing. Coming from a maintenance unit which held, among other things, B-17 Fortresses and Bristol Beaufighters, I didn't take the horrific stories too seriously. The B-17 required some manual dexterity with those quaint three-barred-gate throttles, and the Merlin-engined Beau, in particular, sometimes felt as if it was going sideways when leaving the ground — so the Mosquito, I thought, couldn't be all that more ferocious.

Maybe it *was* a little more prone to swing either way at the slightest provocation; the important thing was to avoid

Viewed from the Cockpit

by H A Taylor

such provocation. On take-off you made sure, leading with the port throttle, that it was rolling straight and that there were ample air loads on the rudder pedals before taking the throttles through the "rated" gate. On landing you made equally sure that there was no drift, or that this had been kicked-off accurately, before closing the throttles; so far as was possible, you avoided badly out-of-wind runways. Differential brake had to be used quickly and firmly at the first sign of deviation from the straight. Nothing could prevent disaster once the Mosquito had really got out of hand after landing — but there seemed to be no reasonable excuse for allowing a swing to develop on take-off when all you had to do was to swallow your pride, close the throttles and start all over again. These throttles, incidentally, were difficult to operate differentially; they were short, very close together and had to be moved with finger-tips and thumb rather than with the palm of the hand. This difficult feature may well have been the cause of some of the take-off accidents.

Pilot's cockpit of the Mosquito NF Mk 30 (one of the later night fighter variants), showing the stick control and the throttles, extreme left.

A late-model Mosquito T Mk III in all-yellow trainer finish; this type shared with the fighter models a flat bullet-proof windscreen.

At No 27 MU, Shawbury, we had, at one period, more Mosquitos than any other type, yet, during a year's flying at the unit, I can remember only one case of a swing which might have been catastrophic. A new test pilot, fresh from operations and "conversion", and with little twin-engined experience, was caught napping on his first flight with a Mosquito. He had spent the better part of an hour familiarising himself with it before coming in for his first landing. In the meantime a freshening wind had veered through 40° or so and the landing "T" had not been moved. His landing, with kicked-off drift, was good and square along the runway. For a few seconds all seemed to be well, though there was a plume of smoke from the heavily braked port wheel — then, like a flash, the Mosquito kicked into the wind and finished its run going backwards on the grass in a cloud of dust and smoke. He taxied in successfully, though both the undercarriage legs (which were a shade on the weak side laterally) were listing 10° or more from the vertical and there were some nasty squealing noises.

The essential harmlessness and good qualities of the Mosquito were demonstrated for me on what happened to be my 13th test flight on the type. When coming in to land with a Mk II fighter I noticed that the airspeed indicator needle was behaving in a very strange way. Eventually, after hurried last-second overshoot action, the needle settled back to the zero equivalent on the scale and stayed there. So, with not much more than half-a-dozen hours of experience to support the operation, and using a rather short

runway, I had to manage without that useful instrument. On the first landing attempt the Mosquito was still floating, throttles closed, when we were a third of the way along the runway. On the second attempt, using a fair amount of power and trying to gauge the speed from aileron reaction, the Mosquito settled in from a three-point attitude as the throttles were closed just beyond the threshold. Obviously we had very little speed in hand and the Mosquito had stalled as soon as the benefits of slipstream-lift had been removed. Later I was to learn that really short landings were best made with power maintained during a fairly slow and gentle check; even from quite high engine-off final approach speeds the powerful elevators could induce a *g*-stall and a heavy premature landing while there was still speed enough for continued untidy flight towards a final definitive three-pointer.

Although the Mosquito had excellent engine-out characteristics and single-engined performance when flying "clean", it could be troublesome, after a real-life failure, if all did not go well on final approach. The lowered undercarriage was a fairly massive drag-producing excrescence and could not be raised rapidly in emergency. Nevertheless, one of our test pilots — after losing an engine far from home and lining up with everything down for a forced landing at a temporarily disused aerodrome — managed a successful go-around when a heavy motor roller, or some similar mobile obstruction, suddenly appeared on the runway.

At one period, when our Mosquitos were overcrowding

Standard armament of the fighter Mosquito, such as the FB Mk VI shown here, comprised four 0·303-in machine guns and four 20-mm cannon in the nose and forward fuselage.

the dispersals, I was told to report on the prospects of flying them into and out of satellite fields where only Battles and Blenheims had previously been parked. After a number of mildly frightening dummy landings and take-offs, I decided that we could no doubt get them in, but might never be able to get them out — except by Queen Mary*. The climb-away, following a take-off which was, in any case, likely to be marginal from those unshorn grass fields, was not bright enough, while the undercarriage was retracting, to ensure adequate clearance over the surrounding trees and other obstructions.

We suffered from another form of overcrowding at Shawbury. The principal unit at the station was an advanced flying training school, whose Oxfords packed the circuit, so it was a special pleasure to be able to get away from it all with a Mosquito II or VI on a cannon-firing test; with a Mk XIII night fighter on an AI (Air Interception) radar test; or with any aeroplane on an anti-aircraft gunnery co-operation flight. For these outings — to the Irish Sea for the cannon tests, and to anywhere you liked for the AI tests — it was nice to have two engines and all that fuel, so you could (without radio) mess about above cloud, and chase random AI targets for a hundred miles or so without getting too worried about fuel-endurance limitations during a sometimes devious map-reading return to base.

We preferred the cannon-firing tests and the A-A flights because, usually, we were then flying solo. One of the prices paid for the outstanding performance of the Mosquito was its cramped accommodation and marginally acceptable crew-escape arrangements. The fuselage — made in two halves like the pod of the pea and with the equipment installed before these were mated — was so narrow that the seats of the two crew members had to be staggered so as to provide acceptable elbow-room. In the fighter versions, entry or exit (by parachute or otherwise) was made through a small door on the starboard side. In the bomber and photographic-reconnaissance variants the way in and out was through a trap-door below the second crew member's seat — involving very careful practice and planning for successful exit in emergency.

Even in normal conditions entry and exit with all versions was mildly difficult; personal loose ends were constantly

---

* *The pet name of the period for the long, low-loading transporters of aeroplanes and bits of aeroplanes.*

getting hooked up on something and, when finally installed, the most phenomenal contortions were required to reach the fuel cocks. The fighter and fighter-bomber versions gave the impression of being more roomy and comfortable than the bomber and PR variants. Their internal volume can have been little, if any, greater; the entirely subjective impression was largely the effect of the better forward view and the absence of the impeding spectacle aileron control. The shallow, raked-vee screen of the bomber, with its anti-misting/icing sandwich construction, no doubt gave an entirely adequate field of forward view, but it didn't *appear* to do so — which merely emphasises the inevitably subjective nature of judgments in such matters. The PR version, with its pressurised cabin, was claustrophobically uncomfortable in a special way and was known by the Hatfield test pilots as "the boiler" — pressure-cookers had not yet been invented.

For the AI tests — which were sometimes combined with the cannon-firing tests in the open sea off the North Wales coast — we tried to operate in pairs, taking turns to close in on each other while the radar operator called out the relative position and range (distance) of the target as indicated on the display. His answers were nearly always right so far as we could judge by visual means, though there were occasionally some curious time lags when the target changed position, and now and again false or negative returns were announced.

Because the pilot couldn't easily see the blips, or whatever, we never learned in what form the information was provided and nobody seemed to be very keen to tell us. Looking back over the better part of 30 years, I can only suppose that we were too busy with our own test-flying problems to wish to know, and treated AI, as we had for long been treating IFF (Identification, Friend or Foe), simply as valuable but mysterious "boffinry".

These radar tests provided some of the most relaxed and pleasurable outings during six years of war — and especially so when no second aircraft was available and we hunted for random anonymous targets. We searched the sky while cruising gently along at zero boost and 2,000 rpm at about 8,000 ft and forgot all about that wood and glue which worried us just a little when on more drastic test work. The curious and possibly fundamental thing about the Mosquito was that, though a "hot" aircraft for its day, it retained a kind of innocence — with a strange resemblance in its handling characteristics to some of its much less pugnacious predecessors from Hatfield. □

Three views of the Mosquito B Mk IV, a standard bomber version, with an additional side view of the NF Mk 30 night fighter.

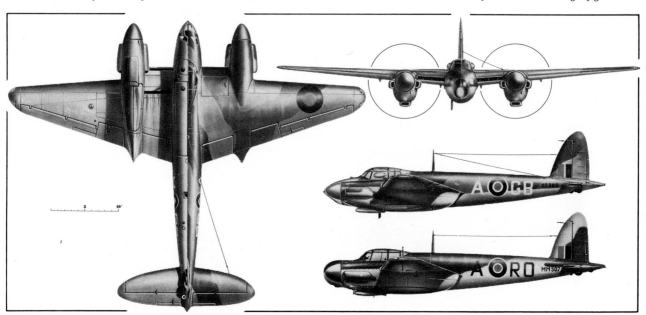

THE FLEDGELING FIGHTER

Thirty-five years ago, at a still relatively insignificant aircraft factory between Augsburg and the little town of Haunstetten in Bavaria, the first production examples of a warplane that was to carry Germany to the forefront of the international fighter scene were nearing completion. Their designer was almost unknown outside Germany's borders, but within a few years his name was to become virtually synonymous with a fighter the equal of any extant and vastly superior to most, the Messerschmitt Bf 109. This warplane was to retain a position in the forefront of its class for three-quarters of a decade; it was to be built in larger numbers than any combat aircraft before or since, and it was to achieve a record for production longevity unlikely ever to be surpassed by a fighter.

This month Macdonald publish AUGSBURG EAGLE*, without a doubt the most detailed account of the development and career of this truly remarkable warplane, and with the permission of the publishers, AIR ENTHUSIAST has adapted from this new book the following account of the first production models of the famous Messerschmitt fighter; the fledgelings from which evolved the eagles that played so large a part in the drama that unfolded in the skies over Europe and North Africa during the years 1939–45.

I N FEBRUARY 1937, the average Augsburger sitting over his *Tassen Kaffee* in the Königsbau, on the Adolf-Hitler Platz, had no idea that, little more than a score of kilometres from the café in which he was seated, an event was taking place of immense significance; an event which, although not immediately apparent even to the youthful *Luftwaffe* of Germany's Third Reich, marked the beginning of German ascendancy in a vitally important category of warplane — the single-seat fighter. On the apron outside the factory hangars of the Bayerische Flugzeugwerke at the little airfield between Augsburg and Haunstetten, the first production example of a new and, from some aspects, radical fighter, the Messerschmitt Bf 109, was about to be handed over to the *Luftwaffe* for its acceptance flight.

The event, which was attended by Robert Ritter von Greim, the *General der Jagdflieger*, and by *Oberst* Ernst Udet, who, on seeing the first prototype of the Bf 109 barely 18 months earlier, had commented, "This machine will never make a fighter", followed a hard-fought contest with another contender for *Luftwaffe* orders, the Heinkel He 112. The contest, which had seen the Bf 109 progress from rank outsider via odds-on favourite to outright winner, had become the subject of discord within both governmental and industrial circles. Each principal contender had had its adherents; both factions had been vocal, and the dissension had been in no way mitigated by the intense personal dislike

that existed between Willy Messerschmitt and Ernst Heinkel. The selection of the Bf 109 had become the subject of much adverse criticism, largely the result of personal prejudices and jealousies, but Messerschmitt's team was convinced that, from the early September day in 1935 when their senior test pilot, Hans D "Bubi" Knoetzsch, had flown the first Bf 109 prototype from that same airfield, it had displayed the hallmark of the thoroughbred. Now it was to face its sternest test — *combat* service! Messerschmitt's fighter was not to be afforded the gradual working up process usually allowed entirely new warplanes; the inevitable teething troubles and wrinkles were to be overcome under true operational conditions.

When the first production example of Messerschmitt's

(At head of page) Messerschmitt Bf 109C-2s fresh from Focke-Wulf's Bremen assembly line in the autumn of 1938, and (below) a Bf 109B-2 at Augsburg-Haunstetten prior to acceptance by the Luftwaffe during the summer of 1937.

*AUGSBURG EAGLE, published by Macdonald & Co (Publishers) Ltd, 49/50 Poland Street, London W1A 2LG, £3·50 (UK only).

A Bf 109B-1, the initial production version of Messerschmitt's fighter, photographed during acceptance trials in 1937.

fighter, the Bf 109B-1, left the Augsburg-Haunstetten assembly line in February 1937, the premier *Luftwaffe* fighter *Geschwader*, JG 132 "*Richthofen*", had been designated the first unit to receive the new combat aircraft, and had been assigned the task of working up the Bf 109 to combat status. It was intended that II *Gruppe* at Jüterbog-Damm would be the first to convert, followed closely by I *Gruppe* at Döberitz. However, the re-equipment of at least part of *Jagdgruppe* 88 in Spain took precedence owing to the mastery over the unit's He 51s being displayed by the Soviet I-15 and I-16. Furthermore, the Spanish conflict provided the *Luftwaffe* with an admirable opportunity to evaluate its new fighter under true combat conditions and, simultaneously, evolve suitable operational tactics. Thus, after a hurried conversion course, II/JG 132 personnel were posted to Spain where Bf 109B-1s began to arrive in April 1937, these being issued to the 2.*Staffel der Jagdgruppe* 88 which relinquished its He 51Bs.

The Bf 109B-1 did not offer any serious problems, despite the appreciably higher performance and very different handling characteristics that it presented pilots versed in the art of flying the unsophisticated He 51 biplane. Some concern was initially expressed over the somewhat alarming tendency of the Bf 109B to drop the port wing during take-off and just before touch-down, but with experience the 2./J 88 pilots found that the careful use of the rudder almost eliminated the wing-dropping tendency.

Noticeable aileron shudder occurred when the slots were opened at high speed, and the phenomena of the high-speed stall was encountered for the first time by many of the pilots, necessitating practise of turns at varying speeds and rates of turn to accustom them to the symptoms of an approaching stall. Coarse use of the controls at high speed sometimes resulted in violent aileron shudder, and the high axis of the fighter produced a tendency to swing pendulum-fashion, a movement having to be counteracted by use of the rudder. However, familiarity with the aircraft and its peculiarities brought confidence in the capabilities of the Bf 109B, and the pilots of 2./J 88 were soon convinced, and not without some justification, that they were flying the best fighter extant.

The oval-section light metal monocoque fuselage of the Bf 109B was manufactured in two halves and joined longitudinally top and bottom, each half being constructed of longitudinal stringers and vertical panels, the latter having flanged edges to form Z-frames which were pierced for the stringers. The single-spar wing was attached to the fuselage at three points, two on spar flanges and one at the leading edge, and the entire trailing edges were hinged, slotted ailerons

outboard and slotted flaps inboard, automatic slots being carried by the outboard leading edges. The tailplane was braced to the fuselage by a single strut on each side, and the main undercarriage members were raised outward into the wings by hydraulic jacks.

Power was provided by a Jumo 210Da engine which, with a two-speed supercharger, 87 octane A2 fuel, and driving a two-bladed wooden Schwarz fixed-pitch airscrew, offered 720 hp at 2,700 rpm for take-off, maximum continuous sea level outputs being 610 and 545 hp at 2,600 and 2,500 rpm respectively. At its rated altitude of 8,860 ft (2 700 m), the Jumo 210Da offered 640 hp for five minutes, and maximum continuous outputs of 575 and 510 hp at 2,600 and 2,500 rpm.

The fuselage tank of 55 Imp gal (250 l) capacity was contoured behind and beneath the pilot's seat, the pilot was provided with a Carl Zeiss *Reflexvisier* C/12C reflector sight, and the intended armament comprised two 7,9-mm MG 17 machine guns mounted on the engine crankcase with their muzzles protruding into blast troughs in the upper nose decking, and a third MG 17 mounted between the cylinder banks and firing through the airscrew hub, all three weapons having 500 rpg. Trials with the Bf 109 V4, V5 and V6 had revealed the fact that, as a result of inadequate cooling, the central MG 17 was prone to seizure after a few bursts, and this was, therefore, omitted from the Bf 109B-1 which was delivered to *Jagdgruppe* 88 with twin-gun armament. Provision was also made for the installation of short-range single-waveband FuG 7 R/T equipment, but this was considered to be a dispensable luxury, and was not fitted to the Bf 109B-1s despatched to Spain.

Blooded over Spain

The Bf 109B-1 was, in fact, considered only as an interim production model of the fighter, resulting from the haste displayed by the *Luftwaffenführungsstab* in bringing the new warplane to service status, and fewer than 30 examples of the initial model had left the Augsburg-Haunstetten assembly line before it was replaced by the Bf 109B-2 which appeared during the early summer of 1937, the initial examples differing from the Bf 109B-1 solely in having the fixed-pitch wooden Schwarz airscrew replaced by a variable-pitch two-bladed Hamilton metal airscrew manufactured under licence by VDM. Retaining the Jumo 210Da engine, the first batch of Bf 109B-2s was immediately shipped to Spain where 1.*Staffel der Jagdgruppe* 88 converted.

The two Bf 109B-equipped *Staffeln* of the *Legion Condor* formed *Grupo* 6 of the Nationalist air arm. The 2.*Staffel* had served as top cover for the He 51s of the 1. and 3.*Staffeln* operating in the close-support rôle during the Brunete offensive, and its Bf 109B-1s had established a clear ascendancy over the opposing I-16s and I-15s. They had also acquitted themselves well as escorts for the bombers of the *Kampfgruppe* 88. The 2.*Staffel* had then been transferred north to participate in the campaign against Santander, and shortly afterwards was joined by the re-equipped 1.*Staffel*, their primary rôle being that of maintaining air superiority, but their missions including offensive sweeps, ground strafing, and bomber escort tasks.

During their offensive sweeps, the Bf 109B-equipped *Staffeln* frequently encountered large formations of Republican fighters, and despite their limited numbers, the Messerschmitt-designed aircraft of 1. and 2./J 88 established a remarkable reputation for the *Luftwaffe's* new standard fighter during the summer and autumn of 1937. The experience gained with them was to prove invaluable in the greater conflict to come. Several of the *Luftwaffe's* future "aces" were "cutting their teeth" with the two Bf 109B-equipped *Staffeln*, and *Leutnant* Wilhelm Balthasar, who was to achieve a measure of fame three years later by destroying an enemy aircraft a day for 21 days, had succeeded in shooting down four Republican I-16 fighters within six minutes during one action.

While the Bf 109 was being "blooded" in action with understandably little publicity over Spain, the Propaganda Ministry in Berlin, while considering it impolitic to publicise the activities of the *Legion Condor*, had every intention of capitalising on the capabilities of the Messerschmitt-designed fighter to raise German aviation prestige abroad, and the 4th International Flying Meeting held at Zürich-Dübendorf between 23 July and 1 August 1937 provided the ideal opportunity to furnish the world with a practical demonstration of the Bf 109's potentialities. The German team included no fewer than five Bf 109s, and none was, in fact, a completely standard production model, comprising the Bf 109 V8 (D-IPLU) and Bf 109 V9 fitted with the fuel-injection Jumo 210Ga engine, a Bf 109B-2 with all armament removed, the Bf 109 V10 (D-ISLU) fitted with the new Daimler-Benz DB 600Aa engine, and the similarly-powered Bf 109 V13 (D-IPKY).

So impressive were the results achieved by the team that several of the competitors withdrew from the contest, and the claims made by the Propaganda Ministry on behalf of the Bf 109 had been fully vindicated. The statement that the fighter was already in *large-scale* service was an exaggeration, however, for apart from the Bf 109Bs serving with the *Legion Condor* in Spain, only I and II *Gruppen* of JG 132 "*Richthofen*" at Döberitz and Jüterbog-Damm, and I *Gruppe* of JG 234 "*Schlageter*" at Cologne included Bf 109Bs in their inventories, and no *Gruppe* was at full strength at the time the Zürich meeting was held.

During the spring and summer of 1937, the pace of development of the Bf 109 had accelerated, and plans had been formulated for the licence manufacture of the new fighter by other companies. The Bf 109 V7 (*Werk-Nr* 881 D-IALY), which had commenced its flight test programme in March 1937 as a prototype for the Bf 109B-2, had been fitted with a VDM-Hamilton variable-pitch metal airscrew driven by a Jumo 210G employing direct fuel injection, a two-stage supercharger, and an automatic boost control. Direct fuel injection enabled the power plant to function equally well in any position, a distinct advantage in a fighter aircraft, and the Jumo 210G offered 700 hp for take-off and 730 hp at 3,280 ft (1 000 m), with 675 hp being available at 12,500 ft (3 800 m). This version of the engine had not been ready for production installation, however, when deliveries of the Bf 109B-2 had begun in the summer of 1937, and the majority of the fighters of this model were to retain the Jumo 210Da, with the Jumo 210G being installed only in the last batches off the line.

The Bf 109 V8, V9 and V10 had all begun life as development aircraft for the Bf 109C, which, by the spring of 1937, was envisaged as the definitive Jumo 210-engined production variant of the fighter. Both the V8 and V9 featured a repositioned oil cooler intake ahead of the radiator bath beneath the nose, and additional auxiliary cooling slots, but the definitive Bf 109C, which dispensed with the latter and introduced a redesigned and deeper radiator bath, standardised on an oil cooler position similar to that employed by the Bf 109B.

As the difficulties with the engine-mounted MG 17 machine gun were still unresolved, the Jumo 210Da-powered Bf 109 V8 introduced a pair of wing-mounted MG 17s installed immediately outboard of the undercarriage wells and demanding few modifications to the wing structure apart from local strengthening. Firing trials led to some stiffening of the inboard wing leading edge, and a minor flutter problem was rectified by balancing the ailerons, but tests were successful, and the installation was adopted as standard for the Bf 109C-1. Armament was later removed, the Jumo 210Da gave place to a Jumo 210Ga and, with these changes,

the Bf 109 V8 had participated in the Zürich meeting in July 1937. The Bf 109 V9 was essentially similar to the V8 in its original form, but the wing-mounted machine guns were replaced by similarly-positioned 20-mm MG FF cannon. The development of this installation was to be somewhat protracted, and, in the event, was not to be standardised until the introduction of the E-series of the fighter.

In November 1937, a fourth *Gruppe*, II/JG 234 "*Schlageter*" at Düsseldorf, began converting to the Bf 109B, collecting its first aircraft on the 13th of that month, and by the end of the year deliveries from the Augsburg-Haunstetten plant were being supplemented by deliveries from the Gerhard Fieseler Werke at Kassel which had initiated production of the Bf 109B-2 under licence. The Erla Maschinenwerk at Leipzig-Heiterblick had reached an advanced stage in tooling for the Bf 109C, and the Focke-Wulf Flugzeugbau at Bremen had also received a contract for the licence production of this model.

Deliveries of the Bf 109C-1 began in the early spring of 1938, this variant having the Jumo 210Ga engine with deeper radiator bath and revised exhaust exits, and four MG 17 machine guns, two in the fuselage with 500 rpg and two in the wings with 420 rpg. As with earlier production variants, some of the first Bf 109C-1s off the line were shipped to Spain for service with the *Legion Condor*, and during the course of the summer, I/JG 132 began to exchange its Bf 109B-1s for the newer model.

A total of 40 Bf 109B-1s and B-2s had been shipped to Spain for use by the *Legion Condor*, and a batch of 12 Bf 109C-1s was now delivered to enable 3.*Staffel der Jagdgruppe* 88 to convert from the He 51 during the course of July 1938. Apart from its heavier armament and more powerful engine, the Bf 109C-1 of 3.*Staffel* differed from the Bf 109B of 1. and 2.*Staffeln* in having FuG 7 R/T equipment, the value of which had by this time been appreciated, both for aerial combat and for efficient liaison with ground forces during close support missions. Simultaneously with the arrival of the Bf 109C-1s for 3.*Staffel*, a Heinkel He 112B-0 was delivered to *Jagdgruppe* 88's operational base near Teruel where it was to be evaluated by leading Spanish Nationalist fighter pilots. García Morato, the most successful of the

(Above and below) A Bf 109B-1 as delivered to 2./J 88 in the spring of 1937. Note pointed spinner and absence of radio mast.

Nationalist pilots, arrived at the base to fly the Heinkel fighter on 27 June 1938, and took the opportunity to also test the Bf 109C-1. As a result of his recommendations, the Nationalist government requested that Bf 109 fighters be supplied for the re-equipment of its *Escuadrillas de Caza*.

Actually, Spanish pilots had already begun conversion to the Bf 109 within the *Legion Condor*. The German High Command's fear that the forthcoming Sudeten *coup* might evoke violent reaction from Britain and France was resulting in the progressive recall to Germany of experienced *Luftwaffe* personnel, and necessitating the gradual substitution of Spanish for German pilots in the ranks of the *Jagdgruppe 88*. Spanish pilots flying Bf 109s actually joined combat during the final phases of the Battle of the Ebro, in October 1938, by which time the three Bf 109 *Staffeln* of *Jagdgruppe 88* were operating from La Cenia, Tarragona. On 23 December 1938, a total of 37 of the 52 Bf 109s sent to Spain for use by the *Legion Condor* remained in the first-line inventory, 32 of these being based at La Cenia and five at León.

Such was the success that had meanwhile attended the Bf 109, and, in consequence, the international acclaim received by (now) Prof Dr-Ing Willy Messerschmitt, that the management of the Bayerische Flugzeugwerke was understandably anxious to capitalise on this fame, and readily agreed to the appointment of Messerschmitt as Chairman and Managing Director. On 11 July 1938 the name of the company was changed to that of Messerschmitt AG.

The year 1938 was an important one for both the *Luftwaffe* and for its new fighter. Production of the Jumo 210-engined Bf 109 was to continue throughout the year, although development was by now concentrated on variants of the basic design powered by the more powerful Daimler-Benz series of 12-cylinder inverted-vee liquid-cooled engines. Difficulties involving the adequate cooling of an engine-mounted MG 17 machine gun had been largely resolved, and the Bf 109C-2 had appeared on the assembly lines, this supplementing the quartette of MG 17s carried by the C-1 with a fifth engine-mounted weapon. Yet a further C-series variant, the Bf 109C-4, was evolved on which the engine-mounted MG 17 was replaced by a 20-mm MG FF/M cannon, but this type was never issued to a service unit, the few examples completed being retained for test and development purposes.

At the time of the Austrian *Anschluss* of 13 March 1938, the *Luftwaffe* possessed 12 *Jagdgruppen* (exluding *Jagdgruppe 88* in Spain with the *Legion Condor*) of which half had converted or were in process of converting to the Bf 109, these being I/JG 131 at Jesau, I/JG 132 at Döberitz, II/JG 132 at Jüterbog-Damm, I/JG 234 at Cologne, II/JG 234 at Düsseldorf, and I/JG 334 at Wiesbaden. Apart from I/JG 136 at Eger-Marienbad which was still flying the obsolete He 51C-2 while awaiting imminent delivery of Bf 109s, the remaining five *Gruppen* — I and II/JG 134 at Dortmund and Werl, I/JG 135 at Aibling, I/JG 137 at Bernburg, and II/JG 334 at Mannheim — were for the most part equipped. with the Arado Ar 68E and 68F.

As tension increased with the approach of the planned recovery of the Sudetenland, a programme was hastily formulated by the *Luftwaffengeneralstab* for the rapid expansion of the fighter force, and on 1 July 1938, no fewer than eight new *Jagdgruppen* officially came into existence. These

---

Bf 109B-2 Specification

Power Plant: One Junkers Jumo 210Da 12-cylinder inverted-vee liquid-cooled engine rated at 720 hp for take-off, 680 hp for five min at 8,860 ft (2 700 m), and 610 hp maximum continuous.

Performance: (At 4,310 lb/1 955 kg) Maximum speed, 255 mph (410 km/h) at sea level, 260 mph (418 km/h) at 3,280 ft (1 000 m), 276 mph at 8,200 ft (2 500 m), 289 mph (465 km/h) at 13,120 ft (4 000 m); normal cruise at 4,740 lb (2 150 kg), 217 mph (350 km/h) at 8,200 ft (2 500 m); maximum range, 430 mls (690 km); time to 19,685 ft (6 000 m), 9·8 min; service ceiling, 26,900 ft (8 200 m).

Weights: Empty, 3,318 lb (1 505 kg); maximum loaded, 4,740 lb (2 150 kg).

Dimensions: Span, 32 ft 4½ in (9,87 m); length, 28 ft 0⅔ in (8,55 m); height 8 ft 0½ in (2,45 m); wing area, 174·053 sq ft (16,17 m²).

Armament: Three 7,9-mm Rheinmetall Borsig MG 17 machine guns with 500 rpg.

Bf 109C-1 Specification

Power Plant: One Junkers Jumo 210Ga 12-cylinder inverted-vee liquid-cooled engine rated at 700 hp for take-off, 730 hp at 3,280 ft (1 000 m), and 675 hp at 12,470 ft (3 800 m).

Performance: (At 4,405 lb/1 998 kg) Maximum speed, 261 mph (420 km/h) at sea level, 292 mph (470 km/h) at 14,765 ft (4 500 m); normal cruise at 5,062 lb (2 296 kg), 214 mph (344 km/h) at 10,170 ft (3 100 m); maximum range, 405 mls (652 km); time to 16,400 ft (5 000 m), 8·75 min; service ceiling, 27,560 ft (8 400 m).

Weights: Empty, 3,522 lb (1 597,5 kg); maximum loaded, 5,062 lb (2 296 kg).

Dimensions: Span, 32 ft 4½ in (9,87 m); length, 28 ft 0⅔ in (8,55 m); height, 8 ft 0½ in (2,45 m); wing area, 174,053 sq ft (16,17 m²).

Armament: Four 7,9-mm Rheinmetall Borsig MG 17 machine guns, the two fuselage-mounted weapons each with 500 rpg and the two wing-mounted weapons each with 420 rpg.

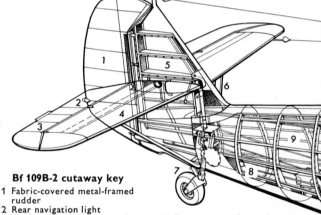

Bf 109B-2 cutaway key

1 Fabric-covered metal-framed rudder
2 Rear navigation light
3 Fabric-covered elevator (+ or −31·6 deg) with ground-adjustable tab (port and starboard)
4 Variable-incidence metal-skinned tailplane (+3 deg to −8 deg travel)
5 Structure of cambered fin
6 Tailplane bracing strut (port and starboard)
7 Castoring non-retractable tailwheel with spring type shock absorber
8 Lift point
9 Monocoque fuselage structure
10 FuG 7 receiver-transmitter pack (range: 28–31 mls/45–50 km air-to-ground and 37–40 mls/60–65 km air-to-air)
11 Radio equipment access door (port)
12 Aerials for FuG 7 R/T
13 Radio mast
14 Oxygen bottle
15 Jettisonable canopy hinging to starboard
16 Contoured tank for 55 Imp gal (250 l) of 87 octane A2
17 Control cables
18 Non-armoured bucket-type seat

19 Emergency undercarriage retraction and tailplane incidence wheels
20 Revi C/12C reflector sight
21 Twin 7,9-mm Rheinmetall Borsig MG 17 machine guns (1,200 rpm rate of fire and 2,477 ft/sec — 755 m/sec muzzle velocity)
22 Control column with offset grip
23 Gun troughs
24 Junkers Jumo 210Da 12-cylinder inverted-vee liquid-cooled engine
25 VDM-Hamilton controllable-pitch metal-bladed airscrew
26 Pitch control mechanism
27 Air intake
28 Coolant oil header tank
29 Engine bearer frame
30 Exhaust outlets
31 Radiator bath
32 Coolant radiator
33 Exhaust air exit trap
34 Hydraulically-operated undercarriage retraction jack
35 Wing spar/fuselage anchorage point
36 Wheel well
37 Mainspar

41

Gruppen (III and IV/JG 132 at Jüterbog-Damm and Werneuchen respectively, IV/JG 134 at Dortmund, II/JG 135 at Aibling, II/JG 137 at Zerbst, I/JG 138 at Aspern, III/JG 234 at Düsseldorf, and III/JG 334 at Mannheim) were hurriedly formed on obsolescent fighters, one of them, I/JG 138, being established with the personnel of the fighter squadrons of the Austrian *Luftstreitkräfte* absorbed by the *Luftwaffe* with the annexation. Another *Gruppe*, III/JG 132 which had been formed on Ar 68Es, was transferred to Fürstenwalde from Jüterbog-Damm shortly after its creation, and converted to the He 112B-0, a batch of 12 of which, being readied for shipment to Japan, having been requisitioned for *Luftwaffe* use.

As at 1 August 1938 the *Luftwaffe*'s first-line inventory included 643 fighters of which fewer than half were Bf 109s, but such was the impetus placed behind production by the hazardous plans of the *Führer* that output was rising more rapidly than the fighters could be absorbed by the *Jagdgruppen*. Arado at Warnemünde had phased into the Bf 109 production programme, which, in addition to Messerschmitt, already included Erla, Fieseler, and Focke-Wulf, and an Ob.d.L. strength return for 19 September 1938 recorded a total of 583 Bf 109s of all types, and of these 510 were serviceable. A substantial number of these fighters had still to be issued to the first-line *Jagdgruppen*, and the Bf 109B was preponderant, despite the carefully fostered belief that the Jumo 210-engined versions of the fighter were being *phased out* in favour of the Daimler-Benz-engined models that had allegedly already supplanted them on the assembly lines. In fact, a serious bottleneck in the supply of the Daimler-Benz engines had necessitated the retention of the Jumo 210-engined Bf 109 in production.

With the scheduled launching imminent of Germany's

first aircraft carrier, then known as "*Träger A*", on 1 August 1938 a shipboard unit, *Trägergruppe* 186, was formed at Kiel-Holtenau. This initially comprised 4.*Trägersturzkampfstaffel*/186 equipped with the Ju 87A dive bomber, and 6. *Trägerjagdstaffel*/186 equipped with the Bf 109B. The latter was to be joined in the summer of the following year by another Bf 109B-equipped *Staffel*, this being 5.*Trägerjagdstaffel*/186, the intention being that both *Staffeln* should eventually re-equip with a specialised shipboard version of the Bf 109 for operation from the *Graf Zeppelin* (as "*Träger A*" was christened when launched on 8 December 1938), but in the event, the *Trägerjagdstaffeln* were destined to be confined to shore-based operations.

On 1 November 1938 a major reorganisation of the *Luftwaffe* took place, and for the first time the fighter component was divided into *leichten* (light) and *schweren* (heavy) *Jagdgruppen*, the latter eventually (from 1 January 1939) becoming *Zerstörergruppen*. The *leichten Jagdgruppen* were I/JG 130 (Jesau), I/JG 131 "*Richthofen*" (Döberitz), I and II/JG 132 "*Schlageter*" (Cologne and Düsseldorf), I and II/JG 133 (Wiesbaden-Erbenheim and Mannheim), I/JG 134 (Aspern), I and II /JG 231 (Bernburg and Zerbst), I/JG 233 at Aibling, I/JG 331 (Mähr Trubau), I and II/JG 333 (Herzogenaurach and Eger-Marienbad), and I/JG 433 (Böblingen). Apart from I/JG 130 which was operating both Bf 109C and the DB 600-powered Bf 109D fighters, and one or two *Gruppen* such as I/JG 133 awaiting re-equipment, all the *leichten Jagdgruppen* were equipped with either the Bf 109B or Bf 109C.

The *schweren Jagdgruppen* comprised I/s.JG 141 (Jüter-

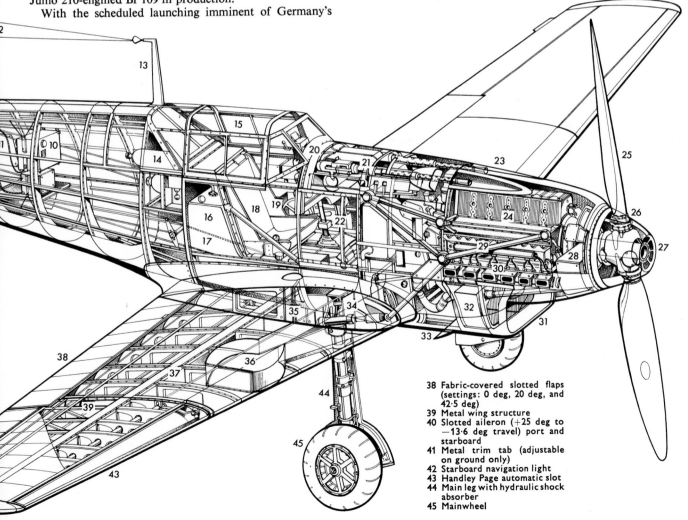

38 Fabric-covered slotted flaps (settings: 0 deg, 20 deg, and 42·5 deg)
39 Metal wing structure
40 Slotted aileron (+25 deg to −13·6 deg travel) port and starboard
41 Metal trim tab (adjustable on ground only)
42 Starboard navigation light
43 Handley Page automatic slot
44 Main leg with hydraulic shock absorber
45 Mainwheel

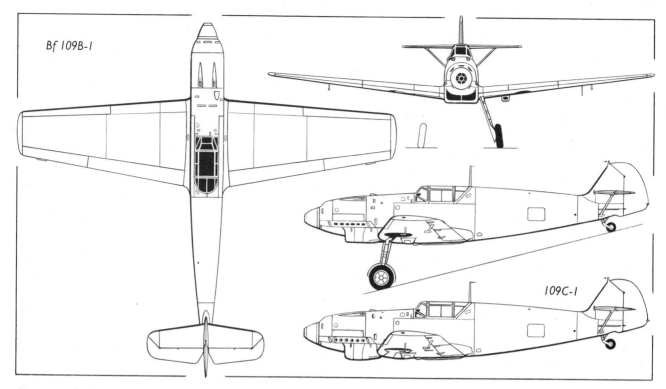

Bf 109B-1

109C-1

bog-Damm) with Bf 109Ds, II/s.JG 141 (Fürstenwalde) with Bf 109Cs, I, II, and III/s.JG 142 (Dortmund, Werl, and Lippstadt) with a mixture of Bf 109Bs, 109Cs, and 109Ds, and I/s.JG 143 (Illesheim) and I/s.JG 144 (Gablingen), both with Bf 109Bs. On 1 January 1939, with the redesignation of the *schweren Jagdgruppen* as *Zerstörergruppen,* JG 141 became ZG 1, the I *Gruppe* commencing conversion to the Messerschmitt Bf 110 shortly afterwards; JG 142 became ZG 26 *"Horst Wessel"*; JG 143 became ZG 52, and JG 144 became ZG 76, a II *Gruppe* being formed with the Bf 109B and the I *Gruppe* converting to the Bf 110.

Whereas total Bf 109 fighter production had barely exceeded 400 machines in 1938 (the entire German single-seat fighter output for the year, which included a small number of Ar 68s and He 112s, having been merely 450 machines), no fewer than 1,091 Bf 109s left the assembly lines between 1 January 1939 and 1 September when the codeword *Ostmarkflug* launched the aerial assault on Poland. This represented an average monthly output of 136·4 Bf 109s for the first eight months of 1939, and the Quartermaster-General's strength return to the *Oberbefehlshaber der Luftwaffe* at the time hostilities commenced included 1,056 Bf 109 fighters of which 946 were serviceable.

The beginning of the end

Throughout the spring and summer of 1939, the *Jagd-staffeln* were feverishly engaged in conversion to the Bf 109E-1 from the older versions of the fighter, and simul-taneously new units were being formed. Some idea of the rapidity with which the Bf 109E-1 was absorbed by the *Luftwaffe's* operational inventory may be gained from the number of *Jagdstaffeln* mounted on this type and included in the order of battle at 1 September 1939, although a sub-stantial proportion of the units were still operating earlier models of the Bf 109. It is of interest to note that on this date, too, five of the seven *Zerstörergruppen* still mounted on Bf 109s were temporarily assigned *Jagdgruppe* designations, these being II/ZG 1 (which became J.Gr. 101), I/ZG 2 (J.Gr. 102), III/ZG 26 (J.Gr. 126), I/ZG 52 (J.Gr. 152), and II/ZG 76 (J.Gr. 176,) the designations of I and II *Gruppen* of ZG 26, which were flying Bf 109Bs and 109Cs, remaining unchanged as their conversion to the Bf 110C was imminent.

Bf 109-equipped units, the sub-type of the fighter with which they were operating and their numerical strengths (serviceable aircraft being indicated in parentheses) under the *Luftflotten* to which they were surbordinated on the eve of hostilities were as follows:

Luftflotte 1: I/JG 1 with 54 (54) Bf 109Es; I/JG 2 with 42 (39) Bf 109Es; 10.(*Nacht*)/JG2 with 9 (9) Bf 109Cs; *Stab* and I/JG 3 with 51 (45) Bf 109Es; I/JG 20 with 21 (20) Bf 109Es and I/JG 21 with 29 (28) Bf 109Cs and 109Es (these two *Gruppen* having originally been scheduled for activation as night fighter units on 1 November but hurriedly pressed into service as day fighter units shortly before hostilities began); J.Gr. 101 (alias II/ZG 1) with 36 (36) Bf 109Bs, and J.Gr. 102 (alias I/ZG 2) with 44 (40) Bf 109Ds.

Luftflotte 2: *Stab* and I/JG 26 with 51 (51) Bf 109Es, II/JG 26 with 48 (44) Bf 109Es; 10.(*Nacht*)/JG 26 with 10 (8) Bf 109Cs; I and II/ZG 26 with 96 (92) Bf 109Bs and 109Ds awaiting conversion to the Bf 110C, and J.Gr. 126 (alias III/ZG 26) with 48 (44) Bf 109Bs and 109Cs.

Luftflotte 3: I/JG 51 with 47 (39) Bf 109Es; I/JG 52 with 39 (34) Bf 109Es; J.Gr. 152 (alias I/ZG 52) with 44 (43) Bf 109Bs; I and II/JG 53 with 51 (39) and 43 (41) Bf 109Es respectively, plus 1. and 2./JG 70 with 24 (24) Bf 109Es, and 1. and 2./JG 71 with 39 (18) Bf 109Cs and 109Es in process of formation as night fighter units.

Luftflotte 4: I/JG 76 with 49 (45) Bf 109Es; I and II/JG 77 with 50 (43) and 50 (36) Bf 109Es respectively, and J.Gr. 176 (alias II/ZG 76) with 40 (39) Bf 109Bs and 109Cs.

In addition, the so-called *Luftwaffe-Lehrdivision* (Instruc-tional Division), which had been formed on 1 November 1938 primarily for the development of operational tactics and techniques, included the *Stab* and I(*Jagd*)/LG 2 with 39 (37) Bf 109Es on strength, and 11.(*Nacht*)/LG 2 with 10 (9) Bf 109Es, while II/JG 186, comprising 5. and 6.*Staffeln* with 24 (24) Bf 109Bs was ostensibly under the control of the *Oberkommando der Marine.*

The Jumo-engined Bf 109Bs and Bf 109Cs were steadily phased out of the first-line units during the first months of World War II, and had been largely relegated to the *Jagdfliegerschulen* by the end of 1939, their last refuge in operational service being with night fighter *Staffeln* such as 10.(*Nacht*)/JG 2 and 10.(*Nacht*)/JG 26. □

FIGHTER A TO Z

(Above) The ANF-Mureaux 114 fighter, described in the previous instalment.

The ANF-Mureaux 170 first prototype (above) and second prototype (below).

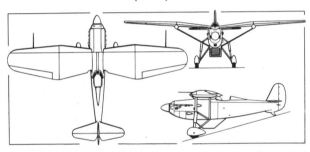

ANF-MUREAUX 170   FRANCE

Designed to meet the requirements of the 1930 single-seat fighter programme which specified the use of a supercharged engine and a maximum speed of at least 217 mph (350 km/h), the Mureaux 170 was powered by a 690 hp Hispano-Suiza 12 Xbrs 12-cylinder liquid-cooled engine and carried an armament of two 7,7-mm MAC (Vickers) machine guns in the wings. Of all-metal construction, the Mureaux 170 was first flown in November 1932, a second prototype, built as a private venture, flying in March 1934. Several changes were made to the size and positioning of the radiator bath, and the fighter was rejected primarily because of the poor visibility from the cockpit for landing. Max speed, 236 mph (380 km/h) at 14,765 ft (4 500 m). Climb to 32,810 ft (10 000 m), 23 min 25 sec. Empty weight, 2,643 lb (1 199 kg). Loaded weight, 3,682 lb (1 670 kg). Span, 37 ft 4 in (11,38 m). Length, 25 ft 11 in (7,90 m). Height, 9 ft 10¼ in (3,00 m). Wing area, 210·54 sq ft (19,56 m²).

ANF-MUREAUX 180   FRANCE

Essentially a two-seat derivative of the Mureaux 170, the Mureaux 180 employed a similar all-metal structure but a frontal radiator was provided for the 690 hp Hispano-Suiza 12 Xbrs engine. The sole prototype was first flown on 10 February 1935, but in April 1935 the original single fin-and-rudder assembly was replaced by twin fins and rudders, and an HS 12 Xcrs engine with provision for a 20-mm Hispano-Suiza cannon firing through the airscrew shaft was installed, engine output being unchanged. Proposed arma-

ment comprised one 20-mm engine-mounted cannon, two wing-mounted 7,7-mm MAC machine guns, and one 7,7-mm gun on a flexible mounting in the rear cockpit. The following figures relate to the Mureaux 180 in its final form. Max speed, 235 mph (379 km/h) at 16,400 ft (5 000 m). Climb to 21,325 ft (6 500 m), 7 min 30 sec. Range, 466 mls (750 km). Empty weight, 2,791 lb (1 266 kg). Loaded weight, 4,306 lb (1 953 kg). Span, 37 ft 4¾ in (11,40 m). Length, 25 ft 8¼ in (7,83 m). Height, 10 ft 8⅜ in (3,26 m). Wing area, 210·54 sq ft (19,56 m²).

ANF-MUREAUX 190   FRANCE

Designed by André Brunet, the Mureaux 190 lightweight single-seat fighter was of all-metal construction and powered by a Salmson 12 Vars 12-cylinder inverted-vee air-cooled engine of 450 hp. Proposed armament comprised an engine-mounted 20-mm cannon and two wing-mounted 7,7-mm machine guns. The prototype Mureaux 190 was flown for the first time in July 1936, but tests were abandoned in 1937 because of the poor reliability of the engine, and a projected development with a retractable undercarriage, the Mureaux 191, was dropped. Max speed, 311 mph (500 km/h) at 13,125 ft (4 000 m), 267 mph (430 km/h) at sea level. Endurance, 2 hr 30 min. Empty weight, 1,874 lb (850 kg). Loaded weight, 2,844 lb (1 290 kg). Span, 27 ft 6 in (8,38 m). Length, 23 ft 7½ in (7,20 m). Height, 9 ft 10 in (3,00 m). Wing area, 107·64 sq ft (10,00 m²).

ANSALDO S.V.A.   ITALY

In the summer of 1916, *Ingegneri* U Savoia and R Verduzio of the *Direzione Tecnica dell' Aeronautica Militare* (Technical Directorate of Military Aviation) began designing a single-seat fighter around the 205 hp SPA 6A six-cylinder liquid-cooled engine. The task of supervising the development and production of the fighter was assigned to the Società Ansaldo, and thus the prototype, first flown on 19 March 1917, was designated S.V.A. (Savoia-Verduzio-Ansaldo). The S.V.A. was a conventional biplane of wooden construction with interplane bracing of the Warren truss type and an armament of two synchronised 7,7-mm Vickers

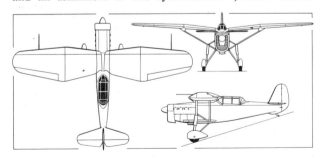

The ANF-Mureaux 180 in original form (above) and final form (below).

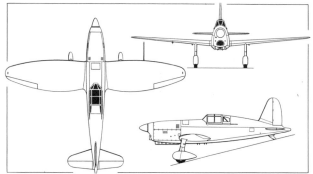

(Above and below) The ANF-Mureaux 190 prototype.

machine guns. The S.V.A. displayed exceptional speed capability but, inherently stable, it was considered to lack the manœuvrability demanded for fighter-versus-fighter combat. However, its excellent range characteristics rendered it suitable for the reconnaissance fighter rôle, and the *Aviazione Militare* decided to adopt the S.V.A. for this task. Deliveries of the initial production version, the S.V.A.2, had meanwhile commenced in the autumn of 1917, this model being assigned to training. Max speed, 137 mph (220 km/h). Time to 9,840 ft (3 000 m), 11 min 20 sec. Endurance, 3 hr. Empty weight, 1,477 lb (670 kg). Loaded weight, 2,100 lb (952 kg). Span, 29 ft 10¼ in (9,10 m). Length, 26 ft 6⅞ in (8,10 m). Height, 8 ft 8⅓ in (2,65 m). Wing area, 260·487 sq ft (24,2 m²).

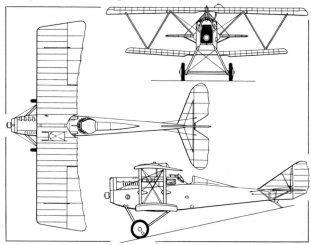

(Above and below) The Ansaldo S.V.A.3 ridotto *short-span interceptor.*

ANSALDO S.V.A.3   ITALY

Built under licence by the AER concern at Orbassano, the S.V.A.3 was a reconnaissance fighter production derivative of the S.V.A. fighter, and essentially similar to the S.V.A.4 built in parallel by the Ansaldo factories at Borzoli and Bolzaneto. In the spring of 1918 a special interceptor version was produced, this having wings of reduced span and area. Known as the S.V.A.3 *ridotto* (reduced), this model was used primarily for airship interception, and although standard armament remained two synchronised 7,7-mm Vickers guns, some examples were fitted with an additional weapon firing upwards at an oblique angle. Power was provided by a semi-supercharged SPA 6A engine of 220 hp. Max speed, 149 mph (240 km/h). Time to 13,125 ft (4 000 m) 13 min. Endurance, 3 hr. Empty weight, 1,470 lb (667 kg). Loaded weight, 1,965 lb (891 kg). Span, 25 ft 5⅛ in (7,75 m). Length, 26 ft 6⅞ in (8,10 m). Height, 8 ft 8⅓ in (2,65 m). Wing area, 236·8 sq ft (22,0 m²).

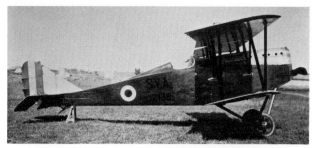

(Above) The Ansaldo S.V.A.4 reconnaissance fighter.

(Above and below) The Ansaldo S.V.A.5 reconnaissance fighter-bomber.

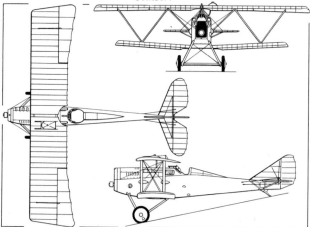

ANSALDO S.V.A.4   ITALY

The S.V.A.4 was the first reconnaissance fighter development of the S.V.A. to be built in substantial quantities. It did not demand an escort in performing reconnaissance missions as it could accept combat with fighters on reasonably equal terms, and break off combat at will by utilising its high maximum speed. It was powered by a 205 hp SPA 6A six-cylinder liquid-cooled engine, and normally carried two synchronised Vickers guns, although the starboard gun was sometimes removed when a reconnaissance camera was carried. The S.V.A.4 entered service with the *Aviazione Militare* early in 1918. Max speed, 134 mph (216 km/h). Time to 9,840 ft (3 000 m), 12 min. Max endurance, 3 hr 35 min. Empty weight, 1,545 lb (701 kg). Loaded weight, 2,150 lb (975 kg). Span 29 ft 10¼ in (9,10 m). Length, 26 ft 6⅞ in (8,10 m). Height, 8 ft 8⅓ in (2,65 m). Wing area, 260·487 sq ft (24,2 m²).

ANSALDO S.V.A.5   ITALY

Built in larger numbers than any other single-seat derivative of the S.V.A., the S.V.A.5 was a reconnaissance-fighter-bomber armed with two 7,7-mm synchronised Vickers machine guns and carrying two reconnaissance cameras or light bombs slung on the fuselage sides on special clips. Initial production examples were powered by the 205 hp SPA 6A but later examples had the semi-supercharged version of this engine rated at 220 hp. The majority of the 1,248 S.V.A. aircraft built during 1917–8 were S.V.A.5s. Max speed, 143 mph (230 km/h). Time to 9,840 ft (3 000 m), 10 min. Normal endurance, 3 hr. Empty weight, 1,500 lb (680 kg). Loaded weight, 2,315 lb (1 050 kg). Span, 29 ft 10¼ in (9,10 m). Length, 26 ft 6⅞ in (8,10 m). Height, 8 ft 8⅓ in (2,65 m). Wing area, 260·487 sq ft (24,2 m²).

The Ansaldo I.S.V.A. single-seat floatplane fighter.

ANSALDO I.S.V.A.   ITALY

A single-seat float fighter version of the S.V.A., the I.S.V.A. (the "I" prefix indicating *Idro* or Water) was built at La Spezia in 1918. Power was provided by a 205 hp SPA 6A engine and armament consisted of two synchronised 7,7-mm Vickers machine guns. A total of 50 I.S.V.A. fighters was manufactured, and these aircraft were used both for the defence of naval bases and coastal reconnaissance. Max speed, 121 mph (195 km/h) at sea level, 112 mph (180 km/h) at 6,560 ft (2 000 m). Endurance, 3 hr. Empty weight, 1,936 lb (878 kg). Loaded weight, 2,425 lb (1 100 kg). Span, 29 ft 10¼ in (9,10 m). Length, 30 ft 2⅓ in (9,30 m). Height, 12 ft 1⅔ in (3,70 m). Wing area, 263·72 sq ft (24,5 m²).

ANSALDO A.1 BALILLA   ITALY

Owing much to the S.V.A., the A.1 Balilla (Hunter) single-seat fighter was flown for the first time in the autumn of 1917, but lacking the agility of its contemporaries it was manufactured in only limited quantities, a total of 166 Balillas being built in 1918. These were confined to home defence tasks, and 75 were supplied to Poland in 1920–1, a further 50 being licence-built by Plage & Laśkiewicz. The Balilla was of wooden construction and carried an armament of two synchronised 7,7-mm Vickers guns. The power plant was either the 205 hp SPA 6A or the 220 hp semi-supercharged version of this engine, while the A.1*bis* was fitted with the 250 hp Isotta-Fraschini V6. The following details apply to the 220 hp A.1. Max speed, 137 mph (220 km/h) at sea level. Time to 9,840 ft (3 000 m), 8 min 30 sec. Endurance, 2 hr 30 min. Empty weight, 1,411 lb (640 kg). Loaded weight, 1,951 lb (885 kg). Span, 25 ft 2⅓ in (7,68 m). Length, 21 ft 3⅞ in (6,50 m). Height, 9 ft 4¼ in (2,85 m). Wing area, 226·04 sq ft (21,0 m²).

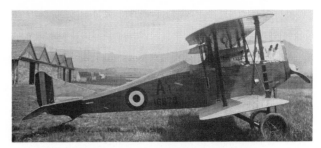

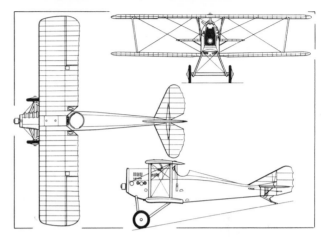

(Above and below) The Ansaldo A.1 Balilla.

ANSALDO A.C.2   ITALY

In 1924 Aeronautica Ansaldo SA acquired manufacturing rights in the Dewoitine D.1 single-seat fighter, assembling one example as the A.C.1. This served as a basis for a modified version of the fighter with marginally smaller overall dimensions, which entered production as the A.C.2. Powered by a 300 hp Hispano-Suiza HS 42–8 12-cylinder liquid-cooled engine and carrying an armament of two synchronised 7,7-mm Vickers guns, the A.C.2 was of metal construction and 112 examples were delivered to the *Regia Aeronautica* during 1925. Max speed, 150 mph (242 km/h) at sea level. Time to 3,280 ft (1 000 m), 2 min 7 sec, to 9,840 ft (3 000 m), 8 min 9 sec. Endurance, 2·6 hr. Empty weight, 1,828 lb (829 kg). Loaded weight, 2,522 lb (1 144 kg). Span, 35 ft 8⅓ in (10,88 m). Length 24 ft 2½ in (7,38 m). Height, 9 ft 1⅞ in (2,79 m). Wing area, 215·28 sq ft (20,0 m²).

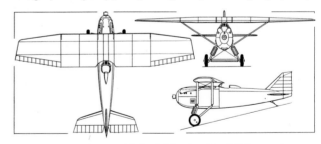

(Above and below) The Ansaldo A.C.2.

PLANE FACTS

Spanish Dragon Rapides

I have recently read an account of the early days of aerial warfare over Spain in August 1936, and the author mentioned that de Havilland Dragon Rapide biplanes were used by the Nationalist side as bombers. I cannot imagine this type serving as a combat airplane, and I shall be glad if you can verify the use of the Dragon Rapide, and, if possible, publish photos of it in Spanish service.

F E Jacobs, San Diego, Calif 92115

During the course of 1935, the Spanish government ordered from de Havilland three examples of the D.H.89M, a military version of the Dragon Rapide commercial transport, for policing duties in Spanish Morocco. The D.H.89M had earlier been evolved to meet specification G.18/35 which had called for a general reconnaissance aircraft for use by Coastal Command, a standard Rapide being fitted with a Mk V Vickers gun in the starboard side of the nose, a Lewis gun on a flexible mounting in the dorsal position, and a weapons bay for two 100-lb (45,3-kg) and four 20-lb (9,0-kg) bombs. The prototype, K4227, carried a crew of three comprising pilot, radio operator/gunner and navigator/bomb aimer, but no orders were forthcoming from the RAF which selected the Avro Anson. The Spanish examples were fitted with a fixed 7,7-mm Vickers E gun in the starboard side of the nose, a Vickers F gun on a flexible mounting in an open dorsal position, and a third Vickers gun which fired through a trap in the floor. A Spanish bomb sight was installed, and a rack beneath the fuselage could carry twelve 26·45-lb (12-kg) bombs.

The three D.H.89Ms were delivered to the

The sole example completed of the Stout ST-1 torpedo-bomber of the early 'twenties.

Spanish government in January 1936, and when the Civil War began, two of these were at bases in territory that proclaimed itself for the Nationalist cause and the third was captured by the Nationalists when it landed at Saragossa on 18 July. The three aircraft were immediately pressed into service for bombing and reconnaissance missions over the Guadarrama, the D.H.89Ms forming a *grupo* commanded by *Capitán* Juan A Ansaldo, and operating from Logroño, Burgos and Saragossa. They saw extensive operational service throughout most of the remainder of 1936 but towards the end of that year were relegated to the transport rôle. Powered by two 200 hp de Havilland Gipsy Six six-cylinder air-cooled engines, the D.H.89M had a maximum speed of 151 mph (243 km/h), cruised at 125 mph (201 km/h), had an initial climb rate of 890 ft/min (17,5 m/sec), and a range of 550 miles (885 km). Empty and loaded weights were 3,368 lb (1 528 kg) and 5,372 lb (2 437 kg).

A Stout effort

I have been trying to track down a photograph of the Stout ST-1 torpedo-bomber circa 1920 without much luck, and although it is something of a long shot, I wonder if your files include such a photograph which could be published in the "Plane Facts" column, together with some details of this elusive aircraft?

Wm E Parks, Chippewa Falls, Wis 54729

One of the first — perhaps *the* first — all-metal monoplanes offered to the US Navy, the Stout ST-1 was designed by William Stout of the Stout Engineering Laboratories to meet a US Navy requirement for a torpedo-bomber suitable for operation from wheel or float undercarriages. The US Navy placed a contract for three aircraft (A-5899, A-5900 and A-5901), one of which was to have been a static test airframe, and flight testing of the only ST-1 completed is believed to have commenced in 1921.

The ST-1 was a two-seat shoulder-wing monoplane powered by two 398 hp Packard V-1237 engines. The fuselage was covered by smooth aluminium sheet and the wings by corrugated aluminium sheet, and both pilot and dorsal gunner were seated in open cockpits. Empty and loaded weights were 6,557 lb (2 977 kg) and 9,817 lb (4 453 kg), and overall dimensions included a span of 60 ft 0 in (18,28 m), a length of 37 ft 0 in (11,28 m), a height of 14 ft 0 in (4,27 m), and a wing area of 790 sq ft (73,39 m²). The ST-1 attained a maximum speed of 110 mph (177 km/h) at which range was 385 miles (619 km), the flight test programme being conducted at Mount Clemens, Michigan, by Edward Stinson. During the first take-off the portside main undercarriage member failed, but after reinforcement of the attachment points the ST-1 prototype was flown 13 times, crashing during the 13th landing. The US Navy contract was subsequently cancelled, and no further examples of the ST-1 were completed.

Close-support Breda

I am trying to compile a history of specialised assault airplanes, but so far have drawn something of a blank with the Italian contribution to this category of warplane. One airplane of which I am particularly anxious to obtain detail is the Breda Ba 65 which was, I understand, employed operationally by the Regia Aeronautica. Can you provide a rundown on the development history and career

Two of the three de Havilland D.H.89M Dragon Rapides used operationally by the Spanish Nationalist air arm, 40-2 "Capitán Vela" (above) and 40-1 "Capitán Pouso" (below).

(Above) A Fiat A 80-powered Breda Ba 65 of the 101ª Squadriglie (5° Stormo), and (below) a Ba 65bis of the Portuguese Arma da Aeronáutica with the Breda "L" dorsal turret and the Isotta-Fraschini-built Gnôme-Rhône 14K engine.

Ba 65 attained a maximum speed of 258 mph (415 km/h) at 16,400 ft (5 000 m) and 217 mph (350 km/h) at sea level. Maximum cruise was 223 mph (360 km/h) at 13,125 ft (4 000 m), and range was 466 miles (750 km) with a 440-lb (200-kg) bomb load. An altitude of 13,125 ft (4 000 m) was attained in 8 min 40 sec, and service ceiling was 25,590 ft (7 800 m). Empty and loaded weights were 5,291 lb (2 400 kg) and 6,504 lb (2 950 kg), and overall dimensions were: span, 39 ft 0½ in (11,90 m), length, 31 ft 6 in (9,60 m), height, 10 ft 11 in (3,33 m), wing area, 252·95 sq ft (23,5 m²).

The Deerhound engine

Some years ago in "Air Pictorial" mention was made of the Armstrong Siddeley Deerhound engine. I shall be grateful if you can publish a photograph and some details of this engine in your "Plane Facts" column.

D Caswell, RAF St Athan, Barry, Glam

The Armstrong Siddeley Deerhound was a 21-cylinder three-row air-cooled radial engine with a two-stage supercharger. The Deerhound was originally planned to provide 1,185 hp for take-off, although this was uprated to 1,350 hp during development, and it was intended to install a pair of Deerhounds in the A.W.39 heavy bomber submitted to meet specification B.1/35. The A.W.39 was a development of the Whitley with generally similar dimensions and layout, although somewhat heavier and featuring a Boulton Paul dorsal turret. Two Deerhound engines were to be buried in the wing leading edges of the A.W.39, resulting in a very clean aeroplane, and design performance included a range of 1,500 miles (2 414 km) with a 5,000-lb (2 268-kg) bomb load at 230 mph (370 km/h) at 15,000 ft (4 570 m). In the event, the B.1/35 contract was finally awarded to the Vickers Warwick, but prior to this the 27th production Whitley II (K7243) had been fitted with a pair of Deerhounds as part of the A.W.39 development programme. The first flight with Deerhounds was performed from Baginton on 6 March 1940 with Flt Lt C K Turner-Hughes at the controls.

of this type, together with a general arrangement drawing and photographs?

John R Preston, Annapolis, Maryland 21412

Evolved from the Breda Ba 64 of 1933, the Ba 65 assault aircraft prototype powered by a Fiat A 80 RC 41 18-cylinder two-row radial engine rated at 1,000 hp for take-off was flown for the first time in 1935. A single-seat all-metal cantilever low-wing monoplane with aft-retracting main undercarriage members, the Ba 65 carried a wing-mounted armament of two 12,7-mm and two 7,7-mm Breda-SAFAT machine guns and provided internal stowage for a 440-lb (200-kg) bomb load. Production began in 1936, the initial model (81 built) having an Isotta-Fraschini-built Gnôme-Rhône 14K 14-cylinder radial of 900 hp, and in August 1937 the 65ª *Squadriglia Assalto* of the 35*mo Gruppo Autonomo Bombardamento Veloce* arrived in Spain with 12 Ba 65s for service with the *Aviazione Legionaria.*

In the meantime, a two-seat version, the Ba 65bis, had been developed and export orders for the Breda assault monoplane had been solicited. Fifteen were ordered in 1937 for the Royal Iraqi Air Force and were delivered in 1938 to equip No 5 Squadron of that service. Thirteen of the aircraft delivered to Iraq were Ba 65bis two-seaters equipped with an hydraulically-operated Breda L dorsal turret mounting a 12,7-mm Breda-SAFAT machine gun. Ten single-seater Ba 65s were delivered to the Soviet Union and 20 to Chile, and 10 examples of the Ba 65bis were ordered for 1939 delivery to Portugal's *Arma da Aeronáutica.* All export Ba 65s were powered by the Isotta-Fraschini engine, but 119 were delivered to the *Regia Aeronautica* in 1938-9 with the Fiat A 80 RC 41, and a total of 154 of the Breda assault aircraft was included in the service's inventory when Italy entered WW II in 1940, this total including a small

number of Ba 65bis two-seaters which were not fitted with the Breda L turret. Owing to the unsatisfactory performance of the Fiat A 80 RC 41 under North African conditions, all Ba 65s with this power plant were re-engined with the 14K.

The Ba 65 was operated by the 101ª and 102ª *Squadriglie* of the 19° *Gruppo* of the 5° *Stormo*, and the 159ª and 160ª *Squadriglie* of the 12° *Gruppo* and the 167ª and 168ª *Squadriglie* of the 16° *Gruppo* of the 50° *Stormo*, but it proved an ineffectual combat aircraft, and those that remained operational in North Africa with components of the 50° *Stormo* played little part in the air war. With the 900 hp Isotta-Fraschini-built Gnôme-Rhône 14K engine, the single-seat

This general arrangement drawing depicts the G-R 14K-powered single-seat Ba 65, the additional side views depicting the G-R 14K-powered Ba 65bis (centre) and the A 80-powered Ba 65bis.

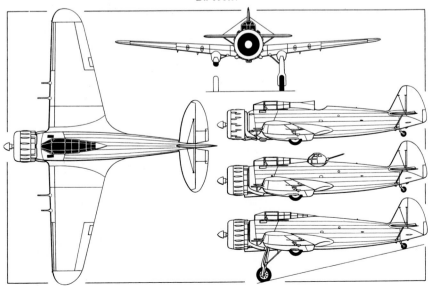

Subsequently, the aircraft was flown on a number of occasions by Flt Lt Eric Greenwood, but most flights terminated in emergency landings resulting from the overheating of the engines. A new type of cowling of reverse-flow type was tested, and this, coupled with an increase in cylinder fin area, resulted in an improvement in cooling, but the rear cylinder bank of the Deerhound still tended to overheat, and this problem had not been resolved when the Whitley test-bed was lost in a crash during take-off — believed to have been the result of pilot error — and with the destruction of virtually all the remaining Deerhound prototype engines when the Armstrong Siddeley factory was bombed further development of this power plant was abandoned.

(Below) The Armstrong Siddeley Deerhound.

Klemm of the 'thirties

During a recent visit to Germany I had the opportunity to look at some of the aeroplanes on the small strip at Nabern, not far from Stuttgart. One was referred to as a Klemm 35 of which I had not previously heard. It was a tandem two-seat monoplane fitted with a small "glasshouse"-type canopy (although I was told that open cockpits are more usual for this type) and a fixed, spatted undercarriage. I would be interested to see some details and perhaps a photo of this type in "Plane Facts".

R J Parkhouse, Harpenden Herts
The tandem two-seat Klemm Kl 35 primary trainer and touring monoplane was first produced by the Leichtflugzeugbau Klemm (later Klemm-Flugzeugbau GmbH) of Böblingen in 1935, the prototype being powered by an 80 hp Hirth HM 60R engine. It was subsequently built in substantial numbers as the Kl 35A and Kl 35AW floatplane with the Hirth HM 60R, as the Kl 35B and BW with the 105 hp Hirth HM 504A, and as the Kl 35D and DW with the 105 hp Hirth HM 504A-2. Some 2,000 examples of the Kl 35D alone were built for flying clubs and for use as a primary trainer by the *Luftwaffe*, and a manufacturing licence was acquired by Sweden where, during 1941-2, it was built in some numbers for use by *Flygvapnet* as a trainer. Some 40 ex-*Flygvapen* Kl 35Ds were later sold to Swedish flying clubs, and at least a dozen of these eventually found their way onto the Federal German register. The example that you saw at Nabern, Mr Parkhouse, was probably one of these.

The Kl 35 was produced both with and without enclosed cockpits, the majority having open cockpits, and was of mixed construction with fabric- and plywood-

An Heinkel He 111H-22 of the Stab/KG 3 with an Fi 103 beneath the starboard wing root.

skinning. The Kl 35D had a maximum speed of 130 mph (210 km/h), a cruising speed of 118 mph (190 km/h), and a range of 413 miles (665 km). An altitude of 3,280 ft (1 000 m) was attained in 5 min, and service ceiling was 14,435 ft (4 400 m). Empty weight was 1,014 lb (460 kg), and loaded weight was 1,653 lb (750 kg), overall dimensions including a span of 34 ft 1½ in (10,40 m), a length of 24 ft 7¼ in (7,50 m), a height of 7 ft 6½ in (2,30 m), and a wing area of 163·5 sq ft (15,20 m²).

Missile-carrying He 111

I would like to have information on the type of Heinkel He 111 which carried the V1, my aim being to make up a plastic model of a bomber of this type carrying the missile.

Charles Galea, Lija, Malta
The principal version of the Heinkel He 111 bomber that carried the Fi 103 (FZG 76), or V1, missile was the H-22, this being an adaptation on the assembly line of the He 111H-21 powered by Jumo 213E-1 engines.

The He 111H-22 usually carried the Fi 103 slung from a rack between the starboard engine nacelle and the fuselage. It was powered by two Junkers Jumo 213E-1 12-cylinder liquid-cooled engines with three-speed two-stage superchargers and induction coolers, these each offering 1,750 hp for take-off and 1,320 hp at 32,000 ft (9 750 m). The electrically-operated dorsal turret mounting a single 13-mm MG 131 machine gun (first introduced by the He 111H-16/R1) was standard. The rack itself was an adaptation of the standard ETC 2000 bomb rack. Apart from the He 111H-22 which left the assembly line as an Fi 103-carrier, a number of He 111H-16s and H-20s were converted for the missile-carrying rôle at a modification centre at Oschatz. The III *Gruppe des Kampfgeschwaders* 3 became operational in the missile-launching rôle late in July 1944, and the accompanying photograph depicts an He 111H-22 (5K+GA) of the *Stab/KG* 3 with an Fi 103 beneath the starboard wing root.

(Above) A Klemm Kl 35D trainer of the Slovak Air Force, and (below) a Swedish-built Kl 35D formerly operated by the Swedish Flygvapen.

VETERAN & VINTAGE

A KINGFISHER COMES HOME

Until this year, only two examples of the Vought Sikorsky OS2U Kingfisher reconnaissance aircraft were on display, survivors of the 1,520 examples built during World War II by Vought and the Naval Aircraft Factory. Today, there are three examples, thanks to the outstanding efforts of the Vought Aeronautics Division of LTV and a number of other organisations and individuals. The new veteran is now on permanent display aboard the USS *North Carolina*, a "memorialised battleship" at Wilmington Harbor, NC.

The story of this particular Kingfisher begins in August 1942. Japan had invaded the Aleutian island chain and America was pouring men and supplies northwards to counter the threat. On 1 August, a squadron of Kingfishers was despatched to Dutch Harbor, Alaska, leaving behind three crews to pick up new aircraft at Floyd Bennett NAB, New York, and proceed to Alaska separately. Flying brand-new Kingfishers, they crossed the continental US in seven days, swopped their fixed landing gear for floats at Sand Point, Washington, and proceeded north. On the morning of 20 August, the flight, led by Lt (jg) Ray G Thorpe, left Coral Harbor, BC, for Ketchikan, Alaska.

After an hour's flight they reached Calvert Island, 300 miles (483 km) north of Vancouver, BC, but they then ran into dense fog and became separated. One Kingfisher, flown by Ensign Mac J Roebuck, was flying at 1,500 ft (457 m) when it crashed into the side of Mt Buxton, tearing off the right wing and main pontoon. Neither Roebuck nor his gunner, Stanley S Goddard, was injured

The recovery begins: a Vertol H-21 of the RCAF is used to lift the remains of the Kingfisher from the side of Mt Buxton where it had lain for 22 years.

and the radio was still working, so Roebuck contacted his flight leader, Thorpe, who called for help from the Royal Canadian Navy.

Roebuck and Goddard made their way down the mountain to the shore, where they were rescued. Shortly after, Roebuck and six RCAF personnel returned to the crash site and salvaged the guns, instrument panels, radios and 450 hp Pratt & Whitney engine from the wrecked Kingfisher.

The salvaged engine was loaded on the wrecked right wing and the "sled" was pulled six miles down the mountainside to the shore, where a US Navy ship picked it

up and brought it back to the US. The rest of the aircraft lay on the tundra of Mt Buxton, exposed to deep snow, corrosive salt air, rain and sun for 22 years. Identities of the pilot and crewman were soon forgotten; after a few years all that was known was that there was a wrecked Kingfisher somewhere on a Birtish Columbia coastal island. No one remembered who had flown it or what had become of them.

In the spring of 1963 the Air Museum of Canada at Calgary, Alberta, heard of the plane's existence. It was suggested to Minister of National Defense Paul Hellyer that salvaging it could serve as a valuable

After spending another six years in storage in the US, the Kingfisher became the subject of a restoration project by retirees of the Vought Aeronautics Division of LTV. After a year's work, it re-emerged in May in the pristine condition shown below, ready to be handed back to the US Navy for permanent display aboard the USS North Carolina.

practice operation for members of the Sea Island-based Search and Rescue Flight, 121 K.U. He authorised the utilisation of RCAF equipment in the salvage operations.

In the spring of 1964, RCAF Vertol H-21 helicopters airlifted the wrecked Kingfisher to Port Hardy on North Vancouver Island's coast. From there a C-130 Hercules of No 435 Squadron ferried it to Calgary. The Air Museum could not reconstruct it, however, so subsequently turned it over to the USS *North Carolina* Battleship Commission, which trucked the pieces 2,700 miles (4 345 km) to Wilmington, NC. The *North Carolina* had carried three Kingfishers throughout World War II, and participated in every major naval engagement of the conflict from Guadalcanal to the surrender of Japan.

Since the Battleship Commission had no facilities to rebuild the aircraft, through the Navy it requested Vought Aeronautics to undertake the restoration. The company agreed and delegated the job to members of its Quarter Century Club who had retired. Many of them, while employed in the experimental department, had worked on the first hand-built Kingfisher in 1939.

After the remains of the Kingfisher — minus the starboard wing, wing floats, guns, instrument panels and with damaged radio — reached Dallas, the group of 14 Vought veterans set to work. Its restoration took just over a year, the completed aircraft in mint condition (but unflyable) being displayed on 13 May at Dallas prior to its transfer for dedication aboard the *North Carolina* on 25 June. Among those participating in the first public showing were the Kingfisher's crew and the formation leader, Ray Thorpe.

Pilot recollections
Recalling the fateful flight of nearly 30 years before, Mac J Roebuck (now a Navy contracting official) said: "I was flying on instruments at 85 to 90 knots (157–167 km/h). I just happened to be in a landing attitude and speed when I hit the mountain. A tremendous boulder ripped the right wing off but we didn't get a big jar or anything. We didn't even have time to get scared. When we hit I actually kept on trying to fly

The Kingfisher is now exhibited on the forward deck of the USS North Carolina *in Wilmington, NC. This "memorialised battleship" is permanently open to the public and is one of three such in the US that now have Kingfishers on show.*

the aircraft for a few minutes. I had been thinking about Goddard in the rear seat and wondering what he was thinking, entrusting his life to a man he hardly knew.

"Being fresh out of flight school, I guess I had only 100 hours of flight time and not much of that instrument time. I really was sweating it out trying to keep the needle in the middle of the ball and the air speed at the proper place.

"After the crash the instruments went wild. Then I heard Thorpe's voice on the radio and I realised that even though I had crashed the radio still worked. So we had time to chat a few minutes and planned how

we were going to get together again." After hiking down the mountain, Roebuck said, they reached a beach and the Canadians picked them up in a boat.

"Flight Leader Thorpe and Sanderson had landed their Kingfishers. They flew on to Alaska with Goddard and I stayed behind, camped in a tent on the ocean beach with the RCAF men to see what could be salvaged from the plane. The Canadians and I made about eight trips up the mountain and carried out the machine guns, instrument panels and radio on our backs. The engine was too heavy so we loaded it on the right wing and pulled it down the mountain till the brush got too thick.

"We then dismantled the engine and carried it the rest of the way to the beach. There we reassembled it and cleaned the machine guns. All this took us two weeks. The crowning blow came later when, after all that work dragging the engine out, I heard the Navy in Seattle had decided the engine was unrepairable and had junked it!"

The two other surviving Kingfishers are aboard the USS *Massachusetts* and the USS *Alabama*. The former exhibit is a loan from the National Air Museum; that on the *Alabama* is one of six supplied to the Mexican Navy during the war and recovered from Mexico as a landplane, having subsequently been converted to a floatplane, in which form the Kingfisher invariably operated when "spotting" from battleships. Other users of the Kingfisher were the Royal Navy, Australia, Uruguay, Chile, Dominican Republic and Cuba. So far as is known, no examples survive in any of these countries. □

Another view of the restored OS2U-2 Kingfisher, one of 1,220 built by Vought Sikorsky during World War II. Another 300 were built by the Naval Aircraft Factory as OSN-1s.

Volume 1 Number 7 December 1971

CONTENTS

AIRSCENE

MILITARY AFFAIRS

AUSTRALIA

Major savings in **defence expenditure** revealed by Defence Minister Fairbairn in the annual report of the Defence Department presented to the House of Representatives on 8 October have been effected by cancellation of the order for 11 Bell AH-1G HueyCobra gunships which would have cost $A13·2m (£6·168m), a reduction of 12 helicopters in current orders for the UH-1H Iroquois saving $A8·8m (£4·11m), and the deferment of an order for six low-cover radar units for the RAAF which were to have cost $A6m (£2·8m). Mr Fairbairn said that the planned introduction of the AH-1G HueyCobra in 1974 had been reassessed in the light of operational experience, developments in the field of close air support, and the Australian withdrawal from Vietnam, and it is understood that the RAAF will shed no tears over this cancellation, having remained unconvinced that the HueyCobra was the answer to its close support requirements. The reduction in the order for the UH-1H Iroquois brings to 30 the planned RAAF acquisition of this type, seven having already been delivered.

At the time of closing for press it was anticipated that the Australian government would confirm the order for **24 Convair (General Dynamics) F-111C** aircraft before Parliament rises on 10 December, but delivery is not likely to be effected before March-April 1973 owing to the extent of the modifications that will be applied throughout next year, and the subsequent IRAN (Inspect and Repair As Necessary) programme to which they will be subjected. They will thus reach the RAAF some 10 years after being ordered, and during this period their cost to Australia will have more than doubled, from $US146m (£60·83m) to $US333m (£138·7m). One result of the F-111C acceptance will be termination of the lease by the RAAF of the 24 F-4E Phantoms currently forming the interim equipment of No 82 Bomber Wing. There would now seem some likelihood, however, that the Phantoms will be retained by the RAAF after the F-111C achieves full operational status in 1974, as the cost of their outright purchase will by then be largely nominal. The Phantoms, if retained, will be converted for the ground attack rôle currently being performed by the Mirage IIIOA, allowing all Mirages to be transferred to the intercept rôle and permitting the introduction of a new interceptor to be deferred until 1980 or thereabouts.

BRUNEI

The Air Wing of the Royal Brunei Malay Regiment (*Askar Melayu Diraja Brunei*) has recently added two **Bell Model 212** Twin-Pac helicopters to its inventory in continuance of its recently-adopted practice of standardising on helicopters of Bell design since the withdrawal from service of the British-supplied Whirlwind Srs 3 ard Wessex Mk 54 helicopters. In addition to the Model 212s, the Air Wing now has two Model 205As and four Model 206A JetRangers which are used for liaison, SAR, communications, patrol and transport duties. One Hawker Siddeley 748 transport was delivered to the Air Wing early this year for the personal use of the Sultan of Brunei.

The first of 12 Shackleton 2s to be converted for the airborne early warning rôle by Hawker Siddeley made its first flight on 30 September and this illustration shows the new radome under the forward fuselage. In January, No 8 Squadron, RAF, is to be formed at Kinloss to operate this version of the aircraft, and will eventually move to Lossiemouth, where RAF AEW operators have been undergoing training by the Fleet Air Arm.

FINLAND

Among the main recommendations of a recent study of military requirements over the period 1972-6, conducted by a parliamentary defence committee headed by Prof Jan-Magnus Jansson, are a substantial **increase in the size** of *Ilmavoimat* coupled with increased equipment procurement for all three armed services. The study recommends an increase of some *Markka* 25m (£2·42m) annually throughout the five-year period for procurement, with equipment appropriations reaching approximately *Marrka* 250m (£24.2m) by the end of the period, and that two more combat squadrons and a reconnaissance unit should be formed within *Ilmavoimat*. The procurement of additional helicopters, and a small number of new transport aircraft, and the replacement of the Saab Safir basic trainer by a more modern licence-built type are recommended. If the study recommendations are accepted, anticipated air defence procurement is estimated at *Markka* 610m (£59·15m) over the five-year period, including base, communications, and antiaircraft equipment, and new ground radar.

FRANCE

The first *Armée de l'Air* unit designated to receive the **Dassault Mirage F1** is the 30ᵉ *Escadre* which has been operating the Sud-Aviation Vautour IIN since 1957. Its *Escadron* 2/30 is scheduled to begin receiving 15 Mirage F1s from the late summer of 1973, *Escadron* 3/30 following suit at the beginning of 1974. The second *Escadre* likely to convert to the Mirage F1 is the 5ᵉ *Escadre* flying Mirage IIICs from Orange, and supplementary orders (those placed to date calling for 85 aircraft) are expected to enable the 12ᵉ *Escadre* at Cambrai to replace its Super Mystère B2s.

The first **Jaguar A** tactical strike fighters are scheduled to enter service with the 1ᵉʳ *Commandement Aérien Tactique* from next year in support of the 2nd French Army Corps based in Germany. The French government has so far placed orders for 59 single-seat Jaguar As and 40 two-seat Jaguar Es, and delivery of the first production example of the latter to the *Armée de l'Air* was imminent at the time of closing for press. The fly-away cost of the Jaguar A is now quoted as NF 22m (£1·6m) which is somewhat more than double that anticipated when the Anglo-French strike fighter was initiated.

FEDERAL GERMANY

Finalisation of the Federal German order for 175 **McDonnell Douglas Phantoms** for the *Luftwaffe* has resulted in the selection of the two-seat F-4F model in preference to

the single-seat F-4E(F) originally proposed specifically to meet the *Luftwaffe*'s requirements. The F-4F is essentially similar to the USAF's F-4E but optimised for the intercept rôle and embodying some of the modifications requested by the *Luftwaffe,* such as the addition of leading-edge slats. Brig Gen Daniel James of the US Department of Defense has stated that the F-4F may be operated by the *Luftwaffe* with a one-man crew. The F-4E(F) originally to have been acquired for the *Luftwaffe* would have had minimal air-to-air capability as a result of the deletion of its Sparrow III AAM system. The prototype F-4F should be flying at St Louis by January 1973, but production deliveries for the *Luftwaffe* are not scheduled to start until early 1974 with completion of deliveries in February 1976.

The German Defence Ministry has decided that only those **F-104G Starfighters** that are scheduled to remain in *Luftwaffe* service into the 'eighties will now receive the new or strengthened wings with which it had previously been planned to equip the entire *Luftwaffe* Starfighter fleet. This decision stems from that concerning the purchase of the F-4F Phantom, the acquisition of which will enable the number of hours flown by the Starfighters during the second half of the 'seventies to be substantially reduced.

The *Luftwaffe* has decided that all of the approximately 200 Dassault-Breguet Dornier **Alpha Jets** which it is expected to take into its inventory from 1976 will now be configured for the ground attack rôle, and, apart from maintaining reserve pilot proficiency, the Alpha Jet will not be used for training. In the *Armée de l'Air* the Alpha Jet will be used solely in the basic training rôle, and at one time the development of a specialised single-seat ground attack version for the *Luftwaffe* was considered, but this would have been heavier, more complex and costlier than the French version, and the *Luftwaffe* has, with a certain amount of reluctance, accepted a minimum-modification variant of the trainer with reduced ground attack mission capability. One of the first two Alpha Jet prototypes will be completed as the *Luftwaffe* ground attack version with five external store stations.

ISRAEL

The Air Section of the Israel Defence Forces, *Heyl Ha'Avir,* failed in its first attempt to intercept two **MiG-23s** engaged on a reconnaissance flight off the Israeli coastline on 10 October. Flown by Soviet pilots on a southerly course some 18 miles (29 km) from the coastal strip between Ashdod and Ashqelon, south of Tel-Aviv, the MiG-23s were picked up by Israeli radar and a flight of F-4E Phantoms attempted an intercept. However, the Israeli aircraft were unable to catch the MiG-23s which disappeared from the radar screens over Port Said.

Hughes Aircraft is to build a computer-based **air defence system** (see page 350) for Israel under an unannounced $13·2m (£5·25m) contract. The system will include two main centres with central data processing, and three remote sites. The main computer will be the Hughes 4118 digital processor, the Israeli computer manufacturer, Elbit, will produce displays under Hughes licence, and Xerox Data Systems of Israel will supply secondary or disc memories.

JAPAN
The Defence Agency will commence the preparation of specifications for a **new ASW helicopter** early in the New Year. It is anticipated that the new helicopter will feature an integrated data computer system and will possess a speed of 200 knots (370 km/h), a combat radius of 230 miles (370 km) plus two hours loiter time, and a gross weight of approximately 22,050 lb (10 000 kg). Five prototypes will be built of which two will be for flight tests, two for static tests and one for tie-down dynamic tests, and flight testing will commence in 1977. It is anticipated that 150-200 helicopters of this type will be required by the Maritime and Ground Self-Defence Forces, the latter using a transport version.

The Air Self-Defence Force is adapting the two prototypes of the F-4EJ Phantom for in-flight launching trials with the indigenous **AAM-2** infra red-homing air-to-air missile. Production deliveries of the AAM-2 to the Air Self-Defence Force are scheduled to commence in 1975.

SAUDI ARABIA
The Saudi Arabian government has signed a $130m (£52m) programme agreement with the US government for the purchase of 20 two-seat **Northrop F-5Bs** and 30 single-seat **F-5Es**, plus automatic ground checkout equipment, spares, mobile training units and various other items of support equipment. The flyaway price of the F-5E was recently quoted by USAF Secretary Robert C Seamans Jr as $1,533,000 (£613,000), but the flyaway price of the Saudi Arabian F-5Es is, because of additional equipment specified, reportedly of the order of $1·7m (£680,000).

SOVIET UNION
A new variable-geometry **air superiority fighter** developed by the Mikoyan bureau and dubbed *Fearless* by NATO is undergoing flight testing in the Soviet Union. According to US sources, the *Fearless* has a thrust-to-weight ratio of the order of 1·2, a combat radius of the order of 345 miles (555 km) and a speed of the order of Mach 2·5. Gross weight is believed to be some 40,000 lb (18 145 kg) and power is provided by two turbojets each having a maximum afterburning thrust of some 24,000 lb (10 885 kg). It may be assumed that the *Fearless* owes much to the earlier single-engined *Flogger* variable-geometry single-engined fighter from the Mikoyan bureau which is currently entering the V-VS inventory. The *Flogger,* somewhat smaller than the *Fearless* with a gross weight of around 28,000 lb (12 700 kg), is believed to be capable of Mach 2·3 with external stores and to have a combat radius of some 600 miles (965 km).

UNITED ARAB REPUBLIC
The Air Force of the United Arab Republic has recently received approximately 20 **Mil Mi-6** heavy assault helicopters from the Soviet Union. Examples of this type of helicopter had previously been supplied to the UAR but all were destroyed in the six-day war.

Reports from Cairo indicate that the current UAR ground attack fighter element equipped with the **Sukhoi Su-7** is in process of expansion with the delivery of additional Su-7s from the Soviet Union. Latest estimates of UAR air strength suggest an inventory of 520 combat aircraft.

UNITED KINGDOM
Three specially-equipped **Hawker Siddeley Nimrods** have been delivered to RAF Wyton for use by No 51 Squadron on electronic reconnaissance missions. They are additional to the 38 Nimrods built as Shackleton replacements for Strike Command and replace three Comet 2s operated by 51 Squadron for many years.

The **Armstrong Whitworth Argosy C Mk 1** is no longer in service as a transport with the RAF, the last squadron to use the type, No 114, having been disbanded at RAF Benson on 8 October. Operational use of the Argosy is at present limited to No 115 Squadron of Strike Command, which is using the E Mk 1 version (see AIR ENTHUSIAST September p172), but a number of the retired transports are to be converted

for service with Training Command as advanced trainers for navigators, air electronics operators and air engineers. In the converted aircraft, engineers will be trained on the existing flight decks, from which some navigation and communication equipment will be removed, while the cabins will be arranged to train navigators and AEOs. Total complement in the training rôle will be 12 students and instructors plus two pilots.

USA
The USAF plans to procure off-the-shelf 15 Fairchild Peacemaker (a licence-built version of the Pilatus Turbo-Porter) and 15 Helio Stallion aircraft for competitive trials prior to the selection of one as a light STOL gunship for use by the South Vietnamese Air Force. This programme, designated **"Credible Chase"**, envisages the use of a "mini gunship" by the South Vietnamese as a standard counter-insurgency weapon, and the specification demands 20-mm cannon armament with side-door firing. It is anticipated that the cost of the selection programme will be $14·5m (£5·8m).

The Convair Aerospace Division of General Dynamics is conducting an extensive aircraft **structural integrity programme** at its Lindbergh Field plant, San Diego, with two F-102 Delta Daggers and an F-106 Delta Dart, the result of which, it is hoped, will enable the USAF and the Air National Guard to increase the certified "life" of these fighter types and thus extend their operational use. The F-102s are being subjected to loads that would be experienced in 28,000 hours of flying time and the F-106 to loads equivalent to 32,000 hours of flying time. This could result in F-102s being re-certified for 7,000 hours and F-106s for 8,000 hours of flight. Both types are currently certified for 4,000 flying hours.

The US Navy now anticipates that of a total buy of 301 **Grumman Tomcat** variable-geometry fighters 179 will be of the F-14B version with the Pratt & Whitney F401-PW-401 turbofans of some 30,000 lb (13 610 kg) thrust. This estimate is based on the belief that the Pratt & Whitney engine will be ready for service use by 1975.

AIRCRAFT AND INDUSTRY

ARGENTINA
Production of **Cessna light aircraft** under licence at the Military Aircraft Factory in Cordoba is expected to total 440 units over the next five years. A renewed and expanded agreement was signed between Cessna and the Argentine Air Force early in October, providing for production of the Model 150, Model A182 and Agwagon 180, for sale in South America. Projected quantities under the five-year plan are 320 Model 150s, 50 Model 182s and 70 Agwagons, with deliveries to start in the first quarter of 1972.

FEDERAL GERMANY
The second prototype of the **VAK 191B,** the V2, was flown for the first time on 2 October, only 22 days after the initial flight of the V1. With Ludwig Obermeier at the controls, the V2 flew at a forward speed of approximately 37 mph (60 km/h) at a height of some 130 ft (40 m), landing after four minutes in the air. The test programme with the first two prototypes is currently continuing, and, at the time of closing for press, the gradual transition to aerodynamic flight was in progress and flight testing of the third prototype, the V3, was imminent.

The Sikorsky S-67 Blackhawk helicopter gunship has recently been undergoing armament tests at the US Army Underhill Firing Range, Vermont. For these trials, it carried a General Electric turret containing a 20-mm cannon, two rocket pods each carrying 19 2·75-in FFARs, and two 300 US gal (1 136 l) drop tanks. The stub wings have a total of six hard points.

FRANCE

Limited manufacture of the **Aérospatiale SN-600 Corvette** is being launched, with a view to having production aircraft available at the end of 1973, when certification is expected. Following the fatal accident suffered by the prototype last April, reportedly as the result of a tailplane stall and subsequent loss of control, two more prototypes are to be built, plus two production models, and the Aérospatiale management has approved ordering of long lead time items for four more production aircraft. The prototype 02, due to fly next December, will have 2,200 lb (998 kg) thrust JT15D-1 engines, like 01, while 03 will fly in March 1973 with 2,310 lb (1 048 kg) JT15D-4 engines, as planned for the production models.

First flight of the **Dassault Falcon 10-02** was made on 15 October from Bordeaux-Merignac. This is a production prototype and is powered by two 3,260 lb st (1 470 kgp) AiResearch TFE-731-2 engines, whereas General Electric CJ610s were fitted in the Falcon 10-01. The latter has now completed more than 70 hrs in 56 flights.

JAPAN

Kawasaki Heavy Industries has been awarded a £300,000 contract by the Maritime Self-Defence Force for preliminary research for the **PX-L** four-turbofan shore-based maritime patrol and ASW aircraft with which it is intended to replace the Kawasaki P-2J. Kawasaki is currently conducting wind tunnel testing with a one-tenth scale model of the proposed 110,230-lb (50 000-kg) aircraft, and is simultaneously designing an integrated data computer system in co-operation with Hitachi, Tokyo-Shibaura and Fujitsu.

Considerable concern has been expressed in Japan regarding production cost overruns, inter-departmental wrangling over finance and inadequate cost-planning which are jeopardising the prospects of the Japanese aircraft industry. The Ministry of Finance maintains that the indigenous **development of military aircraft** has become prohibitively expensive. The Mitsubishi XT-2 trainer, for example, is costing *Yen* 1,200m (£1·45m), or about three times the initial estimate, whereas the Northrop F-5 is available off-the-shelf for some *Yen* 600m (£720,000). The Kawasaki C-1 is also running at some three times the budgeted level, and is now approaching *Yen* 4,000m (£4·84m), whereas the C-130 Hercules can be purchased at some *Yen* 1,500m (£1·81m).

Development of a **turbofan engine** in the 20,000–30,000 lb (9 070–13 605 kg) thrust bracket for commercial use is going ahead in Japan under a nine-year programme. The engine is designated FJR 710/10 and FJR 710/20 (the letters indicating Fan Jet Research and 710 representing 1971) and the basic and detail design work is shared between Ishikawajima–Harima, Kawasaki and Mitsubishi. Phase One of the programme, costing *Yen* 6,700m (£81·2m), will be completed by 1975 and covers testing of a 10,000 lb (4 535 kg) thrust engine.

UNITED KINGDOM

A letter of intent for a **new advanced trainer** to succeed the Gnat was received by Hawker Siddeley in October. The programme reportedly revolves around a fly-away price of £450,000, and although the RAF commitment in terms of quantity has not yet been established, it is expected to cover 175–180 aircraft with the first flight

North American Rockwell's General Aviation Division has added the Aero Commander Model 111 to its range of light aircraft for 1972. Flown for the first time recently, the Model 111 is a fixed undercarriage version of the Model 112, both types being illustrated in this photograph. The Model 111 and 112 are both four-seaters powered by the 180 hp Lycoming engine and production is planned to reach a level of three a day by the end of 1972.

taking place early in 1974 and production deliveries commencing during 1976–7. The specification for the new trainer, which will presumably be based on the HS 1182 proposal, has yet to be finalised, a power plant selected and avionics completely defined, and it is unlikely that a contract will be awarded for some time.

The second **Westland Lynx** made its first flight on 28 September at Yeovil. This aircraft is actually Lynx 00-03 (XW837) and it joins the Lynx 00-01 (XW835) in the flight test programme. Twelve Lynx helicopters have been assigned to the development and each is to be finished in a distinctive colour: 01 and 03 are yellow and red respectively. The first prototype has now flown about 30 hours since the first flight on 21 March, reaching a speed of just over 186 mph (300 km/h) and flying at weights up to about 8,600 lb (3 900 kg) compared with the normal gross of 8,000 lb (3 630 kg).

The future of both the Islander and the Trislander was obscure at the time of going to press, following the appointment of a **receiver for Britten-Norman Ltd.** Despite the success of the Islander and the excellent future prospects for both the Islander and Trislander, the company recently encountered cash-flow problems as a result of a slow-down in Islander sales coupled with the impact of Trislander development costs.

The 1972 **SBAC Flying Display and Exhibition** at Farnborough will take on the character of a truly European show for the first time. The SBAC has decided that the products of companies in ten West European countries (including Great Britain) will be eligible for display; hitherto, only British products or foreign aircraft powered by British engines have been eligible. The latter qualification will continue to apply for exhibits from countries outside the West European bloc, including the USA. Dates of the 1972 event are 4 to 10 September, with public admission on the final three days.

The RAF has now initiated a programme to uprate the Rolls-Royce Bristol Pegasus engines in its Hawker Siddeley Harriers from Pegasus 101 to Pegasus 102 standard. These designations are equivalent, res-

pectively, to the 19,000 lb st (8 618 kgp) Pegasus Pe6 and 20,500 lb st (9 300 kgp) Pegasus Pe10; the modified aircraft are designated **Harrier GR Mk 1A**. An aircraft of this type was flown to Japan (in a Belfast freighter) during October to participate in the International Aerospace Show at Nagoya, taking the place of Hawker Siddeley's Harrier T Mk 52 two-seat demonstrator which had been damaged in a landing accident at Dunsfold. The latter aircraft, with civil registration G-VTOL, had made its first flight on 16 September and is to the latest production two-seat standard with a taller fin.

Flight testing of the **Rolls-Royce RB.199** — the engine for the Panavia MRCA — will begin in the autumn of 1972, in an Avro Vulcan. This is the aircraft used until recently for flight development of the Olympus engine for the Concorde, and it has now been grounded for conversion. The first bench run of the RB.199 was made on 27 September at the Bristol Engine Division.

USA

Boeing has received a contract valued at $10,226,426 (£4,090,570) for the conversion of 96 **B-52G and B-52H** bombers to carry up to 20 SRAMs (Short-Range Attack Missiles) each, eight being mounted on a rotary launcher in the weapons bay and a further six being mounted beneath each wing. Simultaneously Hughes Aircraft is to build 300 forward-looking infra red sets for mounting in twin electro-optical chin turrets on B-52G and B-52H bombers under a contract valued in excess of $30m (£12m). The electro-optical viewing system will enable the crews of the bombers to navigate at low altitudes during day or night.

It is now anticipated that the first **McDonnell Douglas F-15** air superiority fighter will commence its flight test programme in July. A number of significant changes have been made to the F-15 design since the award of the development contract on 23 December 1969, these changes including the provision of a more symmetrical nose radome, the provision of cowl flaps on the engine intakes to enhance directional stability, a 5-in (12,7-cm) aft shift of both the wing and tailplane for reasons of balance, the removal of the ventral fins, and the redesign and increase in height of the vertical tail fins. The 25-mm GAU-7/A rotary cannon currently being developed for eventual installation in the F-15 is expected to offer nearly 10 times the "kill" probability of the 20-mm M-61 owing to its larger projectile, higher muzzle velocity, and improved ballistics.

Although the **North American Rockwell B-1** strategic bomber development programme has been scaled down, most of the primary performance characteristics have not been downgraded. The revised programme calls for a reduction in the number of flight test aircraft from five to three, and in the number of General Electric F101 engines from 40 to 27. The first flight of the B-1 is now scheduled for May 1974, and a production decision will be made a year later, at which time the development programme will have cost $2,300m (£920m). It is now estimated that the total programme cost based on the production of 240 aircraft will be about $11,110m (£4,440m). Take-off performance has been reduced by some 10 per cent to cut costs but most other aspects of the performance remain unchanged. On a typical 3,600-mile (5 795-km) mission, the B-1 will fly at approximately 32,000 ft (9 755 m) and Mach 0·8 with wings extended to the vicinity of hostile territory which would be entered at an altitude of less than 500 ft (150 m) at high subsonic speed with wings swept. Once inside enemy territory the B-1 would employ its high-altitude supersonic capability. If the production go-ahead is received, it is anticipated that the B-1 will be produced at a rate of three-four aircraft per month over a six-year period commencing 1977, but no significant quantity of B-1s is likely to be in service until 1979.

Boeing has obtained certification for the **Advanced 737,** powered by 15,500 lb st (7 030 kgp) JT8D-15 engines and delivery of the first example will be made to Saudi Arabian Airlines in March. Compared with the 737-200 powered by 14,500 lb st (6 577 kgp) JT8D-9 engines, the Advanced 737 will be able to take-off at a 7,000 lb (3 175 kg) greater weight from given runway lengths, or at the same weights from runways 1,500 ft (457 m) shorter. The extra weight allows a full passenger load to be carried 400 naut mls (740 km) farther.

North American Rockwell's range of business aircraft now includes two turbo-prop versions of the original Turbo Commander — the 681B and the 690. The **Turbo Commander 681B** is basically the same as

the Hawk Commander, powered by two 575 shp Garrett AiResearch TPE331-43-BL engines and with a gross weight of 9 400 lb (4 264 kg). The newer **Turbo Commander 690** has 700 shp TPE331-5-251K engines and a gross weight of 9,850 lb (4 468 kg). The span has also been increased by extending the centre section. This is the variant offered to the RAF to satisfy its requirement for a multi-engine pilot trainer and the prototype has been evaluated at the A & AEE, Boscombe Down.

Firm orders for the **Lockheed L.1011 TriStar** now stand at 105, but the number on option for "second buys" has been reduced to 49 through cancellation of the Air Holdings commitment for 29. Air Holdings Ltd originally ordered 50 TriStars as part of the deal to provide British support for the programme in return for Lockheed's selection of the Rolls-Royce RB.211 engine. The aircraft were for re-sale by Air Holdings outside of the US and 21 have been so disposed of (19 to Air Canada and two to Air Jamaica). The Air Holdings commitment to pay progress payments on the other 29 ended, in common with that of other airline customers, when Rolls-Royce collapsed, and the company has decided not to renew the agreement. Payments already made on these aircraft are subject to return to Air Holdings when the aircraft are eventually sold by Lockheed and the aircraft ordered recently by Court Line (two plus an option on three) will be part of the original Air Holdings batch. The fourth L.1011 joined the test programme on 24 October, by which time the first three had flown about 700 hrs. The fifth was to fly in December.

Fairchild Republic Division has won a US Navy Air Systems Command contract to design, construct, test and analyse an advanced helicopter rotor system called a **reverse velocity rotor.** Under a 10-month contract the company will test a scaled-down version of the rotor system at speeds up to 350 knots (649 km/h).

Increased weights have been announced for the **Cessna Citation,** deliveries of which began in October. To take effect after certification next February, the new weights are: max ramp, 11,000 lb (4 990 kg); max take-off, 10,850 lb (4 921 kg); zero fuel, 8,400 lb (3 810 kg) and max landing, 10,400 lb (4 717 kg). An increase of 200 lb (90,7 kg) in usable fuel capacity has also been announced.

Flight testing of a de Havilland CC-115

Buffalo with an **air cushion landing system** (ACLS) is to begin late in 1972 as part of a development programme initiated by the USAF Flight Dynamics Laboratory at Wright Patterson AFB in conjunction with the Canadian government. The ACLS was developed by Bell Aerospace and has been tested on a Lake LA-4 since 1967, this aircraft having demonstrated the ability to operate to and from water, hard runways, mud, snow and ice. The system comprises an inflatable rubberised cell on the underside of the fuselage, in which there are hundreds of small holes; when this cell is inflated, using air from a special auxiliary power system in the aircraft, air escapes through the holes to form an air cushion which supports the aircraft, with a ground pressure of less than 3 lb/sq in (0,21 kg/cm²). The Buffalo is being converted by de Havilland Aircraft of Canada and United Aircraft of Canada is responsible for the auxiliary power system.

First flight of the 50,000 lb st (22 680 kgp) **General Electric CF6-50** turbofan was made in a Boeing B-52 test-bed on 21 September. This new version of the CF6 is destined to power the McDonnell Douglas DC-10 Srs 30 and the Airbus A.300B.

The planned merger between **Piper and Swearingen** was abandoned late in September, Piper having found it impossible to proceed without accepting certain Swearingen commitments which appeared unlikely to prove profitable. Swearingen is reported to have an outstanding commitment with Fairchild for production of 270 sets of wings for Merlins IIIs, and with AiResearch for a similar quantity of aircraft to be furnished, finished and fitted with avionics. The Piper view was that the agreed price for these contracts was too high.

NASA has fitted a **rotating cylinder flap** to a North American Rockwell OV-10A, which is now operating in civil markings (N718NA) at the Ames Research Center. The system comprises cylinders immediately ahead of the aircraft's flaps, which rotate when the flaps are lowered to maintain a better airflow and, therefore, higher lift. The OV-10A has been modified to have Lycoming T55 engines which are interconnected so that either one can drive both propellers and the cylinders.

CIVIL AFFAIRS

BAHAMAS
A new charter airline is in process of formation in the Bahamas, with the name **Trans-Oceanic Airways.** It is expected to start operations by the end of the year from Nassau and Freeport to cities in North America and Europe, using a leased Boeing 707.

CEYLON
Air Ceylon has concluded an agreement with **UTA** whereby the latter company will provide commercial and technical assistance, replacing an earlier agreement with BOAC. Under the agreement, UTA will provide a DC-8 for use in Air Ceylon markings for three years commencing 1 April 1972.

FEDERAL GERMANY
Operations by **Paninternational,** a German IT operator, were suspended early in October on orders of the German government. The company lost a BAC One-Eleven on 6 September (engine failure being caused by contamination of the water-injection

Deliveries of the Hispano HA-220 Super Saeta single-seat light ground attack aircraft to the Ejército del Aire began recently, the first batch of 10 aircraft built against an order for 25 aircraft of this type being seen in this photograph.

system by kerosene). Since that date, according to agency reports, one of the airline's two Boeing 707s has been impounded for alleged non-payment of debts at Munich, and One-Elevens were similarly impounded at Frankfurt and Dubrovnik.

FRANCE
The **Super Guppy 201** arrived at Le Bourget on 29 September on delivery to Airbus Industrie for use in transporting A.300B components. It is being operated by Aeromaritime, the charter associate of UTA. Certification had been obtained on 26 August, and four more examples are under construction by Aero Spacelines. One of the first uses of the Guppy in Europe was to transport the first set of Airbus wings from Chester to Toulouse in November.

A Breguet Deux Pont has been presented to **Air Afrique** by Air France, which recently retired the type from service. It will be used by the company's flying school at Dakar Yoff Airport to give instruction in equipment and engine handling.

INDIA
Air India has set up a non-IATA subsidiary to operate charter flights. Known as **Air India Charters,** it will lease aircraft from the parent company.

NEPAL
Re-equipment plans of **Royal Nepal Airlines Corp** include the acquisition of a DC-9, to operate the Kathmandu-Tokyo service through Calcutta, Bangkok and Hong Kong. Three ex-Australian DC-3s are in course of delivery and two Turbo-Porters are to be obtained.

UNITED KINGDOM
British Air Ferries, the successor of the original Silver City and Channel Air Bridge companies specialising in vehicle ferry services across the English Channel, is reported to be the subject of a take-over bid by **Transmeridian Air Cargo.** The latter company has recently obtained approval for car ferry services with CL-44s from the UK to Switzerland, and would probably use the same aircraft to replace BAF's five Carvairs on the shorter routes.

BEA has finally ended its **Viscount operations** in Germany, the last service being flown out of Templehof Airport, Berlin, for Hanover on 1 November. The airline's German Division is now all-jet equipped and the last Viscount has been transferred to Scottish Airways Division.

USA
The FAA has offered a grant of $229,746 (£91,900) towards the estimated cost of a study into the financial, operational and environmental feasibility of constructing a **V/STOLport** in the New Jersey area. The State of New Jersey will contribute one-third of the total cost of the study, which is the first attempt to develop a master plan for a site suitable for the operation of V/STOL aircraft.

CIVIL CONTRACTS AND SALES

BAC One-Eleven: Court Line has confirmed the purchase of two Srs 500, previously operated by Bahamas Airways. ☐ British Caledonian has sold one Srs 500 to the British Ministry of Defence (Aviation).

Boeing 707: The 707-320B acquired recently by Cathay Pacific is now known to be ex-Northwest Airlines. Two more may follow.

Boeing 737: Pacific Western Airlines in Canada ordered its fifth, an Advanced 737 for April delivery.

Boeing 747: Universal Airlines ordered one 747B for service starting next June. Aircraft was originally built for Braniff, but not delivered. ☐ VIASA will lease one from KLM from next April for once-a-week Caracas-Europe service.

Canadair CL-44D: With effect from November, all Loftleider's aircraft of this type have been transferred to the company's cargo associate, Aircargolux.

Convair 990: Three more have been bought by Spantax from American Airlines to make a total of nine in the fleet; the company is now the largest operator of this type.

Fairchild FH 227: Government of Mexico has bought two for use by the Hydraulics Resources and Public Works departments. Each has a 17-seat executive-style interior.

Fokker F.27: Pertamina, the Indonesian petroleum concern, bought four from All Nippon Airways, which is in process of retiring its fleet of 25.

Fokker F.28: German IT operator Bonair has ordered three Mk 1000s (with a fourth on option) for delivery next March and April. They will be operated by Germanair. ☐ Permina in Indonesia is reported to have ordered one. ☐ During October, Iranair leased one from Fokker-VFW in connection with the celebration of the 2,500th anniversary of the Persian monarchy.

Hawker Siddeley HS 748: Zambia Airways ordered one more, its fourth, for delivery at the end of 1971.

Helicopters: Houston Metro Airlines plan to use one Sikorsky S-58T for a six-month feasibility trial in 1972, linking downtown Houston with the airport. ☐ Hong Kong Air International acquired one Bell 212 for service linking the island with Macau. ☐ Pertamina has acquired four Aérospatiale Pumas, to be operated by Schreiner's Indonesian Air Transport on its behalf. ☐ Bow Industries (formerly Bullock Helicopters) of Calgary, Alberta, has taken delivery of the first Bell 212 for commercial use in Canada.

Light aircraft: The Japanese Civil Aeronautic College at Miyazuki, Kyushu, has ordered 13 Fuji FA-200s for delivery by next March; total sales of the type stand at 157. ☐ Corsario de Aviacao ordered 20 Ipanema agricultural aircraft from Embraer in Rio de Janeiro.

Light twins: All Nippon Airways has taken delivery of three more PA-31 Navajos, making a total of seven, to replace Beech 18s as instrument trainers. ☐ The Zambia Government Communications Flight has bought three PA-23 Aztec Es. ☐ Beech Aircraft has sold five King Air A100s to the Canadian government and two to the Mexican government, all for checking navigation aids and airport facilities. ☐ Aerolineas Centrales de Colombia ordered two Saunders ST-27s.

L 410 Turbolet: Four have been ordered by Slovak Air, a new local operator in Czechoslovakia; the first two entered service in September, bearing the legend Slov-Air, between Prague and Brno.

Lockheed L.1011 TriStar: Court Line of

Luton ordered two for 1973 service, with an option on three more.

McDonnell Douglas DC-8: Air Jamaica is acquiring one -51 second-hand from the manufacturer. ☐ Loftleider has purchased one -63 previously on lease, from Seaboard World.

NAMC YS-11: Perta Air Service, maintained by Indonesian oil company Pertamina to transport personnel and supplies, bought two Series 300.

Yak-40: J F Airlines has signed a letter of intent for one, subject to ARB approval and other conditions. The company operates Scottish Aviation Twin Pioneers between Portsmouth and Jersey and Guernsey.

MILITARY CONTRACTS

Bell OH-58A Kiowa: A contract for 400 additional OH-58A Kiowa helicopters valued at $26,183,948 (£10,473,590) was awarded on 29 July, deliveries against this contract having been scheduled to commence last month (November) and extending production into July 1973. This contract brings to 2,200 the number of Kiowa helicopters ordered by the US Army.

Boeing 737: Five 737-100s are to be acquired from Lufthansa by the *Luftwaffe* as staff and VIP transports, replacing Convair 440s used at present.

Hawker Siddeley Buccaneer: An order for 16 more new Buccaneer S Mk 2Bs is to be placed on behalf of the RAF, it was announced by the Ministry of Defence late in October.

Hawker Siddeley HS 125: It was announced on 15 September that the South African Air Force has ordered a further three HS 125 (Mercurius) light transports at a cost of some £1·5m to replace the three aircraft of this type lost earlier this year.

Kawasaki KH-4: Kawasaki Heavy Industries has delivered seven four-seat KH-4 helicopters to the Thai government during September.

North American Rockwell T-2C Buckeye: The US Naval Air Systems Command has awarded North American Rockwell's Columbus Division a $4·6m (£1·44m) contract for the production of a further 36 T-2C Buckeye trainers to be manufactured during the 1973 calendar year as follow-on deliveries to existing contracts.

North American T-28D-10: The USAF has taken up an option with Fairchild Aircraft Service Division's St Augustine facility for the conversion of a further 22 T-28 trainers to T-28D-10 counter-insurgency aircraft configuration. This is an extension of a contract involving 50 aircraft awarded in November 1970, and value of the contract extension is $1,394,321 (£557,700).

Northrop F-5: The Saudi Arabian Air Force will receive 20 F-5Bs and 30 F-5Es under a $130m (£52m) package contract placed in the USA (see news item in "Military Affairs" on page 339).

Vickers Viscount: A second Type 836 has been acquired for the Muscat and Oman Air Force (see this column in AIR ENTHUSIAST for October). Both aircraft were in service with the RAAF until 1969 and have been refurbished in the UK.

ISRAEL – PREPARING FOR THE NEXT ROUND

The casual visitor to Israel is unlikely to see many obvious signs of the fact that a state of war exists. There is no blackout or curfew to be contended with in Tel-Aviv, Haifa, or any other of the heavily-populated urban areas; there is little open military activity, and the *kibbutznics* tilling their fields and tending their vineyards appear perfectly relaxed. But war between Israel and her Arab neighbours has simmered continuously and boiled over periodically ever since the modern State of Israel was created on 14 May 1948 by a resolution of the United Nations. The Israelis have learned to live with the tension engendered by the constant hostility of their neighbours and its constant threat to their survival as a nation.

Yet thriving though Israel's citizens undoubtedly are, they never forget that they are poised on a knife's edge, and their *esprit de corps,* which has played no small part in enabling Israel to emerge victorious each time the war of nerves and minor border clashes has escalated into a full-scale shooting affair, remains undiminished after 24 years. There are, of course, signs of this very real struggle if the visitor seeks them. On highways almost unbelievable numbers of soldiers are to be seen hitch-hiking to and from their bases, and within a radius of three or four miles of the major bases the *kibbutznics* give place to the *Nahal,* the pioneer settlers who farm high-risk areas instead of giving military service, and whose weapons are only a little less obvious than their farming implements. The visitor relaxing on the beach at, say, Tel-Aviv, cannot but be aware of the constant coming and going of military aircraft or the frequency of sonic-boom thumps as *Heyl Ha'Avir* aircraft continuously exercise over the Mediterranean, while the security consciousness of the average Israeli becomes immediately obvious if the unwary visitor happens to point his camera in the direction of the occasional helicopter!

The Israelis are almost fanatically proud of their air arm, the Air Section of the Israel Defence Forces, or *Tsvah Haganah Le Israel/Heyl Ha'Avir,* for it is improbable that any nation today is more reliant on air power for its continued existence. What is *certain* is the fact that no air arm of such recent creation possesses so dramatic a history. The combination of assiduity, finesse and sheer impertinence displayed by *Heyl Ha'Avir* during the opening phase of the six-day war of 1967 astounded the strategists, and, today, all but a half-decade later, the personnel of this air arm exude

The Israeli authorities recently granted AIR ENTHUSIAST facilities to visit *Heyl Ha'Avir,* the Air Section of the Israel Defence Forces, and we were thus enabled to see something of the changes that have taken place in this air arm, youthful in years but old in experience, since its dazzling demonstration of the effective use of air power five years ago dramatically if temporarily changed the balance of Middle Eastern military power. The following report, which is illustrated with photographs by Stephen Peltz, is based on frank discussions with several high-ranking *Heyl Ha'Avir* officers and with personnel of the squadrons at the three bases that we were permitted to visit, but the Israelis are extremely security conscious, and, in consequence, the officers that we interviewed must remain anonymous, and much has, for obvious reasons, to remain unsaid.

The increased capability of Heyl Ha'Avir *since the six-day war is due in no small part to the acquisition of the Phantom, the F-4E version of which is seen below and immediately above right, with the RF-4E reconnaissance model being illustrated top right. Each of the* Heyl Ha'Avir *Phantom squadrons is reported to include two RF-4Es in its inventory.*

such confidence in their ability to establish once again complete mastery over their opponents when the bell rings for the next round in the conflict, despite the massive resupply and expansion of the neighbouring Arab air forces by the Soviet Union, that it is difficult to imagine them ever being beaten.

At the time of our visit, *Heyl Ha'Avir* was ostensibly operating on a *peacetime* footing, but the officers with whom we talked made plain their belief that the relative quiet of the past 15 months has been a lull before the storm inevitably breaks once again: since August of last year no incidents had been reported until 11 September when one of two Egyptian Sukhoi Su-7s, apparently on a reconnaissance mission, was shot down by Israeli groundfire in the northern sector of the Suez Canal area. Many believe this next round of full-scale fighting to be imminent, and there are detectable signs that *Heyl Ha'Avir* shares this view and is metaphorically girding its loins in preparation.

Introducing a "bantam" bomber

Much has happened in the North African littoral since the six-day war. These few years have seen the Soviet Union become steadily more deeply entrenched, taking over much of the responsibility for the air defence of Egypt, prompted, in the Israeli view, by the manifest inability of the Egyptians to handle effectively the sophisticated hardware lavished upon them. They have also seen the imposition of the French arms embargo and, as a direct result, the extensive "Americanisation" of *Heyl Ha'Avir* from the equipment standpoint. Whereas five years ago the Israeli air arm was equipped almost in its entirety with the products of the French aircraft industry, today combat aircraft of American design and manufacture preponderate in the hardware inventory; a change which, if not of Israeli choosing, is one with which *Heyl Ha'Avir,* though perhaps not the Ministry of Finance, is well satisfied.

Now numerically the most important combat aircraft in the inventory of *Heyl Ha'Avir* is the McDonnell Douglas A-4 Skyhawk; Ed Heinemann's "bantam" bomber which has now taken over the tactical attack rôle for which the Dassault Mirage 5J was originally ordered. The initial Israeli contract placed in 1967 called for 48 examples of the single-seat A-4H and a pair of tandem two-seat TA-4H Skyhawks, these being essentially similar to the US Navy's A-4F and TA-4F apart from having the enlarged and square-tipped vertical tail surfaces associated with provision of a tail drogue 'chute, and what the Israelis consider to be the ineffective US Navy 20-mm Mk 12 cannon replaced by the Israeli-manufactured version of the longer-barrel 30-mm DEFA weapon. One other difference, when the aircraft were originally delivered to Israel, was the omission of the A-4F's dorsal avionics compartment from the A-4H.

Deliveries of the A-4H to *Heyl Ha'Avir* began late in 1967, the first Skyhawk-equipped squadron attaining operational status early in the following year, and indicative of the rapidity with which the Israelis gained confidence in the McDonnell Douglas warplane is the fact that almost immediately upon being cleared for operations this squadron participated in attacks on heavily-defended *El Fatah* terrorist bases on the East bank of the River Jordan.

US governmental approval of the delivery of further Skyhawks to Israel was obtained while deliveries against the initial A-4H contract were in progress, and in 1969 the A-4Hs were supplemented by the first batch of 25 ex-US Navy A-4E Skyhawks, further deliveries having since been made and *Heyl Ha'Avir* now including upwards of a hundred Skyhawks of all versions in its inventory. These equip four squadrons at two different bases, the Skyhawks at one base being all of the A-4H type and those at the other base being A-4Es, and it was the base housing the latter that we were

permitted to see during our visit to *Heyl Ha'Avir.*

We were soon left in no doubt as to the satisfaction with which *Heyl Ha'Avir* views its acquisition of Skyhawks, all the pilots that we interviewed readily extolling the aircraft's virtues. Previously Israel had largely confined her combat aircraft procurement to types capable of fulfilling more than one rôle, the exception being the Vautour IIA attack aircraft, and while the Skyhawk, which is now the primary bombing platform of *Heyl Ha'Avir,* allegedly possesses counter-air capability, provision being made for a pair of Sidewinder AAMs to supplement its twin DEFA cannon, it was tacitly admitted that the A-4 is somewhat on the slow side for air-to-air combat, and would not normally engage enemy fighters unless attacked. "The Skyhawk," we were told, "was given a thorough shake down on operations before the cease-fire of August last year, and although the pilot of a MiG-21MF or other high-performance Soviet fighter now being flown by the Egyptians can undoubtedly break off combat with our aircraft at will, we have no doubt that our Skyhawks, with their outstanding manœuvrability, are fully capable of defending themselves. Prior to the cease-fire one of our Skyhawks had downed a Syrian MiG-19 over the Lebanon with a 2·5-in (6,35-cm) air-to-ground rocket! Another had bagged a MiG-17 with its DEFA cannon."

"The Skyhawk," the squadron commander informed us, "is a relatively undemanding aircraft to fly and easily the most manœuvrable warplane in Middle Eastern skies, not excluding the earlier MiGs." He added that, in his view, the Skyhawk is "one of the best bombing aircraft ever produced for round-the-clock operations. It can carry an 8,200-lb (3 720-kg) ordnance load as compared with the 4,000 lb (1 815 kg) or so of the Mirage IIIC that we used for both strike and air superiority missions during the six-day war." Few problems, he said, had been encountered in phasing this US weapon system into the Israeli inventory, and he evinced particular satisfaction with the navigational and flight-refuelling equipment, and with the general sturdiness of the Skyhawk which enabled it to regain its base after suffering battle damage that would have knocked many current warplanes out of the sky. He instanced the example of an A-4E which, while returning from a deep-penetration mission over Egyptian territory on 29 July last year, was bounced by a MiG-21MF (*Fishbed-J*) and had a wing punctured by an *Atoll* AAM but still succeeded in eluding its pursuer and returning safely to base.

It was during this action, incidentally, that the *Heyl Ha'Avir* had its suspicions confirmed that Soviet pilots were actually flying operational missions. A sizeable force of MiG-21s had scrambled from Inchas — one of the six Egyptian bases then under the partial or complete control of the Russians — to intercept a relatively small force of Skyhawks and Mirage IIICs, driving the Israeli formation back and damaging the A-4. The Israeli pilots reported that the Russian language was being used both by the ground control stations and by the pilots of the fighters that intercepted them, and a Russian-speaking amateur radio operator corroborated these reports and was also instrumental in reporting Soviet procedures. On the following day, 30 July, *Heyl Ha'Avir* sent a small Skyhawk force in the direction of Inchas as bait to draw the Russian-flown interceptors into the air. The bait was swallowed, a sizeable force of MiG-21s being scrambled, and then the trap was sprung, a substantial Israeli force of Mirage IIICs and Phantoms meeting the MiGs over the Egyptian border, south of the city of Suez. In the ensuing mêlée four MiG-21MF interceptors were destroyed — two by Sidewinders launched from Phantoms, one by a Sidewinder launched from a Mirage, and the fourth by cannonfire from another Mirage — without loss to the Israeli force.

During one of the last attacks on the Russo-Egyptian

The F-4E Phantom (above) is now one of the most important weapons in the Israeli inventory, and allegedly equips three squadrons of Heyl Ha'Avir. Some Israeli aircraft, such as that illustrated here, appear to have the barrels of their M-61A-1 rotary cannon extended forward beneath the nose.

Numerically the Bell 205 Iroquois (not Agusta-Bell AB 205 as is often reported) is the most important helicopter of Heyl Ha'Avir, an example of this extremely versatile machine being illustrated right.

The advent of the Sikorsky S-65C-3 (above) in Israeli service has tremendously increased the vertical lift capability of Heyl Ha'Avir. The Nord 2501 Noratlas (below) is used primarily as a paratroop transport but, on occasions, also serves in the target-towing rôle.

The Sikorsky S-65C-3 heavy-lift helicopter, one of the Heyl Ha'Avir examples of which is illustrated left, has now been in Israeli service for some two years, making a spectacular operational début on 26 December 1969 when it was used in the capture of an entire "Barlock" mobile radar installation.

missile sites a few days before the ceasefire, a Skyhawk received a direct hit in the aft fuselage from an anti-aircraft shell. The pilot of the Skyhawk succeeded in retaining control of the aircraft but told his wingman that he thought that he would have to eject. The wingman flew aft and below the damaged Skyhawk, which was emitting a lot of smoke and losing fuel, and reported to his companion that he thought his aircraft could just about reach a forward emergency strip in the Sinai if it did not break up in the air first! The Israeli pilot stayed with his aircraft and succeeded in putting it down on the forward strip. Recounting the story, he told us: "I jumped from the cockpit to take a look at the damage. Much of the rear fuselage was missing, the tailpipe was mangled and the tail assembly seemed to be hanging in shreds. I passed out cold!" Replacement aft fuselage, tailpipe and tail assembly were flown into the strip by helicopter, and two days later the Skyhawk was safely back in the overhaul hanger of its home base.

Between 7 January last year, when Heyl Ha'Avir increased the missions flown against Egyptian targets and launched a series of deep penetration attacks on objectives more than 10 miles (16 km) behind the ceasefire line along the Suez, and seven months later when the ceasefire of 8 August began, eight A-4 Skyhawks were lost, mostly victims of SA-2 missiles, but these have been replaced on a one-for-one basis by the USA, apart from the deliveries of supplementary batches of aircraft, and Heyl Ha'Avir is now engaged in retrofitting its Skyhawks with new navigation and attack systems which are partly housed by a fairing "hump" aft of the cockpit similar to that first introduced on the A-4F.

At the Skyhawk base that we visited A-4Es both with and without the dorsal avionics compartment were to be seen.

Affording a substantial improvement in weapons delivery accuracy, the new equipment includes a Lear Siegler inertial navigation system using a Singer-General Precision stabilised platform and driving an Elliott Automation head-up display. We understand that the most recent batch of 18 A-4Es for Heyl Ha'Avir had been fitted with the Lear Siegler weapons delivery package prior to arrival in Israel, but our enquiry as to the substance behind US reports that this latest batch of Skyhawks was equipped for the Walleye television-guided glide bomb elicited the reply from the Skyhawk squadron commander, admittedly accompanied by something of a twinkle of the eye: "Walleye? Never heard of it!"

More expensive hardware

The McDonnell Douglas F-4E Phantom, which, at a unit cost of the order of $4·5m (£1·83m), is about the most expensive item of hardware ever purchased for Heyl Ha'Avir, began to enter the Israeli inventory a little more than two years ago, in September 1969, some 120 Israeli personnel having previously undergone conversion training in the USA. The initial Israeli contract called for 50 F-4E and six RF-4E Phantoms, which, according to US sources, were flown to Israel by USAF and USMC reservists under contract to McDonnell Douglas. The aircraft were flown across the Atlantic to the Azores, flight-refuelled by USAF KC-135 tankers en route, and from the Azores following a route designed to take them across as few countries as possible, touching down in Italy or Greece, and then making a final

Still the principal Israeli air superiority fighter, the Dassault Mirage IIICJ is marginally less manœuvrable than the MiG-21 but Heyl Ha'Avir believes that the combination of features embodied by this French fighter render it superior to its principal antagonist in Middle Eastern skies.

The Aérospatiale SA 321K Super Frelon was the largest helicopter in the Israeli inventory until the arrival of the S-65C-3. Heyl Ha'Avir originally planned to acquire 24 helicopters of this type, but the French embargo was imposed after the delivery of nine Super Frelons.

stop in Cyprus where the USAF insignia was removed and replaced by Israeli markings. Understandably, the Israelis would neither confirm nor deny this method of delivery.

The F-4E Phantom was first committed to operations by *Heyl Ha'Avir* on 7 January 1970 in a strike against an SA-2 missile site at Dahashur, a few miles south of Heluan, and saw considerable use prior to the ceasefire, by which time nine had been lost, mostly to surface-to-air missiles. Five of the Phantoms were destroyed between 29 July and 7 August when *Heyl Ha'Avir* pounded the missile sites so heavily that the service used nearly its entire available bomb supply. Of the other Phantoms lost, one was destroyed by Syrian groundfire east of the Golan Heights on 2 April 1970, and two others were lost to non-combat causes. Phantom attrition has since been made good by new deliveries on a one-for-one basis and, as in the case of the Skyhawk, additional batches of aircraft have been supplied to Israel, bringing the total number of Phantoms taken into the inventory of *Heyl Ha'Avir* to something of the order of 80 aircraft, but one of the several matters on which our hosts refused to be specific was that of aeroplane quantities.

It would seem that *Heyl Ha'Avir* is now operating three squadrons of Phantoms primarily for the strike mission, each squadron being assigned two of the RF-4Es for reconnaissance. The F-4E has hitherto been operated in Israeli service virtually entirely in the strike rôle, though, as previously recounted, it did come to grips with the MiG-21 in air-to-air combat on 30 July 1970. As one officer put it: "As long as we are opposed by the MiG-21 the Mirage IIIC is more than competent to handle the air superiority rôle,

and, in any case, the Phantom is too expensive an item of equipment to commit to a dogfight without good reason!" Nevertheless, although the majority of the F-4E Phantoms currently flying with *Heyl Ha'Avir* are configured for the strike mission, they all possess a secondary intercept rôle in which they can tote a quartet of Sparrow AAMs and a similar number of Sidewinders. Furthermore, the last half-score or so F-4Es delivered during the summer of last year against the initial order had, in fact, been re-configured to incorporate an optical gun sight to render them more effective in the intercept mode.

How the Phantom matches up to the Soviet-flown MiG-23 (*Foxbat*) interceptor capable of Mach 3·0 dash at 80,000 feet (24 385 m) is, understandably, the sixty-four dollar question, and has provided a subject for much speculation in *Heyl Ha'Avir* since the first flights of this advanced warplane from Cairo West were monitored late in March. The consensus of opinion of those Israeli pilots with

Two of the older but still very useful French combat aircraft in the Israeli inventory are the Super Mystère B2 (above right) and the Mystère IVA (below). Each equips one squadron of Heyl Ha'Avir, *and these fighters are likely to remain in service for some considerable time to come.*

A substantial number of Douglas C-47s remains in the Israeli transport inventory as the "maids-of-all-work" of Heyl Ha'Avir, *and at the present time no plans exist for a replacement for this type which is scheduled to soldier on well into the mid 'seventies.*

Cessna U 206C Super Skywagons, one of which is seen right, have been in Heyl Ha' Avir *service for some two years. These serve in the utility rôle and are equipped with a detachable glass-fibre Cargo-Pack which incorporates loading doors on the side and at the rear. The Super Skywagon is used primarily for light transport and liaison tasks, but can also serve in the casualty-evacuation rôle.*

The Piper Super Cub has served Heyl Ha' Avir *for many years in the primary training, liaison and spotting rôles. Although the numbers in Israeli service have now diminished, it still fulfils grading and spotting tasks, and an example of this type is illustrated left.*

The ex-US Navy A-4E Skyhawks, which have supplemented the initial deliveries of the A-4H version of this highly-successful little warplane, are now being fitted with the "avionics hump" which will eventually be added to all Israeli Skyhawks. An A-4E modified by the addition of the "hump" is illustrated below.

which we discussed the challenge presented by the MiG-23 is that this warplane is a purely defensive interceptor aimed primarily at combating the next generation of strategic bombers, such as North American Rockwell's B-1A. On the basis of available information it would seem to enjoy a markedly superior performance to the Phantom at extreme altitudes, but to descend to lower altitudes and engage the lighter and more manœuvrable McDonnell Douglas fighter could well be disastrous for the MiG-23. To achieve altitude performance, Mikoyan's fighter would seem to have sacrificed performance aspects in which the Phantom excells and is likely to find itself at a serious disadvantage below 30,000 ft (9 145 m). To oppose a conventional strike going in at between 15,000 and 20,000 feet (4 570 and 6 095 m), the MiG-23 would have to rely on its superior speed to come down from the altitudes at which it was designed to operate and make a single stern attack. This would place the Soviet fighter in the low-altitude environment where the Phantom is supreme.

Numerically now only third in importance in the Israeli combat aircraft inventory, the Dassault Mirage IIICJ is today confined primarily to the air superiority rôle, and still equips three squadrons of *Heyl Ha'Avir,* although attrition over the years has reduced to something under 60 the number of aircraft of this type on strength. The fact that so many Mirages remain represents no mean achievement when it is recalled that the 72 fighters of this type (plus three two-seat Mirage IIIBJs for type conversion) were procured during 1963-6, and that for a considerable proportion of the intervening years these excellent Dassault aircraft have borne the brunt of much of the fighting to which *Heyl Ha'Avir* has been committed.

Innovation and a modicum of ingenuity have kept Israel's Mirages in top flying condition despite the French embargo on the supply of spares and components. Thanks to Israel Aircraft Industries and the Bet-Shemesh Engine concern, *Heyl Ha'Avir* is now largely independent of France for spares for the service's French-manufactured aircraft, and stocks of those spares that Israel *cannot* manufacture are, we were assured, sufficient for several more years.

No plans currently exist to re-engine the entire Israeli Mirage fleet with the General Electric J79 turbojet, as has been suggested from time to time in both the popular and technical western press, and *Heyl Ha'Avir* has not been directly concerned with the experimental adaptation by Israel Aircraft Industries of a Mirage airframe to take a J79-GE-17 under the programme name of *Salvo.* The modified Mirage has been flying since 19 October 1970, and although a problem involving unacceptably high heat levels with afterburner has been resolved, and the J79-powered aircraft has displayed significant improvements in take-off and climb performance coupled with substantially reduced specific fuel consumption, the effort demanded by a major conversion programme is not considered justified in view of the fact that, as we were told, "The SNECMA Atar currently installed is expected to perform perfectly adequately and provide no spares problem throughout the remaining and not inconsiderable anticipated airframe service life of the Mirage".

No comment was forthcoming regarding the new J79-powered strike fighter which has allegedly been developed and built by Israel Aircraft Industries under the reported programme name of *Black Curtain.* This aircraft is said to be based on the Mirage III airframe but to have markedly superior external weapons capability and to have commenced its test programme during the late summer.

Other French combat aircraft remaining in the first-line strength of *Heyl Ha'Avir* and likely to do so for some time yet to come comprise one squadron of Mystère IVA fighter-bombers, the survivors of 60 acquired in 1956, which recently completed 15 years in Israeli service; one squadron of barely less venerable Super-Mystère B2 fighters with 12 years of service behind them, and one squadron of single-seat Vautour IIA long-range attack aircraft, which, together with several two-seat Vautour IINs modified for electronic reconnaisance and countermeasures tasks, will probably be phased out over the next year. The real old man of Middle Eastern skies is, of course, the now aged but still sprightly Dassault Ouragan which, first acquired by *Heyl Ha'Avir* in 1955, has now been relegated to second-line and training tasks with one squadron.

Accent on airborne mobility

The airborne mobility tactics employed so effectively by the Israeli forces during the six-day war have been progressively developed over the past few years by *Heyl Ha'Avir*, and much effort has been placed behind the expansion and modernisation of the helicopter force. All the Sikorsky S-58s have now been retired, some of these having been sold back to Sikorsky for conversion to turbine-powered S-58Ts, and these have been replaced by Bell Model 205 Iroquois helicopters, some 25 having been delivered to equip two squadrons. However, the most recent addition to the helicopter inventory is the Sikorsky S-65C-3, basically similar to the US Marine Corps' CH-53D heavy assault transport, and sporting armour, armament, and jettisonable auxiliary fuel tanks.

The S-65C-3 began to reach Israel during 1969, and 10 are now in service, but its existence in the Israeli inventory was not *officially* revealed until the Independence Day Parade on 11 May, although this rotorcraft had made its rather spectacular début on 26 December 1969 when it was used to lift a commando team to Ras Ghareb, about 115 miles (185 km) from Suez, to capture, dismantle and return to Israel a complete Soviet-built P-12 *Barlock* mobile radar installation which was being used to detect low-flying aircraft. Two vans containing radar and generators, and weighing three and four tons respectively, were carried back across the Gulf by the S-65C-3s, and subsequently transported to the Weizmann Institute at Rehovoth, near Tel-Aviv. Four ranking Egyptian Air Force officers were later reported to have been executed as a result of this audacious raid.

Prior to the arrival of the S-65C-3 in Israel, the largest helicopter in the inventory of *Heyl Ha'Avir* was the SA 321K Super Frelon. Six of these French helicopters had been delivered to Israel prior to the six-day war, an initial quantity of 12 having been ordered, but only a further three arrived before an embargo was imposed on the delivery of the remaining Super Frelons, and a supplementary order for an additional 12 was cancelled. The Super Frelon, too, has had its share of spectacular missions, such as that early in 1968 when three helicopters of this type carried Israeli commandos 200 miles (320 km) from the Red Sea into Egyptian territory to attack the Quena Dam and the Nag Hammadi hydro-electric power station in the Upper Nile Valley. With the Bell 205 and S-65C-3 squadrons, the Super Frelon unit forms a composite wing, and other helicopters currently in Israeli use include four Alouette IIs, and a number of Bell 47G-3Bs for the primary training rôle.

Increasing emphasis is being placed on transport support, and there are now about a dozen ex-Pan American Boeing 377 Stratocruisers and ex-USAF KC-97s operating as a unit from Lod. All of these aircraft have been extensively modified and, in some cases, virtually rebuilt by Israel Aircraft Industries, and fulfil several different functions. At the time of our visit two equipped for electronic countermeasures tasks, three modified as swing-tail freighters, three operating in the straight transport rôle, and two equipped as flight refuelling tankers were to be seen. Confirmation that *Heyl Ha'Avir* included a flight refuelling

The Sud-Aviation Vautour IIA long-range attack aircraft shares with the Skyhawk and Phantom the main punch of Heyl Ha'Avir *but is likely to be phased out of the inventory over the next year or so.*

tanker component first came on 16 July 1970 when one of these elderly Boeing transports fitted with refuelling pods outboard of the engines and with two A-4H Skyhawks attached participated in the Air Force Graduation Day display. Both the Skyhawks and the Phantoms serving with *Heyl Ha'Avir* are equipped for aerial refuelling, and although the Boeing tankers are relatively slow, they endow the strike component of the Israeli air arm with a substantial increment of range and place targets in the west and south of Egypt within Israeli reach. One of the Boeings — reportedly one of those equipped for ECM tasks — was lost, incidentally, to an Egyptian SAM on 16 September while flying 15 miles (24 km) inside Israeli-occupied Sinai.

For paratroop transportation the Nord 2501 Noratlas is used, and about 30 of these are on strength. Originally 12 of these transports were acquired from France, but these have been supplemented over the years by similar ex-*Luftwaffe* aircraft and would seem likely to soldier on in Israeli service until at least the late 'seventies, no requirement for a successor currently existing. A substantial number of Douglas C-47s, mostly ex-*Armée de l'Air* aircraft, also remain with the Israeli transport component, serving in the same unit with the Boeing 377s and KC-97s at Lod. The officers with whom we talked discounted reports that plans exist to replace the C-47 with the indigenous Arava 201, one high-ranking officer commenting: "The C-47 is irreplaceable and, in any case, is good for quite a few years yet! We have placed no order for the Arava and will certainly not order this aircraft until a genuine military prototype is available for evaluation. If this meets our requirements then we will be interested, but it is not our task to finance the development work of Israel Aircraft Industries!" Some upgrading of the logistic support capability of *Heyl Ha'Avir* is expected to result from the acquisition of several Lockheed C-130 Hercules transports, but none of these had arrived in Israel by the time that we left.

The principal enemy
At the time of our visit to *Heyl Ha'Avir* a decision concerning a new computer-controlled air defence system was believed imminent. Proposals for an optimum computerised network were submitted by Hughes Aircraft and the Systems Development Corporation after a 200-man evalua-

tion of Israeli needs had been conducted last year, and Hughes Aircraft, which has already sold some $10m worth of ALQ-71/ALQ-72 jamming pods and related electronic countermeasures equipment to Israel, has offered to build a plant in Israel for the manufacture of much of the equipment needed for the system which will include the Westinghouse TPS-43 three-dimensional radar purchased by the Israelis separately. Elbit, the Israeli computer manufacturer, is to build displays for the system under Hughes licence, and preliminary site preparations for the system have allegedly begun. Airborne avionics used by *Heyl Ha'Avir* includes the ALT-27 electronic jammer, which, built by the Amecom Division of Litton, is similar to that used by the US Navy's EKA-3C Skywarrior and the USAF's EB-66E Destroyer ECM aircraft.

Weapons that may be added to Israel's armoury include the Shrike anti-radar missile employed so effectively by the USAF in Vietnam, and the Walleye television-guided glide bomb, but as was stressed on more than one occasion during our visit, Israel's pocket is not bottomless. With a population of barely more than three millions, her 1971-2 defence budget of 5,193 million Israeli pounds ($1,483,700,000) represents almost $500 per capita, or something of the order of 27 per cent of the gross national product, by far the largest percentage of any nation in the world!

Headed up by 46-year-old Major General Mordecai "Motke" Hod, who assumed command of *Heyl Ha'Avir* shortly before the six-day war of 1967, the Israeli air arm today possesses something like treble the striking power that it had five years ago, although, numerically, its size has changed little, but there is little danger of complacency. As one senior officer put it: "Although the Arabs undoubtedly learned much from the six-day war, there is little evidence that the capability of their pilots has improved to any great extent. Of course, Egypt's programme of providing bomb-proof shelters for the bulk of her combat aircraft will render them more difficult for us to get at on the ground, but it makes them no more effective in the air. No, our *principal* enemy is not the aircraft that we are likely to encounter in combat but the many batteries of radar-controlled anti-aircraft guns and anti-aircraft missiles that now oppose us. But past experience has taught us that we can get through anything if we use the right tactics." □

One of the "straight transport" versions of the Boeing Stratocruiser that currently provide Heyl Ha'Avir *with its principal heavy logistic support capability. One of these elderly Boeings flying over the Sinai was knocked down by an Egyptian SAM in September.*

Il-28

a Soviet "Canberra"

THROUGHOUT the 'fifties, two aircraft, the MiG-15 and the Il-28, symbolised more than any others the amazingly rapid postwar advance of Soviet combat aeroplane design and manufacturing technology. But they were not only symbolic of the speed with which the Soviet aircraft industry was drawing abreast of western standards; they had become instruments of Soviet foreign policy. Whatever the country towards which Soviet aid was directed, the MiG-15 and Il-28 for its air force followed as a natural sequence of events. Thus, these first-generation turbojet-powered warplanes had attained quite remarkable ubiquity, both serving with a score or more air arms.

It is in the nature of things that a fighter loses its currency more rapidly than does the contemporary bomber, and when the 'sixties dawned the MiG-15 had lost most of its significance, both militarily and politically. The Il-28, on the other hand, although obsolescent by the standards of the day, still possessed some importance as late as 1966 when the mere presence of some bombers of this type on the island of Hainan, in the Gulf of Tonking, within striking distance of the South Vietnamese capital of Saigon, evoked considerable concern. Four years earlier, in December 1962, the Il-28 had gained the distinction of becoming probably the only war-plane ever to make the world's headlines as *deck cargo* when crated bombers of this type were shipped out of Cuba aboard the freighters *Krasnograd*, *Kasimov* and *Okhotsk*.

For many years the Il-28 equipped the entire light tactical bombing element of the Soviet Air Forces, the *Voenno-vozdushniye Sily* (V-VS), as well as many of their tactical reconnaissance units, providing the backbone of the so-called Frontal Aviation, or *Frontovaya Aviatsiya* (FA), until supplanted in the mid 'sixties by such types as the Yak-28 (Brewer). It equipped a substantial proportion of the shore-based anti-shipping and torpedo-bomber units of the Naval Air Force, the *Aviatsiya Voenno-morskovo Flota* (AV-MF), and it remains in service today with the air forces of nearly a dozen nations, 22 years after joining the ranks of the V-VS. This extraordinary longevity results not from inspired basic design, as in the Canberra with which the Il-28 is so frequently compared, but from a thoroughly sound if un-ambitious approach, using conventional engineering and state-of-the-art aerodynamics, the end product being a competent and versatile warplane with a useful potency.

Flown in prototype form nine months before its British equivalent, the English Electric Canberra, the Il-28 was destined to be referred to frequently as the "Soviet Canberra", but the Canberra was, in fact, a somewhat heavier warplane capable of lifting larger offensive loads, and placing greater emphasis on aerodynamic cleanliness and low wing loading which combined to produce a faster, more manœuvrable and higher-flying aircraft. The only true comparisons which could be drawn between British and Soviet bombers concerned time scale and operational ubiquity.

The Il-28 was actually conceived in 1947, more than a year later than the Canberra, in the design bureau of Sergei Vladimirovich Ilyushin whose name was primarily associated with heavily-armoured assault aircraft, the *shturmoviki*. Ilyushin's team had already designed a turbojet-driven bomber, the Il-22, which, when the Il-28 began its journey across the drawing boards, had still to make its first flight. The Il-22 was a shoulder-wing monoplane powered by four TR-1 axial-flow turbojets designed by Arkhip Lyulka and rated at 2,866 lb (1 300 kg) thrust, but predicted performance was unspectacular as was to be confirmed during the sole prototype's relatively short test programme which was to commence on 24 July 1947, and meanwhile appreciably higher performances were being demanded.

The principal obstacle in the path of attaining such performances as were now mandatory was lack of sufficiently powerful and reliable turbojets. As wartime amic-ability between eastern and western allies had translated to postwar frigidity, the Soviet Union had made intensive efforts to assimilate its heritage of German data on advanced design and manufacturing techniques, but by mid-1946 it had become obvious that despite the German legacy power plant technology was lagging far behind that of the West. Work on indigenous turbojets held out little immediate prospect of resulting in sufficiently powerful engines to produce the aircraft performances now requested, and the situation had given rise to grave concern until, at a *Min-aviaprom* meeting at which Iosif Stalin castigated senior members of the Ministry and power plant design bureau leaders for their failure to produce turbojets of sufficient power to meet the demands of the airframe designers, it was suggested that an attempt be made to purchase suitable engines from the UK.

Stalin is on record as having replied to this suggestion:

Although Finland's air arm, Suomen Ilmavoimat, never acquired the Ilyushin Il-28 for operational tasks, two examples of this aircraft were obtained for the target-towing rôle, these being flown by the Transport Squadron, or Kuljetus laivue, at Utti, which also possesses a target-towing commitment, and one of these is illustrated above.

"What fool would sell *us* his secrets!" He found it difficult to believe that Rolls-Royce's Derwent and Nene were no longer considered to *be* secret, but authorised immediate negotiation for their purchase, and within six months virtually every major airframe design bureau in the Soviet Union had initiated design studies for combat aircraft utilising the British power plants, including that of Sergei Ilyushin which had begun the design of a light tactical bomber powered by two Nenes to meet the latest V-VS specification.

The Ilyushin bureau found itself in competition with that of the veteran bomber designer Andrei N Tupolev which was also endeavouring to meet the V-VS specification with a twin-Nene bomber, while the bureau of Pavel Sukhoi attempted to obtain the same results with a bomber which, designated Su-10, employed an uprated version of Arkhip Lyulka's first production turbojet, the TR-1A rated at 3,307 lb (1 500 kg) thrust, four power plants of this type being accommodated in two wing nacelles, each housing one TR-1A superimposed above another.

Proven features

Although the trend in the West was to omit tail defensive armament from jet bombers, relying on their speed to evade interception, the V-VS requirement stipulated, apart from the broad performance parameters and alternative war-loads, the provision of both rear defence and a fixed forward-firing armament. The Ilyushin aircraft, which was subsequently to receive the designation Il-28, was a straightforward design using proven aerodynamics, its only distinctive feature being the combination of straight wing and swept tail surfaces, the relatively thin unswept wing which was of TsAGI SR-5S profile offering both long range and moderately high cruising speed while avoiding the aerodynamic development problems such as tip-stalling and "Dutch roll" at that time associated with swept wings, and the swept tail surfaces ensuring the retention of longitudinal and directional control at the design diving speed. The wing, the structure of which was carried straight through from tip to tip, was set high on the fuselage to avoid interfering with the weapons bay, and the engines were underslung so that the thrust line passed through the CG and thus avoided trim changes with power settings. The main undercarriage units, which had a track of 24 ft 3⅓ in (7,40 m), were attached to the engine nacelle structures, their legs rotating through 90 deg during retraction in order to lie flat beneath the jetpipes.

The engines were cantilevered well forward from the wing in order to keep the maximum depth of the engines ahead of the mainwheels, acceptable aircraft balance being obtained by positioning the radio, batteries, air-conditioning units, and other equipment in the extreme rear fuselage to supplement the weight of the Il-K6 tail turret with its twin 23-mm NR-23 cannon which had been introduced in the design to provide the stipulated aft defence. Fuel capacity for a total of 1,738 Imp gal (7 900 l) was provided by three forward fuselage tanks, two aft fuselage tanks, integral wing tanks, and auxiliary wingtip tanks. Provision was made for the installation of a single vertical AFA 33/20, 33/75-50 or 33/100 camera beneath the rearmost tank in the forward fuselage, and two fixed forward-firing 23-mm NR-23 cannon with 85 rpg were mounted on either side of the nosewheel bay.

Accommodation was provided for three crew members. The pilot, who was provided with a gyro gunsight for use with the fixed cannon, was well-positioned for all-round

The Indonesian air arm, Angkatan Udara Republik Indonesia, received some 20 Il-28s from the Soviet Union during the Sukarno régime, but since the establishment of an anti-Communist western-orientated government in 1966, no spares for the Il-28s have been forthcoming, and these aircraft, which were operated by No 21 Squadron of the Operations Command, have been grounded. The general arrangement tone drawing on the opposite page depicts an Il-28 of the Air Force of the Chinese People's Republic.

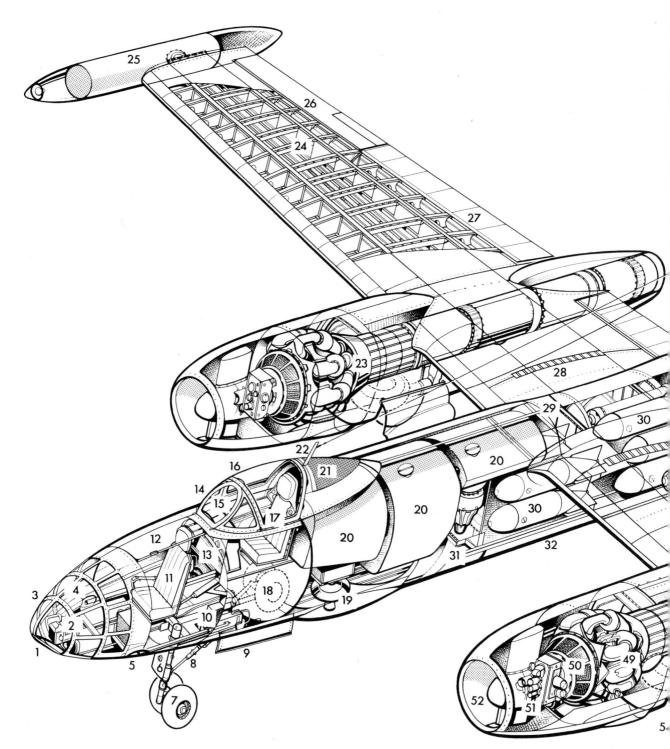

Ilyushin Il-28 cutaway key

1 Optical aiming flat
2 Bombsight
3 Nose glazing
4 Starboard 23-mm NR-23 cannon
5 Port cannon muzzle
6 Nosewheel flap with landing light
7 Twin-nosewheel unit
8 Nosewheel retraction jack
9 Nosewheel door
10 Port 23-mm NR-23 cannon
11 Navigator/bombardier's ejector seat
12 Jettisonable entry hatch
13 Instrument console
14 Flat windscreen
15 Gyro gunsight
16 Hinged canopy

17 Pilot's ejector seat
18 Nosewheel well
19 Ground-mapping radar
20 Three forward-fuselage fuel tanks
21 Dielectric panel
22 HF aerial mast
23 Starboard Klimov VK-1 turbojet
24 Two-spar wing construction
25 Starboard auxiliary tiptank
26 Starboard aileron
27 Starboard flap
28 Wing centre section/outer panel attachment point
29 Wing carry-through centre section
30 Weapons bay
31 Ventral camera port
32 Weapons-bay doors
33 JATO rocket

34 Two aft-fuselage fuel tanks
35 Aerial
36 Fixed-incidence tailplane
37 Elevator
38 Tailfin construction
39 De-icing air vents
40 Rudder
41 Rudder trim tab
42 Radio operator/gunner's compartment
43 Twin 23-mm NR-23 cannon
44 Radio operator/gunner's seat
45 Tail warning radar
46 Ventral entry/escape hatch
47 R/T equipment, batteries, etc
48 Tailpipe
49 Port VK-1 combustion chambers
50 Engine air intakes

51 Starter motor/gearbox drives
52 Bifurcated intake
53 Mainwheel well
54 Offset mainwheel door
55 Mainwheel
56 Mainwheel retraction mechanism
57 Wing skinning
58 Port flaps
59 Aileron trim tab
60 Port aileron
61 Port auxiliary tiptank
62 Fuel filler cap
63 Port navigation light

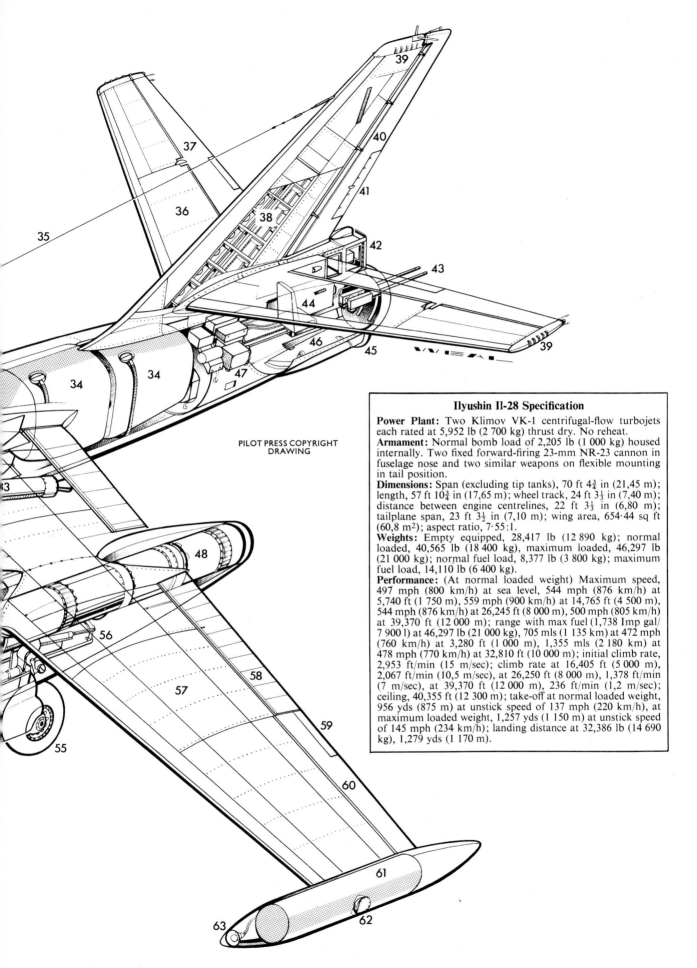

PILOT PRESS COPYRIGHT
DRAWING

Ilyushin Il-28 Specification

Power Plant: Two Klimov VK-1 centrifugal-flow turbojets each rated at 5,952 lb (2 700 kg) thrust dry. No reheat.

Armament: Normal bomb load of 2,205 lb (1 000 kg) housed internally. Two fixed forward-firing 23-mm NR-23 cannon in fuselage nose and two similar weapons on flexible mounting in tail position.

Dimensions: Span (excluding tip tanks), 70 ft 4¾ in (21,45 m); length, 57 ft 10¾ in (17,65 m); wheel track, 24 ft 3⅓ in (7,40 m); distance between engine centrelines, 22 ft 3⅓ in (6,80 m); tailplane span, 23 ft 3½ in (7,10 m); wing area, 654·44 sq ft (60,8 m²); aspect ratio, 7·55:1.

Weights: Empty equipped, 28,417 lb (12 890 kg); normal loaded, 40,565 lb (18 400 kg), maximum loaded, 46,297 lb (21 000 kg); normal fuel load, 8,377 lb (3 800 kg); maximum fuel load, 14,110 lb (6 400 kg).

Performance: (At normal loaded weight) Maximum speed, 497 mph (800 km/h) at sea level, 544 mph (876 km/h) at 5,740 ft (1 750 m), 559 mph (900 km/h) at 14,765 ft (4 500 m), 544 mph (876 km/h) at 26,245 ft (8 000 m), 500 mph (805 km/h) at 39,370 ft (12 000 m); range with max fuel (1,738 Imp gal/7 900 l) at 46,297 lb (21 000 kg), 705 mls (1 135 km) at 472 mph (760 km/h) at 3,280 ft (1 000 m), 1,355 mls (2 180 km) at 478 mph (770 km/h) at 32,810 ft (10 000 m); initial climb rate, 2,953 ft/min (15 m/sec); climb rate at 16,405 ft (5 000 m), 2,067 ft/min (10,5 m/sec), at 26,250 ft (8 000 m), 1,378 ft/min (7 m/sec), at 39,370 ft (12 000 m), 236 ft/min (1,2 m/sec); ceiling, 40,355 ft (12 300 m); take-off at normal loaded weight, 956 yds (875 m) at unstick speed of 137 mph (220 km/h), at maximum loaded weight, 1,257 yds (1 150 m) at unstick speed of 145 mph (234 km/h); landing distance at 32,386 lb (14 690 kg), 1,279 yds (1 170 m).

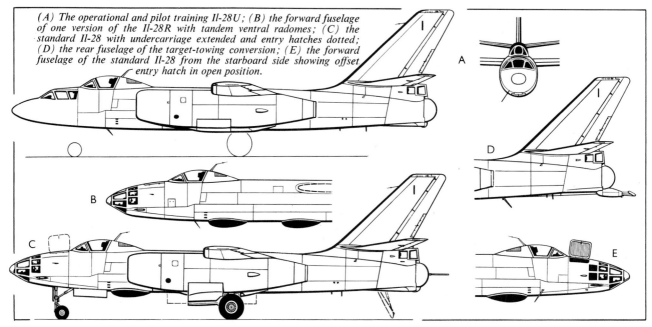

(A) The operational and pilot training Il-28U; (B) the forward fuselage of one version of the Il-28R with tandem ventral radomes; (C) the standard Il-28 with undercarriage extended and entry hatches dotted; (D) the rear fuselage of the target-towing conversion; (E) the forward fuselage of the standard Il-28 from the starboard side showing offset entry hatch in open position.

view beneath a starboard-hinging jettisonable canopy; the navigator/bombardier was positioned forward, below and slightly to port of the pilot, access to his compartment being provided by a portside-hinged heavily-padded metal hatch, and the radio-operator, who also served as tail gunner, was seated in the extreme rear fuselage. Both pilot and navigator/bombardier were provided with ejection seats, the radio-operator escaping through a trap in the floor of his compartment. Provision was made for the installation of a PSB-N ground-mapping radar for navigation and blind strikes, but the primary bomb sight was a modification of the M-9 Norden sight acquired under lease-lend during WW II.

The competitive bomber design from the Tupolev bureau, which was primarily the work of A A Arkhangel'ski, was a somewhat larger, heavier and more complex aircraft embodying remotely-controlled gun barbettes for aft defence. Initiated under the design bureau designation of Tu-72, it carried a crew of four, and before metal was to be cut on this aircraft, latent doubts on the part of Arkhangel'ski regarding the reliability of its intended power plants were to lead to the decision to introduce a third turbojet in the extreme rear fuselage. an intake for this engine being provided above the fuselage as an extension of the vertical tail surfaces.

Iosif Stalin constantly urged speed on the part of the Tupolev and Ilyushin bureaux, allocating the highest priority to the tactical jet bomber programme, and in the spring of 1947 personally ordered work to commence on prototypes of both contenders for production orders. The first Tupolev prototype, the Tu-73, was flown for the first time on 29 October 1947, had completed its factory test programme on 14 June 1948, and had been cleared to commence its State test programme before the first prototype of the Il-28 had been rolled out of the experimental assembly shop. A second prototype of the Tupolev bomber, the Tu-78, had also joined the flight test programme before, on 8 August 1948, the Il-28 was flown for the first time with the Ilyushin bureau's chief test pilot, Vladimir K Kokkinaki, at the controls, the third competitive tactical bomber design, the Sukhoi Su-10, having meanwhile been abandoned.

Quantity production

Powered by the initial Soviet production derivative of the Rolls-Royce Nene, the RD-45, the prototype Il-28 proved to possess pleasant flying characteristics, was reasonably

manœuvrable, and promised to provide a stable bombing platform. Its normal internal bomb load was 2,205 lb (1 000 kg) which could be carried over a maximum range of 1,367 miles (2 200 km) cruising at 466 mph (750 km/h) at 32,810 ft (10 000 m), and its field performance was acceptable. It met the V-VS specification in virtually every respect, but so too did the competing Tu-78 which offered substantially greater range and appreciably more internal weapons capacity. However, the tail-mounted auxiliary turbojet and barbette-mounted defensive armament of the latter were not favoured, and the Tupolev offering was less agile than the Il-28, and possessed an inferior field performance.

To settle the matter of which bomber should be ordered into production for the V-VS, the C-in-C, Marshal of Aviation Konstantin A Vershinin, ordered the two bombers to be evaluated side by side by three different crews, all crews flying both aircraft types. All three crews favoured the Il-28 which was accordingly ordered into immediate large-scale production. The Tu-78, rejected by the V-VS but favoured by the AV-MF owing to its greater range and weapon-carrying capability, was ordered into limited production for the re-equipment of the shore-based naval anti-shipping, torpedo-bomber and reconnaissance units.

When the order for full production of the Il-28 was given, three prototypes had been completed, and Iosif Stalin was most insistent that 25 of the new bombers be available to participate in the 1950 May Day fly-past, barely more than 12 months ahead. Several factories were assigned to the Il-28 manufacturing programme, and the Ilyushin bureau devoted maximum effort to productionising the bomber, evolving a system whereby the wings, tail surfaces and fuselage were split down their horizontal axes and manufactured in halves, an admittedly space-consuming arrangement which effectively reduced actual unit manufacturing time at a cost of 1·5 per cent increase in structural weight. That this system achieved the desired result is evidenced by the fact that the 1950 May Day celebrations included the appearance of a formation of 25 Il-28 bombers led by Lt Col A A Anpilov. Later, this form of manufacture was refined, and some weight was saved by the adoption of large mechanically-milled panels.

By the late summer of 1950, when the first FA *polki* had relinquished their piston-engined Tu-2s and were engaged in the not entirely trouble-free task of working up on the Il-28, the efforts of the Ilyushin bureau and the TsAGI (*Tsentralny Aerogidrodinamichesky Institut* — Central Aero and Hydrodynamic Institute) had resulted in a number of minor aerodynamic and major structural refinements, and the initial Soviet version of the Nene, the RD-45, had given place to Klimov's improved development, the VK-1 of 5,952 lb (2 700 kg) thrust. By the standards of the day, the Il-28 possessed a very respectable performance if somewhat short on internal weapon capacity. At its normal loaded weight of 40,565 lb (18 400 kg), which included 8,377 lb (3 800 kg) of fuel and a 2,205-lb (1 000-kg) bomb load, it took-off within 956 yards (875 m), unsticking at 137 mph (220 km/h), had an initial climb rate of 2,953 ft/min (15 m/sec), and attained a maximum speed at military power (11,560 rpm) of 559 mph (900 km/h) at 14,765 ft (4 500 m), cruising at 472 mph (760 km/h) at this altitude. With maximum fuel of 1,738 Imp gal (7 900 l) and the same bomb load, the Il-28 attained its maximum range of 1,355 miles (2 180 km) at 478 mph (770 km/h) and 32,810 ft (10 000 m). This performance was much in accordance with Western intelligence assessments of the period based on the 12 per cent thickness/chord ratio of the SR-5S wing profile which offered a critical Mach in excess of 0·8 under zero lift conditions.

With large-scale production spread between a number of factories, several variants of the basic design began to appear in the early 'fifties, the first of these being the Il-28U operational and pilot training version in which the nose glazing was deleted, together with the ground-mapping radar housing, and a second cockpit with a full set of controls for the pupil pilot was inserted ahead and below the standard cockpit. Most Il-28-equipped *polki* were issued with two or three examples of this version for pilot checks, etc. The Il-28U was followed by the Il-28R for tactical reconnaissance, provision being made for the insertion of alternative packs containing optical or electronic sensors in the weapons bay, and some examples having an additional radome under the centre fuselage. Simultaneously, a torpedo-bombing model for the AV-MF, the Il-28T, was

introduced to supplement the larger Tu-14T, this being able to carry two short torpedoes and having modified avionics.

At a relatively early stage in the Il-28's service career, this light tactical bomber began to appear in the armouries of China, Czechoslovakia, and Poland, the first-mentioned receiving more than 500 examples from the Soviet Union and subsequently manufacturing the Il-28 under licence, and as the years passed the number of national insignia carried by the Ilyushin light tactical bomber steadily increased until more than a score of air forces listed the Il-28 in their inventories. Early in 1956, *Aeroflot* acquired several demilitarised examples of the Il-28 to which civil registrations were applied, and which, bearing for some obscure reason the designation Il-20, were used to carry newspaper matrices from Moscow to Sverdlovsk, Novosibirsk, and elsewhere to enable newspapers to be published in those places simultaneously with the publication of the Moscow editions. As the Il-28 gave place in the first-line FA units to later types many were converted for various second-line tasks, including target-towing and meteorological flights.

Today, 21 years after the Il-28 made its public début in the May Day 1950 fly-past, this first-generation Ilyushin bomber remains in service with a number of air arms. Rendered obsolete in the context of major conflagrations by its speed and ceiling, it is still a useful weapon for light shipping attack and for "policing" operations, and it is likely to be several years before the last air arm to include the Il-28 in its inventory finally phases out this reliable and generally well-liked warplane. Nobody in East or West is likely to allege that the Il-28 was technically an outstanding aeroplane for its day, but if straightforward of design, it has proved itself worthy of inclusion in that most useful breed of military aircraft which combine freedom from aerodynamic and structural problems with a useful performance, and, furthermore, belong to that increasingly exclusive club of commercially successful military aircraft. □

Still serving in some numbers with the AV-MF, the Il-28T is illustrated by this recent photograph.

IN PRINT

"Early Aviation at Farnborough"
by Percy B Walker
Macdonald & Co, London, £6.50
284 pp, 7¼ in by 9¾ in, illustrated
SUBTITLED "Balloons, Kites and Airships", this is the first volume in an intended larger work devoted to a history of the Royal Aircraft Establishment. Its author spent much of his working life at the RAE, finally as head of the Aircraft Structures Department and Special Consultant to the Director of the RAE, and is therefore specially qualified both to research his subject and to interpret the findings of his research.

This volume deals very fully with the period from 1899 to the beginning of 1910. During this time, the establishment at Farnborough was known as the Balloon Factory, and its activities in building balloons and, later, airships are fully described, as well as its involvement with S F Cody and his kites. But this is not only a history of the Factory itself; rather, it is an account of British military aviation before the aeroplane appeared on the scene. Although the subject matter of the book may make it of somewhat limited interest, Dr Walker has made a valuable and lucid contribution to history with this account and further volumes in the series are awaited with interest.

"Hawker Aircraft Since 1920"
by Francis K Mason
Putnam & Co, Ltd, London, £4.20
496 pp, 5¼ in by 8½ in, illustrated
REVISED second edition of one of the first volumes in Putnam's series of British aircraft company histories. The new material deals principally with the Harrier and with the more recent Hunter exports. There is a useful breakdown of the latter, listing mark numbers from 50 to 79.

"Guide to Approach and Landing
at Heathrow Airport"
Published by BEA and IAL, 60p
JOINTLY prepared by BEA and International Aeradio, this guide will be of primary interest to visitors to London Heathrow, and to those living in the vicinity of the airport. The wallet contains three items: a colourful map of the airport with, on the reverse side, a layman's guide to approach procedures; a description of the aircraft stacks around Heathrow; and a copy of the Aerad Flight Guide Chart for Heathrow.

Copies of the Guide can be obtained from Mr V Windett, International Aeradio Ltd, Hayes Road, Southall, Middlesex, price 60p including postage.

"The Debden Eagles"
by Garry L Fry
Available outside American Continent from W E Hersant Ltd, London, £3.25 post paid
102 pp, 8½ in by 11 in, illustrated
AMONG the Fighter Groups of the USAAF that served in the UK during World War II, the 4th FG was of special interest, since it was formed by absorbing the RAF's three

Tiger Squadrons — units manned by American volunteer pilots. From August 1942 to May 1945, the 4th flew from Debden, claiming a total of 1,016 enemy aircraft destroyed.

This volume provides a history of the Group from the time it was activated in the USAAF to the time of its return to the USA. Primarily this is achieved in pictures, which amply capture the atmosphere of a fighter base at war. Salient points in the history are briefly enumerated in a written introduction.

"Igor Sikorsky"
by Frank J Delear
Dodd, Mead & Co, New York, $4.95
272 pp, 6 in by 9 in, illustrated
THE subtitle of this biography is "His Three Careers in Aviation", drawing attention to the three distinct phases of aircraft design by one of aviation's greatest surviving pioneers. The first phase was his work in Russia from 1910 to 1918, when, after a false start with an attempted helicopter, he designed and built a series of biplanes culminating in the *Ilya Mouromets* four-engined bombers, largest aircraft of their day. Following emigration to America, Sikorsky's second career encompassed the production of a well-known series of flying-

Three stages in the evolution of the Sikorsky VS-300, the first practical helicopter developed by Igor Sikorsky. The pictures show, top to bottom, the VS-300 on its first flight on 14 September 1939; after the addition of outrigged tail rotors in 1940 and with a single horizontal tail rotor in 1941.

boats as well as some less successful land-planes. Finally, in 1939 he launched into the third and most important phase of his design career with the development of one of the first truly successful helicopters. Since that time, the company that bears Sikorsky's name has produced only rotary-winged aircraft, and the Sikorsky formula is still today the basis for virtually all other successful helicopter designs.

The three careers are described in this volume with obvious affection by Frank Delear, who has been Sikorsky Aircraft's public relations manager for most of the helicopter period. Primarily a biography of the man, the book also contains plenty of information about his aircraft and many of the illustrations are drawn from Igor Sikorsky's personal collection and other private sources.

"The Black Eagle"
by John Peer Nugent
Steain and Day, New York, $6.95
191 pp, 5½ in by 8¼ in
THIS is a "story of" type biography of Hubert Fauntleroy Julian, the "Black Eagle" — "of Harlem" during the 1920s and "of Abyssinia" for an instant in the 1930s, and a sagacious arms merchant of the 1950s and 1960s.

Born into a rather substantial family in Trinidad in 1897, Julian was sent to England as a young man, emigrated to Canada where he learned to fly in 1919, and then went to the United States, which has been his home base since. The book describes his barnstorming career around New York's Harlem and his brief command of the Abyssinian Air Force (three aeroplanes), in 1930, which was cut short by his offending Haile Selassie; his flying for the religious mystic Father Divine and his abortive return to Abyssinia after the Italian invasion of 1935. He was also a mercenary in the Finnish War of 1940. But it is clear from the text that Julian was less an airman than a con man, although a very, very colourful (no pun intended) one.

During World War II, however, he was too old for the USAAF's all-Negro 99th Fighter Squadron and he served as a private in the Army. Real war, in any case, was not Julian's cup of tea. But after the war he flourished again, selling guns to Guatemala, Batista's Cuba, and the Dominican Republic; and he tried to procure P-51s for Castro's Cuba and B-26s for Katanga, although without success.

This is a lively story well told, although not without its errors. The first Negro aviator was not Herbert Julian. As far as can be determined it was the American-born Eugene Bullard (1894-1961), who flew with the *Escadrille Lafayette* in 1916, and the reader is advised to bring a few pinches of salt with him to be sprinkled on various other passages. Nevertheless, after such a long and noisy career scattered through a half-century of newspaper columns, it is nice to have Julian's stormy life finally done up tidily in a hard binding. — RKS

THE NOT-SO-ANCIENT MARINER

D ELIVERY of the 2,500th McDonnell Douglas A-4 Skyhawk recently places this diminutive warplane firmly in the front rank of combat aircraft produced since World War II, while confirmation of planned production at least until 1974 will ensure that the Heinemann-designed light tactical strike aircraft has one of the longest unbroken production runs of any aircraft in its class. Skyhawks are currently in service with four air arms in addition to the US Navy and Marine Corps, and have built up an impressive record of successful operational deployment in Vietnam and Israel.

At a time when multi-rôle aircraft win easy political acclaim because of the real or imagined reductions in budget expenditure they make possible, the Skyhawk is an interesting and highly effective example of a diametrically different design approach, in which a single operational rôle — close support attack — was pursued without compromise. Consequently, the A-4 has been operated in virtually one rôle throughout its 15-year service life, yet, as the statistics in the opening paragraph indicate, it has been produced in far greater quantities than many aircraft types that have been evolved in a variety of different rôle variants.

The design of the Skyhawk was initiated during 1952 by a project team headed by Ed Heinemann (then chief engineer of the Douglas Aircraft plant at El Segundo, California), as an outgrowth of earlier studies directed towards a lightweight fighter. The period was one when many designers were growing alarmed by the rapid growth in airframe weights (attributable both to structure and to equipment). Although increases in jet engine thrusts more than matched the high gross weights that resulted from this trend, airframe performance in terms of take-off and landing speeds and manœuvrability inevitably suffered. The philosophy of lightweight fighter designs propounded by the late W E W Petter in producing the original Folland Gnat was echoed by Douglas proposals to the US Navy early in 1952.

The Navy suggested, however, that Heinemann's team should apply these design principles to an attack aircraft rather than a fighter, experience in the Korean War having shown the need for an aircraft to replace the piston-engined A-1 Skyraider whilst retaining the same strike capability. An opportunity to fund the project was presented to the Navy later in 1952, when the Douglas A2D Skyshark was cancelled through difficulties with the paired Allison XT40-A-6 turboprop engine and contraprop.

The objective set for Douglas Aircraft was to produce a lightweight, high performance carrier-based day attack aeroplane capable of dive bombing, interdiction and close support missions while carrying conventional or nuclear weapons and operating against sea or land targets with or without fighter escort in a hostile environment. Target gross weight set for the new design was 15,000 lb (6 800 kg), only half the weight that the Navy had previously assumed would be reached by a jet replacement for the A-1.

Douglas drew upon the series of earlier simplification studies to design the Skyhawk, which evolved as the smallest combat jet aircraft operated by the USN. Somewhat unconventional was the use of a delta wing planform together with an orthodox tail unit, but structurally the design was simple and straightforward. By rigorously confining the overall dimensions to the maximum acceptable by the flight deck lifts of modern US carriers, Heinemann did away with the

Latest production version of the McDonnell Douglas Skyhawk is the A-4M, shown (above) in the markings of the first Marine Corps unit to fly the type (VMA-324) and (below) awaiting delivery. Among the new features are the square-tipped fin, angled refuelling probe, drag 'chute and deepened cockpit canopy.

A McDonnell Douglas TA-4F, two-seat version of the Skyhawk, in the colourful markings of Commander Readiness, Attack Carrier Air Wing Twelve, home-based at NAS North Island, California.

need for wing folding (thus saving weight); another example of simplification was provided by retracting the under-carriage legs forward, so that in the event of a system failure, they would fall free and lock under airstream pressure, making an emergency system unnecessary.

This approach to the design problems resulted in an aircraft with an empty equipped weight of only about 8,000 lb (3 630 kg), less than half the maximum gross weight which, in the initial production version, reached 20,000 lb (9 070 kg). As would be expected, cost as well as weight benefited from the designers' efforts to keep the new attack aircraft simple. Today, the latest version of the Skyhawk, the A-4M, is reported to be available fully equipped for less than $1,500,000 (£625,000) — a highly competitive figure.

Marine Corps choice

The low cost of the A-4M was one of the factors leading to selection of the A-4M over the Vought A-7 Corsair II by the US Marine Corps in 1969. Production of single-seat Skyhawks for the US Navy ended in 1968, since which time it has been USN policy to phase the A-7 into service as a replacement for the A-4. The Marine Corps, however, reversed its original decision to join the Navy in the A-7 programme and decided, instead, to retain the A-4 in its inventory, replacing earlier models with the A-4M which was developed specially to meet Marine Corps requirements.

Use of an uprated engine, plus a number of other new features, makes the A-4M the most potent of all the Skyhawk variants and one having a considerable performance advantage over the A-4F, the final Navy version. By comparison with the latter, the A-4M differs in having a Pratt & Whitney J52-P-408A turbojet with a rated thrust of 11,200 lb (5 080 kg), some 20 per cent more than the previous model. This gives it better manœuvrability, rate of climb, acceleration and take-off performance, for a small reduction in range.

Specifically, the improvement in manœuvrability at Mach 0·75 at sea level, with external stores, is more than 100 per cent, from 1·5g to 3·2g; the improvement in rate of climb is 50 per cent, from 5,620 ft/min (2,9 m/sec) to 8,440 ft/min (4,3 m/sec), and the increase in longitudinal acceleration during dive bombing recovery is 23 per cent. Maximum speed at sea level, with 4,000 lb (1 814 kg) external load, is increased by 45 knots (83,4 km/h) to 560 knots (1 038 km/h) and the take-off distance on a standard day is reduced from 3,720 ft (1 134 m) to 2,700 ft (823 m), with even more spectacular reductions at higher temperatures.

The J52-P-408A engine in the A-4M is fitted with smokeless burner cans, similar to those now being introduced on commercial jet engines, and this feature virtually eliminates the tell-tale smoke trail left by earlier A-4s. The new engine requires greater mass flow during take-off, and the intake area on the A-4M has therefore been increased by about 7 per cent, by widening each inlet 2½ in (6 cm). Addition of two-position inlet guide vanes and a new constant speed drive/generator in the nose bullet fairing makes the P-408A engine 1½ in (3,7 cm) longer than the P-8A version. It also operates at higher turbine inlet temperatures.

Other changes in the A-4M are an increase in electrical generating capacity, self-contained electrical starter, doubling of the internal armament capacity to a total of 400 rounds for the two 20-mm cannon in the wing roots; installation of a ribbon-type drag 'chute in a fairing just aft of the arrester hook; increased fin area; redesigned flight refuelling probe; enlarged cockpit canopy and a new fixed gun sight. A Douglas Escapac I-C-3 rocket ejection seat with zero-zero capability is fitted, as in earlier Skyhawks.

The canopy redesign raises and widens the one-piece upward-hinged main canopy and introduces a new rectangular windshield; as a result, the angle of vision downwards ahead of the aircraft is increased from 16° to 18°, and to each side, from 46° to 55°. The refuelling probe is now angled out to the starboard side of the aircraft, instead of extending forwards, to prevent interference between the probe and the nose-mounted radar, which may be of a more advanced type in due course.

Both the enlarged fin and the drag 'chute were first developed for export models of the Skyhawk (described later) and have now been adopted for a US service version for the first time. The fin change, to contain an antenna, results in a marked angular appearance for the fin tip and provides a ready means of distinguishing the A-4M from the A-4F.

Advanced avionics and nav/attack equipment has been planned for the A-4M, although development of some of these items has lagged behind aircraft production and early A-4Ms were flown with the same standard of avionics as fitted in the A-4F. Basic equipment includes a Bendix automatic flight control system; ARC-52 UHF radio transceiver; ARA-50 UHF direction finder; APX-64 IFF equipment; ARC-115 VHF/FM radio transceiver; APG-53A terrain clearance radar; ASN-41/WOS-600/APN-153 (V) Doppler/inertial navigation system and Elliott Type 546 head-up display. Space provision is made for ARR-69 auxiliary UHF radio receiver, ARN-52 TACAN and APN-141 radar altimeter. Classified ECM equipment is carried, with antennae on the tail and in a small chin fairing on the nose.

A laser target designator system has been under development for the A-4M at the Naval Weapons Center, China

Lake, California. This system illuminates selected targets with a laser beam to provide a suitable signal source for specially-equipped homing bombs or rockets.

The A-4M has five underwing stores stations, including one on the fuselage centreline, and can carry and deliver all air-to-ground weapons at present in the Naval inventory, as well as carrying all types of modern tactical armament. Capacities of the external stations are 3,575 lb (1 620 kg) on the centreline, 2,240 lb (1 015 kg) on the inboard wing position and 570 lb (258 kg) outboard. Standard built-in armament comprises two Mk12 20-mm guns with a rate of fire of 1,000 rounds/min. The guns are located in the wing just beneath the air intakes and 200 rounds are carried for each gun.

The centreline and two inboard wing pylons can be used to supplement the A-4M's internal fuel capacity, which comprises 240 US gal (905 l) in a self-sealing fuselage tank and 560 US gal (2 364 l) in the integral wing tanks. The fuselage station can carry a 150-, 300- or 400-US gal (568-, 1 137- or 1 514-l) tank, while the two wing stations can accommodate up to 300 US gal (1 137 l) each. Thus, the

maximum possible fuel load is 1,800 US gal (6 812 l) which, with standard reserves, gives the A-4M a ferry range of 1,785 naut miles (3 307 km). A buddy refuelling pod can be carried on the centreline.

Flying a combat close support mission with a 4,000-lb (1 814-kg) bomb load (Mk 81 and Mk 82 bombs), the A-4M can fly a distance of 150 naut miles (278 km) and then loiter for 65 minutes at 5,000 ft (1 524 m) before returning. For a lo-lo-lo strike mission, a 2,000 lb (907 kg) bomb load can be carried just over 300 naut miles (556 km) with an allowance of 5 minutes over the target; in this case, 600 US gal (2 274 l) of external fuel is carried. For a hi-lo-lo-hi mission, the radius in this configuration increases to 450 naut miles (834 km), with the last 100 naut miles (185 km) into and out of the target flown at low level.

Basic operating weight of the A-4M is 12,280 lb (5 570 kg) with gun armament and ammunition and pilot. The normal gross weight of 24,500 lb (11 113 kg) leaves a margin of 12,220 lb (5 543 kg) for fuel and external ordinance. Maximum external load on the five pylons is 9,155 lb (4 153 kg), in which case the weight available for internal fuel is 3,065 lb

(Above) An A-4E in Israeli colours, this being one of a batch of ex-USN Skyhawks recently delivered to Israel to supplement the earlier purchase of A-4Hs. Features of the latter, shown in the tone drawing (below), include the drag 'chute and square-tipped fin. Both types have 30-mm DEFA guns in the wings and are in course of being fitted with dorsal avionics packs in Israel.

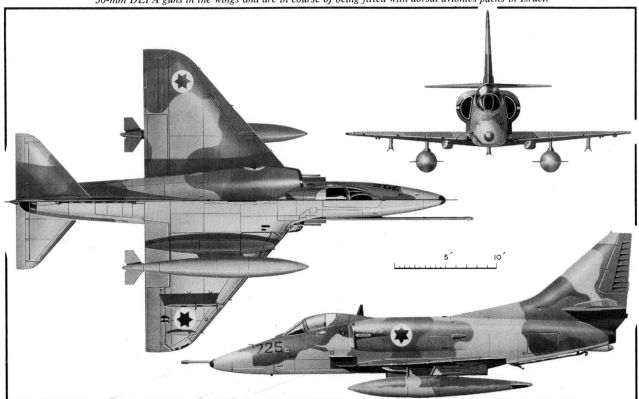

To date, Argentina has received 50 ex-USN A-4Bs, with another 20 to be delivered at the end of this year. The first batch of 25 were in natural metal finish as shown below but the second batch have a green-and-brown camouflage pattern, illustrated above on two A-4Bs demonstrating the buddy-pack refuelling facility.

(1 655 kg), equivalent to 470 US gal (1 771 l) of JP-4. Maximum internal fuel load of 800 US gal (3 028 l) weighs 5,202 lb (2 360 kg).

The A-4M was first flown from Palmdale, where the Skyhawk final assembly line is now located, on 10 April 1970. Deliveries to the Marine Corps began just over a year later, on 16 April 1971, when four A-4Ms were handed over to VMA-324 at MCAS Beaufort, South Carolina. By that time, four other A-4Ms were already engaged in BIS (Bureau of Inspection and Survey) trials at the NATC Patuxent River. The fifth operational A-4M delivery, on 20 April, was made the subject of formal ceremonies at NAF Washington marking the completion of 2,500 Skyhawks. About 100 more A-4Ms and TA-4Js are believed to be on order at present, apart from possible additional export orders, for such countries as Israel and Switzerland.

Export models

To date, about 200 Skyhawks have been exported to four nations, some of these being refurbished US Navy aircraft and the others new-built. First overseas deliveries were made in October 1966 and comprised 25 ex-USN A-4Bs refurbished for the Argentine Air Force (Fuerza Aérea Argentina) and used to equip the IV Grupo Caza-Bombardero at Villa Reynolds in the San Luis Province. These aircraft, the first of which flew at Tulsa after overhaul on 31 December 1965, were joined in 1969 by a second batch of 25 A-4Bs, distinguished from the earlier batch in that they were camouflaged rather than having plain metal finish. Twenty more A-4Bs have now been acquired by the Argentine Navy to equip the I and II Escuadrones de Ataque which provide detachments for service aboard the aircraft carrier 25 de Mayo.

In 1967, Australia became the first overseas nation to acquire new Skyhawks with delivery of eight A-4Gs and two two-seat TA-4Gs. These aircraft were similar to the US Navy A-4E and TA-4F respectively; the single-seaters were used to equip No 805 Squadron aboard HMAS Melbourne while the two-seaters serve with No 724 Squadron. A second batch, of similar quantity but comprising refurbished and re-engined US Navy aircraft brought up to the same stan-

dards, was delivered to the RAN earlier this year, the single-seaters being for service aboard HMAS Sydney.

Also in 1967, delivery began of a new version of the Skyhawk, the A-4H, to Israel. This variant was also based on the A-4E, but introduced several new features to the specific requirements of the Israel Defence Forces/Air Section (Tsvah Haganah Le Israel/Heyl Ha'Avir), including a drag 'chute installation and two 30-mm DEFA 552/553 cannon in place of the 20-mm Mk 12s, with 150 rpg. This was also the first Skyhawk version to feature the slightly larger, square-tipped fin and rudder used on the A-4M. An initial batch of 48 aircraft was delivered to this standard, but subsequent deliveries have brought the total of Israeli Skyhawks to about 100, including some two-seat TA-4Hs. Among the more recent deliveries are ex-US Navy A-4Es and some of these aircraft are now flying in Israel with the A-4F-type avionics pack. Israel is reported to be modifying all Skyhawks to this configuration, using kits obtained from America.

Fourth and most recent Skyhawk customer was the Royal New Zealand Air Force, which acquired, early in 1970, eight A-4Ks and four two-seat TA-4Ks. Based on the A-4F, the A-4K introduced new features to RNZAF specification, including a drag 'chute and changed radio. Arrester gear was retained, for use with runway arrester equipment, and these aircraft also featured the squared-off fin. The RNZAF Skyhawks equip No 75 Squadron, the TA-4Ks serving with No 14 Squadron for operational conversion duties.

The Skyhawk has been on offer to the Swiss Air Force to meet its requirement for a Venom replacement and was reported to have been rated second to the Vought A-7 in the latest round of technical evaluations. Low cost is a major factor in favour of the A-4M in this respect, but it has a less sophisticated nav/attack system than the A-7, and carries a smaller offensive load (see "Pirates Ashore", AIR ENTHUSIAST August 1971), but it has now reportedly been eliminated from the contest for reasons not connected with the Skyhawk's capabilities.

US Navy models

Although the Skyhawk no longer figures in US Navy procurement plans as a combat aircraft, the great majority of the 2,500 delivered so far have been for USN operational use and the A-4 has been responsible for a major share of all Navy combat sorties flown over Vietnam in the last ten years. As related at the beginning of this feature, the Skyhawk was originally designed to a US Navy specification, and the first contract covering the design activity was placed on 21 June 1952. Before the end of the same year, contracts were awarded covering 20 aircraft of which the first was to be designated XA4D-1 and the remainder A4D-1.

First flight of the prototype XA4D-1, powered by a 7,200 lb (3 266 kg) thrust Wright J65-W-2 engine (license-built Armstrong Siddeley Sapphire) was made on 22 June 1954, and further aircraft followed before the end of the year. The first carrier landing was made on 12 September 1955 aboard the USS *Ticonderoga*. The third Skyhawk was used in October 1955, during the course of its test programme, to set a 500-km closed circuit speed record of 695.163 mph (1 119 km/h). Deliveries to the first operational unit, VA-72, (*The Hawks*) began on 26 October 1956, this unit being already familiar with the type as it served also as the service test unit for the Fleet Indoctrination Program.

Following VA-72, which was based at NAS Quonset Point, RI, and attached to the Atlantic Fleet, VA-93 (*The Blue Blazers*) became the first operational Skyhawk squadron with the Pacific Fleet, and in January 1957, VMA-224 became the first Marine Corps unit to fly the A4D-1.

Production of the A4D-1 totalled 166 including prototypes and test models, and was sufficient to equip all Navy and Marine attack squadrons designated to fly jet aircraft. Production aircraft were powered by the 7,700 lb (3 493 kg) thrust J65-W-4 or -4B engine.

A few months after fleet introduction of the A4D-1 began, a new version, the A4D-2, made its first flight, on 26 March 1956. Powered by a 7,700 lb (3 493 kg) J65-W-16A engine, the A4D-2 featured flight refuelling capability, with a long probe extending along the starboard side of the fuselage; the so-called "tadpole" rudder which comprised a single central skin with external stiffening ribs; improved bomb delivery system and automatic dead reckoning navigation computer; a dual hydraulic system and provision for carrying Bullpup air-to-ground missiles. Structural strengthening of the fuselage and a new tailplane increased the manœuvre limit to 7g at maximum speed. Deliveries of this version began in September 1957 to the Marine unit VMA-211, with VA-12 becoming the first Navy unit to fly the type in February 1958. Production totalled 542.

Limited all-weather capability was added to the Skyhawk in its next version, the A4D-2N. The major innovations in this model included use of the AJB-3 all-attitude reference and loft-bombing system, TPQ-10 ground control blind bombing system, an angle-of-attack indicating system, APG-53 terrain clearance radar in a lengthened nose, an improved gun sight and automatic flight control system. A Douglas Escapac rocket ejection seat was fitted and other small improvements were introduced. First flight was made on 21 August 1958 and introduction into service began in March 1960 at the hands of Marine Squadron VMA-225 at Cherry Point, NC. First Navy unit was VA-192, equipped in May 1960. Production totalled 638.

The fourth major Skyhawk variant appeared as the A4D-5, flown on 12 July 1961. This type represented a complete modernisation of the Skyhawk, reflecting a shift in basic philosophy towards limited war and close support operations with emphasis placed upon the carriage of large conventional weapon loads and air-to-ground missiles. The major changes were introduction of a Pratt & Whitney J52-P-6 turbojet (a version of the civil JT8B), rated at 8,500 lb (3 856 kg), and the addition of two more underwing stores stations, making five in all. Provision was made for JATO, and a swivelling nosewheel was introduced, as well as jet blast rain removal for the windscreen and other new equipment. The bombing system was updated to AJB-3A standard, a Mk 9 bomb director was added for toss bombing, and Doppler radar, TACAN and radar altimeter were installed.

Before deliveries of the A4D-5 could begin, the Department of Defense introduced its new tri-service aircraft designation system, as a result of which the Skyhawk became the A-4 rather than A4D. Versions redesignated retrospectively were A-4A (A4D-1), A-4B (A4D-2), A-4C (A4D-2N)

and A-4E (A4D-5). The A-4D designation was skipped to avoid confusion with the earlier A4D. Deliveries of the new A-4E began in December 1962 with VA-23 (the *Black Knights*) becoming the first unit operational on the type. Production ended with 499 built.

A number of further design improvements were proposed for the basic Skyhawk, such as a rocket seat with zero-zero capability, nose-wheel steering for crosswind taxying and wing spoilers to serve primarily as lift dumpers, also with the purpose of improving the landing performance in a crosswind. All these were adopted by the USN, in the first instance, for a two-seat operational training version of the Skyhawk, ordered in 1964. To accommodate the second cockpit, the overall fuselage length was increased by 28 in (71 cm) and the modification resulted in a reduction in internal fuel capacity and therefore endurance. This apart, however, the two-seat Skyhawk, powered by the 9,300 lb (4 218 kg) J52-P-8A engine, retained the full operational and weapon capability of the single-seater.

Two prototypes were built with the designation TA-4E, first flights being made on 30 June and 2 August 1965 respectively. By the time production two-seaters began to appear, however, the US Navy had ordered a new series of single-seaters designated A-4F with most of the new features that had been adopted in the TA-4E. Consequently, the production trainer was designated TA-4F, the first example being flown in April 1966, with first delivery to VA-125 at NAS Lemoore on 19 May 1966.

When the first single-seat A-4F flew on 31 August 1966, it looked externally similar to the A-4E, the principal differences being the J52-P-8A engine, zero-zero ejection seat, nosewheel steering and wing spoilers. Production models, deliveries of which began on 20 June 1967, were readily distinguishable, however, by the large dorsal fairing extending between the cockpit and the fin and containing the aircraft's avionics equipment.

A second training version of the Skyhawk was acquired by the Navy following a decision early in 1968 to adopt the

(Above) The Royal New Zealand Air Force has eight A-4Ks and four two-seat TA-4Ks in service. The A-4Ks have most of the external features of the A-4M apart from the deepened cockpit canopy. (Below) The Royal Australian Navy's A-4G (background) and TA-4G two-seaters are externally similar to the A-4E and TA-4F respectively.

McDonnell Douglas A-4M Specification

Power Plant: One Pratt & Whitney J52-P-408A turbojet rated at 11,200 lb (5 443 kg) static thrust. No afterburner.
Weights: Empty, 10,465 lb (4 747 kg); basic operating weight (zero fuel, zero external ordnance), 12,280 lb (5 570 kg); max internal fuel weight, 5,202 lb (2 360 kg); max external load, 9,155 lb (4 153 kg); max take-off weight, 24,500 lb (11 113 kg).
Performace: Maximum speed, clean, 582 knots (1 078 km/h) at sea level, Mach 0.94 at 25,000 ft (7 620 m); take-off distance at 23,000 lb (10 433 kg) gross weight, 2,700 ft (8 230 m) at 15°C, 3,160 ft (963 m) at 32°C, 3,860 ft (1 177 m) at 40°C; rate of climb, 8,440 ft/min (4,3 m/sec) at sea level, 2,500 ft/min (1,27 m/sec) at 25,000 ft; tactical radius, close air support mission, 4,000 lb (1 814 kg) bomb load, 290 naut mls (538 km); ferry range, 1,785 naut mls (3 307 km).
Dimensions: Span, 27 ft 6 in (8,38 m); length, 40 ft 3¼ in (12,27 m); height, 15 ft 0 in (4,57 m); tailplane span, 11 ft 3½ in (3,44 m); undercarriage track, 7 ft 9½ in (2,38 m); gross wing area, 260 sq ft (24,16 m²); sweepback at quarter chord 33.2°.

The McDonnell D

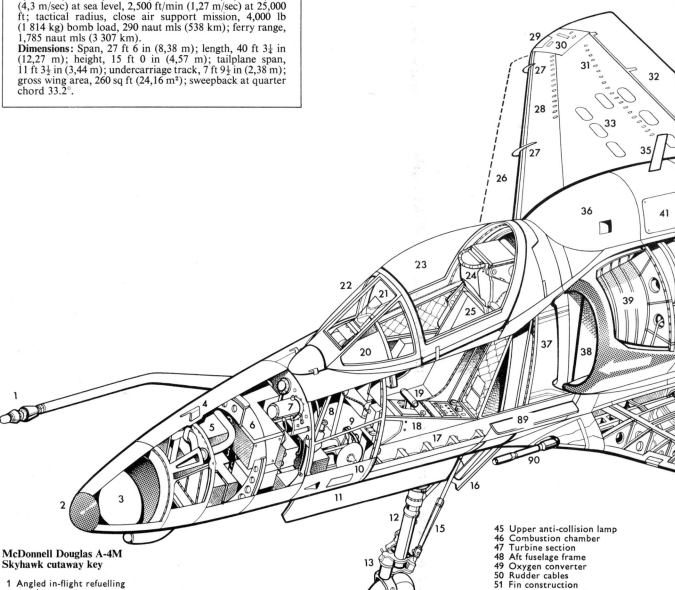

McDonnell Douglas A-4M Skyhawk cutaway key

 1 Angled in-flight refuelling probe
 2 Di-electric nose cone
 3 APG-53A radar scanner
 4 Pitot head
 5 Electronics pack
 6 Avionics pack
 7 Cockpit forward bulkhead
 8 Control column
 9 Rudder pedal
10 Internal armour plate
11 Single nosewheel door
12 Oleo leg
13 Steering cylinder
14 Nosewheel
15 Shortening link
16 Retraction jack
17 Integral armour area
18 Port instrument console
19 Throttle control lever
20 Instrument panel shroud

21 Fixed or lead computing gunsight
22 Bullet resistant rectangular windscreen
23 Cockpit canopy
24 Headrest
25 Zero-zero ejection seat
26 Leading-edge slat (open position)
27 Starboard flow fences
28 Leading-edge slat
29 Starboard navigation lamp
30 Aerials
31 Vortex generators

32 Aerodynamically-balanced aileron
33 Wing inspection panels
34 Split flap
35 Antenna
36 Dorsal avionics pack
37 240-US gal (909-l) self-sealing fuel cell behind cockpit
38 Engine intake
39 Intake trunking
40 J52-P-408A engine
41 Inspection panel
42 Power supply amplifier
43 Compass adapter
44 Engine firewall

45 Upper anti-collision lamp
46 Combustion chamber
47 Turbine section
48 Aft fuselage frame
49 Oxygen converter
50 Rudder cables
51 Fin construction
52 Tip of fin containing antenna
53 Externally-braced rudder construction
54 Rudder hinge post
55 Rudder control power unit
56 Trim tab
57 Rear navigation lamp
58 Tailplane incidence control power unit
59 Elevator actuator
60 Elevator
61 Tailplane
62 Submerged tailpipe outlet
63 Drag 'chute housing
64 Tailpipe
65 Port air brake
66 Arrester hook
67 Arrester hook actuating mechanism
68 Gyro platform

ouglas *A-4M Skyhawk*

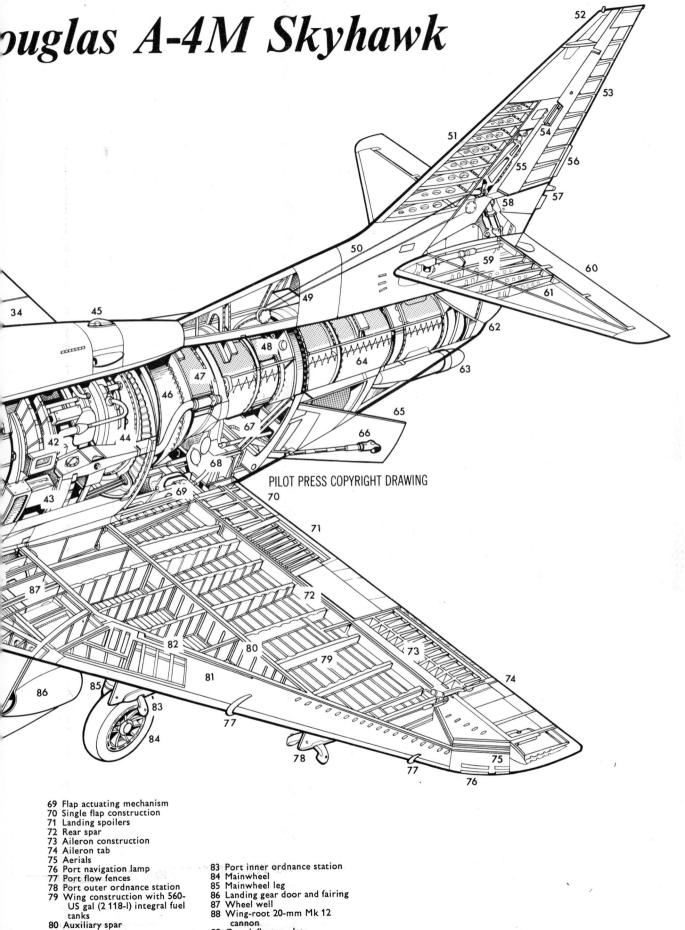

PILOT PRESS COPYRIGHT DRAWING

69 Flap actuating mechanism
70 Single flap construction
71 Landing spoilers
72 Rear spar
73 Aileron construction
74 Aileron tab
75 Aerials
76 Port navigation lamp
77 Port flow fences
78 Port outer ordnance station
79 Wing construction with 560-
   US gal (2 118-l) integral fuel
   tanks
80 Auxiliary spar
81 Leading-edge slat
82 Front spar

83 Port inner ordnance station
84 Mainwheel
85 Mainwheel leg
86 Landing gear door and fairing
87 Wheel well
88 Wing-root 20-mm Mk 12
   cannon
89 Gun deflector plate
90 Cannon muzzle

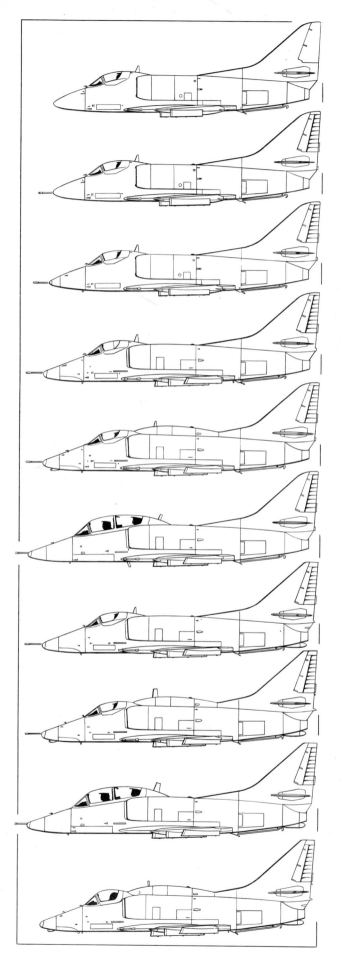

type as a replacement for TF-9J Cougar advanced trainers in service with Advanced Air Training Command. For this rôle, at the Navy basic jet training facilities at Kingsville (VT-21) and Beeville, Texas (VT-22), the aircraft was simplified by deletion of the air-to-air and air-to-ground weapon launch equipment. Having a lower gross weight, the TA-4J, as the new version was designated, needed only the lower-rated J52-P-6 engine, but in most other respects it was identical with the TA-4F. First flight was made in May 1969 and production is scheduled to continue alongside the A-4M into 1974.

As the earlier model Skyhawks were retired from operational service, a number were assigned to Naval Air Reserve units, and for this purpose, some A-4Cs have been modified up to A-4F standard, with the dorsal avionics pack and other features. When so modified, they are designated A-4L. The A-4As and A-4Bs have not been modified in this way.

Although the A-4 has operated almost exclusively in the ground attack rôle, its manœuvrability ("like an agile cat" says one pilot) gives it good air-to-air combat capability. In an unusual episode in the Skyhawks' history, an A-4 flown by Lt T R Swartz, USN, is credited with the destruction of a MiG-17 in air-to-air combat over North Vietnam. Lt Swartz was attacking two MiGs on the ground at Kep airfield when one of his comrades called him to say he had two MiGs at six o'clock (on his tail).

"I spotted the attacking aircraft," he subsequently reported, "and put my A-4 into a high barrel roll, dropping in behind the MiGs. From this markedly advantageous position, I fired several air-to-ground rockets at the number two MiG and then got another call that there was a MiG at my six o'clock again. I was not able to see my rockets hit as I bent my A-4 hard, checking for the suspected third MiG." However, his wingman saw one of the enemy aircraft hit the ground, and Lt Swartz's kill was confirmed. Subsequently, at least two other MiG fighters have been downed by Israeli Skyhawks.

The Skyhawk has also demonstrated considerable ability to take combat damage and make it back home. A typical incident, described by a pilot from VA-94, occurred over Vinh in North Vietnam, when the aircraft was attacking a well-defended target. "I was flying as the fourth man in the slot. Right after my roll-in, my plane was hit in the left wing; it inverted completely. All I could think about was the downtown dance in Vinh and how I was going to be a participant.

"It took a bit of strain to right the plane, but the Skyhawk was still up there and would not give in. I could see from the reflection on the ground that I was streaming a trail of fuel, making a good target. An A-3 Skywarrior tanker picked me up at the coastline and pumped fuel to the engine all the way to the ship where even with a hole blasted in the wing, I made a nice, easy landing."

Maintenance of the A-4 is aided by the grouping of the systems in easily accessible areas. Its small size also makes for easy accessibility. Navy experience has shown an average of 12·8 maintenance man hours per flight hour, with about 35 flight hours per month. Continued operation of the Skyhawk is planned well into the 1980s even if no further production contracts are placed, making certain that the aircraft dubbed in its early days as "Heinemann's Hot Rod" will become one of the few combat types to achieve a quarter century of service. □

VFW 614

The recent first flight of the VFW 614 marks a triumph over years of adversity . . . but its future remains clouded by uncertainties

MONDAY, 19 July 1971, dawned brilliantly in Bremen, West Germany, with bright sun and blue skies giving a boost to the spirits of the VFW-Fokker flight test team at the company's airfield. Excitement was in the air, and not surprisingly, for ten long and difficult years spent developing Germany's first jet airliner were to be culminated this day with the inaugural flight of the VFW 614 short-haul twin-jet transport. Few aeroplanes in the past decade have had to overcome so many set-backs in the course of their evolution; the determination with which the VFW team has tackled every event threatening potential disaster must surely augur well for this interesting newcomer in the civil airliner marketplace.

Whilst the name of the VFW 614's manufacturer — Vereinigte Flugtechnische Werke-Fokker GmbH — may sound strange to most ears, the new transport's ancestry is impeccable. In fact, from the very same spot where the VFW 614 made its first flight, one of its famous forebears, the Focke-Wulf Fw 200 Condor, also became airborne for the first time almost exactly 34 years before, on 27 July 1937. The Condor was also an innovation in the civil transport world, the outcome of discussions between the former Focke-Wulf company (of which Prof Dipl-Ing Kurt Tank was then technical director) and Deutsche Lufthansa (headed by Dr Stüssel). Kurt Tank's concept of a sleek, low-wing four-engined monoplane for transatlantic operations was in the forefront of contemporary airliner design, preceding similar American types by a considerable margin.

Three prototypes of the Fw 200 were laid down, as well as an initial production batch of nine, and little time was lost in getting the first prototype into the air, only about 15 months elapsing from the initial discussion between Focke-Wulf and DLH, and a few days more than a year from the signature of a contract. The original Focke-Wulf company is now one of the merged partners in VFW-Fokker, but the launching of the VFW 614 proved less simple than that of its illustrious ancestor, and ten years elapsed from the first discussion of the design to first flight. Like the Fw 200, the VFW 614 programme uses three prototypes and the aircraft is one of considerable importance to the future of the German aircraft industry. Far from having transatlantic range, however, the latest product to take flight from Bremen is designed for operations over very short stage lengths, and can be regarded in the general category of a "DC-3 replacement".

Even in this respect, the VFW 614 has undergone considerable change since its design was originated. In the words of the manufacturer, it was "designed . . . as an aircraft for developing countries . . . (but) developed into a luxury jet for regional traffic in the 1970's". Not only the concept changed, however: in the decade between project design and first flight, the parent company has survived two intricate mergers, the engine chosen originally has been cancelled, the second choice of powerplant has proved little more auspicious in view of the Rolls-Royce difficulties and the whole programme has had to survive one major financial crisis only to be confronted with another at the present time. Through all these difficulties, the design team headed by Dr Rolf Stüssel has pursued its objective of producing the first German jet airliner with single-minded purposefulness.

The need for a small, low-cost transport aircraft for use in the underdeveloped areas of the world was attracting considerable attention in the early 'sixties, especially in the USA, where various reports and studies had attempted to define the market need for a "DC-3 replacement". Project studies for such an aircraft were started in Germany within the Entwicklungsring Nord, a grouping of three aircraft companies — Weser Flugzeugbau GmbH, Focke-Wulf GmbH and Hamburger Flugzeugbau — located in the north of Germany. At this point in time, the German aircraft industry was still seeking to re-establish itself after the enforced termination of all design and production activities in 1945 and the dispersion of the design teams. One consequence of the situation that had prevailed in the immediate post-war years was that many engineers formerly engaged in the German aircraft industry were dispersed throughout the world, allowing the home-based industry to keep in touch, at least informally, with some of the more advanced work going on in America and elsewhere.

This unofficial "grapevine" was instrumental in the

selection of engines for the ERNO project when some former Junkers engineers, then employed by Lycoming, visited the Weser plant in Bremen. Lycoming plans to develop a commercial turbofan engine based on the military T55 turboprop were discussed and the Weser team realised that the new engine would make an excellent power plant for the proposed feeder-liner. The engine, designated PLF 1B, was in the 5,000 lb st (2 268 kgp) power bracket and offered a good thrust/weight ratio, low noise levels and a low sfc.

Design evolution

Around two of these engines, the project took shape during 1962 with the designation E 614 (fourth project originated in 1961), Weser being the principal instigators within the ERNO group. The major design objectives were to produce an aircraft of about 12 tons with a useful load of some 5 tons, which could operate from grass runways and would have a range of about 620 miles (1 000 km). Emphasis was to be placed on large capacity and ease of loading, as well as rugged construction and ease of maintenance. The favoured

layout used a low, unswept wing of relatively high aspect ratio, on a circular-section fuselage the entire nose-section of which hinged sideways behind the flight deck to facilitate straight-in loading. Ventral stairs were also planned, for passenger access to the rear cabin while freight went in and out at the front.

The most unusual feature of the design was the location of the engines on pylons *above* each wing, a position chosen primarily to avoid the risk of ingesting stores and debris when operating from grass or dirt strips. Underwing pods would have required a longer undercarriage, as well as bringing risk of engine damage, while the then-in-vogue rear-fuselage mounting incurred unacceptable weight penalties. The original E.614 design also featured a T-tail arrangement, despite the short moment arm provided by the stubby fuselage: at the time this feature was adopted, the deep stall danger of the T-tail had not been tragically demonstrated by the BAC One-Eleven.

First details of the project were made public, as the Weser WFG 614, at the 1963 Paris Air Show, where the renascent German aircraft industry appeared with a clutch of futuristic designs reminiscent of the hey-days of 1945. In the welter of unorthodox models on the German stands at Le Bourget, it would have been easy to dismiss the WFG 614 as just another pipe dream, but the merits of the design did not go unnoticed by potential customers, who were impressed by the promised characteristics, including a take-off distance (FAA requirements) of 2,790 ft (850 m), landing distance of 1,350 ft (412 m), cruising speed of 278 knots (515 km/h), range of 155 miles (250 km) with a maximum payload of 9,900 lb (4 500 kg) and gross weight of 25,350 lb (11 500 kg). The engines were 5,220 lb st (2 350 kgp) Lycoming PLF

(Above) An early model of the Weser WFG 614 showing the T-tail and hinged nose. (Below) Head-on, plan and side (bottom right) views of the VFW 614 as completed and flown, and side views (right, top to bottom) of the original WFG 614 and two intermediate stages in the evolution of the VFW 614.

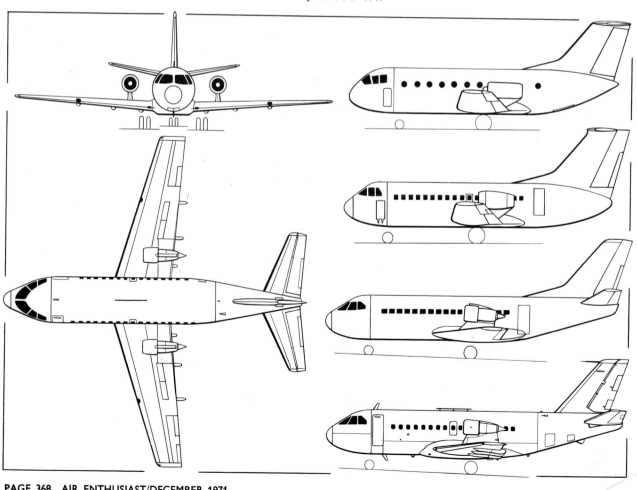

1B-2 turbofans and up to 40 passengers could be accommodated, four abreast at a pitch of 29 in (73 cm).

The designers now found themselves in a classic "chicken-and-egg" situation. Airlines which might very well be prepared to buy the WFG 614 needed an assurance that the project would go ahead, and firm delivery dates, before committing themselves to a contract. But ERNO had no money with which to launch even prototype construction, and was unlikely to be able to interest possible backers, including the West German government, without a more positive indication of orders. Before this vicious circle of indecision could be broken the project had to face and overcome other difficulties.

On 1 January 1964, the process of rationalising the German aircraft industry, which had too many companies chasing too little work, was taken a stage further when Weser Flugzeugbau merged with Focke-Wulf to form VFW, and the ERNO group was reorganised to exclude aircraft design and development. The links between Weser and Hamburger thus ended, but in 1965 Ernst Heinkel Flugzeugbau GmbH joined VFW, which thus brought together three of the oldest-established German aircraft manufacturers in a single entity employing a work force of about 10,000.

Through 1964, design work was continued on the feeder-liner, now designated VFW 614, on a low priority and at company expense, the activity being concerned primarily with keeping the project up-to-date in line with changing market requirements. Early in 1965, the project suffered a further severe blow when Lycoming decided to abandon work on the PLF 1 because the US Army discontinued funding for the uprated T55 which formed the basis for its development. This left the VFW 614 without a power plant and led the designers to intensive activity at the 1965 Paris Air Show in the search for an alternative. Initial discussions took place at Le Bourget between VFW, Bristol Siddeley and SNECMA around the possible development by the two engine companies of a civil version of the M45F, the engine then being developed with joint Anglo-French government funding to power the AFVG.

Designated M45H, the new engine had a thrust of 7,715 lb (3 500 kg) and considerable redesign of the VFW 614 was necessary to permit its use. By this time, however, the attraction of the developing countries as the principal market for the VFW 614 had waned, and contemporary market research, particularly in South America, Australia and the Far East pointed to the need for a more sophisticated DC-3 replacement to operate as a feeder-liner on established routes. At the end of 1965, the original concept was formally dropped by VFW and the whole basis of the project was reorientated to meet this newer requirement.

Technical description

As the VFW 614 was reshaped, little remained of the original but the overwing engine location. The wing was given 15 deg of sweepback, the swing nose and ventral staircase were eventually abandoned, and a low tailplane position took the place of the original T-tail. Maximum accommodation increased to 44 in a lengthened fuselage, and the gross weight went up to 38,000 lb (17 236 kg). With the more powerful engines and swept wing, the cruising speed went up to 400 knots (741 km/h) but the take-off performance was degraded, balanced field length becoming 3,760 ft (1 146 m) at max weight. The range with 8,200 lb (3 720 kg) payload was 410 st miles (660 km).

Structurally, the VFW 614 was designed to fail-safe criteria, with high-strength aluminium alloys used to provide optimum fatigue characteristics. The wing design was a compromise between the needs of high speed cruising and low speed control without recourse to high lift devices.

Two views of the VFW 614 G1 showing the overwing engine installation, the integral stairs to the passenger cabin and the large flap guide shrouds. The nose probe is for flight test only.

Structurally, it was built around a continuous two-spar torsion box of which the centre section is assembled integral with the fuselage. Fuel tanks were incorporated in the outer wing panels, the centre section containing the wells for the main wheels. In the fashion of the Boeing 737, the VFW 614 dispensed with wheel well fairing doors over half of each well.

Control surfaces comprised ailerons and rudder with trim tabs; elevators with servo tabs; variable incidence tailplane; double-slotted Fowler-type flaps and two-section spoilers on the top surface of each wing to serve as air brakes and lift dumpers. Basic flight controls were mechanically operated. A single channel (with emergency back-up) 3,000 psi (211 kg/cm²) hydraulic system supplied nosewheel steering, undercarriage operation, spoilers, flaps and brakes. The tailplane was electrically operated.

An APU was fitted to provide on-ground electric supply and air conditioning, and compressed air for self-contained engine starting. The VFW 614 was designed to have a double air-cycle machine type air-conditioning system, cabin pressurisation by hot bleed air from the engine compressors, and AC brushless generators driven by the engine constant speed drive. A liquid de-icing system was adopted for the wing and tail leading edges; engine nose-cowl de-icing was by hot air and windscreen de-icing by electric current.

Cockpit layout of the VFW 614 was designed for two-crew operation, with a folding supernumerary seat provided, facing the centre console incorporating the radio panels. The uninterrupted constant diameter cabin length of 31 ft 1 in (9,783 m) provided four-abreast seating with a centre aisle 18 in (45,7 cm) wide, and accommodates 44 passengers at a seat pitch of 31 in (78 cm), with a small galley and a toilet at the rear. Main access to the cabin was by means of an integral stairway in the forward door, and a compartment for carry-on baggage was provided adjacent to this entry. Two underfloor cargo compartments had a combined volume of 121·1 cu ft (3,43 m³).

Spreading the risk

Airline interest remained steady, but financing a production programme was still the major problem. In the summer of 1966, the German government finally agreed in principle that the VFW 614 qualified for official support in the form of a loan of up to 60 per cent of the total German share of the development costs.

In order to keep the other 40 per cent within limits that VFW could afford from its own resources, the company had embarked upon a series of possible risk-and-production-sharing arrangements. The first of these, originating in August 1965, was a consortium of VFW, Fokker and Shorts. This proposal, so far as Shorts was concerned, came to an end late in 1965 when, following the recommendations of the Plowden Report on the future of the British aircraft industry, the UK government decided that the Belfast company should not undertake development work on new aircraft such as the VFW 614. Briefly in 1966, again at the instigation of the British government, Hawker Siddeley joined the VFW 614 consortium, but left it again in February because it thought the market needed an aircraft of HS.144/Fokker F.28 size, larger than the VFW 614.

Fokker's proposed 30 per cent participation in the project was limited during 1966, at Dutch government insistence, to 20 per cent. To find additional backers, VFW began discussions with WMD/Siat in Germany; with SABCA in Belgium and with Douglas in the US. The German company, now a division of Messerschmitt/Bölkow/Blohm, entered the programme to the extent of a 9 per cent share of development costs. SABCA took a 2 per cent share, while Avions Fairey in Belgium was also brought in with a 2 per cent share to meet a Belgian government condition that more than one company should share in the work in order to qualify for government financial help. The Fokker share of development costs became 15 per cent and Rolls-Royce had a 3 per cent share in respect of engine pod production. Thus, of the total development costs, 22 per cent were shared by the UK, Netherlands and Belgium, with the Dutch and Belgian governments bearing between 60 and 80 per cent of each of the national commitments. These final arrangements were not concluded easily or quickly, however.

Discussions with Douglas were terminated in the spring of 1967 following the McDonnell Douglas merger. In their place, talks began with North American and for the remainder of the year, VFW enjoyed a close technical co-operation with the American company. NAA conducted a

wind tunnel test programme in its own facilities, joint-market research studies were conducted, and an integrated North American/VFW team made a series of visits to airlines in North and South America, and the Far East. However, this arrangement was also terminated early in 1968 after the North American Rockwell merger and a decision not to set up a second production line for the VFW 614 in America, as previously planned.

By April 1968, VFW had reached the point where further work on the 614 could only continue if a new basis for financing the launching costs could be found. Urgent discussions with the German government produced a solution whereby the latter increased its share to 80 per cent of the total German contribution, and made this contribution repayable from eventual sales, rather than a simple loan as before. This new agreement finally allowed the VFW management to give the go-ahead for detailed design, production and testing of five airframes, including two for static and fatigue tests. This milestone was reached in July 1968, and the decision was rewarded before the end of the year with the first airline option, for five examples, placed by Sterling Airways.

VFW-614 cutaway key
1 Weather radar scanner
2 Forward pressure bulkhead
3 Pilot's rudder pedals
4 Rudder control linkage
5 Elevator control linkage
6 Elevator trim-wheel
7 Pilot's seat and control column
8 Co-pilot's seat and control column
9 Throttle levers
10 Starboard instrument bank
11 Radio compartment
12 VHF blade aerial
13 Baggage compartment gate
14 Fire extinguisher bottles
15 Oxygen bottles
16 Battery
17 Taxying light
18 Nosewheel doors
19 Nose gear pivot

20 Twin steerable nosewheels
21 Retractable airstairs
22 Passenger entrance lobby
23 Baggage compartment
24 Foremost passenger seats (40-seater configuration)
25 Foremost cabin window (17 port, 15 starboard)
26 Forward under-floor baggage bay
27 Cabin air-conditioning equipment
28 Air-conditioning intake
29 Main wing spar
30 Fuselage mainframe attachment to wingspar
31 Emergency exit (port and starboard)
32 Fuselage main-frame (two)
33 Main gear retraction jack
34 Main gear pivot
35 Radius rod hinge
36 Twin mainwheels
37 Main gear outer door
38 Main engine bearer
39 R-R/SNECMA M.45H pylon-mounted turbofan engine

VFW 614 Specification

Power Plant: Two Rolls-Royce/SNECMA M45H-01 turbofans. Static thrust, each, 7,760 lb (3 520 kg).
Weights: Operating weight empty, 26,896 lb (12 200 kg); max payload, 8,598 lb (3 900 kg); max take-off and landing weight, 41,006 lb (18 600 kg); max zero fuel weight, 35,494 lb (16 100 kg).
Performance: Balanced take-off field length at max weight 3,897 ft (1 188 m); max level speed, 457 mph (735 km/h) at 21,000 ft (6 400 m); max cruising speed, 449 mph (722 km/h) at 25,000 ft (7 620 m); range with max payload (FAR Pt 121 reserves), 391 miles (630 km); range with max fuel (same reserves), 1,146 miles (1 845 km); landing distance required 3,766 ft (1 145 m).
Dimensions: Span, 70 ft 6½ in (21,50 m); length, 67 ft 7 in (20,60 m); height, 25 ft 8 in (7,84 m); wheeltrack, 12 ft 9½ in (3,90 m); wheelbase, 23 ft 0¼ in (7,02 m); sweepback 15 deg at quarter chord; wing dihedral, 3 deg; gross wing area, 689 sq ft (64,00 m²).

With the merger in 1969 between VFW and Fokker, to form the first international aerospace company in Europe, the two major partners in the VFW 614 came together, becoming responsible for 84 per cent of the development programme. Further significant stages were reached at the end of 1969, when the go-ahead was given for ordering long lead-time items for production aircraft, and in June 1970, when full series production was approved. By the time the first flight was made in July 1971, the company held 26 options from nine customers.

Engine problems

To ensure development of the engine, the German government agreed to fund 50 per cent of the development cost of the M45H, Bristol Siddeley and SNECMA being responsible

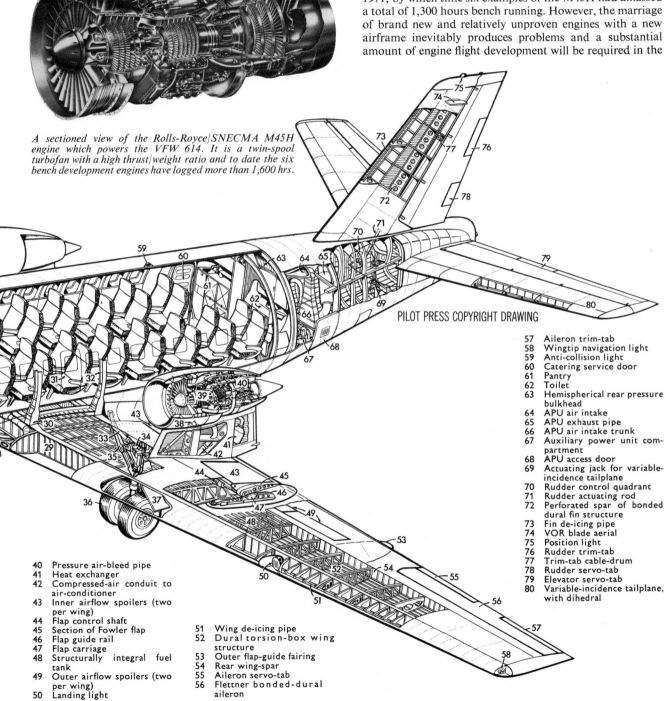

for the other half. Agreements to this effect were signed in July 1967, only a few months before the French decision to abandon the AFVG resulted in the military forerunner of the M45H being cancelled. Since that time, the M45H has been developed solely for the VFW 614 and up to the present time there is no other firm application planned for the engine. In the autumn of 1968, Bristol Siddeley merged with Rolls-Royce and the latter company assumed responsibility for the British share of the M45H programme.

The Rolls-Royce collapse in February 1971 again placed the VFW 614's power plant in jeopardy. Not only were there delays in the delivery of flight engines but future production of the M45H was threatened unless Germany agreed to an additional contribution to development costs and accepted a higher unit cost.

The first two engines for installation in the prototype VFW 614 were delivered to Bremen in February and March 1971, by which time six examples of the M45H had amassed a total of 1,300 hours bench running. However, the marriage of brand new and relatively unproven engines with a new airframe inevitably produces problems and a substantial amount of engine flight development will be required in the

A sectioned view of the Rolls-Royce/SNECMA M45H engine which powers the VFW 614. It is a twin-spool turbofan with a high thrust/weight ratio and to date the six bench development engines have logged more than 1,600 hrs.

PILOT PRESS COPYRIGHT DRAWING

40 Pressure air-bleed pipe
41 Heat exchanger
42 Compressed-air conduit to air-conditioner
43 Inner airflow spoilers (two per wing)
44 Flap control shaft
45 Section of Fowler flap
46 Flap guide rail
47 Flap carriage
48 Structurally integral fuel tank
49 Outer airflow spoilers (two per wing)
50 Landing light

51 Wing de-icing pipe
52 Dural torsion-box wing structure
53 Outer flap-guide fairing
54 Rear wing-spar
55 Aileron servo-tab
56 Flettner bonded-dural aileron

57 Aileron trim-tab
58 Wingtip navigation light
59 Anti-collision light
60 Catering service door
61 Pantry
62 Toilet
63 Hemispherical rear pressure bulkhead
64 APU air intake
65 APU exhaust pipe
66 APU air intake trunk
67 Auxiliary power unit compartment
68 APU access door
69 Actuating jack for variable-incidence tailplane
70 Rudder control quadrant
71 Rudder actuating rod
72 Perforated spar of bonded dural fin structure
73 Fin de-icing pipe
74 VOR blade aerial
75 Position light
76 Rudder trim-tab
77 Trim-tab cable-drum
78 Rudder servo-tab
79 Elevator servo-tab
80 Variable-incidence tailplane, with dihedral

VFW 614s before certification. Fifteen flight engines are being built for the prototype VFW 614s in addition to six bench development engines.

The pace of flight testing of the prototypes, at least in these early stages, is to a large extent controlled by engine availability and serviceability. The first flight of VFW 614 G1 (the first prototype) was delayed by about two months on this score, and only 24 hours before the first flight, new fan units were flown to Bremen for installation in the prototype engines. G2 and G3 were scheduled to fly in September and December and it is not yet certain what effect the engine situation will have on these dates or upon planned certification to German and American standards by early 1973. By that time, the three aircraft are expected to have flown a total of 2,500 hours.

Testing of airframe GS, the static test specimen, began at Lemwerder in March 1971 and dynamic testing of the specimen GD has also begun. The airframe production programme is scheduled to provide the first completed production aircraft, G4, in August 1972, with four more available for delivery by the end of that year. Thereafter, production can continue at a rate of two to four a month according to demand.

The announced options for the VFW 614 comprise Sterling Airways, 5; Filipinas Orient Airways, 2; Bavaria Fluggesellschaft, 3; General Air, 2; TABA, 2; Yemen Airlines, 3; Cimber Air, 3; Société Travail Aerien (Algeria), 2; Spanish Air Ministry, 1; unspecified, 3.

Division of the production effort differs somewhat from the percentages in which development costs are shared, and is as follows: VFW-Fokker, Bremen, 50 per cent (front and main fuselage); MBB (Siat), Munich, 14 per cent (rear fuselage and tail unit); Fokker-VFW, Dordrecht, 22 per cent (mainplanes and engine pylons); SABCA, Gosselies, 5 per cent (landing flaps); Fairey, Gosselies, 5 per cent (ailerons and spoilers); Rolls-Royce (sub-contracted to Short Bros, Belfast), 4 per cent (engine pods). The main and nose undercarriage units are supplied by Dowty-Rotol.

VFW studies have indicated that the market for a small jet transport in the 614 category is as high as 1,700 aircraft by 1980, and the German company hopes to gain about 20 per cent of this, with 350–400 aeroplanes sold. Breakeven, including repayment of the government loan, comes at about half this figure, on the basis of a planned unit cost of about £1,050,000. The launching costs (which include production tooling but not the materials for production aircraft) are reported to be about £33·5m for the airframe and £35m for the engine, with another £10m now required for the engine.

To finance series production of the VFW 614, the German government and the regional councils in Germany have provided additional support. On 30 November 1970, German banks formed a consortium in order to make the necessary funds — estimated at DM 200m — available to VFW, and most of these loans are guaranteed by the government. Marketing of the new aircraft is in the hands of Fokker-VFW International, a primarily-Dutch organisation based in Amsterdam. This arrangement gives the VFW 614 the benefit of a highly experienced and successful sales team, but also throws into stark relief one of the several problems casting a cloud over the aircraft's future — its competitive situation vis-à-vis the Fokker F.28.

All concerned with the programme admit that there is an overlap between these two types, and that sales of both may be affected by availability of the other. A gradual shift in emphasis by Fokker-VFW towards the larger F.28 Mk 2000 might be expected for this reason, while there will probably be no early attempt to stretch the VFW 614 in passenger capacity, which would bring it closer in specification to the F.28 Mk 1000. In the period during which the VFW 614 has been under development other projects have come and gone — for instance, the Lear Model 40, the Fairchild Hiller FH 327 and FH 228, and the Hawker Siddeley HS.136. None of these has reached prototype construction, so that the major competition for the German type now comes, somewhat unexpectedly, from the Soviet Union, in the tri-motored shape of the Yak-40.

Aviaexport estimates of the potential market for the Yak-40 are as bullish as those of VFW, and there are signs that this aircraft will be the subject of the most powerful attempt by the Soviet Union yet to break into Western markets. For the energetic VFW team at Bremen under Dr Rolf Stüssel, the threat of Soviet competition is just one more problem to be overcome. As the foregoing account shows, the VFW 614 has already encountered and survived enough setbacks to have discouraged a less determined company: if it does not now succeed, it will not be for want of perseverance. □   FGS

The VFW-Fokker "stable" of civil aircraft comprises, left to right, the Fokker F.27 Friendship, the VFW 614 and the Fokker F.28 Fellowship. As explained above, the two last-mentioned types are so close to each other in size and performance as to cause some embarrassment to the VFW-Fokker marketing teams.

TALKBACK

Speed of the Widow

I WAS very pleased to read your article on the P-61 Black Widow (June issue). My interest in the P-61 is that of a former Black Widow pilot, a member of the 422nd Night Fighter Squadron in the UK, France, Belgium and Germany.

The flying characteristics of the Black Widow were very well covered in your article. I believe, however, that the stated final approach speed of 120 mph (193 km/h) was high. If my memory serves me right, with full flaps extended we flew the final approach at 105 mph (169 km/h). Excessive floating resulted before touch-down at higher approach speeds. The outstanding characteristic was the flight stability of this aircraft at slow speeds, even down to 75 mph (121 km/h).

Tadas J Spelis
Lt Col, USAF, Retd
Citrus Heights, Calif, USA

D XXI details

I HAVE READ with interest the article on the Fokker D XXI (August issue) which I found to be excellent. You may be interested to know that Fokker tested a modified wing — the so-called E-wing — on one of the Finnish fighters. It had more dihedral and was more tapered than the normal D XXI wing, with considerable wash-out towards the tips. During the first tests with the D XXI the gun access panel in the upper wing surface blew off on a couple of occasions as a result of the build-up of pressure in the gun bay. We overcame this problem by modifying the forged locks.

In your description of the camouflage of the Dutch D XXI there would seem to be a small misunderstanding. The reddish-brown used for the underside was also used in the camouflage scheme of the wing upper surfaces and fuselage, together with light beige and olive green. There were, therefore, only three colours used in the finish of the D XXI, reddish-brown, beige and olive green. The prototype D XXI was, incidentally, a brownish colour overall in accordance with the NEI specification, and not overall olive green as has been depicted in some publications, this being the scheme for home-based LVA aircraft prior to the adoption of camouflage.

H A Somberg
"Fokker-VFW" N V
Schipol-Oost
Netherlands

Fighter omissions?

AS ONE who collects everything on military airplanes, let me say that your feature "Fighter A to Z" is the finest! I keep close watch on the alphabetical chronology, and I have already found some omissions, but please do not get the idea that I am being overly critical. On the contrary, I am anxious to help if I can. There are three aircraft that I think should have been included. These are the Aeronautica Umbra MB-902 long-range fighter, the Aichi AM-15 experimental fighter apparently owing much to the Boeing P-26, and the Aichi AM-25 (S1A1) 18-*Shi* night fighter, a prototype of which was constructed before the end of World War II.

Bernhard Klein
Midland Park
NJ 07432, USA

We stated clearly in our introduction to the first instalment of the "Fighter A to Z" that this encyclopædia of fighter aircraft is confined to aircraft built and *flown. The three types that you list, Mr Klein, do not qualify for inclusion as they were not flown. The Aeronautica Umbra MB-902 designed by Ing Bellomo was actually built in prototype form in 1942 but was destroyed before flight testing could begin. It was unusual in having two Daimler-Benz DB 605 engines buried in the fuselage and driving twin contra-props mounted outboard on the wings via extension shafts. The Aichi AM-15 shore-based fighter project was abandoned while still on the drawing boards, while the first prototype of the AM-25 (S1A1) was destroyed by air attack when 95 per cent complete, and the second prototype was only 90 per cent complete when hostilities terminated. The S1A1 Denko (Light Bolt) was to have been a twin-engined two-seat night fighter with a remotely-controlled dorsal gun barbette and two Nakajima NK9K-S Homare 22 engines.*

I believe that I have discovered a fighter that you have omitted from your "Fighter A to Z" series. This is the Italian Adamoli-Cattani fighter of 1918. I know that this type was built as a friend of mine in Italy has sent me some cuttings of photographs of this fighter reproduced in the journal of the Italian Aviation Research Branch of Air Britain.

Douglas Robertson
Aberdeen
Scotland

The Adamoli-Cattani fighter was built, Mr Robertson, but it was not *flown! The Adamoli-Cattani fighter was designed by Enea Cattani who was in charge of the Project and Development Department of the Pomilio Company. A single-seat biplane of mixed construction powered by a 200 hp Le Rhône rotary engine, it employed variable-camber wing leading-edges for lateral control instead of the more normal ailerons, and armament comprised two 7,7-mm machine guns. Overall dimensions included a span of 21 ft 1⅞ in (6,45 m), a length of 15 ft 9 in (4,80 m), and a wing area of 189·98 sq ft (17,65 m²), and empty and loaded weights were 1,036 lb (470 kg) and 1,488 lb (675 kg) respectively. The Adamoli-Cattani fighter was completed in 1918, but only static testing had been completed when World War I terminated and further development of the aircraft was abandoned.*

(Above) The so-called E-wing fitted on one of the Finnish Fokker D XXIs, as described in the letter from Mr Somberg printed alongside. (Below) The Adamoli-Cattani fighter, which was built in 1918 but abandoned before flight testing could begin; see accompanying letter from Mr Robertson.

THE UBIQUITOUS HAWK

Continuing the story of the Curtiss Hawk 75 family, the first part of which appeared in AIR ENTHUSIAST for November. Operational use in America and France forms the subject of this instalment.

AFTER official evaluation of the first Curtiss Y1P-36 at Wright Field in June 1937, the US Army Air Corps decided to make the design one of the types on which to base its long-awaited expansion. As a result, the contract placed with Curtiss on 7 July 1937 was for no fewer than 210 P-36s, the largest single contract for fighters placed in the USA since the end of World War I. Its introduction into service, however, was not to be achieved without difficulty.

The production P-36A was essentially similar to the Y1P-36 evaluation model apart from having a fully-rated R-1830-13 Twin Wasp engine driving a Curtiss Electric constant-speed airscrew, and an armament of one 0·5-in (12,7-mm) and one 0·3-in (7,62-mm) calibre Browning gun with 200 and 500 rounds of ammunition respectively. With the exception of the fabric covering of the movable control surfaces, the aircraft was entirely of metal construction. The three-spar wing, which was of NACA 2215 aerofoil section at the root and 2209 at the tip, had 24ST Alclad skinning reinforced with 24ST bulb angle extrusions, and was built in two pieces and joined on the centreline. The fuselage had similar stressed skinning reinforced by transverse bulkheads, stiffeners and longitudinal stringers. Normal fuel capacity was 87 Imp gal (395 l) but provision was made for an overload capacity of 135 Imp gal (613 l).

Empty and normal loaded weights were 4,567 lb (2 071 kg) and 5,470 lb (2 481 kg), overload weight being 6,010 lb (2 726 kg), and performance included a maximum speed of 300 mph (483 km/h) at 10,000 ft (3 050 m), normal range being 825 miles (1 328 km) at 270 mph (434 km/h) at that altitude. Initial climb rate was 3,400 ft/min (17,27 m/sec), an altitude of 15,000 ft (4 572 m) was reached in 4·8 minutes, and service ceiling was 33,000 ft (10 060 m).

The first production P-36A (38-001) was delivered from Buffalo to Wright Field in April 1938, but severe skin buckling in the vicinity of the undercarriage wells immediately became apparent, dictating increased skin thicknesses and reinforcing webs, but these failed to eradicate the

problem entirely. Engine exhaust difficulties and some weaknesses in the fuselage structure were also encountered. The 20th Pursuit Group at Barksdale Field, comprising the 55th, 77th and 79th Pursuit Squadrons, had been designated the first recipient of the P-36A and had relinquished most of its P-26As in anticipation of the delivery of the new Curtiss fighter, but despite both production line and field fixes, the P-36As were grounded again and again. At one time the 20th was down to *six* serviceable aircraft, and even these had to be flown under limitations, aerobatics and combat manœuvres being restricted and pilots being forbidden to exceed 250 mph (402 km/h).

The 1st Pursuit Group at Selfridge Field, Michigan, consisting of the 17th, 27th and 94th Pursuit Squadrons, had also been scheduled for 1938 conversion to the P-36A, but was forced to await the outcome of the strenuous efforts being made at Buffalo to wring out the fighter's troubles, and in the event, the 94th Squadron was the only component of the Group to receive any P-36As during the course of the year, operating a handful of the Curtiss fighters alongside Seversky P-35s. A few were also taken into the inventory of the 27th Squadron early in 1939, but neither squadron was destined to receive a full complement of P-36As, their numbers being made up with P-35s, while the 17th Squadron flew solely the latter type.

By the time that Curtiss-Wright completed the AAC order on 30 April 1939, the P-36 had also joined the 8th Pursuit Group at Langley Field, Virginia, its component squadrons being the 33rd, 35th and 36th, but some two-thirds of the P-36s accepted by the AAC were still grounded pending modification!

Meanwhile, the 10th P-36A airframe (38-010) had been fitted with a 12-cylinder liquid-cooled Allison V-1710-19 engine and redesignated XP-40, flying in this form in October 1938, while in the following month the 20th P-36A airframe (38-020) had been flown with an R-1830-25 engine, affording 1,100 hp for take-off, as the P-36B. This aircraft attained 313 mph (504 km/h) but was subsequently reconverted to P-36A standards. Yet another P-36A airframe (38-004) was retained at Buffalo to evaluate the possibility of reducing drag by fitting an extension shaft to an R-1830-31 engine and enclosing this in a lengthened high-inlet-velocity cowling. Flown in March 1939 as the XP-42, this aircraft was subsequently to be fitted with various types of cowl flaps, and short-nose high- and low-inlet-velocity cowlings with and without cooling fans.

The last 30 airframes built against the original order for 210 machines were completed as P-36Cs (38-181 to -210) with increased armament. Curtiss-Wright had earlier offered to make provision for a 0·3-in (7,62-mm) Colt-Browning in each outer wing panel of the 85th P-36A airframe (38-085), and the success of this installation had resulted in its adoption for the last 30 aircraft at a cost of $1,168 per unit. The P-36C also featured an R-1830-17 (S1C3-G) engine rated at 1,200 hp for take-off and 1,050 hp at 7,500 ft (2 286 m). These changes had been ordered on 26 January 1939, and a distinctive feature of the P-36C was the cartridge case retainer boxes which protruded beneath the wings. Despite the drag of these boxes and the increased weight — empty and loaded weights being raised to 4,620 lb (2 095 kg) and 5,800 lb (2 631 kg) respectively — maximum speed was increased to 311 mph (500 km/h), although range was reduced to 600 miles (966 km) at 270 mph (434 km/h).

By the time the last P-36 for the AAC rolled off the Buffalo assembly line, the manufacturer had made 81 changes in the fighter, some of these being of a major nature, and several aircraft were being employed to test equipment, power plant and armament innovations. The first Y1P-36 (37-068) had been evaluated with two-bladed contra-props, the first such installation on a US aircraft, and a company-owned demonstrator had been fitted with a turbo-supercharged R-1830-SC2-G engine as the Model 75R. The turbo-supercharger was fitted beneath the nose just aft of the engine cowling and the inter-cooler was mounted beneath the trailing edge of the wing. Empty and normal loaded weights were raised to 5,074 lb (2 302 kg) and 6,163 lb (2 795 kg) respectively, and during trials a maximum speed of 330 mph (531 km/h) was attained at 15,000 ft (4 570 m), but the poor reliability and complexity of the installation motivated against AAC acceptance of a turbo-supercharged variant of the fighter, and after Wright Field trials the aircraft was returned to Buffalo where it was re-engined with a Wright R-1820 and, as NX22028, reverted to its rôle as a civil demonstrator.

Armament trials were performed with several modified P-36As, these including the 174th aircraft (38-174) which was withdrawn from squadron service early in 1939 and fitted with four belt-fed Colt-Browning machine guns of 0·3-in (7,62-mm) calibre in the wings with 500 rpg. Simultaneously, the fuselage-mounted armament was changed to a pair of 0·5-in (12,7-mm) guns with 200 rpg, and the modified aircraft received the designation XP-36D. Another armament test-bed was provided by the 147th P-36A (38-147), which, with new outer wing panels each housing four 0·3-in (7,62-mm) guns with 500 rpg, became the XP-36E. The fuselage-mounted 0·5-in (12,7-mm) gun was retained but rendered inoperable. The 172nd P-36A (38-172) served as the XP-36F to evaluate the 23-mm Danish Madsen automatic cannon similar to that specified for the Thai Hawk 75N. The XP-36F retained the standard fuselage-mounted armament, and the cannon were installed in underwing fairings with 100 rounds per gun, but maximum weight rose to 6,850 lb (3 107 kg) and maximum attainable speed fell to 265 mph (426 km/h).

By the time that the USA entered WW II in December 1941, the P-36 was regarded as obsolescent, and had been largely supplanted in first-line Army Air Force (as the Army

Air Corps had become) units by a warplane that it had sired, the P-40. At home the P-36 had been assigned the task of combat training, the 35th and 36th Pursuit Groups activated on 1 February 1940 at Moffett Field, California, and Langley Field, Virginia, respectively, training on this type before converting to P-39s, P-40s and P-38s. Despite its acknowledged obsolescence, however, the P-36 was still in limited first-line service when Japanese forces struck at Pearl Harbour.

At Albrook Field in the Panama Canal Zone, the 16th Pursuit Group comprising the 24th, 29th and 43rd Squadrons was operating P-36s alongside P-26s, and this Group had been joined at the same base by the 32nd Pursuit Group which, activated on 1 January 1941, consisted of the 51st, 52nd and 53rd Squadrons flying the same mix of P-36s and P-26s. During February 1941, 20 crated P-36s reached Elmendorf, Alaska, from where they were flown by the 23rd Pursuit Squadron, and at the same time, 31 P-36s arrived in Hawaii from San Diego aboard the carrier USS *Enterprise*. At the time of the arrival of these P-36s, the fighter defence of Hawaii was provided primarily by the 18th Pursuit Group at Wheeler Field, its component squadrons being the 6th and 19th with P-26s and the 78th with P-36s, and the additional fighters were assigned to the 15th Pursuit Group which had been activated at Wheeler Field on 1 December 1940.

A total of 39 P-36 fighters was available in the Hawaiian Islands at the time of the Japanese assault, most of these being on the strength of the 15th Pursuit Group which was in process of conversion to the P-40. Of the three component squadrons, the 45th had completed conversion, the 47th was still in process of conversion, and the 46th was still entirely equipped with the P-36. Most of the Group's equipment was destroyed on the ground by the first wave of the Japanese attack, but several P-36s and P-40s of the 47th Squadron took-off from Haleiwa between 0815 and 1000 hours, and at 0850 hours four of the 46th Squadron's P-36s took-off on the first retaliatory sortie, intercepting a formation of nine Japanese aircraft of which two were destroyed, thus gaining the first AAF "kills" of the Pacific War.

In European skies

Slightly more than two years earlier, and on the other side of the globe, the Curtiss fighter had gained the first Allied aerial victories in European skies. On 8 September 1939, the *Groupe de Chasse* II/4 equipped with an export version

(Above) A Curtiss P-36A in the markings of the 55th Pursuit Squadron, which, as part of the 20th Pursuit Group at Barksdale Field, was one of the first units to receive the Curtiss fighter in 1938. (Below) A Curtiss P-36C in the olive drab finish re-adopted by the USAAF, together with yellow numerals, in 1940. (Opposite page) A P-36C in temporary "War Games" camouflage; the wing guns and cartridge case retainer boxes can be seen.

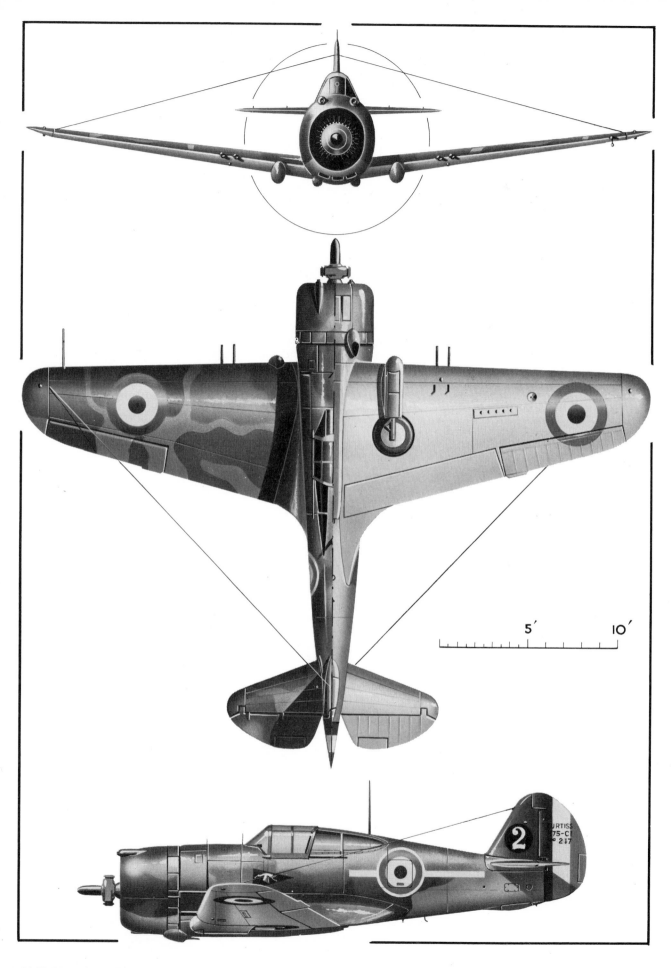

of the P-36 had succeeded in destroying two Messerschmitt Bf 109s. Numerically second only in importance to the indigenous Morane-Saulnier 406 in the fighter inventory of France's *Armée de l'Air*, the Curtiss fighter had begun to enter French service during the previous March, and when hostilities had commenced it had been included on the strength of the *Groupes de Chasse* I/4, II/4, I/5 and II/5, each *Groupe* theoretically comprising 18 aircraft, although, in fact, their strength varied between 20 and 25 aircraft.

In February 1938, two months before the first P-36A for the US Army Air Corps had rolled off the Buffalo assembly line, the French government had entered into negotiations with the Curtiss-Wright Corporation for the supply of 300 fighters of the Hawk 75A type which the American concern had offered to the *Armée de l'Air*. The Hawk 75A was essentially an export version of the P-36, and was being offered with either the Pratt & Whitney Twin Wasp or the Wright Cyclone engine. However, the unit price of Fr 2,365,000 being demanded by Curtiss-Wright was considered exorbitant by the French — the unit price of the Morane-Saulnier 406 was only Fr 1,314,000 — and the proposed delivery schedule commencing in March 1939 with 20 aircraft and continuing at a rate of 30 aircraft monthly was considered totally unacceptable. Furthermore, the US Army Air Corps had made it manifestly obvious that it disapproved of the sale of the Curtiss fighter to France, basing its disapproval on the proven inability of Curtiss-Wright to meet the contractual dates with the delivery of the P-36A.

Nevertheless, German rearmament was rendering the expansion and modernisation of the *Armée de l'Air* a matter of the utmost urgency, and while the French were of the opinion that the American manufacturer was attempting to take advantage of the situation, they persisted with the negotiations, and as a result of direct intervention on the part of President Roosevelt, one of the leading French test pilots, Michel Détroyat, was permitted to fly a Y1P-36 at Wright Field during March 1938. Détroyat submitted a thoroughly enthusiastic report, and Curtiss-Wright suggested that more acceptable delivery schedules could be offered if the French government would consider financing the construction and equipping of supplementary assembly facilities.

The unit price remained an obstacle, however, and on 28 April 1938, the *Comité du Matériel* decided to adjourn a decision until completion of the official trials of the prototype Bloch MB-150, the quoted price of the production version of which at Fr 1,237,000 was barely more than half that demanded for the Hawk 75A. The MB-150 had enjoyed a singularly inauspicious test career and had been subjected to a succession of modifications for almost two years, but at last seemed to have overcome the most serious of its teething troubles. Unfortunately, at this point it was realised that the structure of the MB-150 was unsuited for the rapid mass-production techniques that were now being demanded, and would, in consequence, have to undergo a complete structural redesign.

The reworking of the MB-150 was obviously a time-consuming process, and time was against the *Armée de l'Air*. In consequence, on 17 May 1938, the Minister for Air, Guy La Chambre, announced that, price notwithstanding, the Hawk 75A would be purchased for the *Armée de l'Air*, and a French purchasing mission headed by *Ingénieur Général* Champsaur was instructed to sign contracts for 173 Pratt & Whitney R-1830 Twin Wasp engines and 100 Hawk 75A-1 airframes. The contracts stipulated that the first Hawk

Opposite page: Hawk 75A-3 (No 217) flown by the leading French "Ace" of 1939, Lt Marin la Meslee (20 "kills"), 1st Escadrille Groupe de Chasse I/5 based at Rabat, Morocco, in spring 1941. (Drawing reproduced by courtesy of BPC Publishing Ltd)

75A-1, which was to be produced with machine tools and jigs purchased by the French government, should be flown at Buffalo by 25 November 1938, and that the 100th aircraft should be delivered by 10 April 1939.

The majority of the Hawk 75A-1s were to be shipped in disassembled form, assembly on a production line basis being undertaken by the Société Nationale de Constructions Aéronautiques du Centre (SNCAC) at Bourges. The first two aircraft were completed and flown at Buffalo early in December 1938, only a few days after the contractual date, acceptance tests only being conducted by *Capitaine* Viguier from the *Centre d'Essais du Matériel Aérien*, and the first Hawk 75A-1s (actually the fourth and fifth off the line) reached France on 24 December when the SS *Paris* docked at Le Havre. From Le Havre they were flown in January by Curtiss-Wright pilots to Bourges where they were to serve as pattern aircraft.

Fourteen further Hawk 75A-1s followed for *Armée de l'Air* trials, but all subsequent machines were shipped in disassembled form, assembly by the SNCAC commencing in February 1939. The task of assembling the fighters was performed at Bourges with quite remarkable rapidity. The *Armée de l'Air* had received 16 Hawk 75A-1s in February, including the aircraft that had arrived in assembled condition, and 25 were delivered from Bourges during March with 41 following during April, and all remaining aircraft on the initial contract having been assembled and delivered by 12 May.

The rapidity with which the SNCAC assembled the fighters was emulated by the *Armée de l'Air* in the speed with which it placed the Hawk 75A-1 in service, for both the 4e and 5e *Escadres de Chasse* had initiated conversion from the Dewoitine 500 and 501 during March and April, and by 1 July 1939, the 4e *Escadre* had 54 of the Curtiss fighters on strength while the 5e *Escadre* had 41 in its inventory. The conversion programme had not, of course, passed entirely without incident, one Hawk 75A having been force-landed when an over-speeding airscrew had caused its engine to overheat, and another having been destroyed with the loss of its pilot as a result of a flat spin developing during aerobatic trials with full tanks.

The Hawk 75A-1 was powered by an R-1830-SC-G Twin Wasp with an international rating of 900 hp at 12,000 ft (3 657 m) and affording 950 hp for take-off. Armament comprised four 7,5-mm FN-Browning Mle 38 machine guns, two mounted in the upper decking of the fuselage nose and two mounted in the wings, the former having 600 rpg and the latter 500 rpg. Like the P-36A, the Hawk 75A-1 had provision for a total of 135 Imp gal (613 l) of fuel in overload condition, this being distributed between a forward tank (35 Imp gal/158 l), two underfloor tanks (27 Imp gal/125 l and 25 Imp gal/113 l), and an aft tank (48 Imp gal/217 l). Apart from the altitude indicator all instruments were metric calibrated, a modified seat was introduced to accommodate the French Lemercier back-parachute, and a throttle which operated "French-fashion" (ie, in the reverse direction to the throttles of British or US machines) was fitted. French equipment included a Baille-Lemaire gun sight, a Radio-Industrie-537 radio, and a Munerelle oxygen system.

Following the placing of the initial French order for the Hawk 75A in May 1938, an option had been taken on a further 100 airframes, this being translated into a firm order on 8 March 1939 to ensure continued production on the French assembly line at Buffalo, and at the same time an order was placed for a further batch of 150 R-1830 engines. The second batch of aircraft differed from the initial batch in having provision for an additional FN-Browning machine gun in each wing, some structural reinforcement of the rear fuselage, and the minor modifications necessary to permit

(Above) A Curtiss Hawk 75A-3 in the markings of the 2e Escadrille *of GC I/5. As will be related in the third instalment of this feature, this was one of two Hawk 75* Groupes *which, based in Morocco, fought against the Allies during the Operation* Torch *landings in North Africa in November 1942. (Below) The fourth Curtiss Hawk 75A-1 for the* Armée de l'Air *during a demonstration at Le Bourget early in 1939.*

interchangeability between the R-1830-SC-G and slightly more powerful R-1830-SC3-G, the latter affording 1,050 hp at 2,700 rpm for take-off.

With these changes the aircraft was designated Hawk 75A-2 by its manufacturers, and the first examples of this version of the fighter were disembarked at Le Havre at the end of May 1939. Thirty-two Hawk 75A-2s were accepted by the *Armée de l'Air* during July, 36 during August, and 29 in September. The first 40 of these were basically similar to the A-1 in both power plant and armament, the R-1830-SC3-G being, in fact, tested at the *Centre d'Essais du Matériel Aérien* in the 20th Hawk 75A-2, and the first aircraft to fly with the six-gun armament being the 41st A-2. The first Hawk 75A-2 to have both the uprated engine and the increased armament was actually the 48th off the Buffalo line.

A comparison with the Spitfire

Considerable interest in the Curtiss fighter had been aroused in Britain as a result of brief handling trials of an example of this aircraft by an RAF pilot in France. The Hawk 75A appeared to possess remarkably good controls, and the ailerons in particular were apparently light at high speeds in marked contrast to those of the early Spitfire which had been described as "almost immovable at speeds above 300 mph (483 km/h)". The Royal Aircraft Establishment therefore requested the loan of a Curtiss from the French government, borrowing the 88th production Hawk 75A-2 which was used for comparative trials with a Spitfire I

(K9944) between 29 December 1939 and 13 January 1940, particular attention being paid to lateral control at high speeds.

The Hawk 75A-2 was flown with aft tank empty at a loaded weight of 6,025 lb (2 733 kg), and the three pilots that flew the fighter at the RAE were unanimous in their praise for its exceptional handling characteristics and beautifully harmonised controls. In a diving attack at 400 mph (644 km/h), the Hawk 75A-2 was far superior to the Spitfire I owing to its lighter ailerons, and in a "dogfight" at 250 mph (402 km/h) the Hawk 75A was again the superior machine because its elevator control was not over-sensitive and all-round view was better, but the Spitfire could break off combat at will owing to its very much higher top speed. In a dive at 400 mph (644 km/h) the Spitfire pilot, exerting all his strength, could apply no more than one-fifth aileron because the stick forces became excessive, whereas the pilot of the Curtiss fighter could apply three-quarter aileron.

When the Spitfire dived on the Hawk, both aircraft travelling at 350-400 mph (560-645 km/h), the Curtiss fighter's pilot could avoid his opponent by applying his ailerons quickly, banking and turning rapidly. The Spitfire could not follow the Hawk round, and, in consequence, overshot its target. In the case of the Hawk diving on the Spitfire, the pilot of the American fighter could follow his British opponent round with ease until the Spitfire's superior speed enabled it to pull away. The superior manœuvrability of the Hawk was ascribed mainly to the over-sensitiveness of the Spitfire's elevator which resulted in some difficulty in accurately controlling the *g* in a tight turn, the pilot tending to over-correct, resulting in the danger of an inadvertent stall, the Hawk's lighter ailerons, and its better all-round cockpit view.

The Spitfire I used for the trials was fitted with the early two-pitch airscrew, and the Hawk 75A displayed appreciably superior take-off and climb characteristics. The swing on take-off was smaller and more easily corrected than on the British fighter, and during the climb the Hawk's controls were more effective. However, the American fighter proved to be rather slow in picking up speed in a dive, rendering the Spitfire the more suitable machine of the two for intercepting high-speed bombers. □

(To be concluded next month)

FLYING THE DOUGLAS TWINS

QUITE the most pleasant war-time American aeroplane in my experience was the Douglas DB-7 Boston/Havoc. It was stable, handled well and easily, and was almost unique of its kind in that the controls were light, powerful and well-matched in a manner to which we had become accustomed in most British aeroplanes. The DB-7 provided, for many pilots, a first experience of the great advantages of the nose-wheel layout, of built-in control locks and — except for the many who had been brought up on the North American Harvard trainer — of pedal-operated brakes.

For the newcomer, especially, these three features were, so to speak, complementary. The level attitude and good all-round forward view made fast taxying both possible and safe; you just drove it along as if it were a motor-car with wings, instead of weaving clumsily along at a snail's pace. The control locks — which were engaged by pulling, under very strong disengagement spring-loading, a hook affair from the dashboard and attaching it to the control column — kept the rudder pedals rigid so that the brake pedals could be operated accurately and evenly. Without its use the inexperienced driver found brake operation distinctly awkward as the pedals flopped about under varying toe and air pressures, and the castoring nosewheel tended to turn violently from one full-lock to the other. We soon learned to use the control locks and to keep them in place until (or even after) turning into wind for take-off — with a check of control freedom as one of the last items in the pre-departure drill.

The first DB-7 I flew, in July 1940, was one of a batch originally intended for Belgium, but diverted and assembled and tested at Speke Airport, Liverpool. My job was to deliver it to the maintenance unit at St Athan in South Wales. Luckily, I had flown a few aeroplanes with what were then known as tricycle undercarriages. Though these were pre-war light aircraft, such as the Dutch Scheldemusch, the American Stearman-Hammond, the General Aircraft Cygnet and an experimentally equipped de Havilland Hornet Moth, they had taught me something of nosewheel characteristics and at least disabused me of any belief that you could just fly them into the gound and shown me that some sort of hold-off was advisable or even essential. I had also flown a few other American aeroplanes — so I knew something about their instrumentation, their inertia starters, and their lack of aids to engine handling such as automatic boost control. Nevertheless, I was a somewhat worried man as I settled into the cockpit after climbing in through the long sideways-opening roof hatch and searched for the various essential knobs and gauges.

Having found (labelled in French) such things as the inertia-starter spin-up and mesh switches, the undercarriage and flap lever (all below on the left) and the dual, upper and lower, cooling-gill controls (below on the right), I discovered that the fuel gauge — an ingenious switch-over affair for half-a-dozen tanks — was registering in litres and that the boost gauges were registering not in inches of mercury or pounds per square inch, but in things called *pièzes*. Fortunately the dials carried limitation marks and I'd been told (though didn't expect to remember) the approximate conversion factor. I was bothered by all those fuel tanks and their on-off and balance cocks — but was told to leave them as they were. Just the same I was to waste my time during the hour's flight to St Athan in vain attempts to get the tanks to empty in some rational way.

But all this was forgotten when the Twin Wasps had been run up, the chocks removed, the brakes released and the Boston I was rolling across the apron. This, I thought, was what all taxying should be like — with easy control and a clear view ahead of the whole aerodrome and its parked aeroplanes. No doubt this experience of a new era of flying was the cause of a momentary lapse during take-off, when, feeling more than the usual acceleration and pressure, and

Viewed from the Cockpit

by H A Taylor

An ex-Belgian-contract Douglas DB-7, used by the RAF as a Boston I, distinguished by the short nacelles for the Twin Wasp engines.

glancing at those *pièzes* dials, I realised that the permissible maximum boosts were being exceeded. The final approach at St Athan was, too, a bit on the fast side and the brakes had to be pressed rather too hard and often — which didn't do the greensward much good.

Nearly three years later I was to learn a little more about DB-7s when, as an RAF test pilot, I was posted to No 29 MU, High Ercall, which "held", at that time, a dozen or more Boston IIIs and A-20 Havoc Is and IIs, some of which had already done duty as Turbinlite interceptors operating with Hurricanes from the same station. One of the things I learned was that aerodynamic drag can (or could then) be nearly as effective as brakes for slowing an aircraft down provided that it has a reasonably innocuous stall. In other words, it does not necessarily produce the shortest run — even with a nosewheel aircraft, on which such a manœuvre

The cockpit of a USAAF Douglas A-20G-15, showing the gun sight and gun firing button (on the control wheel) for the forward firing guns.

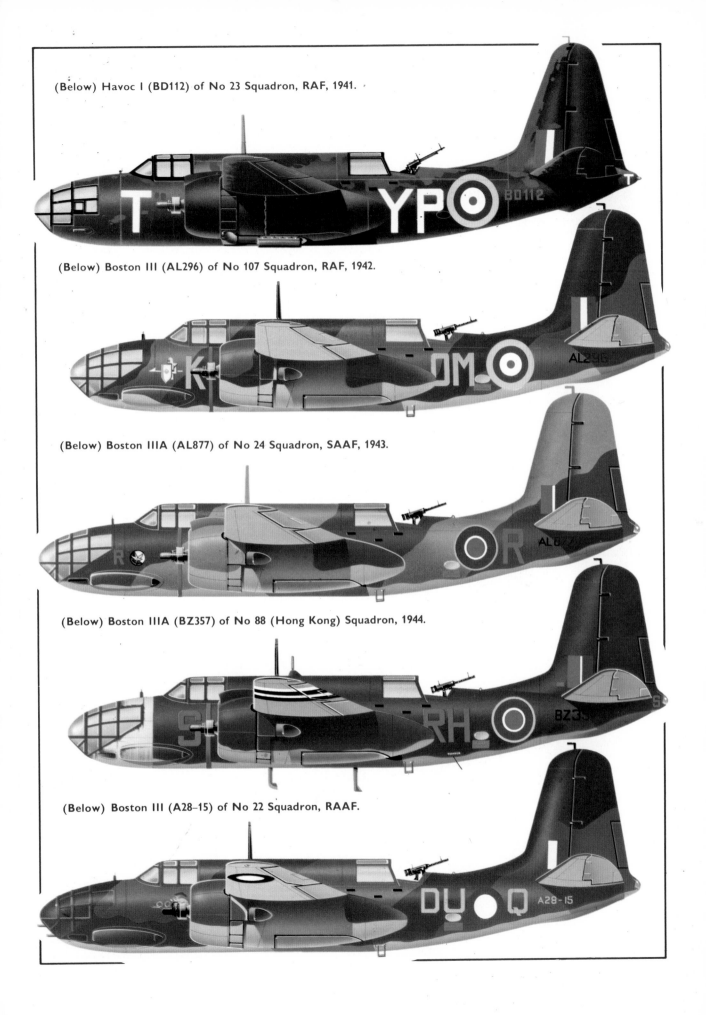

(Below) Havoc I (BD112) of No 23 Squadron, RAF, 1941.

(Below) Boston III (AL296) of No 107 Squadron, RAF, 1942.

(Below) Boston IIIA (AL877) of No 24 Squadron, SAAF, 1943.

(Below) Boston IIIA (BZ357) of No 88 (Hong Kong) Squadron, 1944.

(Below) Boston III (A28–15) of No 22 Squadron, RAAF.

is practicable — to fling the thing on the ground and stamp hard on the brakes (even assuming that these will accept such treatment). The discovery of this minor law of aviation nature was forced on me following hydraulic failure on Boston III W8259. The early DB-7s had no emergency undercarriage-lowering system apart from the handpump; later versions had an air bottle and separate lines to the undercarriage jacks. The event also demonstrated, quite incidentally, the good low-speed handling of the Boston.

After take-off on this routine test the undercarriage had taken an unconscionable time in retracting and could only be locked-up with help from the handpump. For some long-forgotten reason I didn't equate the incident with serious hydraulic trouble — though I may have thought that an engine-driven pump had failed — and carried on with the test. Twenty minutes later, when checking undercarriage and flaps before entering the landing circuit, I realised that things were far from normal. Selection of undercarriage "down" produced no visible results and the handpump had no feeling of pressure. However, working away at it — occasionally with a comforting bit of resistance — the green locked-down lights eventually came up. Little or nothing in

the way of flap movement could be induced with continued pumping against no resistance — so it seemed that the hydraulic fluid must have fled away. Goodness knows what my fitter/foreman-passenger, out of communication as he was in the aft gunner's position, thought was going on during the successive circuits as the undercarriage was being pumped down. It was just as well, perhaps, that he couldn't know about the situation because he would also have realised that without hydraulics there would be no brakes either.

So, making a wide circuit to line up for a protracted, straight-in powered approach to the shortish into-wind runway, I prepared to make a flapless and brakeless landing. As the speed was reduced to leave 5–10 mph in hand above the known flaps-up stalling speed — which I thought would be an adequate margin at our relatively low weight and with so much slipstream lift — the attitude was very nose-high and I was afraid that the tail would touch first. To make sure that it didn't, I let the Boston drop on its main wheels before hauling back on the column to obtain as much drag as possible while elevator control remained. The landing run was certainly much shorter than I expected it to be, but

Douglas Boston III cutaway key

1 Bomb-aimer's flat clear-vision window
2 Bombsight
3 Bomb-aimer's emergency escape hatch
4 Gunsight bead
5 Bomb-aimer's seat
6 Ports for 0·303-in (7,62-mm) Browning machine-guns (two each side)
7 External fairings for outer guns

8 Bomb-aimer's entrance hatch
9 Rearwards-retracting nosewheel
10 Nosewheel doors
11 Control column
12 Instrument panel
13 Throttle quadrant
14 Pilot's seat
15 Compass bowl
16 Pilot's entrance and escape hatch (hinged to starboard)
17 Foremost starboard bomb cluster
18 Foremost port bomb cluster
19 Port bomb-bay door
20 Hamilton-Standard three-blade airscrew

21 Wright Double-Cyclone GR-2600 twin-row radial engine
22 Inboard self-sealing fuel tank (136 US gal; 515 l — each wing)
23 Wing-spar/fuselage anchorage
24 Aerial mast
25 D/F loop
26 Radio equipment bay
27 Rearmost bomb-stowage
28 Oil tank

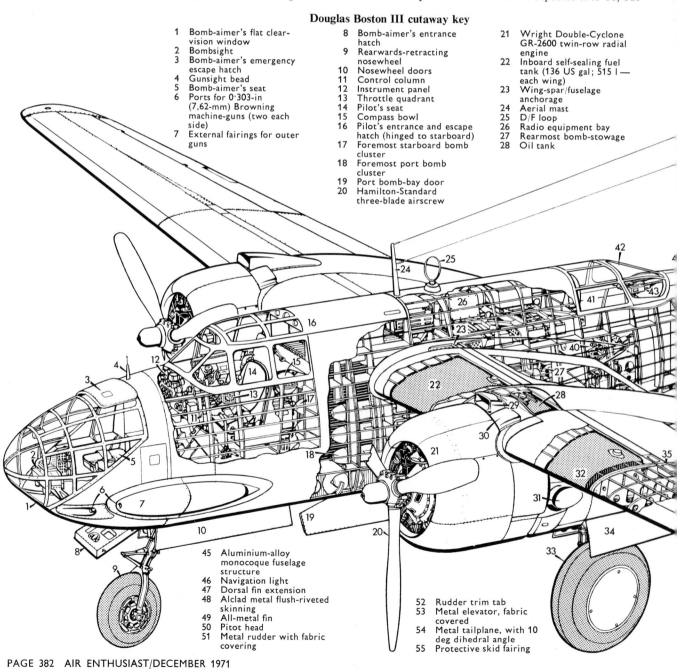

45 Aluminium-alloy monocoque fuselage structure
46 Navigation light
47 Dorsal fin extension
48 Alclad metal flush-riveted skinning
49 All-metal fin
50 Pitot head
51 Metal rudder with fabric covering

52 Rudder trim tab
53 Metal elevator, fabric covered
54 Metal tailplane, with 10 deg dihedral angle
55 Protective skid fairing

without brakes there was no means of finally stopping the thing, which continued to bowl gently along under the tick-over thrust from the two Double Cyclones. Eventually, with the end of the runway coming up, I used a burst of port engine to turn off the paving and into the mud and pulled back the mixture controls to "idle cut-off".

The engine-out controllability of the Boston was adequately (if rather recklessly) demonstrated by one of our test pilots who — after an engine failure and wishing to alert flying control to the need for giving him runway priority — flew, with the propeller of the dead engine feathered, backwards and forwards at near zero feet past the watch office until someone grasped the significance of his manœuvres and rang up the chequered runway control box and fired a couple of green Very cartridges to show the test pilot that all was now understood and forgiven.

As with other nosewheel aircraft, the Boston/Havoc had to be positively pulled (or trimmed) off the ground when such-and-such an airspeed had been reached. A long-ago conversion course at the RAF's Central Flying School had taught me that the use of flap for take-off shortened the run but spoiled the climb-away and engine-out performance —

so I never used them on any aeroplane except when the "book" said you must or when the primary need was to get airborne in the shortest possible distance when, as at satellite fields, the run was strictly limited or the ground very rough. This is mentioned because of a minor peculiarity of the Boston after take-off. As the undercarriage was coming up it tended to sink back on to the ground and had to be held in the air by brute force. The impression was almost exactly similar to that when flaps are raised and on my first flight I thought that I had accidentally lowered the flaps during the cockpit check and that they had been blown up as speed increased. The effect may possibly have had something to do with a shift of centre of gravity, or centre of pressure, as the undercarriage retracted, but, remembered in retrospect, it was probably no more than the result of leaving the ground cushion.

The DB-7 was one of the few aeroplanes which saw the war through more or less from start to finish — first as the RAF's multi-gun Havoc night fighter and Turbinlite interceptor, and later, when the skies over the Continent were less menacing, as the usually fighter-escorted Boston day bomber. □

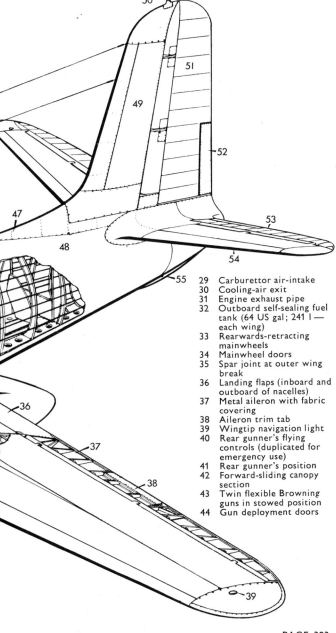

29 Carburettor air-intake
30 Cooling-air exit
31 Engine exhaust pipe
32 Outboard self-sealing fuel tank (64 US gal; 241 l — each wing)
33 Rearwards-retracting mainwheels
34 Mainwheel doors
35 Spar joint at outer wing break
36 Landing flaps (inboard and outboard of nacelles)
37 Metal aileron with fabric covering
38 Aileron trim tab
39 Wingtip navigation light
40 Rear gunner's flying controls (duplicated for emergency use)
41 Rear gunner's position
42 Forward-sliding canopy section
43 Twin flexible Browning guns in stowed position
44 Gun deployment doors

(Above) A pre-delivery picture of a British-contract Boston III and (below) a Havoc II after conversion in Britain to have a 12-gun nose and early AI equipment with aerials on the nose and wing.

Another variant of the Douglas A-20 family used by the RAF was the A-20J Boston IV shown (above) with ventral ferry tank and (below) with underwing bombs.

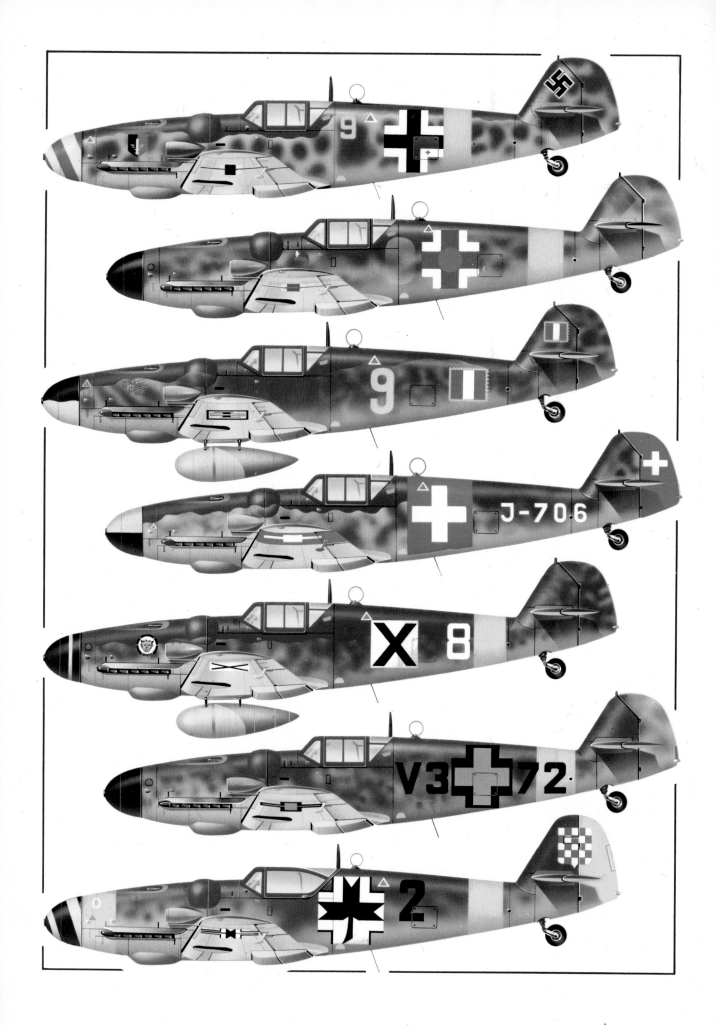

Transparently obvious

NOTHING, but nothing, mars the appearance of a model more than a poor set of transparencies! Crudely-finished or semi-opaque cockpit canopies and large "transparent" areas have proved the ruination of uncountable models that, in every other respect, approached perfection, and have, we suspect, resulted in many a staunch modeller turning to less wearing pastimes such as big game hunting or bullfighting. Transparencies are the particular *bête noire* of the conversion addict. All other features of a conversion can normally be finished — usually with paint — in such a way as to hide minor flaws and imperfections, but transparencies must stand naked and unashamed. Not only is any shortcoming patently obvious; it can also be magnified by unkind lighting.

Many and varied are the expedients adopted by modellers to overcome the "transparency peril". Some use vacuum-forming equipment, but this has the drawbacks of being expensive and not readily available outside the USA. Others labour for hours over the carving of male and female moulds from wood, and then shaping the transparent part from acetate sheet softened by heat, while there are those possessing the patience demanded to build up a canopy from layer upon layer of clear dope applied to a waxed wooden form.

It may come as a surprise to these modellers, and cause some teeth-gnashing at the thought of past hours of unnecessary labour, to learn that such complex and time-consuming effort is quite unnecessary; that canopies may be shaped in exactly the same fashion as other plastic parts provided that the necessary care is taken. There are, as may be suspected, a few special points to be taken into account, however. The transparent plastic used for model canopies has somewhat different characteristics to those of the plastic used for other components of a kit. It is harder and more brittle, and, in consequence, is prone to fatigue failure, the canopy suddenly crazing

Opposite page, top to bottom: Bf 109G-6 of I/JG 52 at Leipzig (Rumania), summer 1944; Bf 109G-6 of the 14th Slovakian Fighter Squadron, Crimea, spring 1943; Bf 109G-6 of II° Gruppo Caccia Terrestre, 3ª Squadriglia "Diavoli", Villafranca, Verona, October 1944; Bf 109G-6 of Fliegerkompagnie 7 of the Swiss Fliegertruppe, autumn, 1944; Bf 109G-6 of the Bulgarian 6th Fighter Regiment, Wrasdebna, April 1944; Bf 109G-6 of the Hungarian 102 Independent Fighter Group, summer 1944; Bf 109G-10/U4 of the kroat.Jagdstaffel, Eichwalde, November 1944.

or shattering, usually without warning and often when reshaping is approaching completion.

The prevention of such disasters is comparatively easy. All that is needed is a roughly-carved former fitting sufficiently tightly inside the canopy to prevent flexing when it is being filed. By the use of such a former, virtually any canopy may be filed to shape without risk of shattering, and very thin sections may be produced with ease. Once the canopy has taken on the desired shape it has, of course, to be returned to its original clarity. This is simply a matter of sanding with very fine emery and polishing until all scratches are removed. Some modellers coat their canopies with varnish, but this practice is not recommended. Not only is it unnecessary if polishing has been sufficiently thorough, but the varnish tends to yellow with age, and with the passage of time the model will be sporting an ugly, semi-opaque canopy to the infinite detriment of its appearance.

One of the most tedious of all "transparency jobs" is that of providing a long row of cabin windows for a model of an airliner. This is often done by cutting out and fitting into place up to a hundred tiny pieces of acetate sheet. Happily, this chore is quite unnecessary. All that is required is to fill each window aperture with a blob of white emulsion glue, which, in a few hours, will dry out transparent. Owing to surface tension, it will also dry out slightly concave, and to fill the concavity and also to protect the fine glue film, a drop of Fuel Proofer should be applied to each window. When this has dried the model will feature a row of clear and strong cabin windows.

This month's colour subject

As long as military aircraft exist, the Messerschmitt Bf 109 fighter will be remembered. Its claims to fame are numerous, quite apart from the fact that it was manufactured in larger quantities than any fighter before or since, and it has always exercised a fascination for the modeller, and this fascination is unquestionably engendered in no small way by the infinite variety of marking and finish that may be applied to models of this famous warplane. Built in immensely greater numbers than any sub-type of the Bf 109 was, of course, the *Gustav* — the Bf 109G, and our colour page this month, which reveals some of the interesting insignia that this fighter carried in service, comprises a selection of colour profiles reproduced from *The Augsburg Eagle — The Story of the Messerschmitt 109* by kind permission of the publishers of this new book, Macdonald & Co (Publishers) Limited. This book contains 24 pages of full-colour illustrations by John Weal which

depict examples of the Bf 109 in the finishes and markings applied throughout its long *Luftwaffe* career as well as those applied by the many other air arms with which this distinguished warplane served.

Among several quite good kits are those to 1/72nd scale by Airfix and to 1/70th scale by Hawk, the latter being the more accurate but appreciably cruder than the former. The principal shortcoming of the Airfix kit is presented by the nose which is both incorrectly shaped and too short, and to put this right it is necessary to insert a section 3/32 in (2,38 mm) wide at the rear of the cowling, which is no simple task. Perhaps the best answer for a Bf 109G model in the 1/72nd range is to convert Frog's very good Bf 109F kit to *Gustav* standards which is easier than correcting the Airfix kit. Revell issues a generally good kit of the Bf 109G to 1/32nd scale, and those with a penchant for the larger model will undoubtedly choose this, but there is an obvious gap to be filled by a kit of a *Gustav* to 1/48th scale, and there are rumours that Fujimi intends to produce just such a kit. There are many modellers who will be delighted if these rumours prove to be fact.

Shipboard welter weight

North American Rockwell's formidable RA-5C Vigilante, the largest and heaviest warplane ever to look upon an aircraft carrier as home, has been well served by the plastic kit industry. Two good kits of this immense aeroplane have been on the market for some while, and now these are joined by a third, this time from Airfix. This, too, is in most respects a very good kit, being neatly moulded in white plastic of good quality, accurate in outline, and finely detailed, but the fit of its component parts could be so much better, demanding filing and filling in various places, notably at the rear end of the fuselage and on the wing inserts.

It would seem that Airfix has at last taken heed of the volume of criticism directed against its decal sheets from various directions, not excluding this column, for the decals accompanying this kit are excellent. They provide the very colourful markings of a Vigilante of RVAH-14 operating from the USS *John F Kennedy,* are extremely well printed, and correctly gloss-finished. For a comparatively simple model such as this the instruction sheet is adequate, and at its UK price of 50p, Airfix's Vigilante kit is good value for money.

A variety of new decals

Companies the world over are issuing specialist decals, but with only one or two exceptions, the UK has not so far been outstanding in this field, and we are particularly glad, therefore, to welcome the arrival of a new company, Modelmark, which, if its future releases maintain the standard established by its first issue, should find a ready market.

Modelmark's first sheet offers markings to 1/72nd scale for a Messerschmitt Bf 109E-

7/U2, a Messerschmitt Bf 110C, and a Messerschmitt Me 262A. These decals, which are very well printed and matt finished, will make attractive adornments for any models depicting the previously-mentioned aircraft, but they should be closely trimmed as the film, in some cases, spreads some distance beyond the designs. Accompanying the decal sheet is a really first-class sheet of instructions which provides excellent drawings and descriptions, not only appertaining to the markings and their positioning, but to the camouflage schemes, interior colours, and so on, of the aircraft that sported them. These decals are available from Modelmark (8 Michell Avenue, Redbridge, Essex) at 25p, including postage, and the company advises us that its next release will feature decals for the McDonnell F-101 Voodoo and the Republic F-105 Thunderchief.

The latest three decal sheets received from Max Abt of France (Nos 120, 121 and 122) are certainly off-beat, as, all to 1/72nd scale, they cover a Zero-Sen in surrender markings, the last P-40 to fly from Bataan, and vicious-looking "shark's teeth" applied to a Hellcat belonging to VF-27 aboard USS *Princeton*. As is usual with Abt sheets, these are beautifully printed and matt-finished. The mass of light blue bullet-hole patches for the P-40 are of the press-on type, and, not requiring soaking, are an excellent idea, but as the film on which they are printed is exceedingly glossy, it is essential that the patches themselves are cut out with extreme accuracy.

Latin Lightning

Italian aircraft of the WW II period have suffered some neglect at the hands of the plastic kit manufacturers who, irrespective of their nationality, have for the most part elected to concentrate on the warplane progeny of Britain, Germany, Japan and the USA. From time to time, however, the occasional kit of an aircraft operated by the *Regia Aeronautica* appears but only thanks to — appropriately enough — the Italian Artiplast concern and the few other of Italy's kit manufacturers. The latest Artiplast product to appear on the market is a kit of the Macchi C.200 Saetta to 1/50th scale.

Artiplast kits are seldom faultless, and the Saetta — which can be translated as Lightning or Arrow (the weapons of Jupiter) — is no exception, although it does display a marked improvement in quality over *some* of this company's earlier products. It is generally accurate in outline, and there is plenty of fine, if somewhat irregular, surface detail, while the component parts are moulded in silver plastic of good quality. These parts have more than their fair share of flash, however, and their fit leaves much room for improvement. Much filling is demanded, particularly around the root joints of the wing and tailplanes, while the cockpit canopy, which does not fit any too well and calls for careful filing if it is to take up its correct position, is somewhat opaque and demands much polishing if the very good cockpit detail is not to be obscured.

While the component parts of this kit are open to criticism, the decal sheet is not. Semi-matt finished and extremely well

A welcome addition to the range of Regia Aeronautica *aircraft kits available is the Macchi C.200 Saetta recently issued by Artiplast. The example of the Saetta illustrated by this photo is an early production series aircraft of the 374° Squadriglia flying over Grottaglie in November 1940. The "Asso di Bastoni" emblem of the 152° Gruppo can be seen on the rear fuselage. This aircraft is finished in the temperate scheme of green with a darker brownish-green dapple.*

printed, it provides the markings for no fewer than six aircraft, four of which are illustrated by the instruction sheet. The latter, printed in four languages, is adequate, but the exploded-view drawing should not be followed in its entirety. It is impossible to assemble the cowling correctly if the drawing is followed but, fortunately, the text makes the assembly procedure clear. Despite our criticisms, this is basically a good kit and well worth the trouble of getting. The Saetta was a fine aeroplane and its praises are all too rarely sung.

Vac-u-form kits from VHF

VHF Supplies Limited has recently introduced a range of Vac-u-form kits of aircraft not as yet available in *normal* kit form. Following the company's normal policy, most of the models so far issued represent commercial aircraft, and four kits have recently been received for review, these being the Short Skyvan and the Hawker Siddeley 748 to 1/72nd scale, and the Lockheed Super Constellation and Boeing Stratocruiser to 1/144th scale. As is usual with Vac-u-form kits, the component parts must be cut from thin sheet styrene, air-screws, undercarriages, etc, being added from normal commercial kits.

These are *not* models for the novice, considerable skill having to be exercised in their assembly. The manufacturer recommends the use of wood strengthening inside wings and tailplanes, and this advice should *definitely* be followed. Surface detail may be added by means of a sharp-pointed scriber or blade drawn alongside a steel rule, and when assembled by a skilled modeller, these kits can result in pleasing miniature replicas of aircraft otherwise unavailable in kit form, and, indeed, indistinguishable from models assembled from normal kits. The instructions accompanying these Vac-u-form kits are reasonably comprehensive, but it would be of much value to the modeller if the manufacturer recommended suitable airscrews, etc. As things are, some difficulty is inevitably experienced in working out what components from which kits should be used — in the case of the HS 748 the airscrews, pilots' canopy and undercarriage of the Airfix Friendship may be used.

The difficulties inherent in their assembly and completion apart, these kits are well worth buying by those possessing the necessary skill and wish to include the aircraft previously listed in their collections.

Dornier and Grumman from Frog

We have found it necessary to criticise some recent Frog kits rather harshly; some of this company's kits have fallen far short of the standards established by their predecessors. This decline would now seem to have been arrested by the Dornier Do 17Z and Hellcat now issued as these, happily, revert to the standards of former years. Both are accurate and neatly moulded to 1/72nd scale, and their component parts fit together well, though the Hellcat is the better of the two in this respect as some of the joints of the Do 17Z demand some work from the modeller. Both display a mass of fine straight-line surface detail and have excellent thin, crystal-clear transparencies. The superbly-detailed engines provided with both kits are particularly worthy of mention, though the power plant provided for the Hellcat is in error in one respect: the two rows of cylinders line up incorrectly, those comprising the rear row lining directly behind those of the front row instead of being staggered. Fortunately, this error is simple to correct.

The decal sheet of the Do 17Z is excellent, offering markings for an aircraft of I/KG 3 operating over the UK in 1940 and for an aircraft of Finland's PLeLv 46. The decal sheet accompanying the Hellcat kit, which offers the markings of two British-operated aircraft, one from No 800 Squadron operating in home waters and one from No 1839 Squadron flying from HMS *Indomitable* in the Indian Ocean, is definitely weak on colour, and there is an ugly black line separating the red and blue portions of the roundels of the No 800 Squadron aircraft. It may be that this weakness in the colour is deliberate, representing an effort to simulate the weathered effect sported by the markings of many naval aircraft, but if so, the result is not very successful.

Both kits include the usual all-drawing style instruction sheets which serve their purpose adequately. These are certainly excellent products and are to be recommended highly, being the best 1/72nd scale representations of their subjects on the market today, and at their UK prices of 47p for the Do 17Z and 20p for the Hellcat they offer good value. □   W R MATTHEWS

RECENTLY ISSUED KITS			
Company	Type	Scale	Price
Frog	Dornier Do 17Z	1/72	—
Otaki	Lockheed C-5A Galaxy	1/144	—
Fujimi	Spitfire Mk VB	1/50	350 Yen
MCCNE	Yak-40	1/125	—

PLANE FACTS

Pander's Hunter

Can you publish details, a photograph and a general arrangement drawing of an aircraft, built in Holland, which participated in the 1934 race to Australia under the name "Panderjager"? I believe that it was damaged or destroyed somewhere in India.

M B Dessaur, Bat-Yam 59644, Israel

The Pander S.IV Panderjager (Pander Hunter) which participated in the London-Melbourne race of 1934 was built by H Pander & Zonen in the Netherlands as a fast long-range mailplane for carrying mail between Amsterdam and Batavia in the Netherlands East Indies, and was flown for the first time on 6 October 1933 at Schiphol with Capt G J Geyssendorffer of KLM at the controls. Initially known as the Postjager (Mail Hunter), the S.IV was registered PH-OST and was an extremely clean low-wing monoplane of wooden construction powered by three 420 hp Wright R-975-E2 Whirlwind nine-cylinder air-cooled radial engines. A total of 462 Imp gal (2 100 l) of fuel was housed in six wing tanks, and accommodation was provided for a crew of three comprising pilot and co-pilot seated side-by-side and a radio operator immediately aft. To the rear of the flight deck was the freight hold which was intended to accommodate 1,102 lb (500 kg) of mail.

It was intended to make the first mail flight between Amsterdam and the East Indies in December 1933, and with 600 lb (272 kg) of mail, the S.IV took-off from Schiphol on the 9th of that month, but the starboard engine stopped after losing oil pressure over southern Italy, necessitating a landing at Grottaglie, near Brindisi. As it was found that a replacement engine was

The photograph of the Heinkel He 112 V5 which accompanied Harold Thiele's letter concerning He 112 prototypes used in the experimental rocket propulsion programme of the mid 'thirties.

called for, the mail was transferred to another aircraft, the S.IV continuing its journey to Batavia on 28 December and arriving there three days later, the total flying time from Amsterdam to Batavia being 45 hours. The S.IV returned to the Netherlands in January 1934, and, renamed Panderjager, was entered for the London-Melbourne race by a private Dutch syndicate. The S.IV reached Mildenhall, the departure airfield, on 16 October 1934, and on 20 October took-off for Melbourne. Unfortunately, 36 hours after leaving Mildenhall and during the landing approach at Bamrauli, Allahabad, India, one of the mainwheels jammed in the retracted position. The aircraft was landed on one mainwheel but suffered relatively light damage. Repairs were begun immediately, and the S.IV took-off to continue its journey to Melbourne on 26 October. However, just as the aircraft left the runway one wheel struck a tractor, the aircraft crashed

and was completely burned out.

The Pander S.IV had a maximum speed of 224 mph (360 km/h), a maximum cruise of 186 mph (300 km/h), and a service ceiling of 19,850 ft (6 050 m). An altitude of 7,218 ft (2 200 m) was attained in 7 min 2 sec. Empty and loaded weights were 7,055 lb (3 200 kg) and 12,566 lb (5 700 m), and overall dimensions included a span of 53 ft 5½ in (16,60 m), a length of 41 ft 0 in (12,50 m), a height of 10 ft 9⅞ in (3,30 m), and a wing area of 495·14 sq ft (46,0 m²).

Rocket-boosted Heinkel

The liquid-fuel rocket experiments carried out by Wernher von Braun and others with the He 112 which culminated in the first rocket-propelled flight during the summer of 1937 have been somewhat neglected by historians. According to Heinkel's memoirs (Sturmisches Leben), two fuselages and two complete He 112 prototypes were used for these experiments, one of the latter being destroyed by an explosion during ground tests. According to Das Buch der Deutschen Luftfahrttechnik *one of the two prototypes was the V5, while* Warplanes of the Third Reich *gives no details of these rocket experiments, but the civil registrations of all prototypes are mentioned with the exception of the V5 and the V11.*

I am sending with this letter a photograph of an He 112 which appears to be similar to the V3, V4, V5 and V6, but its registration which would seem to be either D-IIZO or IIZQ is not mentioned in Warplanes. *Could it be the V5? Can you tell me which prototype is depicted by this photograph and which He 112 prototypes were used for the rocket experiments?*

Harold Thiele,
79 Ulm-Söflingen, Germany

There would seem to be little doubt, *Herr* Thiele, that the A-series He 112 prototype depicted by your photograph (either D-IIZO or D-IIZQ) is the He 112 V5 as the registrations of the V3, V4 and V6 are known (D-IDMO, D-IZMY and D-ISJY), and this prototype *did* participate in the rocket propulsion experiments, being the example that was destroyed on the ground at Neuhardenburg in March 1937 as a result of an explosion while the pilot, Erich Warsitz, was preparing to take-off for the initial test flight. Although we have been unable to confirm positively which He 112 prototype

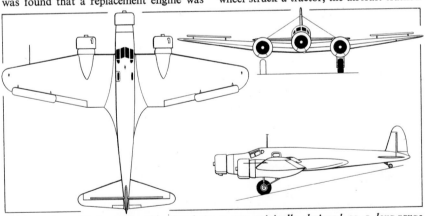

The Pander S.IV Postjager alias Panderjager was originally designed as a long-range mailplane for use by KLM and participated in the 1934 London-Melbourne Race.

fuselages had been used in the test programme prior to this accident, we believe
that these were the V1 (D-IADO) and V3
(D-IDMO), both of which were destroyed
by explosions during the course of the
rocket experiments at Kummersdorf. The
V2 (D-IHGE) had earlier crashed during
spinning trials, and the V4 (D-IZMY) was
evaluated in Spain by the *Legion Condor*,
and some considerable time after the first
three He 112 airframes participating in the
experimental rocket programme had been
destroyed, this particular prototype was
demonstrated at Zürich-Dübendorf. As no
other He 112 prototypes had been completed
and flown at the time of the Kummersdorf
accidents, a process of elimination leaves
only the V1 and V3!

Concerning the fourth He 112 used in the
experiments, the testing of which began in
April 1937 to culminate in the first flight
solely on the power of a liquid-fuel rocket
motor, this is known to have been a DB
600-powered prototype and was *presumably*
the He 112 V8 (D-IRXO) which was the last
A-series airframe and was completed as a
test-bed for the Daimler-Benz DB 600Aa
engine. This preceded the He 112 V7
(D-IKIK) which was the first B-series
airframe and also DB 600 powered and was
still serving in the engine development rôle
late in 1937, while the He 112 V9 (D-IGSI)
did not fly until July 1937 and, in any case,
was sold to Hungary in 1940.

The last Stinson tri-motor

*I am researching the Stinson A low-wing
trimotor. Perhaps you could publish some
information in a forthcoming issue.*

Don Hartsig,
Cicero, Illinois 60650, USA

The Stinson Model A 10-passenger airliner
powered by three 260 hp Lycoming R-680-5
radial air-cooled engines was unveiled late
in 1934 at a unit price of $37,500. It had a
fabric-covered welded chrome-molybdenum
steel-tube fuselage, a fabric-covered steel-
tube wing, electrically-operated all-metal
flaps, and electrically-operated fully-
retractable main undercarriage members.
The Model A was adopted by several
airlines, including Delta and American, but
was only suitable for short-haul operations,
particularly in undeveloped areas, and did
not offer effective competition for the new

(Above) The first Miles M.20 in its original markings; it later carried the serial number
AX834. (Below) The second M.20, to naval requirements with a modified undercarriage and
catapult points under the centre section.

generation of commercial transports, such
as the Boeing 247 and Douglas DC-2. In
consequence, relatively few were built.

The Model A attained a maximum speed
of 180 mph (290 km/h) and cruised at
170 mph (273 km/h) at 7,000 ft (2 135 m).
Initial climb rate was 1,000 ft/min (19,68
m/sec), service ceiling was 17,000 ft (5 180
m), and cruising range was 640 miles
(1 030 km). Empty weight was 7,200 lb
(3 266 kg), loaded weight was 10,200 lb
(4 627 kg), and overall dimensions included
a span of 60 ft 0 in (18,29 m), a length of
37 ft 0 in (11,28 m), a height of 12 ft 0 in
(3,66 m), and a wing area of 500 sq ft
(46,45 m2).

Miles "emergency" fighter

*I believe that Miles Aircraft built a fighter
aircraft in the short space of nine months
somewhere around 1940. I think that it had a
Napier Sabre engine and a spatted undercarriage. I will be very interested to learn
further details of this plane, and shall be glad
if you can tell me if the Napier engine was
ever used in a production aircraft.*

A G Wilson, Southmeed,
Bristol BS10 5JN

The Miles fighter to which you refer, Mr
Wilson, was the M.20, but this was powered
by a 1,460 hp Rolls-Royce Merlin XX 12-
cylinder vee liquid-cooled engine and *not* by
the Napier Sabre 24-cylinder horizontal-H
engine, the production installations of the
latter being the Hawker Typhoon and
Tempest V and VI.

The M.20 was conceived during the
summer of 1940 as a utility fighter which,
without forfeiting performance, could be
manufactured rapidly in large quantities.
Phillips and Powis Aircraft tendered the
proposal to the then Minister of Aircraft
Production, Lord Beaverbrook, who immediately gave authority to proceed with
the project. Specification F.19/40 was
written around the proposal, and Walter
Capley was placed in charge of the design.
Within 65 days (*not* nine months) the first
prototype M.20 (AX834) had been designed,
built and flown! All features of the design
were subordinated to speed of production,
standard Master advanced trainer components were employed wherever possible,
all hydraulics were eliminated, and a fixed
spatted undercarriage was adopted. The

*One of the relatively small number of Stinson Model A 10-passenger transports built during
1936-7.*

airframe comprised a two-spar wooden wing with plywood and fabric skinning, a wooden semi-monocoque fuselage, and a fabric- and ply-covered wooden tail unit. The Merlin XX "power egg" was interchangeable with that of the Beaufighter II, provision was made for the installation of 12 0·303-in (7,7-mm) Browning machine guns in the wings (although, in the event, only eight were fitted), and wing tanks housed 154 Imp gal (700 l) of fuel. The "bubble"-type canopy adopted was one of the first of its kind.

The initial flight was performed by Tommy Rose on 14 September 1940, but the shortage of Spitfires and Hurricanes, to safeguard against which the M.20 had been designed, did not occur and the Miles fighter was not ordered into production. The first prototype suffered some damage during the winter of 1940 and was written off. Hugh V Kennedy, Miles' assistant chief test pilot, had landed at Woodley but when he applied the brakes the wheels locked, and the M.20 ran through the boundary fence and ended up in a sandpit. A second prototype, the M.20 Mk II (DR616), was completed in April 1941 to meet the requirements of specification N.1/41 which called for a shipboard fighter. Apart from some redesign of the undercarriage fairings and radiator bath contours, the application of a more pointed airscrew spinner, and provision for an arrester hook and catapult spools, the second prototype was identical to its predecessor. However, no production order was forthcoming.

Although the M.20 could attain 350 mph (563 km/h) in level flight at 20,600 ft (6 280 m) in lightly loaded condition, maximum speed at this altitude with full armament and fuel was 333 mph (536 km/h), and 288 mph (463 km/h) at sea level. Economic cruise was 275 mph (442 km/h) at which range was 550 mls (885 km), but at long-range cruise of 167 mph (268 km/h) at 9,000 ft (2 743 m) an absolute range of 870 mls (1 400 km) was attainable. Initial climb rate was 3,200 ft/min (16,26 m/sec), service and absolute ceilings were 31,400 ft (9 570 m) and 35,500 ft (10 820 m) respectively, and stalling speeds were (flaps up) 101 mph (162 km/h) and (flaps down) 80

The first prototype of the Bücker Bü 133 Jungmeister powered by a Hirth HM 6 air-cooled engine.

mph (129 km/h). Empty and loaded weights were 5,870 lb (2 662 kg) and 7,758 lb (3 519 kg), and overall dimensions were: span, 34 ft 7 in (10,54 m), length, 30 ft 1 in (9,17 m), height, 12 ft 6 in (3,81 m), wing area, 234 sq ft (21,74 m²).

Hirth-powered Young Champion

The Bücker Jungmeister is well known as a radial-engined aircraft but I can find little reference to the extremely elegant Hirth inline powered version. Perhaps you can shed some light on this variant of the Jungmeister?
Michael Godwin, Leicester
The Bü 131 Jungmeister (Young Champion) produced by the Bücker Flugzeugbau GmbH of Berlin-Rangsdorf was originally offered with the 140 hp Hirth HM 6 (Bü 133A) or 160 hp Hirth HM 506 (Bü 133B) six-cylinder inline inverted air-cooled engines, or 160 hp Siemens Sh 14A-4 (Bü 133C) seven-cylinder radial engine, and the first prototype of this single-seat aerobatic bi-plane (D-EVEO) which flew in 1935 was powered by the Hirth HM 6, the second (D-EAKE) having the Siemens Sh 14A. In the event, the HM 6-powered Bü 133A progressed no further than the prototype stage, and although a manufacturing licence for the HM 506-powered Bü 133B was

acquired by the Spanish Construcciones Aeronauticas SA and pattern aircraft were supplied, the Sh 14A-powered Bü 133C was the version eventually built. All production by the parent company was confined to the radial-engined model.

The HM 6-powered Bü 133A attained a maximum speed of 143 mph (230 km/h), cruised at 124 mph (200 km/h), and landed at 50 mph (80 km/h). An altitude of 3,280 ft (1 000 m) was attained in 3 min, 6,560 ft (2 000 m) in 6·7 min, and 9,840 ft (3 000 m) in 11 min, and maximum range was 310 miles (500 km). Empty and loaded weights were 904 lb (410 kg) and 1,290 lb (585 kg) respectively. Overall dimensions were: span, 21 ft 7⅞ in (6,60 m), length, 20 ft 2⅛ in (6,15 m), height, 10 ft 8 in (2,25 m), wing area, 129·167 sq ft (12,0 m²).

Jet-flap investigation

Can you please publish information on the Hunting H.126 jet-flap research aircraft? What was its fate?
D A Bennett, Yateley, Camberley, Surrey
The Hunting H.126 (XN714) was built to the requirements of specification ER.189D to flight test the jet-flap principle in which a turbojet's exhaust efflux is ducted to the trailing edge of the aircraft's wing and ejected through a narrow slit. As well as providing a means of propulsion, the efflux can be deflected downward to form a "jet-flap" of high-velocity gas resulting in lift coefficients of 10 or more. The H.126 was powered by a single Bristol Siddeley Orpheus turbojet, the greater part of the jet efflux from which was ducted into the wings and ejected as a thin gaseous sheet over the flaps which acted as jetstream deflectors. The stream followed the flaps as they were lowered, resulting in an appreciable increase in lift. The H.126 was flown for the first time on 26 March 1963, and subsequently fulfilled an extensive flight test programme at the RAE Bedford. In the spring of 1969 it was flown to the USA to undergo wind-tunnel tests at Ames Research Laboratory. It has since returned to Bedford and is unlikely to fly again.

A single-seater, the H.126 was of all-metal construction, and its overall dimensions included a span of 45 ft 4 in (13,82 m), a length including nose probe of 50 ft 2 in (15,29 m), and a height of 15 ft 6 in (4,72 m).

The Hunting H.126 built to test the jet-flap principle and which was tested by NASA in 1969/70 and is now at RAE Bedford awaiting a decision on its future disposal.

The production version of the Ansaldo A.C.3.

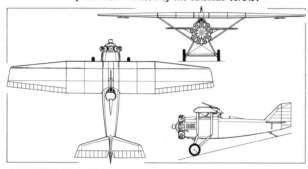

ANSALDO A.C.3   ITALY

Based on the Dewoitine D.9, the A.C.3 differed primarily in having slightly increased wing span and area and a marginally reduced overall length. Powered by a 420 hp Gnôme-Rhône Jupiter IV nine-cylinder radial, the prototype was flown early in 1926 and a total of 150 A.C.3s was delivered between September 1926 and April 1927. Armament normally comprised two fuselage-mounted and two wing-mounted 7,62-mm Darne machine guns, but the latter were sometimes replaced by a single gun above the wing centre section mounted to fire upward at an oblique angle. During the 'thirties the A.C.3s were employed in the assault rôle, and were finally phased out in the summer of 1938. Max speed, 153 mph (247 km/h) at sea level. Time to 3,280 ft (1 000 m), 1 min 41 sec, to 9,840 ft (3 000 m), 6 min 11 sec. Endurance, 2·83 hr. Empty weight, 2,114 lb (959 kg). Loaded weight, 2,981 lb (1 352 kg). Span, 41 ft $11\frac{7}{8}$ in (12,80 m). Length, 23 ft $10\frac{3}{4}$ in (7,28 m). Height, 9 ft $7\frac{1}{3}$ in (2,93 m). Wing area, 269·1 sq ft (25,0 m2).

ANSALDO A.C.4   ITALY

The A.C.4 was a direct development of the A.C.2 from which it differed primarily in having a 410 hp Fiat A.20 engine. Possessing a similar armament to the earlier fighter, the A.C.4 was flown in 1927 but only one prototype was built. Max speed, 157 mph (253 km/h) at sea level. Time to

The sole prototype of the Ansaldo A.C.4.

3,280 ft (1 000 m), 2 min 12 sec, to 9,840 ft (3 000 m), 7 min 38 sec. Endurance, 2·33 hr. Empty weight, 2,227 lb (1 010 kg). Loaded weight, 2,879 lb (1 306 kg). Span, 35 ft $8\frac{1}{3}$ in (10,88 m). Length, 24 ft $2\frac{1}{2}$ in (7,38 m). Height, 9 ft $1\frac{7}{8}$ in (2,79 m). Wing area, 215·28 sq ft (20,0 m2).

ARADO SD I   GERMANY

The first fighter to be built by the Arado Handelsgesellschaft of Warnemünde, the SD I single-seat sesquiplane designed by Ing Walter Rethel, was flown for the first time in 1927, having been built clandestinely to the requirements of the *Reichswehrministerium*. It featured a welded steel-tube fuselage with wooden wings and plywood and fabric skinning, and power was provided by a 425 hp Gnôme-Rhône Jupiter 9-cylinder radial engine. Armament comprised two 7,9-mm machine guns. The SD I failed to meet RWM requirements, and only two prototypes were completed. Max speed, 171 mph (275 km/h) at 16,405 ft (5 000 m), 152 mph (345 km/h) at sea level. Empty weight, 1,874 lb (850 kg). Loaded weight, 2,712 lb (1 230 kg). Span, 27 ft $6\frac{3}{4}$ in (8,40 m). Length, 22 ft $1\frac{3}{4}$ in (6,75 m).

(Above) The sole prototype of Arado's first fighter, the SD I.

(Right) The nose of the Arado SD II, the basic design of which eventually provided the Luftwaffe's first indigenous fighter, the Ar 64.

ARADO SD II   GERMANY

Apart from a common designer and similar construction, the SD II single-seat fighter which appeared in 1929 bore no relationship to the SD I. Powered by a Siemens-built Jupiter VI geared 9-cylinder radial driving a three-bladed wooden airscrew and rated at 530 hp for take-off, the SD II was intended to carry an armament of two 7,9-mm machine guns, and only one prototype was completed and flown. Max speed, 146 mph (235 km/h) at 16,405 ft (5 000 m), 138 mph (222 km/h) at sea level. Empty weight, 3,186 lb (1 445 kg). Loaded weight, 3,902 lb (1 770 kg). Span, 32 ft $5\frac{3}{4}$ in (9,90 m). Length, 24 ft $4\frac{1}{3}$ in (7,40 m).

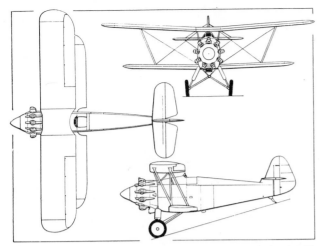

The Arado SD III which was developed in parallel with the SD II.

ARADO SD III — GERMANY

Evolved in parallel with the SD II and flown in 1929, the SD III designed by Ing Walter Rethel differed primarily in having a direct-drive Siemens-built Jupiter 9-cylinder radial rated at 490 hp and driving a two-bladed fixed-pitch wooden airscrew. Armament comprised two 7,9-mm machine guns and only one prototype was built, both this and the SD II serving as development aircraft for the Ar 64. Max speed, 140 mph (225 km/h) at 13,125 ft (4 000 m), 132 mph (212 km/h) at sea level. Span, 32 ft 5¾ in (9,90 m). Length, 25 ft 5⅛ in (7,75 m).

ARADO Ar 64 — GERMANY

The Ar 64 single-seat fighter which appeared in 1930 was essentially a derivative of the SD II and SD III, and was produced by the Arado Handelsgesellschaft as a result of an invitation received in 1929 from the RWM to develop a successor for the Fokker D XIII fighters then in use at the

The photograph (above) depicts the Ar 64c and the general arrangement drawing (below) illustrates the Ar 64d.

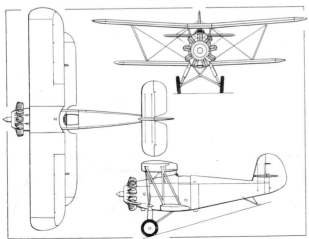

clandestine German flying training school at Lipezk in the Soviet Union. The first prototype, the Ar 64a, was powered by a 530 hp Siemens-built Jupiter VI radial driving a four-bladed wooden airscrew, and was, like earlier Arado fighters, of mixed construction with a welded steel-tube fuselage and wooden wings. This was followed by two examples of the Ar 64b, which, powered by a BMW VI 6,3 12-cylinder liquid-cooled engine rated at 640 hp and driving two-bladed airscrews, were evaluated at Lipezk in 1931. The Ar 64c was similarly powered to the Ar 64a but embodied some structural changes, and the initial production model was the Ar 64d with enlarged vertical tail surfaces, a revised undercarriage, and a Jupiter VI driving a four-bladed airscrew, this being followed by the Ar 64e which differed in having a two-bladed airscrew. Some 20 Ar 64d and 64e fighters were built and were used at the DVS (*Deutsche Verkehrsfliegerschule*) at Schleissheim, armament being two 7,9-mm guns. (Ar 64d): Max speed, 155 mph (250 km/h) at 16,405 ft (5 000 m). Empty weight, 2,667 lb (1 210 kg). Loaded weight, 3,704 lb (1 680 kg). Span, 32 ft 5¾ in (9,90 m). Length, 25 ft 7⅞ in (7,82 m).

Both photograph above and general arrangement drawing below depict the final production version of the Ar 65, the Ar 65f.

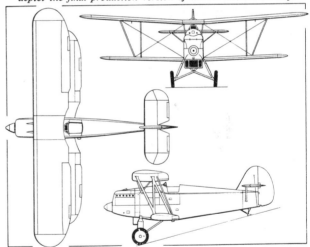

ARADO Ar 65 — GERMANY

Developed as a successor to the Ar 64, the Ar 65 appeared in 1931 with a 750 hp BMW VI 7,3 12-cylinder liquid-cooled engine and an armament of two 7,9-mm MG 08/15 machine guns. Three prototypes, the Ar 65a, 65b and 65c which embodied equipment changes and differed in minor structural details, were followed by the initial production model, the Ar 65d, in 1933, this being armed with two 7,9-mm MG 17 machine guns with 500 rpg. Minor modifications resulted in the Ar 65e and 65f production models which had the vertical fuselage magazine for six 22-lb (10-kg) bombs deleted. Production continued until early 1936 and 85 examples of these models were built. (Ar 65e): Max speed, 186 mph (300 km/h). Cruise, 153 mph (246 km/h). Initial climb, 2,086 ft/min (10,6 m/sec). Empty weight, 3,329 lb (1 510 kg). Loaded weight, 4,255 lb (1 930 kg). Span, 36 ft 9 in (11,20 m). Length, 27 ft 6¾ in (8,40 m). Height, 11 ft 2¾ in (3,42 m). Wing area, 322·92 sq ft (30,0 m²).

Remembering Reginald Mitchell

FORTY years ago, on 13 September 1931, Britain won the coveted Schneider Trophy outright with the Supermarine S.6B racing seaplane, designed by Reginald Mitchell. The Trophy had first been contested in 1913 and was regarded for the next 18 years as a major prize for aviation development. In these 18 years, the winning speed increased from 45 mph to 340 mph (71 km/h to 547 km/h) and the fact that the 1931 winner outperformed contemporary RAF fighters despite the encumbrance of large floats is some indication of the stimulus that the contest provided for aircraft designers. In Mitchell's case, his experience in building a series of racing seaplanes provided the basis for the design of the Spitfire.

The Schneider-winning S.6B S1595 is now displayed in the Science Museum, South Kensington, London. Its sister ship, S1596, is in the hands of the Southampton Junior Chamber of Commerce and is to provide the central exhibit of a proposed Mitchell Memorial Museum in that city. This S.6B was used, on the same day that the Schneider Trophy was won, to set a World Speed Record of 379 mph (610 km/h). A few days later it sank after capsizing when landing. A few years ago it was refurbished by Vickers Armstrongs.

The Museum will also include a Spitfire I, X4590, in the markings of 609 Squadron, and a series of models depicting various marks of Spitfire and Mitchell's earlier designs, together with various other mementoes. It is hoped eventually to include a library of reference material to Mitchell's life and his aircraft.

Site of the Mitchell Memorial Museum will be alongside the Royal Pier in Southampton, facing Southampton Water and Calshot where many of Mitchell's products operated. Funds are still needed if a hoped-for 1972 opening is to be achieved, and donations can be sent to the Financial Organiser, Mitchell Museum, 53 Bugle Street, Southampton.

Another Hind recovered

A HAWKER HIND which crashed in New Zealand 30 years ago has recently been rediscovered and it is hoped that it will be fully restored to exhibition standard. This particular aircraft was being flown by an

(Above) The Supermarine S.6B S1596, intended for permanent display in the Mitchell Memorial Museum (see alongside), during one of its rare appearances. At this time (1967) the spinner was incorrectly painted black, but has since been corrected.

(Below) Reader Roland Poehlmann of Erlangen, Germany, provides this picture in response to our request for news of 40-year-old veterans still flying. D-ELIS is a Comte AC-4B (c/n 34), first registered in September 1930 as CH-262 and flown later as HB-IKO before being sold to Germany. A 140 hp AS Genet Major now replaces the original ADC Cirrus Hermes. Owned by Herr F Dioszeghi, the AC-4B is based at Bamberg, has a red finish with silver cowl and carries advertising slogans for Motorola Sprechfunk.

RNZAF pilot, G C Stewart, when it went down in thick bush country in the Tararua Range. The pilot was unhurt and is still living in New Zealand.

A search party sent to the scene of the crash decided that recovery was impracticable. An attempt (fortunately unsuccessful) was made to destroy the Hind by igniting a flare in the fuel tank, but the fabric covering was destroyed to prevent other pilots from thinking they had discovered a newly-crashed aircraft if they

spotted it from the air. Parachute and instruments were recovered.

Several attempts had been made to rediscover the crash site, and success came recently to a 13-strong party including members of the Aviation Historical Society of New Zealand. Even then, its discovery was almost accidental, being made by one searcher when he had briefly lost contact with the main party. The Rolls-Royce Kestrel is reported to be in good condition, and the oil tank still contained some oil.

(Below) The sole Tiger Moth Seaplane in Britain, G-AIVW, obtained a renewed C of A during August. The picture (right) shows 'VW refuelling at Aldeburgh en route for a demonstration at the Oulton Broad regatta on 30 August. The D.H.82C floatplane (left), CF-FUG, was used by Tom Dow of Thunder Bay, Ontario, to attend the recent EAA Fly-In at Oshkosh, Wisconsin, and is shown on nearby Lake Winnebago.

AIR Enthusiast VOLUME ONE

ILLUSTRATIONS

Photographs — half-tone and (c) colour

Line illustrations and (c/a) cutaway drawings

Tone illustrations and (c) colour drawings